Lib: J. H. Rolston

QD505
H995

Lib: J. H. Rolston

Hydrogen Effects in Catalysis

CHEMICAL INDUSTRIES

A Series of Reference Books and Textbooks

Consulting Editor
HEINZ HEINEMANN
*Heinz Heinemann, Inc.,
Berkeley, California*

Volume 1: Fluid Catalytic Cracking with Zeolite Catalysts,
Paul B. Venuto and E. Thomas Habib, Jr.

Volume 2: Ethylene: Keystone to the Petrochemical Industry,
Ludwig Kniel, Olaf Winter, and Karl Stork

Volume 3: The Chemistry and Technology of Petroleum,
James G. Speight

Volume 4: The Desulfurization of Heavy Oils and Residua,
James G. Speight

Volume 5: Catalysis of Organic Reactions,
edited by William R. Moser

Volume 6: Acetylene-Based Chemicals from Coal and Other Natural Resources, *Robert J. Tedeschi*

Volume 7: Chemically Resistant Masonry,
Walter Lee Sheppard, Jr.

Volume 8: Compressors and Expanders: Selection and Application for the Process Industry, *Heinz P. Bloch, Joseph A. Cameron, Frank M. Danowski, Jr., Ralph James, Jr., Judson S. Swearingen, and Marilyn E. Weightman*

Volume 9: Metering Pumps: Selection and Application,
James P. Poynton

Volume 10: Hydrocarbons from Methanol,
Clarence D. Chang

Volume 11: Foam Flotation: Theory and Applications,
Ann N. Clarke and David J. Wilson

Volume 12: The Chemistry and Technology of Coal,
James G. Speight

Volume 13: Pneumatic and Hydraulic Conveying of Solids,
O. A. Williams

Volume 14: Catalyst Manufacture: Laboratory and Commercial Preparations, *Alvin B. Stiles*

Volume 15: Characterization of Heterogeneous Catalysts,
edited by Francis Delannay

Volume 16: BASIC Programs for Chemical Engineering Design,
James H. Weber

Volume 17: Catalyst Poisoning,
L. Louis Hegedus and Robert W. McCabe

Volume 18: Catalysis of Organic Reactions,
edited by John R. Kosak

Volume 19: Adsorption Technology: A Step-by-Step Approach to Process Evaluation and Application, *edited by Frank L. Slejko*

Volume 20: Deactivation and Poisoning of Catalysts,
edited by Jacques Oudar and Henry Wise

Volume 21: Catalysis and Surface Science: Developments in Chemicals from Methanol, Hydrotreating of Hydrocarbons, Catalyst Preparation, Monomers and Polymers, Photocatalysis and Photovoltaics
edited by Heinz Heinemann and Gabor A. Somorjai

Volume 22: Catalysis of Organic Reactions,
edited by Robert L. Augustine

Volume 23: Modern Control Techniques for the Processing Industries, *T. H. Tsai, J. W. Lane, and C. S. Lin*

Volume 24: Temperature-Programmed Reduction for Solid Materials Characterization, *Alan Jones and Brian McNicol*

Volume 25: Catalytic Cracking: Catalysts, Chemistry, and Kinetics, *Bohdan W. Wojciechowski and Avelino Corma*

Volume 26: Chemical Reaction and Reactor Engineering, *edited by J. J. Carberry and A. Varma*

Volume 27: Filtration: Principles and Practices, second edition, *edited by Michael J. Matteson and Clyde Orr*

Volume 28: Corrosion Mechanisms, *edited by Florian Mansfeld*

Volume 29: Catalysis and Surface Properties of Liquid Metals and Alloys, *Yoshisada Ogino*

Volume 30: Catalyst Deactivation, *edited by Eugene E. Petersen and Alexis T. Bell*

Volume 31: Hydrogen Effects in Catalysis: Fundamentals and Practical Applications, *edited by Zoltán Paál and P. G. Menon*

Additional Volumes in Preparation

Hydrogen Effects in Catalysis

Fundamentals and Practical Applications

edited by

Zoltán Paál
Institute of Isotopes of the
Hungarian Academy of Sciences
Budapest, Hungary

P. G. Menon
Dow Chemical (Nederland) B.V.
Terneuzen, The Netherlands
Chalmers University of Technology
Gothenburg, Sweden

Marcel Dekker, Inc.   New York and Basel

Library of Congress Cataloging in Publication Data

Hydrogen effects in catalysis.

  (Chemical industries ; v. 31)
  Includes index.
  1. Catalysis. 2. Hydrogen. 3. Surface chemistry.
I. Paál, Zoltán. II. Menon, P. G.
(P. Govind). III. Series.
QD505.H93   1988   541.3'95   87-24599
ISBN 0-8247-7774-3

Copyright   1988 by MARCEL DEKKER, INC.   All Rights Reserved

Neither this book nor any part may be reproduced or transmitted in any form or by any means, electronic or mechanical, including photocopying, microfilming, and recording, or by any information storage and retrieval system, without permission in writing from the publisher.

MARCEL DEKKER, INC.
270 Madison Avenue, New York, New York 10016

Current printing (last digit):
10 9 8 7 6 5 4 3 2 1

PRINTED IN THE UNITED STATES OF AMERICA

Preface

It is now being increasingly recognized that catalytically active surface sites may (or should) contain atoms different from that of the catalyst material. Several attempts have been made to define working catalysts as "catalytic systems" rather than clean substances. Boudart wrote as early as 1975 that "the definition of a working catalyst must include all the surface species of the catalyst in its working state": the catalyst itself, added promoters, chance impurities, and components arising from the reactant(s). However, the exact information on this working state still "remains an elusive goal of research in catalysis" (M. Boudart, *J. Vac. Sci. Technol.*, *12*, 329 [1975]).

One should not think that the additives responsible for catalytic action are necessarily present in minute amounts. G. K. Chesterton in one of his Father Brown stories describes a case when four reliable and observant persons testify that nobody entered a house where and when a murder was committed. Father Brown's simple wisdom is necessary to realize that a "mentally invisible man" must be held responsible—a man who in a "striking and showy costume" entered the house "under eight human eyes"—namely a postman. "Nobody ever notices postmen, somehow," he concludes, "yet they have passions like other men" (G. K. Chesterton, The Invisible Man. In *The Innocence of Father Brown*, Penguin, Harmondsworth [1950], p. 92).

Hydrogen, especially in metal catalysis, can be compared to this "mentally invisible" postman. Its presence (usually in ample amounts) has been taken for granted and, so far, most studies have not tried to dig deep and look at whether it participates in creating catalytically active surfaces, and if so, how. Unlike postmen, hydrogen does not wear a "striking and showy costume": on the contrary, it is a rather difficult species to detect by most of the regular techniques of surface science. So we may conclude that nobody ever notices hydrogen; yet it has an influence on catalytic systems as other additives. Not that this influence would be murderous to catalysis; experience shows just the contrary—but alas! "mentally invisible" benefactors escape notice much easier than villains.

The interest in hydrogen increases in various fields of science and technology, especially its prospective use as an absolutely "clean" energy source (see also Chapter 23). Several changes in the chemisorptive and catalytic properties of metals brought about by hydrogen have been reviewed recently by us elsewhere (Z. Paál and P. G. Menon, *Catal. Rev.-Sci. Eng.*, *25*, 229 [1983]). The time and context now are considered proper for an attempt to devote a full book to hydrogen effects in catalysis. Such a book should necessarily cover a very wide range of subjects. Throughout the present compilation, a vertical treatment has been attempted which covers hydrogen effects in catalysis in the broadest sense, from surface science to industrial applications. This is reflected by the five main parts of this work.

Part I deals with phenomena observed with well-defined catalyst surfaces—the typical surface science approach. Most of the methods applied here are indirect ones. These are summarized in Chapter 1. Chapter 2 deals with vibrational spectroscopy, representing a rare family of techniques suitable for gaining direct information from surface hydrogen. The results can be summarized in terms of postulating a "hydrogen fog" *on* metal surfaces which can be almost as much delocalized as the "electron cloud" *in* metals.

Physico-chemical methods suitable for characterizing practical catalysts have been collected in Part II. Traditional techniques supplying indirect information such as adsorption heat measurements, surface titrations, and temperature-programmed studies are presented first. The description and results of less common techniques such as magnetic measurements, infrared spectroscopy, and neutron diffraction follow, the last two being able to supply direct information on hydrogen. Their vast possibilities are still dormant. Also included is a chapter on electrochemical methods, which are able to give exclusive results on what happens on the surface proper.

Perhaps the climax of the book is Part III, dealing with the role of hydrogen to produce active working catalyst surfaces. The complex nature of catalysis science is truly reflected in its five chapters. One or more of the phenomena treated here (e.g., sintering, dissolution of hydrogen, and eventual hydride formation—all solid-state transformations) or the two particular aspects related to catalysis, namely hydrogen spillover and strong metal-support interactions, may often take place in the lifetime of any catalyst, be it the few grams in a laboratory test or the several tons in an industrial reactor. In Chapter 12 it is correctly remarked that hydrogen spillover may not be tremendously significant by itself, but as a first crucial step in a sequence it may be very important. The same is more or less true for other effects treated in Part III and, perhaps, in the whole book.

Real catalytic chemistry comes in Part IV, where hydrogen effects on various reactions are treated. These include those when hydrogen is an "astoichiometric" component, i.e., when it does not participate in the stoichiometric equation of the reaction. Most typical examples of reactions are hydrocarbon transformations on metals. That is the reason that these reactions were selected to illustrate the general kinetics of hydrogen effects. In Chapter 15, rather uncommon effects of hydrogen are revealed. Such delicacies as comparing low- and high-pressure reactions on single crystals follow: hydrogen effects may help construct that long-awaited bridge between single-crystal work and real catalysts. Both hydrogenation and skeletal reactions of hydrocarbons offer a number of examples of hydrogen effects. Still, attention has to be drawn to Chapter 19, where a largely unexplored field has been opened up: the catalytic transformations of oxygenated organic compounds. Although a host of phenomena are highlighted here, these compounds still represent a minority of organic compounds. The present trends toward striving at producing special chemicals in increasing numbers certainly will benefit from hydrogen effects in selectivity regulation. The last two chapters of this part illustrate that hydrogen effects are not confined to metal catalysts; oxide and sulfide catalysts—though less explored at present—also exhibit some of them.

The last part (Part V, Technological Implications) can necessarily give only an incomplete excerpt on what hydrogen can do in industrial catalysis.

Preface

Again, metal catalysts and hydrocarbon reactants represent most widely studied systems, as indicated by the three chapters dealing with this subject. Apart from industrial hydrogenations, such important processes as those in catalytic reforming and such novelties as hydrogen diffusion through metal catalysts are dealt with. There are indications that even an important process like ammonia synthesis can also exhibit typical hydrogen effects; these have, however, not been explored to an extent that a separate chapter could be devoted to them. The last two chapters also divert attention from metallic systems: hydrogen transfer reactions, so important in up-to-date industrial processes using zeolite catalysts, have been treated adequately. The last chapter calls attention to hydrogen effects in Ziegler-Natta and other olefin polymerization, an area with which most people in (heterogeneous) catalysis research and development are usually not very familiar. Just as with metallic and oxide catalysts, here, too, hydrogen plays a double role: on the one hand, it is part of the catalyst system itself; on the other hand, it is a powerful probe to investigate the nature of active sites in olefin polymerization catalysts (because it can respond so selectively and specifically to the different polymerization centers).

The book contains altogether 28 chapters, written by 38 authors from 10 different countries. A beautiful example, indeed, of international cooperation, typical for catalysis as for any other specialized field of science. The editors want to thank here all the other 36 authors—colleagues and friends, whose whole-hearted cooperation has made this venture possible and successful. We also thank Dr. Heinz Heinemann, Consulting Editor of the Chemical Industries series, for his advice and helpful comments at various stages in the compilation of this work.

Compiling this book was undertaken with a double goal in mind: to obtain state-of-the-art reviews on the phenomena collectively called "hydrogen effects" and to draw the attention of the catalysis community to the importance of these phenomena both in the science and technology of catalysis. Hence this book does not close or summarize the subject; instead, it is meant to introduce the various facets of this subject and stimulate further studies in this field.

Zoltán Paál
P. G. Menon

Contributors

Calvin H. Bartholomew Head, BYU Catalysis Laboratory; Associate Director, Advanced Combustion Engineering Research Center; and Professor of Chemical Engineering, Brigham Young University, Provo, Utah

Mihály Bartók Professor of Organic Chemistry, József Attila University, Szeged, Hungary

Brian E. Bent* Department of Chemistry, University of California at Berkeley, Berkeley, California

Jean-Paul Bournonville Chemical Engineer, Kinetics and Catalysis Department, Institut Français du Pétrole, Rueil-Malmaison, France

Robbie Burch Reader, Catalysis Research Group and Chemistry Department, University of Reading, Reading, Berkshire, England

N. Y. Chen Senior Scientist/Research Advisor, Princeton Research Laboratory, Mobil Research and Development Corporation, Princeton, New Jersey

Klaus R. Christmann Professor of Physical Chemistry, Chemistry Department, Freie Universität Berlin, Berlin, Federal Republic of Germany

William Curtis Conner, Jr. Associate Professor, Department of Chemical Engineering, University of Massachusetts, Amherst, Massachusetts

Duncan R. Coupland Principal Scientist, Engineering Applications Group, Catalytic Systems Division—Engineered Products, Johnson Matthey, Royston, Hertfordshire, England

József Engelhardt Senior Scientific Research Associate, Department of Hydrocarbon Catalysis, Central Research Institute for Chemistry of the Hungarian Academy of Sciences, Budapest, Hungary

---

Current affiliation: Postdoctoral Fellow, AT&T Bell Laboratories, Murray Hill, New Jersey

Alfred Frennet Senior Research Associate, Unité de recherches sur la catalyse, Université Libre de Bruxelles, Brussels, Belgium

Jean-Pierre Franck R & D Project Manager for Catalytic Reforming, Kinetics and Catalysis Department, Institut Français du Pétrole, Rueil-Malmaison, France

J. W. Geus Professor of Inorganic Chemistry, University of Utrecht, Utrecht, The Netherlands

W. O. Haag Senior Scientist, Princeton Research Laboratory, Mobil Research & Development Corporation, Princeton, New Jersey

Richard D. Kelley Chemist, Surface Science Division, National Bureau of Standards, Gaithersburg, Maryland

Tamás Mallát Senior Research Fellow, Department of Organic Chemical Technology, Technical University of Budapest, Budapest, Hungary

C. Mathew Mate[*] Department of Physics, University of California at Berkeley, Berkeley, California

Tibor Máthé Associate Professor and Senior Research Fellow, Institute for Organic Chemical Technology, Hungarian Academy of Sciences, Budapest, Hungary

P. Govind Menon Specialist for Catalysis, Research & Development Department, Dow Chemical (Nederland) B.V., Terneuzen, The Netherlands; and Research Professor, Department of Engineering Chemistry, Chalmers University of Technology, Gothenburg, Sweden

Árpád Molnár Associate Professor, Department of Organic Chemistry, József Attila University, Szeged, Hungary

Richard B. Moyes Senior Lecturer, Chemistry Department, University of Hull, Hull, North Humberside, England

Zoltán Paál Professor, Head of Department of Catalysis, Institute of Isotopes of the Hungarian Academy of Sciences, Budapest, Hungary

Wacława Palczewska Professor of Chemistry, Department of Catalysis on Metals, Institute of Physical Chemistry, Polish Academy of Sciences, Warsaw, Poland

József Petró University Professor, Department of Organic Chemical Technology, Technical University of Budapest, Budapest, Hungary

John Philpott Special Projects Manager, Catalytic Systems Division–Engineered Products, Johnson Matthey, Royston, Hertfordshire, England

---

[*]*Current affiliation*: Postdoctoral Fellow, IBM Almaden Research Center, San Jose, California

Contributors

Éva Polyánszky Assistant Lecturer, Department of Organic Chemical Technology, Technical University of Budapest, Budapest, Hungary

Eli Ruckenstein Distinguished Professor, Department of Chemical Engineering, State University of New York at Buffalo, Buffalo, New York

Nils-Herman Schöön Professor of Chemical Reaction Engineering, Chalmers University of Technology, Gothenburg, Sweden

Shashikant Senior Research Officer, Research Centre, Indian Petrochemicals Corporation, Ltd., Vadodara, Gujarat, India

Swaminathan Sivaram Senior R & D Manager, Research Centre, Indian Petrochemicals Corporation, Ltd., Vadodara, Gujarat, India

Gerard V. Smith Director of Molecular Science Program and Professor of Chemistry and Biochemistry, Southern Illinois University at Carbondale, Carbondale, Illinois

Gabor A. Somorjai Professor of Chemistry, University of California at Berkeley, Berkeley, California

Poondi R. Srinivasan Senior Research Officer, Research Centre, Indian Petrochemicals Corporation, Ltd., Vadodara, Gujarat, India

Iruvanti Sushumna Research Associate, Department of Chemical Engineering, State University of New York at Buffalo, Buffalo, New York

Tibor Szilágyi Scientific Research Associate, Institute of Isotopes of the Hungarian Academy of Sciences, Budapest, Hungary

Terrence J. Udovic Chemical Engineer, Institute for Materials Science and Engineering, National Bureau of Standards, Gaithersburg, Maryland

József Valyon Senior Scientific Research Associate, Department of Hydrocarbon Catalysis, Central Research Institute for Chemistry of the Hungarian Academy of Sciences, Budapest, Hungary

Francisco Zaera Professor of Chemistry, University of California at Riverside, Riverside, California

Contents

Preface	iii
Contributors	vii

PART I SURFACE CHEMISTRY OF HYDROGEN ON METALS

1. Hydrogen Sorption on Pure Metal Surfaces 3
   Klaus R. Christmann

2. Vibrational Spectroscopy of Hydrogen Adsorbed on Metal Surfaces 57
   C. Mathew Mate, Brian E. Bent, and Gabor A. Somorjai

PART II CHARACTERIZATION OF SURFACE HYDROGEN ON CATALYSTS

3. Energetics of Hydrogen Adsorption on Porous and Supported Metals 85
   J. W. Geus

4. Hydrogen as a Tool for Characterization of Catalyst Surfaces by Chemisorption, Gas Titration, and Temperature-Programmed Techniques 117
   P. G. Menon

5. Hydrogen Adsorption on Supported Cobalt, Iron, and Nickel 139
   Calvin H. Bartholomew

6. Neutron Scattering Studies of Hydrogen in Catalysts 167
   Terrence J. Udovic and Richard D. Kelley

7. Infrared Spectroscopy of Adsorbed Hydrogen 183
   Tibor Szilágyi

8. Effects of Adsorption of Hydrogen on the Magnetic and Electrical Properties of Metals 195
   J. W. Geus

9. Electrochemical Investigation of Surface Hydrogen on Metal Catalysts 225
   József Petró, Tamás Mallát, Éva Polyánszky, and Tibor Máthé

PART III HYDROGEN AS A PARTNER IN PRODUCING ACTIVE CATALYSTS

10. Hydrogen Effects in Sintering of Supported Metal Catalysts 259
    Eli Ruckenstein and Iruvanti Sushumna

11. Hydrogen-Induced Sintering of Unsupported Metal Catalysts 293
    Zoltán Paál

12. Spillover of Hydrogen 311
    William Curtis Conner, Jr.

13. Strong Metal-Support Interactions 347
    Robbie Burch

14. Catalytic Properties of Metal Hydrides 373
    Wacława Palczewska

PART IV HYDROGEN EFFECTS IN CATALYTIC REACTIONS

15. General Kinetics of Hydrogen Effects: Hydrocarbon Transformations over Metals as Model Reactions 399
    Alfred Frennet

16. Role of Hydrogen in Low- and High-Pressure Hydrocarbon Reactions 425
    Francisco Zaera and Gabor A. Somorjai

17. Hydrogen Effects in Skeletal Reactions of Hydrocarbons over Metal Catalysts 449
    Zoltán Paál

18. Hydrogen Effects in Organic Hydrogenations 499
    Árpád Molnár and Gerard V. Smith

19. Hydrogen Effects in Catalytic Transformations of Oxygenated Carbon Compounds on Metals 521
    Mihály Bartók

20. Role of Hydrogen in CO Hydrogenation 543
    Calvin H. Bartholomew

21. Effect of Hydrogen on the Activity of Oxide Catalysts 565
    József Engelhardt and József Valyon

22. Reactivity of Hydrogen in Sulfide Catalysts 583
    Richard B. Moyes

PART V TECHNOLOGICAL IMPLICATIONS

23. Hydrogen Effects in Industrial Catalysis — 611
    P. G. Menon

24. Selectivity in Competitive Hydrogenation Reactions — 621
    Nils-Herman Schöön

25. Hydrogen and Catalytic Reforming — 653
    Jean-Paul Bournonville and Jean-Pierre Franck

26. Metal Membranes for Hydrogen Diffusion and Catalysis — 679
    John Philpott and Duncan R. Coupland

27. Hydrogen Transfer in Catalysis on Zeolites — 695
    N. Y. Chen and W. O. Haag

28. Hydrogen Effects in Olefin Polymerization Catalysts — 723
    Poondi R. Srinivasan, Shashikant, and
    Swaminathan Sivaram

Index — 747

Hydrogen Effects in Catalysis

1
SURFACE CHEMISTRY OF HYDROGEN ON METALS

1 | Hydrogen Sorption on Pure Metal Surfaces

KLAUS R. CHRISTMANN

Freie Universität Berlin, Berlin, Federal Republic of Germany

I.	INTRODUCTION	3
II.	ADSORPTION OF HYDROGEN ON METAL SURFACES	4
	A. Mechanism and Energetics of Hydrogen Adsorption	4
	B. Dynamics of Adsorption	23
	C. Coadsorption of Hydrogen and Other Gaseous Adsorbates	34
III.	ABSORPTION OF HYDROGEN: SURFACE EFFECTS	40
	A. Equilibrium: Surface-Bulk	40
	B. Subsurface Hydrogen	42
	C. Influence of Hydrogen Sorption on the Lattice Parameters	44
IV.	REACTION OF HYDROGEN WITH METAL SURFACES	45
	A. Conditions for Hydride Formation	45
	B. Surface Reconstruction Induced by Hydrogen	46
V.	CONCLUSIONS	50
	REFERENCES	51

I. INTRODUCTION

During the past 20 years the dominant role of hydrogen in heterogeneous catalysis, energy storage technology, and metallurgy has become widely evident. Consequently, surface scientists have become more and more interested in the corresponding elementary processes that occur on or within the surfaces and that involve the hydrogen atom as a reactant or product.

There are at least four different processes that govern the interaction of gaseous hydrogen (or its isotope deuterium) with metal surfaces: (a) the *adsorption* of the diatomic molecule $H_2(D_2)$ on the surface, (b) the formation of H *atoms* on the surface (dissociation), (c) the migration of the H atoms through the near-surface region into the bulk of the metal crystal (*absorption/dissolution*), and (d) the possible *reaction* between the H atom and the metal atom(s) to form a surface or bulk hydride compound.

Adsorption is always the first and often the rate-limiting step in the interaction and is therefore of particular importance for a detailed understanding of a heterogeneous surface reaction. This is why we give this subject extensive treatment here. The absorption may also be of great technical interest, particularly in the hydrogen storage area. Although for physical reasons the adsorption and absorption phenomena are closely related, research in the latter field has developed to a practically independent science. Within the framework of this chapter, we can, therefore, deal only briefly with absorption processes, for example, in the context of H on Pd (which metal is known to dissolve hydrogen readily) and otherwise simply refer to the numerous available monographs. Details of absorption processes can also be found in Chapter 14.

The reaction of H atoms with solid surfaces is crucial in materials science and metallurgy (hydrogen embrittlement and fracture). Concerning the interaction between hydrogen and titanium, niobium, vanadium, zirconium, and some other transition metals, the formation of bulk metal hydride takes place readily even if the hydrogen is in the molecular state; *atomic* hydrogen strongly reacts with alkali and alkaline earth metals to form the respective hydride compound. Again, due to the complexity of this matter, there will be only a very brief treatment in this chapter. However, recent investigations seem to demonstrate some importance of the preceding step of the hydride formation in the field of heterogeneous catalysis, namely, the H-induced reconstruction of metal surfaces. This may easily occur in the course of a chemical surface reaction and will often lead to structural changes of the solid surface, especially at elevated temperatures and reactant pressures (i.e., under catalytical conditions). Some examples are presented in this area; the consequences of these processes in the field of catalyst sintering are discussed in Chapters 10 and 11.

This chapter is divided into three sections, each of which contains some of the most relevant literature references. There will be selection of data, in that model studies on well-defined single-crystal surfaces will be given preference, despite the still existing "gap" between these model studies and real catalysis [1].

II. ADSORPTION OF HYDROGEN ON METAL SURFACES

A. Mechanism and Energetics of Hydrogen Adsorption

It is a reasonable approach to distinguish the various cases of hydrogen interaction with metal surfaces with respect to the energy involved. In this sense we have (a) the weak (molecular) interaction based on physisorptive (van der Waals) interaction forces with binding energies of the order of 10 to 20 kJ mol^{-1}. Examples include (although there is a lack of experimental data for single-crystal surfaces) hydrogen on Al, Au, and Ag. The latter system has been investigated by Demuth's group [2] using a Ag(111) surface, and it was shown that *molecular* hydrogen exists on the surface, which desorbs around 20 K; there was no hint of any dissociation under these conditions. There is a simple way to illustrate the van der Waals interactions in the simple-one-dimensional potential energy diagram according to Lennard-Jones [3]. As Fig. 1a shows, the physisorptive interaction leads to a shallow minimum far outside the surface in which the trapped H_2 molecule will be held. The situation changes completely (b) as soon as the electronic configuration of the metal (i.e., the position of its d band) enables the H_2 molecule to dissociate into the atoms: The energy gain due to the overlap of the hydrogenic 1s wavefunctions with sp or d functions of the metal atoms overcompensates the binding energy of the H_2 molecule, which immediately causes its spontaneous dissociation and the formation of two stable hydrogen-metal bonds. This is depicted in Fig. 1a by the dashed lines: The *chemisorbed*

H atom is strongly held in a deep (80 to 120 kJ mol^{-1}) potential well relatively close to the surface. Evidently, the detailed pathway of the dissociation depends sensitively on the relative position of the physisorptive and chemisorptive potential energy curves. If the crossover point of the two curves is below the zero-energy level, we have the case of spontaneous ("nonactivated") dissociation; if it is above E = 0, there is an activation energy barrier, E*, built up which prevents a rapid spontaneous dissociation of H_2 molecules impinging on the surface with thermal energy. Excitations using supersonic molecular beams can, however, substantially increase the dissociation probability and hence the uptake of chemisorbed hydrogen. These cases of activated and nonactivated adsorption are illustrated in Fig. 1.

As an example, the activated adsorption experiments by Balooch et al. [4] using Cu single crystals and a molecular beam of H_2 (D_2) have to be mentioned. In these hydrogen adsorption was detected for certain kinetic energies of the incoming H_2 molecules. Even more instructive than the one-dimensional potential energy diagram in illustrating the dissociation process is the two-dimensional representation shown in Fig. 1b. This figure is based on theoretical work by Nørskov dealing with the dissociation of hydrogen on a Mg(0001) surface [5]. The dissociation of the molecule (i.e., the variation of the H-H bond length x) was calculated as a function of the distance of the molecule from the surface, y. In Fig. 1b, y is plotted against x, and the reaction coordinate is indicated by a dashed line. It can be seen that there exists an entrance channel [in which the molecular bond is not stretched (constant x)], followed by an activation barrier as the molecule approaches the surface. As y gets smaller, x *increases* until beyond point P (located in the exit channel) the molecule can be regarded as dissociated into the individual atoms. It should be noted that the activation barrier is not necessarily located in the exit channel but may also be in the entrance channel. This would correspond to the existence of a *molecular precursor* state which may (with some caution) be interpreted as a physisorbed H_2 molecule.

For hydrogen interaction with transition metal surfaces (which are used predominantly in heterogeneous catalysis), nonactivated adsorption is the rule. This is indicated by high sticking probabilities (rapid uptake of hydrogen), between 0.1 and 1.0 [6]. Small sticking coefficients, on the other hand, may sometimes indicate the existence of a small activation barrier in the adsorption path. As demonstrated recently by Robota et al. [7] there is such a barrier (height ca. 9 kJ mol^{-1}) for hydrogen chemisorption on the Ni(111) surface which can be overcome by thermally excited H_2 molecules. It is interesting at this point to look at the face specificity: On the crystallographically more "open" Ni(110) surface there is no such barrier; the sticking probability for H is close to 1.0 [8].

A great deal of work has been directed toward the determination and calculation of heats of adsorption of hydrogen and H-Me binding energies. For dissociative adsorption, the H-Me bond energy can be obtained simply from the relation (which follows directly from Fig. 1a)

$$E_{Me-H} = \frac{1}{2}(E_{diss} + E_{ad}) \qquad (1)$$

in which E_{diss} denotes the dissociation energy of the H_2 molecule (432 kJ mol^{-1}) and E_{ad} the heat of adsorption (kJ mol^{-1}) as determined in the experiment. As will be shown later, the values for the heat of adsorption typically range between 60 and 120 kJ mol^{-1} for hydrogen on transition metal surfaces. Accordingly, the numbers for E_{Me-H} lie in the range 200 to 300 kJ mol^{-1} and thus represent energies typical for chemical bonds. It has to be borne in mind that these values refer to a situation in which a single H_2 molecule inter-

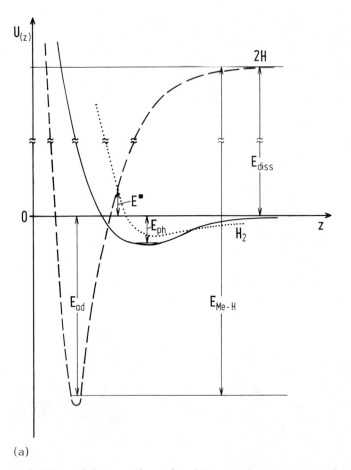

(a)

FIGURE 1 (a) One-dimensional Lennard-Jones potential energy diagram for adsorption of a diatomic molecule (H_2). z denotes the reaction coordinate (i.e., the distance perpendicular to the surface plane). Solid line, physisorption of a H_2 molecule; dashed line, chemisorption of a predissociated H atom; dotted curve, physisorption potential energy curve leading to an activated atomic adsorption.

acts with a clean single-crystal surface. Higher surface concentrations of chemisorbed hydrogen in most cases lead to a more or less pronounced decrease in the Me-H bond energies due to competing mutual interactions of the adsorbed particles. Therefore, one has to distinguish *two* regimes of interaction: the "zero-coverage" limit (H coverage θ = number of adsorbed particles/number of possible adsorption sites $\to$ 0) and the "multiparticle" limit ($0 < \theta < 1$).

Zero-Coverage Limit (Single-Particle Interaction)

When considering the physical and chemical aspects of single-particle interaction, at least five questions come to mind:

1. What is the chemical nature of the interaction; that is, does the H atom form a covalent or more ionic bond with a significant amount of charge transfer, or does the bond even have metallic character? This question

1. *Hydrogen Sorption on Pure Metal Surfaces*

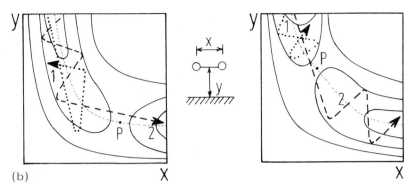

(b) Two-dimensional representation of the potential energy surface of the H_2 molecule interacting with a metal surface. Various trajectories are indicated: (1) denotes a reflection trajectory without chemisorption; (2) a trajectory that leads to dissociative adsorption. Note that the saddle point can be located either in the exit channel (left-hand side) or in the entrance channel (right-hand side). [(b) After Ref. 5.]

includes the frequently discussed problem of the polarization of adsorbed hydrogen.
2. What is the strength of the Me-H bond; that is, what amount of energy is required to break this bond to either have the hydrogens recombined to H_2 or to have them reacted with other surface species?
3. Is there any influence of the a priori heterogeneity of the substrate surface? Does an H atom chemisorbed on a defect site (kink, ledge, corner, etc.) experience a higher or lower binding energy than that of the "flat" portions of the surface? May it be that these defect sites facilitate the necessary dissociation process?
4. What are the absolute location and configuration of the chemisorbed H atom (i.e., the geometry of the H-Me complex); what are the bond lengths, bond angles, and so on?
5. How many substrate (metal) atoms participate (are required) in the bonding of one H atom? Is the concept of a localized complex ("surface molecule") as proposed by Newns [9] a more appropriate description than the idea of embedding the H atom in a delocalized surface electron band (which may be responsible for ensemble and ligand effects frequently reported in heterogeneous catalysis)?

This brief list of possible questions does not even include the problems mentioned before, namely, hydrogen absorption or surface reconstruction/ reaction, which represent additional complications. From the wealth of experimental data most of these questions can now be answered fairly satisfactorily.

1. It is generally accepted that the hydrogen-metal interaction leads preferentially to *covalent* (atomic) H-Me bonds with only small amounts of charge transfer. Compared, for example, to the halogen or alkali metal adsorption, the change in the surface work function $\Delta\phi$, is a factor of 10 smaller; the derived dipole moments range between 0.05 and 0.08 D [10], whereby on most surfaces an increase in ϕ ($\Delta\phi > 0$) is the rule (i.e., the surface dipole has its negative end outward). For comparison, $\Delta\phi$ decreases of up to 3 eV have been found with alkali metals, corresponding to dipole moments of ca. 6 D. Some hydrogen-induced work function changes are listed in Table 1.

TABLE 1 Some Selected Values of the Hydrogen-Induced Work Function Changes

System	Work function change (mV) at saturation	Adsorption temperature, T_{ad} (K)	Ref.
H/Ni(100)	+160	300	11
	+100/50	100	12
H/Ni(111)	+195	300	11
		100	13
H/Ni(110)	+530	120	14
	+510	120	8
H/Pd(100)	+200	120	15
H/Pd(111)	+180	300	16
H/Pd(110)	+320	300	16
	+350	300	17
H/Fe(110)	−85	150	18
H/Fe(100)	+65	150	18
H/Fe(111)	+220	150	18
H/Ru(0001)	+25/−50	173	19
	+25/−150	273	19
H/Rh(110)	+930	80	20
H/Pt(111)	−300	150	21
H/W(100)	+900	300	22

From those rare cases in which the absolute configuration of a hydrogen adsorption complex has been evaluated (H/Ni(111) [13]), a charge transfer of about 1/10 e_0 could be derived. Yet this number is crude only since contributions to $\Delta\phi$ of structural changes in the surface caused by the adsorption (relaxation or reconstruction) will become significant. These contributions cannot simply be delineated. An example of an experimental $\Delta\phi(\Theta_H)$ curve is given later (Fig. 26).

Attempts have been made to correlate the charge located on an adsorbed hydrogen atom (as reflected by the observed work function change) with the nature (geometry) of the adsorption site. An example is perhaps provided by the adsorption of hydrogen on stepped Pt(997) surfaces [23]. Precise work function and hydrogen coverage measurements indicated that besides the H atoms adsorbed on the flat terraces, two differently held hydrogen species could be distinguished in and near the steps of the Pt surface. The situation is depicted in Fig. 2. Part (a) shows the observed work function change as a function of hydrogen coverage, whereas part (b) gives a schematic structure model of the two possible locations of the hydrogen atoms adsorbed in the step region. These two types of hydrogen sites might lead to locally different H-Me dipole moments (i.e., give rise to a site-dependent polarization of the chemisorbed hydrogen). While the general validity of this idea was supported by Baró and Ibach in their electron loss spectroscopy work on the

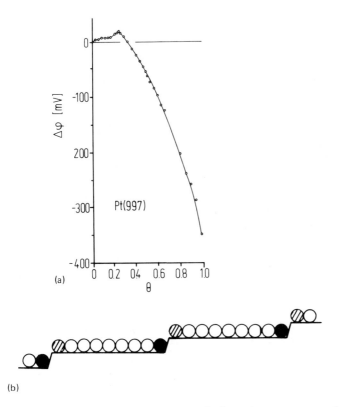

FIGURE 2 (a) Change of the work function, $\Delta\phi$, as a function of the hydrogen coverage, θ_{H_2}, for a stepped Pt(997) surface. (b) Structure model for the Pt(997) surface (cut perpendicular through the surface plane). Two different adsorption sites for hydrogen near the steps are indicated by hatched and dark circles. [(a), After Ref. 23.]

same [H/Pt(997)] system, they found only one "extra" step site sufficient to explain their experimental data [24].

At this point it is advisable to sound a warning: As mentioned earlier, the experimentally observed work function changes are by no means 1:1 related to the local geometry and/or electronic structure of an adsorption site. It must be borne in mind that the work function of a surface is a complicated physical property which generally cannot simply be split up into its two contributions: namely, the surface dipole (influenced by, among others, the adsorbed particles) and the chemical potential μ of the bulk (which reacts very sensitively upon the location of the atoms of the surface region). As pointed out in more detail in Section III.C, the adsorbed H atom, owing to its small size, may significantly perturb the surface electronic structure and can easily induce relaxation, if not reconstruction effects of the metal itself. In this case, however, the chemical potential of the metal will be affected and hence its work function contribution. It is self-evident that in this case the hydrogen-induced work function change does not simply monitor the dipole moment of the H-Me bond. Thus it seems somewhat dangerous to identify, for example, "s-type" and "r-type" hydrogen adsorption sites as theoretically predicted by Toya [25] with hydrogen-induced work function changes of different polarity. Nevertheless, there is some evidence from recent studies of the hydrogen adsorption around 100 K that a more weakly held (but still atomic) hydrogen species can cause a different sign of the work function change. Such an effect was proposed some years ago by Eley and Pearson [26] for hydrogen adsorption on Pd(100) (on which surface, however, no different signs of the hydrogen-induced work function changes could be detected [15]). Examples are hydrogen on Ni(111) [13], Ni(100) [12], and Ru(001) [19]. It may, however, also be possible that the more weakly bound H atoms produce a work function change with the *same* sign as the strongly bound species. Examples are hydrogen on Ni(110) [8], Pd(110) [17], or Rh(110) [20]. Whether or not a particular work function change is always correlated with an adsorption site of distinct local geometry is not an easy question to answer. The available data rather suggest an *electronic* origin for the work function change. All these effects occur at elevated hydrogen coverages where both strongly and weakly held hydrogen species exist on the surface. For small hydrogen coverages, on the other hand, there is no hint of an a priori heterogeneity for hydrogen adsorption on defect-free low-index metal surfaces. This is also the reason why the frequently observed ordered hydrogen phases exhibit a pronounced lattice gas behavior (i.e., *equivalent* adsorption sites are statistically occupied by H atoms). Any heterogeneity is only introduced by lateral mutual interactions between the adsorbed particles as their density is increased. The electronic interaction between adsorbed hydrogen and a metal surface also shows up in the ultraviolet photoelectron (UP) spectra. Fairly broad resonance levels ca. 5 to 8 eV below the Fermi edge E_f are characteristic for hydrogen adsorption.

This "extra" emission arises from overlap of the hydrogenic 1s wavefunction with metallic sp orbitals. However, the d electron bands are also affected by hydrogen chemisorption, as the marked suppression of metallic d-state density right at the Fermi edge indicates. An example taken from a UV-photoemission study of hydrogen on Ni, Pd, and Pt surfaces by Demuth [27] is presented in Fig. 3. Often, the hydrogen-induced resonance state is so wide and uncharacteristic that it may easily be overlooked, particularly if hydrogen is coadsorbed with surface species that produce strong emission features (CO). Comparatively sharp UP spectral peaks may be observed if a reconstruction of the metal surface is driven by adsorbed hydrogen. The reconstructed 1 × 2 phase of hydrogen on Ni(110) is an example of this type. A sharp "extra" emission peak about 1 eV below E_f is associated with the 1 × 2 reconstructed phase [28], which is discussed in more detail in Section IV.B.

2. As pointed out earlier the strength of the chemisorptive bond of hydrogen to metals, E_{Me-H}, ranges from 200 to 300 kJ mol^{-1}. Several review articles contain listings of E_{Me-H} and E_{ad} values, respectively: for example, the work by Knor [29]. In Table 2 we present some values for E_{Me-H} as obtained from single-crystal work. Also given are experimental values for the initial heat of adsorption, E_{ad}, as derived from isosteric heat or thermal desorption measurements. Where available [30] the bond energies of the corresponding diatomic metal hydride molecules are also given in Table 2. It can be seen from the table that the chemisorptive Me-H bonds follow roughly the trend of the bond energies of the hydride molecules. Moreover, the energies of hydrogen on the noble metals Cu, Ag, and Au are substantially lower than those of the typical transition metals Ni, Pd, and Pt. This again elucidates the important role of the d electrons in hydrogen adsorption: With Ni, Pd, or Pt the d band has its maximum close to the Fermi edge, whereas the noble metals exhibit their d-band maximum around 2 to 3 eV below E_f. Melius et al. [42] have put forth a theory in which they explain the role of the d electrons in the bonding. Apparently, the d functions are able to reorganize themselves so as to facilitate the dissociation process of the H_2 molecule.

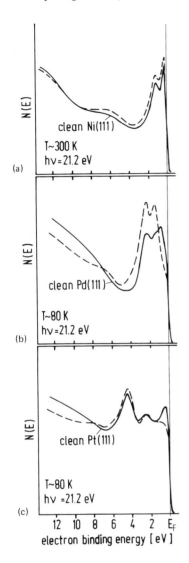

FIGURE 3 Photoelectron energy distribution curves for clean and hydrogen-covered (111) surfaces of Ni, Pd, and Pt at 300 K and 80 K, respectively. The respective H_2 exposures are (a) 6 L H_2 on Ni(111), (b) 1 L H_2 on Pd(111), and (c) 3 L H on Pt(111). (After Ref. 27.)

If the heat of adsorption ($\hat{=}$ the depth of the potential energy well) is supplied to the system, the trapped H atoms can leave their binding sites and may recombine to form the H_2 molecule, thereby releasing the heat of dissociation. This will immediately lead to desorption of the molecular entity. However, if reactive intermediate species are present on the surface (e.g., unsaturated hydrocarbons), the H atoms may rapidly be trapped by the intermediate and thus prevented from recombination and desorption.

3. The presence of surface defects (crystallographic defects or impurity atoms) may extremely sensitively influence the adsorption of hydrogen, whereby energetic effects have to be distinguished from kinetic processes in which only the rate of adsorption is affected. As noted earlier, we have systematically investigated the influence of crystallographic defects on the binding energy of adsorbed hydrogen by using a Pt(997) surface which consists of terraces of (111) orientation nine atoms wide separated by monoatomic steps in the (111) direction [23]. The adsorption of hydrogen was followed using low-energy electron diffraction (LEED), thermal desorption,

TABLE 2 Some Selected Values for Initial Heats of Adsorption ($E_{ad,o}$), Me-H Binding Energy (E_{Me-H}), and Bond Strength of Diatomic Hydride Molecules (E_{hy})

Metal surface	$E_{ad,o}$ (kJ mol^{-1})	E_{Me-H} (kJ mol^{-1})	E_{hy} kJ mol^{-1})	References
Fe(110)	109	271		18
Fe(100)	100	265		18
Fe(111)	88	260		18
Ni(100)	95	264	251	11, 12
Ni(111)	95	264		13, 11
Ni(110)	90	261		11, 28
Cu(311)	39	236	276	31
Ru(0001)	80	256		32
	120	276		19
Rh(111)	78	255		33
Rh(110)	77	255		20
Pd(100)	99	266		15
Pd(110)	102	267		16, 34
Pd(111)	88	260		16, 35
Ag(111)	ca. 15	241	222	2
W(100)	134	283		36
W(110)	138	285		36, 37
W(111)	155	294		36, 37
Pt(111)	42	237	347	21
	77	255		38, 39
	75	254		40
Au			310	
Ir(110)	77	255		41

Source: Ref. 9.

and work function measurements. It could be shown that the very first H atoms adsorb preferentially on and near the steps, thereby confirming the general trend that H atoms prefer binding in a highly coordinated site [10]. On the average the heat of adsorption on these sites is about 10 kJ mol^{-1} higher than on the terrace sites characteristic of the flat (111) surface. This holds fairly independently of the metal surface and its crystallographic orientation. In Fig. 4 we show the coverage dependence (the work function change was used as a coverage monitor) of the isosteric heat of adsorption of hydrogen on a Ni(100) surface [11], which clearly exhibits an initially higher adsorption energy, owing to the presence of about 10% defect sites.

Although the effect of impurity atoms such as carbon or sulfur becomes really pronounced at higher hydrogen coverages, the initial heat of adsorption

1. *Hydrogen Sorption on Pure Metal Surfaces*

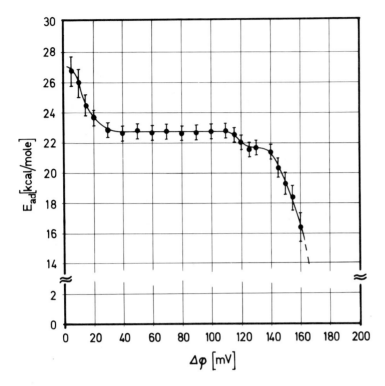

FIGURE 4 Isosteric heat of adsorption (kcal mol^{-1}) for hydrogen on a Ni(100) surface as a function of the hydrogen-induced work function change (which is a direct measure of the surface concentration of hydrogen). Note the increased E_{ad} value at small coverages, which reflects adsorption on surface defect sites. (After Ref. 11.)

is also affected. In addition, the coadsorption of gases such as carbon monoxide may lower the heat of adsorption of hydrogen by several kJ mol^{-1} as shown for Ni(110)/H + CO by Schober [43] and Penka [44]. This subject is covered in more detail in Section II.C.

As far as kinetics is concerned, the effect of defects or impurities becomes really striking. For Pt(997) we found an increase in the initial sticking probability by almost a factor of 5 [23], owing to the presence of the steps. Quite recently, Poelsema et al. [38,39] reinvestigated the Pt(111)/H system by means of He ion scattering and found the presence of steps crucial for the dissociation and sticking of the H_2 molecules. This observation fits well into the picture, whereafter the sticking probability of hydrogen on flat Pt(111) is very low [21]. The particular catalytic activity of stepped surfaces was nicely and convincingly demonstrated by angle-dependent molecular beam measurements by Somorjai's group [45]. It was found that the rate of the H_2-D_2 isotope exchange reaction (which requires the dissociation step)

$$H_2 + D_2 \rightarrow 2HD \tag{2}$$

$$H_2 + 2^* \rightarrow 2H_{ad} \quad (^* \text{denotes an empty adsorption site}) \tag{3}$$

$$D_2 + 2^* \rightarrow 2D_{ad} \tag{4}$$

$$2H_{ad} + 2D_{ad} \rightarrow 2HD + 4^* \tag{5}$$

depended sensitively on how the molecular beam was directed with respect to the azimuth of the surface. If the beam was directed toward the steps, a high HD production rate was observed, whereas a beam direction along the steps gave a minimum reaction rate. The implications of this behavior for catalytic applications are self-evident: The high sticking probability on defect sites may easily cause high local concentrations of reactive species associated with step sites, and catalytic reactions may preferentially run at and near those sites. This also provides a simple explanation for the enhanced catalytic activity of highly dispersed materials. On the other hand, as soon as the active sites become blocked by irreversible adsorption of foreign atoms (C, P, S) the overall reaction rate will be lowered and the catalyst material then regarded as deactivated or poisoned. From many of our previous gas adsorption studies involving CO, O_2, NO, or large organic molecules (CH_3OH, C_6H_6, etc.) it appears that hydrogen adsorption is particularly sensitive to defects. The very small H atom can be regarded as an excellent probe for defects and impurity concentrations on a surface.

4. This leads directly to the fourth question raised earlier, that involving the local binding geometry of a H-Me adsorption complex (see also Fig. 1 of Chapter 2). Fortunately, there were several structure determinations for H on metal adsorption systems using LEED or ion channeling techniques. Table 3 compiles the recent LEED structure analyses. The site coordination number and the metal-hydrogen bond length for several Ni/H and Fe/H chemisorption systems are presented, together with the original references. The essence of the data is twofold. First, as mentioned earlier, the H atom usually "prefers" a highly coordinated site (so far the famous H/W(100) system seems to be the only case in which the H atoms are located in twofold (bridge) sites [49]). Second, the Me-H bond length ranges between a largest value of 2.0 Å observed with Pd(110) and a smallest value of 1.64 Å found with Ni(110). These bond distances are substantially smaller than those reported for O, N, CO, or NO on transition metal surfaces. For CO on Pd(100), for example, a bond length of 1.93 Å has been determined [50]. High-resolution electron energy loss spectroscopy (HREELS) can also be employed to provide information about the local geometry of an adsorption site. Measurements of this type revealed the fourfold hollow site for the systems hydrogen on Ni(100) [51] and hydrogen on Pd(100) [52]. For surfaces with hexagonal symmetry, the threefold site is usually occupied by hydrogen, as demonstrated for Ni(111) by Ibach and Bruchmann [53], for Ru(0001) by Conrad et al. [54] and Barteau et al. [55], and for Pt(111) by Baro et al. [56] and Lee et al. [57]. Additional details about the local

TABLE 3 Binding Sites and Bond Lengths of Hydrogen Adsorbed on Transition Metal Surfaces as Determined by Low-Energy Diffraction (LEED)

Surface	Coordination of binding site	Me-H bond length (Å)	Ref.
Ni(111)	Threefold ("honeycomb")	1.84 (± 0.06)	13
Ni(110)	Threefold	1.64 (± 0.1)	46
Fe(110)	Threefold	1.75 (± 0.05)	47
Pd(110)	Threefold	2.0 (± 0.1)	48

1. Hydrogen Sorption on Pure Metal Surfaces

geometry of the hydrogen adsorption site as, for example, probed by the EELS technique are provided in Chapter 2.

5. A very important problem is how many metal atoms are involved in the bonding of one H atom, which is not very easy to address experimentally. Qualitatively, the generally observed decrease in the heat of adsorption of hydrogen with increasing coverage may not only indicate the operation of repulsive mutual interactions but may also point to the participation of an *ensemble* of metal atoms in the chemisorptive bond. More quantitative information about the ensemble size can be obtained from measurement of the uptake of strongly chemisorbed hydrogen on so-called "bimetallic" surfaces, consisting of one inert and one active component for hydrogen adsorption (e.g., systems such as Cu-Ru, Cu-Ni, or Au-Ru). Figure 5 presents an example of how the amount of strongly chemisorbed hydrogen on a Ru(0001) surface decreases with increasing surface concentration of copper on that surface [58]. An analysis based on statistical theory following the argumentation of Burton and Hyman [59] reveals at least four to five adjacent Ru atoms necessary for an unperturbed adsorption of hydrogen. This number, n, is in good agreement with reports by Yu et al. [60], who derived an ensemble size of n = 4 for hydrogen chemisorption on Cu-Ni alloys. Very recent results from our own laboratory using the hydrogen-on-Au-Ru system [61] confirm the general trend that a metal atom ensemble has to consist of approximately four atoms in order to bond a H atom with a strength characteristic of the pure metal surface. This result also casts some light on the first question raised in this section. Apparently, the theoretical approach of the "surface molecule" does not represent a very appropriate description of the real situation. Again, the consequences for "real" catalysis are important. Our

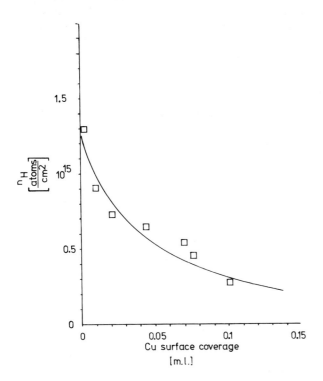

FIGURE 5 Number of H atoms chemisorbed on a Ru(0001) surface at 300 K, as a function of the surface concentration of copper atoms. (After Ref. 58.)

results with the Cu-Ru system [32, 58] not only explain the catalytic investigations by Sinfelt's group [62] but also cause one to question those studies in which the number of active surface sites of a catalyst material was "titrated' using hydrogen chemisorption. This approach is valid only if there exists a 1:1 site blocking ratio, which is not the rule for dissociative hydrogen chemisorption (see also Chapters 4 and 5).

Multiparticle Interaction

The situation of the potential energy depicted in Fig. 1 fails to describe the reality if the adsorption of a whole ensemble of H atoms is considered, as is usually the case. A very important experimental observation in this context is that chemisorbed gases frequently form phases with long-range order which give rise to "extra" features in LEED. This holds for hydrogen-adsorbed phases, too, although complications may arise from the high mobility of adsorbed H atoms and from the weak scattering power for low-energy electrons. The formation of an ordered overlayer with lattice spacings that differ from those of the metal lattice clearly indicates an induced site heterogeneity in which certain adsorption sites are given preference over others. The reason for this behavior is illustrated by means of Fig. 6, which shows the variation in potential energy in the x,y direction (i.e., *parallel* to the surface) for three different cases: (a) single-particle adsorption, (b) adsorption of four adjacent particles exerting *repulsive* interaction forces on each other, and (c) adsorption of four adjacent particles exerting *attractive* forces on each other. It is evident from this one-dimensional model that the mutual (pairwise) interaction forces between the adparticles sort of *modulate* the periodic potential of the surface so as to create sites with higher or lower binding energy for the adsorbate atom. This means that the H-Me binding energy, E_{Me-H}, and therefore the heat of adsorption, E_{ad}, will be influenced by both the concentration and the configuration of the adsorbed particles. Again, as in the zero-coverage approach, several questions can be put forth that provide a better understanding of the principal difficulties:

1. What is the chemical origin of the lateral interactions (which we shall henceforth denote by the symbol ω)?
2. What is the magnitude of these interaction forces, what is their sign (repulsive or attractive), and how do they decay as the distance between two interacting H atoms varies? Are there many-particle interactions (three-body interactions, etc.)?
3. Which is the critical coverage for the lateral interactions to occur? (This is related to question 2.)
4. What is the maximum possible coverage of hydrogen in the adsorbed layer? This actually implies the minimum possible distance between two adjacent H Atoms.
5. Is there any formation of long-range periodicity under the given conditions of coverage and temperature? (This question involves the entire field of phases and phase transitions, including the statistical treatment.)

Many experimental and theoretical investigations have addressed these questions. Experimentally, measurements of the heat of adsorption as a function of hydrogen coverage as well as of the phase transition behavior of hydrogen phases with long-range order are most suited to obtaining information on the magnitude and sign of the lateral interaction forces.

1. Clearly, quantum mechanics has to be invoked to explain the origin of the lateral interactions between adjacent adparticles. Two different cases

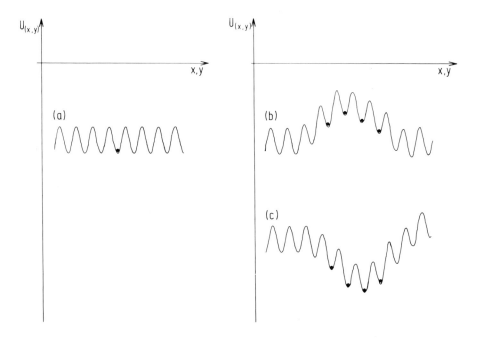

FIGURE 6 Lateral variation of the potential energy of a metal surface: (a) single-particle interaction; (b) multiparticle interaction of repulsive character; (c) multiparticle interaction with attractive character.

have to be distinguished: (a) There may be a direct overlap (interference) of atomic or molecular orbitals of the adsorbed particles, which usually leads to repulsive interactions. These are called "direct interactions." They play a role in the adsorption of larger atoms (O, N, Ar, Xe) and of molecules (CO, NO, NH_3, etc.), with the consequence that a sort of "site blocking" occurs. For example, it is known from many investigations of CO adsorption that the closest mutual distance between adjacent CO molecules in the adsorbed state is about 3 Å—completely in line with the known van der Waals radii of the gas-phase CO molecule. (b) As soon as more than the one or two metal atoms underneath the chemisorbed particle are involved in the chemisorptive bond, the adsorption of another particle in the direct vicinity of the given adatom requires that the electronic charge of the metal atoms underneath is shared between the two chemisorptive bonds. This simply means that the strength of the one bond is affected by the vicinity of the other. Since these interaction forces are actually mediated or conducted *through* the metal surface, they are called "through-bond" interactions or "*indirect*" interactions. There has been a large number of quantum mechanical calculations concerning this subject; details can be found in the literature [63-68]. For the special case of hydrogen adsorption we have to recall the small size of the adsorbed H atom: Before any direct repulsions due to orbital interpenetrations can become effective, the indirect interactions dominate. The small H atom has to be regarded as a strong perturbation of the electronic structure of the metal; theoretical calculations have shown [66] that there may not be either repulsive or attractive forces *alone* but also—dependent on the distance—they may oscillate, thus promoting certain adsorption sites and inhibiting others.

2. The magnitude of the interaction forces may be estimated from the rule of thumb that they are about one-tenth of the heat of adsorption. For the

common transition metal surfaces this means that ω ranges between 5 and 10 kJ mol^{-1}. As mentioned earlier, this information can be obtained, for example, from the coverage dependence of the heat of adsorption. As soon as the lateral interactions become noticeable, the heat of adsorption drops (in the case of repulsions) or increases (in the case of attractions). Repulsive interactions are the rule at higher coverages, and there are many examples that show a drop of $E_{ad}(\theta_H)$ when the hydrogen coverage is increased beyond a critical value. Figure 7 shows some selected examples taken from previous measurements of the coverage dependence of the isosteric heat of adsorption. Note that E_{ad} is a differential molar quantity. The system hydrogen on Ni(111) is a particularly suitable case to consider in this context. Exactly at a coverage of 0.5 monolayer, θ = 0.5, all H atoms are located in a honeycomb-like hexagonal network (Fig. 8) which produces a c2 × 2 structure in LEED. The (to the first approximation *equivalent*) threefold adsorption sites are occupied such that each H atom is surrounded by three first neighbors at a distance of 2.87 Å. Clearly, the addition of any new H atom beyond θ_H = 0.5 will affect and perturb this ordered configuration significantly; in other words, pronounced repulsive mutual interaction forces will destroy the long-range order very effectively. In the $E_{ad}(\theta)$ relation (Fig. 7) this shows up as a drop in the heat of adsorption around θ_H = 0.5. These repulsions will finally dominate the entire adsorption process and lead to a dramatic decrease of E_{ad} into a regime where the adsorption is so weak that no *chemisorption* can occur anymore. For hydrogen on Ni(111) the maximum attainable coverage is about 0.8 monolayer [13], although, in principle, there are more "possible" adsorption sites on Ni(111). On Pd(100), where a c2 × 2 hydrogen phase is formed [15], a hydrogen coverage of almost 1 monolayer can be reached until the heat of adsorption drops. In Fig. 9 we present the corresponding structure models for the c2 × 2 phase at θ_H = 0.5 (a), and for the 1 × 1 phase at

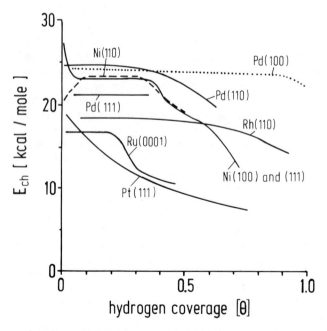

FIGURE 7 Compilation of the coverage dependence of the heat of adsorption for various metal-hydrogen chemisorption systems.

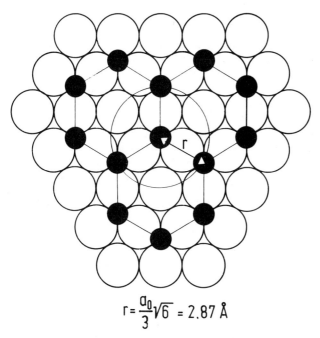

$$r = \frac{a_0}{3}\sqrt{6} = 2.87 \text{ Å}$$

FIGURE 8 Structure model of the c2 × 2-2H phase observed below 270 K for hydrogen adsorption on Ni(111). Indicated is the closest mutual H-H distance, r_{H-H}.

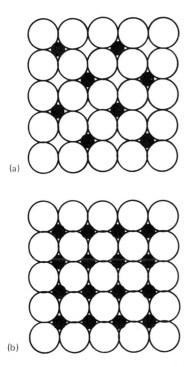

FIGURE 9 Structure model for the two hydrogen phases observed below 260 K on a P(100) surface; (a) c2 × 2 phase at $\theta_H = 0.5$ monolayer; (b) 1 × 1 phase at $\theta_H = 1.0$ monolayer.

$\theta_H = 1.0$ (b), in which *all* fourfold sites are occupied by H atoms. The Pd(100)/H is also a good example for the occurrence of attractive interaction forces. However, this information cannot be achieved simply from isosteric heat measurements. Instead the phase transition behavior of the c2 × 2 phase has to be investigated experimentally and interpreted theoretically. For the sake of completeness, a brief summary is given here; for details, readers are referred to the original work [69-71]. If a Pd(100) surface is exposed to H_2 at a temperature below 260 K, LEED "extra" spots of a c2 × 2 pattern appear which reach their maximum intensity at 1/2-monolayer coverage; thereafter they decay again and disappear around $\theta_H = 0.9$. As can be shown [68], the intensity of a LEED spot can be taken as a good measure of the long-range order within an overlayer. This intensity samples all H atoms which are located in the "right" adsorption sites of Fig. 9a. It is evident that only at T = 0 K are all H atoms located in the minima of the potential energy curve of Fig. 6; higher temperatures introduce thermal mobility so as to enable a certain fraction of the H atoms to migrate or "hop" to adjacent adsorption sites. At a certain temperature, the *critical* temperature T_c, long-range order no longer exists; the phase has suffered an order-disorder transition. This critical temperature sensitively depends on the magnitude of the lateral interaction energies ω. The statistical mechanical treatment (with the formalism of the two-dimensional Ising model applied [71] to the square-lattice geometry) yields the following relation between T_c and ω [68]: $\omega = 1.76\ k_B T_c$. From this the magnitude may be estimated for the Pd(100)/H system with $T_c = 260$ K to be about 5.5 kJ mol^{-1}, which is a crude number only since a detailed study of the coverage dependence of the transition temperature actually reveals that the phase transition cannot be described

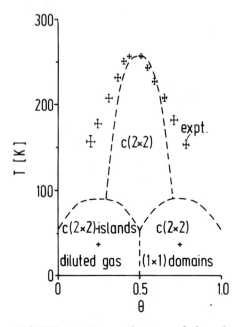

FIGURE 10 Phase diagram of the c2 × 2 structure observed with H on Pd(100) at $\theta_H = 0.5$ monolayer. The experimental data points are marked with their error bars; the dashed lines refer to theoretical predictions by Binder and Landau [70]. Note the extended width of the experimental phase diagram.

TABLE 4 Ordered Chemisorbed Hydrogen Phases on Metal Surfaces Without Surface Reconstruction

Metal surface	Phase	Hydrogen coverage (monolayers)	Critical temp., T_c (K)	Refs.
Fe(110)	2 × 1	0.5	245	73
	3 × 1	0.67	265	73
Ni(111)	c2 × 2-2H	0.50	270	13
Ni(110)	2 × 3-1D	0.33		8
	c2 × 4	0.50		8
	c2 × 6	0.65		8, 74
	c2 × 6	0.83		74
	2 × 1-2H	1.0	≤ 200	8, 14, 74
Pd(100)	c2 × 2	0.50	260	15, 87
Pd(110)	2 × 1-2H	1.0		17
Pd(111)	p$\sqrt{3}$ × $\sqrt{3}$R 30°	0.33	~80	75
	c$\sqrt{3}$ × $\sqrt{3}$R 30°	0.67	~100	75
Rh(110)	p1 × 3	0.33	~140	20
	p1 × 2	0.50	~180	20
	1 × 3-2H	0.67	~220	20
	1 × 2-2H	1.5	~190	20
	1 × 1-2H	2.0	<180	20
Co(10$\bar{1}$0)	c2 × 4	0.5	270	76
	2 × 1-2H	1.0	290	76

sufficiently well by the Ising formalism [15, 69, 70]. The complete *phase diagram* of the c2 × 2 structure (Fig. 10) was obtained by measuring the fractional-order LEED intensity as a function of temperature for various (constant) hydrogen coverages. Compared to the theoretical predictions [70], this phase diagram has an extended width; that is, ordered islands of c2 × 2 are formed at too-low coverages. This is always indicative of the operation of *attractive* interaction forces, which cause an island formation or a clustering of particles at a very low coverage. An increasing number of hydrogen adsorption systems exhibit phases with long-range order [72]. These cases are listed in Table 4.

3. The third question—as to the critical coverage for lateral interactions to occur—has already been answered, but it is interesting to point out that for certain hydrogen adsorption systems (e.g., Ni(110)/H [8, 74]), the mutual interactions become observable even at coverages as low as 0.1 monolayer and then lead to structures with extremely large unit cells. In the case of Ni(110) the formation of a c2 × 6 structure has been reported by Rieder and Engel [77]. In this phase, an adsorption site about six lattice spacings away is affected by the adsorption of a given H atom. The model is shown in Fig. 11. This behavior again demonstrates the *long-range* character of the chemisorptive Me-H bond: Apparently an isolated adsorbed H atom can affect the adsorption properties of surface atoms which are 10 Å away. This fits well into the picture of the delocalized nature of the Me-H bond.

4. From all the ordered hydrogen phases known so far and listed in Table 4, the impression arises that two H atoms cannot be brought closer

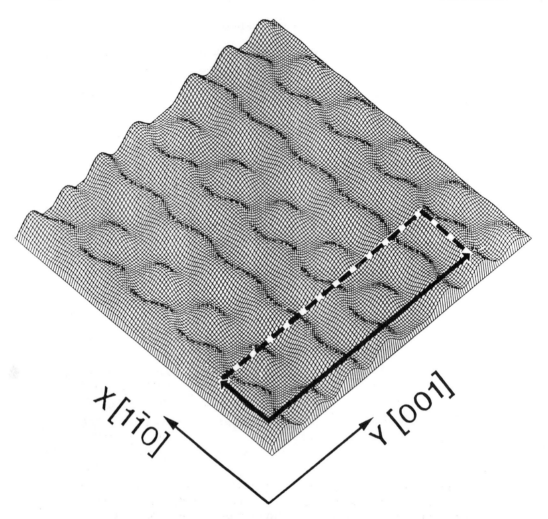

FIGURE 11 Perspective view of the best-fit corrugation function for the H c2 × 6 phase observed by He diffraction from Ni(110). The unit cell is indicated. (After Ref. 77; reproduced with permission.)

together than about 2.7 Å, as in the 1 × 1 phase on Pd(100). However, very recent investigations in our laboratory at 70 K on the H/Rh(110) system revealed the surprising result that after prolonged exposure to H_2, a 1 × 1 phase is formed which contains *two* nominal hydrogen monolayers [20] in which adjacent H atoms have a mutual distance of approximately only 2 Å. Of course, this phase is merely weakly bound in the chemisorbed state and becomes destroyed at 120 K by thermal desorption. So, taking all the information together, the maximum possible density of hydrogen adsorbed phases is about 2×10^{15} atoms cm^{-2} [for the extreme case of the Rh(110) surface]; a more "normal" value would be about 1.5×10^{15} atoms cm^{-2}.

5. Ordered adsorbed overlayers represent interesting model systems for statistical mechanics. The simplest concept to calculate or predict the configuration of an ensemble of interacting particles is the lattice gas model with pairwise interactions only. Adsorbed hydrogen phases frequently obey the conditions of a lattice gas and have thus been used extensively in Monte

Carlo calculations (see, e.g., Ref. 78). However, apart from a theoretical calculation of interaction parameters, it is believed that the implications on catalytic phenomena are not significant and it is not deemed useful to enter this field, which is covered in many excellent review articles (see, e.g., Ref. 79).

So far, the interaction of the metal surface with an ensemble of particles has been restricted to the case in which the adsorption does not affect the *structure* of the metal itself. However, as precise LEED structure determinations of clean adsorbate-covered surfaces demonstrate, the adsorption of gases frequently gives rise to minor or major displacements of the topmost metal atoms from their equilibrium positions (relaxation and reconstruction, respectively). This may not only affect the heat of adsorption and thus the equilibrium coverage under certain pressure and temperature conditions but may also strongly influence the dynamics of adsorption, as pointed out further below [80].

Clearly, relaxation and reconstruction processes represent the tendency of an adsorption system to lower its total free energy. In this sense a clean single-crystal surface is always an energetically unfavorable system because for the atoms on top, one-half of the neighbors are missing. This "clean" system usually lowers its total energy in that the perpendicular distance between the first and second layers deviates from the typical bulk value; normally, contraction is observed. However, as soon as adsorbate atoms saturate the free valencies of the topmost atoms, this contraction is removed. Strongly interacting gases (O_2, N_2, H_2) in some cases even lead to larger displacements of the surface atoms (reconstruction), a situation that may end with the formation of a surface compound (oxide, nitride, hydride). This is why we discuss these phenomena in greater detail in Section IV.

B. Dynamics of Adsorption

Rate of Adsorption

From the kinetic-molecular theory of gases the number of particles incident per unit time on a surface with unit area is expressed as the flux ϕ:

$$\phi = \frac{dn}{dt} = P(2\pi m k_B T)^{1/2} \tag{6}$$

where P is the gas pressure, n the particle number, and k_B is Boltzmann's constant. As in a chemical reaction, not all of the impinging particles remain chemisorbed on the surface, but a major fraction may be reflected directly into the gas phase. The ratio between the particles that have actually "stuck" to the surface and the incident particles is denoted as the sticking coefficient. This sticking probability usually depends strongly on the number of adsorbed particles (coverage θ), that is, $s = s_0 f(\theta)$, with s_0 being defined as the initial sticking probability. The analysis of the function $s(\theta)$ leads to certain conclusions about the mechanism of adsorption, whereas the value of s_0 (which lies between 0 and 1) contains all the information about the mechanism of trapping of a particle on the surface. With the sticking probability introduced, the rate of adsorption takes the general form

$$\frac{dn_{ad}}{dt} = s_0\, f(\theta_H)\, \exp\left(-\frac{E^*_{ad}}{kT}\right) \cdot \phi \quad \frac{\text{particles}}{m^2 \cdot s} \tag{7}$$

(E^*_{ad} denotes the activation energy for adsorption), which for nonactivated adsorption simplifies to the expression

$$\frac{dn_{ad}}{dt} = s_0 \, f(\theta_H) \, \Phi \quad \frac{\text{particles}}{m^2 \cdot s} \tag{8}$$

For hydrogen molecules H_2 we obtain for Φ at 300 K the following relation to the gas pressure, p_{H_2}:

$$\Phi_{H_2} = 1.0717 \times 10^{23} p_{H_2} (Pa) \quad \frac{\text{particles}}{m^2 \cdot s} \tag{9}$$

If the H_2 molecule would stick and dissociate with unity probability, an atomic hydrogen monolayer (ca. 10^{19} atoms m^{-2}) would be built up after approximately 50 s if the hydrogen pressure is taken as 10^{-6} Pa.

Before we discuss the sticking probability-coverage function (which is very important in heterogeneous catalysis since it governs the rate of uptake of hydrogen under high-pressure conditions) it is worthwhile to expand somewhat on the interpretation of the initial sticking coefficient, s_0. The question of whether an incident particle sticks (i.e., remains on the surface for much longer than just one vibrational period in the potential energy well) is crucially dependent on the shape (softness and depth) of the potential, the total kinetic energy of the incident particle, its mass, and of course, its chemical nature. In principle, the particle has to transfer its kinetic energy to the surface; similarly, the heat of adsorption released in the chemisorption reaction must be dissipated by the surface. As pointed out elsewhere [81], there are at least two possible channels for this energy accommodation: Excitation of metallic phonons (then the crystal lattice acts as a heat bath) or the creation of electron-hole pairs near the Fermi surface of the metal (electronic excitations). According to Brenig's model (see Ref. 82 and the references therein) the phonon coupling mechanism should be particularly effective for incident particles with a high atomic mass, whereas light atoms or molecules such as H_2 are not expected to transfer their kinetic energy effectively to the surface by this mechanism. Experiments have been reported in the literature that tried to distinguish between these two mechanisms; the results are controversial. Whereas Strongin et al. [83] find evidence for the electron-hole-pair mechanism by means of uptake and photoemission measurements on a niobium-palladium bilayer, recent molecular beam scattering experiments utilizing H_2 and D_2 by Ertl's group [7] do not support the electron-hole mechanism. Rather, differences in the elastic and direct inelastic scattering behavior of H_2 and D_2 (which differ in mass by a factor of 2) support the phonon excitation mechanism [84]. The absolute numbers of the sticking probabilities s_0 range from 0.01 up to 1.0; numbers significantly smaller than 1.0 dominate in the literature. Table 5 gives some selected examples of s_0 for some H-Me adsorption systems. A more detailed list of s_0 values is given in the article by Morris et al. [37].

If the general trend of the numbers of Table 5 is compared with a similar list of s_0 numbers, such as those for carbon monoxide, it is quite striking that on the average the hydrogen sticking probabilities are markedly smaller than those for CO. It is quite obvious that the requirement for dissociation of a diatomic molecule such as H_2 introduces another probability factor; in other words, a certain spatial orientation of the incident molecule may be required for an easy dissociation path. Even if the initial sticking probability is very high, there is, for dissociative chemisorption, a fairly rapid decrease in available ("free") adsorption sites with increasing gas exposure. In the simplest case this decrease can be described by the function $f(\theta) = (1 - \theta)^2$, since each dissociation and adsorption event consumes two empty sites. There

TABLE 5 Some Selected Values of Initial Sticking Probabilities, s_0, for Hydrogen on Metal Surfaces

Metal surface	Initial sticking coefficient, s_0	Temperature (K)	Refs.
Ni(111)	0.1	150-300	7, 13
Ni(110)	0.96	120	7, 8
Ni(100)	0.06-0.1	150	12
Pd(111)	0.5	80	35, 75
Pd(110)	0.7	120	17, 34
Pd(100)	0.5	120	15, 85
Fe(110)	0.16	150	18
Ru(0001)	0.2-0.3	120	19, 32
W(100)	0.17-1.0	300	37
W(110)	0.3-0.45	300	37
Pt(111)	0.1	150	21
Pt(997)	0.4	150	23
Ir(110)	1.0 (β_2 state)	130	41
	0.007 (β_1 state)	130	41
Rh(110)	1.0	85	20
Co(10$\bar{1}$0)	1.0	120	76

are some examples in the literature where this relation has been verified [18]. In many cases, however, the function $f(\theta_H)$ is more complicated, even if the influence of surface defects is neglected. The reason for this is the existence of a weakly bound molecular state at a greater distance from the surface, the so-called "precursor" state. This means that an adsorbing H_2 molecule will first pass or enter this van der Waals-like state before it can enter the chemisorption potential energy well, according to the reaction scheme

$$[H_2]_g \underset{k_{-1}}{\overset{k_1}{\rightleftharpoons}} [H_2]_{prec} \underset{k_{-2}}{\overset{k_2}{\rightleftharpoons}} 2[H]_{ad} \qquad (10)$$

and the rate of uptake of atomic hydrogen actually depends on the ratio of the rate constants k_1/k_2. k_{-1} can be associated with the reflection of a H_2 molecule from the precursor state; and k_{-2} will govern the rate of desorption from the chemisorbed state. It has to be kept in mind that owing to the validity of the *detailed balancing* or microscopy reversibility concept, states passed during the *adsorption* are also passed during *desorption*. Many attempts have been made to model the precursor state and the corresponding adsorption/desorption kinetics. The most famous model was developed by Kisliuk [86], who derived the rate expression on the basis of the function

$$f(\theta) = \frac{(1-\theta)^2}{1 - \theta(1-K) + \theta^2 s_0} \qquad (11)$$

The constant K contains the probabilities for desorption from a precursor site with a chemisorbed atom underneath divided by the probabilities of desorption from a precursor with an empty site underneath plus that of adsorption into the chemisorbed state. These coverage functions are plotted in Fig. 12 for $s_0 = 1.0$ and varying the constant K. K = 1 means that no precursor exists. An experimental example is provided by the Pd(100)/H system as demonstrated in Fig. 13; clearly, the $s(\theta)$ function exhibits the typical convex chape characteristic of a precursor state. Also, hydrogen on Ni(110) [8], Pd(110) [17], and Rh(110) [20] give evidence of mobile precursor states.

Qualitatively, the existence of a precursor state shows up by a sticking probability which is more or less constant over an appreciable coverage range. Apparently, the incident particle is trapped in the mobile precursor state for a long enough time to search successfully for an empty adsorption site in which it will finally be trapped. In terms of quantum mechanical interaction [81,87], a precursor state represents a molecule with little more than physisorption energy, which nevertheless influences the charge density of the metal surface in its vicinity.

From the viewpoint of heterogeneous catalysis, the possible existence of weakly bound precursor states may be of great significance. Particularly at higher reactant pressures those states may become filled and their population rate determining, respectively. The $s(\theta)$ function may be even more complicated in cases where structural changes in the metal substrate occur under the influence of the adsorbing gas which then sensitively influence the sticking probability. A good example is the reconstruction of the Pt(100) surface during the adsorption and reaction of CO and O_2. The two possible structures of the Pt(100) surface exhibit marked differences in the sticking probability of oxygen. It was demonstrated in various exciting studies by Ertl and collaborators [88] that kinetic instabilities and even oscillations may then occur on Pt(100) for the reaction $CO + \frac{1}{2}O_2 = CO_2$. Although an equally striking case has not been found so far for H_2 adsorption, there are some pronounced effects found with the Ni(110)/H system, which clearly show the relation between surface structure and sticking probability [89]. As described

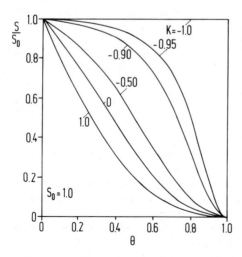

FIGURE 12 Plot of the Kisliuk function, $\frac{S}{S_0} = (1 - \theta)^2 [1 - \theta(1 - K) + \theta^2 s_0]^{-1}$ for the case $s_0 = 1$ and various (indicated) values for K, according to a second-order precursor state model. (After Ref. 86.)

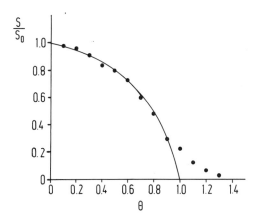

FIGURE 13 Experimental example for a hydrogen precursor state: s/s_0 for hydrogen adsorption on Pd(100) as a function of θ_H. The data points are connected by a line fit according to a first-order precursor state model. (After Ref. 15.)

in more detail in Section IV.B, there is a reconstructive phase transformation with Ni(110) in the presence of hydrogen at T = 120 K in which a nonreconstructed 2 × 1-2H phase transforms to a reconstructed 1 × 2 phase. In this phase paired rows of Ni surface atoms are formed along the [110] direction as soon as θ_H exceeds 1 monolayer. In the $s(\theta_H)$ curve shown in Fig. 14, this gives rise to a plateau around $\theta_H = 1.0$.

Rate of Thermal Desorption

The kinetics of thermal desorption of particles from a homogeneous surface is frequently described by the Wigner-Polanyi equation:

$$-\frac{dn_{ad}}{dt} = k n_{ad}^x \qquad (12)$$

where k is a rate constant and x is the reaction order. In terms of a simple Arrhenius model, we obtain for the temperature dependence of k,

$$k(T) = \nu_{(x)} \exp\left(-\frac{E_d^*}{kT}\right) \qquad (13)$$

where ν denotes the "frequency factor," which is discussed further below, and E_d^* stands for the activation energy of desorption, which for nonactivated adsorption, equals the depth (E_{ad}) of the chemisorption potential well of Fig. 1. Although it is impossible within the scope of this chapter to consider all the aspects and models developed to analyze thermal desorption data nor to get into the quantum statistical theory, some very brief remarks ought to be made.

In principle, thermal desorption is a very complicated process since it comprises the transfer of the thermal energy of the solid (phonons) to the adsorbed molecule. There is, of course, a whole "ladder" of vibrational states to which the bound particle can be excited before it will leave the adsorption potential. However, for most purposes of phenomenological thermal desorption spectroscopy a few simple models are quite appropriate (e.g., the Redhead model [90]). Within its (certainly crude) approximation, both E_d and

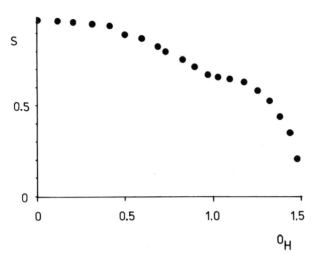

FIGURE 14 Structure sensitivity of the sticking probability-coverage function: hydrogen on Ni(110) at T = 120 K is given as an example. Note the plateau region around θ_H = 1 monolayer.

the preexponential ν are assumed coverage independent, which fails to describe the situation correctly at higher coverages (e.g., when mutual particle interactions come into play). Nevertheless, the Redhead model is quite helpful in getting preliminary information about the activation energy for desorption, which is the most interesting quantity. A more advanced approach that considers the coverage dependences of E_d^* and ν, known as line-shape analysis, has been described by King [91] and Bauer et al. [92].

It is obvious that the mechanism of desorption strongly governs the shape of a thermal desorption peak (the cases considered here consist of a linearly temperature-ramped crystal mounted in a vacuum system with infinite pumping speed for the desorbing gas). Here we have to distinguish between processes of zero, first, and second order (x = 0, 1, and 2, respectively). For desorption from a *constant number* of desorption centers, a zero order is always observed; one-point (molecular) desorption leads to a first-order process, whereas a thermal desorption in which the recombination of individual particles to the desorbing molecule is rate limiting obeys a second-order kinetics. Therefore, the desorption of molecular hydrogen should clearly be second order, and indeed, this is most commonly observed, for example with the system H/Ru(0001), as displayed in Fig. 15. Sometimes, however, there are pronounced deviations from this "normal" desorption path. Again, the Ni(110)/H system may be taken as an example. We have already mentioned the reconstructed 1 × 2 phase (which saturates at θ_H = 1.5 monolayers). This phase is thermally unstable and decomposes spontaneously around T = 220 K [93], thereby loosing about 0.5 monolayer of adsorbed hydrogen within a very short period. In this process, the recombination of individual atoms to H_2 molecules is no longer rate determining, but the dissolution of 1 × 2 islands from their perimeters. This causes a reaction order which is close but not equal to zero, which means that there is an exponential increase of the rate of desorption which, at a certain temperature, suddenly drops to zero. This behavior can be seen from Fig. 16, in which a family of thermal desorption curves is shown for a Ni(110) surface, covered at 120 K with increasing amounts of hydrogen. Note the sharp α state associated with breakdown of the 1 × 2 phase. Sometimes similar thermal desorption peaks

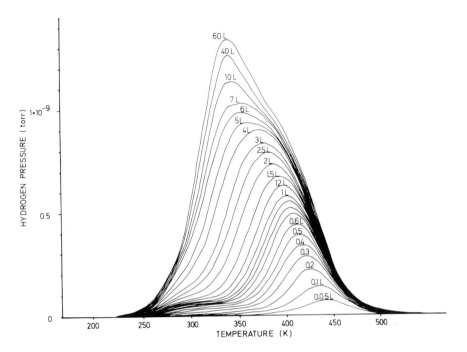

FIGURE 15 Example for a second-order kinetics in the thermal desorption reaction. Shown is a family of TPD curves obtained with the hydrogen-on-Ru(0001) adsorption system. The temperature shift of the peak maximum toward lower temperatures with increasing coverage indicates a second-order rate process. (After Ref. 32.)

occur in the course of structural phase transformations (in which adsorption sites are annihilated) or during surface reactions (e.g., decomposition reactions), as demonstrated some time ago by Falconer and Madix [94].

At any rate, the activation energy follows fairly directly from the temperature position of the thermal desorption peak. The frequency factors are somewhat more difficult to obtain in an experiment. In order to understand the meaning of ν, it is again helpful to utilize a simple theory, namely, transition state theory, which has been proven to describe the path of a thermal desorption reaction quite well. For more detailed considerations the article by Laidler [95] is recommended. In transition state theory (TST) ν is expressed

$$\nu = \frac{kT}{h} \frac{f^{\ddagger}}{f^2_{ad}} \kappa_d \tag{14}$$

where kT/h is the universal frequency factor 6.25×10^{12} s^{-1} at 300 K; $f^{\ddagger}$ the partition function of the desorption complex [(the H_2 molecule "ready" for desorption), i.e., of the gaseous molecule with 1 vibrational degree of freedom (= kT/h) removed]; f_{ad} the partition function of the adsorbed H atoms; and κ_d the transmission (tunneling) factor (close to 0.1 for H_2). f_{ad} may be split up into the contributions from the two-dimensional translational motion and from the vibration v of the adsorbed atom with respect to the solid surface. $f^{\ddagger}$, on the other hand, contains the normal translational, rotational, and vibrational partition functions. With the molecular parameters known, one can fairly easily calculate the contributions of the various parti-

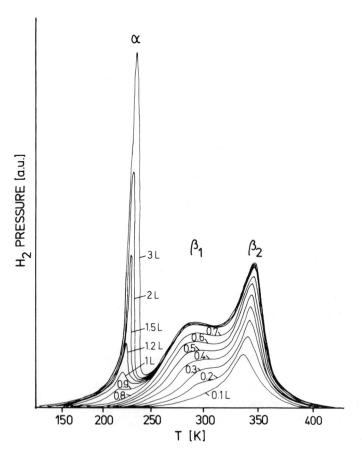

FIGURE 16 Series of thermal desorption curves obtained from a Ni(110) surface after various hydrogen exposures at T = 120 K. The sharp α state is directly related to the occurrence of the 1 × 2 reconstructed surface phase. Note the very narrow width of the α state and the unusual desorption kinetics (peak shift toward *higher* temperatures with increasing coverage). (After Ref. 28.)

tion functions to f_{ad} at a given temperature. Here one has to distinguish between a *mobile* and an *immobile* adsorbate. In the first case, f_{ad} is simply equal to the number of available surface sites, N_s, because there exist N_s possibilities to distribute a given particle on a surface with N_s sites. For completely mobile adsorption, on the other hand, f_{ad} is equal to the partition function of the two-dimensional translation: $f_{ad} = 2\pi m k T/h^2$. This approximation holds as long as mutual particle interactions are not significant; at higher coverages the vibrational and configurational states of the adsorbed layer may easily change; that is, the preexponential factor will become coverage dependent. Also, the occurrence of precursor states in the adsorption path will be felt in the desorption. That this is the case was shown some years ago by King [96] and Gorte and Schmidt [97].

Numerically, the frequency factors for hydrogen desorption range between 10^{-5} and 10^0, as can be seen from Table 6. The dimension of ν_2 is cm² (H atoms · s)$^{-1}$; simple TST yields for *immobile* H_2 adsorption a value of 5×10^{-1} cm² (H atom · s)$^{-1}$ and for *mobile* adsorption about 1×10^{-3} cm² (H atom · s)$^{-1}$.

TABLE 6 Experimentally Determined Frequency Factors, ν_2, for H_2 Desorption from Metal Surfaces

Metal surface	Frequency factor [cm^2 (H atom s)$^{-1}$]	Refs.
Ni(100)	8×10^{-2}	11
	2.5×10^{-1}	144
	3×10^0	12
Ni(111)	2×10^{-1}	11
	2.3×10^{-2}	13
Pd(100)	1×10^{-2}	15, 85
Pd(111)	1.3×10^{-1}	35
Ru(0001)	4×10^{-3}	32
	5×10^{-2} (state 2)	19
	3×10^0 (state 1)	19
Pt(111)	3×10^{-9}	21
	10^{-3}	98
Pt(997)	6×10^{-8}	23
	10^{-2}	38
Ir(110)	1.5×10^{-2} (β_2)	41
	2×10^{-7} (β_1)	41
Rh(111)	1.2×10^{-3}	33

Table 6 shows that (apart from some exceptions) TST describes the reality fairly well.

Although this section is devoted to desorption *kinetics*, it may be worthwhile here to add some remarks on the energetic quantity, E_d^*, of Eq. (13) because it is, aside from the "kinetic" parameter ν, the most important quantity to be evaluated in a TPD experiment. As mentioned in Section II.B, the isosteric heat of adsorption, E_{ad}, can be identified with E_d^* if no activation barrier for adsorption is involved (i.e., $E_{ad}^* = 0$). In a thermal desorption experiment the existence of various "states" of adsorbed hydrogen is most easily indicated by the corresponding TPD *peaks*. In that sense weakly held H atoms require a small thermal energy to be desorbed and therefore produce a TD peak at low temperatures, whereas tightly bound atoms leave the surface only after an appreciable thermal energy is supplied to the crystal and thus appear at the high-temperature end of the TPD spectrum. A convincing example is provided by the Rh(110)/H system, which is illustrated in Fig. 17: A "superhigh-density" hydrogen phase of 2×10^{15} atoms cm^{-2} (with strong repulsive forces between the adsorbed H atoms) desorbs around 130 K (α_1 state) ($E_d^* = 33$ kJ mol^{-1}), whereas a less densely packed phase (1.5×10^{15} atoms cm^{-2}) desorbs only around 230 K (α_2 state). Lower concentrations of adsorbed hydrogen on Rh(110) ($\theta < 1.0$) give rise to the "normal" second-order desorption peak around 300 K (β state) [20].

The existence of various hydrogen binding states does not always show up so clearly in the TPD spectra. It is quite common that the second state (induced by lateral interactions) appears only as a shoulder on the low-T side of the main TPD peak. This is the case, for example, for hydrogen on Pt(111); a typical set of TPD spectra is reproduced in Fig. 18.

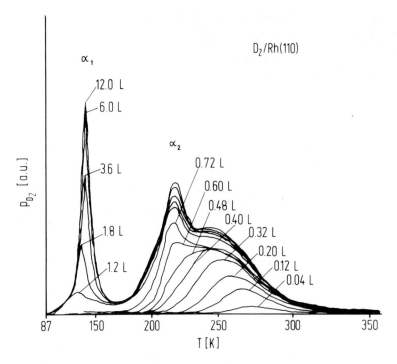

FIGURE 17 Set of thermal desorption spectra obtained after increasing amounts of exposed deuterium on a Rh(110) surface at T = 85 K. The sharp states are closely related to the adsorbate structure. (After Ref. 20.)

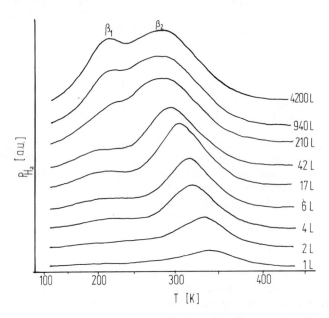

FIGURE 18 Thermal desorption spectra of hydrogen from a clean Pt(111) surface which was exposed to increasing amounts of hydrogen gas at T = 120 K. The peak shift toward lower temperatures with increasing coverages indicates a second-order desorption mechanism. The shoulder at high coverages is due to the operation of repulsive mutual H-H interactions. (From Ref. 21.)

Again, a warning may be added here as to the significance of E_d^* values derived from TPD measurements: These numbers are *always* tied to certain *kinetic* models of the desorption process; depending on the applied model quite different values for E_d^* and ν may be obtained. Again, H on Pt(111) may serve as an example: Assuming a strictly second-order desorption kinetics, Christmann et al. [21] arrived at $E_d^* = 40$ kJ mol^{-1} and $\nu \approx 10^{-8}$ cm^2/(molecule · s)$^{-1}$. If, however, a first-order desorption was admitted, as by McCabe and Schmidt [98], more reasonable numbers ($E_d^* = 70$ kJ mol^{-1} and $\nu \approx 10^{-2}$ cm^2/(molecule · s)$^{-1}$) were evaluated.

Finally, we should mention a recent work by Bowker et al. [37] in which a wealth of hydrogen TPD spectra are reproduced for a variety of hydrogen on metal systems; this may serve as a further source of references.

Surface Diffusion of Hydrogen

Probably one of the first attempts to investigate the surface diffusion of adsorbed hydrogen was carried out by Gomer's group [99], who watched a platinum field-emission tip, half of which was covered with adsorbed hydrogen. It was found in these experiments that the hydrogen front was constantly moving and expanding even at temperatures as low as 150 K. Since then more refined (mostly field-emission) measurements of the surface diffusion have been made (e.g., by Difoggio and Gomer with W(110) [100]). Commonly, small activation energy barriers for surface diffusion are observed if the "normal" mechanism applies. For W(110)/H, E_{diff}^* values around 18 kJ mol^{-1} were reported [100]; the diffusion coefficients, D, were given as 5×10^{-5} cm^2 s^{-1}. However, just for H on W(110), DiFoggio and Gomer found that a different diffusion path is chosen by hydrogen below 140 K, namely, *tunneling*. This process is nearly not activated and occurs at a much faster rate. It appears from Gomer's investigation that tunneling processes will run preferentially at very low temperatures. Since catalytic reactions are probably not carried out below 200 K, we shall deal briefly with surface diffusion of hydrogen according to the Arrhenius equation:

$$D = D_0 \exp\left(-\frac{E_{diff}^*}{kT}\right) \tag{15}$$

where D is the diffusion coefficient at T as defined by Fick's law in terms of the time dependence of the concentration and the concentration gradient:

$$D = -\frac{(\partial N/\partial t)_x}{(\partial N/\partial x)_t} \tag{16}$$

Thus E_{diff}^* is obtained from measurements of D at different temperatures. D, on the other hand, may be accessible from FEM measurements by making use of the relation

$$\bar{x} = (dt)^{1/2} \tag{17}$$

where $\bar{x}$ is the mean distance traveled after time t. To understand the mechanism of surface diffusion, it is worthwhile to refer to Fig. 6 again. Surface diffusion is nothing but a sequence of hopping events of H atoms from one potential minimum to the next. Since this is an activated process, supply with thermal energy will increase the number of such events until the entire layer is mobile. The residence time of a particle in a potential minimum is then given by

$$\tau' = \tau_0' \exp\left(\frac{E^*_{diff}}{kT}\right) \tag{18}$$

with E^*_{diff} estimated to be 16 kJ mol^{-1} (general rule: $E_{diff} \simeq 1/10$ of E_{ad}) and $\tau_0' \simeq 10^{-13}$ s, the residence time of a H atom in a certain site is only 10^{-10} s at 300 K but 2×10^{-5} s at 100 K. This extremely high mobility agrees very well with the experimental observation, whereafter the long-range order within hydrogen adsorbed phases [e.g., the c2 × 2 structure on Ni(111)] is formed almost instantaneously after the appropriate coverage conditions have been adjusted. Even the very complicated unit cells of the c2 × 6 structure formed on Ni(110) are immediately visible even at 100 K.

There has been a concept proposed [13] and pursued later by Nørskov and coworkers [101] to describe the motion of an adsorbed H atom or a proton, H$^+$, parallel to a corrugated surface by means of a proton band model in which the protons are regarded as delocalized. At higher vibrational excitations, the narrow vibrational states broaden and exhibit reasonable overlap, which may also account for an extremely rapid motion of a proton parallel to the surface. A more extensive report on that subject is given in Chapter 2. From the catalytic viewpoint it is relevant to keep in mind that adsorbed H atoms tend to migrate across two-dimensional surfaces very rapidly; it is therefore not imaginable that there exists any surface reaction in which the hydrogen diffusion could represent the rate-limiting step. More detailed considerations about surface diffusion in general can be found in the review article by Morris et al. [37] mentioned earlier.

C. Coadsorption of Hydrogen and Other Gaseous Adsorbates

For the sake of understanding the mechanism of heterogeneous catalytic reactions it is extremely important to consider also the interactions between *different* adsorbed species; this involves the field of catalyst poisoning (inhibition), promoter activity, synergistic effects, and so on. Again, one promising route to obtaining relevant information here is provided by coadsorption studies of reversibly bound gaseous species on single-crystal surfaces. The simplest case involving hydrogen is certainly the aforementioned H_2-D_2 isotope exchange reaction, which may be used to probe the catalytic activity of a surface or to prove whether hydrogen adsorbs dissociatively or molecularly. More complex coadsorption systems are $CO + H_2$, $N_2 + H_2$, or $O_2 + H_2$, and before entering into any details it seems worthwhile to provide some more general remarks about the mechanism of coadsorption.

There are basically two cases that may be distinguished when dealing with coadsorption of two different species A and B: (a) competitive adsorption, and (b) cooperative adsorption. The governing factor is the interaction between the species, ε, three types of which must be distinguished: ε_{A-A}, ε_{B-B}, and ε_{A-B}. In case (a) the interactions between equal particles, ε_{A-A} and ε_{B-B}, dominate over ε_{A-B} and a sort of segregation occurs between A and B. Case (b) represents the two-dimensional analog of a three-dimensional regular mixed phase with the interaction between different species, ε_{A-B}, exceeding either ε_{A-A} or ε_{B-B}. The situation is illustrated in Fig. 19. In the *first* case, species A tends to form large domains of type A, as does species B. It depends on the enthalpy of adsorption, $\Delta H_{ad}^{(A)}$ and $\Delta H_{ad}^{(B)}$, respectively, which species wins in the competition for the adsorption sites of the surface. Therefore, displacement effects may play a dominant role. Of course, the temperature influence may be of major importance here since kinetic (diffusion) barriers often prevent the system from reaching thermodynamic equilibrium.

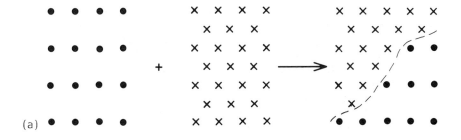

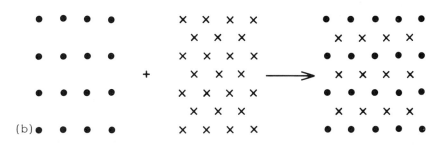

FIGURE 19 (a) Example for the case of competitive adsorption of two different species, A and B, forming patches with their own characteristic structure. (b) Example for cooperative adsorption. Species A and B form a new phase with long-range order different from that of either of the "pure" A and B structures.

If $\Delta H_{ad}^{(A)} \simeq \Delta H_{ad}^{(B)}$, a situation is obtained in which large islands of species A and B coexist on the surface, each having its own characteristic adsorbate structure. The LEED pattern of such an adsorbed layer then consists of a superposition of the structure elements of the respective "pure" phases. An example is the coadsorption of oxygen and carbon monoxide on a Pd(111) surface at room temperature as investigated by Ertl and Koch [102]. On clean Pd(111), CO_{ad} alone forms a $(\sqrt{3} \times \sqrt{3})R30°$ structure, whereas O_{ad} causes a (2×2) structure. The diffraction pattern of the coadsorbed species clearly is a superposition of both spot systems.

The other case (which is chemically much more interesting) is characterized by an intimate mixture (e.g., on the atomic scale) of the coadsorbed species. As again illustrated by Fig. 19b, the species A and B form a somewhat regular solution leading to unit cells with a new internal structure. Therefore, the structure amplitudes of those units differ from the pure phases A or B and give rise to new diffraction patterns not observed with either species alone. One clear example here is the coadsorption of $CO + H_2$ on a Pd(110) surface as investigated by Conrad et al. [103]. With the coadsorbed layer a new (1×3) LEED structure was observed that is indicative of the formation of individual CO_{ad}-H_{ad}-Pd complexes with long-range order.

It is quite evident that the mechanism and rate of a heterogeneous chemical reaction will depend very sensitively on the lateral distribution of the reactants on the surface. With competitive adsorption the reaction will preferentially run at or near the grain boundaries of the different domains (islands) of type A or type B. Conversely, if the reactants already coexist in intimately mixed form, no diffusion processes will be rate limiting but only the activation barrier to reach the transition state and to form the product. There are various examples of both types of coadsorption, as described next; we concentrate on coadsorption phenomena of hydrogen with carbon monoxide, nitrogen, and oxygen. Within the framework of this chapter the quite interesting phenomena of coadsorption of H_2 and C, S, and so on, on metal surfaces cannot be covered, although they have some impact on the catalytic activity of these materials (e.g., in the methanation reaction on nickel as pointed out by Kelley and Goodman [104].

Hydrogen and Carbon Monoxide

The coadsorption of these gases is of fundamental interest in view of the Fischer-Tropsch reaction:

$$CO + H_2 \longrightarrow \begin{array}{l} C_xH_yO_z \\ C_xH_y + H_2O \end{array} \qquad (19)$$

which may lead to either oxygenated or hydrocarbon species. Which route of reaction is actually followed by the reactants depends sensitively on the properties of the catalyst material (structure, composition, dispersion, promoter additives, etc.). It has been the aim of surface scientists for many years to model these reactions under ultrahigh-vacuum (UHV) conditions. Although at these low pressures thermodynamics does not favor any compound formation, there have been numerous attempts to coadsorb CO and H_2 and to search for surface complexes between these two species which might be regarded as precursors or intermediates of the product species. A review of this "surface science approach" has recently been given by Bonzel and Krebs [105]. As a more specific example, the extensive studies by White and co-workers on the Ni(100)/CO + H_2 system [106-108] are described briefly here. With this system, evidence was obtained for the formation of a CO···H surface complex (based on previous observations by Goodman et al. [109]) by TDS [106], UPS [106], LEED [107], and HREELS [108]. Most revealing are the thermal desorption results, which are displayed in Fig. 20. If hydrogen was preadsorbed and CO postadsorbed, a Σ state with an unusually narrow width appeared in the TPD spectra of both hydrogen and carbon monoxide around 210 to 220 K. Furthermore, coadsorbed CO slightly shifts the entire hydrogen desorption spectrum toward lower (ca. 20 K) temperatures. The Σ species forms in fairly large quantities and is interpreted as a surface complex between CO and H. It involves a hydrogen species that exhibits a surface binding energy of ca. 10 kcal mol^{-1} less than "normally" adsorbed hydrogen. Ultraviolet photoemission, however, reveals that no new H—CO or CO—H bond is formed. Nevertheless, some evidence was obtained of a CO species with a somewhat extended bond length with respect to the metal when hydrogen was coadsorbed. The situation is certainly quite complicated since the *sequence* of the adsorption also has a dramatic influence on the observed phenomena: The Σ state is completely suppressed if CO is adsorbed first. Obviously, the CO molecule irreversibly blocks sites required for hydrogen chemisorption or dissociation.

The same conclusion was reached by Benziger and Madix [110], who investigated the coadsorption of CO and H_2 on Fe(100). They observed a common hydrogen and carbon monoxide state at 475 K only when H_2 was preadsorbed, whereas CO adsorption blocked the subsequent dissociative adsorption of hydrogen.

1. Hydrogen Sorption on Pure Metal Surfaces

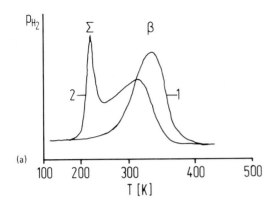

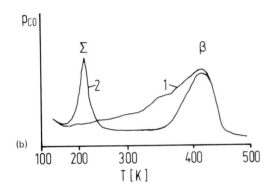

FIGURE 20 Thermal desorption spectra of H_2 and CO from a Ni(100) surface. (a) Hydrogen desorption. Curve (1) was obtained after a dosage of 17 L of H at 98 K on clean Ni(100); Curve (2) shows the hydrogen desorption after 17 L of CO was postadsorbed. (b) CO desorption. Curve (1) describes the CO desorption after 17 L of CO exposure to clean Ni(100); curve (2) the CO desorption after preadsorption of 17 L of H_2. Note the sharp Σ state that appears around 210 to 220 K in *both* coadsorption spectra. (After Ref. 106.)

The occurrence of cooperative adsorption between CO and H_2 can nicely be established by LEED experiments. As already mentioned, Conrad et al. found a new 1 × 3 LEED phase on Pd(110) formed by CO and H_2 together but not by either of these adsorbates alone. Again, hydrogen had to be preadsorbed to obtain the cooperative structure [103]. For CO + H_2 on Ni(100), White and co-workers [107] studied temperature dependences of the adsorbate structure by means of LEED and observed a sequence of various phases for the coadsorbed system: A c(2 × 2) up to 138 K, a c($\sqrt{2} \times \sqrt{2}$)R45° phase between 140 and 210 K, followed by a disordered phase between 210 and 280 K. Clearly, the c($\sqrt{2} \times \sqrt{2}$)R45° structure could be associated with the H_2 and CO Σ states since their desorption made the LEED pattern disappear.

On Ni(110) + CO + H_2, Behm and Penka [111] found similar interesting cooperative effects which involve the structure of the Ni substrate. Interestingly, CO exposure to the hydrogen-saturated (1 × 2) reconstructed Ni(110) surface at 120 K *removes* the surface reconstruction without any noticeable H_2 desorption. At the same time a c(2 × 2) phase is formed which contains 0.5 monolayer of CO. Above T = 150 K two other ordered phases form whereby the corresponding phase transitions are accompanied by characteristic sharp TPD maxima and work function changes.

Besides these examples of cooperative adsorption between H_2 and CO there are reports in which rather competitive effects have been observed. One example is again provided by investigations by White's group, namely, for CO + H_2 interacting with a Ru(0001) surface [112]. Both gases block adsorption sites of the respective other species (H_2 less strongly), and strong repulsive interactions between CO_{ad} and H_{ad} provide an extensive crowding of the CO molecules on the surface in the presence of hydrogen. The results are interpreted in terms of a segregation of H_2 + CO on the Ru surface.

Hydrogen and Nitrogen

This coadsorption system is of interest because of the well-known ammonia synthesis reaction,

$$N_2 + 3H_2 \rightarrow 2NH_3 \tag{20}$$

which is practically carried out over promoted Fe catalysts, under high pressures and at elevated temperatures, in the Haber-Bosch process. The exact modeling of this catalytic reaction under UHV conditions is difficult (if not impossible) because at the high reaction temperatures noticeable surface coverages can be achieved only at high pressures (in the torr regime) where the common spectroscopic tools can no longer be used. This is why the reverse reaction, namely, the decomposition of NH_3 into H_2 and N_2, is studied to elucidate the NH_3 formation reaction. The idea behind this is that the reaction sequence (including the formation of NH_x intermediates) follows the principle of microscopic reversibility, and steps detected in the decomposition path should also play a part in the synthesis reaction. Pioneering work in that area done by Ertl and his group is documented in numerous publications [113].

There are rather few studies concerned with actual coadsorption between H and N on well-defined single-crystal surfaces. It appeared from Ertl's investigation that the dissociation of the N molecule to give N_{ad} is a very slow process and thus probably rate determining in NH_3 formation. A number of studies were devoted to the question of whether the coadsorption of hydrogen could accelerate or retard N_{ad} formation. Whereas some authors concluded that indeed the presence of H_{ad} was favorable for N_2 dissociation [114,115], more recent studies revealed rather a negative influence of the hydrogen coadsorption on the rate of N_2 dissociation [116]. In this study study it was confirmed that hydrogen could be dissociatively adsorbed even in the presence of preadsorbed *molecular* nitrogen on a Fe surface, whereas the reverse process, namely the adsorption of molecular N_2, is inhibited by preadsorbed atomic hydrogen. This molecular $N_{2(ad)}$, however, is a necessary precursor for the formation of N_{ad}, and hence it is clear that preadsorbed hydrogen is not beneficial for the buildup of this species. This is illustrated by Fig. 21, in which the increase of $[N_{ad}]$ on Fe(111) is plotted against the N_2 exposure for the two cases without and with hydrogen in the gas phase. In the latter case hydrogen is competing with N_2 for the empty adsorption sites, and it is evident from Fig. 21 that there is an appreciably slower uptake of N_{ad} in the presence of hydrogen.

Vice versa, there is a strong reduction of adsorbed atomic hydrogen if N_{ad} is preadsorbed. It is believed that the strength of the individual H-Me, N_2-Me, and N-Me bonds governs the stability of the respective surface species. Clearly, this strength increases in the order $N_{2(ad)}$-, $H_{(ad)}$-, and $N_{(ad)}$- on a metal surface, in particular on the Fe(111) surface [116]. Generally, an

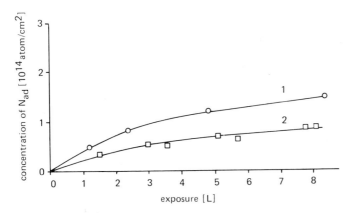

FIGURE 21 Increase of the surface coverage of atomic nitrogen N_{ad} on Fe(111) surface at 430 K as a function of the N_2 gas exposure. Curve (1) was obtained with a clean Fe(111) surface, while curve (2) reflects the situation after simultaneous H_2 exposure using the same gas pressure. (After Ref. 116.)

inhibiting effect of preadsorbed hydrogen on the formation of N_{ad} and hence of NH_3 can be stated. This is also supported by studies of NH_3 formation over Ru catalysts [117].

Whereas coadsorption studies could in no case reveal any formation of NH or NH_2 intermediate under UHV conditions, these species were discovered during decomposition of ammonia by UPS [118], SIMS [119], and HREELS [120], which confirms previous assumptions by Ertl and co-workers that the mechanism of NH_3 formation involves the stepwise hydrogenation of adsorbed N_{ad} [113].

Hydrogen and Oxygen

The heterogeneously catalyzed oxidation of hydrogen can also be studied by coadsorption of H_2 and O_2 on clean, well-defined metal surfaces. From the numerous studies in that field (which are reviewed, for example, in the article by Norton [121]) only a few examples will be given here. Again, the most active surfaces with respect to water formation are among the transition metals (especially group VIII metals). In their coadsorption studies Fisher and Gland [122] exposed a clean Pt(111) surface at 100 K first to oxygen (which, in the interesting coverage and temperature range, adsorbs dissociatively) up to a coverage of 0.25 and then to hydrogen. No water formation was detected by means of photoelectron spectroscopy. At a slightly higher temperature ($\Delta T \sim 20$ K), however, adsorbed H_2O was indeed formed, and an activation energy of that reaction of ca. 35 kJ mol^{-1} was calculated. It was also shown that the rate-limiting step in that reaction was $O_{ad} + H_{ad} \rightarrow (OH)_{ad}$. Once $(OH)_{ad}$ was present, water could be formed readily by the fast reaction step $OH_{(ad)} + H_{(ad)} = H_2O_{(ad)}$. To our knowledge there is no extensive study of the structural properties of coadsorbed hydrogen and oxygen on platinum-group metal surfaces.

Whether competitive or cooperative adsorption will occur can therefore not be answered at present. It is, however, clear that preadsorbed oxygen will inhibit the dissociative adsorption of hydrogen, whereas preadsorbed hydrogen will have only a minor effect on subsequent oxygen adsorption. An extreme case is provided by metal surfaces which can easily be oxidized by adsorbed oxygen (e.g., nickel(110), (100), or (111) [123]). On these surfaces (provided that the oxygen coverage is high enough, i.e., > 0.5 monolayer) first O-induced reconstruction and later oxide formation may lead to a completely different reaction mechanism involving hydrogen interaction with a surface oxide. Benninghoven et al. [124] investigated the coadsorption of hydrogen and oxygen on polycrystalline nickel foils by means of secondary ion mass spectroscopy (SIMS) and thermal desorption techniques and obtained the following results. Oxygen exposure to a nickel surface covered with hydrogen led to gradual removal of the adsorbed hydrogen layer simply by displacement effects (the O-Me bond is at least a factor of 3 greater than the respective H-Me bond). Continuous exposure of oxygen at this stage caused incorporation of O into the surface and subsequent Ni-oxide formation. The intriguing result of the reverse adsorption sequence (i.e., oxygen adsorption first, followed by hydrogen exposure) was that OH_{ad} groups were formed on the surface with different binding energies. Heating of these hydroxyl species resulted in the disappearance of the OH species and the simultaneous desorption of water (H_2O) molecules. Since Benninghoven's data were obtained with polycrystalline material, it appears that these oxygen-hydrogen interactions are rather structure insensitive, which would be expected for the case of competitive adsorption.

All in all, the certainly quite important coadsorption phenomena could only be touched in this section; it is hoped, however, that the material presented has given the reader some idea of the problems involved in the field.

III. ABSORPTION OF HYDROGEN: SURFACE EFFECTS

A. Equilibrium: Surface-Bulk

As mentioned in Section I, it is not our intention to present an exhaustive description of the absorption and bulk effects of hydrogen in metals. The interested reader will find the basic principles as well as more details and specificities in Ref. 125 and 126.

With regard to the surface phenomena on materials known to dissolve hydrogen readily (e.g., Ti, V, Zr, Nb, or Pd), complications may arise from the fact that H atoms may penetrate the phase boundary and enter the bulk crystal, and vice versa. The reaction sequence for the simple case of equilibrium between the surface and the bulk hydrogen population was formulated for Pd by Wagner [127] as follows:

$$[H_2]_{gas} + 2[*] \underset{k''}{\overset{k'}{\rightleftarrows}} 2[H]_{ad} \tag{21}$$

$$[H]_{ad} \underset{k''''}{\overset{k'''}{\rightleftarrows}} [H]_{Me} + [*] \tag{22}$$

with [*] denoting an empty adsorption site and $[H]_{Me}$ a H atom dissolved in the bulk. Of course, this is a very simple thermodynamic description which does not consider any kinetic limitations (activation barriers) or any specific differences between various bulk hydrogen states which are known to exist.

Much work has been directed toward the interaction of hydrogen with bulk metals, experimentally (e.g., Ref. 125) and theoretically [128]. Concerning the chemical state of hydrogen in metals three classes may be distinguished:

1. The first involves a mere solution of hydrogen in the metal crystal, whereby no stoichiometric hydride compound is formed and the H atoms are located in interstitial sites. By varying the pressure and temperature, almost

any bulk concentration of hydrogen within the solution capability of the metal can be achieved. In addition to the "normal" lattice plane of the surface through which the H atoms enter the bulk, there may be easier penetration channels, such as lattice distortions caused by impurity atoms or surface reconstructions, grain boundaries, or defects (dislocations). This matter has been considered briefly elsewhere [129].

2. There is a reaction between hydrogen and the bulk metal leading to metallic (stoichiometric) hydrides the composition of which does not change if pressure and temperature are varied within certain limits. Bulk hydrides exist for Pd, Ti, Nb, Zr, and some other metals. The hydrogens are located in tetrahedral or octahedral sites; an overview has been given by Burch [130]. Chapter 4 is devoted to the catalytic properties of bulk hydrides and related phenomena.

3. There may be a selective chemical reaction between the permeating or dissolved hydrogen and a nonmetallic alloy or impurity constituent of the bulk (C, P, S) which may lead to gaseous reaction products (CH_4, PH_3, SH_2) inside the crystal. These products will leave the bulk via grain boundaries or at least put a strain on the metal crystal lattice.

In this section only the first process of simple hydrogen absorption is considered. If we neglect for the moment all kinetic barriers that definitely exist, the amount of dissolved hydrogen will be governed by the free energy of the system, in which the heat of dissolution is an important ingredient. McLellan and Oates [131] have published a variety of experimental values for the heat of solution for hydrogen in metals, ΔH_{sol}; for elements such as Ca, Sc, Ti, V, Pd, or Zr *negative* values ranging from -0.2 eV to almost 1 eV have been reported, whereas for elements such as Cr, Mn, Fe, Co, Ni, Ru, Rh, or Pt, positive values up to 0.6 eV are characteristic. This is in line with the solubility of hydrogen in those metals, which is actually very small for catalytically relevant metals such as Fe, Co, Ni, Ru, Rh, or Pt. Only Pd represents an important exception. Model investigations using Pd(110) single crystals by Conrad et al. [16] revealed that after prolonged exposures of hydrogen at room temperature there was an appreciable uptake of bulk hydrogen, which extended to fairly deep regions of the crystal. This bulk hydrogen was slowly released into the gas phase when the sample was subjected to a temperature-programmed thermal desorption experiment. This is shown in Fig. 22. The first sharp TPD state corresponds to the desorption of the surface hydrogen, while the broad desorption state around 500 to 600 K indicates H atoms that have moved through the bulk to the surface (which requires many thermally activated steps) and have left the surface at these elevated temperatures. Another manifestation of the importance of hydrogen bulk-surface equilibria is provided by molecular beam experiments by Engel and Kuipers [132] using the Pd(111)/H (D) system. These authors investigated the H_2-D_2 isotope exchange reaction, which is the most elementary surface reaction that involves the breaking and formation of new chemical bonds. Although this reaction actually takes place as a Langmuir-Hinshelwood reaction *on* a surface only, dissolved hydrogen (deuterium) can have a dramatic influence on the reaction rate, in that the surface-near bulk region provides a reservoir of reactive species. Due to the relatively small solubility of hydrogen in most metals, an effect of this type is normally overlooked. With Pd, however, it may become quite significant, as Fig. 23 demonstrates. This figure shows a digitally averaged wave form for HD formed on a Pd(111) surface in a modulated (chopping frequency 63 Hz) molecular beam experiment at 693 K. The amount of hydrogen and deuterium taken up from the beam which reacts off as HD is an order of magnitude greater than that which can be chemisorbed in the outermost layer. This documents that most of the H_2 or D_2 is absorbed (probably in near-

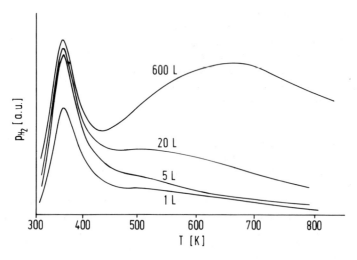

FIGURE 22 Thermal desorption spectra from a Pd(110) surface after various H_2 exposures from 1 to 600 L. The broad peak between 200 and 600°C reflects H atoms that have diffused through the near-surface bulk region of the crystal. (After Ref. 16.)

surface regions) rather than adsorbed. The large, temperature-independent relaxation time in the HD formation is indicative of a diffusion-controlled reaction.

B. Subsurface Hydrogen

Unfortunately, the situation described above is even more complicated, as recent investigations of chemisorption on Pd and Ru surfaces demonstrate. It is possible that there exists another potential for hydrogen trapping between the chemisorbed and absorbed hydrogen, called the "subsurface hydrogen." The population of the corresponding state has to be in equilibrium with both the surface and the bulk hydrogen. Actually, the subsurface hydrogen may be thought of as H atoms located directly underneath the topmost layer of metal atoms, which can, however, very easily move to the surface and are therefore very difficult to delineate from the actual *surface hydrogen* phase. LEED and molecular beam as well as photoemission studies of recent years revealed some insight into the surface-subsurface-bulk exchange kinetics, although the processes are not understood in detail [133,134].

The potential energy diagram, including all possible hydrogen states on and in the surface as well as in the bulk, is reproduced in Fig. 24. The right-hand part of this figure is identical to Fig. 1 and simply describes the dissociative chemisorption potential for H_2 molecules arriving from the gas phase. Below the surface, however, which is indicated by a hatched region, there exists no steep repulsive potential, only an activation barrier of a few kJ/mol until the H atom can be trapped in the subsurface potential (the depth of which depends, of course, on the population of the chemisorbed sites). From this subsurface potential, E_{ss}, the H atoms can move farther into the interior of the bulk once they have overcome a second activation barrier, which separates the subsurface potential from the bulk (dissolution) potential. The potential energy diagram, taken from our own work on a Pd(110) surface [135], represents just one example; similar curves have been

1. *Hydrogen Sorption on Pure Metal Surfaces*   43

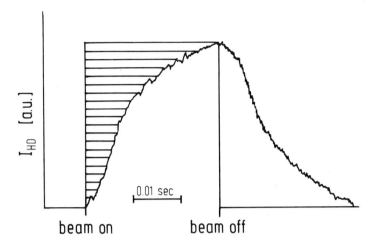

FIGURE 23 Digitally averaged waveform for HD produced on a Pd(111) surface at 693 K. The shaded area reflects the amount of adsorbed HD. The chopping frequency was 63 Hz. (After Ref. 132.)

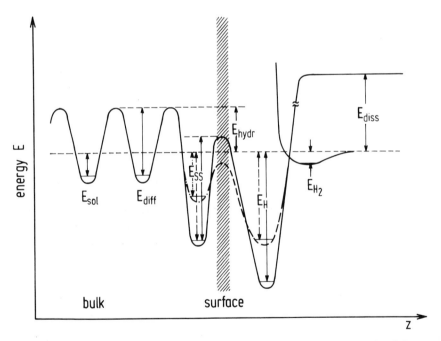

FIGURE 24 Potential energy diagram for adsorption, "sorption" into so-called "subsurface sites," and bulk absorption for Pd(110). The symbols have the following meaning: E_{sol}, heat of solution; E_{diff}, activation energy for bulk diffusion; E_{ss}, activation energy for desorption from the H subsurface sites; E_{hydr}, activation energy for hydrogenation; E_H, activation energy for desorption from the chemisorbed state; E_{H_2}, adsorption energy of the physisorbed H_2; E_{diss}, heat of dissociation of the H_2 molecule. The dashed line indicates the potential energy situation after the surface has been reconstructed by 1.5 monolayers of hydrogen. Clearly, the activation barrier for population of the subsurface sites is removed by reconstruction.

reported in the literature [130]. There has also been a wealth of theoretical calculations of thermodynamics and kinetics of the absorption of hydrogen into subsurface and bulk states, to which readers are referred for further details [134,136]. Experimentally, subsurface hydrogen very likely occurs with such materials as Ti, Nb, Zr, and of course, Pd, whereby the uptake kinetics depends strongly on the orientation of the respective metal face and on pressure and temperature. Here again we mention the investigations by Engel and Kuipers [132], which clearly demonstrate that, particularly at high pressures, the bulk and especially the subsurface region of a crystal may become loaded with hydrogen and can easily act as a buffering source of hydrogen. The role of the substrate face orientation in the population of subsurface sites is crucial. From our investigations with Pd surfaces of orientation (100), (110), and (111), it turned out that especially the (110) surface provided subsurface sites even at 100 K and hydrogen pressure in the 10^{-6} Pa regime, whereas higher pressures and temperatures were necessary to populate such sites with the more densely packed planes.

In contrast to UV photoemission reports by Eberhardt et al. [137], who found evidence for subsurface hydrogen even with the "dense" (111) planes of Ni and Pt, we could clearly rule out this subsurface species for Ni(111) [13]. For Pt(111), ion channeling work by Norton et al. [138] also indicated the *absence* of subsurface hydrogen, although they found some evidence for a small surface relaxation due to hydrogen chemisorption. Quite recently, a similar issue has come up with hydrogen on Ru(0001) where the possibility of subsurface hydrogen species was discussed by Feulner and Menzel [19] and Yates et al. [139]. We believe, however, that penetration of hydrogen through this hexagonal close-packed surface to subsurface sites is rather unlikely. Clearly, precise LEED investigations have to be performed to resolve this problem unambiguously.

C. Influence of Hydrogen Sorption on the Lattice Parameters

If hydrogen is adsorbed at 300 K onto a Pd(111) surface, no "extra" spots are observed in the LEED pattern. This indicates the absence of a chemisorbed hydrogen layer with long-range periodicity. Nevertheless, if the intensities of the integral-order LEED beams of the Pd 1 × 1 structure are measured with a Faraday cup, rather dramatic changes can be observed [140]. In the low-energy region between 20 and 150 eV there is a continuous change in the entire structure of the I-V curves, in that clean surface peaks disappear and new peaks come up as the hydrogen adsorbs (Fig. 25). These alterations certainly have to be attributed to changes in the lattice spacings of the topmost layer(s), probably due to the population of subsurface sites. At higher electron energies (150 to 400 eV) a pronounced shift of the position of the primary Bragg maxima toward lower energies indicates a certain expansion of the Pd surface lattice, which was determined to be around 2% [140].

Similar effects of slight peak shifts of Bragg maxima have been observed with Pt(111) surfaces covered by hydrogen [21]. Apparently, prolonged exposure to hydrogen causes a slight "surface relaxation," which means that in the direction normal to the surface, the lattice spacings change somewhat as the hydrogen chemisorbs; however, no changes occur in the lattice spacings *parallel* to the surface (i.e., no superstructure can be seen by LEED). What has to be kept in mind is the fact that many stable metal surfaces loose their stability somewhat in the presence of a hydrogen atmosphere: The small H atom can more easily "squeeze" into the lattice, particularly of Pd, Ti, Nb, or Zr, not to mention alloys of La with Mg or Nb. The result is not only an uptake of hydrogen in the surface region but also a slight distortion of the

1. Hydrogen Sorption on Pure Metal Surfaces

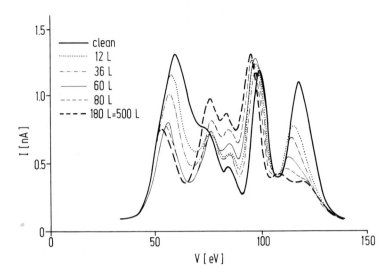

FIGURE 25 Variation of the LEED intensity of the (0,0) beam of a Pd(111) surface with the hydrogen exposure at 300 K. The electron energy range from 40 to 140 eV is shown, which exhibits the most pronounced alterations.

metal lattice, which in view of catalytic applications, may increase the activity of the respective metal, but may also cause an easier chemical attack by reactive intermediates or impurity atoms (catalyst poisons). A hydrogen-rich surface layer furthermore promotes sintering (recrystallization) of metal particles in dispersed catalysts, particularly if this layer persists at high temperatures (which is likely in the presence of high H_2 pressures).

IV. REACTION OF HYDROGEN WITH METAL SURFACES

A. Conditions for Hydride Formation

The forementioned remarks are directly related to the subject of this final section. It is quite clear that the process of relaxation is a first step toward a chemical *reaction* between hydrogen and the metal leading to hydride compounds. These hydrides are very well known to exist for a variety of metals (e.g., for Li, Na, Ca, Sr, Ti, V, Cr, Cu, Pd, Zr, etc.). However, the chemical nature of these hydrides may be quite different. There exist hydrides with ionic (saltlike) character (e.g., NaH) and there are hydrides which have metallic properties (e.g., PdH). Also, covalent hydrides are known (e.g., GaH_3, AlH_3). An excellent overview of primary solid hydrides is given by Gibb [141] (see also Chapter 14). Following Gibb [141] the stability of a hydride and therefore also the tendency of its formation depend strongly on how many d electrons exist in the metal. Metals rich in d electrons (located on the right-hand side of the periodic table) exhibit strong metallic bonds. There is little reason for them to form hydrides (except Pd, for different reasons), because the overlap of these d wavefunctions with the hydrogenic 1s wavefunction does not lead to a sufficient energy gain; rather, it costs energy. On the other hand, as we move to the left-hand side of the periodic table, there is a lack of d electrons, and a metal system can gain energy by forming chemical bonds to H atoms so as to react to a hydride. In the *surface region* of metal crystals, the situation may deviate substantially from that of the bulk metal crystal. Owing to the lack of neighbors, the metallic bond of the surface atoms is weaker than that of the bulk atoms, and it may in some cases be advantageous for a metal surface to form (locally) a surface hydride.

Difficulties exist if such surface hydride is to be distinguished from chemisorbed hydrogen since there are few differences in the chemical binding situation (i.e., the strength of a chemisorptive bond is comparable to that of a hydridic bond). An important point here is the so-called "decohesion" [142] of the metal which accompanies the hydride formation. The metal atoms involved in the hydride molecule become decoupled from their bulk neighbors to some extent, so that indeed, a surface complex arises.

B. Surface Reconstruction Induced by Hydrogen

As we pointed out several times earlier, chemisorbed H atoms may very well induce noticeable lateral displacements of metal surface atoms, which leads to new, "reconstructed" phases. This reconstruction may be considered as a significant first step toward the formation of a surface or bulk hydride, especially in cases where the reconstruction of the metal surface requires the adsorption of *many* H atoms [e.g., with Ni(110) or Pd(110)]. A common feature of hydrogen-induced reconstruction processes is that surface atoms tend to form densely packed complexes *in the surface*. A pairing of atoms to form dimers (as in the W(100)/H system [143]) or two from paired rows (as with the systems H/Ni(110) [28] and H/Pd(110) [17]) is quite pronounced, even if the metal atoms have to leave the sites that formerly provided a high coordination with respect to the neighboring metal atoms. Table 7 compiles the hydrogen-metal adsorption systems for which surface reconstruction phenomena have been observed. Since these reconstruction phenomena are relevant for heterogeneously catalyzed surface reactions, it is deemed useful to describe the problems in some detail. Particularly suited for this is the H/Ni(110) system, not only because it is probably one of the most intensely studied adsorption systems, but because it provides examples of almost any surface process that can be initiated by adsorbed hydrogen [148].

TABLE 7 Hydrogen-Metal Adsorption Systems Which Undergo Hydrogen-Induced Reconstruction

Metal surface	Temperature range of reconstruction (K)	Reconstructed phase	Hydrogen coverage (monolayers)	Refs.
Ni(110)	≤200	1×2	1	8
Ni(110)	>220	"Streak"	0.1	89
Pd(110)	120	1×2	1	17
Pd(110)	≥300	Streaky 1×2	<1	17
Mo(100)	>150	$c2 \times 2$ + complex phases	0.5-2	146
W(100)	>150	$c2 \times 2$ + complex phases	0.5-2	22
W(110)	300	Nonprimitive 2×2	0.5	145

1. Hydrogen Sorption on Pure Metal Surfaces

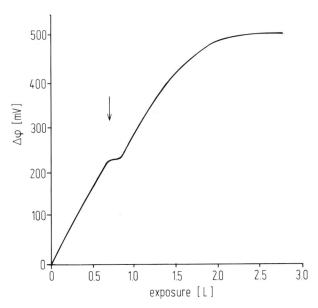

FIGURE 26 H-induced work function change of a Ni(110) surface as a function of the H_2 exposure at 120 K. The arrow indicates the coverage region in which the surface structure changes from 2 × 1-2H to 1 × 2 (reconstructed). The saturation coverage is 1.5 monolayers and is reached after 2.5 L.

In Fig. 26 we display the work function change, $\Delta\phi$, induced by hydrogen adsorption at T = 120 K as a function of the H_2 exposure. An interesting phenomenon is the plateau around 0.7 L, which is associated with a hydrogen coverage close to one monolayer. Figure 27 shows the coverage dependence of the LEED intensity of various hydrogen-induced "extra" spots, namely, of the final lattice gas phase, a 2 × 1-2H, which does *not* yet involve a surface reconstruction, and of the subsequent 1 × 2 phase which is reconstructed and in which the Ni atoms form pairs in the [001] direction along the [1̄10] direction. A structure model for this phase transformation is given in Fig. 28. Also given in that figure is the side view (i.e., a cut perpendicular to the surface plane), so that the actual coordination of the Ni and H atoms can be seen. From this figure it is quite evident that if the topmost Ni atoms are displaced so as to touch one another in the paired row configuration, these atoms have to be lifted and supported by H atoms that have squeezed between the first and second Ni lattice planes. This process is energetically favorable only after a "critical" density of adsorbed H atoms has been reached, namely, just one monolayer. The corresponding structural reorganization not only leads to new surface periodicities but also has consequences for the kinetics of adsorption. As mentioned briefly in Section II.B, there is a change in the sticking probability around 1-monolayer coverage when the 1 × 2 reconstructed phase starts to form. This is reflected directly in the work function curve of Fig. 26. The uptake of H atoms in the unreconstructed 2 × 1-2H phase comes to an end, as indicated by the saturating first branch of the $\Delta\phi$ curve, and only after the reconstruction to the 1 × 2 phase has occurred can the surface adsorb additional H atoms, leading to the second part of the $\Delta\phi$ curve. The question as to the driving force of the 1 × 2 reconstruction can, at present, not be answered exhaustively. From the chemical point of view the possible formation of a Ni surface hydride has to

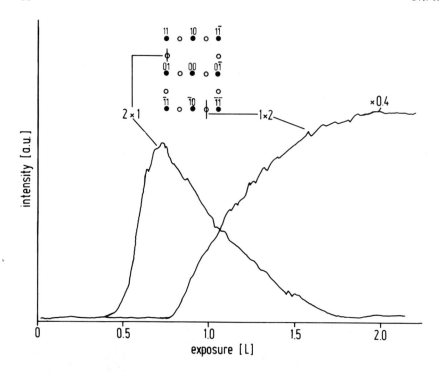

FIGURE 27 Exposure dependence of the LEED intensity of a 2 × 1-2H and of a 1 × 2 "extra" spot of a Ni(110) surface (see inset LEED pattern) at T = 120 K. The intensity maximum of the 2 × 1-2H structure reflects the hydrogen coverage of 1 monolayer.

be taken into account, for the following reasons. As pointed out before, Ni(110) is a relatively "open" surface, which could provide the prerequisites for the stability of a surface hydride (although the existence of a bulk Ni hydride is questionable [141]). Moreover there is a new electronic state associated with the formation of the 1 × 2 phase, as angle-resolved and energy-dependent UV photoemission measurements showed [147]. This could well be a hydridic state. Finally, the reconstructed 1 × 2 phase is thermally unstable and decomposes abruptly around T = 220 K, thereby giving rise to the very sharp thermal desorption state of Fig. 16, which obeys a fractional-order kinetics (which is unusual for recombinative H_2 desorption). This strange desorption kinetics could be indicative of the decomposition of a surface hydride. Interestingly, there is another phase transformation associated with this decomposition of the 1 × 2 (hydride?) phase: The two-dimensionally ordered 1 × 2 phase transforms into a merely one-dimensionally ordered "streak" phase which is also reconstructed. A possible structure model is reproduced in Fig. 29, from which it can be seen that the nice correlation of the paired rows in the [001] direction is entirely lost. This streak phase is the only stable phase at temperatures above 220 K. This has the implication that catalytic reactions involving hydrogen and carried out over Ni-containing catalysts might also involve this streak phase, provided that Ni microfacets with (110) orientation are thermodynamically stable. There is another consequence, which has to be viewed somewhat more generally: It may very well happen in the course of a catalytic reaction that a given surface changes its orientation at certain

1. *Hydrogen Sorption on Pure Metal Surfaces*

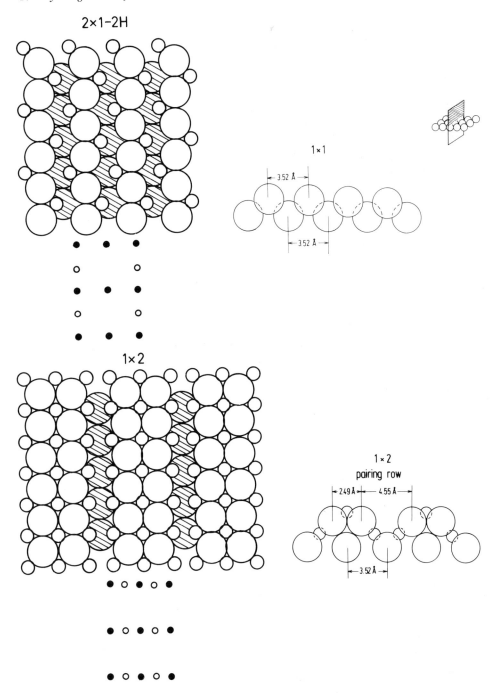

FIGURE 28 Structure models of the unreconstructed 2 × 1-2H and of the reconstructed (pairing row) 1 × 2 phase. The small circles represent the adsorbed H atoms, the shaded circles the second-layer Ni atoms. Also given in the figure are the side views of both phases and the corresponding LEED patterns. Quite recently, a LEED structure analysis [149] revealed, in addition to the row pairing, a second-layer buckling in which the second-layer Ni atoms are alternatley displaced in the z direction by about 0.25 Å. The mean displacement of the top Ni atoms is approxiamately 0.4 Å in the x, y direction.

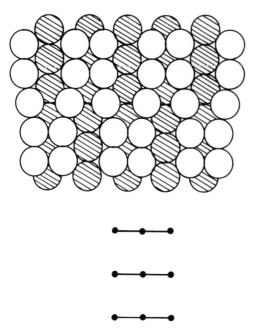

FIGURE 29 Tentative structure model of the "streak" phase observed with a Ni(110) surface upon hydrogen exposure at room temperature or by thermal decomposition of the 1 × 2 reconstructed phase at 220 K. Also given in the figure is the streaky LEED pattern observed experimentally.

conditions of hydrogen pressure (leading to the related surface coverage) and temperature (which might induce thermally activated decomposition reactions of surface complexes). This could lead to particularly reactive surface phases or to a buildup of very high effective surface pressures of reactants (e.g., hydrogen).

V. CONCLUSIONS

Hydrogen (H_2) is a chemically quite reactive gas which adsorbs dissociatively on most transition metal surfaces with heats of chemisorption between 60 and 120 kJ mol^{-1}, which corresponds to metal-hydrogen binding energies of 230 to 270 kJ mol^{-1}. The preferential adsorption site has a high coordination; that is, threefold sites are favored on digonal, trigonal, and hexagonal surfaces, while fourfold sites are the rule for tetragonal surfaces. There is ample evidence that the strong chemisorption of one H atom requires ensembles of several (four to seven) adjacent metal atoms (ensemble effect). The Me-H binding energy is sensitively affected as soon as these ensembles become diluted by chemically different (and less active) atoms. At higher coverages indirect mutual interactions come into play which are in most cases repulsive. They lead to a substantial lowering of the initial metal-hydrogen binding energy, a phenomenon called induced surface heterogeneity. As a consequence, weakly held (and chemically reactive) hydrogen species are formed.

With regard to the dynamics of hydrogen adsorption, it is noteworthy that the sticking probability of hydrogen is, on the average, lower than 1.0, probably owing to spatial requirements at the moment of the impact of the molecule. Defect sites may be very important in the energy accommodation and dissociation process, in that an atomically "rough" surface has a particular promoting effect. The kinetics of adsorption and desorption often follow second-order kinetics. However, an increasing number of systems have been reported in which the adsorption and desorption of hydrogen occurs via a mobile precursor state leads to marked deviations from the "normal" second-order behavior.

Adsorbed atomic hydrogen is very mobile compared to adsorbed oxygen or carbon monoxide; there are reports of a H atom or proton undergoing tunneling in certain temperature regimes. The very small size of a H atom enables it to enter a metal surface fairly deeply and to introduce a strong perturbation in the electronic structure of the surface. Normally, this will only cause a relaxation of the surface lattice. Sometimes, however, noticeable displacements of metal atoms from their equilibrium positions also result, called "surface reconstruction." This may be viewed as a reaction leading to absorption in the bulk or even a reaction with the metal atoms (hydride formation). Crystallographically more "open" surfaces are particularly susceptible to these processes. Of course, metals such as Ti, Cr, V, Zr, Pd, La, and some others act per se as absorbing materials for hydrogen, whereby intermediate subsurface states may facilitate the transition from the chemisorbed to the adsorbed state.

ACKNOWLEDGMENTS

This work was financially supported by the Deutsche Forschungsgemeinschaft (DFG) through SFB 6. Stimulating discussions with H. J. Brocksch, D. Tomanek, F. Rys, and L. D. Roelofs as well as technical assistance by K. Schubert and R. Cames are gratefully acknowledged.

REFERENCES

1. H. P. Bonzel, *Surf. Sci.*, 69:239 (1977).
2. J. E. Demuth, D. Schmeisser, and Ph. Avouris, *Phys. Rev. Lett.*, 47: 1166 (1981); Ph. Avouris, D. Schmeisser, and J. E. Demuth, *Phys. Rev. Lett.*, 48:199 (1982).
3. J. E. Lennard-Jones, *Trans. Faraday Soc.*, 28:28 (1932).
4. M. Balooch, M. J. Cardillo, D. R. Miller, and R. E. Stickney, *Surf. Sci.*, 46:358 (1974).
5. J. K. Nørskov, A. Houmøller, P. K. Johansson, and B. Lundqvist, *Phys. Rev. Lett.*, 46:257 (1981).
6. K. Christmann and G. Ertl, in *Catalyst Characterization Science* (M. L. Deviney and J. L. Gland, eds.), ACS Symposium Series 288, American Chemical Society, Washington, D.C. (1985), p. 222.
7. H. J. Robota, W. Vielhaber, M. C. Lin, J. Segner, and G. Ertl, *Surf. Sci.*, 155:101 (1985).
8. V. Penka, K. Christmann, and G. Ertl, *Surf. Sci.*, 136:307 (1984).
9. D. M. Newns, *Phys. Rev.*, 178:1123 (1969).
10. K. Christmann, *Surf. Sci. Rep.*, to be published.
11. K. Christmann, O. Schober, G. Ertl, and M. Neumann, *J. Chem. Phys.*, 60:4528 (1974).

12. K. Christmann, Z. Naturforsch. Teil, A 34:22 (1979).
13. K. Christmann, R. J. Behm, G. Ertl, M. A. van Hove, and W. H. Weinberg, J. Chem. Phys., 70:4168 (1979).
14. T. N. Taylor and P. J. Estrup, J. Vac. Sci. Technol., 11:244 (1974).
15. R. J. Behm, K. Christmann, and G. Ertl, Surf. Sci., 99:320 (1980).
16. H. Conrad, G. Ertl, and E. E. Latta, Surf. Sci., 41:435 (1974).
17. M. G. Cattania, V. Penka, R. J. Behm, K. Christmann, and G. Ertl, Surf. Sci., 126:382 (1983).
18. F. Bozso, G. Ertl, M. Grunze, and M. Weiss, Appl. Surf. Sci., 1:103 (1977).
19. P. Feulner and D. Menzel, Surf. Sci., 154:465 (1985).
20. K. Christmann, M. Ehsasi, J. H. Block, and W. Hirschwald, Chem. Phys. Lett., 131:192 (1986).
21. K. Christmann, G. Ertl, and T. Pignet, Surf. Sci., 54:365 (1975).
22. P. J. Estrup and J. Anderson, J. Chem. Phys., 45:2254 (1966).
23. K. Christmann and G. Ertl, Suf. Sci., 60:365 (1976).
24. H. Baró and H. Ibach, Surf. Sci., 92:237 (1980).
25. T. Toya, J. Res. Inst. Catal. Hokkaido Univ., 10:236 (1962).
26. D. D. Eley and E. J. Pearson, J. Chem. Soc. Faraday Trans. 1, 72 (1975).
27. J. E. Demuth, Surf. Sci., 65:369 (1977).
28. K. Christmann, F. Chehab, V. Penka, and G. Ertl, Surf. Sci., 152/153:356 (1985).
29. Z. Knor, in Catalysis, Science and Technology, Vol. 3 (J. R. Anderson and M. Boudart, eds.), Springer, Berlin (1982), p. 231ff.
30. A. G. Goyden, Dissociation Energies and Spectra of Diatomic Molecules, Chapman & Hall, London (1968).
31. J. Prichard, T. Catterick, and A. K. Gupta, Surf. Sci., 53:1 (1975).
32. H. Shimizu, K. Christmann, and G. Ertl, J. Catal., 61:412 (1980).
33. J. T. Yates, Jr., P. A. Thiel, and W. H. Weinberg, Surf. Sci., 84:427 (1979).
34. M. G. Cattania, K. Christmann, V. Penka, and G. Ertl, Gazz. Chim. Ital., 113:433 (1983).
35. H. Conrad, Ph.D. thesis, University of Munich (1976).
36. P. W. Tamm and L. D. Schmidt, J. Chem. Phys., 51:5352 (1969); 52:1150 (1970); 54:4775 (1971).
37. M. A. Morris, M. Bowker, and D. A. King, in Simple Processes at the Gas-Solid Interface, Comprehensive Chemical Kinetics Series, Vol. 19 (C. H. Bamford, C. F. H. Tipper, and R. G. Compton, eds.), Elsevier, Amsterdam (1984), Chap. 1
38. (a) B. Poelsema, R. L. Palmer, G. Mechtersheimer, and G. Comsa, Surf. Sci., 117:60 (1982); (b) B. J. J. Koeleman, S. T. de Zwart, A. L. Boers, B. Poelsema, and L. K. Verheij, Nucl. Instrum. Methods, 218:225 (1983).
39. B. Poelsema, L. K. Verheij, and G. Comsa, Surf. Sci., 152/153:496 (1985).
40. K. Wangemann, J. Rüstig, and K. Christmann, Verh. Dtsch. Phys. Ges., 4:935 (1985) and to be published.
41. D. E. Ibbotson, T. S. Wittrig, and W. H. Weinberg, J. Chem. Phys. 72:4885 (1980).
42. C. F. Melius, J. W. Moskowitz, A. P. Mortola, M. B. Baillie, and M. A. Ratner, Surf. Sci., 59:279 (1976).
43. O. Schober, unpublished result.
44. V. Penka, Ph.D. thesis, University of Munich (1985).

45. (a) R. J. Gale, M. Salmeron, and G. A. Somorjai, *Phys. Rev. Lett.*, *38*:1027 (1977); (b) M. Salmeron, R. J. Gale, and G. A. Somorjai, *J. Chem. Phys.*, *67*:5324 (1977).
46. W. Reimer, R. J. Behm, G. Ertl, and V. Penka, *Verh. Dtsch. Phys. Ges.*, *4*:933 (1985) and to be published.
47. W. Moritz, R. Imbihl, R. J. Behm, G. Ertl, and T. Matsushima, *J. Chem. Phys.*, *83*:1959 (1985).
48. M. Skottke, R. J. Behm, G. Ertl, W. Moritz, and V. Penka, *Verh. Dtsch. Phys. Ges.*, *5*:1382 (1986) and *J. Chem. Phys.*, in press.
49. M. R. Barnes and R. F. Willis, *Phys. Rev. Lett.*, *41*:1729 (1978).
50. R. J. Behm, K. Christmann, G. Ertl, and M. A. van Hove, *J. Chem. Phys.*, *73*:2984 (1980).
51. S. Andersson, *Chem. Phys. Lett.*, *55*:185 (1978).
52. (a) C. Nyberg and C. G. Tengstål, *Solid State Commun.*, *44*:251 (1982); (b) *Phys. Rev. Lett.*, *50*:1680 (1983).
53. H. Ibach and H. Bruchmann, *Phys. Rev. Lett.*, *44*:36 (1980).
54. H. Conrad, R. Scala, W. Stenzel, and R. Unwin, *J. Chem. Phys.*, *81*:6371 (1983).
55. M. A. Barteau, J. Q. Broughton, and D. Menzel, *Surf. Sci.*, *133*:443 (1983).
56. A. M. Baro, H. Ibach, and H. D. Bruchmann, *Surf. Sci.*, *88*:384 (1979).
57. J. Lee, J. P. Cowin, and L. Wharton, *Surf. Sci.*, *130*:1 (1983).
58. K. Christmann and G. Ertl, *J. Mol. Catal.*, *25*:31 (1984).
59. J. J. Burton and E. Hyman, *J. Catal.*, *37*:114 (1975).
60. K. Y. Yu, D. T. Ling, and W. E. Spicer, *J. Catal.*, *44*:373 (1976).
61. C. Harendt, K. Christmann, W. Hirschwald, and J. C. Vickerman, *Surf. Sci.*, *165*:413 (1986).
62. J. H. Sinfelt, Y. L. Lam, J. A. Cusamano, and A. E. Barnett, *J. Catal.*, *42*:227 (1976).
63. J. Koutecky, *Trans. Faraday Soc.*, *54*:1038 (1958).
64. T. B. Grimley and M. Torrini, *J. Phys.*, *C6*:868 (1973).
65. T. L. Einstein and J. R. Schrieffer, *Phys. Rev.*, *B7*:3629 (1973).
66. T. L. Einstein, *Surf. Sci.*, *84*:L497 (1979).
67. K. H. Lau and W. Kohn, *Surf. Sci.*, *75*:69 (1978).
68. G. Doyen, G. Ertl, and M. Plancher, *J. Chem. Phys.*, *62*:2957 (1975).
69. K. Binder and D. P. Landau, *Surf. Sci.*, *61*:577 (1976).
70. K. Binder and D. P. Landau, *Surf. Sci.*, *108*:503 (1981).
71. (a) E. Ising, *Z. Phys.*, *31*:253 (1925); (b) L. Onsager, *Phys. Rev.*, *65*:117 (1944); (c) T. D. Lee and C. N. Yang, *Phys. Rev.*, *87*:410 (1952).
72. K. Christmann, *Ber. Bunsenges. Phys. Chem.*, *90*:307 (1986).
73. R. Imbihl, R. J. Behm, K. Christmann, G. Ertl, and T. Matsushima, *Surf. Sci.*, *117*:257 (1982).
74. K. H. Rieder, *Phys. Rev.*, *B27*:7799 (1983).
75. T. E. Felter and R. H. Stulen, *J. Vac. Sci. Technol.*, *A3*:1566 (1985).
76. K. Christmann and K. H. Ernst, in preparation.
77. K. H. Rieder and T. Engel, *Phys. Rev. Lett.*, *43*:373 (1979); *45*:824 (1980).
78. K. Binder (ed.), *Monte Carlo Methods in Statistical Physics*, Springer, Berlin (1979).
79. (a) L. D. Roelofs, *Appl. Surf. Sci.*, *11/12*:425 (1982); (b) L. D. Roelofs, in *Chemistry and Physics of Solid Surfaces*, Vol. 4 (R. Vanselow and R. Howe, eds.), Springer, Berlin (1982), p. 219.
80. K. Christmann, *Bull. Soc. Chim. Fr.*, *3*:288 (1985).
81. J. K. Nørskov, *J. Vac. Sci. Technol.*, *18*:420 (1981).

82. W. Brenig, *Z. Phys. B*, *48*:127 (1982).
83. M. El-Batanouny, M. Strongin, G. P. Williams, and J. Colbert, *Phys. Rev. Lett.*, *46*:269 (1981).
84. G. Doyen, *Surf. Sci.*, *89*:238 (1979).
85. R. J. Behm, Ph.D. thesis, University of Munich (1980).
86. P. Kisliuk, *J. Phys. Chem. Solids*, *3*:95 (1957); :78 (1958).
87. K. H. Rieder and W. Stocker, *Surf. Sci.*, *138*:139 (1984).
88. (a) G. Ertl, P. R. Norton, and J. Rüstig, *Phys. Rev. Lett.*, *49*:177 (1982); (b) M. P. Cox, G. Ertl, R. Imbihl, and J. Rüstig, *Surf. Sci.*, *134*:L517 (1983); (c) G. Ertl, *Ber. Bunsenges. Phys. Chem.*, *90*:284 (1986).
89. K. Christmann, R. J. Behm, V. Penka, and G. Ertl, in preparation.
90. P. Redhead, *Vacuum*, *12*:203 (1962).
91. D. A. King, *Surf. Sci.*, *47*:384 (1975).
92. E. Bauer, F. Boczek, H. Poppa, and G. Todd, *Surf. Sci.*, *53*:877 (1975).
93. K. Christmann, V. Penka, F. Chehab, G. Ertl, and R. J. Behm, *Solid State Commun.*, *51*:487 (1984).
94. J. L. Falconer and R. J. Madix, *Surf. Sci.*, *46*:473 (1974).
95. K. J. Laidler, *J. Phys. Colloid Chem.*, *53*:712 (1949).
96. D. A. King, *Surf. Sci.*, *64*:43 (1977).
97. R. Gorte and L. D. Schmidt, *Surf. Sci.*, *76*:559 (1978).
98. R. W. McCabe and L. D. Schmidt, *Surf. Sci.*, *65*:189 (1977).
99. R. Gomer, R. Wortmann, and R. Lundy, *J. Chem. Phys.*, *27*:1099 (1957).
100. R. DiFoggio and R. Gomer, *Phys. Rev. Lett.*, *44*:1258 (1980).
101. M. J. Puska, R. M. Nieminen, M. Manninen, B. Chakraborty, S. Holloway, and J. K. Nørskov, *Phys. Rev. Lett.*, *51*:1081 (1983).
102. G. Ertl and J. Koch, in *Adsorption-Desorption Phenomena* (F. Ricca, ed.), Academic Press, New York (1972), p. 345.
103. H. Conrad, G. Ertl, and E. E. Latta, *J. Catal.*, *35*:363 (1974).
104. R. D. Kelley and D. W. Goodman, *Surf. Sci.*, *123*:L743 (1982).
105. H. P. Bonzel and H. J. Krebs, *Surf. Sci.*, *117*:639 (1982).
106. B. E. Koel, D. E. Peebles, and J. M. White, *Surf. Sci.*, *107*:L367 (1981).
107. B. E. Koel, D. E. Peebles, and J. M. White, *Surf. Sci.*, *125*:L87 (1983).
108. G. E. Mitchell, J. L. Gland, and J. M. White, *Surf. Sci.*, *131*:167 (1983).
109. D. W. Goodman, J. T. Yates, Jr., and T. E. Madey, *Surf. Sci.*, *93*:L135 (1980).
110. J. B. Benziger and R. J. Madix, *Surf. Sci.*, *115*:279 (1982).
111. V. Penka, R. J. Behm, and G. Ertl, *Verh. Dtsch. Phys. Ges.*, *4*:921 (1985) and to be published.
112. D. E. Peebles, J. A. Schreifels, and J. M. White, *Surf. Sci.*, *116*:117 (1982).
113. (a) G. Ertl, *Catal. Rev.*, *21*:201 (1980); (b) G. Ertl, in *Catalysis, Science and Technology*, Vol. 4 (J. R. Anderson and M. Boudart, eds.), Springer, Berlin (1983), p. 257ff.
114. K. Tamaru, *Trans. Faraday Soc.*, *59*:979 (1963).
115. M. Grunze, F. Bozso, G. Ertl, and M. Weiss, *Appl. Surf. Sci.*, *1*:241 (1978).
116. G. Ertl, M. Huber, S. B. Lee, Z. Paál, and M. Weiss, *Appl. Surf. Sci.*, *8*:373 (1981).
117. H. Amariglio, *Symposium on Nitrogen Fixation*, Tokyo (1980).
118. M. Weiss, G. Ertl, and F. Nitschké, *Appl. Surf. Sci.*, *2*:614 (1979).
119. M. Drechsler, H. Hoinkes, H. Kaarmann, H. Wilsch, G. Ertl, and M. Weiss, *Appl. Surf. Sci.*, *3*:217 (1979).

120. J. Küppers, personal communication.
121. P. R. Norton, in *The Chemical Physics of Surfaces and Heterogeneous Catalysis*, Vol. 4 (D. P. Woodruff and D. A. King, eds.), Elsevier, Amsterdam (1982), p. 27.
122. G. B. Fisher, J. L. Gland, and S. J. Schmieg, *J. Vac. Sci. Technol.*, 20:518 (1982).
123. (a) L. D. Roelofs, A. R. Kortan, T. L. Einstein, and R. L. Park, *Phys. Rev. Lett.*, 46:1465 (1981); (b) D. E. Taylor and R. L. Park, *Proceedings of the International Conference on Phase Transitions on Surfaces*, University of Maine at Orono, Maine (1981), p. 45.
124. A. Benninghoven, P. Beckmann, K. H. Müller, and M. Schemmer, *Surf. Sci.*, 89:701 (1979).
125. G. Alefeld and J. Völkl (eds.), *Hydrogen in Metals*, Vols. 1 and 2, Springer, Berlin (1978).
126. F. A. Lewis, *The Palladium-Hydrogen System*, Academic Press, London (1967).
127. C. Wagner, *Z. Phys. Chem. Abt. A*, 159:459 (1932).
128. (a) J. K. Nørskov, F. Besenbacher, J. Bøttiger, B. B. Nielsen, and A. A. Pisarev, *Phys. Rev. Lett.*, 49:1420 (1982); (b) M. S. Daw and M. I. Baskes, *Phys. Rev. Lett.*, 50:1285 (1983).
129. K. Christmann, in *Atomistics of Fracture* (R. M. Latanision and J. R. Pickens, eds.), Plenum Press, New York (1983), p. 363.
130. R. Burch in *Chemical Physics of Solids and Their Surfaces* (M. W. Roberts and J. M. Thomas, eds.), Royal Society of Chemistry, London (1980), p. 1.
131. R. B. McLellan and W. A. Oates, *Acta Metall.*, 21:181 (1973).
132. T. Engel and H. Kuipers, *Surf. Sci.*, 90:162 (1979).
133. M. Lagos, *Surf. Sci.*, 122:L601 (1982).
134. J. W. Davenport, G. J. Dienes, and R. A. Johnson, *Phys. Rev.*, B25:2165 (1982).
135. R. J. Behm, V. Penka, M. G. Cattania, K. Christmann, and G. Ertl, *J. Chem. Phys.*, 78:7486 (1983).
136. J. K. Nørskov, *Phys. Rev.*, B26:2875 (1982).
137. W. Eberhardt, F. Greuter, and E. W. Plummer, *Phys. Rev. Lett.*, 46:1085 (1981).
138. J. A. Davies, D. P. Jackson, P. R. Norton, D. E. Posner, and W. N. Unertl, *Solid State Commun.*, 34:41 (1980).
139. J. T. Yates, Jr., C. H. F. Peden, J. E. Houston, and D. W. Goodman, *Surf. Sci.*, 160:37 (1985).
140. K. Christmann, G. Ertl, and O. Schober, *Surf. Sci.* 40:61 (1973).
141. T. R. P. Gibb, Jr., *Prog. Inorg. Chem.*, 3:315 (1962).
142. D. G. Pettifor, in *Atomistics of Fracture* (R. M. Latanision and J. R. Pickens, eds.), Plenum Press, New York (1983), p. 281.
143. (a) M. K. Debe and D. A. King, *Surf. Sci.*, 81:193 (1979); (b) D. A. King and G. Thomas, *Surf. Sci.*, 92:201 (1980).
144. J. Lapujoulade and K. S. Neil, *Surf. Sci.*, 35:288 (1972); *J. Chem. Phys.*, 57:3535 (1972).
145. W. Chung, S. C. Ying, and P. J. Estrup, *Phys. Rev. Lett.*, 56:749 (1986).
146. T. E. Felter, R. A. Barker, and P. J. Estrup, *Phys. Rev. Lett.*, 38:1138 (1977).
147. K. Christmann, H. Kuhlenbeck, M. Ehsasi, F. Chehab, H. Saalfeld, and M. Neumann, in preparation.
148. K. H. Rieder and W. Stocker, *Surf. Sci.*, 164:55 (1985).
149. G. Kleinle, R. J. Behm, G. Ertl, W. Moritz, and V. Penka, *Phys. Rev. Lett.*, 58:148 (1987).

2 | Vibrational Spectroscopy of Hydrogen Adsorbed on Metal Surfaces

C. MATHEW MATE*, BRIAN E. BENT[†], AND GABOR A. SOMORJAI

University of California at Berkeley, Berkeley, California

I.	INTRODUCTION	57
II.	HIGH-RESOLUTION ELECTRON ENERGY LOSS SPECTROSCOPY OF HYDROGEN ADSORBED ON METAL SINGLE-CRYSTAL SURFACES IN ULTRAHIGH VACUUM	59
	A. Dipole Scattering	60
	B. Impact Scattering	62
	C. Site Determination by HREELS	65
	D. Delocalized Hydrogen	68
	E. Trends in the HREEL Spectra	72
III.	VIBRATIONAL SPECTROSCOPY OF HYDROGEN ADSORBED ON HIGH-SURFACE-AREA METALS AT ATMOSPHERIC PRESSURE	75
	A. Incoherent Inelastic Neutron Scattering	75
	B. Infrared and Raman Spectroscopy	76
IV.	CONCLUSIONS	78
	REFERENCES	79

I. INTRODUCTION

Vibrational spectroscopy is essentially the study of atomic motion. A bound atom, either in a molecule or to a surface, undergoes quantized vibrational motion, and the vibrational frequencies are measurable by vibrational spectroscopy.

In surface reactions, atoms have to move from site to site as well as break bonds and form new ones, so a knowledge of atomic motion is an important part of any atomic scale understanding of the chemical processes. Further, since the vibrational frequencies reflect the symmetry of atomic motion and the nature of interatomic potentials, vibrational spectroscopy provides an excellent insight into the bonding environment of atoms and molecules.

Current affiliations:
*IBM Almaden Research Center, San Jose, California
[†]AT&T Bell Laboratories, Murray Hill, New Jersey

TABLE 1 Vibrational Spectroscopies Used to Measure Vibrational Frequencies of Atoms and Molecules Adsorbed on Surfaces

Vibrational spectroscopy	Principle	Samples	Resolution (cm^{-1})	Spectral range (cm^{-1})	Refs.
High-resolution electron energy loss spectroscopy (HREELS)	Inelastic scattering of low-energy (1–150 eV) electrons	Single crystals, thin films, and foils in UHV	30–90	100–5000	5
Incoherent inelastic neutron scattering	Incoherent inelastic scattering of thermal neutrons	50–100 g of powder	5–50	4–4000	6
Raman spectroscopy and surface-enhanced Raman spectroscopy (SERS)	Inelastic scattering of photons of visible light	100 mg of powder, single crystals, electrodes	1–10	200–5000	7, 8, 1
Reflection absorption infrared spectroscopy (RAIRS)	Absorption of infrared radiation detected in the reflected beam	Foils, single crystals	1–10	700–5000	9
Transmission absorption infrared spectroscopy (TAIRS)	Absorption of infrared radiation detected in the transmitted beam	10–100 mg of pressed powder, solution	1–10	400–5000	10, 1–3
Infrared emission spectroscopy (IES)	Detection of the emitted blackbody radiation from vibrating molecules	Foils, single crystals, zeolite on a Au wire	1–10	400–4000	11, 1
Inelastic electron tunneling spectroscopy (IETS)	Inelastic tunneling of electrons between metals through an oxide layer containing the sample	1–10 mg of Al or other metal oxide (20 Å thick, supported on metal film)	10–50	10–5000	12
Photoacoustic spectroscopy (PAS)	Vibrational excitation with pulsed light source and detection of the sound waves generated	100 mg of powder or metal film	1–10	400–5000	13

2. Vibrational Spectroscopy of Hydrogen on Metals

Over the last 25 years a number of vibrational spectroscopies have achieved sufficient sensitivity to measure vibrations of adsorbates on surfaces [1-13]; these techniques are listed and described in Table 1. In this chapter we examine the vibrational spectroscopies used to study hydrogen adsorption on metal surfaces, which are the first five spectroscopies listed in Table 1: HREELS, IINS, RAIRS, TAIRS, and Raman spectroscopy.

First, we review what is presently known from vibrational spectroscopy about adsorption of hydrogen in ultrahigh vacuum (UHV) on well-defined single-crystal surfaces. Here the preferred technique of investigation is high-resolution electron energy loss spectroscopy (HREELS). Next, we review the adsorption of hydrogen at atmospheric pressure on samples that have a high surface-to-volume ratio and that are similar to the metal surfaces of heterogeneous catalysts used in industry. Here the results are primarily from incoherent inelastic neutron scattering, infrared spectroscopy, and Raman spectroscopy.

II. HIGH-RESOLUTION ELECTRON ENERGY LOSS SPECTROSCOPY OF HYDROGEN ADSORBED ON METAL SINGLE-CRYSTAL SURFACES IN ULTRAHIGH VACUUM

Much of our atomic-scale understanding of how hydrogen interacts with metal surfaces has come from experiments studying hydrogen chemisorption on single-crystal surfaces under ultrahigh-vacuum conditions. Ultrahigh vacuum (i.e., gas pressures of less than 10^{-8} torr) is necessary to ensure that a substantial quantity of gases other than hydrogen do not adsorb on the metal surface during the course of an experiment. By studying hydrogen on atomically clean single-crystal surfaces, one can learn how hydrogen interacts with well-characterized metal surfaces, an important starting point toward an eventual understanding of how hydrogen behaves on the less-well-defined surfaces present on high-surface-area catalysts.

Single-crystal surfaces make it possible to study the bonding of hydrogen on surfaces with different types of adsorption sites. Figure 1 shows some of the low-Miller-index crystal surfaces of face-centered cubic (fcc), body-centered cubic (bcc), and hexagonal close-packed (hcp) metals; top sites, bridge sites, and three- and fourfold hollow sites are indicated.

The most important vibrational spectroscopy currently used to study hydrogen adsorbed on single-crystal surfaces is high-resolution electron energy loss spectroscopy (HREELS). Most HREEL spectrometers have a high sensitivity to adsorbates and obtain spectra over the entire vibrational frequency range (100 to 5000 cm^{-1}) in less than 20 min, which accounts for the widespread use of HREELS to study all types of adsorbates on single-crystal surfaces. These attributes are particularly important for studying adsorbed hydrogen, since hydrogen is difficult to detect with many experimental probes. HREELS is able to detect a few percent of a monolayer of hydrogen adsorbed on a single-crystal surface. Also, for adsorbed hydrogen, a large frequency range is necessary, since the vibrational frequencies observed can range from 400 cm^{-1} for atomic deuterium chemisorbed on hexagonally close-packed metal surfaces [14] to 4159 cm^{-1} for the H-H stretching mode of molecularly adsorbed hydrogen [15].

Figure 2 shows a HREEL spectrometer used by the authors, which is typical of spectrometers in current use. The monochromator produces a beam of low-energy electrons (1 to 10 eV) which is focused on the sample. The energy and angle of the scattered electrons are then determined by the analyzer. Most (99.9%) of the electrons incident on a metal single-crystal surface are reflected elastically without any change in energy, but a fraction

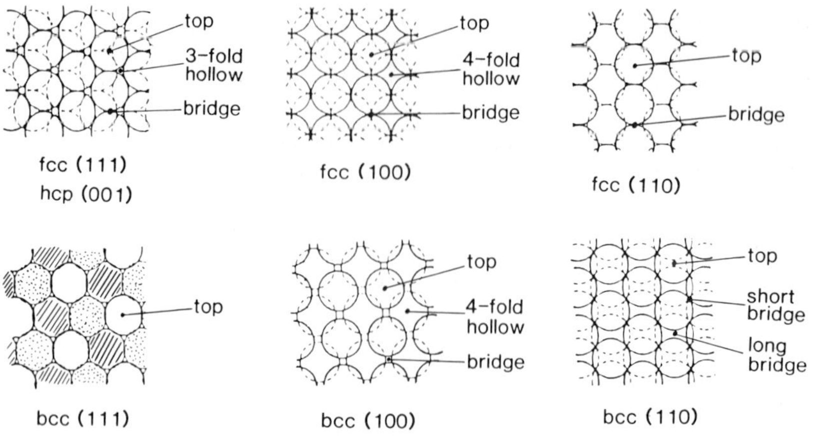

FIGURE 1 Top views of the atomic arrangement in the first two or three layers of the common low-Miller-index faces of face-centered cubic (fcc), body-centered cubic (bcc), and hexagonal close-packed (hcp) crystals. Thin-lined atoms are behind the plane of thick-lined atoms. For the "open" bcc(111) crystal face, second-layer atoms are dotted and third layer atoms are slashed. High-symmetry binding sites are indicated.

of the electrons scatter inelastically; that is, they lose energy by exciting a vibrational mode of an adsorbate or of some other surface excitation. For vibrational excitations, the energy lost by the exciting electron is relatively small compared to the energy of the incident electron beam, so the spectrometer must be able to operate at fairly high resolution, typically less than 10 meV or 80 cm^{-1} (1 meV = 8.0655 cm^{-1}), in order to distinguish the weak loss peaks from the intense peak of elastically scattered electrons; hence the name of the technique—high-resolution electron energy loss spectroscopy. A large number of reviews [16] and several excellent books [5] have been written on the design of HREEL spectrometers and their application to the study of atoms and molecules adsorbed on surfaces.

In 1967, Propst and Piper reported the first vibrational spectra, obtained by HREELS, of adsorbates on a metal surface, W(100) [17]. Among the gases adsorbed in this experiment on the W(100) surface was hydrogen. Since then the study of hydrogen adsorbed on W(100) by HREELS has played a central role both in the development of HREELS and in studies of adsorbed hydrogen. We use this system to illustrate how vibrational frequencies measured by HREELS are used to determine adsorption sites on a well-defined metal surface. Making this assignment requires first an understanding of the mechanisms of vibrational excitation by electron scattering from adsorbate-covered surfaces [18-29]. According to our present views, these mechanisms can be conveniently divided into two categories: a long-range scattering process called dipole scattering and a short-range scattering process called impact scattering.

A. Dipole Scattering

Dipole scattering is similar in nature to the vibrational excitation mechanism in infrared (IR) vibrational spectroscopy. The long-range Coulomb field of

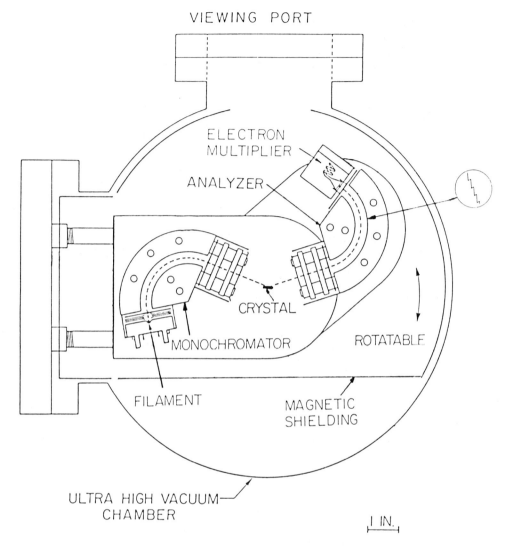

FIGURE 2 Schematic diagram of a typical high-resolution electron energy loss spectrometer. The energy dispersive elements are 127° cylindrical sectors.

the incident electrons interacts with the "dynamic dipole" of a vibrating adsorbate just as the electric field of the incident photons in IR spectroscopy interacts with the dynamic dipole of a molecule. This coupling between the electron and the dynamic dipole moment of the adsorbate enables the electron to lose a quantum of energy by exciting an adsorbate vibrational mode.

An important characteristic of dipole scattering arises from the physical nature of the Coulomb field of the incident electron at the metal surface. Electric fields can only have a component perpendicular to a metal surface since parallel components are screened by the metal. Therefore, only the component of the dynamic dipole moment perpendicular to the surface can couple with incoming electrons. This results in what is usually called the "surface dipole selection rule," which states that only vibrations which have a net dynamical dipole moment perpendicular to the surface can scatter electrons via the dipole scattering process.

Another characteristic of dipole scattering, which enables one to distinguish between dipole and impact scattering processes, is the angular dependence of the scattering intensity; the dipole scattering intensity falls off rapidly for scattering angles away from the specularly reflected electron beam. The angular half-width of the dipole scattering intensity is on the order of $\hbar\omega_0/2E_I$, where ω_0 is the frequency of the vibration and E_I is the incident electron energy. In typical HREELS experiments, the dipole scattered electrons are concentrated within several degrees of the specular beam.

A good example of dipole scattering is the 130-meV (1050-cm^{-1}) loss that appears in the HREEL spectra for a saturation coverage of hydrogen atoms on the unreconstructed surface of W(100). As shown in Fig. 3, there is an intense loss at 130 meV for spectra obtained in the specular direction. Figure 4 shows that the intensity of this loss decreases with sample rotation as the analyzer collects electrons scattered at angles away from specular direction. This angular dependence of the scattered intensity allows us to identify the 130-meV loss with a dipole active vibrational mode; then, using the surface dipole selection rule, we can conclude that the 130-meV loss corresponds to the vibration of hydrogen perpendicular to the surface.

B. Impact Scattering

Figure 3 also shows that at off-specular collection angles, several other weak losses are visible in the HREEL spectrum. These losses are the result of vibrational modes parallel to the surface being excited by impact scattering. During the impact scattering process, the incident electrons lose energy by interacting directly with the atomic potentials while the electrons are within a few angstroms of the surface. During this very short-range interaction, the electrons essentially become "impacted" in the surface atoms for a short period of time, allowing vibrational modes oriented parallel to the surface, as well as those oriented perpendicular, to be excited by the electrons. Figure 4 shows that the losses from the parallel vibrational modes of hydrogen atoms on the W(100) surface have a broad angular distribution, characteristic of impact scattering and unlike the narrow angular distribution of dipole scattering.

This short-range impact scattering is physically a much more complicated process than the long-range dipole scattering process. The complicated nature of impact scattering is well illustrated by the HREEL spectra for hydrogen adsorbed on Rh(111) [30]. For this system the loss peaks in the vibrational spectra show interesting intensity variations with changes in the incident electron beam energy. Figure 5 shows the HREEL spectra for saturation coverage of hydrogen on Rh(111) taken at several different incident electron energies, while Fig. 6 shows how the intensities of the elastic peak and of the three main losses in the spectra vary as a function of incident electron energy. All three loss peaks have maximum intensity at ca. 4.7 eV, while the elastic peak shows maximum intensity at ca. 2 eV. Further, this maximum intensity at 4.7 eV is due to impact scattering as determined by measuring the angular distribution of these peak intensities at different incident beam energies. For incident energies less than about 3 eV, the loss intensities fall off dramatically at angles away from the specular direction, indicating dipole scattering as the main scattering mechanism in this energy range. Above 3 eV, however, the angular distribution of the loss intensities is fairly broad, implicating impact scattering as the main scattering mechanism at these energies. The loss intensity maxima at ca. 4.7 eV are attributed to a "resonant" impact scattering off the atomic poten-

2. Vibrational Spectroscopy of Hydrogen on Metals 63

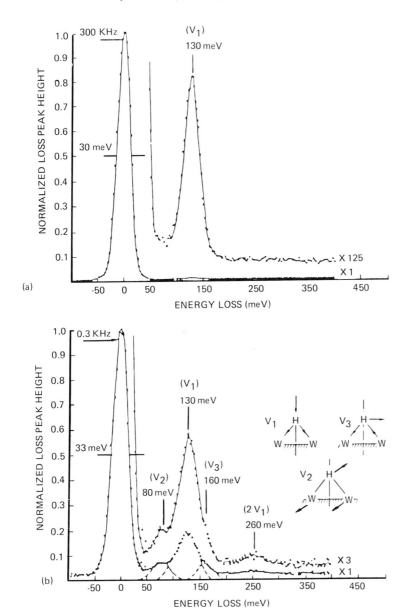

FIGURE 3 Normalized electron energy loss spectra for a saturation coverage of hydrogen chemisorbed on W(100) for $\theta_i = 23°$ incident angle and an impact energy $E_0 = 9.65$ eV showing the different vibrational losses observable in (a) the specular beam direction and (b) the +17° off-specular direction. (From Ref. 21; used with permission.)

tials. Although resonant impact scattering is poorly understood at present, resonant electron scattering from gas-phase atoms and molecules has been studied extensively where it is thought to involve the temporary capture of the incident electron to form a short-lived negative ion [31].

Even though the mechanism of impact scattering is not as well understood as that for dipole scattering, a detailed knowledge is fortunately usually not

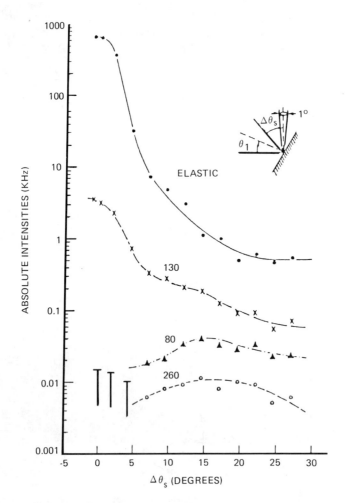

FIGURE 4 Absolute intensities of the energy loss peaks in Fig. 3 as a function of collection angle away from the specular direction $\Delta\theta_S$. The scattering geometry is shown in the upper inset. (From Ref. 21; used with permission.)

required in order to interpret HREEL spectra; in most cases, it suffices to know only whether a particular mode has some dipole scattering contribution and consequently, a net dynamic dipole perpendicular to the surface. Impact scattering does, however, have some useful selection rules which apply in certain cases [32]:

1. The inelastic scattering intensity vanishes for a vibration that is odd with respect to a mirror plane symmetry if trajectories of both the incident and scattered electrons lie in this symmetry plane.
2. The inelastic scattering intensity vanishes in the specular direction for a vibration that is odd with respect to a mirror plane symmetry if the plane containing the trajectories of the incident and scattered electrons is perpendicular to the mirror plane and the surface.
3. The inelastic scattering intensity vanishes in the specular direction for a vibration that is odd with respect to a twofold rotation symmetry

2. Vibrational Spectroscopy of Hydrogen on Metals

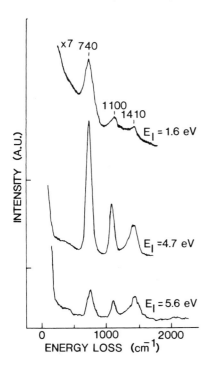

FIGURE 5 High-resolution electron energy loss vibrational spectra taken, in the specular direction, at three different incident electron beam energies for H_2 dissociatively adsorbed on Rh(111) at 77 K. The intensity variations are the result of "resonant" impact scattering of the electrons at an incident beam energy of ca. 4.7 eV.

C. Site Determination by HREELS

These selection rules for impact scattering, together with the surface dipole selection rule, help determine the symmetry of the hydrogen adsorption site [21, 23, 24, 26, 29]. To illustrate this point, we again use the case of hydrogen adsorbed on the W(100) crystal surface. As shown in Fig. 2, a total of four losses are observed, with energies of 80, 130, 160, and 260 meV, resulting from dissociatively adsorbed hydrogen. Since the 260-meV loss has twice the energy of the 130-meV loss, it is probably an overtone or multiple loss of the 130-meV loss. The remaining three losses (80, 130, and 160 meV) come from the motion of hydrogen atoms on the surface.

If we assume that the hydrogen atoms are well localized on the surface, these three losses correspond to the three vibrational modes of hydrogen atoms bound at a particular type of site or sites. An atom bound to a surface has three "normal modes" of vibration, since a free atom has three degrees of freedom. At least one of these vibrational modes must involve motion of the atom perpendicular to the surface. Since only the 130-meV mode is dipole active, one can conclude that there is only one type of adsorption site. The 80- and 160-meV modes are not dipole active, so they correspond to vibrations parallel to the surface, and since the 80- and 160-meV losses are not seen in the specular direction (Fig. 3a), they must satisfy one of the selection rules of impact scattering. Of the possible point groups for a hydrogen atom on W(100)–C_{4v}, C_{2v}, C_s, and C_1–only for C_{2v} can adsorbed hydrogen

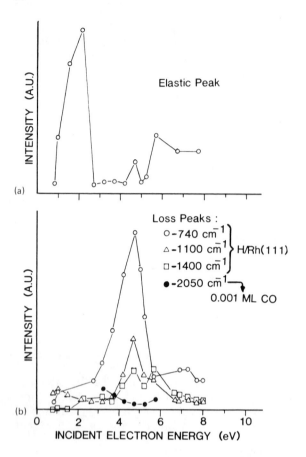

FIGURE 6 Absolute intensities in the specular direction of the elastic peak (a) and of the energy loss peaks (b) in Fig. 5 as a function of incident electron beam energy for hydrogen on Rh(111) at 77 K. Note that both the elastic peak and the energy loss peak for 0.001 monolayer of coadsorbed CO are at low intensity for E = 4.7 eV, where the hydrogen energy loss peaks are strongly enhanced by "resonant" impact scattering.

have two nondegenerate, nondipole active vibration modes of the correct symmetry to satisfy the impact scattering selection rule. In this case, once we know the symmetry of the adsorption site, we also know the adsorption site, since a bridge site (as shown in Fig. 1) is the only site on the W(100) surface with C_{2v} symmetry.

For H adsorbed on the W(100) surface, one can also determine bond lengths and bond angles from the vibrational frequencies by making the following assumptions: (a) the mass of the metal atoms is infinite, (b) the bonds between the hydrogen and the metal atoms can be approximated by springs, and (c) the angle-bending force constant is much smaller than the bond-stretching constant. If ν_{sym} is the vibrational frequency perpendicular to the surface and ν_{as} is the vibrational frequency parallel to the surface along the M-H-M bond for hydrogen bound at a bridge site, then the M-H-M interbond angle, α, is related to the vibrational frequencies by $\nu_{as}/\nu_{sym} = \tan(\alpha/2)$. This theoretical relationship between frequencies and bond angles has been applied to bridge-bonded hydrogen atoms in metal hydrides [33] as

2. Vibrational Spectroscopy of Hydrogen on Metals

shown in Fig. 7, which plots the ν_{as}/ν_s ratio measured by IR spectroscopy versus the $\tan(\alpha/2)$ determined by x-ray or neutron diffraction. The correlation between the theoretical and experimental angles is quite good. Furthermore, for cluster hydrides, once the M-H-M bond angle is determined from the ν_{as}/ν_s ratio, one can determine the metal-hydrogen bond distance if the metal-metal separation is known.

Willis has applied this correlation for metal cluster hydrides to surfaces to determine bond angles and bond lengths for hydrogen on W(100) [23]. The ν_{as}/ν_s ratios for hydrogen on unreconstructed p(1 × 1) W(100) and on reconstructed c(2 × 2) W(100) are also plotted in Fig. 7. Figure 8 shows the bonding geometries determined from Fig. 7 for hydrogen atoms on W(100). It is noteworthy that a specific surface site (bridge site, for example) does not have "characteristic" vibrational frequencies. Within a given adsorption site, small changes in bond lengths and bond angles have a large effect on the measured vibrational frequencies, as demonstrated in Fig. 7 by the variation from 1 to 2 in the ν_{as}/ν_s ratio for bridge-bonded hydrides.

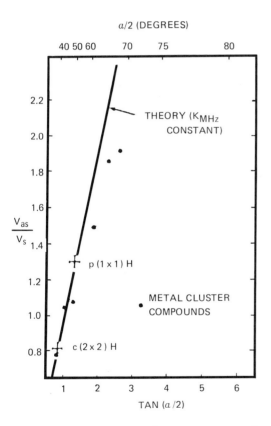

FIGURE 7 Plot of the M-H bond stretching frequency ratio (ν_{as}/ν_s) observed in the infrared spectra of bridged metal hydrides versus the tangent of half the interbond M-H-M angle, α. The ν_{as}/ν_s ratio determined by HREELS for chemisorbed, bridge-bonded hydrogen on the unreconstructed [p(1 × 1)H] and reconstructed [c(2 × 2)H] W(100) surfaces is plotted on the theoretical fit [$\tan(\alpha/2) = \nu_{as}/\nu_s$ ($k_{MH} \simeq 108$ N m^{-1})] to determine the surface bonding geometry shown in Fig. 8. (From Ref. 23; used with permission.)

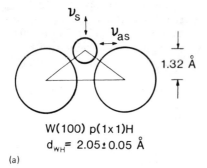

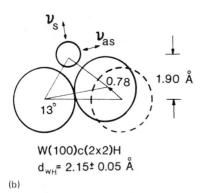

FIGURE 8 Side view of the bridge-bonding geometry of hydrogen on (a) unreconstructed [p(1 × 1)H] W(100) and (b) reconstructed [c(2 × 2)H] W(100) as determined by comparison to metal hydride clusters as shown in Fig. 7. (From Ref. 23.)

D. Delocalized Hydrogen

For hydrogen adsorbed on W(100), it is fairly straightforward to determine the bonding geometry using HREELS.[†] For hydrogen adsorbed on other metal surfaces this is not always possible, for several reasons: (a) The selection rules for impact scattering do not apply in every experimental situation; (b) hydrogen is a weak scatterer of electrons, so often a vibrational mode is too low in intensity to be detectable; and (c) the hydrogen atoms may not be well localized at an adsorption site, in which case the atomic motion no longer approximates that of a simple harmonic oscillator.

The third effect is somewhat unexpected and has only recently been discussed in the literature [30, 35-37]. This effect arises because vibrating hydrogen has a large displacement amplitude and because the potential energy of hydrogen on a surface does not go to infinity parallel to the surface as it does, to a good approximation, perpendicular to the surface. Instead, a finite potential barrier exists between the energetically favorable adsorption sites. If the ground-state vibrational energy is on the same order as the barrier height, it is possible for hydrogen to tunnel from one site to another. Also, the energy of the excited vibrational states can be larger than the height of the barrier potential, in which case it is no longer valid to think

---

[†]However, a recent publication [34] indicates that a new assignment of the HREEL spectrum for hydrogen on W(100) will be proposed in a future publication.

of small vibrations about an equilibrium position; instead, one should think of the atom moving across the surface with a certain momentum and with its motion perturbed by the periodic potential of the surface. The phenomenon is analogous to electrons in solids, where their motion is described by energy bands.

For electrons in solids, there are two physical models which are commonly used to explain the formation of energy bands and band gaps. One model, called the free-electron model, treats the electrons in the solid as a free-electron gas and the periodic potential of the lattice as a perturbation on the free-electron energies. The second model, called the tight-binding method, approximates the electron states as the atomic orbitals of the free atoms. In the second model, the bands occur from the overlap of orbitals of neighboring atoms. Although one can show that the two models are equivalent ways of looking at the same physical problem, one usually obtains a better intuitive understanding of how electrons behave in a solid by using the free-electron model when the periodic potential seen by the electrons is small compared to their energy and by using the tight-binding method when the electron emergies are smaller than the maximum potential energy between atoms.

Therefore, for the lowest energy bands of hydrogen adsorbed on metal surfaces, the tight-binding method would provide the best understanding of the nature of these bands. In this method one first determines the wavefunctions that are solutions of the Schrödinger equation for a potential well centered over a particular site. Since the wavefunctions still have a finite value at distances away from the center of the site, a fraction of each wavefunction will overlap to a certain extent with the wavefunctions of neighboring sites. The wavefunctions of neighboring sites are degenerate in energy with those of the original sites before the overlap is taken into account since the potential wells of the neighboring sites are the same as the original site; however, the overlap of wavefunctions lifts the degeneracies, resulting in the formation of energy bands.

For hydrogen adsorbed on W(100), the separation distance between neighboring bridge sites is fairly large (3.16 Å), so the overlap of wavefunctions is probably small. Therefore, the low-energy bands of hydrogen on W(100) should be fairly narrow, and the wavefunctions well localized at bridge sites. Consequently, even for such "localized" hydrogen atoms, the previous analysis would be valid.

However, other types of crystal faces can have adsorption sites with substantially smaller separation distances. Consequently, the overlap of even the ground-state wavefunctions on these surfaces can be significant, leading to broad energy bands. In this case, the hydrogen atoms are not well localized at adsorption sites but instead, can tunnel from site to site.

Recently, Puska et al. [36,37] have calculated the low-energy position wavefunctions for hydrogen adsorbed on the Ni(100), Ni(111), and Ni(110) surfaces where the distance between hollow sites is, respectively, 2.49 Å, 1.44 Å, and 1.44 Å. These calculations use the effective-medium approximation of density functional theory to calculate the potential energy of hydrogen on these surfaces. The Schrödinger equation is then solved numerically to determine the wavefunctions. Figure 9a shows the potential energy, the A_1 wavefunctions, and the densities for hydrogen chemisorbed on the Ni(100) surface; Fig. 10 shows the corresponding band structure for the A_1 wavefunctions of hydrogen chemisorbed on the Ni(100) surface. The lowest-energy wavefunction (Fig. 9c) is fairly well localized at the hollow sites, but the wavefunctions of the first two excited bands (Fig. 9e and g) are substantially less localized. The wavefunctions of the excited bands correspond to motion both perpendicular and parallel to the surface. The coupling between

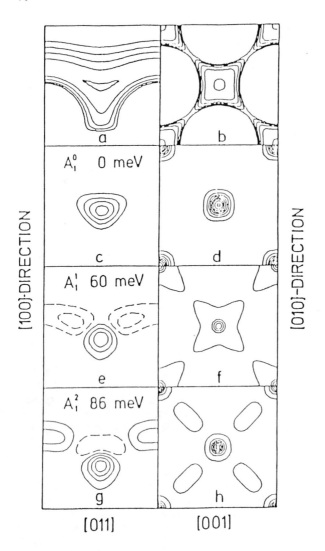

FIGURE 9 Potential, A_1 wavefunctions, and densities for hydrogen chemisorbed on the Ni(100) surface. The left panel shows the potential and wavefunctions in a vertical plane along the <110> direction through the fourfold center position where the potential has its minimum. The lengths of the cuts are the Ni nearest-neighbor distance ($4.7a_0$) in both the parallel and perpendicular directions. At the top of the right panel, the potential is shown in a cut parallel to the surface through the absolute minimum. Underneath are shown the hydrogen densities, integrated perpendicular to the surface in the same parallel cut. In the right panel the cuts are one Ni lattice constant ($6.65a_0$) in each direction. All wavefunctions are evaluated at Γ. (From Ref. 36; used with permission.)

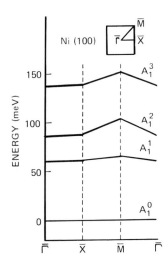

FIGURE 10 The band structure for hydrogen chemisorbed on the Ni(100) surface shown along the high-symmetry directions indicated in the inset. Only the states belonging to the A_1 representation of the C_{4v} point group are shown. The zero of energy is the ground-state energy (-2.6 eV) at the Γ point. This includes a zero-point energy of 0.1 eV. In the inset the Brillouin zone has been rotated 45° relative to the convention used in Fig. 9. (From Ref. 36; used with permission.)

the perpendicular and parallel motion results from the anharmonicity of the potential well, as shown in Fig. 9b, over the distance of the zero-point motion of the hydrogen atom.

Table 2 lists the calculated band centers and bandwidths for hydrogen on Ni(111). The bandwidths are larger than those calculated for Ni(100) because the smaller separation distances between hollow sites on this surface results in more overlap of neighboring wavefunctions. Since the lowest-energy band is fairly narrow (ca. 4 meV), this band will be fully occupied, even at liquid-nitrogen temperatures, but the higher-energy band will be only sparsely occupied at this temperature. Consequently, the losses observed in HREELS for hydrogen adsorbed on this surface would correspond to transitions from all parts of the Brillouin zone of the first band to higher-energy bands; the transitions are "vertical," since for HREEL spectra obtained in the specular direction $\Delta k_\parallel = 0$.

Table 2 also lists the experimentally observed values for transitions between bands for hydrogen adsorbed on several hexagonally close-packed surfaces. Off-specular measurements indicate that the 820- and 1140-cm^{-1} transitions of hydrogen on Ru(0001); the 750-, 1100-, and 1430-cm^{-1} transitions of hydrogen on Rh(111); and the 550-cm^{-1} transition on Pt(111) are dipole active, so these excitations correspond to $A_1^0 \rightarrow A_1^n$ transitions. These frequencies have been assigned by us, somewhat arbitrarily, to transitions that are closest to those predicted theoretically for hydrogen on Ni(111). Many of the transitions that are predicted to occur have not been observed, most likely due to very low excitation probabilities. For hydrogen adsorbed on Rh(111), all the low-energy transitions have been observed by choosing incident beam energies that enhance the various excitation probabilities.

The observation of all the low-energy transitions for the motion of hydrogen on the Rh(111) surface is strong experimental evidence that the hydrogen

TABLE 2 Calculated Band Centers (Bandwidths; cm^{-1}) for Hydrogen on Ni(111)[a]

A_1^0	A_1^1	A_1^2	E^1	E^2
16 (32)	597 (347)	1113 (323)	323 (210)	1095 (468)

Experimentally measured transition energies by HREELS (cm^{-1}) and heats of adsorption (kcal mol^{-1}) for hydrogen adsorbed on hexagonally close-packed surfaces:

Metal surface	Refs.	$A_1 \to A_1^1$	$A_1^0 \to A_1^2$	$A_1^0 \to A_1^3$	$A_1^0 \to E^1$	$A_1^0 \to E^2$	Heat of adsorption
Ni(111)	25, 38	710	1120	–	–	–	23
Ru(001)	39	690 (low θ)	1070	–	–	–	29
		820 (high θ)	1140	1550			
Rh(111)	30, 40	750	1100	1430	400	1280	19
Pt(111)	14, 41	550	1230	–	–	–	9.5

[a]From Refs. 36 and 37.

atoms are fairly delocalized on this surface (see also Chapter 4). Further, when deuterium is adsorbed instead of hydrogen on Rh(111), the band positions and bandwidths undergo isotopic shifts that do not correspond to those predicted by a simple harmonic oscillator model; however, the shifts do fit a model where the motion parallel to the surface is that of a free atom perturbed by a periodic potential [30].

E. Trends in the HREEL Spectra

Except for HREEL spectra of H_2 physisorbed on Cu and Ag at 10 K [15], HREEL spectra of H_2 adsorbed on transition metal surfaces from 70 to 300 K show dissociative adsorption of H_2. Tables 2 and 3 summarize the vibrational frequencies observed for hydrogen chemisorbed on metal single-crystal surfaces. No vibrational frequencies for hydrogen bonding below the surface (e.g. between the first and second layers of metal atoms) have been reported. In most cases, the vibrational frequencies in Tables 2 and 3 are relatively insensitive to coverage, varying by at most 25 cm^{-1} as a result of lateral interactions [49a]. However, on Ni(110), W(100), and Mo(100) the vibrational frequencies for adsorbed hydrogen change dramatically with coverage as a result of surface reconstructions. The metal atoms on these surfaces are induced by hydrogen adsorption at certain coverages to adopt new equilibrium positions. This change in metal surface geometry, as we show below, greatly affects the H-atom bonding geometry and thus the vibrational frequencies.

Assigning the observed vibrational frequencies in Tables 2 and 3 and determining the adsorbed state of the hydrogen atoms is not straightforward. It appears that H atoms on the hexagonally close-packed surfaces tend to be delocalized parallel to the surface as supported by the assignments in Table 1 and their agreement with the calculations for Ni(111) [36,37]. Since this delocalization results from the overlap of the wavefunctions of H atoms in adjacent sites, surfaces with larger corrugation or with a greater distance between favored adsorption sites may bond H atoms to a greater degree at

TABLE 3 Frequencies Observed by HREELS for Adsorbed Hydrogen

Metal surface	Refs.	Frequencies (cm^{-1})	Proposed adsorption site	Heat of adsorption (kcal mol^{-1})
W(100)	23, 42	440, 960, 1215 (low coverage, reconstructed)	Bridge	32
		680, 1000, 1280 (high coverage, unreconstructed)	Bridge	
W(110)	43, 45	645, 775, 1290	Long bridge	33
W(111)	42, 46	1290	Top	37
Fe(110)	47, 48	880, 1060	Short bridge	26
Mo(100)	48	1125, 1240–1260 (low coverage, reconstructed)	Several bridge sites	24
		555, 1030, 1125 (high coverage, unreconstructed)	Bridge	
Ni(100)	38, 43	595	Fourfold hollow	23
Ni(110)	38, 44	650, 1060 (low coverage, unreconstructed)	Bridge and threefold hollow	22
		610, 940 (high coverage, reconstructed)	Several bridge sites	
Pd(100)	49	515	Fourfold hollow	21
Pt[6(111) × (111)]	50	500, 1130, 1270	Threefold hollow, and bridge	12

localized sites. The bonding of H atoms in bridge sites on W(100) is a good example of this localized adsorption.

For hydrogen chemisorbed on other surfaces, the interpretation of the observed vibrational frequencies is less clear. The most common approach is to use a localized, valence-force field interpretation, such as was used to describe hydrogen on W(100). Table 3 includes proposed sites for hydrogen adsorption based on this assumption. It is not clear, however, whether the wavefunctions are localized enough on these surfaces for this approach to be valid. For example, on Ni(100) the only observed loss (at 74 meV, 595 cm^{-1}) was attributed by Andersson [43] to a vibration of the H atoms perpendicular to the surface in a fourfold hollow site. However, this frequency also agrees reasonably well with the calculated energy of transition between the A_1^0 and A_1^1 bands (62 meV) for delocalized hydrogen on Ni(100) [36,37]. Even though the calculations of Puska et al. predict five transitions in the range of frequencies studied, the observation of only one loss for hydrogen on the Ni(100) surface may be due to low excitation probabilities for these other transitions.

It should be noted that even for "delocalized" hydrogen adsorption, certain surface sites will have a higher probability of occupation than others, as shown in the right panel of Fig. 9. For example, dynamical low-energy electron diffraction (LEED) calculations have determined that H atoms bond in threefold hollow sites on Ni(111) [35]. The M-H bond length of 1.84 Å determined by LEED compares favorably with that of 2.05 Å for hydrogen on W(100) as determined by HREELS [23], since the covalent radius of W is 0.15 Å longer than that of Ni.

Despite the difficulties in interpreting the HREEL spectra of adsorbed H atoms, some general observations are possible. First, the vibration of a hydrogen atom perpendicular to the surface shifts to lower frequency with increased coordination to the metal [51]. Thus $\omega_\perp$ for top site > bridge site > threefold hollow site > fourfold hollow site. Second, the observed vibrational frequencies are highly sensitive to the local bonding geometry [51], as shown in Fig. 7 for bridge-bonded H atoms in hydrides. Third, the local bonding geometry of H atoms, and thus the observed vibrational frequencies, are sensitive to the geometry of the metal. This can be seen by comparing the vibrational frequencies observed for hydrogen on the (111), (100), and (110) faces of Ni and W. For these two metals it appears that higher corrugation of the surface favors lower coordination of the H atom. For example, on the rough W(111) surface, hydrogen bonds to top sites versus bridge sites on the smoother W(100) and W(110) surfaces and, on the bumpy Ni(110) surface, hydrogen bonds to bridge sites versus three- and fourfold hollow sites on Ni(111) and Ni(100). Finally, metals toward the left of the transition metal series (W, Mo, Fe) seem to favor lower coordination of H atoms than do those on the right (Pt, Pd, Ni).

It is interesting to try to correlate the vibrational frequencies with other physical parameters of adsorbed hydrogen. One parameter that has been measured for hydrogen adsorbed on many single-crystal metal surfaces is the heat of adsorption. Tables 2 and 3 list the heats of adsorption of hydrogen on metal surfaces together with the observed vibrational frequencies. Only the heats of adsorption for low hydrogen coverages are listed since, at high coverages, lateral interactions become important and can complicate the interpretation of the data. For hexagonally close-packed surfaces, the heats of adsorption vary dramatically, from 9.5 kcal mol^{-1} for hydrogen on Pt(111) to 29 kcal mol^{-1} for hydrogen on Ru(001), even though the vibrational frequencies of chemisorbed hydrogen are similar on these surfaces. This suggests that the vibrational modes are more sensitive to the structure of the crystal face than to the strength of the hydrogen-metal bond. Further evidence for this can be found by comparing different crystal faces of the same metal; the vibrational spectra are quite different for hydrogen adsorbed on the different crystal faces of nickel and tungsten, even though the heats of adsorption do not change dramatically from one crystal face to another on these metals.

III. VIBRATIONAL SPECTROSCOPY OF HYDROGEN ADSORBED ON HIGH-SURFACE-AREA METALS AT ATMOSPHERIC PRESSURE

While vacuum conditions are necessary for electron spectroscopies such as HREELS, several techniques—incoherent inelastic neutron scattering, transmission-adsorption infrared spectroscopy, reflection-adsorption infrared spectroscopy, and Raman spectroscopy—are able to do vibrational spectroscopy of the metal/gas interface at high gas pressures. Since these probes interact only weakly with atoms and molecules, it is often necessary to use samples with a large surface-to-volume ratio in order to be sensitive to surface adsorbates. Using samples with a large surface-to-volume ratio is not necessarily a disadvantage since the same types of samples are used in heterogeneous catalysis; so by these techniques we are able to study hydrogen under conditions similar to those present in catalysis.

Although Chapters 6 and 7 deal in detail with neutron scattering and IR spectroscopy of dispersed catalysts, respectively, in what follows we describe these techniques briefly and review the results obtained for hydrogen adsorbed on metal surfaces.

A. Incoherent Inelastic Neutron Scattering

In a neutron scattering experiment, the neutrons scatter by interacting with the nuclei of the atoms in the sample. As was the case for electrons in HREELS, a fraction of the neutrons lose energy when they scatter, by exciting vibrational modes of the sample. Unlike HREELS, where the electrons do not penetrate significantly beyond the surface, neutrons can easily transverse a sample; so to be useful as a surface probe, the adsorbate must have a cross section which is substantially larger than that of the substrate. Fortunately, the incoherent inelastic cross section of hydrogen is at least 20 times greater than that of any other common element, so it is possible to obtain reasonable scattered intensity from hydrogen-containing adsorbates. Several good reviews [6] have been written on neutron scattering as well as its application to the study of adsorbates.

To date, only a few types of metallic surfaces have been used in neutron scattering experiments, due to the difficulties in obtaining samples with a surface that is both clean and has a large area. So far, the adsorption of hydrogen has been studied by incoherent inelastic neutron scattering (IINS) on Raney nickel [52] and on platinum [53] and palladium blacks [54]. Table 4 gives a summary of the inelastic neutron scattering results obtained for hydrogen adsorption on Raney nickel and platinum black with those obtained by HREELS for hydrogen on Ni(100), Ni(111), Ni(110), Pd(100), and Pt(111) shows that most of the HREELS frequencies are also observed in the neutron scattering experiments. This is probably due to hydrogen adsorption on the same type of crystal faces in the high-surface-area samples. However, more modes are observed by IINS than by HREELS; these could be due to adsorption on different crystal surfaces and on imperfections such as steps or kinks, or could be due to modes that are weak scatterers by HREELS (i.e., modes parallel to the surface) but that are relatively intense by IINS.

TABLE 4 Frequencies (cm^{-1}) Observed by IINS for Adsorbed Hydrogen

Raney nickel [52]			
Low coverage	High coverage	Platinum black [53]	Palladium black [54]
		2000-2250	
		1696	
1080	1100	1296	
	940	936	916
780		856	823
	600-640	616	
	400	512	

Since the surfaces used in IINS are less well defined than the single-crystal surfaces used in HREELS, it is difficult to make a definite assignment of the observed frequencies. Most researchers have interpreted the IINS data for hydrogen adsorption in terms of localized harmonic oscillators at multicoordinated sites. As was the case in interpreting HREEL spectra, however, hydrogen may not always be well approximated by a localized harmonic oscillator, but such an approximation is probably more valid for high-surface-area metals, which have more defects and a larger surface corrugation than "smooth" single-crystal surfaces. Further, most of the values for the transition energies calculated by Puska et al. [36,37] for delocalized hydrogen on Ni(100) and Ni(111) are in poor agreement with those observed in the neutron scattering experiments, implicating a more localized adsorption on these high-surface-area metals.

B. Infrared and Raman Spectroscopy

These spectroscopies, like IINS, offer both better spectral resolution than HREELS and the ability to study gas adsorption at atmospheric pressures on high-surface-area catalysts. IR and Raman spectroscopies are also less costly and more readily available than thermal neutron sources. However, because of the small dynamic dipole moment and small polarizability of metal-hydrogen bonds, measurement of the vibrational spectrum for chemisorbed hydrogen by IR or by Raman spectroscopy is difficult. Only a few metal/hydrogen systems have been studied, to date, and transmission absorption infrared spectroscopy (TAIRS) has been used in all but a few of these studies.

In TAIRS, an infrared beam is transmitted through the sample and its adsorption as a function of frequency is measured before and after the introduction of gas into the sample cell. The differences in the adsorption spectra gives the vibrational frequencies of the adsorbed species. To use this spectroscopy, it is important to have high-surface-area samples so that the spectra have a significant contribution from the surface. Also, it is important to have samples that have good transmission in the frequency range of interest. However, even with the most infrared transparent samples, the spectral range is limited to 1000 to 5000 cm^{-1}. For these reasons, TAIRS studies of adsorbed hydrogen have been limited to metal powders or metals dispersed on oxide supports that give high IR transmission and that have a high surface area.

Reflection absorption infrared spectroscopy (RAIRS) is similar to TAIRS except that the absorption of infrared radiation is detected by the change in intensity of a reflected IR beam when an adsorbate is present on the reflecting surface. Samples must therefore be reflective and are generally polycrystalline foils or single crystals. The spectral range is limited by the transmission of the infrared windows to frequencies above 400 cm^{-1}. Recently, vibrational spectra of hydrogen adsorbed on a single-crystal surface [W(100)] in ultrahigh vacuum have been obtained using RAIS [34]; in the future, this technique promises to be an important competitor to HREELS for obtaining vibrational spectra of hydrogen adsorbed on single-crystal surfaces. An excellent review of this spectroscopy has been written by Hoffmann [9].

The vibrational spectrum in Raman spectroscopy (reviewed in Refs. 7 and 8) is recorded by measuring the inelastic scattering of visible light. Raman spectroscopy has an even lower cross section for vibrational excitation of adsorbed hydrogen than does IR spectroscopy, so high-surface-area samples like those for TAIRS are generally used. However, because visible rather than infrared light is used, supports are transparent to lower frequencies, and spectra down to 200 cm^{-1} can be obtained.

Table 5 gives a summary of the TAIRS, RAIRS, and Raman spectroscopy data for hydrogen adsorbed on high-surface-area metals and metals dispersed on oxide supports. All the frequencies observed are fairly high (>1800 cm^{-1}), in contrast to the lower values observed for hydrogen adsorption on unsupported metals. However, the lower-frequency modes (<1000 cm^{-1}) lie outside the obtainable spectral range and are probably present but not observed. The high-frequency modes (which are distinguished from adsorbed CO and other contaminants by measuring the frequency shift for deuterium) are attributed to hydrogen atoms bonded to single metal atoms (i.e., bonded at a "top site"). These frequencies are consistent with those for terminally bound hydrogen atoms in metal hydride clusters (1900 to 2250 cm^{-1}). As discussed earlier, bridge and hollow sites are thought to be the more favorable hydrogen adsorption sites for hydrogen on low-surface-area metals.

TABLE 5 Frequencies Observed by TAIRS, RAIRS, and Raman Spectroscopy for Hydrogen Adsorbed on Dispersed Metal Surfaces

Metal/support	Spectroscopy	Ref.	ν_{M-H} (cm^{-1})	ν_{M-D} (cm^{-1})
Pt/Al$_2$O$_3$	TAIRS	55	2120, ~2060	1520, 1490
Ir/Al$_2$O$_3$	TAIRS	55	2120, ~2050	1520, 1490
Ni/Al$_2$O$_3$	TAIRS	55	1880	1360
Fe, Co, Ni, Rh, Pd, and Ir/Al$_2$O$_3$	TAIRS	55	1850-1940	—
Ni/SiO$_2$	Raman	56	2028, 1999 1600, 950 725, 692	1410, 1380 1150, 740 —, —
Pd film	RAIR	57	760, 880	—, —

Indeed, the RAIRS study of H adsorbed on a Pd film shows only lower-frequency peaks characteristic of a bridge or hollow site, while the Raman spectroscopy study of H adsorbed on Ni/SiO$_2$ shows both high($>$1200 cm^{-1})- and low ($<$1200 cm^{-1})-frequency peaks. This confirms that both top-site hydrogen atoms and multiply bonded hydrogen atoms are present on high-surface-area dispersed metals, but the latter sites are not detectable by TAIRS.

IV. CONCLUSIONS

Vibrational spectra of hydrogen adsorbed on metal surfaces show that on both single-crystal and high-surface-area metal surfaces, hydrogen is dissociatively adsorbed. The exact nature of the adsorbed hydrogen atoms is difficult to determine from vibrational spectroscopy alone, but some general trends have been found. The adsorbed state of H atoms is sensitive to both the chemical nature of the metal and the geometry of the metal. It appears that hydrogen atoms bond preferentially at bridge and hollow sites on "flat" single crystals and thin films, but bonding at top sites also occurs on "rougher" high-surface-area metals. The observed vibrational frequencies depend more on the structure of the crystal face than on the strength of the metal-hydrogen bond as measured by heats of adsorption. Also, the vibrational frequencies are quite sensitive to changes in bond lengths and bond angles, so that analogous surface sites on different metal surfaces have different vibrational frequencies.

While the vibrational frequencies and bonding geometries of hydrogen atoms on metal surfaces appear in many cases to be analogous to those of metal hydride clusters, recent vibrational spectra and calculations for H atoms on "smooth" metal surfaces indicate a unique form of hydrogen adsorbed on surfaces—"delocalized hydrogen." This delocalized adsorption is more pronounced for hydrogen atoms than for other adatoms, because hydrogen's low mass results in a large "zero-point" motion. Since much of the reactivity of surface hydrogen undoubtedly occurs from thermally excited atoms, which are particularly delocalized, perhaps one should not think of these atoms as having a well-defined bonding geometry; maybe, just as one thinks of "electron clouds" for electrons, one should think of a "hydrogen cloud" or, more descriptively, a "hydrogen fog" that covers the surface and is responsible for the reactivity of adsorbed hydrogen atoms.

Experimental techniques for doing vibrational spectroscopy of adsorbates at surfaces are continually being developed and improved. Recent improvements in infrared and Raman spectroscopy should make it possible to extend the vibrational spectroscopy data base for adsorbates on single-crystal surfaces in ultrahigh vacuum to low- and high-surface-area metals at high pressures. New directions are now possible, including the study of the reactivity of different types of adsorbed hydrogen by isotopic labeling [58], and determination of the effect of coadsorbed hydrogen on the bonding of other adsorbates [59]. HREEL spectra on single-crystal surfaces in UHV after high-pressure gas exposures using isotope labeling show promise in following the reactivity of adsorbed hydrogen, and studies have shown that adsorbed H atoms may force changes in the adsorption site of coadsorbed species. Clever use of vibrational spectroscopy, together with a variety of well-chosen catalytic chemical reactions on different surfaces, will hopefully enable us to develop an atomic-scale understanding of the role of hydrogen in heterogeneous catalysis.

ACKNOWLEDGMENTS

We thank Dr. M. A. Van Hove and Professor L. M. Falicov for helpful discussions in connection with this work. The work was supported by the Director, Office of Energy Research, Office of Basic Energy Sciences, Materials Sciences Division of the U.S. Department of Energy under Contract DE-A003-76SF00098.

REFERENCES

1. G. T. Haller, *Catal. Rev.-Sci. Eng.*, 23(4):477 (1981).
2. W. N. Delgass, G. L. Haller, R. Kellerman, and J. H. Lunsford, *Spectroscopy in Heterogenous Catalysis*, Academic Press, New York (1979).
3. M. L. Hair, *Infrared Spectroscopy in Surface Chemistry*, Marcel Dekker, New York (1967).
4. C. B. Duke, *J. Electron Spectrosc. Relat. Phenom.*, 29:1 (1983).
5. H. Ibach and D. L. Mills, *Electron Energy Loss Spectroscopy and Surface Vibrations*, Academic Press, New York (1982).
6. P. G. Hall and C. J. Wright, in *Chemical Physics of Solids and Their Surfaces*, Vol. 17 (M. W. Roberts and J. M. Thomas, eds.), The Chemical Society, Burlington House, London (1978).
7. R. P. Cooney, G. Curthoys, and N. T. Tam, *Adv. Catal.*, 24:293 (1975).
8. R. K. Chang and T. E. Furtak (eds.), *Surface Enhanced Raman Scattering*, Plenum Press, New York (1982).
9. F. M. Hoffmann, *Surf. Sci. Rep.*, 3(2/3):107 (1983).
10. R. P. Eischens and W. A. Pliskin, *Adv. Catal.*, 10:1 (1958).
11. S. Chiang, R. G. Tobin, and P. L. Richards, *Phys. Rev. Lett.*, 49(23):648 (1984).
12. R. M. Kroeker and P. K. Hansma, *Catal. Rev.-Sci. Eng.*, 23(4):553 (1981).
13. F. Trager, H. Coufal, and T. J. Chuang, *Phys. Rev. Lett.*, 49(23):1720 (1982).
14. A. M. Baro, H. Ibach, and H. D. Bruchmann, *Surf. Sci.*, 88:384 (1979).
15. J. E. Demuth, D. Schmeisser, and Ph. Avouris, *Phys. Rev. Lett.*, 47:1166 (1981); Ph. Avouris, D. Schmeisser, and J. E. Demuth, *Phys. Rev. Lett.*, 48:199 (1982); S. Andersson and J. Harris, *Phys. Rev. Lett.*, 48:545 (1982).
16. H. Froitzheim, in *Electron Spectroscopy for Surface Analysis, Topics in Current Physics*, Vol. 4 (H. Ibach, ed.), Springer-Verlag, Berlin (1977), pp. 205-250; H. Ibach, H. Hopster, and B. Sexton, *Appl. Surf. Sci.*, 1:1 (1977); B. E. Koel and G. A. Somorjai, in *Catalysis: Science and Technology*, Vol. 38 (J. R. Anderson and M. Boudart, eds.), Springer-Verlag, New York (1986); B. A. Sexton, *Appl. Phys.*, A26:1 (1981); Ph. Avouris and J. E. Demuth, *Ann. Rev. Phys. Chem.*, 35:49 (1984).
17. F. M. Propst and T. C. Piper, *J. Vac. Sci. Technol.*, 4:53 (1967).
18. H. Froitzheim, H. Ibach, and S. Lehwald, *Phys. Rev. Lett.*, 36:1549 (1976).
19. C. Backx, B. Feuerbacher, B. Fitton, and R. F. Willis, *Phys. Lett.*, A60:145 (1977).
20. A. Adnot and J.-D. Carette, *Phys. Rev. Lett.*, 39:209 (1977).
21. W. Ho, R. F. Willis, and E. W. Plummer, *Phys. Rev. Lett.*, 40:1463 (1978).
22. M. R. Barnes and R. F. Willis, *Phys. Rev. Lett.*, 41:1729 (1978).

23. R. F. Willis, *Surf. Sci.*, *89*:457 (1979).
24. R. F. Willis, W. Ho, E. W. Plummer, *Surf. Sci.*, *80*:593 (1979).
25. W. Ho, N. J. DiNardo, and E. W. Plummer, *J. Vac. Sci. Technol.*, *17*: 134 (1980).
26. W. Ho, R. F. Willis, and E. W. Plummer, *Phys. Rev.*, *B21*:4202 (1980).
27. B. M. Hall, S. Y. Tong, and D. L. Mills, *Phys. Rev. Lett.*, *50*:1277 (1983).
28. E. F. J. Didham, W. Allison, and R. F. Willis, *Surf. Sci.*, *126*:219 (1983).
29. S. R. Bare, P. Hofmann, M. Surman, and D. A. King, *J. Electron Spectrosc. Relat. Phenom.* *29*:265 (1983).
30. C. M. Mate and G. A. Somorjai, *Phys. Rev.*, *B34*:7417 (1986).
31. G. J. Schulz, *Rev. Mod. Phys.*, *45*:378 (1973).
32. H. Ibach and D. L. Mills, *Electron Energy Loss Spectroscopy and Surface Vibrations*, Academic Press, New York (1982), pp. 116-120.
33. M. W. Howard, U. A. Jayasooriya, S. F. A. Kettle, D. B. Powell, and N. Sheppard, *Chem. Commun.*, *929*:18 (1979).
34. Y. J. Chabal, *Phys. Rev. Lett.*, *55*(8):845 (1985); refers to a future publication by J. P. Woods and J. L. Erskine.
35. K. Christmann, R. J. Behm, G. Ertl, M. A. Van Hove, and W. H. Weinberg, *J. Chem. Phys.*, *70*:4168 (1979).
36. M. J. Puska, R. M. Nieminen, M. Manninen, B. Chakraborty, S. Holloway, and J. K. Norskov, *Phys. Rev. Lett.*, *51*:1081 (1983).
37. M. J. Puska and R. M. Nieminen, *Surf. Sci.*, *157*:413 (1985).
38. K. Christmann, O. Schober, G. Ertl, and M. Neumann, *J. Chem. Phys.*, *60*:4528 (1974).
39. M. A. Barteau, J. Q. Broughton, and D. Menzel, *Surf. Sci.*, *133*:443 (1983); H. Conrad, R. Scala, W. Stenzel, and R. Unwin, *J. Chem. Phys.*, *81*:6371 (1984); P. Feulner and D. Menzel, *Surf. Sci.*, *154*:465 (1985).
40. J. T. Yates, Jr., P. A. Thiel, and W. H. Weinberg, *Surf. Sci.*, *84*:427 (1979).
41. K. Christmann, G. Ertl, and T. Pignet, *Surf. Sci.*, *54*:365 (1976).
42. P. W. Tamm and L. D. Schmidt, *J. Chem. Phys.*, *54*:4475 (1971).
43. S. Andersson, *Chem. Phys. Lett.*, *55*:185 (1978).
44. N. J. DiNardo and E. W. Plummer, *Surf. Sci.*, *150*:89 (1985); L. Olle and A. M. Baro, *Surf. Sci.*, *137*:607 (1984); M. Nishijima, S. Masuda, M. Jo, and M. Onchi, *J. Electron Spectrosc. Relat. Phenom.*, *29*:273 (1983).
45. G. B. Blanchet, N. J. DiNardo, and E. W. Plummer, *Surf. Sci.*, *118*:496 (1982).
46. C. Backx, B. Feuerbacher, B. Fitton, and R. F. Willis, *Surf. Sci.*, *63*:193 (1977).
47. A. M. Baro and W. Erley, *Surf. Sci.*, *112*:L759 (1981); F. Bozso, G. Ertl, M. Grunze, and M. Weiss, *Appl. Surf. Sci.*, *1*:103 (1977).
48. F. Zaera, E. B. Kollin, and J. L. Gland, *Surf. Sci.*, *166*:L149 (1986).
49. C. Nyberg and C. G. Tengstal, *Phys. Rev. Lett.*, *50*(21):1680 (1983); C. Nyberg and C. G. Tengstal, *Surf. Sci.*, *126*:163 (1983); K. Christmann and J. E. Demuth, *J. Chem. Phys.*, *76*(12):6318 (1982).
50. A. M. Baro and H. Ibach, *Surf. Sci.*, *92*:237 (1980).
51. P. Nordlander, S. Holloway, and J. K. Norskov, *Surf. Sci.*, *136*:59 (1984).
52. (a) H. Jobic and A. Renouprez, *J. Chem. Soc. Faraday Trans. 1*, *80*:991 (1984); (b) A. Renouprez, P. Fouilloux, G. Coudurier, D. Tocchetti, and R. Stockmeyer, *Trans. Faraday Soc.*, *73*:1 (1977); (c) R. Stockmeyer, H. M. Stortnik, I. Natkaniec, and J. Mayer, *Ber. Bunsenges.*

Phys. Chem., 84:79 (1980); (d) R. D. Kelley, J. J. Rush, and T. E. Madey, Chem. Phys. Lett., 66:159 (1979).
53. J. Howard, T. C. Waddington, and C. J. Wright, J. Chem. Phys., 64: 3897 (1976); J. Howard, T. C. Waddington, and C. J. Wright, Neutron Inelastic Scattering 1977, Vol. 2, IAEA, Vienna (1978), p. 499.
54. J. Howard, T. C. Waddington, and C. J. Wright, Chem. Phys. Lett., 56: 258 (1978).
55. U. A. Jayasooriya, M. A. Chesters, M. W. Howard, S. F. A. Kettle, D. B. Powell, and N. Sheppard, Surf. Sci., 93:526 (1980).
56. W. Krasser and A. J. Renouprez, J. Raman Spectrosc., 8(2):92 (1979).
57. I. Ratajczykowa, Surf. Sci., 48:549 (1975).
58. B. E. Koel, B. E. Bent, and G. A. Somorjai, Surf. Sci., 146:211 (1984).
59. P. A. Thiel and W. H. Weinberg, J. Chem. Phys., 73(8):4081 (1980); G. E. Mitchell, J. L. Gland, and J. M. White, Surf. Sci., 131:167 (1983); R. D. Kelley, J. J. Rush, and T. E. Madey, Chem. Phys. Lett., 66: 159 (1979).

II
CHARACTERIZATION OF SURFACE HYDROGEN ON CATALYSTS

ℭ | Energetics of Hydrogen Adsorption on Porous and Supported Metals

J. W. GEUS

University of Utrecht, Utrecht, The Netherlands

I.	INTRODUCTION	85
II.	FORMATION OF BULK HYDRIDES AND ADSORPTION OF HYDROGEN	86
III.	EXPERIMENTAL DETERMINATION OF THE EXTENT OF HYDROGEN ADSORPTION	93
IV.	EXTENT OF ADSORPTION OF HYDROGEN BY PURE METALS	95
V.	ADSORPTION OF HYDROGEN BY SUPPORTED METALS	99
VI.	CONCLUSIONS	113
	REFERENCES	113

I. INTRODUCTION

Many important technical reactions involve catalytic hydrogenations using metallic catalysts. Sabattier and Senderens developed catalytic hydrogenation at the turn of the century [1]. It would therefore be expected that catalytic hydrogenation has been dealt with exhaustively. However, many details of catalytic hydrogenations, especially selective hydrogenations, are still obscure, and a more profound understanding of the fundamental atomic or molecular processes proceeding on the surface of metallic hydrogenation catalysts is greatly desired. A first step toward such an understanding is a broad survey of the amount and properties of adsorbed hydrogen.

The adsorption of hydrogen is also being used to determine the free metal surface area of supported catalysts. This determination is very important in elucidating the mechanism of catalytic reactions. Many fundamental studies on the catalytic properties of metals are based on turnover numbers [2]. These numbers are based on metal surface areas that are generally calculated from the extent of hydrogen adsorption on previously reduced and degassed catalysts. Important characteristics of supported catalysts, such as structure sensitivity [3-6], are thus based on the extent of hydrogen adsorption and the assumptions used in the calculation of the metal surface area from the experimental data.

This chapter deals with the adsorption of hydrogen on porous and supported metals. In contrast to experiments on single crystals and filaments, transport of hydrogen into pores present in metal specimens or in supported catalysts can appreciably affect the results [7,8]. Since in many technically important instances porous or supported metals are utilized, it is very relevant to survey the adsorption of hydrogen on these materials and the properties of the hydrogen thus adsorbed. It is especially important to apply the results obtained on single crystals and filaments to polycrystalline porous and supported metals.

The chapter begins with a survey of the ability of elements to react to bulk hydrides in relation to their position in the periodic table and the ability to dissociatively adsorb hydrogen. In addition to Christmann's contribution (Chapter 1), some remarks about the transition between adsorption and absorption of hydrogen will be made. Subsequently, both activated and nonactivated adsorption of hydrogen are dealt with.

In Section III we consider the experimental determination of the extent of hydrogen adsorption. Besides the conventional volumetric procedure, some other techniques to assess the amount of hydrogen adsorbed have been developed. In Section IV we discuss the extent of hydrogen adsorption on various metals as measured on porous metal specimens. The hydrogen adsorption on these unsupported metals can be related to the BET (Brunauer-Emmett-Teller) surface area, which allows a more direct assessment of the number of hydrogen atoms adsorbed per metal surface atom.

Experimental results on the adsorption of hydrogen by supported metals are described in Section V. A number of features to be considered in a reliable determination of the amount of hydrogen adsorbed per reduced metal atom are discussed.

II. FORMATION OF BULK HYDRIDES AND ADSORPTION OF HYDROGEN

Hydrogen forms chemical bonds with almost all elements in the periodic table. Depending on the difference in electronegativity, hydrogen forms saltlike compounds, in which the hydrogen is negatively charged, and metallic or covalent hydrides, where the charge on the hydrogen or the metal atoms is small [9,10]. Although bare protons may be important in the transport of hydrogen through metals, the presence of protons at stable positions in metals is unlikely [9]. Here we concentrate on the metallic and saltlike hydrides. As can be expected from the low electronegativities, the elements of groups IA and IIA of the periodic table form saltlike solid hydrides. While the hydrides of elements of group III have properties intermediate between metallic and saltlike, the elements of groups IV and V, together with Cr and U of group VI, can react to metallic bulk hydrides. Palladium occupies a peculiar position; it is the only metal of group VIII that reacts to a bulk hydride at low hydrogen pressures. Of this group nickel also forms a bulk hydride, but only at elevated pressures (about 4000 bar) [11]. The properties of copper hydrides are still poorly known.

With the formation of bulk hydrides a distinction has to be made between the reaction to a separate hydride phase and a mere solution of hydrogen into the metal. In the foregoing criterion regarding the formation of hydrides we have assumed the formation of a solid hydride, not the solution of generally limited amounts of hydrogen into the metal. Although with small amounts of hydrogen taken up, a new solid phase is not formed, a separate solid phase coexisting with the original metal containing a small amount of dissolved

3. Energetics of Hydrogen Adsorption

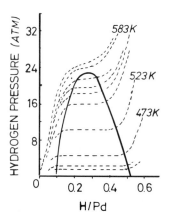

FIGURE 1 Absorption isotherms for hydrogen in palladium. The horizontal parts of the isotherms correspond to the equilibrium between the α- and β-hydride. The β-hydride has a much higher content of hydrogen, whereas the α phase contains a small amount of mobile hydrogen.

hydrogen is observed with metals exhibiting the formation of bulk hydrides. The presence of an additional phase is evident from the fact that the hydrogen pressure displays a constant plateau at a fixed temperature as long as the two solid phases are present. Since the energy of hydrogen atoms accommodated into the hydride structure is lower, but the entropy is lower as well, an equilibrium analogous to that exhibited with condensation-evaporation equilibria is set up (Fig. 1).

Since in the formation of bulk hydrides the hydrogen atoms have to be accommodated between the metal atoms, it is obvious that adsorption of hydrogen can proceed more easily than the formation of bulk hydrides. Adsorption of hydrogen is thus exhibited by a much wider range of metals than the formation of bulk hydrides. As a criterion for hydrogen adsorption we take nonactivated adsorption of about a monolayer of hydrogen. A survey of the literature shows that the metals of groups IIA to VIII display the hydrogen adsorption described above (Fig. 2). The elements of groups IB and IIB do not adsorb nonactivatedly about a monolayer of hydrogen [12-15]. Copper, an element of group IB, is especially interesting, since copper exhibits activity in both dehydrogenation and hydrogenation reactions.

The foregoing criterion for the adsorption of hydrogen makes demands on both the thermodynamics and kinetics of the adsorption. Since the dissociation energy of molecular hydrogen is 436 kJ, the adsorption energy of hydrogen atoms must be above about 220 kJ to compensate for the loss of entropy and thus to render adsorption thermodynamically stable. Starting from atomic hydrogen, adsorption does not involve dissociation of molecular hydrogen. It has been observed that exposure of clean metal surfaces to atomic hydrogen initially always leads to the adsorption of hydrogen atoms. Accordingly, the adsorption of hydrogen from gaseous molecular hydrogen merely indicates that the adsorption energy is above about 220 kJ. The fact that transition metals adsorb a monolayer of hydrogen atoms and, for example, Ib metals do not, indicates that the adsorption bond with transition metals is stronger, but does not imply that the Ib metals are not adsorbing hydrogen atoms.

Whether the hydrogen taken up from gaseous hydrogen atoms is stable depends on the adsorption energy and the mobility of the adatoms [16]. Provided that the hydrogen atoms are sufficiently mobile, atoms adsorbed

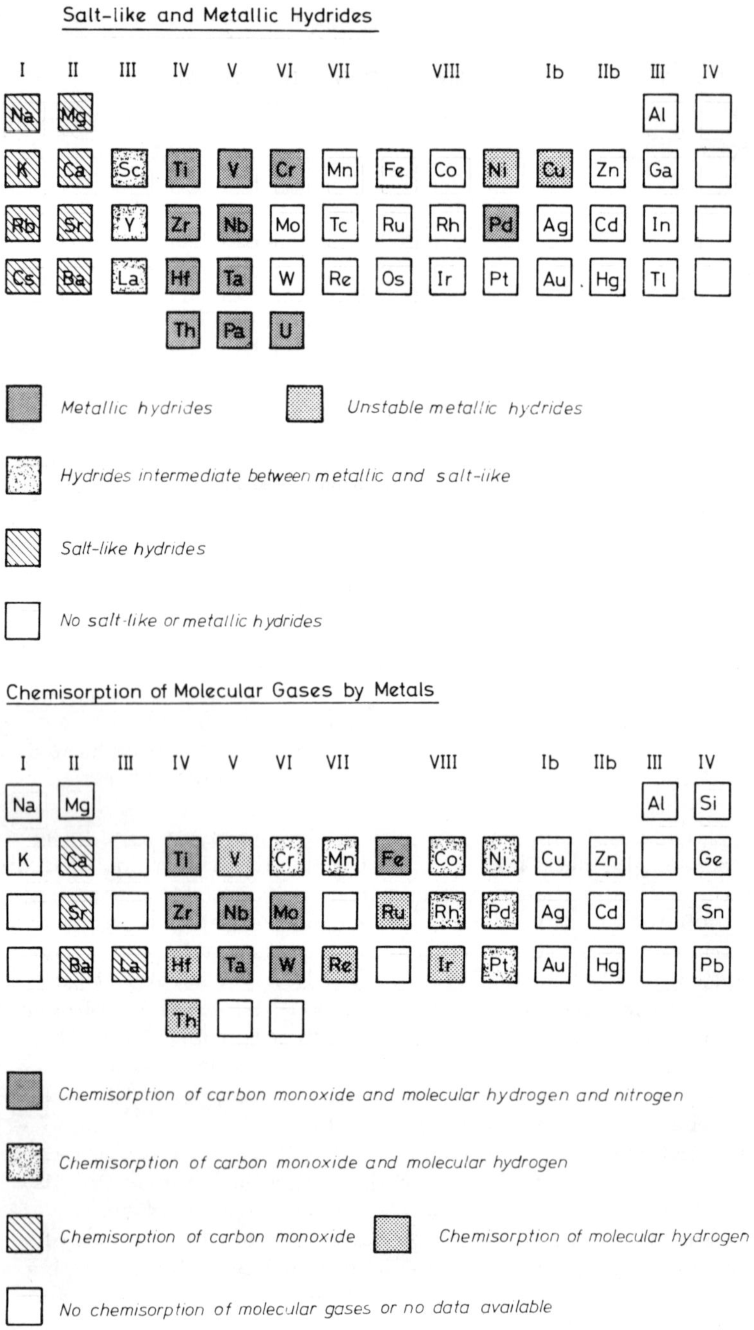

FIGURE 2 Survey of the elements forming bulk hydrides and those exhibiting nonactivated adsorption of about a monolayer of hydrogen atoms.

with an energy below about 220 kJ will desorb as hydrogen molecules. At temperatures where the hydrogen atoms are not mobile, hydrogen atoms bonded with an energy below about 220 kJ also remain adsorbed. When at rising temperatures mobility of the adsorbed hydrogen sets in, rapid desorption of the adatoms bonded with an energy below 220 kJ proceeds. The mobility of adsorbed hydrogen atoms can also be substantially decreased by adsorbed poisoning atoms or molecules. It has been established that adsorbed sulfur atoms can strongly lower the mobility of hydrogen atoms and thus impede desorption of molecular hydrogen. These experiments were carried out with hydrogen atoms produced electrochemically [17]. It is possible also that nondissociatively adsorbed carbon monoxide can substantially decrease the surface mobility of adsorbed hydrogen and thus prevent complete desorption of hydrogen.

With adsorption of hydrogen both dissociation of the hydrogen molecule and modification of the metal surface involving weakening of intermetallic bonds or changing the positions of metal surface atoms have to be considered. Both the modification of the metal surface structure and the dissociation of the hydrogen molecules can affect the kinetics of the dissociative adsorption of hydrogen. Generally, the dissociation of molecular hydrogen is considered only. Since adsorption of atomic hydrogen thus calls for rupture of the strong chemical bond between the hydrogen atoms, it can be expected that the kinetics of the dissociative adsorption strongly determines the extent of hydrogen adsorption.

Accordingly, in the old literature the adsorption of hydrogen on many metal surfaces is stated to be activated; this implies that a considerable thermal energy is required for an incident hydrogen molecule to pass over the energy barrier to the adsorbed atomic state (Fig. 3). As a result, the extent of adsorption of hydrogen passes through a maximum as the temperature is raised (Fig. 4). However, almost all of the older results have been measured on contaminated metal surfaces. Field-emission experiments [18,19] especially have shown adsorption of hydrogen, even at 10 K, not to be activated with many metals provided that the metal surface is (almost) completely clean. Hence the interaction of hydrogen with clean metal surface is not activated, unlike the reaction of molecular hydrogen with adsorbed impurities present in the old work. Generally, activated adsorption has been observed with metal specimens the surfaces of which were not sufficiently freed from oxygen in the previous reduction treatment or were oxidized during evacuation subsequent to the reduction.

However, porous vapor-deposited metal films thrown in ultrahigh vacuum onto cooled substrates also exhibit a maximum in the adsorption isobar [20-22]. Nevertheless, the surfaces of carefully thrown metal films are doubtless clean. It has been shown that the maximum in the extent of adsorption with the temperature is due to the transport of hydrogen atoms into narrow pores in vapor-deposited metal films. To cover metal surfaces of narrow pores, surface migration of hydrogen atoms, and thus surface mobility, are required.

Copper is the only metal for which activated adsorption of hydrogen seems to be well established [15]. Since the energy of adsorption of hydrogen on copper is rather small, the equilibrium coverage at higher temperatures is low. Since more elevated temperatures are required to establish dissociative adsorption of hydrogen on copper, the extent of adsorption remains small. Nevertheless, the ability of copper to dissociate hydrogen at higher temperatures accounts for the activity in hydrogenation reactions. With hydrogenation over copper catalysts, generally high hydrogen pressures are required to establish a sufficiently high coverage of hydrogen atoms. Therefore, many catalytic hydrogenations over copper catalysts call for hydrogen pressures from 100 to 300 bar. In Chapter 1, Christmann dealt with the effect of the

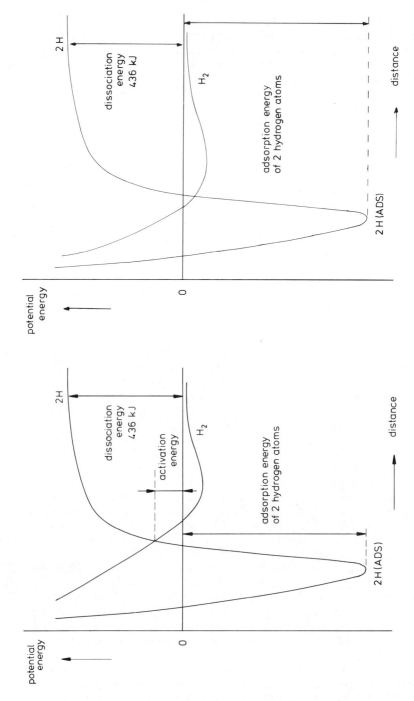

FIGURE 3 Potential energy of a hydrogen molecule as a function of the distance to a metal surface able to adsorb hydrogen dissociatively. *Left*: activated absorption; *right*: nonactivated adsorption.

3. Energetics of Hydrogen Adsorption

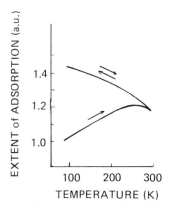

FIGURE 4 Adsorption isobar of hydrogen on a vapor-deposited nickel film. When the temperature of a film exposed to hydrogen at 77 K is raised to 300 K, the amount of hydrogen taken up passes through a maximum. Recooling to 77 K brings about an appreciable increase in the amount adsorbed. The upper branch of the adsorption isobar is reversible.

structure of the metal surface on the dissociative adsorption of hydrogen. Also with (supported) copper particles, there are indications that dissociative adsorption of hydrogen can proceed more rapidly on defects on the copper surface.

Ehrlich [23] rationalized the absence of activated adsorption of hydrogen on most clean transition metals by considering the activation energy for surface migration as being only 10 to 20% of the adsorption energy. The relatively low activation energy for surface migration indicates that these metal surfaces contain sites at a mutual distance corresponding to the distance of the hydrogen atoms in a hydrogen molecule where both atoms of a molecule can be bonded strongly. If one hydrogen atom is placed in a stable site, the other atom can be bonded in a site the adsorption energy of which is only 10 to 20% smaller. A considerable drop in the adsorption energy by the small distance of the two hydrogen atoms is not envisioned. In Chapter 1 Christmann deals with small energy barriers to dissociative adsorption of hydrogen and with the effect of the surface structure of the metal. Arguing that the adsorption energy of hydrogen increases with the number of metal atoms that can be contacted per hydrogen adatom, he explains the preferential adsorption of hydrogen on step sites by higher adsorption energy. A deeper well of the adsorbed hydrogen atoms in Fig. 3 also lowers the energy maximum through which the hydrogen atoms have to pass on dissociative adsorption. Accordingly, Poelsema et al. [24] have obtained evidence that the rate of dissociative adsorption of hydrogen on Pt (111) is related to the number of steps present on the surface. The local structure of the metal surface can affect not only the final state of adsorbed hydrogen, but also the transition state. As is evident from calculations on the reaction

$$H_2 + D = HD + H$$

the reaction can proceed only if the incoming deuterium atom is approaching the hydrogen molecule almost precisely along the molecular axis. The reaction does not take place when the deuterium atom is colliding with the hydrogen molecule along another direction (Fig. 5). Dissociation of hydrogen can therefore be expected to proceed most smoothly when an adsorbed hydrogen molecule can approach a metal surface atom along its molecular axis. Since the polarizability of hydrogen is larger perpendicular to the molecular axis, the hydrogen molecules will preferably be adsorbed with the axis parallel to the metal surface. Since the metal atoms of a step will be approached axially by a hydrogen molecule migrating over the surface, the rate of dissociative adsorption can be much larger at steps. Moreover, hydrogen atoms adsorbed into step sites are capable of assisting in the dissociation of hydrogen. This explains why the presence of steps raises the rate of adsorption not only into step sites, but all over the metal surface.

It has been established that the surfaces of the group Ib metals copper, silver, and gold are very smooth. Using ion scattering, Algra et al. [25] have observed that stepped copper surfaces are smoothed by penetration of metal edge atoms in between the atoms present underneath. Earlier, Rhead [26] argued that the small anisotropy of the surface energy of copper indicates very smooth surfaces. Therefore, it may be questioned whether the activated adsorption of hydrogen by copper, especially, is due to the absence of a significant number of surface defects or to the atomic properties of copper. The fact that even steps at copper surfaces have an atomically very smooth structure renders bonding of hydrogen atoms to many copper atoms impossible. Alternatively, an approach to metal surface atoms as indicated in Fig. 5 cannot proceed on copper surfaces.

As dealt with in Chapter 8, the perturbation of the electronic structure of metals in the adsorption of hydrogen and the reaction to bulk hydrides is usually small. The most important effect is that the interaction between the metal atoms is strongly decreased in the reaction to bulk hydrides, as has to be expected. Consequently, an activated perturbation of the surface structure of the metal cannot be expected with most crystallographic surfaces.

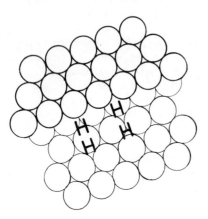

FIGURE 5 Dissociation of molecular hydrogen at steps on a metal surface. *Top*: cross-sectional picture of a step with an adsorbed hydrogen molecule. *Bottom*: two hydrogen molecules adsorbed on a (111) terrace in contact with a step. The metal atoms in the step can lead to a linear chain of three atoms that leads to dissociation of hydrogen, analogous to the reaction of molecular "protium" with a deuterium atom.

3. Energetics of Hydrogen Adsorption

A relatively open surface such as the (110) surface of face-centered cubic (fcc) metals has been shown to assume a considerably different structure on the adsorption of hydrogen [27,28]. The fact that the different structure is established only at temperatures above about 200 K shows that the perturbation of the metal surface calls for some activation [28-30]. Since the effect on the surface structure is exhibited even at relatively low temperatures, the energy required is relatively small. Penetration of hydrogen atoms in between metal surface atoms leads to a structure resembling the structure of the corresponding bulk hydride. Accordingly, nucleation of bulk hydride can proceed easily at crystallographic surfaces of an open atomic structure. When the presence of adsorbed atoms or molecules strongly decreases the surface mobility of hydrogen adatoms and thus the desorption of hydrogen molecules, the virtual hydrogen pressure corresponding to the coverage of hydrogen atoms can be considerable. The elevated thermodynamic hydrogen pressure can readily lead to formation of bulk hydrides stable only at high hydrogen pressures.

A small effect on the metal structure cannot be expected with the alkali metals. The difference in electronegativity causes the valency electron of the metal to be transferred almost completely to the hydrogen atom. As a result, the intermetallic bonding, and hence the metal structure, will be completely destroyed and changed into an ionic structure. In view of the considerable effect on the metal structure, the activation energy not only involves breaking of the bond between the two atoms of a hydrogen molecule, but also disrupture of the intermetallic bond. As a result, the adsorption of molecular hydrogen is activated. At 300°C or higher, reaction of alkali metals with molecular hydrogen is not restricted to adsorption, but the interaction also leads to the formation of bulk hydride. Anderson et al. [31] showed that the activation energy is determined by the dissociation of hydrogen. Atomic hydrogen reacts with alkali metals at room temperature to form bulk hydride.

In addition to the difference in electronegativity, the fact that the transition metals have more valency electrons causes the metallic structure of these metals to be affected much less by the interaction of hydrogen. The hydrides of these metals therefore retain their metallic properties, which is apparent most clearly from the hydrides exhibiting an electrical resistance that increases with temperature. Consequently, the adsorption of molecular hydrogen can proceed rather smoothly on these metals.

III. EXPERIMENTAL DETERMINATION OF THE EXTENT OF HYDROGEN ADSORPTION

To get an absolute measure of the extent of adsorption of hydrogen per unit surface area of metal, comparison of the extent of physical adsorption of, for example, xenon with that of the chemical adsorption of hydrogen is indispensable. Although objections can be raised against the theory underlying the BET equation, its reliability has been well established over the years. The extent of physical adsorption can therefore accurately provide the surface area. Measurement of the hydrogen adsorption onto the same surface as measured by physical adsorption calls for specimens containing metal surfaces only. Supported metals, where the support is always taking up an appreciable fraction of the BET surface area, cannot be utilized. Consequently, either vapor-deposited metal film thrown onto cooled substrate or nonsupported metal powders have to be used.

The deposition of porous metal films exhibiting a surface area of 1000 to 10,000 cm^2 has been dealt with extensively [32]. Provided that the filaments from which the film are being thrown are previously carefully cleaned and deposition of oxides onto the substrate onto which the film is to be thrown is prevented, metal films with clean surfaces can be obtained. Using mercury cutoffs or greaseless valves instead of the usual stopcocks, contamination of the surface of the film during adsorption measurements can be avoided. Preparation of metal powders with clean surfaces is difficult. Since unsupported metals rapidly sinter at elevated temperatures, cleaning by reduction at high temperatures cannot be done. Therefore, reduction for prolonged periods of time is required.

Supported metals have to be reduced analogously to metal powders, but now the temperatures can be more elevated. After reduction the adsorbed hydrogen must be removed by evacuation at high temperatures. Two factors are of paramount importance with the evacuation: the pumping speed at the specimens to be measured and the release of water vapor from the generally extensive surface area of the support. Usually, to restrict the "dead volume" of the apparatus as much as possible, stopcocks with small bores are utilized, which lower tremendously the pumping speed at the specimen. Consequently, long evacuation times are required to remove the hydrogen from the sample. With supported catalysts, which are often highly porous, transport of the adsorbed hydrogen through the pores can also call for prolonged evacuation times. Nevertheless, evacuation times of 1 h are often used, which are generally totally inadequate to remove the adsorbed hydrogen completely from the system. However, evacuation for prolonged periods of time leading to complete removal of the hydrogen out of the porous system can easily lead to oxidation of the metal surface by water vapor released from the surface of the support.

The extent of adsorption of hydrogen can be measured in three ways: volumetrically, by means of a flow system, and by exchange with deuterium (see also Chapter 4). The volumetric procedure does not call for much explanation. At present the difficult manipulation of mercury manometers and McLeod gauges can be avoided. Modern pressure transducers or Bourdon-type manometers (Texas Instruments) enable us to record the pressures easily.

To avoid vacuum equipment, the use of a thermal conductivity cell with a flow system has been proposed. An inert gas, such as argon or nitrogen, is passed through the catalyst and pulses of hydrogen are added to the flow. The takeup of the pulses provides the amount of hydrogen totally adsorbed. A serious drawback of this procedure is that the partial pressure of the hydrogen is not well defined in these experiments. As the adsorption of hydrogen at temperatures of 23°C or higher depends strongly on the hydrogen pressure, the takeup of hydrogen thus obtained is ambiguous.

With noble metals a better procedure has been proposed. According to this procedure a flow of inert gas containing, say, 5% of hydrogen is passed over the catalyst. Pulses of carbon monoxide are added to the gas flow. On metals where carbon monoxide can bring about desorption of adsorbed hydrogen, evolution of hydrogen results that can be accurately measured by the thermal conductivity cell. Since evacuation of adsorbed hydrogen remaining after the reduction is not required, this procedure is much faster than the usual volumetric method. However, this method depends on the ability of carbon monoxide to produce merely desorption of hydrogen. With metals such as platinum and palladium, penetration of hydrogen into the subsurface layer of the metal has been inferred mainly from temperature-programmed desorption studies on monocrystalline specimens [33-35]. With nickel single crystals, slow desorption of adsorbed hydrogen has been found [36-40]. Silica-supported nickel particles, however, did not show a significant desorption of hydrogen on exposure to carbon monoxide.

3. Energetics of Hydrogen Adsorption

An elegant procedure to measure the adsorption of hydrogen has been developed by Scholten [40a]. He cooled the reduced sample to be measured down to 195 K without previous evacuation. At 195 K the sample is evacuated to remove the hydrogen from the gas phase. Subsequently, the adsorbed hydrogen is exchanged with deuterium and the content of H ("protium") is determined. From the amount of protium, the extent of adsorption can readily be calculated.

IV. EXTENT OF ADSORPTION OF HYDROGEN BY PURE METALS

As stated above, vapor-deposited metal films are very well suited for measuring the extent of hydrogen adsorption per unit area of metal surface. A difficulty is that the extent of hydrogen adsorption on different crystallographic surfaces cannot be assessed by using metal films. The fractions of different crystallographic planes in the surface of the metal films can be only crudely estimated from electron-microscopic investigation of the film. It therefore seems to be much more obvious to measure the extent of adsorption on single-crystal surfaces using ultrahigh-vacuum techniques.

However, it is difficult to establish the amount of hydrogen adsorbed at rather elevated pressures in an ultrahigh-vacuum system. Usually, the maximum intensity of a LEED pattern corresponding to a known stoichiometry of adsorbed species is determined as a function of the pressure. On the surface thus characterized, the Auger spectrum of the adsorbed species is measured or, preferably, the ellipsometric parameters. The intensity of the Auger electrons or the ellipsometric parameters are calibrated accordingly. However, hydrogen adsorption not often leads to LEED patterns different from that of the clean surface. Only recently a number of metal surfaces was found to exhibit LEED patterns due to ordered hydrogen overlayers at low temperatures [41-44]. Moreover, the electrons used in taking the LEED pattern can bring about dissociation of molecular hydrogen and thus raise the extent of adsorption. Another complication is that the adsorption of hydrogen is not evident from an Auger electron spectrum nor from an effect on the ellipsometric parameters. The only way to assess the adsorption of hydrogen is to measure the effect on the work function or to carry out (temperature-programmed) thermal desorption. Owing to desorption from the shanks of the single crystals, however, the amount of hydrogen desorbing cannot give reliable information about the extent of adsorption. Moreover, the sample has to be evacuated before a temperature-programmed desorption run. Accordingly, measurements at high pressures cannot be carried out. Another complicating factor is that the roughness factor of single crystals is usually difficult to establish. We therefore feel that we have to rely on data measured on vapor-deposited metal films.

We will first consider the adsorption of hydrogen on nickel, for which the most extensive body of experimental evidence is available (see also Chapter 5). The heat of adsorption of hydrogen continuously drops as a function of coverage. As a result, the hydrogen coverage will depend on the pressure if the temperature is not low. Kavtaradze [45,46] investigated the adsorption of hydrogen on nickel films at a range of temperatures and hydrogen pressures. He observed that the amount of hydrogen adsorbed at 77 K depends hardly at all on the hydrogen pressure. Raising the pressure from 10^{-7} torr to 5×10^{-3} torr increases the extent of adsorption by less than 3%. After heating the film under hydrogen to room temperature and recooling to 77 K, the film was brought to 200 K. It was observed that at a constant pressure of about 10^{-2} torr about 5% of the hydrogen desorbs at 200 K. Hence it can be concluded that the hydrogen coverage does not depend significantly on the hydrogen pressure at temperatures of 200 K or lower. Raising the temperature to 300 K, however, causes about 20% of the hydrogen to be desorbed. At this temperature about 50% of the hydrogen can be removed by evacuation. Germer and MacRae [47], studying single crystals of nickel, observed that hydrogen could be removed completely by evacuation at room temperature for prolonged periods of time. The slower desorption of hydrogen from porous metal films is due to the slow transport through the pores of the film. Kavtaradze established that at 473 K the adsorbed hydrogen could be removed completely by pumping. The strong dependence of the extent of adsorption on the hydrogen pressure at higher temperatures was also found by Schuit and de Boer [48] for nickel-on-silica catalysts and by Gundry and Tompkins [21] and by Rideal and Sweet [49] for nickel films. Frennet and Wells [50] have established for supported platinum particles that the extent of hydrogen adsorption at not too low temperatures does not saturate. The adsorption increases continuously with hydrogen pressure.

Consequently, the most reliable value for hydrogen adsorption is obtained by measuring the extent of adsorption at temperatures below 200 K, where the adsorption does not depend strongly on the hydrogen pressure and extrapolation to some standard hydrogen pressure is not required. Most measurements on vapor-deposited metal films have been done at relatively low maximum pressures. Since measurements on supported catalysts must be carried out at more elevated hydrogen pressures, as will be explained below, an unequivocal comparison asks for the results on vapor-deposited films measured at low temperatures. As mentioned above, the adsorption of hydrogen by vapor-deposited metal films measured at a constant hydrogen pressure exhibits an irreversible maximum when the nickel film having adsorbed hydrogen at 77 K was heated to 300 K and subsequently cooled to 77 K. After having brought the film at 273 K and again to 77 K, the extent of adsorption has generally increased by 20 to 40% [21, 51-53]. Knor and Ponec [54] extensively studied the maximum in the amount adsorbed; their experimental results are summarized in Fig. 4.

After much discussion it has become clear that the maximum in the amount of hydrogen adsorption exhibited in Fig. 4 is brought about by transport through the narrow pores in nickel film. As observed by Kooy and Nieuwenhuizen [55] for vapor-deposited aluminum films and by Geus [32] for iron and nickel films, vapor-deposited films contain a large number of narrow pores. The axis of the pores is oriented parallel to the direction of the incident metal atoms. The presence of narrow pores in vapor-deposited metal films thrown onto a cooled substrate can also be inferred from the BET surface area being proportional to the weight of the film and being much larger than the geometrical surface area. The surface of very narrow pores cannot be covered by gas molecules incident from the gas phase. Migration of adsorbed atoms or molecules over the metal surface is required to cover the surface of these narrow pores. Gomer and co-workers [18] observed in field-emission experiments on nickel tips that adsorbed hydrogen atoms become mobile at about 200 K. Accordingly, the surface of very narrow pores cannot be covered at temperatures below 200 K, whereas at this temperature hydrogen atoms migrating over the surface can reach the metal surface situated at very narrow pores. As has to be expected, the maximum of the adsorption isobar is observed at 200 K, the temperature where the adsorbed atoms become mobile. At higher temperatures the equilibrium shifts to lower coverages, whereas at lower temperatures after the establishment of equilibrium the extent of adsorption increases as the temperature decreases.

3. Energetics of Hydrogen Adsorption

TABLE 1 Extent of Hydrogen Adsorption on Vapor-Deposited Nickel Films at 77 K

Physically adsorbed gas	Ref.	Ratio, monolayer volume/H_2 adsorbed
Krypton	45	0.83
	56	0.855
	57	0.79
	93	0.77
	58	0.99 ± 0.06
Methane	56	0.78
n-Butane	56	0.505
Neon	56	1.01-1.45

The extent of hydrogen adsorption has to be related to that of physical adsorption. Now it is not certain that the molecules being physically adsorbed, such as krypton, xenon, and methane, are able to penetrate into pores in which hydrogen cannot enter at 77 K. If they cannot, we can use the extent of hydrogen adsorption measured at 77 K, whereas with penetration of physically adsorbed molecules into narrow pores, the extent of adsorption measured at 77 K must be raised by 20 to 40%.

The values for the ratio of hydrogen adsorbed at 77 K to the monolayer volume of physically adsorbed gas are collected in Table 1. With the exception of Brennan and Hayes' result [58], the data for krypton of the various authors are in remarkably good agreement, especially if we consider the difficulties in the preparation of metal films with reproducible properties. We want to restrict the discussion to the data for krypton, for which the most results are available. Omitting Brennan and Hayes' ratio, the values of the other authors are between 0.77 and 0.855. Before we can derive the extent of adsorption of hydrogen per unit surface area of nickel, we have to establish the surface area taken up by a physically adsorbed krypton molecule. Table 2 summarizes the surface areas taken up per physically adsorbed krypton atom determined by a number of authors. Accepting a surface area of 0.21 nm^2 for a physically adsorbed krypton molecule, the surface area taken up by a hydrogen atom is 0.081 to 0.090 nm^2. These values correspond to a number of adsorbed hydrogen atoms of 1.11 to 1.24 × 10^{15} cm^{-2}.

TABLE 2 Surface Area Occupied by Physically Adsorbed Krypton Molecules (Values Determined Experimentally)

Refs.	Surface area (10^{-16} cm^2)	
	Based on P_0 liquid	Based on P_0 solid
59	19.5 ± 2.0	—
60	21.0 ± 1.0	22.5 ± 1.5
61, 62	21.6	21.0 ± 3.0
57	—	21.0

TABLE 3 Number of Metal Surface Atoms in Nickel Surfaces

Crystallographic plane	Number of metal atoms (10^{15} cm^{-2})
(111)	1.87
(100)	1.61
(110)	1.14

The number of metal surface atoms in the most stable crystallographic planes of nickel are collected in Table 3. When we take as a rough estimate for the ratio of the (111), (100), and (110) planes in the surface of vapor-deposited nickel films 2:2:1, a mean number of metal atoms per square centimeter of 1.62×10^{15} is calculated. If we consider the additional adsorption of 20 to 40% of hydrogen onto the surfaces of very narrow pores and we assume that the surfaces of these pores can be covered by the physically adsorbed molecules, about 1.33 to 1.74×10^{15} hydrogen atoms per square centimeter are adsorbed. The data agree well with the coverage quoted by Christmann (Chapter 1), that is, 0.8 monolayer of hydrogen atoms, usually adsorbed at saturation.

For nickel powders Roberts and Sykes [63] published an extensive investigation of the extent of hydrogen adsorption. With a powder reduced for a prolonged period of time these authors observed a ratio of the krypton monolayer to the amount of hydrogen taken up at 90 K that was slightly higher than the value measured on films, viz. 1.0 as against 0.8 for vapor-deposited metal films. This would indicate a lower adsorption of hydrogen and hence a slight contamination. Geus and Scholten [40a] measured the extent of hydrogen adsorption for a nickel powder prepared by precipitation with ammonia from a nickel nitrate solution. The powder was reduced for 20 h at 250°C. Since the adsorption of hydrogen was determined by exchange with deuterium, evacuation at elevated temperatures was not required. For the ratio of hydrogen adsorbed at 195 K to physically adsorbed methane, values of 0.73 and 0.85 were obtained for a powder of surface area 1.5 and 2.3 m^2 per gram of metal, respectively. These values compare well with the value of 0.78 measured by Beeck and Ritchie [56] on vapor-deposited nickel films. These authors quoted a ratio of 0.78 (see Table 2). Livingston's [64] value for the surface area of a physically adsorbed methane molecule, 0.16 nm^2, leads to a number of adsorbed hydrogen atoms of 1.47 to 1.71×10^{15} cm^{-2}. Again these values agree quite well with the number of nickel surface atoms to be expected, provided that each nickel surface atom adsorbs one hydrogen atom.

For iron, for which the experimental evidence is less extensive, analogous values have been measured. Thus Zwietering et al. [65] and Geus [66] measured at 77 K an extent of adsorption of 6×10^{14} molecules per square centimeter. An average number of iron surface atoms of 1.5×10^{15} leads to a coverage of 0.8. This is of the same order of magnitude as found with nickel. The surface area of 0.224 nm^2 per xenon atom adsorbed in the monolayer was assumed.

Geus et al. [67] extensively measured the extent of hydrogen adsorption by vapor-deposited tungsten films. With five films the extent of hydrogen adsorption at 77 K was found to be 1.2 to 1.3×10^{15} hydrogen atoms per square centimeter. The BET surface area of the films was also measured by physical adsorption of xenon. The surface area per xenon molecule was again taken to be 0.224 nm^2. Anderson and Baker [68] also arrived at a higher

3. Energetics of Hydrogen Adsorption

extent of adsorption with vapor-deposited tungsten films: 1.5×10^{15} cm^{-2}. Assuming the same proportion of closely packed crystallographic planes in the surface of the tungsten films, the mean number of surface atoms is 1.2×10^{15}. Hence the coverage of tungsten is higher. In Chapter 8 we bring forward evidence that vapor-deposited tungsten films have atomically rougher crystallographic planes. Accordingly, hydrogen atoms can penetrate between the surface atoms more easily. Adsorption of subsurface and surface hydrogen causes the amount of hydrogen taken up at 77 K per square centimeter of metal surface to be relatively high.

The foregoing results indicate that the extent of adsorption of hydrogen is of the order of 0.8 to 1.0 monolayer. This extent of adsorption has been measured at 77 K. At temperatures above about 200 K the extent of hydrogen adsorption starts to depend rather strongly on pressure. More elevated pressures are required to establish the coverage of 0.8 to 1.0. Often, the extent of adsorption is calculated by extrapolation of the experimental points to zero pressure. In this way any physical adsorption of hydrogen should be eliminated. The experimental results on films and porous unsupported metal specimens suggest, however, that the extent of adsorption at higher temperatures (e.g., 296 K) is significantly lower at zero pressure than below 200 K, where the extent of adsorption does not depend strongly on the hydrogen pressure.

V. ADSORPTION OF HYDROGEN BY SUPPORTED METALS

With supported metals the extent of hydrogen adsorption per gram of reduced metal has to be determined. Therefore, in addition to the extent of adsorption, the degree of reduction has to be established. Assessment of the extent of adsorption per gram of metal allows calculation of the dispersion: the number of metal surface atoms over the total number of metal atoms.

While vapor-deposited metal films and sintered unsupported metal powders display surfaces on the order of about 0.5 to 5 m^2 g^{-1}, the BET surface area of supported metals can be much larger, 100 to 500 m^2 g^{-1}. Consequently, the pore volume in supported catalysts is much greater and the pores are much more extended. As a result, the hydrogen pressures admitted to supported metals have to be much higher. Transport of hydrogen into long, narrow pores proceeds slowly. To arrive at a reasonable surface coverage without the need to extend the measurement over a week, hydrogen has to be admitted to the sample at a pressure of about 10 torr or higher [69]. To be sure that the hydrogen is penetrating the entire pore system, some workers always measure the extent of hydrogen adsorption on nickel at a hydrogen pressure of 1 bar. Although most workers do not use such elevated hydrogen pressures, hydrogen pressures with supported metals must be considerably higher than with, for example, vapor-deposited films, where the pores have lengths on the order of, at most, 1 μm. Supported metals, on the other hand, exhibit pore lengths of up to 1 mm.

The extended porous system of many supported metals also makes it almost impossible to have uniform coverage of the total metal surface. Unless the sticking probability of hydrogen is strongly diminished by considerably raising the temperature, the hydrogen admitted to supported catalysts will initially cover the surfaces of the metal particles present at the outermost regions of the porous particles of the supported catalyst. Since the heat of adsorption of hydrogen drops with the coverage and the sticking probability decreases even more rapidly, hydrogen adsorption on supported metal catalysts is more favorable than, for example, the adsorption of carbon monoxide or the sorption of molecular oxygen. Nevertheless, a nonuniform distribution of hydrogen over the metal particles has to be envisioned. With pores of a dimension that Poiseuille flow can proceed instead of Knudsen diffusion, the flow of hydrogen into the pores can be rapid provided that the pores are empty and a pressure difference can be established over the pore length. When the pores are previously filled with an inert gas (e.g., in the flow procedure of determining the extent of hydrogen adsorption), transport is much slower and the distribution of hydrogen, or of adsorbed CO with the desorption of hydrogen, is much less uniform. Therefore, it is advisable to use the smallest possible catalyst particles and to prevent the small particles from being carried away with the gas flow during evacuation.

Another complication with supported catalysts can be adsorption of hydrogen onto the surface of the support, which can be much more extended than the metal surface. The support can take up large amounts of hydrogen especially at low temperatures. In special experiments Geus and Eurlings have established that at about 120 K and hydrogen pressures of about 10 torr, the support takes up much hydrogen. Magnetic measurements have also demonstrated that most of the hydrogen admitted to supported catalysts at, for example, 77 K does not reach the metal particles in the more interior parts of the catalyst particles [70]. Since the adsorption is also much slower at low temperatures, hydrogen is usually admitted at room temperature. When the extent of adsorption has to be determined at lower temperatures, the supported catalyst is cooled down after equilibration at room temperature. When the surface area of the support is much larger than that of the active metal, as with supported platinum catalysts containing 0.5 to 1.0 wt % Pt, some workers prefer to measure the hydrogen adsorption at 200°C to restrict adsorption of hydrogen onto the support as much as possible.

The generally highly porous support also presents problems with the evacuation of the catalyst. The transport of hydrogen out of the porous system can proceed much more slowly than that out of a system containing no pores or only wide pores. With metals liable to be oxidized by water, the usually large amounts of water adsorbed on the surface of the support can also cause difficulty. As stated above, evacuation for prolonged periods of time is required to remove the hydrogen from the supported catalysts. During the evacuation, desorption of water and thus reoxidation can proceed [71]. It is therefore advisable to use a high pumping speed and small catalyst particles to be able to reduce the evacuation period as much as possible.

As stated above, the degree of reduction also has to be determined. It is most reliable to determine the degree of reduction of the sample onto which the adsorption of hydrogen is being measured. The extent of reduction can be determined from (a) the hydrogen consumption during the reduction, (b) the consumption of iodine or bromine after reaction with the reduced catalyst, (c) the amount of hydrogen evolved on dissolution in acids, (d) the saturation magnetization, or (e) the increase in weight of the catalyst after oxidation.

Measurement of the consumption of hydrogen during the reduction can be carried out in the apparatus where the extent of adsorption is determined. However, the reduction often runs very slowly, which leads to a small decrease in the hydrogen content of the gas flowing over the catalyst. Therefore, an accurate measurement with a small drift of the base line is required. Chemical determination of the degree of reduction can be done after measuring the adsorption of hydrogen by the sample. This can be done on the same catalyst sample for which the extent of hydrogen adsorption has been measured. This also holds true for the determination of the amount of hydrogen evolved on dissolving the reduced catalyst into an acid solution [72]. As a matter of fact, this procedure can only be applied with metals such as nickel, which dissolve in acid, with the formation of hydrogen. A difficulty with this procedure is that all adsorbed hydrogen must have been removed previously. Measurement of the saturation magnetization can provide accurate data about the degree of reduction. This procedure can be used only with ferromagnetic metals: iron, cobalt, and nickel. It is possible to carry out the measurement with the sample in the measuring cell in which the extent of hydrogen adsorption is measured. It is also possible, after measuring the extent of hydrogen adsorption, to put the sample into a vessel in a glove box and subsequently put the tightly closed vessel into the apparatus for measuring the magnetization. Especially with small nickel particles, the saturation magnetization has to be determined by extrapolation of data measured at low temperatures and elevated magnetic field strengths. Finally, the weight of the sample can be recorded during reduction. Since with the reduction of supported metal catalysts both desorption of water and formation of water by the reduction proceeds, it is difficult to assess the degree of reduction by measuring the decrease in weight during the reduction. When the metal particles are small and thus can be completely oxidized to a stoichiometric oxide, the degree of reduction can be accurately determined by measuring the increase in weight on reoxidation of the catalyst. With iron and nickel catalysts this procedure has provided excellent results. To measure the takeup of hydrogen gravimetrically is difficult in view of the low weight of hydrogen. Therefore, the gravimetric procedure is difficult to carry out on the sample used in the determination of the hydrogen adsorption.

In Section I we mentioned that measurement of the extent of hydrogen adsorption by supported metal catalysts is carried out, first of all, for determining the free metal surface area per unit volume or per unit weight of catalyst. The free metal surface area indicates any significant sintering of the metal particles. Moreover, the dispersion of the metal can be calculated and the turnover numbers. The dispersion and turnover numbers can indicate whether a catalytic reaction is surface sensitive (demanding) or structure insensitive (facile). With structure-sensitive reactions it is often not desirable to utilize very small metal particles. In spite of the large free metal surface area the activity per unit surface area of the metal is so low or the selectivity so bad, that larger metal particles are exhibiting a better catalytic performance.

Additionally, information about the interaction with the support and the presence of grain boundaries in the catalytically active metal particles can be obtained. When the free metal surface as calculated from the extent of hydrogen adsorption is significantly smaller than that calculated from, say, the mean size of the particles as determined from x-ray line broadening or from electron micrographs, for example, a significant contact area with the support can be expected. When the particle size as calculated from the x-ray line broadening is much smaller than that resulting from the extent of hydrogen adsorption, the presence of grain boundaries is indicated.

To be sure that the extent of hydrogen adsorption is proportional to the surface area of the metal particles, some trivial problems must be resolved, the first being the presence of contaminants at the metal surface affecting the adsorption of hydrogen. We mentioned above the reaction of hydrogen with adsorbed contaminants being activated and proceeding at higher temperatures only. With nickel the effect of presorbed oxygen on the subsequent reaction with hydrogen has been studied [73-78]. With severely oxidized nickel surfaces, reaction with hydrogen proceeds only at high temperatures. However, when not more than about a monolayer of oxygen is taken up, hydrogen molecules can still dissociate on the metal surface. Dissociation of hydrogen leads to active hydrogen atoms being able to react with adsorbed oxygen at room temperature. Although reaction of the thus adsorbed oxygen proceeds more slowly than adsorption of hydrogen onto a clean nickel surface, transport through the pores of a porous system may obscure the relatively slow reaction of the hydrogen admitted. Besides the presence of impurities on the metal surface capable of reacting with hydrogen, blocking impurities can also be present on the surface. Instances are sulfur atoms, carbon layers, and silica or titania moieties. Adsorbed blocking impurities decrease the extent of hydrogen adsorption. Spillover effects, on the other hand, may lead to too large an extent of hydrogen adsorption. The reactive hydrogen atoms resulting from the dissociation of hydrogen on the free metal surface are mobile at not too low temperatures. At still higher temperatures the hydrogen atoms may migrate to the surface of the support. Reaction with oxidic supports has been postulated, leading to OH groups that give rise to desorption of water at high temperatures. Such problems are discussed in detail in the following chapters.

A final fundamental problem is the surface sensitivity of the extent of hydrogen adsorption. As catalytic reactions can be surface sensitive, the extent of hydrogen adsorption may also be surface sensitive. It is possible that the extent of hydrogen adsorption per unit surface area of metal with very small metal particles or with metal particles exposing atomically rough surfaces is different from that with more extended, smooth metal surfaces. To assess the extent of adsorption of hydrogen dependent on the atomic structure and the dimensions of the adsorbing metal surfaces, the dispersion (i.e., the fraction of metal atoms present in the metal surface) has to be determined. The dispersion of supported metal catalyst can be determined from x-ray line broadening, high-resolution electron microscopy, and EXAFS. With supported nickel particles, the particle-size distribution can be calculated from the magnetization measured as a function of field strength and temperature. The dispersion of supported metal particles has been determined for nickel on silica, platinum on silica and on platinum, and rhodium and iridium particles on silica, alumina, and titania supports.

For nickel an extensive body of experimental results is available. As dealt with above, the extent of hydrogen adsorption on unsupported reduced nickel powders and on vapor-deposited nickel films has been measured together with the BET surface area. Consequently, the extent of adsorption on extended, atomically smooth nickel surfaces is known rather accurately. The extent of hydrogen adsorption on supported nickel particles has also been studied extensively. Schuit and de Boer [48] measured the hydrogen adsorption on silica-supported nickel particles at a range of pressures and temperatures. Schuit and van Reijen [79] derived particle-size distributions for nickel-on-silica particles from magnetic measurements. The amount of hydrogen adsorbed at 195 K and 100 mmHg turned out to be appreciably lower than expected from the particle-size distribution. The results of these authors suggest that the unexpectedly low extent of hydrogen adsorption is confined to very small nickel particles. The initial slope of the magnetization curve at 300 K correlates very well with the amount of hydrogen adsorbed. The initial slope of the magnetization curve is proportional to the size of the larger particles in the nickel catalysts. The amount of hydrogen taken up per gram of nickel is inversely proportional to the size of the nickel particles. That with larger nickel particles the initial slope does not increase further, whereas the extent of hydrogen adsorption continues to drop, may be due to the higher magnetic anisotropy of larger nickel particles. The higher anisotropy prevents the magnetization of the larger particles to line up with a magnetizing field. Schuit and van Reijen ascribed the relatively low extent of hydrogen adsorption to the fact that part of the nickel particles are embedded in the silica support. Because of this the surface of the nickel particles is not (completely) accessible for hydrogen. Especially with coprecipitated catalysts that contain a large amount of nickel hydrosilicate

3. Energetics of Hydrogen Adsorption

before reduction, it is quite reasonable that the embedding involves the small nickel particles preferentially.

The results of Dietz and Selwood [69] exhibit the same relatively low adsorption of hydrogen by nickel-on-silica particles. However, this is not apparent from their paper, owing to an arithmetical error. From the magnetization at 4.2 K and high magnetic field strengths exhibited by a coprecipitated nickel-on-silica catalyst, these authors calculated a mean particle radius of 1 nm. They considered this value to be an upper limit for the mean particle radius. A mean particle radius of 1 nm corresponds to a nickel surface area of 337 m^2 per gram of nickel, not to 31 m^2 per gram. A surface area occupied per adsorbed hydrogen atom of about 6.2×10^{-16} cm^2 leads to a monolayer of about 100 mL (STP) of hydrogen per gram of nickel. Dietz and Selwood [69] measured an adsorption of only 24 mL (STP) per gram of nickel. This extent of adsorption corresponds to a nickel surface area of about 80 m^2 per gram of nickel and not of 43 m^2 per gram as mentioned by the authors. Nevertheless, the extent of hydrogen adsorption measured experimentally is much smaller than the nickel surface area calculated from the independently determined mean particle size.

When the average size of the nickel particles is calculated from x-ray line broadening, an unexpectedly low hydrogen adsorption is again observed. First, the extensive and carefully conducted work by Coenen and Linsen [72,80] on nickel-on-silica catalysts has to be mentioned. The authors determined the extent of reduction from the amount of hydrogen evolved on reacting the reduced and degassed catalysts with sulfuric acid. The broadening of the (200) reflection was used to calculate the mean particle size. While Schuit and van Reijen [79] assumed an embedding of small nickel particles in the silica, Coenen and Linsen [80] propose a considerably contact area between the nickel particles and the silica support. They showed that the nickel (111) plane fits well on the antigorite, the nickel hydrosilicate lattice. They argue that the strong interaction between the two lattices leads to the small nickel particles assuming a hemispherical shape. The equatorial plane of the hemispheres contacting the silica cannot adsorb hydrogen. As a result, the adsorption of hydrogen is lower than expected.

To survey the implication of the shape of the nickel particles being hemispherical, we calculated the diameter of a completely spherical particle, of a hemispherical particle, and of a cube from the dimension calculated from the x-ray line broadening using Scherrer's formula. The dimension D_W resulting from the x-ray line broadening is

$$D_W = \sqrt[3]{V_p}$$

where V_p is the particle volume. This dimension is equal to the edge length, D_c, of cubic particles. Hence

$$D_c = D_W$$

The equivalent diameter D_s of completely spherical particles is

$$D_s = D_W \sqrt[3]{\frac{6}{\pi}}$$
$$= 1.241 \, D_W$$

For a hemispherical particle the relation for the diameter D_{hs} is

$$D_{hs} = D_w \sqrt[3]{\frac{12}{\pi}}$$

$$= 1.563 \, D_w$$

If we assume the same proportion of 2:2:1 for the (111), (100), and the (110) planes as used above, the mean number of nickel surface atoms is 1.62×10^{15} cm^{-1}, and the mean surface area is 6.17×10^{-16} cm^2 per nickel surface atom. From this mean surface area the dispersion, the fraction of the total number of nickel atoms present in the surface, can be calculated from the mean dimension of nickel particles of spherical, hemispherical, or cubic shape. With the hemispherical particles the nickel atoms present in the equatorial plane are excluded and with the cubic particles those in one of the faces that is supposed to be in contact with the support.

With very small nickel particles a significant fraction of the surface atoms will be present at edges and corners. These atoms are taking up a larger surface area than nickel atoms in closely packed extended nickel surfaces. With a cubo-octaeder of dimensions 1 nm, for instance, the mean surface area per nickel atom is 12×10^{-16} cm^2, whereas the mean surface area of a nickel atom in extended (111) and (100) planes is only 5.6×10^{-16} cm^2. If the surface area per nickel surface atom is larger, the dispersion calculated from the dimension obtained by x-ray line broadening will be lower, since the surface will contain a smaller number of nickel atoms. Nevertheless, the dispersion of small cubo-octaeders is still considerable; a cubo-octaeder of a diameter of 1.86 nm exhibits a dispersion of 0.55 and a cubo-octaeder of a diameter of 2.75 nm of 0.43.

Linsen [72] investigated impregnated and coprecipitated nickel-on-silica catalysts. The impregnated catalysts were prepared by ball-milling of the support together with $Ni(NO_3)_2 \cdot 6H_2O$. Subsequently, the intimate mixtures were kept for 4 days at 75°C in stoppered bottles. At this temperature a solution of $Ni(NO_3)_2$ in its crystal water results. Linsen's procedure leads to an attack of the silica support by nickel ions. This author also studied nickel-on-silica catalysts prepared by coprecipitation. Both types of catalysts exhibit a strong interaction of the reduced nickel and the support. Linsen measured the amount of hydrogen taken up at a pressure of 760 mmHg after 16 h.

Table 4 shows Linsen's results for silica-supported nickel catalysts. Linsen reduced the impregnated catalysts at 300°C, where he observed a degree of reduction of 0.985. The coprecipitated catalysts, NiWAG and ViALG, which are more difficult to reduce, were reduced at 500°C. The dispersion has been calculated from the extent of hydrogen adsorption assuming that one hydrogen atom is taken up per nickel surface atom. The dispersion has also been calculated for spherical, cubic, and hemispherical particles. It can be seen that the dispersion calculated for spherical and cubic particles is significantly higher than that measured by hydrogen adsorption. The dispersion calculated for hemispherical particles agrees reasonably well with the extent of hydrogen adsorption.

Coenen and Linsen [80] measured the x-ray line broadening for a number of nickel-on-silica catalysts that were partially poisoned by sulfur. From the sulfur content and the extent of hydrogen adsorption the mean crystallite size was calculated, which was compared with the crystallite size from the x-ray line broadening assuming a hemispherical shape. The dimensions thus obtained are compared in Table 5. It can be seen that again the agreement is satisfactory although the x-ray line broadening results are often too small for relatively small nickel particles.

TABLE 4 Linsen's Data for Extent of Hydrogen Adsorption and Weight-Mean Particle Size from X-Ray Line Broadening

Catalyst	Ni content (wt %)	Hydrogen adsorption (mL H_2/g Ni)	Dispersion (%)	Weight-mean diameter (nm)	Dispersion (%)			Ratio dispersion	
					Sphere	Cube	Hemisphere	Sphere/H_2	Hemisphere/H_2
De 01	5.0	17.5	9.2	5.4	15.9	16.4	12.6	1.7	1.4
De 02	9.7	16.8	8.8	5.9	14.5	15.0	11.5	1.6	1.3
De 03	16.9	13.7	7.2	8.4	10.2	10.5	8.1	1.4	1.1
De 04	22.7	11.3	5.9	11.0	7.8	8.0	6.2	1.3	1.0
De 05	27.95	11.2	5.9	8.8	9.7	10.1	7.7	1.7	1.3
De 06	33.0	11.2	5.9	9.5	9.0	9.3	7.2	1.5	1.2
De 16	34.1	20.3	10.6	8.6	10.0	10.3	7.9	0.9	0.7
De 26	34.0	16.3	8.5	7.1	12.1	12.5	9.6	1.4	1.1
cog 1	9.3	6.9	3.6	17.5	4.9	5.1	3.9	1.4	1.1
cog 2	17.0	5.8	3.0	20.6	4.2	4.3	3.3	1.4	1.1
cog 3	23.5	4.8	2.5	20.9	4.1	4.2	3.3	1.6	1.3
NiWAG	66.4	12.9	6.8	10.2	8.4	8.7	6.7	1.2	1.0
NiALG	64.0	15.7	8.2	9.0	9.5	9.8	7.6	1.2	0.9

Source: Ref. 72.

TABLE 5 Results for Sulfur-Containing Nickel-on-Silica Catalysts: Crystallite Size from Nickel Surface Area and X-Ray Line Broadening

Catalyst	Nickel content (wt %)	Nickel surface area (m²/g Ni)	Crystallite size		Ratio
			Adsorption (nm)	X-ray (nm)	
P 3	31	111	3.9	3.2	1.2
P 93	43	109	4.0	3.8	1.05
P 1607	60	90	4.8	4.5	1.1
P 60	38	82	5.3	4.0	1.3
P 94	55	77	5.7	5.9	1.0
P 21	63	74	5.8	5.7	1.0
P 85	62	66	6.5	6.8	1.0
P 84	62	55	7.8	6.3	1.2
P 95	70	54	7.9	6.9	1.15

Source: Ref. 80.

Geus and Eurlings [40a] determined the extent of hydrogen adsorption at 195 K and the X-ray line broadening of a number of nickel-on-silica catalysts produced by deposition-precipitation. The extent of hydrogen adsorption at 195 K was measured by exchange with deuterium, as dealt with above. Also, with these catalysts prepared by deposition-precipitation at about 90°C the silica support is attacked. The reaction of the silica is apparent from the increase in the BET surface area. Accordingly, a strong interaction of the reduced nickel with the silica has also to be expected with these catalysts. As can be seen in Table 6, the results show the same trend as in the data of Coenen and Linsen [80]—the hemispherical model is providing fair agreement. The still relatively low extent of hydrogen adsorption may be due to the reduction not being complete, the larger surface area taken up per nickel atom in small nickel particles, and possibly embedding of very small nickel particles in the support. The very small extent of hydrogen adsorption observed by Dietz and Selwood [69] for tiny nickel-on-silica particles also points to embedding of the nickel particles in the support.

Mustard and Bartholomew [81] were the first to determine the particle-size distribution of supported nickel particles from electron micrographs quantitatively. They criticized earlier work on nickel-on-silica catalysts in view of the lack of a quantitative assessment of the particle-size distribution from the electron micrographs [82-84]. Former results on nickel-on-alumina catalysts, where the extent of hydrogen adsorption and the x-ray line broadening of the nickel was determined, were rejected, since the estimates of the nickel surface area from the x-ray line broadening varied by as much as a factor of 2 to 3. Moreover, the nickel surface areas obtained from hydrogen adsorption and x-ray line broadening also differed by a factor of 2 to 3. Mustard and Bartholomew studied nickel supported on silica, alumina, and titania carriers. The catalysts were prepared by a routine impregnation procedure, while two nickel-on-silica catalysts prepared by deposition-precipitation were also investigated. The authors compared the extent of hydrogen adsorption with the nickel crystallite size as apparent from high-resolution electron micrographs and from x-ray line broadening. The uptake of hydrogen was

3. Energetics of Hydrogen Adsorption

TABLE 6 Dispersion of Deposition-Precipitated Nickel-on-Silica Catalysts from Hydrogen Adsorption and X-Ray Line Broadening

Ni content (wt %)	Hydrogen adsorption (mL H_2/g Ni)	Dispersion (%)	Weight-mean diameter (nm)
11.1	57.9	30.3	1.9
37.2	50.0	26.2	2.3
40.9	22.6	11.8	4.0
37.4	26.1	13.7	4.8
39.7	24.7	12.9	3.0
58.0	20.6	10.8	7.0

Nickel content (wt %)	Dispersion (%)			Ratio dispersion	
	Sphere	Cube	Hemisphere	Sphere/H_2	Hemisphere/H_2
11.1	45.1	46.6	35.8	1.5	1.2
37.2	37.2	38.5	29.6	1.4	1.1
40.9	21.4	22.1	17.0	1.8	1.4
37.4	17.8	18.4	14.2	1.3	1.0
39.7	28.5	29.5	22.7	2.2	1.8
58.0	12.2	12.6	9.7	1.1	0.9

Source: Data from Ref. 40a.

determined by extrapolation of the straight-line part of the isotherm to zero pressure. The adsorption of hydrogen by nickel-on-titania was anomalous due to SMSI (strong metal-support interaction) effects.

To analyze Mustard and Bartholomew's results we will utilize the same procedure as above, comparing the dispersions obtained from the various experimental techniques. While Mustard and Bartholomew use 6.75×10^{-16} cm^2 as the mean surface area per nickel atom, we stick to the value 6.17×10^{-16} cm^2. The difference is due to another proportion of (111), (100), and (110) crystallographic planes assumed to be in the surface of the supported nickel particles. Since it can be assumed that with the impregnated catalysts the interaction with the silica is much weaker than with the catalysts prepared by deposition-precipitation, we will consider the two types of silica-supported nickel catalysts separately.

Table 7 shows the extensive data. To get different particle-size distributions, the catalyst containing 13.5 wt % of nickel was treated at 923, 973, and 1023 K in hydrogen. The size of the nickel particles in the freshly reduced catalysts was too small to be determined reliably by x-ray line broadening. As can be seen at the bottom of Table 7 the extent of hydrogen adsorption for the sintered catalyst agrees well with the dispersion calculated from x-ray line broadening assuming hemispherical particles. It can hence be concluded that in spite of the different procedures used in assessing the extent of

TABLE 7 Mustard and Bartholomew's Results on Nickel-on-Silica Catalysts Prepared by Deposition-Precipitation

Nickel content (wt %)	Pretreatment Time (h)	Pretreatment Temp. (K)	Dispersion from H_2 adsorption (%)	Weight-mean diameter E.M. (nm)	Weight-mean diameter X-ray (nm)
3.6	Fresh		37	2.7	—
13.5	Fresh		41	2.9	—
	50	923	22	4.5	2.8
	50	973	15	7.1	5.4
	50	1023	15	8.0	4.8

Nickel content (wt %)	Dispersion electron microscopy (%) Sphere	Cube	Hemisphere	Ratio Sphere/H_2	Hemisphere/H_2
3.6	32	34	26	0.9	0.7
13.5	30	31	23	0.7	0.6
	19	20	15	0.9	0.7
	12	12	10	0.8	0.7
	11	11	9	0.7	0.6

Nickel content (wt %)	Dispersion x-ray (%) Sphere	Cube	Hemisphere	Ratio Sphere/H_2	Hemisphere/H_2
3.6	—	—	—	—	—
13.5	—	—	—	—	—
	31	32	24	1.4	1.1
	16	16	13	1.1	0.9
	18	18	14	1.2	0.9

Source: Ref. 81.

hydrogen adsorption, the results of Linsen and Coenen, Geus and Eurlings, and Mustard and Bartholomew point consistently to a significant contact area of the nickel particles with the silica support, which can be accounted for by, for example, a hemispherical shape of the nickel particles. The weight-mean particle size as evaluated from electron micrographs leads to a slightly smaller nickel surface area. This must be due to the fact that small nickel particles are almost completely oxidized to badly crystallized nickel oxide on exposure to atmospheric air. The small nickel oxide particles not exhibiting Bragg reflection are difficult to distinguish from the silica support. Fumed silica supports also display very small amorphous particles in high-resolution electron micrographs. Consequently, the mean particle size as determined from the electron micrographs is slightly too low, due to the omission of a fraction of very small nickel particles. As in the reduced catalyst, these particles are also adsorbing hydrogen, the nickel surface area as calculated from the extent of hydrogen adsorption is slightly larger. Kuo et al. [85] determined the particle-size distribution of a coprecipitated nickel-on-silica

3. Energetics of Hydrogen Adsorption

catalyst using x-ray single profile analysis and electron microscopy. They claim to have obtained results in excellent agreement by both techniques. Also, the extent of hydrogen adsorption is stated to have agreed with the average particle size as calculated from the extent of hydrogen adsorption.

The results for the impregnated nickel-on-silica catalysts are collected in Table 8. The x-ray line broadening of the 2.7 wt % nickel catalyst could not be determined. Van Hardeveld [86] was the first to establish that thermal pretreatment of nickel catalysts prepared by impregnation with, say, nickel nitrate strongly affects the particle-size distribution obtained after reduction. When an impregnated nickel catalyst is freshly reduced, a very broad particle-size distribution results. Very small and much larger nickel particles are present in the reduced catalyst. Previous calcination leads to a much more uniform size of larger particles. As can be seen in Table 8, both the catalyst containing 2.7 and that containing 15.0 wt % of nickel exhibit a much larger mean particle size after having previously been calcined. This again confirms van Hardeveld's observations. The effect of the particle-size distribution is apparent from the bottom of the table, where the dispersions calculated from the x-ray line broadening are compared with those calculated from the extent of hydrogen adsorption. X-ray line broadening leads to a mean diameter

TABLE 8 Mustard and Bartholomew's Results on Impregnated Nickel-on-Silica Catalysts

Nickel content (wt %)	Pretreatment Time (h)	Pretreatment Temp. (K)	Dispersion from H_2 adsorption (%)	Weight-mean diameter E.M. (nm)	Weight-mean diameter X-ray (nm)
2.7	Fresh		51	2.9	—
	3	573	16	11.0	—
15.0	Fresh		19	9.4	12.0
	22	773	5.8	24.0	19.0

Nickel content (wt %)	Dispersion electron microscopy (%) Sphere	Cube	Hemisphere	Ratio Sphere/H_2	Hemisphere/H_2
2.7	30	30	23	0.6	0.5
	7.8	8.0	6.2	0.5	0.4
15.0	9.1	9.4	7.2	0.5	0.4
	3.6	3.7	2.8	0.6	0.5

Nickel content (wt %)	Dispersion x-ray (%) Sphere	Cube	Hemisphere	Ratio Sphere/H_2	Hemisphere/H_2
2.7	—	—	—	—	—
	—	—	—	—	—
15.0	7.1	7.4	5.7	0.4	0.3
	4.5	4.7	3.6	0.8	0.6

Source: Ref. 81.

which is much more dominated by the larger nickel particles present in the catalyst than by the hydrogen adsorption. As a result, the more narrow particle-size distribution of the previously calcined catalyst leads to a much smaller discrepancy with the dispersion calculated from the hydrogen adsorption than the very broad distribution of the freshly reduced catalyst. The larger nickel particles are more likely to be included in the assessment of the particle-size distribution from electron micrographs than are the small particles. As stated above, the completely oxidized particles do not scatter the incident electrons appreciably more than do silica particles of the same size that are also present in fumed silicas. This brings about the smaller dispersions calculated from the weight-mean particle size shown in Table 8.

Mustard and Bartholomew's results for nickel-on-alumina catalysts are shown in Table 9. The authors mentioned that the (200) peak, the only prominent peak that was not obscured by a peak due to alumina, was weak in the catalyst containing 15 wt % of nickel and moderately strong in the catalyst containing 23 wt % of nickel. The mean particle size of the catalyst containing 23 wt % of nickel could thus be determined reliably from line broadening. As can be seen in Table 9, the dispersion calculated from the weight-mean particle size calculated from x-ray line broadening is slightly smaller than that calculated from the extent of hydrogen adsorption. This is to be expected. Apparently, the weakness of the peak due to (200) nickel with the catalyst containing 15 wt % of nickel leads to a mean particle size which is much too small. With the evaluation of the electron micrographs of this catalyst, areas of sufficient contrast between the nickel particles and the acicular support were selected. The agreement between the dispersion calculated from the electron micrographs is about the same as with the nickel-on-silica catalysts prepared by deposition-precipitation.

The foregoing extensive and carefully measured data show that interaction of the support with nickel particles can affect the contact area and thus the extent of hydrogen adsorption. Also, embedding of nickel particles can proceed, especially with coprecipitated catalysts. With nickel-on-silica catalysts a relatively large contact area and hence a relatively small extent of hydrogen adsorption has been observed consistently provided that the conditions were favorable to establish a strong interaction. Since the shape of supported metal particles determines the proportion of different crystallographic planes exposed and thus with surface-sensitive reactions the catalytic properties, it is important to pursue Mustard and Bartholomew's approach. A combination of different techniques can provide information about the interaction with the support and the shape of the metal particles. Magnetic measurements may be quite useful in this respect. The requirement to carry out measurements on samples from the same batch having been subjected to the same pretreatment is evident from data by Richardson and Dubus [87]. The latter authors observed a mean particle size of 3.0 nm for a catalyst containing 17 wt % of nickel reduced for 15 h at 400°C. The percentage reduction was 47%. With a nickel catalyst also prepared by deposition-precipitation containing 13.5 wt % of nickel, Mustard and Bartholomew measured a mean diameter of the nickel particles of 2.9 nm after reduction for 14 to 16 h at 450°C. However, the percentage of reduction was now 95%. In spite of the large difference in percentage reduction the mean size of the nickel particles thus does not differ significantly.

Whereas the extent of hydrogen adsorption on silica-supported nickel particles can be lower than expected from the mean size of the nickel particles, the adsorption of hydrogen by platinum, rhodium, and iridium particles has been found to be considerably larger than expected. The mean particle size of the supported metal particles has been determined by electron microscopy and EXAFS with these metals.

3. Energetics of Hydrogen Adsorption

TABLE 9 Mustard and Bartholomew's Results on Impregnated Nickel-on-Alumina Catalysts

Nickel content (wt %)	Pretreatment Time (h)	Pretreatment Temp. (K)	Dispersion from H_2 adsorption (%)	Weight-mean diameter E.M. (nm)	Weight-mean diameter X-ray (nm)
15.0	Fresh		17.5	4.6	—
	72	1023	9.5	11.0	4.6
	13	1023	9.5	12.0	4.9
23.0	Fresh		16.0	6.4	6.8

Nickel content (wt %)	Dispersion electron microscopy (%) Sphere	Cube	Hemisphere	Ratio Sphere/H_2	Hemisphere/H_2
15.0	18.6	19.3	14.8	1.1	0.8
	7.8	8.1	6.2	0.8	0.7
	7.1	7.4	5.7	0.7	0.6
23.0	13.4	13.8	10.6	0.8	0.7

Nickel content (wt %)	Dispersion x-ray (%) Sphere	Cube	Hemisphere	Ratio Sphere/H_2	Hemisphere/H_2
15.0	—	—	—	—	—
	18.6	19.2	14.8	2.0	1.6
	17.5	18.1	13.9	1.8	1.5
23.0	12.6	13.0	10.0	0.8	0.6

Source: Ref. 81.

An extensive investigation on silica-supported platinum particles has been carried out by the Eurocatalysis group. A silica-supported platinum catalyst containing 6.3 wt % of platinum was investigated in a number of laboratories. The precipitated silica support did not exhibit small silica particles. The size of the platinum particles has been measured mainly electron microscopically [88]. The difference in contrast between the amorphous silica and the platinum particles was excellent. The results were confirmed by x-ray line broadening. The consistent results indicate an extent of adsorption which is definitely higher than one hydrogen atom per metal surface atom. The size of the supported platinum particles is of the order of 2 nm from which a dispersion of about 60% was calculated assuming hemispherical or cubic particles. The extent of hydrogen adsorption is of the order of 64 mL (STP) per gram of platinum, which corresponds to 1.116 hydrogen atoms per platinum atom [50]. Since the dispersion as apparent from electron micrographs and x-ray line broadening is only 60%, the extent of hydrogen adsorption is much larger than expected.

First, it can be questioned whether x-ray line broadening and electron microscopy can measure very small clusters of platinum atoms. One of the reasons to choose silica as a support was that amorphous silica does not display a diffraction pattern. As a result, even very small platinum particles can rather easily be distinguished in a high-resolution electron microscope. Also, the x-ray diffraction pattern is not obscured by diffraction from the support. It is therefore not likely that many smaller platinum particles are present in the catalyst and have escaped detection.

The relatively high extent of hydrogen adsorption may be due to spillover of hydrogen to the support, the takeup of more hydrogen than one hydrogen atom per platinum surface atom, and reaction of hydrogen with oxygen present at the interface of the metal particles with the support.

EXAFS measurements on the catalyst indicated that evacuation of the previously reduced catalyst leads to the formation of platinum oxygen bonds. The platinum-platinum nearest-neighbor peak was reduced by about a factor of 2. It has therefore been supposed that the following reactions proceed at the interface with the support:

H_2 + (Pt-O-Si) = -Pt-Pt- + 2HO-Si

H_2 + (Si-O-Pt) = H-Pt- + HO-Si

These reactions are assumed to be reversible. At about 550 K hydrogen taken up in the reactions is being desorbed. Hydrogen desorbing at 380 and 240 K is assumed to be adsorbed on metallic platinum. Additionally, about 5% of the hydrogen taken up is assumed to spill over onto the support.

However, other EXAFS experiments with noble metal catalysts did not indicate the formation of metal-oxygen bonds on evacuation of reduced catalysts. Thus Kip et al. [89] investigated supported rhodium, iridium, and platinum catalyst by EXAFS. The metals had been applied onto silica, alumina, and titania supports. The mean size of the metal was accurately determined by EXAFS measurements. No increase in the number of metal-oxygen bonds was observed after evacuation of previously reduced catalysts. From the EXAFS data the mean coordination number to other metal atoms was calculated. As expected, the adsorbed hydrogen-to-metal atomic ratio increased with decreasing coordination number. However, the hydrogen-to-metal atomic ratio even rose to above 1 with platinum and to above 2 with rhodium and iridium. With iridium the hydrogen-to-metal atomic ratio was higher than with rhodium, which was in turn higher than found with platinum at an equal coordination number of metal atoms. The atomic ratio depended only on the nature of the metal and the mean coordination number, not on the nature of the support. Therefore, the authors rejected a reaction at the interface with the support.

Penetration of hydrogen atoms into the metal surface accompanied by adsorption of hydrogen atoms onto the surface, leading to adsorption beyond a monolayer, was not indicated by the EXAFS data. For small metal particles, penetration of hydrogen atoms between the metal surface atoms would have led to a change in the metal-metal bond, which was not observed. Kip et al. therefore assume that the metal atoms at the edges and corners, which are relatively numerous with small particles, can adsorb more than one hydrogen atom. Organometallic compounds of tungsten, platinum, rhodium, and iridium show that bonding of more than one hydrogen atom to a metal atom in a stable compound is possible. The relatively high number of corner and edge atoms with small metal particles thus could lead to a high atomic ratio of adsorbed hydrogen-to-metal surface atoms.

Earlier, Mignolet [90] considered the adsorption of hydrogen on extended closely packed metal surfaces in excess of a 1:1 ratio. This author considered the sites on a fcc (111) and a (100) surface. The distance of the two identical triangular sites on the (111) surface is 0.58D, where D is the diameter of the metal atoms. An adsorbed hydrogen atom diameter of 0.19 nm leads to a ratio of 0.69 for a Pt(111), since the diameter of a platinum atom is 0.277 nm. A compression of the adsorbed hydrogen atoms is therefore required to take up a layer exhibiting a stoichiometry of 2, that is, two hydrogen atoms per metal surface atom. With platinum (0.277 nm), the compression is much less than with nickel (0.249 nm), where the ratio is 0.76, which is appreciably higher than the ratio of the distance of the sites, 0.58D. Mignolet therefore assumed that with Pt more hydrogen could be taken up, leading to compression of the adsorbed hydrogen layer. Following Mignolet, the adsorption in excess of a 1:1 ratio can also be due to adsorption of more than one hydrogen atom onto flat, closely packed metal surfaces.

A rather remarkable particle-size effect has been reported in the literature regarding the extent of adsorption on iron particles [91]. Large iron particles, such as those present in commercial ammonia catalysts, exhibit normal adsorption of hydrogen provided that the catalysts have been carefully reduced. Small supported iron particles, on the other hand, display only activated adsorption of hydrogen, despite an extensive reduction procedure. Work on iron single crystals has demonstrated that it is extremely difficult to remove the last monolayer of oxygen from all iron surfaces except the Fe(110) surface. Since small iron particles have no surfaces of the low reactivity of extended (110) surfaces, the last monolayer of oxygen cannot be removed by hydrogen reduction. Adsorption of hydrogen on the oxygen-covered surface is activated.

VI. CONCLUSIONS

An accurate assessment of the free metal surface area and the structure of the surfaces of supported metal particles from the extent of hydrogen adsorption is still not completely unambiguous. Although the extent of hydrogen adsorption on unsupported metal specimens has been established with results that agree fairly well, the effects of particle size, shape, surface structure, and interaction with the support on the extent of hydrogen have not yet been completely elucidated. The relatively low extent of hydrogen adsorption on silica-supported nickel particles, the elevated adsorption onto platinum, rhodium, and iridium particles, and the difficulty of removing the oxygen monolayer on some iron surfaces are not completely resolved. Nevertheless, present work in which the size of the metal particles is determined accurately by an independent method, such as electron microscopy, magnetic measurements, or EXAFS, is leading to results that allow more reliable determination of the free metal surface area. The work on single crystals, which provides data about the ordering of adsorbed hydrogen, the possible presence of subsurface hydrogen, and the abundancy of differently adsorbed hydrogen, is also very helpful.

Although measurement of the extent of hydrogen adsorption has already been carried out for decades, and while many samples are being measured each year, there are still a number of experimental details to be determined. The importance of careful reduction and degassing has already been mentioned. The temperature and pressure at which the extent of hydrogen adsorption have to be determined are still being debated.

REFERENCES

1. H. S. Taylor, *J. Am. Chem. Soc.*, 66:1615 (1944).
2. R. L. Burwell, Jr. and M. Boudart, in *Investigation of Rates and Mechanisms of Reactions*, Part 1 (E. S. Lewis, ed.), Wiley, New York (1974), p. 700.

3. M. Boudart, A. Aldag, J. E. Benson, N. A. Dougharty, and C. G. Hawkins, *J. Catal.*, *6*:92 (1966).
4. M. Boudart, in *Interactions on Metal Surfaces* (R. Gomer, ed.), Springer-Verlag, Berlin (1975), p. 275.
5. M. Boudart, in *Physical Chemistry: An Advanced Treatise*, Vol. 7 (H. Eyring, W. Jost, and D. Henderson, eds.), Academic Press, New York (1975), Chap. 7.
6. G. C. Bond, in *Electronic Structure and Reactivity of Metal Surfaces* (E. G. Derouane and A. A. Lucas, eds.), Plenum Press, New York (1976), p. 523.
7. C. N. Satterfield, *Mass Transfer in Heterogeneous Catalysis*, MIT Press, Cambridge, Mass. (1970).
8. R. Aris, *The Mathematical Theory of Diffusion and Reaction in Permeable Catalysts*, Vols. 1 and 2, Clarendon Press, Oxford (1975).
9. T. R. P. Gibb, Jr., in *Progress in Inorganic Chemistry*, *3*:315 (1962).
10. W. M. Mueller, J. P. Blackledge, and G. G. Libowitz (eds.), *Metal Hydrides*, Academic Press, New York (1968).
11. B. Baranowski, in *Festkörperchemie* (V. Bodyrev and K. Meyer, eds.), VEB, Leipzig (1973), p. 364.
12. F. Bloyart, L. D'Or, and J. C. P. Mignolet, *J. Chim. Phys. (Fr.)*, *54*: 74 (1957).
13. R. V. Culver, J. Pritchard, and F. C. Tompkins, in *Proceedings of the 2nd Congress on Surface Activity*, Vol. 2, Butterworth, Wobrun, Mass. (1957), p. 243.
14. J. Pritchard and F. C. Tompkins, *Trans. Faraday Soc.*, *56*:540 (1960).
15. J. Pritchard, *Trans. Faraday Soc.*, *59*:437 (1963).
16. V. Ponec, Z. Knor, and S. Cerny, *J. Catal.*, *4*:485 (1965).
17. B. Baranowski and M. Smialowski, *Z. Phys. Chem.*, *45*:206 (1965).
18. R. Gomer, R. Wortman, and R. Lundy, *J. Chem. Phys.*, *26*:1147 (1957).
19. R. Wortman, R. Gomer, and R. Lundy, *J. Chem. Phys.*, *27*:1099 (1957).
20. P. M. Gundry and F. C. Tompkins, *Trans. Faraday Soc.*, *52*:1609 (1956).
21. P. M. Gundry and F. C. Tompkins, *Trans. Faraday Soc.*, *53*:218 (1957).
22. Z. Knor and V. Ponec, *Collect. Czech. Chem. Commun.*, *26*:37 (1961).
23. G. Ehrlich, in *Structure and Properties of Thin Films* (C. A. Neugebauer, J. B. Newkirk, and D. A. Vermilyea, eds.), Wiley, New York (1959), p. 423.
24. B. Poelsema, L. K. Verhey, and G. Comsa, *Surf. Sci.* *152/153*:496 (1985).
25. A. J. Algra, S. B. Luitjens, E. P. Th. M. Suurmeijer, and A. L. Boers, *Appl. Surf. Sci.*, *10*:273 (1982).
26. G. E. Rhead, in *Electronic Structure and Reactivity of Metal Surfaces* (E. G. Derouane and A. A. Lucas, eds.), Plenum Press, New York (1976), p. 229.
27. G. J. R. Jones, J. H. Onuferko, D. P. Woodruff, and B. W. Holland, *Surf. Sci.*, *147*:1 (1984).
28. K. Christmann, F. Chehab, V. Penka, and G. Ertl, *Surf. Sci.*, *152/153*: 356 (1985).
29. K. Griffith, P. R. Norton, J. A. Davies, W. N. Unertl, and T. E. Jackman, *Surf. Sci.*, *152/153*:374 (1985).
30. L. Olles and A. M. Baro, *Surf. Sci.*, *137*:607 (1984).
31. J. R. Anderson, I. M. Ritchie, and M. W. Roberts, *Nature*, *227*:704 (1970).
32. J. W. Geus, in *Chemisorption and Reactions on Metallic Films*, Vol. 1 (J. R. Anderson, ed.), Academic Press, New York (1971), p. 129.
33. M. P. Kiskinova and G. M. Bliznakov, *Surf. Sci.*, *123*:61 (1982).
34. Nguyen Van Hieu and J. H. Craig, Jr., *Surf. Sci.*, *160*:483 (1985).
35. J. Ratajczykowa, *Surf. Sci.*, *172*:691 (1986).

3. Energetics of Hydrogen Adsorption

36. H. Conrad, G. Ertl, J. Kuppers, and E. E. Latta, in *Proceedings of the 6th International Congress on Catalysis, London 1976* (G. C. Bond, P. B. Wells, and F. C. Tompkins, eds.), Chemical Society, London (1977), p. 427.
37. D. E. Peebles, J. R. Creighton, D. N. Belton, and J. M. White, *J. Catal.*, 80:482 (1983).
38. N. D. S. Canning and M. A. Chesters, *Surf. Sci.*, 175:811 (1986).
39. B. E. Koel, D. E. Peebles, and J. M. White, *Surf. Sci.*, 125:709 (1983).
40. B. E. Koel, D. E. Peebles, and J. M. White, *Surf. Sci.*, 125:739 (1983).
40a. Unpublished results, Central Laboratory DSM, The Netherlands.
41. R. Imbihl, R. J. Behm, K. Christmann, G. Ertl, and T. Matsushima, *Surf. Sci.*, 117:257 (1982).
42. T. E. Velter and R. H. Stulen, *J. Vac. Sci. Technol.*, A3:1566 (1985).
43. R. J. Behm, K. Christmann, and G. Ertl, *Surf. Sci.*, 99:320 (1980).
44. K. Christmann, R. J. Behm, G. Ertl, M. A. van Hove, and W. H. Weinberg, *J. Chem. Phys.*, 70:4186 (1979).
45. N. N. Kavtaradze, *Dokl. Akad. Nauk SSSR*, 123:498 (1958).
46. N. N. Kavtaradze, *Z. Phys. Chem.*, 28:376 (1961).
47. L. H. Germer and A. U. MacRae, *J. Chem. Phys.*, 37:1382 (1962).
48. G. C. A. Schuit and N. H. de Boer, *Recl. Trav. Chim. Pays-Bas*, 72:909 (1953).
49. E. K. Rideal and F. G. Sweet, *Proc. R. Soc. London*, A257:291 (1960).
50. A. Frennet and P. B. Wells, *Appl. Catal.*, 18:243 (1985).
51. D. F. Klemperer and F. S. Stone, *Proc. R. Soc. London*, A243:375 (1957).
52. D. D. Eley and P. R. Norton, *Proc. R. Soc. London*, A314:319 (1970).
53. G. Wedler and F. J. Bröcker, *Surf. Sci.*, 26:454 (1971).
54. Z. Knor and V. Ponec, *Collect. Czech. Chem. Commun.*, 26:37 (1961).
55. C. Kooy and J. M. Nieuwenhuizen, in *Basic Problems in Thin Film Physics* (R. Niedermayer and H. Mayer, eds.), Vandenhoeck und Ruprecht, Göttingen (1966), p. 181.
56. O. Beeck and A. W. Ritchie, *Discuss. Faraday Soc.*, 8:159 (1950).
57. Z. Knor and V. Ponec, *Collect. Czech. Chem. Commun.*, 26:961 (1961).
58. D. Brennan and F. H. Hayes, *Trans. Faraday Soc.*, 60:589 (1964).
59. R. A. Beebe, J. B. Beckwith, and J. M. Honig, *J. Am. Chem. Soc.*, 67:1554 (1945).
60. R. T. Davis, T. W. DeWitt, and P. H. Emmett, *J. Phys. Colloid Chem.*, 51:1246 (1947).
61. J. Wollock and B. L. Harris, *Ind. Eng. Chem.*, 42:1347 (1950).
62. R. A. W. Haul, *Angew. Chem.*, 68:238 (1956).
63. M. W. Roberts and K. W. Sykes, *Trans. Faraday Soc.*, 54:548 (1958).
64. H. K. Livingston, *J. Colloid Sci.*, 16:181 (1948).
65. P. Zwietering, H. T. L. Koks, and C. van Heerden, *J. Phys. Chem. Solids*, 11:18 (1959).
66. J. W. Geus, in *Chemisorption and Reactions on Metallic Films*, Vol. 1 (J. R. Anderson, ed.), Academic Press, New York (1971), p. 327.
67. J. W. Geus, H. L. T. Koks, and P. Zwietering, *J. Catal.*, 2:274 (1963).
68. J. R. Anderson and B. G. Baker, *J. Phys. Chem.*, 66:482 (1962).
69. R. E. Dietz and P. W. Selwood, *J. Chem. Phys.*, 35:270 (1961).
70. P. W. Selwood, *Chemisorption and Magnetization*, Academic Press, New York (1975), p. 85.
71. G. C. A. Schuit and N. H. de Boer, *Recl. Trav. Chim. Pays-Bas*, 70:1067 (1951).
72. B. G. Linsen, thesis, Technical University of Delft (1964).
73. C. M. Quinn and M. W. Roberts, *Trans. Faraday Soc.*, 60:899 (1964).
74. C. M. Quinn and M. W. Roberts, *Trans. Faraday Soc.*, 61:1775 (1965).
75. M. W. Roberts and B. R. Wells, *Trans. Faraday Soc.*, 62:1608 (1966).

76. M. W. Roberts and B. R. Wells, *Discuss. Faraday Soc.*, *41*:162 (1966).
77. V. Ponec and Z. Knor, *Collect. Czech. Chem. Commun.*, *27*:1443 (1962).
78. Z. Knor and V. Ponec, *Collect. Czech. Chem. Commun.*, *31*:1172 (1966).
79. G. C. A. Schuit and L. L. van Reijen, *Adv. Catal.*, *10*:242 (1958).
80. J. W. E. Coenen and B. G. Linsen, in *Physical and Chemical Aspects of Adsorbents and Catalysts* (B. G. Linsen, ed.), Academic Press, New York (1970), p. 471.
81. D. G. Mustard and C. H. Bartholomew, *J. Catal.*, *67*:186 (1981).
82. R. van Hardeveld and A. van Montfoort, *Surf. Sci.*, *4*:396 (1966).
83. R. van Hardeveld and F. Hartog, *Adv. Catal.*, *22*:75 (1972).
84. C. S. Brooks and G. L. M. Christopher, *J. Catal.*, *10*:211 (1968).
85. H. K. Kuo, P. Ganesan, and R. J. De Angelis, *J. Catal.*, *64*:303 (1980).
86. R. van Hardeveld, in *Proceedings of the 3rd International Congress on Catalysis, Amsterdam 1964* (G. C. A. Schuit, W. M. H. Sachtler, and P. Zwietering, eds.), North-Holland, Amsterdam (1965).
87. J. T. Richardson and R. J. Dubus, *J. Catal.*, *54*:207 (1978).
88. J. W. Geus and P. B. Wells, *Appl. Catal.*, *18*:231 (1985).
89. B. J. Kip, F. B. M. Duivenvoorde, D. C. Koningsberger, and R. Prins, to be published.
90. W. M. Mueller, J. P. Blackledge, and G. G. Libowitz (eds.), *Metal Hydrides*, Academic Press, New York, (1968), p. 651.
91. J. C. P. Mignolet, "Etude théorique et expérimentale de quelques problèmes d'adsorption," Université de Liège (1958).
92. A. J. H. M. Kock and J. W. Geus, *Prog. Surf. Sci.*, *20*:229 (1985).
93. W. A. Crossland and J. Pritchard, *Surf. Sci.*, *2*:217 (1964).

| Hydrogen as a Tool for Characterization of Catalyst Surfaces by Chemisorption, Gas Titration, and Temperature-Programmed Techniques

P. G. MENON

Dow Chemical (Nederland) B.V., Terneuzen, The Netherlands; and Chalmers University of Technology, Gothenburg, Sweden

I.	INTRODUCTION	117
II.	HYDROGEN CHEMISORPTION	119
III.	SURFACE TITRATIONS WITH H_2	119
	A. Monometallic and Bimetallic Reforming Catalysts	119
	B. Pt-Ru	121
	C. Pd-Ni	123
	D. Supported Ag	123
	E. Superiority of Surface Titration over Chemisorption in Cases of SMSI	124
IV.	TEMPERATURE-PROGRAMMED TECHNIQUES	125
	A. TPD	125
	B. TPR	130
	C. TPSR	131
	D. Fingerprinting the Surface of a Used Catalyst	133
V.	CONCLUDING REMARKS	136
	REFERENCES	136

I. INTRODUCTION

For measurement of specific metal area or metal dispersion in supported metal catalysts, the chemisorption method is today perhaps the most useful single tool. This method was introduced in the mid-1930s in the pioneering work of Emmett and Brunauer [1] on composite iron catalysts for ammonia synthesis. A more direct stimulus for the method was the advent of catalytic reforming in the 1950s, with only 0.3 to 0.6 wt % of the precious metal Pt dispersed on an alumina support. It was very necessary to know how well this precious metal was dispersed on the support. After all, only the exposed Pt atoms on the surface can come in contact with the hydrocarbon molecules; the Pt atoms inside the metal crystallites are inaccessible to the reactant molecules and hence just wasted Pt. Furthermore, the performance of present-day reforming catalysts is so good that every exposed Pt atom on the catalyst surface effectively catalyzes the reactions of over 1 billion hydrocarbon molecules during the expected lifetime of a few years for the catalyst. For optimum performance of the bifunctional reforming catalyst, a delicate balance also has to be maintained between the Pt-metal function and the acid function of the alumina. Hence measurement of Pt dispersion became a routine fingerprint for the reforming catalyst, not only for comparison of activities of different catalysts but also as a periodic check on the performance of the 8 to 20 metric tons of catalyst in a typical reformer plant. This problem was further highlighted when supported Pt catalysts soon found other important applications, such as xylene isomerization in the manufacture of dimethyl terephthatate or DMT, the intermediate for polyester; benzene hydrogenation for nylon manufacture; paraffin isomerization, as an octane-number boosting step in refineries; selective dehydrogenation of n-paraffins to n-olefins to make linear alkylbenzenes for biodegradable detergents; and during the last decade, for purification of automotive exhaust emissions.

Other supported metal catalysts in use include those for fat hydrogenation (Ni); hydrogenation/dehydrogenation (Pt, Pd, Ni); dehydration, methanation, and steam reforming (Ni); epoxidation (Ag); and so on. The oil and energy crisis dating from 1973 and the revised interest in coal chemistry, Fischer-Tropsch synthesis, and the new C_1 chemistry have provided the latest incentive to the study of supported metal catalysts. Measurement of specific metal surface in supported metal catalysts has now become a sine qua non for any work on such catalysts.

Expensive chemisorption automats have become fairly common today in industrial laboratories and richer university catalysis laboratories. Conventional volumetric, gravimetric, or gas-chromatographic methods of chemisorption are used by others to obtain similar results with quite comparable reliability and accuracy.

Surface gas titration techniques are particularly suitable in cases where the first hydrogen chemisorption may not be very reproducible, for a variety of reasons. Temperature-programmed desorption (TPD) and reduction (TPR) studies are very relevant for the normal performance of a catalyst. TPD of H_2 from a catalyst gives a characteristic spectrum of the catalyst-hydrogen interaction as a function of temperature, including the temperature ranges of interest for the catalytic reaction. TPR reveals the reduction-oxidation or redox properties of the catalyst and hence gives a direct indication of whether or not the adsorption and catalytic properties are associated with a redox mechanism involving the catalyst. The reducibility of an active component of a catalyst depends very much on the strength of its interaction with other components in the catalyst or with the catalyst support. Hence TPR can often provide a direct insight into such interactions between the active component and the support. Temperature-programmed surface reaction (TPSR) is a powerful technique to detect reactive intermediates on the catalyst surface as well as to characterize and differentiate the different types of carbon (loosely called "coke") on the catalyst under actual reaction conditions.

The purpose of this chapter is to focus attention on the detailed characterization and fingerprinting of catalysts possible nowadays using hydrogen, and on more recent highlights in this area. Typical examples will be cited only to illustrate the power of the techniques. Since detailed reviews on these techniques are readily available in recent literature, no attempt will be made here for comprehensive coverage of all interesting work in this field. The work on single crystals and evaporated metal films has been covered in Chapters 1 and 3. Hence the discussion here is confined to supported metal catalysts of the type usually used in industry.

II. HYDROGEN CHEMISORPTION

Hydrogen chemisorption is undoubtedly the most commonly used technique for the measurement of dispersion of Pt and Ni in supported catalysts. Its advantages are that (a) the chemisorption is relatively simple, (b) the physical adsorption on the metal or support is negligible, and (c) practically no adsorption occurs on oxide catalysts. As against these, the disadvantages are (a) possibility of dissolution or formation of hydrides, for example, with Pd (see Chapter 14); (b) sensitivity of hydrogen chemisorption to impurities on the surface; and (c) sometimes very misleading results, for example, when TiO_2-supported metals are subjected to reduction above 500°C [cf. the recent controversy on strong metal-support interaction (SMSI) discussed by Burch in Chapter 13].

Detailed sketches and descriptions of static volumetric adsorption equipment are available in the monographs of Hayward and Trapnell [2], Ponec et al. [3] and Anderson [4]. Lemaitre et al. [5] have discussed the measurement of metal dispersion by chemisorption, electron microscopy, and x-ray techniques; they also give a critical evaluation of the methodology to be followed for assessing the accuracy of dispersion measurements. Various aspects of chemisorption on supported metal catalysts have been reviewed by Ponec et al. [3], Anderson [4], Moss [6], and Ferrauto [7]. Scholten et al. [8] have presented a detailed review on surface characterization of supported and unsupported hydrogenation catalysts. This review is especially addressed to "organic chemists who make use of supported-metal hydrogenation catalysts, and who are unfamiliar with surface characterization techniques." Hence it can be strongly recommended as an excellent introduction for newcomers in the area of metal catalysts. A short review on the characterization of catalysts by reactive gas titration has been given elsewhere [9]. Chemisorption of H_2 on supported Fe, Co, and Ni catalysts is discussed in Chapter 5.

III. SURFACE TITRATIONS WITH H_2

A. Monometallic and Bimetallic Reforming Catalysts

Reactive surface gas titrations were introduced into catalysis by Benson and Boudart [10] in 1965 to titrate Pt surface in Pt/Al_2O_3 reforming catalysts:

$Pt + (1/2)H_2 \rightarrow Pt\text{---}H$ (H chemisorption, HC)

$Pt + (1/2)O_2 \rightarrow Pt\text{---}O$ (O chemisorption, OC)

$Pt\text{---}O + (3/2)H_2 \rightarrow Pt\text{---}H + H_2O$ (H titration, HT)

$2Pt\text{---}H + (3/2)O_2 \rightarrow 2Pt\text{---}O + H_2O$ (O titration, OT)

Stoichiometry HC:OC:HT   1:1:3

Many conflicting stoichiometries were reported for this surface titration during the next 12 years, as shown in Table 1. Prasad et al. [11] suggested that much of this controversy was perhaps caused because every research group used the very first hydrogen chemisorption on their freshly reduced catalyst as the basis for all calculations. But this first hydrogen chemisorption itself was often not reproducible because the surface was still in a rather metastable condition after the reduction. They found [11] that if the calculations were based on hydrogen titration, after a few H_2-O_2 cycles at room temperature, the stoichiometry was *always* 1:1:3, independent of the pretreatment

TABLE 1 Some Conflicting Surface Stoichiometries Reported for Hydrogen Chemisorption (HC):Oxygen Chemisorption (OC):Hydrogen Titration (HT) of Supported Platinum Catalysts

Year	Ref.[a]	HC:OC:HT	Remarks/observation
1965	Benson and Boudart (1)	1:1:3	
1967	Mears and Hansford (2)	2:1:4	
1970	Wilson and Hall (3)	2:1:4 to 2:1:3	Depends on sintering
1972	Freel (4)	2:1:4 to 1:1:3	Depends on pretreatment
1975	Netzer and Gruber (5)	1:0.5:2 to 1:1:3	Depends on temperature
1972	Basset et al. (6)	1:1:3	Only at p_{H_2} <1 torr
1970	Barbaux et al. (7)	2:1:4	At 10 torr
1970	Vannice et al. (8)	1:1:3	Even at 100 torr
1973	Menon et al. (9)	1:1:3	O_2-H_2 cycles at 20°C are necessary
1976	Prasad and Menon (10)	1:1:3	Special precautions needed
1978	Carabello et al. (11)	1:1.5:—	Careful reduction needed
1978	Prasad et al. (12)	1:1:3	Always, provided that two conditions are fulfilled

[a]References: (1) J. Catal., 4:704 (1965); (2) J. Catal., 9:125 (1967); (3) J. Catal., 17:190 (1970); (4) J. Catal., 25:149 (1972); (5) Z. Phys. Chem. (Neue Folge), 96:25 (1975); (6) Proceedings of the 5th International Congress on Catalysis (1972), p. 915; (7) J. Chim. Phys., 67:1035 (1970); (8) J. Catal., 16:348 (1970); (9) J. Catal., 29:188 (1973); (10) J. Catal., 44:314 (1976); (11) J. Catal., 52:507 (1978); (12) J. Catal., 52:515 (1978).

of the catalyst and independent of sintering or Pt crystallite size. The need to wait for a few O_2-H_2 cycles to obtain reproducible results had been observed earlier by Menon et al. [12]. It looks as if the heterogeneity of the freshly reduced catalyst surface is smoothed out or annealed after one or more gas titrations, thereby producing a more "homogeneous" surface on which the H_2-O_2 titrations can be carried out with a definite stoichiometry. From subsequent studies on the effect of reduction temperature on the properties of supported Pt catalysts, Menon and Froment [13-16] have proposed that the peculiar behavior or nonreproducibility of a freshly reduced supported Pt catalyst may be due to the residual hydrogen retained by the reduced metal. These and other aspects of hydrogen effects in metal catalysts have been reviewed in detail by Paál and Menon [17].

The advent of the bimetallic Pt-Re/Al_2O_3 (Chevron) reforming catalyst in 1969 made it imperative to be able to measure the Pt and Re dispersions separately in such bimetallic systems. Freel [18] and Menon et al. [12] adopted the H_2-O_2 titration technique to the much faster gas-chromatographic pulse technique, which was then extended [12] to the Pt-Re/Al_2O_3 bimetallic catalyst as follows. After calcination and reduction at 500°C and cooling in H_2 to room temperature:

4. Hydrogen for Catalyst Characterization

1. First oxygen pulsing gave chemisorption by Pt + Re.
2. Subsequent reduction at 293 K converted Pt···O to Pt···H.
3. Renewed oxygen pulsing gave chemisorption by Pt only.
4. The difference between steps 1 and 3 gave the chemisorption by Re only.

The equipment for such gas titrations [12] is quite simple, consisting mainly of a gas chromatograph with a TC detector and a samping valve.

Special mention may be made of three applications of the aforementioned H_2-O_2 titration technique for obtaining insight into three different aspects of the bimetallic reforming catalyst:

1. For alumina-supported bimetallic catalysts of Pt with Re, Ir, or Ru, it is important to differentiate between two extreme cases: where the two metals exist as separate particles and where they coexist as bimetallic aggregates or alloys of constant composition. Charcosset [19] has pointed out that in repeated or cyclic H_2-O_2 titrations at room temperature, the first and second O_2 titrations will give equal values only when all of the Re is in the form of Pt-Re alloy. The value for the second O_2 titration will be lower when a part of the Re atoms on the surfaces is in a free or unalloyed state (because the chemisorbed O on these free atoms cannot react with H_2 at room temperature).
   In this way, depending on the activation procedure for the bimetallic catalysts, either almost all the surface Re atoms can be in an alloyed form or a significant part of them can be in free Re particles. The gas titration is the simplest technique available at present to detect these free Re particles on the surface.
2. Isaacs and Petersen [20] have employed the surface titration technique, together with temperature-programmed reduction, to investigate the extent of alloy formation between Pt and Re when subjected to various pretreatments.
3. Barbier et al. [21] have quite recently confirmed the foregoing stoichiometry for the H_2-O_2 titration. They have also extended the titration technique to get a direct measure of hydrogen spillover from the metal to the alumina at typical reforming temperatures.

The method based on the differences in reactivity of chemisorbed oxygen toward hydrogen, or vice versa, which works so well for characterizing Pt-Re/Al_2O_3 catalysts, fails in the case of Pt-Ir/Al_2O_3. The reducibility of oxygen chemisorbed on Ir does not differ enough from that of Pt [22]. This is best seen in the TPR profile of Pt-Ir/Al_2O_3: a single sharp reduction peak at 508 K, halfway between the individual reduction peaks for Pt at 473 K and for Ir at 523 K.

B. Pt-Ru

For the bimetallic Pt-Ru/Al_2O_3 catalyst, Blanchard et al. [23] have employed a temperature-programmed titration (TPT) technique to determine Pt and Ru dispersions separately. The oxygen chemisorbed on Ru atoms on the surface of Ru-Pt alloy particles was found to be much more reducible than the oxygen chemisorbed on pure Ru. Heating the catalyst in a stream of 5% H_2 in argon would show the ready consumption of H_2 at two different temperature ranges as followed by a TC detector of a gas chromatograph (Fig. 1). The chemisorption of O_2 and H_2 and H_2-TPT of chemisorbed oxygen can be used to determine not only the overall metallic (Pt + Ru) area but also the fraction of the metallic

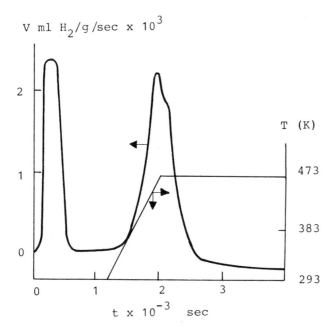

FIGURE 1 Temperature-programmed titration of chemisorbed O with hydrogen on Pt-Ru/Al_2O_3. First peak, Pt and Pt-Ru clusters; second peak, free Ru. (From Ref. 23.)

area of pure Ru. These results could be checked by x-ray emission analysis combined with transmission electron microscopy, which indeed showed the presence of a small fraction of Ru in Pt particles of about 1.5 nm size.

One peculiar difficulty in extending gas titration techniques to supported metallic catalyst systems deserves special mention. The chemisorption stoichiometry of a gas for a metal on a carrier can sometimes change when a second metal is present, particularly if bimetallic clusters or alloys can be formed. This has been shown quite recently by Chakrabarty et al. [24] for Pt-Ru-bimetallic catalysts supported on silica (Table 2). On Ru/SiO_2, the chemisorption of both H_2 and CO gave identical dispersion values for Ru. When the 2% metal loading was changed from pure Ru to Pt:Ru atomic ratios of 20:80 or 50:50, the total metal dispersion value from CO chemisorption was only half of that from H_2 chemisorption. It looks as if CO chemisorption on the bimetallic catalyst is in the bridged form, while in the monometallic

TABLE 2 Chemisorption Data for 2% (Pt + Ru)/SiO_2 Catalysts

Catalyst composition, Pt:Ru	Metal dispersion (%) from chemisorption of:	
	H_2	CO
0:100	12	12
20:80	27.6	13.7
50:50	16.8	8.0

Source: Ref. 24.

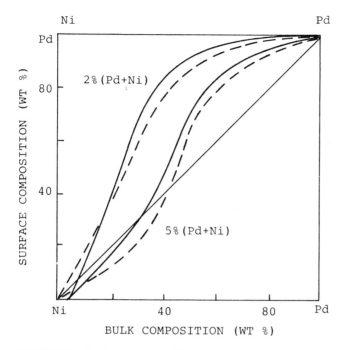

FIGURE 2 Surface composition of Pd-Ni/Al_2O_3. Solid line, from H_2-O_2 titrations; dashed line, from TPR. (From Ref. 25.)

Ru catalyst it is in a linear form. This will have to be checked and confirmed by infrared spectra. But this shows that hydrogen chemisorption is a superior and more reliable primary standard in such metal dispersion measurements.

C. Pd-Ni

Both Pd and Ni have similar chemisorption characteristics toward common adsorbates such as H_2, O_2, and CO. Hence direct chemisorption cannot be used to determine surface composition of Pd-Ni/Al_2O_3 catalysts. The oxidized surfaces of Ni and Pd, however, have quite different reactivities toward H_2 at room temperature. Paryjczak et al. [25] took advantage of this fact and hence could estimate Pd and Ni metal surfaces separately by resorting to H_2-O_2 titrations and temperature-programmed reduction (TPR). The oxidized Pd surface could be titrated with H_2 at room temperature, but not the oxidized Ni surface. Hence the specific surface of Pd could be determined by H_2-O_2 titrations, while the TPR method gave an independent estimate of Ni on the composite surface. The surface compositions determined by these two independent techniques agree quite well, as shown in Fig. 2.

D. Supported Ag

Supported Ag is a unique catalyst for the selective oxidation of ethylene to ethylene oxide. Measurement of Ag dispersion on such a catalyst is usually done by chemisorption of O_2, or better, by reactive chemisorption of N_2O with the Ag atoms on the surface, as proposed by Scholten et al. [26]. Recently, a reactive gas titration of chemisorbed O on Ag with H_2 at 443 K has been proposed [27].

$$Ag_sO + H_2 \rightarrow Ag_s + H_2O$$

Direct chemisorption of oxygen on Ag_S and reactive surface oxidation of Ag_S with N_2O both gave quite comparable results in the H_2 titration. The advantages claimed for this titration method are (a) rapid equilibration; (b) increased sensitivity; (c) no oxygen contamination effects; (d) distinguishes any irreversible oxygen uptake on the support from that on the metal; and (e) very reproducible stoichiometry, $HT/OC \approx 2.0$, not dependent on pretreatment (contrast the so-called SMSI effects on chemisorption of H_2 on TiO_2-supported noble metal catalysts).

E. Superiority of Surface Titration over Chemisorption in Cases of SMSI

Direct chemisorption of H_2 on TiO_2-supported metal catalysts reduced at high temperatures (>773 K) gives a misleadingly low value, due to the so-called SMSI effect (see Chapter 13). Menon and Froment [13-16] have reported that much of this lost H-chemisorption capacity of the metal after HTR can be recovered by a few O_2-H_2 cycles at room temperature. Menon et al. [12] had proposed in 1973 that such O_2-H_2 cycles at 293 K constitute an essential precondition for obtaining reproducible results in surface gas titrations of alumina-supported Pt and Pt-Re reforming catalysts. A detailed explanation for the effect of the O_2-H_2 cycles has been given elsewhere [14,17]. (See also Chapter 13 for more recent applications of O_2-H_2 cycles in surface titrations.)

Table 3 gives recent data of Herrmann [28] for the uptake of gases during H_2 and O_2 chemisorptions (HC and OC) and H_2 and O_2 reciprocal titrations (HT and OT) on Pt/TiO_2 and Rh/TiO_2 catalysts when subjected to a low-temperature reduction (LTR, at 473 K) or a high-temperature reduction (HTR, at 773 K). While HC after HTR gives very low values, OC, OT, and HT give comparable values after HTR and LTR. The titration stoichiometry HC:OC:HT:OT is very nearly 1:1:3:1.5, except for HC after HTR (in the "SMSI state"). On the basis of such data, Herrmann [28] has proposed an inhibition factor R for the suppression of HC after HTR:

$$R = \frac{\text{number of } H_2 \text{ molecules chemisorbed after LTR}}{\text{number of } H_2 \text{ molecules chemisorbed after HTR}}$$

Typical values of R for some TiO_2-supported catalysts are shown in Table 4.

TABLE 3 Uptake of H_2 and O_2 During Chemisorption and H_2-O_2 Reciprocal Titrations (μmol/g Catalyst)

Sequence	5% Pt/TiO_2		5% Rh/TiO_2	
	LTR	HTR	LTR	HTR
H chemisorption	42	6	67	24
O titration	65	56	112	100
H titration	126	105	224	201
O chemisorption	45	37	106	88
H titration	134	114	298	244
O titration	66	53	122	112

Source: Ref. 28.

TABLE 4 Inhibition Factor R for TiO_2-Supported Catalysts

Catalyst	Metal loading (wt %)	Crystallite side (nm)	R
Pt/TiO_2	5	2	7
	0.5	2	∞
Ni/TiO_2	5	13.5	∞
Rh/TiO_2	5	3.5	2.8

Source: Ref. 28.

IV. TEMPERATURE-PROGRAMMED TECHNIQUES

A. TPD

Temperature-programmed desorption (TPD) of H_2 from catalysts, developed in 1963 by Amenomiya and Cvetanovic [29a-d], has become a very handy and popular technique for catalyst characterization during the last two decades. Excellent reviews on the principles, experimental details, and applications of TPD by the foregoing authors [29b-d], Smutek et al. [30], and quite recently by Falconer and Schwartz [31], Scholten et al. [8] and Lemaitre [32] are readily available. Hence these aspects of TPD will not be repeated here.

Special mention may be made of the advantages of a multipurpose experimental unit (Fig. 3) in which chemisorption, surface gas titration, temperature-programmed desorption, combustion (of coke on catalyst), and so on, can all be carried out on one and the same sample of catalyst [9]. Basically, such a unit consists of the TC detector of a gas chromatograph (GC), a sampling valve, and a quartz microreactor. Different gases can be used as a continuous carrier stream for the GC, or for pulsing through the sampling valve and the microreactor (0.2 to 1 g of catalyst). Sophistications possible nowadays include a thermal mass-flow controller to maintain constant flow even when the flow resistance increases with increasing temperature, a computer-driver automatic gas-sampling valve to send identical gas pulses at fixed intervals, computer integration of GC peaks, a quadruple mass spectrometer with a mass selector and a multipen recorder to monitor each desorbing chemical species separately and continuously, and so on. It is very important to use only a glass or quartz reactor, not a stainless-steel reactor, in all TPD/TPR work. Otherwise, the uptake of H_2 by the metal at medium temperatures and its desorption at high temperatures can sometimes yield highly magnified and distorted TPD/TPR curves. This can easily be checked by running a blank experiment with an empty stainless steel reactor.

In general, TPD is a very useful technique for obtaining information on the nature of the interaction between adsorbed species and a catalyst surface. Four basic types of information can be collected from TPD studies: (a) the number of various adsorbed species, (b) the approximate population of various adsorbed states, (c) bond energies between adsorbate and surface, and (d) the possible chemical form of the adsorbed species on the catalyst surface. The number of peaks and their area provide information on types (a) and (b). Peak temperatures and their variation with the heating rate support data with respect to type (c), whereas desorption kinetics give indications concerning type (d). Many examples for all these may be seen in the reviews cited above.

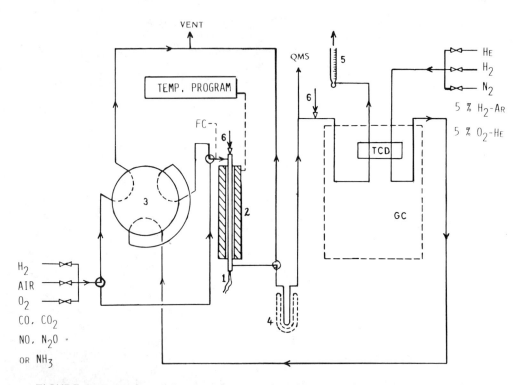

FIGURE 3 Experimental setup for gas titration, TPD/TPR/TPSR, and so on, on one and the same catalyst sample. 1, Thermocouple; 2, reactor; 3, sampling valve; 4, cold trap; 5, soap-bubble flow meter; 6, injection port; FC, thermal mass flow controller; QMS, quadrupole mass spectrometer. (From Ref. 9.)

Two very thorough and comprehensive books on TPD of hydrogen on group VIII metals were published by Popova and co-workers [33a, b] in 1979-1980. They reported TPD curves for various amounts of hydrogen sorbed on every group VIII metal, together with desorption orders and desorption activation energy values [33a]. Detailed fingerprints of TPD curves for every metal are given therein; the reactivity of various types of adsorbed hydrogen is also discussed there. The special value of the more recent book by Popova [33b] is that it gives characteristic TPD curves for bimetallic supported catalysts such as 2% Pd-Ir/Al_2O_3 and 4% Pd-Pt/Al_2O_3 in the whole composition range (five intermediate compositions in both cases). For example, the desorption temperature of the lower-temperature hydrogen peak remained constant (about 320 K) with Pd-Ir atomic ratios between 1:0 and 6:4; similarly, the high-temperature peak remained constant at about 700 K. But at a composition of Pd-Ir 4:6, the lower peak was shifted abruptly from 320 K to about 470 K: further increase in the Ir content affected it only marginally (ca. 500 K at Pd-Ir 1:9). The high-temperature peak remained unaffected during all of the foregoing changes in composition. In the Pd-Pt system, the low-temperature peak disappeared at Pd-Pt atomic ratio 0.4:1; here the addition of even a little Pt (Pd-Pt 1:0.027) already changed the shape of the high-temperature peak. Data on such systems as Pd-Rh, Pd-Os, Pt-Re, Pt-Sn, Pt-Os, and Pt-Ir, all supported on Al_2O_3, Ni-Fe/Cr_2O_3, and so on, are also given in the form of TPD spectra. Hence these two books serve as an interesting and concise source of information on the monometallic and bimetallic supported catalyst systems.

4. Hydrogen for Catalyst Characterization

The adsorption of H_2 on disperse metals reveals in several cases properties similar to those found with single crystals, but the temperatures of the TPD spectra are usually higher. The chemisorption of H_2 on Pt black in the range 77 to 670 K was investigated by Tsuchiya et al. [29d] using the TPD technique. The appearance of four peaks in the TPD spectrum with their maxima at 170, 250, 350, and 570 K indicates at least four different states (designated as α, β, γ, and δ) of chemisorbed hydrogen on Pt in this range. These states have been tentatively assigned the following adsorption forms:

```
H
|
H     H——H    H         H
|     |  |    |        / \
Pt    Pt Pt   Pt      Pt  Pt
```

In the case of alumina-supported Pt catalysts, Aben et al. [34] found that the desorption peaks represented at least three different species of H_2. Recent work on TPD of H_2 from supported Fe, Co, and Ni catalysts are discussed in detail in Chapter 5.

In the context of the recent controversy on strong metal-support interaction (SMSI; discussed by Burch in Chapter 13) the TPD technique has been useful in throwing light on the phenomena occurring on high-temperature reduction (HTR) of supported Pt and other metallic catalysts. Menon and Froment [13-16] found that in the case of Pt supported on Al_2O_3, SiO_2, and TiO_2, and even for unsupported Pt black, the main H_2-desorption peak during TPD is shifted to higher temperatures by 100 to 200 K when the catalyst was reduced at 773 to 873 K, instead of at 673 K (Figs. 4 and 5). On the basis of these and other results, Menon and Froment [13-16] propose that the residual hydrogen, retained in the catalysts after heat treatment and cooling in H_2 to room temperature, may be mainly responsible for much of the SMSI effect.

In the case of TiO_2-supported catalysts, there is an additional complication due to the easy reducibility of the support TiO_2 itself. This was shown [14, 15] by conducting TPD experiments using 5% H_2-Ar as the carrier gas instead of N_2, thereby enabling ready readsorption, or any other uptake of H_2, on the catalyst wherever possible during the progressive rise in temperature of the catalyst. For the catalyst reduced at 470 K, a profuse desorption of H_2 occurs up to about 570 K, which is then followed by a re-uptake of H_2 (or a further reduction of the catalyst) in the range 570 to 770 K (Fig. 6). This re-uptake resolves into two separate peaks around 650 and 770 K as the initial reduction temperature of the catalyst is raised to 570 or 670 K. After reduction at 770 K the re-uptake of H_2 is no longer noticeable. Whatever change has been occurring in the earlier cases at 670 to 770 K has already occurred here during the reduction itself at 770 K. A comparative experiment on the carrier TiO_2 (Fig. 4, top) shows that the reduction peak at about 770 K is characteristic for TiO_2; apparently this occurs in the case of Pt-TiO_2 also in the 5% H_2/Ar stream unless the initial reduction is carried out at 770 K. The migration of reduced TiO_x species on to the top of the metal crystallites has been established recently by XPS/AES techniques (see Chapter 13).

Nagy et al. [35] studied the sorption of hydrogen on an unsupported sintered Pt catalyst by means of electrochemical, volumetric, and thermodesorption (TPD) techniques. They showed that the specific hydrogen sorption depended on the thermal conditions of adsorption. If the saturation of the catalyst with H_2 is carried out at 273 to 288 K, the amount of adsorbed H_2 corresponds to one H atom per surface Pt atom, as is usually observed. If the catalyst is cooled down in H_2 from a higher temperature, the amount

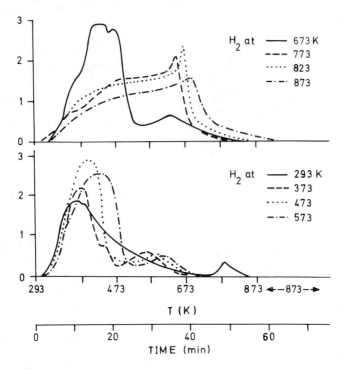

FIGURE 4 TPD of H_2 from 0.6% Pt/Al_2O_3 reforming catalyst, exposed to H_2 at various temperatures for 1 h and cooled in H_2 to 293 K. Catalyst taken 0.84 g; N_2 flow 17 cm^3 min^{-1}; heating rate 10 K min^{-1}; GC attenuation 4×. (From Ref. 13.)

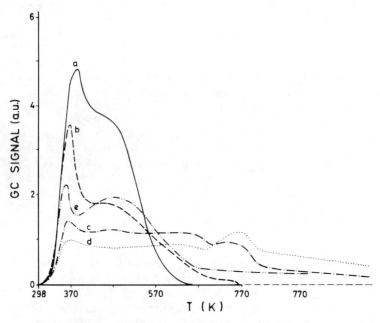

FIGURE 5 TPD of H_2 from 2% Pt/TiO_2 catalyst reduced at 473, 573, 673, 773 K, and finally at 473 K (curves a, b, c, d, and e, respectively). TPD conditions as for Fig. 4. (From Ref. 14.)

4. *Hydrogen for Catalyst Characterization* 129

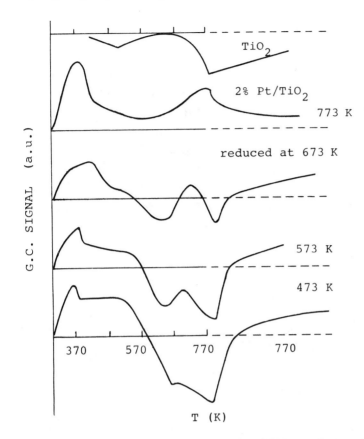

FIGURE 6 Desorption of H_2 from 2% Pt/TiO_2 catalyst and re-uptake of H_2 by it during programmed heating (10 K min^{-1}) in 5% H_2-Ar stream. Signal above the baseline indicates a net desorption and below it a re-uptake of H_2. The catalyst was pre-reduced at the temperatures indicated. The reduction of TiO_2 itself is indicated at the top. (From Ref. 14.)

of H_2 retained by the catalyst could be four to five times larger than in the former case, as shown in Table 5. The hypothesis put forward by Nagy et al. [35] for the interesting results of Table 5 is that H atoms can penetrate the topmost lattice layer of Pt and thus reach the Pt in the second, third, and lower layers. This could be the origin of the residual hydrogen in freshly reduced catalysts, proposed by Menon and Froment [13-16] as the main cause for the discrepancy in surface titration stoichiometries and for the peculiar properties of supported metal catalysts and even unsupported Pt black, when exposed to a higher-temperature reduction, as discussed above and in Section III.A.

Interesting correlations can sometimes be drawn between hydrogen species on the catalyst surface resulting in a peculiar peak during TPD and the activity of the catalyst. Thus Aben et al. [34] found a correlation between the low-temperature H_2 desorption peak at about 253 K in the TPD spectra of Pt/Al_2O_3 catalysts and their hydrogenation activity for benzene at 323 K. Similarly, Menon and Froment [13-14] have reported a direct correlation between the profuse desorption of H_2 in the region 323 to 573 K during TPD and the hydrogenolysis activity for *n*-pentane or *n*-hexane on Pt/Al_2O_3, Pt/SiO_2, and Pt/TiO_2 catalysts. For similar cases, see Section III.A of Chapter 18.

TABLE 5 Sorption of H_2 by Pt: Results from Three Different Methods

Method		μmol H_2/g Pt	Remarks
Electrochemical	298	1.27	Saturation with H_2 at 293 K; desorption measurements
Volumetric	273	1.68	Saturation with H_2 at 273 K
		6.30	Saturation in H_2 by slow cooling from 873 K to 273 K
		5.94	Amount of H_2 not removable by pumping after the last run
TPD	298-823	7.05	Saturation in H_2 by slow cooling from 823 K, then desorption

Source: Ref. 35.

B. TPR

The work on TPR has been covered in fair detail in a recent review by Hurst et al. [36]. Hence only a few highlights will be cited here. The analytical capability of TPR was illustrated by Jenkins [37], who showed a decrease in reducibility (Fig. 7) as the coordinative unsaturation of Pt decreases from $PtCl_2$ to $PtCl_4$, H_2PtCl_6, $Pt(NH_3)Cl_2$, and K_2PtCl_6. Important differences can also be observed in TPR profiles when the same quantity (0.5 wt %) of Pt is deposited on a carrier like alumina either by impregnation or by ion exchange (Fig. 8). Such fingerprinting by TPR gives a fascinating insight into the nature of the resulting metal catalyst, its metal dispersion (on the same sample, measured in the same apparatus by O_2-H_2 titration after the

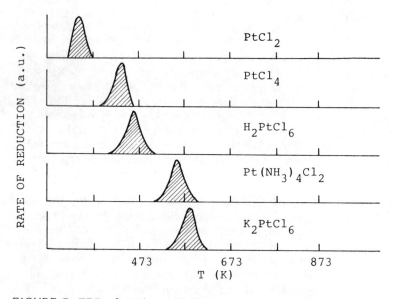

FIGURE 7 TPR of various platinum compounds, showing the decrease in reducibility as the coordinative unsaturation of Pt decreases. (From Ref. 37.)

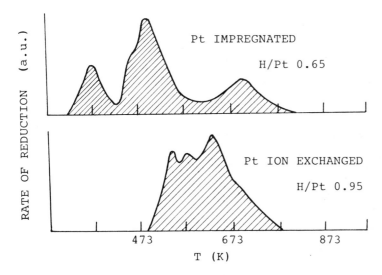

FIGURE 8 Differences in TPR profile when 0.5% Pt is deposited on Al_2O_3 by impregnation or ion exchange. (From Ref. 37.)

TPR run), the metal-support interaction, and so on, when drying and calcination temperature, catalyst support, metal loading, and other preparative variables for the catalyst are changed. In the case of Pt-Re/Al_2O_3 reforming catalyst, TPR studies by Wagstaff and Prins [38a] have shown that in the reduced catalyst Pt and Re are in an alloyed state, but they readily segregate into separate phases on calcination of the catalyst above 773 K. Prins and co-workers have extended such studies to several other systems, the most recent being Co-Rh catalysts supported on alumina [38b], titania [38c], and silica [38d].

TPR studies sometimes reveal interesting correlations between the reducibility of an oxide catalyst and its activity for a particular catalytic reaction. Thus, for various MoO_3-γAl_2O_3 and WO_3-γAl_2O_3 catalysts, Thomas and Moulijn [39] have found a positive correlation between the reducibility of the catalyst and their initial reaction rates for the metathesis of propylene (Fig. 9). The reducibility in this context is expressed by the average reduction temperature at which half of the oxide species has been reduced under standard TPR conditions. It has to be emphasized, however, that excessive reduction of the catalysts invariably occurs with the olefin under metathesis conditions over a longer process time (time on stream). This leads to the creation of other catalytic species on the surface, catalyzing unwanted side reactions such as polymerization and coke formation, and ultimately to rapid catalyst deactivation. Reduction of Re^{7+} to an intermediate (Re^{6+} or Re^{5+} ?), but not much lower Re^{4+} or Re^0, oxidation state is also very important for Re_2O_7/Al_2O_3 metathesis catalysts, as has been reported by Xu Xiaoding [40].

C. TPSR

Reactions of hydrocarbons and of CO on catalysts at high temperatures are often accompanied by deposition of carbonaceous residues on the catalyst surface. Such carbon on the catalyst may be a reactive intermediate, or it can also be an unwanted by-product ("coke"), deactivating the catalyst directly in some cases, and only very slowly and indirectly in other cases. Hence the reactivity of the carbonaceous species toward gases such as H_2,

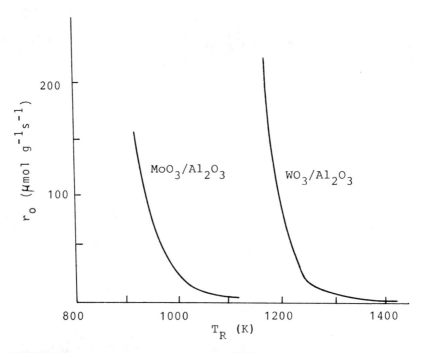

FIGURE 9 Correlation between reducibility of supported MoO_3 and WO_3 and their initial reaction rates for metathesis of propylene. (From Ref. 39.)

O_2, and steam can be used to differentiate or characterize such species. Temperature-programmed surface reaction (TPSR) is a convenient method to use to do this [41]. In TPSR, the "coked" catalyst is heated up at a constant linear rate in a stream of the reactive gas and the products formed are analyzed continuously by an on-line gas chromatograph or mass spectrometer.

Carbon, deposited from CO disproportionation on Ni/Al_2O_3 catalysts, reacts very differently toward H_2 to form CH_4 as the temperature is raised progressively. Thus McCarty and Wise [42] found peaks of CH_4 at 325, 470 ± 20, and 680 ± 30 K for carbon species on the surface, designated as α'-, α-, and β-carbon, respectively (Fig. 10). With C_2H_4 as a carbon source instead of CO, the carbon on the surface has different reactivity patterns toward H_2 during TPSR [43]. The reactivity also changes with the temperature of exposure of the catalyst to C_2H_4. The carbon species (α' and α) which react with H_2 at lower temperatures are nowadays designated as chemisorbed carbon, and the higher-temperature (β) species as carbidic carbon. Studies using Auger electron spectroscopy [44,45] and x-ray photoelectron spectroscopy [46] have confirmed the carbidic nature of this β-carbon, while scanning and transmission electron microscopy studies of coke on the catalyst formed from hydrocarbons at still higher temperature show [47,48] filamentous (δ') and encapsulating carbon (δ, 960 ± 15 K). In a very recent study using TPSR and multiple reflectance infrared spectroscopy, Hayes et al. [49] report the existence of three types of carbonaceous species during methanation on Ni catalysts. Of these, the two species active at normal methanation temperatures are a surface nickel carbide and a carbon triply bonded to one Ni site, thus giving rise to two possible molecular routes for methanation.

4. Hydrogen for Catalyst Characterization

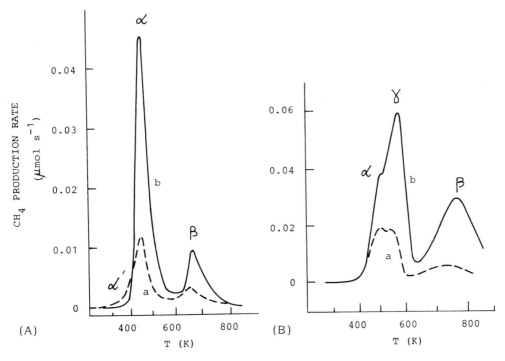

FIGURE 10 Detection of carbon species on Ni/Al$_2$O$_3$ catalyst by TPSR with H$_2$. Heating rate approximately 1 K/min. (A) Carbon deposition by exposure to CO at 550 K; relative carbon deposit: (a) 1.19; (b) 3.14. (B) Carbon deposited by ethylene decomposition at 575 K; relative carbon deposit: (a) 3.7; (b) 11.1. [(A) From Ref. 42; (B) from Ref. 43.]

Carbonaceous surface species with varying amounts of hydrogen and with or without oxygen have been reported on a variety of catalysts for such reactions as methanol and Fischer-Tropsch synthesis, steam reforming of hydrocarbon, decomposition of formic acid, and water-gas shift reaction (see Fig. 2 in Chapter 23). One way to identify and count such reactive species on the catalyst surface under actual reaction conditions is to send another reactant which can react with these surface species and sweep the surface clean. This type of reactive scavenging of catalyst surfaces has been applied very successfully by Ekerdt and Bell [50] and Bell [51]. Recently, Tau and Bennett [52] have studied an active Fe/TiO$_2$ surface used in Fischer-Tropsch synthesis by titrating it with H$_2$, O$_2$, or D$_2$. Their results show that during the reaction the surface is covered with about 46 µmol per gram of CH and large quantities of carbidic carbon. During hydrogenation of ethylene, on the other hand, there are only CH species on the surface. Thus the type and content of hydrogen of the reactive carbonaceous intermediate on the catalyst surface have a decisive influence in determining the conversion and product selectivity in catalyzed organic reactions.

D. Fingerprinting the Surface of a Used Catalyst

Detailed characterization of the surface composition of used industrial catalysts is sometimes possible by a combination of chemisorption, surface titration, and

TPD techniques, supplemented by electron probe microanalysis (EPMA), and other techniques. A typical example of this is the recent work of Duprez et al. [53] on Rh/Al_2O_3 catalysts used in steam dealkylation/reforming at 733 K.

$$C_6H_5CH_3 + 2H_2O \rightarrow C_6H_6 + CO_2 + 3H_2$$

These coked and sulfided catalysts were fingerprinted using two different procedures, both giving essentially the same results:

1. Selective oxidation of carbonaceous deposits by O_2 pulses at 593 K allowed practically all the sulfur to remain chemisorbed on the metal. Chemisorption values of H_2 (specific for free Rh metal surface) before and after the selective oxidation yield the fractions of free metal surface (Rh_f), of coked surface, and of sulfided surface (total Rh surface = $Rh_f + Rh_c + Rh_s$).
2. Oxidation by O_2 pulses at 723 K results in carbon gasification and sulfur oxidation into sulfate with the simultaneous displacement of the sulfur (as sulfate) from the Rh metal surface to the alumina surface. On high-temperature reduction, the sulfate was reduced to sulfur, which migrated back to the Rh metal surface from the support surface.
    (a) Initial hydrogen chemisorption (HC) on spent catalyst = Rh_f
    (b) HC after oxidation at 723 K = total Rh surface = $Rh_f + Rh_s + Rh_c$
    (c) HC after reduction = $Rh_f + Rh_c$
    Hence Rh_c = (c) - (a) and Rh_s = (b) - (c).

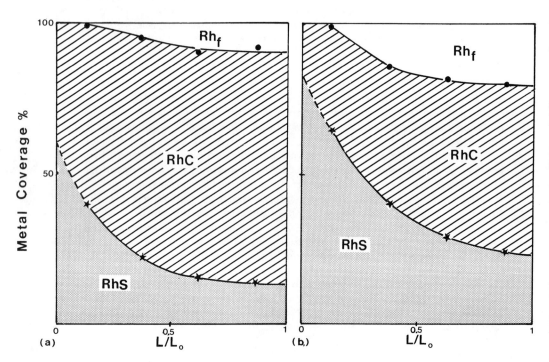

FIGURE 11 Distribution of coke and sulfur in the catalyst bed after toluene-steam reactions on Rh/Al_2O_3 catalyst at 733 K, using toluene with different S contents: (a) 5 ppm S, run for 27 h; (b) 50 ppm S, run for 4 h. (From Ref. 53.)

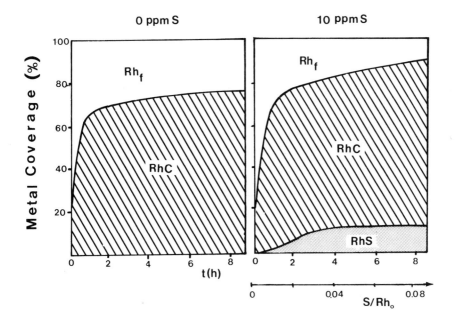
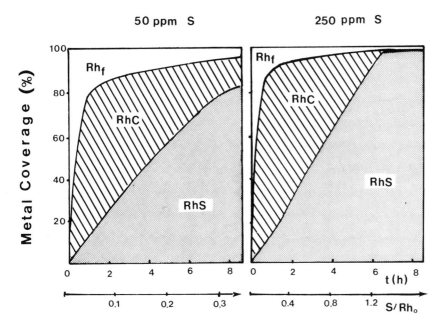

FIGURE 12 Variations of free, coked, and sulfided rhodium surface distributions with time on stream for various S contents in toluene. Toluene-steam reactions on Rh/Al_2O_3 catalyst at 733 K, with mol ratio H_2O/toluene = 6. (From Ref. 53.)

Using procedure 2 on catalyst samples taken from different bed heights in the reactor, profiles or maps of coke and S distribution along the bed could be drawn at different process times (times on stream) and for different S levels in the feed. Figure 11 shows the distribution of coke and S in the catalyst bed after toluene-steam reactions with a toluene feed containing 5 ppm S for 27 h and 50 ppm S for 4 h (catalyst exposed to a total of 9 and 13 μmol S, respectively). EPMA studies on catalyst samples taken from the reactor after the 50-ppm S study showed that (a) in the very first moments of the reaction, S adsorbs on the exterior of the pellets located at the bed inlet and progressively invades the interior of the pellet, and simultaneously, the areas located downstream in the catalyst bed, and (b) S remains always confined on or in the vicinity of the Rh metal particles.

Distributions of coke and S on the entire catalyst bed as functions of process time and S content in the feed are shown in Fig. 12. These surface-concentration profiles reveal that (a) the proportion of free metal surface (Rh_f) decreases rapidly with time whatever the S content in the feed; (b) between and 0 and 1 h on-stream, the catalyst deactivation is mainly due to coke deposition; (c) beyond 1 h, S progressively spreads onto the initially coke-covered surface; and (d) S has a tendency to induce a very high coke deposition.

V. CONCLUDING REMARKS

Specific chemisorption of hydrogen, directly as such, in its modified form as in TPD and O_2-H_2 surface gas titrations, and in a more reactive form as in TPR, TPSR, or reactive scavenging, is a powerful experimental tool for the characterization of catalyst surfaces. These techniques provide not only a measure of the total metal and separate metal dispersions, but also valuable information on the extent and type of metal-support interaction, cluster or alloy formation, the role of promoters, oxidation states of the metal on the surface, surface enrichment, spillover phenomena, and so on. The distinctive advantages of the foregoing techniques involving H_2 are relatively simple and often self-made equipment, much lower cost, and quicker measurements, which enable these techniques to be within the reach of practically every catalysis laboratory today. Their disadvantages include the lack of specificity of chemisorption in some cases, the complexity due to different forms of chemisorption of H_2, surface enrichment sometimes caused by chemisorption itself, slow or time-dependent chemisorption, and strong metal-support interaction (SMSI). Fortunately, these limitations can often be recognized as peculiarities of the particular catalyst-adsorbate system, enabling one to resort to complementary techniques for such specific cases. Except in such restricted cases, a wealth of information can usually be obtained from hydrogen chemisorption or reactive hydrogen titration of composite catalyst surfaces.

REFERENCES

1. P. H. Emmett and S. Brunauer, *J. Am. Chem. Soc.*, 59:310 (1937); S. Brunauer and P. H. Emmett, *J. Am. Chem. Soc.*, 62:1732 (1940).
2. D. O. Hayward and B. M. W. Trapnell, *Chemisorption*, Butterworth, London (1964).
3. V. Ponec, Z. Knor, and Z. Czerny, *Adsorption on Solids*, Butterworth, London (1974).

4. J. R. Anderson, *Structure of Metallic Catalysts*, Academic Press, New York (1975).
5. J. L. Lemaitre, P. G. Menon, and F. Delannay, in *The Characterization of Heterogeneous Catalysts* (F. Delannay, ed.), Marcel Dekker, New York (1984), p. 299.
6. R. L. Moss, in *Experimental Methods in Catalytic Research*, Vol. 2 (R. B. Anderson and P. T. Dawson, eds.), Academic Press, New York (1976).
7. R. J. Ferrauto, *AIChE Symp. Ser.*, 70(143):9 (1974).
8. J. J. F. Scholten, A. P. Pijpers, and A. M. L. Hustings, *Catal. Rev.-Sci. Eng.*, 27:151 (1985).
9. P. G. Menon, in *Catalyst Deactivation* (E. E. Petersen and A. T. Bell, eds.), Marcel Dekker, New York, 1987.
10. S. W. Benson and M. Boudart, *J. Catal.*, 4:706 (1965).
11. J. Prasad, K. R. Murthy, and P. G. Menon, *J. Catal.*, 52:515 (1978).
12. P. G. Menon, J. Sieders, J. Streefkerk, and G. J. M. van Keulen, *J. Catal.*, 29:188 (1973).
13. P. G. Menon and G. F. Froment, *J. Catal.*, 59:138 (1979).
14. P. G. Menon and G. F. Froment, *Appl. Catal.*, 1:31 (1981).
15. P. G. Menon and G. F. Froment, in *Metal-Support and Metal-Additive Effects in Catalysis* (B. Imelik et al., eds.), Elsevier, Amsterdam (1982), p. 171.
16. P. G. Menon and G. F. Froment, *Acta Chim. Acad. Sci. Hung.*, 111:631 (1982).
17. Z. Paál and P. G. Menon, *Catal. Rev.-Sci. Eng.*, 25:229 (1983); see also P. G. Menon, in *Advances in Catalysis Science and Technology* (T. S. R. Prasada Rao, ed.), Wiley Eastern, New Delhi (1985), pp. PL1-15.
18. J. Freel, *J. Catal.*, 25:149 (1972).
19. H. Charcosset, *Platinum Met. Rev.*, 23:18 (1979).
20. B. H. Isaacs and E. E. Petersen, *J. Catal.*, 77:43 (1982); 85:1, 8 (1984).
21. J. Barbier, H. Charcosset, G. de Periera, and J. Riviere, *Appl. Catal.*, 1:71 (1981).
22. L. Tournayan, H. Charcosset, R. Fretty, C. Leclercq, P. Turlier, J. Barbier, and G. Leclercq, *Thermochim. Acta*, 27:95 (1978); see also N. Wagstaff and R. Prins, *J. Catal.*, 59:446 (1979).
23. G. Blanchard, H. Charcosset, H. Dexpert, E. Freund, C. Leclercq, and G. Martino, *J. Catal.*, 70:168 (1981).
24. D. K. Chakrabarty, K. Mohana Rao, A. A. Desai, and P. K. Basu, in *Advances in Catalysis Science and Technology* (T. S. R. Prasada Rao, ed.), Wiley Eastern, New Delhi (1985), p. 199.
25. T. Paryjczak, J. M. Farbotko, and J. Goralski, *J. Catal.*, 88:228 (1984).
26. J. J. F. Scholten, J. A. Konvalinka, and F. W. Beeckman, *J. Catal.*, 28:209 (1973).
27. S. R. Seyedmonir, D. E. Strohmayer, G. L. Geoffry, M. A. Vannice, H. W. Young, and J. W. Linowski, *J. Catal.*, 87:424 (1984).
28. J. M. Herrmann, *J. Catal.*, 89:404 (1984).
29. (a) Y. Amenomiya and R. J. Cvetanovic, *J. Phys. Chem.*, 67:144 (1963); (b) Y. Amenomiya and R. J. Cvetanovic, *Adv. Catal.*, 17:103 (1967); (c) Y. Amenomiya and R. J. Cvetanovic, *Catal. Rev.*, 6:21 (1972); (d) S. Tsuchiya, Y. Amenomiya, and R. J. Cvetanovic, *J. Catal.*, 19:245 (1970).
30. M. Smutek, S. Cerny, and F. Busek, *Adv. Catal.*, 24:353 (1975).
31. J. L. Falconer, and J. A. Schwarz, *Catal. Rev.-Sci. Eng.*, 25:141 (1983).
32. J. L. Lemaitre, in *Characterization of Heterogeneous Catalysts* (F. Delannay, ed.), Marcel Dekker, New York (1984), p. 29.

33. (a) N. M. Popova, L. V. Babenkova, and G. A. Savelyeva, *Adsorption and Interaction of the Simplest Gases with Group VIII Metals* (in Russian), Nauka Kazakhskoi, Alma-Ata (1979); (b) N. M. Popova, *Effect of Support and Metal Structure on Gas Adsorption* (in Russian), Nauka Kazakhskoi, Alma-Ata (1980).
34. P. C. Aben, R. H. van der Eijk, and J. M. Oelderik, *5th International Congress on Catalysis*, Palm Beach, Paper 48 (1972).
35. F. Nagy, D. Móger, M. Hegedus, G. Y. Mink, and S. Szabó, *Acta Chim. Acad. Sci. Hung.*, 100:211 (1979); S. Szabó, D. Móger, M. Hegedus, and F. Nagy, *React. Kinet. Catal. Lett.*, 6:89 (1976).
36. N. W. Hurst, S. J. Gentry, A. Jones, and B. D. McNicol, *Catal. Rev.-Sci. Eng.*, 24:233 (1982).
37. J. W. Jenkins, *Preprints Sixth Canadian Symposium on Catalysis*, Ottawa (1979).
38. (a) N. Wagstaff and R. Prins, *J. Catal.*, 59:434 (1979); (b) H. F. van't Blik and R. Prins, *J. Catal.*, 97:188 (1986); (c) J. H. A. Martens, H. F. J. van't Blik, and R. Prins, *J. Catal.*, 97:200 (1986); (d) H. F. J. van't Blik, D. C. Konigsberger, and R. Prins, *J. Catal.*, 97:210 (1986).
39. R. Thomas and J. A. Moulijn, *J. Mol. Catal.*, 15:157 (1982).
40. Xu Xiaoding, Ph.D. thesis, University of Amsterdam (1985).
41. H. Wise, J. G. McCarty, and J. Oudar, in *Catalyst Deactivation and Poisoning* (J. Oudar and H. Wise, eds.), Marcel Dekker, New York (1985), p. 1.
42. J. G. McCarty and H. Wise, *J. Catal.*, 57:406 (1979).
43. J. G. McCarty, P. Y. Hou, D. Sheridan, and H. Wise, in *American Chemical Society Symposium Series* (L. F. Albright and R. T. K. Baker, eds.), American Chemical Society, Washington, D.C. (1982).
44. J. P. Coad and J. C. Riviere, *Surf. Sci.*, 25:609 (1971).
45. D. W. Goodman, R. D. Kelley, T. E. Madey, and J. M. White, *J. Catal.*, 64:479 (1980).
46. J. de Deken, P. G. Menon, G. F. Froment, and G. Haemers, *J. Catal.*, 70:225 (1981).
47. J. R. Rostrup-Nielsen, and D. L. Trimm, *J. Catal.*, 48:155 (1977).
48. H. H. Gierlick, M. Fremery, A. Skov, and J. R. Rostrup-Nielsen, in *Catalyst Deactivation* (B. Delmon and G. F. Froment, eds.), Elsevier, Amsterdam (1980), p. 459.
49. R. E. Hayes, R. J. Ward, and K. E. Hayes, *Appl. Catal.*, 20:123 (1986).
50. J. G. Ekerdt, and A. T. Bell, *J. Catal.*, 62:19 (1980).
51. A. T. Bell, *Catal. Rev.-Sci. Eng.*, 23:203 (1981).
52. L. M. Tau, and C. O. Bennett, *J. Catal.*, 89:327 (1984).
53. D. Duprez, M. Mendez, and J. Little, *Appl. Catal.*, 27:145 (1986).

5 | Hydrogen Adsorption on Supported Cobalt, Iron, and Nickel

CALVIN H. BARTHOLOMEW

Brigham Young University, Provo, Utah

I.	INTRODUCTION	139
II.	KINETICS AND ENERGETICS OF ADSORPTION	140
	A. Calorimetric and TPD Studies of Hydrogen Interaction with Supported Nickel	140
	B. Calorimetric and TPD Studies of Hydrogen Interaction with Supported Cobalt and Iron	150
III.	REVERSIBILITIES AND STOICHIOMETRIES OF ADSORPTION	154
	A. Reversible and Irreversible Adsorption	154
	B. Adsorption Stoichiometries	158
	C. Measurement of Monolayer Adsorption Capacity	160
IV.	SUMMARY	162
	REFERENCES	163

I. INTRODUCTION

Some of the concepts and fundamentals relating to the kinetics, energetics, and stoichiometries of adsorption of hydrogen on metals have been addressed in Chapters 1, 3, and 4, and methods for characterizing metal surfaces and catalysts using chemisorption titration and TPD were discussed in Chapter 4. Accordingly, it will not be necessary to reintroduce these basic concepts and fundamentals. Rather, they will be used as a foundation for discussing the complexities of hydrogen adsorption on supported, base-metal catalyst systems—especially supported cobalt, iron, and nickel, with emphasis on nickel.

Chapter 3 focused on the assessment of surface area of supported active metals from the extent of hydrogen adsorption. The difficulty of assessing monolayer hydrogen adsorption on metals was emphasized. Evidence was presented for less than monolayer adsorption on some metals, and this was interpreted in terms of "limited accessibility" of gaseous hydrogen to the metal surface. This represents one of the important schools of thought regarding this phenomenon.

In this chapter the evidence for nonstoichiometric adsorption of hydrogen on supported metals is discussed from a different point of view, and the results are interpreted in terms of (a) reversibility of adsorption and (b) changes in the kinetics and energetics of hydrogen adsorption due to the presence of support moieties or metal oxides on the metal surface. This represents a second important school of thought regarding the nonstoichiometry of hydrogen adsorption, thus providing a balanced point of view on this subject.

We also address several other important issues:

1. How are the kinetics, energetics, and stoichiometries of hydrogen adsorption on metals influenced by (a) the important physical and chemical properties of a supported catalyst system: for example, metal dispersion (percent metal exposed) and extent of reduction to the metal, (b) the interaction between metal and support, (c) promoters, and (d) preparation variables?
2. What is meant by "reversible" or "irreversible" adsorption of hydrogen? What types of adsorbed species correspond to the different "adsorption states" observed by temperature-programmed desorption (TPD)?
3. What is activated adsorption? How does it affect the measurement of metal surface area?
4. What are the best methods for measuring monolayer hydrogen adsorption on supported base metals? What are the limitations of these methods?
5. What complications are introduced by unreduced metal oxide phases in estimating dispersion of base metal/support systems, and how can they be treated?

Adsorption/desorption kinetics and energetics will be treated first, since an understanding of these phenomena is prerequisite to the discussion of reversibilities and stoichiometries, topics which follow in that order. It should be emphasized that it is not possible in this short chapter to review all of the pertinent literature. Rather, an effort has been made to discuss representative recent literature.

II. KINETICS AND ENERGETICS OF ADSORPTION

A. Calorimetric and TPD Studies of Hydrogen Interaction with Supported Nickel

Early calorimetric studies of hydrogen adsorption on nickel involved metallic films [1-6]; the author is not aware of any published calorimetric studies of hydrogen adsorption on supported nickel. While numerous TPD studies of hydrogen from single-crystal nickel have been reported [1,7-12], relatively few TPD studies of hydrogen from supported nickel have been published [13-18]. Four of these studies involving high-surface-area catalysts [15,16] or decorated nickel films [17,18] were conducted in the clear absence of transport influences [19-23] and in the case of high-surface-area samples took into account readsorption of hydrogen on the metal crystallites [15,16] or nickel films [17,18] were conducted in the clear absence of transport influences [19-23] and in the case of high-surface-area samples took into account readsorption of hydrogen on the metal crystallites [15,16,20-22], thus providing quantitative kinetic data. The results of these studies are summarized and compared with data from single-crystal TPD and calorimetric film studies in Table 1. Figs. 1 to 7 show TPD spectra of nickel supported on silica, titania, and alumina. To facilitate the discussion of the results in these figures and

TABLE 1 Atomic TPD States and Heats of Adsorption of Hydrogen on Nickel Surfaces

Surface	Temperature maxima (K) of atomic states observed by TPD			Desorption order	$-\Delta H^a$ (kJ mol^{-1})	Refs.
	β_1	β_2	γ			
Ni(100)	300-325	350-400	—	2	96^b (β_2)	10,11
Ni(110)	—	350-360	—	1	90^b (β_2)	10
Ni(111)	320-340	360-395	—	2	96^b (β_2)	10
Polycrystalline Ni	—	—	—		75-130^c	1
	—	410 (θ = 0.5)	—		87^d	16
24% Ni/SiO$_2$	334 (θ = 1)	450 (θ = 1)	620 (θ = 1)		61^d, 85^d, 123^d	14
50% Ni/SiO$_2$	—	450-575	—	2	89^e	15
	—	425 (θ = 0.5)	—	2	87^e	16
10% Ni/SiO$_2$	—	415 (θ = 0.5)	—	2	82^e	16
3% Ni/SiO$_2$	—	410 (θ = 0.5)	—	—	82^d	16
14% Ni/Al$_2$O$_3$	—	335 (θ = 0.5)	605 (θ = 0.5)	—	70^d, 125^d	16
0.7 ML^f Al$_2$O$_3$/Ni	275	350-375	450	—	—	17
10% Ni/TiO$_2$	—	375 (θ = 0.5)	520 (θ = 0.5)	—	84^d, 105^d	16
1 nm Ni/TiO$_2$	—	425^g	500^g	—	—	18
0.3 ML^f TiO$_2$/Ni	—	360	480	2	75, 108	17

^{a}Heat of adsorption or activation energy.
^{b}Activation energies from flash desorption.
^{c}Heats of adsorption by calorimetry.
^{d}Heats of adsorption estimated from desorption peak temperatures according to method of Konvalinka [14], which accounts for readsorption.
^{e}Heats of adsorption using desorption rate isotherms extrapolated to low coverage and assuming readsorption in quasi-equilibrium.
^{f}ML = units of monolayer coverage.
^{g}Reduced in hydrogen at 820 K.

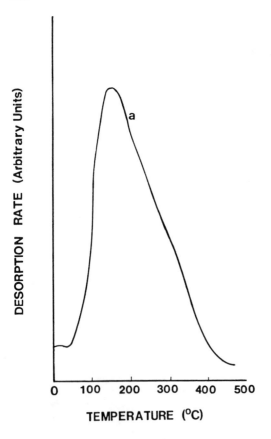

FIGURE 1 Hydrogen TPD spectrum for 10% Ni/silica after adsorption at 300 K. (From Ref. 16.)

in Table 1 and to provide a foundation for discussion of material in sections to follow, a brief review of the fundamentals of hydrogen adsorption on single-crystal nickel is useful.

Two atomic states are observed for hydrogen adsorption on Ni(100), Ni(110), and Ni(111), which desorb in the range 180 to 450 K [10,12]. The β_2 state forms at low coverages, while the β_1 state forms after β_2 at higher coverages (see Fig. 16 in Chapter 1). The existence of these two states is normally interpreted in terms of a through-the-lattice adsorbate-adsorbate repulsion at higher coverages which decreases the binding energy for the β_2 state [10]. Figure 16 in Chapter 1 shows that the maximum rate of desorption for the β_1 state occurs near room temperature; accordingly, the β_1 state is easily desorbed at room temperature [10] and will therefore not be observed in TPD studies in which hydrogen was adsorbed at room temperature followed by purging in inert gas (e.g., the studies in Refs. 15 and 16). According to Christmann et al. [11b], hydrogen adsorption is completely reversible at room temperature. Saturation coverage of hydrogen at 120 to 180 K (β_1 and β_2 states) corresponds to 0.9 to 1.0 monolayer [11b,12], that is, a H/Ni_s ratio of essentially unity. Finally, it is clear that there are only small differences in the binding energies for hydrogen on the three lowest-index planes of nickel (see Table 1); in other words, the energetics of adsorption are little affected by surface structure, at least on the three lowest-index planes.

The data in Table 1 reveal some important similarities and differences for the kinetics and energetics of hydrogen adsorption on supported nickel relative to single-crystal nickel. Based on recent quantitative TPD studies [15,16], hydrogen adsorption on Ni/silica at room temperature involving half-monolayer coverage is very similar to that on single-crystal surfaces, as evidenced by the appearance of a single β_2 peak having a desorption order of 2 and a heat of adsorption (82 to 89 kJ mol^{-1}) very nearly the same as for the single-crystal surfaces (90 to 96 kJ mol^{-1}). The shift to higher temperatures for the peak maximum for the β_2 state on Ni/silica relative to single-crystal nickel is most probably a result of readsorption on the supported catalysts [20-22].

The TPD spectrum for 130-K adsorption to greater than a monolayer of hydrogen on a 24% Ni/silica (Fig. 2) consists of mainly two low-temperature peaks (weakly adsorbed α states) and three large peaks above room temperature corresponding to moderately or strongly adsorbed hydrogen. The peak at 334 K with an adsorption heat of 61 kJ mol^{-1} can be assigned to the β_1 state and that at 450 K (heat of adsorption of 85 kJ mol^{-1}) to the β_2 state. However, the observation of the γ state reported at 620 K (heat of adsorption of 123 kJ mol^{-1}) for Ni/silica is unique to this study; this could be an artifact of the somewhat arbitrary deconvolution of the complex TPD spectrum (see Fig. 2) or a result of the preparation procedure involving precipitation and calcination—the latter treatment possibly causing a shift in adsorption states to higher temperatures [24] (see also the discussion of iron catalysts below). The assignment of this high-temperature state by Konvalinka et al. [14] to edge and corner sites on small crystallites in their sample seems unlikely in view of the structure insensitivity of hydrogen adsorption discussed above, but this possibility cannot be ruled out altogether.

The adsorption of hydrogen on Ni/titania and Ni/alumina is clearly more complicated than that on single-crystal nickel or nickel/silica (see Table 1 and Figs. 3 to 7). In both of these systems, two strongly bound states, β_2 and γ, are observed after hydrogen adsorption at 300 K (Figs. 3 and 5); the high-temperature γ state is the predominant peak in both of these catalysts reduced at 673 and 725 K, respectively. Moreover, the appearance of a dis-

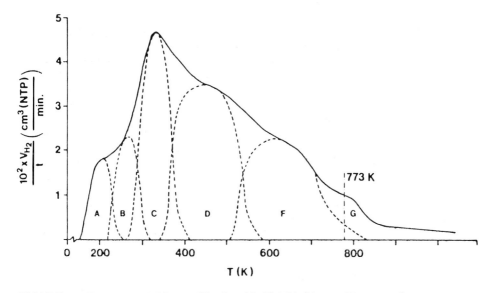

FIGURE 2 Hydrogen TPD profile for 24.3% Ni/silica. (From Ref. 14.)

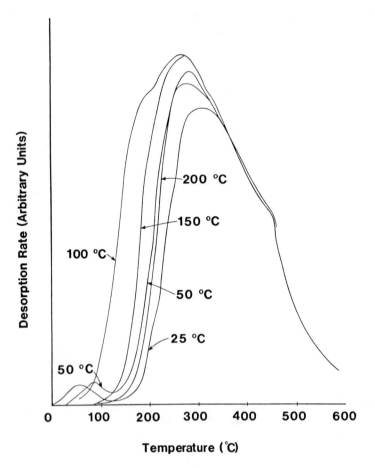

FIGURE 3 Hydrogen TPD spectra for 14% Ni/alumina at different adsorption temperatures. (From Ref. 16.)

tinct shoulder at about 700 K (450°C) in the spectra for hydrogen desorption from 14% Ni/alumina (Fig. 3) provides evidence for the existence of γ_1 and γ_2 states at approximately 600 and 700 K, respectively. One or both of these states apparently involves activated adsorption, since the TPD spectral area increases with increasing adsorption temperature. Raupp and Dumesic [17] observed similarly that alumina on nickel (70% of a monolayer, prepared by sputtering aluminum on the surface of a nickel film, followed by oxidation and reduction) caused the appearance of a new γ state at 400 to 500 K after hydrogen adsorption at 300 K but not after adsorption at 150 K (see Fig. 4). In other words, there was an activation barrier for adsorption into the strongest adsorption state. Thus, adsorption of hydrogen on the alumina-coated nickel film was very similar to that of the 14% Ni/alumina.

Adsorption of hydrogen on nickel is generally assumed to be nonactivated, that is, to have nearly zero activation energy and thus no kinetic barrier to adsorption [25]. Indeed, Roberts [26] observed an activation energy for adsorption of hydrogen on nickel wire of only 1.7 kJ mol^{-1}. Moreover, the kinetics of adsorption/desorption from hydrogen on single-crystal nickel [7-12] and Ni/silica [15,16] are consistent with this assumption; indeed, the amount adsorbed decreases with increasing temperature above about 120 K for these systems, since adsorption and desorption are in a near equilibrium, which under these conditions favors desorption. However, in the case of 14% Ni/alumina (Fig. 3) the activation energy for hydrogen adsorption is significant (i.e., 10 kJ mol^{-1}) [16]. The higher activation barrier to hydrogen adsorption on Ni/alumina and the appearance of new adsorption states at higher binding energies is probably best explained by the presence of un- reduced nickel and aluminum oxide species on the nickel surface or in intimate contact with small nickel crystallites [17, 18, 24, 27, 28]. It is well known that nickel supported on alumina is more difficult to reduce to the metallic state than nickel on silica [29, 30]. The greater apparent adsorption strength of these sites may be due partially to steric hindrance by these contaminant species of hydrogen atoms during surface diffusion and recombination [17],

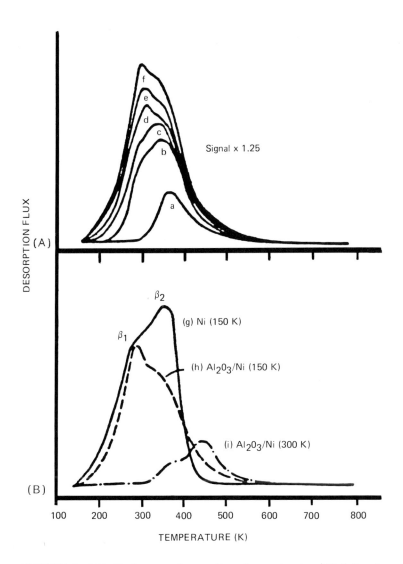

FIGURE 4 (A) Hydrogen desorption from alumina/Ni following different hydrogen exposures at 150 K: (a) 0.3; (b) 1.0; (c) 3.0; (d) 5.0; (e) 13; (f) 30 L. Alumina coverage was 0.7 ml. (B) Hydrogen desorption from saturation coverage on (g) clean Ni dosed at 150 K, (h) alumina/Ni dosed at 150 K, and (i) alumina/Ni dosed at 300 K. (From Ref. 17.)

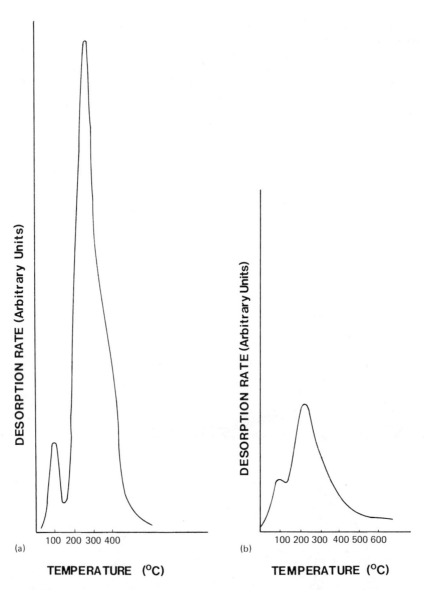

FIGURE 5 Hydrogen TPD spectra for 10% Ni/titania: (a) after reduction at 400°C; (b) after successive reductions of 1 h each at 400, 500, 600, and 700°C. Desorption rates are on the same scale for both runs. (From Ref. 16.)

while the increase in activation energy for adsorption occurs as a result of local interactions of nickel sites with neighboring nickel or aluminum oxide species [17]. The presence of other adlayer species, such as carbon, copper, and potassium, has been shown to induce activation energy barriers for hydrogen adsorption and in the case of potassium increase binding energies, although carbon and copper species decrease binding energies [17,24,31].

The behavior of Ni/titania for hydrogen adsorption is similar to that of Ni/alumina, although the observed effects of support are more dramatic and more sensitive to reduction temperature in the former system. In the case of a 10% Ni/titania catalyst reduced initially at 673 K, there is a dramatic decrease in the intensity and area of the beta and gamma peaks with increasing reduction temperature (see Fig. 5). Raupp and Dumesic [18] found that hydrogen adsorption on a 1-nm nickel film overlaying a titania film was very similar to that on single-crystal nickel (see Fig. 6a); however, as the nickel film/titania film was heated to progressively higher temperatures in hydrogen, the saturation coverage decreased while the peak maximum shifted to higher binding energies (Fig. 6b). It should be emphasized that the observation of these high-temperature states required extended exposures to hydrogen at 300 K. Since Weatherbee and Bartholomew exposed their 10% Ni/titania to pulses of hydrogen for only short time periods, they did not observe these high-temperature states.

The decrease in saturation hydrogen coverage and shift to higher binding energies for the Ni/titania system can be attributed to decoration of the nickel surface with reduced TiO_x species [17,18]. Indeed, Raupp and Dumesic [17] found that titania on a nickel film behaved similarly to Ni/titania; as the titania coverage was increased progressively to about two-thirds of a monolayer, the hydrogen TPD spectrum transformed from one typical of single-crystal nickel with β_1 and β_2 peaks at 300 and 400 K, respectively, to one involving two

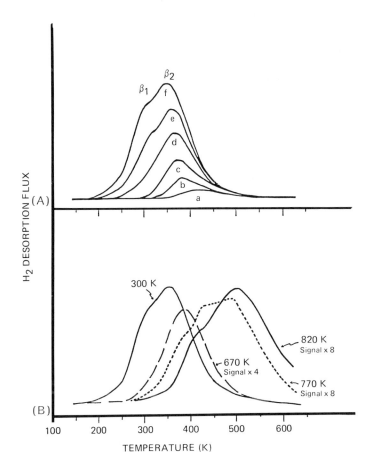

FIGURE 6 (A) Hydrogen desorption from 1 nm of Ni on titania following different hydrogen exposures at 150 K: (a) 0.4, (b) 1.0, (c) 3.0, (d) 6.0, (e) 12, and (f) 20 L. (B) Effect of heating in low-pressure hydrogen at different temperatures (indicated) on desorption from saturation coverage. Sample dosed with hydrogen at 300 K followed by cooling in hydrogen to 150 K. (From Ref. 18.)

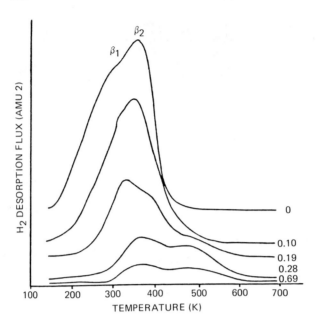

FIGURE 7 Effect of varying TiO_x coverage on hydrogen thermal desorption from nickel following saturation hydrogen exposure at 150 K. (From Ref. 17.)

TABLE 2 Activation Energies and Heats of Adsorption and Desorption for H_2/Cobalt

Catalyst	E_{Aa} (kJ mol^{-1})[a]	E_{Ad} (kJ mol^{-1})[b]	$-\Delta H_a$[c]	Order of desorption	Ref.
Unsupported Co	5.8	151	145 ± 10	2	34
3% Co/SiO$_2$	43		—	—	34
10% Co/SiO$_2$	18	168	145 ± 7	2	34
10% Co/Al$_2$O$_3$	39	144	105	2	34
4% Co/TiO$_2$ (rutile)		88/138			32
Reduced Co (calorimetric)			105		37
Co films (surface potential/TPD)		42/79			38
Single-crystal Co(0001)		67		2	39

[a] Activation energy for adsorption of H_2.
[b] Activation energy for desorption of H_2.
[c] Heat of adsorption, $-\Delta H_a = E_{Ad} - E_{Aa}$.

states of higher binding energy but of lower occupancy (see Fig. 7). Moreover, upon adsorption of hydrogen at 300 K on the samples of higher TiO_x coverage, a new activated state was observed at higher binding energies, which was attributed to adsorption on sites adjacent to the TiO_x species or to spillover onto the titania support [17]. The observed increased binding energy for the new states observed on nickel decorated with titania might be a consequence of direct bonding of hydrogen adatoms to oxygens associated with surface titania species [17], of a steric hindrance to hydrogen adatom migration and recombination [17], or localized electronic effects in the vicinity of TiO_x species.

Together the results for Ni/alumina and Ni/titania systems reveal a pattern that might well explain two important problems in catalysis: (a) metal support interactions in conventional high-surface-area catalysts [17] and (b) "hydrogen effects" discussed in earlier and later chapters and which are sometimes manifest by strongly bound states of hydrogen on these catalysts after reduction at high temperatures. As pointed out by Raupp and Dumesic [17], the underlying cause of "metal-support interactions" or "strong metal-support interactions" involving either so-called "reducible" or "unreducible" supports may be the "contamination or decoration" of the surfaces of metal crystallites by "moieties of the support material." These decorating species, in turn, introduce new hydrogen adsorption states having significantly higher binding energies and activation energies for adsorption. These "high-temperature states" can significantly alter the behavior of these supported catalyst systems in chemisorption titrations involving hydrogen and in many important catalytic reactions involving hydrogen as a reactant.

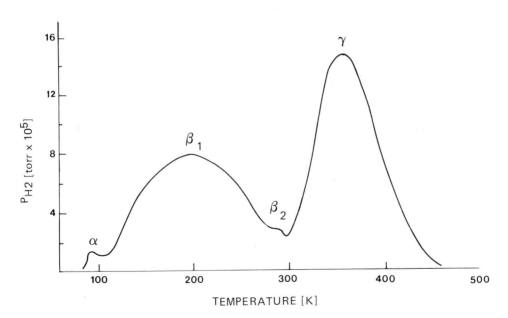

FIGURE 8 Spectrum of thermal desorption of hydrogen from cobalt film after adsorption at 78 K; final equilibrium hydrogen pressure 2.1×10^{-2} torr. (From Ref. 38.)

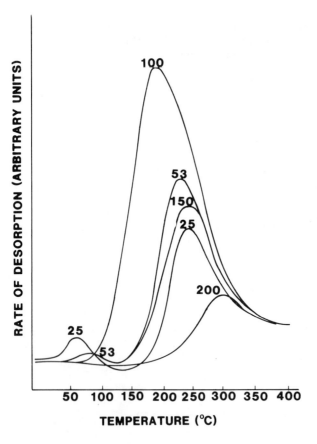

FIGURE 9 Hydrogen TPD spectra for 10% Co/silica as a function of adsorption temperature. (From Ref. 34.)

B. Calorimetric and TPD Studies of Hydrogen Interaction with Supported Cobalt and Iron

As in the case of nickel, previous calorimetric studies of hydrogen adsorption on cobalt and iron were limited to unsupported powders, filaments, or films [1]. Published TPD studies of hydrogen interaction with supported Co [32-34] and Fe [24,35,36] include only six studies from three laboratories. Quantitative kinetic and heats of adsorption data, obtained in only two of the studies of cobalt [32,34], are summarized and compared with calorimetric [37] and film/single-crystal TPD [38,39] data in Table 2. Hydrogen desorption spectra from a cobalt film [38] and from cobalt supported on silica [34] are shown in Figs. 8 and 9.

The hydrogen desorption spectrum for unsupported cobalt (Fig. 8) consists of four states: α, β_1, β_2, and γ. The α state corresponds to adsorption of hydrogen in molecular form, while the β_1, β_2, and γ states correspond to one atomic electropositively polarized form, β^+, and two electronegatively polarized states, β_s^- and β^- [38]. The activation energy for the β^+ state is reported to be 8.8 kJ mol^{-1}, while those for the β^- states are 41.8 and 79 kJ mol^{-1}, respectively [38].

The desorption spectrum for 10% Co/silica after hydrogen adsorption at 298 K (Fig. 9) consists of two states, β_2 and γ, having desorption maxima at 323 and 513 K, respectively. At higher adsorption temperatures (e.g.,

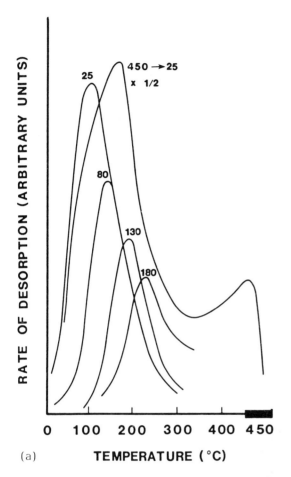

FIGURE 10 (a) TPD spectra of hydrogen from 15% Fe/silica (dried 60 to 80°C prior to reduction) as a function of adsorption temperature. Solid bar indicates temperature hold at 450°C. (b) TPD spectra of hydrogen from 15% Fe/silica (dried 100°C prior to reduction) as a function of adsorption temperature, (c) TPD spectra of hydrogen from 15% Fe/3% K/silica as a function of adsorption of temperature. Solid bar indicates temperature hold at 450°C. [(b) and (c) From Ref. 24.]

326 and 373 K) the population of the β_2 state decreases while that of the γ state increases very significantly, indicating that this higher-temperature state is activated. Maximum adsorption capacity is observed at about 373 K. This highly activated adsorption state for hydrogen is observed for unsupported cobalt and cobalt supported on silica, alumina, titania, and carbon [34]. In Co/MgO and Co/ZSM-5 systems the kinetic limitations are so severe that adsorption is not observed even at elevated temperature except after long periods of exposure to hydrogen [34]. Activation energies for *adsorption* (Table 2) range from 5.8 kJ mol^{-1} for unsupported cobalt to 39 kJ/mol for 10% Co/alumina and 43 kJ/mol for 3% Co/silica; adsorption activation energies increase with decreasing loading and in the order Co, Co/silica, Co/alumina (if loading is held constant in the supported systems). This order is the same as that for increasing degree of interaction of cobalt with the support.

Activation energies for desorption from the γ state, obtained in studies of cobalt, Co/silica, Co/alumina, and Co/titania in which activated adsorption

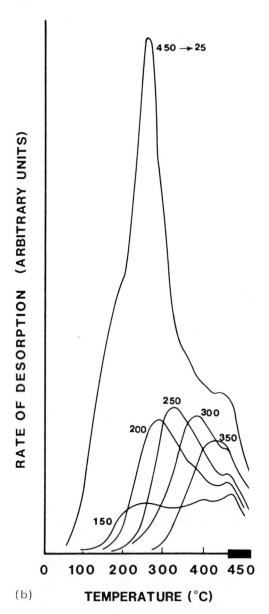

(b)

FIGURE 10 (continued)

was accounted for [32,34], range from 138 to 168 kJ mol^{-1} (Table 2). The The E_d values (Table 2) of 79 and 76 kJ mol^{-1} reported for a Co film [38] and single-crystal cobalt [39] are suspect because a good portion of the activated state was excluded from study due to kinetic limitations.

While the activation energies for desorption do not vary greatly with support, significant differences are observed in the heats of adsorption (Table 2). Indeed, the heat of adsorption for 10% Co/alumina of 105 kJ mol^{-1} is clearly very significantly lower than for either 10% Co/silica or unsupported cobalt (145 kJ mol^{-1} for both).

The significantly different adsorption enthalpy for Co/alumina is due to its much higher activation energy, which, as in the case of nickel, can probably be ascribed to the presence of cobalt or aluminum oxide species located

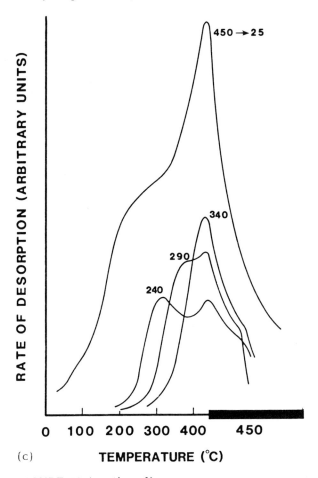

(c)

FIGURE 10 (continued)

either on the cobalt crystallite surface or in intimate contact with cobalt crystallites.

In addition to being influenced by metal-support interactions, the kinetics and energetics of hydrogen adsorption on base metals are influenced by promoters such as potassium [24, 31, 40], other additives such as carbon, copper, and sulfur [27, 31, 40], and by catalyst preparation/pretreatment methods [24]. Effects of calcination pretreatment and potassium promoter on the adsorption of hydrogen on Fe/silica according to Weatherbee et al. [24] are illustrated in Fig. 10.

It is evident from Fig. 10a that hydrogen adsorption on a 15% Fe/silica catalyst precalcined at the lower temperature (60 to 80°C) prior to reduction for 36 h in hydrogen is only moderately activated (i.e., a significant quantity adsorbs at 25°C and the amount adsorbed *decreases* with increasing adsorption temperature); nevertheless, the amount adsorbed upon cooling from 450°C to 25°C in hydrogen is greater by more than a factor of 2 than that adsorbed by the pulse method at 25°C. However, hydrogen adsorption on the catalyst precalcined at the higher temperature (100°C) is considerably more activated; that is, negligible hydrogen adsorbs during exposure to pulses of hydrogen at 25 to 100°C (Fig. 10b). Moreover, the amount adsorbed at 150°C is small and adsorption capacity *increases* with increasing adsorption temperature.

It is evident that the desorption peaks are shifted to higher binding energy with increasing precalcination temperature. Weatherbee et al. [24] and Rankin and Bartholomew [36] have attributed these observed increases in adsorption activation and binding energies with increasing precalcination temperature to either (a) contamination of the iron surface by support species or (b) the presence of iron oxide either on or in contact with iron metal crystallites. That iron supported on oxide carriers such as alumina, magnesia, and silica cannot be completely reduced to the metallic state and that oxides of iron interact strongly with these supports to form highly stable solid solutions is well documented [24,36,41-45].

TPD spectra for 15% Fe/3% K/silica in Fig. 10c when compared with those for 15% Fe/silica in Fig. 9a provide evidence that potassium addition likewise increases adsorption activation and binding energies of hydrogen on iron. In addition, a new adsorption state with a maximum at about 400°C is observed (see Fig. 10c). These significant changes in the kinetics and energetics of hydrogen adsorption are explained by localized electronic perturbations of the metal surface by the potassium additives [31,40].

Since promoters and pretreatment methods can separately greatly influence the kinetics and energetics of hydrogen adsorption, it should be interesting to consider the effects of simultaneous promotion and pretreatment. The hydrogen TPD spectrum for a 15% Fe/3% K/silica catalyst precalcined before reduction at 200°C was found to contain no adsorption states below 400°C [24,36]; in other words, the combined effect of potassium promotion and precalcination is to dramatically increase adsorption activation and binding energies. It will be shown in a later chapter that the substantial change in the hydrogen adsorption properties of this catalyst dramatically affects its activity and selectivity properties in CO hydrogenation.

III. REVERSIBILITIES AND STOICHIOMETRIES OF ADSORPTION

A. Reversible and Irreversible Adsorption

From the TPD data presented in Sections I and II, it is evident that hydrogen adsorption on cobalt, iron, and nickel is reversible over a wide range of conditions; that is, chemisorbed, atomic hydrogen can be desorbed at least in part by evacuation or purging at temperatures above about 200 K. For example, Figs. 2 and 8 reveal the existence of atomic hydrogen states of low binding energy on nickel and cobalt which are easily desorbed at temperatures near ambient (200 to 300 K). The high reversibility of hydrogen adsorption has important implications in regard to proposed techniques for measuring monolayer hydrogen capacities (discussed in Section III.C).

It should be emphasized that the term "reversible" or "irreversible" applied to adsorption has operational meaning only. That is, adsorption of hydrogen on nickel at 120 to 150 K is for all practical purposes irreversible since rates of desorption are negligible (see Fig. 16 in Chapter 1); however, at 300 K, the β_1 state of hydrogen on nickel is reversible since it is easily desorbed, while the β_2 state is irreversible, that is, not desorbable at measurable rates at this temperature.

Besides TPD studies there have been several previous investigations that provide information on the reversibility of hydrogen adsorption on supported cobalt, iron, and nickel catalysts [36,46-49]. These studies, summarized in Table 3, provide evidence that the reversibility of hydrogen adsorption is a function of dispersion, support, extent of reduction to the metal, and metal loading. This is reasonable in view of the TPD data presented in the earlier

TABLE 3 Studies of Reversible Hydrogen Adsorption on Supported Cobalt, Iron, and Nickel Catalysts

Ref.	Catalysts	Effects investigated	Observations and comments
46	4-40% Ni/silica	Effects of nickel content and crystallite size on irreversible uptake	Reversibility increases with increasing crystallite size and nickel content.
47	3, 9% Ni/alumina, 14% Ni/alumina	Pressure dependence of isotherm, Fraction of monolayer desorbing at 300 K	Isotherm pressure dependence increases with decreasing loading. 40% of monolayer desorbs at 300 K.
48	20-30% Ni/silica, 7% Ni/alumina	Effects of temperature, pressure, support, unreduced nickel, and crystallite size	Reversible fraction increases with increasing time and pressure. Reversibility increases with increasing temperature, decreasing extent of reduction, increasing crystallite size, and with change of support from silica to alumina.
49	1-15% Co-alumina, 3 and 10% Co on silica, titania, C, and MgO	Effects of support, cobalt content, preparation, and extent of reduction at 300 K	Percent reversibility ranges from 15 to 90%. Reversibility increases with decreasing extent of reduction and in the order Co/silica, Co/alumina, Co/titania, Co/C, Co/MgO. Reversibility is greater for catalysts prepared by ppt. relative to impregnation and for catalysts of lower loading.
36	15% Fe/silica, 15% Fe/0.2-3% K/silica	Effects of potassium on reversibility at 300 K	Potassium addition increases percentage reversibility from about 13% to 25-30%.

sections showing that the number of binding states, temperatures for desorption maxima, and binding energies are functions of these same catalyst variables.

The results summarized in Table 3 indicate that the percent reversibility of hydrogen adsorbed at 300 K ranges from 10 to 90% and that the extent of reversibility generally increases with increasing temperature of adsorption, crystallite size, and metal loading, and with decreasing extent of reduction. The percent reversibility apparently also increases with increasing interaction of support and metal (i.e., in the order M/silica, M/alumina, M/titania, M/carbon, M/magnesia) and with potassium promotion. Hydrogen adsorption is more reversible on catalysts prepared by controlled pH-precipitation relative to those prepared by impregnation. Most of these trends are consistent with and can be explained by the observed shifts in hydrogen desorption states as a result of changes in these catalyst properties which can lead to higher or lower rates of desorption at ambient temperature (see the discussion in Section II).

However, other explanations for the observed changes in adsorption reversibility have been proposed which should be mentioned. But the discussion of these other possibilities should be prefaced by a brief review of the facts surrounding the behavior of hydrogen adsorption on supported metals. Nickel supported on alumina or silica provides a representative example.

Hydrogen adsorbs on supported nickel at room temperature and pressures of 100 to 400 torr in three stages [47,48,50,51]: (a) much (50 to 80%) of the hydrogen is adsorbed rapidly (within 5 min), (b) a significant quantity is adsorbed within the next 30 to 45 min at which point in time near-equilibrium is achieved, and (c) a small quantity (typically 3 to 5% of a monolayer) is taken up very slowly over the next few hours.

Previously reported isotherms [48,51] are consistent with a Langmuir model involving dissociative adsorption on strong sites followed by adsorption on weaker sites [48]. Reversible adsorption on weaker sites would account for the observed increasing adsorption capacity with increasing pressure observed [48,51]. Indeed, Richardson and Cale [48] were able to fit their isotherms for Ni/silica with a high degree of confidence to an equation of the form

$$N_A = N_O + bP_H^{0.5} \qquad (1)$$

where N_A is the number of moles adsorbed at a given pressure, N_O the moles adsorbed at zero pressure (in other words, strongly adsorbed hydrogen), and b is a constant that reflects the degree of reversibility. The parameter b increases with decreasing nickel content and decreasing extent of reduction and is larger for Ni/alumina than for Ni/silica [48,51].

While adsorption on sites of at least two different binding energies clearly explains the observed pressure dependence and hence the reversibility of hydrogen adsorption, there are a number of other explanations that have been proposed in previous studies [48,52,53], most of which are summarized in Table 4. Of these, only absorption into the metal can be entirely discounted, since permeation rates into nickel are negligibly small [52]. Most of the others involve phenomena observed occasionally and/or which contribute only in small measure to reversibility. For example, the effects of spillover leading to adsorption on the support or unreduced nickel can be neglected in procedures involving a short (e.g., 45 min) equilibration, since spillover to the oxide phases is a slow, multihour process [17,18,53]; accordingly, the very slow hydrogen uptake observed in phase 3 after the initial 45-min equilibration is most probably due to spillover. In most supported base-metal systems, problems of surface contamination and/or surface inaccessibility can be avoided or minimized through careful preparation and pretreatment. The TPD data in Section II showed that multilayer or "predissociative" adsorption does not occur above about 120 to 150 K. The amount of hydrogen that adsorbs strongly on supports under typical conditions is negligible [54], while the contribution of physically adsorbed hydrogen can be eliminated through proper extrapolation of the data (see Sections III.B and III.C). Thus, all of the proposed contributions to reversible hydrogen adsorption in Table 4 can be discounted with the exception of phenomenon 1, involving different adsorption states or strengths. It is the most reasonable and likely explanation because: (a) it is the only one that can, by itself, explain the data [i.e., it provides a very good fit of the isotherm data using Eq. (1)],

TABLE 4 Phenomena Proposed to Explain Reversible Adsorption on Nickel

Phenomenon	Basis for explanation	Confirming or discounting evidence
1. Decreasing heats with increasing coverage	Lower binding energies at higher coverages lead to reversible adsorption.	This model fits experimental data very well. Supported by TPD and magnetic data.
2. Adsorption on support or nickel oxide	Although adsorption on supports or NiO is negligible, Ni crystallites may create new adsorption sites. Physical adsorption.	Magnetic data indicate that adsorption occurs on the metal, not on the support or nickel oxide. Physical adsorption can be corrected for by extrapolation to $P = 0$.
3. Spillover of atomic hydrogen onto support or nickel oxide.	Hydrogen atoms can diffuse from metal to support.	Large body of evidence confirms occurrence of spillover, but process is slow (requires many hours at 300 K).
4. Reaction with adsorbed contaminants such as oxygen	Reaction of hydrogen with adsorbed oxygen has been observed [56].	This type of contamination can be avoided. Reaction is not reversible [56].
5. Decoration of metal crystallites by reduced species from the support (so-called "SMSI" effect)	Adsorption on reduced support species or support "skin" located on the surface of metal crystallites could create new sites for reversible adsorption.	Evidence for contamination of metal by reduced support species is well documented. Effects can be minimized by careful preparation. Adsorption on these species is very slow [17, 18].
6. Surface inaccessibility	"Support skin" or pore trapping causes weak, reversible adsorption.	Same as 5 above. No concrete evidence for "pore trapping."
7. Multilayer adsorption	Greater-than-monolayer adsorption on bridge or hole positions or predissociative molecular adsorption at high coverages.	TPD studies [7-18] indicate that this occurs only at $T < 150$ K.
8. Absorption into the bulk metal	Penetration into bulk metals could be reversible and slow.	Measured permeation rates into Ni are negligible [52].

Source: Refs. 47 to 51.

and (b) it is strongly supported by evidence from TPD and magnetic measurements. The TPD evidence, discussed in Section II, is entirely consistent with the view of different adsorption states or sites of different energies. Magnetic studies [48,55] reveal that changes in the magnetic moment of nickel continue to occur with increasing hydrogen pressure, even up to hundreds of atmospheres, indicating that reversible hydrogen adsorption takes place on nickel crystallites rather than on the support, support contaminant, or unreduced nickel.

B. Adsorption Stoichiometries

Adsorption Stoichiometries of Hydrogen on Supported Nickel

There is reliable evidence from the earlier literature that hydrogen adsorbs on single-crystal nickel [11,12] and polycrystalline nickel [55,56] with a stoichiometry of one hydrogen atom per nickel surface atom. For example, Pannell et al. [56] obtained a H/Ni_s *ratio of 1.1* by comparing the surface area of a high-purity nickel powder calculated *from the total hydrogen uptake at 300 K* (based on an equally weighted average of the planer site densities for the three lowest index planes) with surface areas determined from argon and nitrogen BET measurements (assuming argon and nitrogen areas of 0.169 and 0.162 nm^2/molecule, respectively).

Bartholomew and Pannell [47] investigated the stoichiometries of hydrogen and CO adsorptions on alumina- and silica-supported nickel catalysts having a range of metal loadings, dispersions, and extents of reduction. Their data provide evidence that (a) *room-temperature hydrogen adsorption occurs on Ni/alumina and Ni/silica catalysts of moderate or high loading with a stoichiometry of one hydrogen atom per surface nickel atom, if the total adsorption (reversible and irreversible) is included*, and (b) less than a monolayer of hydrogen is adsorbed (H/Ni_s < 1) on Ni/alumina catalyst of low loading (<3 wt %). The first conclusion is supported by the good to very good agreement of particle-size estimates from hydrogen adsorption (assuming that $H/Ni_s = 1$), x-ray diffraction (XRD), and transmission electron microscopy (TEM) for 14 and 23% Ni/alumina and 13.5 and 15% Ni/silica. Bartholomew and Pannell concluded that the suppression of hydrogen adsorption in catalysts of low loading was due to metal-support interactions. Based on the data presented in Section II, it would be reasonable to postulate that strongly activated adsorption and/or contamination of the metal crystallites with support-derived oxides was responsible for this very significant inhibition of hydrogen adsorption.

It should be emphasized that the estimation of metal crystallite size from hydrogen uptake requires that assumptions be made regarding metal crystallite morphology and location of the unreduced metal oxide. Bartholomew and Pannell [47] assumed the metal to be present as spherical nickel crystallites, while unreduced nickel oxide was assumed to exist in separate thin layers in intimate contact with the support (see Fig. 11). The presence of dense, spherical crystallites of nickel in their catalysts was confirmed by TEM [57]. A stable, hardly-reducible surface nickel aluminate phase in the form of a monolayer on the alumina support is consistent with recently reported ESCA and ISS data [30]. Thus the model used by Bartholomew and Pannell [47] appears to be consistent with recent morphological information.

Several studies [57-61] provide evidence that significantly less than a monolayer of hydrogen is adsorbed on Ni/titania and that the H/Ni_s ratio decreases with increasing reduction temperature and decreasing nickel loading. The recent evidence is most consistent with blocking of adsorption by reduced support species, TiO_x, that migrate onto crystallites during reduction [17,18].

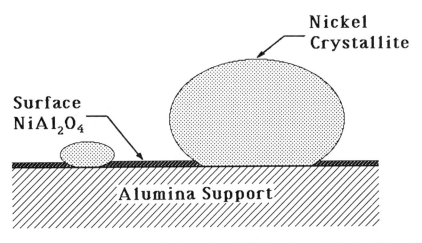

FIGURE 11 Model of partially reduced Ni/alumina catalyst. (From Ref. 47.)

However, Xuan-Zhen et al. [53] observed a H/Ni$_s$ ratio of 1 after adsorbing hydrogen on Ni/titania for 15 to 20 h; this was attributed to spillover of atomic hydrogen from the metal to the reduced TiO$_x$ or to the support in contact with the crystallites.

The phenomenon of "limited accessibility" of gaseous hydrogen to the metal surface was introduced and discussed in Chapter 3. Thus the discussion of this topic here will be brief and focused on the question of how this phenomenon affects adsorption stoichiometry and hence the measurement of surface area.

Access of hydrogen to the surface of a metal crystallite may be affected by (a) the degree to which the metal crystallite contacts of "wets" the support; (b) the presence on the metal crystallite of contaminants such as carbon, sulfur, or reduced support species; and (c) the collapse of pores accompanied by encapsulation of crystallites during sintering at high temperatures. That reduced support species can migrate onto the surface of Ni/titania, Ni/alumina, and Ni/silica catalysts during high-temperature reduction is confirmed from several recent investigations [17,18,62], although much higher reduction temperatures are required to achieve this effects in the Ni/alumina and Ni/silica systems compared to Ni/titania. Recent studies [63-65] also confirm that sintering at high temperatures of Ni/alumina and Ni/silica leads to partial collapse of the micropore structure with encapsulation of nickel crystallites. Nevertheless, the data of Bartholomew and Pannell [47] indicate that *hydrogen is essentially completely accessible to the nickel surface in Ni/alumina and Ni/silica catalysts of moderate loading, reduced at 725 K*, and which are prepared from high-purity supports. While several studies (Chapter 3, Refs. 48 and 63) have reported limited accessibility of hydrogen for unsintered, moderately-loaded nickel catalysts, these results must be questioned, since in most of these studies only strongly adsorbed hydrogen was accounted for; from the previous discussion of reversibility it is clear that this would have led to large (30 to 90%) errors in the measurement of saturation coverage. For example, Desai and Richardson [63] used a flow technique that measured only the strongly held hydrogen, while Richardson and Cale [48] used Eq. (1) to subtract out the contribution of reversible adsorption. Thus much of the previous work reporting inaccessibility needs to be revisited.

Adsorption Stoichiometries of Hydrogen on Supported Cobalt and Iron

Hydrogen adsorption measurements on supported cobalt and iron are complicated by high adsorption activation energies, which severely limit the rate of adsorption at room temperature. In fact, negligible hydrogen were observed to adsorb on 3 to 9% Co/ZSM-5 catalysts at 300 K even after a 45-min equilibration, although 10 to 100 times as much adsorbed at 100°C [34].

Reuel and Bartholomew [49] obviated the problem of activated adsorption and *observed monolayer hydrogen adsorption with a stoichiometry of one hydrogen atom per cobalt surface atom on unsupported cobalt, 3 and 10% Co/silica, and 10 to 15% Co/alumina catalysts* reduced at 623 to 673 K *by measuring total hydrogen adsorption at 373 to 423 K*. The basis of their observation was good to very good agreement of average crystallite diameters estimated from hydrogen adsorption (assuming that $H/Co_s = 1$), XRD, and TEM. However, in the case of Co/titania, the crystallite diameter estimated from hydrogen adsorption was a factor of 2 higher than that estimated by the other techniques; this poor agreement is attributed to decoration of the metal crystallites by reduced TiO_x species as observed in the Ni/titania system [17,18].

In their investigation of hydrogen adsorption on a 15% Fe/silica catalyst reduced at 725 K, Rankin and Bartholomew [36] found good agreement among crystallite sizes estimated from hydrogen adsorption (assuming that $H/Fe_s = 1$), XRD and TEM, if the total hydrogen adsorption capacity was measured by cooling the sample in a predetermined quantity of hydrogen from 673 K to 300 K followed by isotherm measurements at 300 K. Jones et al. [66] observed similar good agreement among crystallite diameters of 10% Fe/carbon measured by the same three techniques and using a similar procedure for measuring hydrogen adsorption capacity. Thus it is safe to conclude that hydrogen adsorbs with a $H/M_s = 1$ stoichiometry on conventional cobalt and iron catalysts of moderate or high loading if the total uptake is measured at elevated temperatures according to the procedures discussed above.

C. Measurement of Monolayer Adsorption Capacity

The measurement of monolayer hydrogen adsorption capacity of supported metals is useful in providing estimates of metal surface area, average crystallite diameter, and the number of potential catalytic sites available for reaction. The *ideal method* for measuring monolayer adsorption capacity would involve *reproducible, convenient, and rapid* measurements using apparatus that is easily and economically fabricated; *it must account for hydrogen that is either reversibly or irreversibly chemisorbed on the metal while excluding chemical and physical adsorptions on the support*. Finally, it *should provide a means of circumventing the kinetic limitations of highly activated adsorption*. Of course, the measurement of an adsorbed monolayer can be realized only *in the absence of surface contaminants* (e.g., carbon, sulfur, and/or reduced support moieties).

Based on the discussions in Sections III.A and III.B, it is clear that *hydrogen adsorbs in a well-defined monolayer on supported cobalt, iron, and nickel* with a stoichiometry of one hydrogen atom per metal surface atom *if the measurement includes the total (reversible and irreversible) quantity chemisorbed*. Reproducible, convenient, static volumetric methods that have been developed and demonstrated for measuring monolayer hydrogen adsorption capacity on supported nickel, cobalt, and iron catalysts are summarized in Table 5. These methods are modifications of the standard procedure for measuring hydrogen chemisorption on Pt developed by Committee D-32 of ASTM [67]; the most significant differences are the procedures developed to circumvent the limitations of highly activated adsorption by means of a high-temperature adsorption in the case of cobalt and the cooling in hydrogen from just below the reduction temperature in the case of iron, the latter procedure having first been suggested by Amelse et al. [68].

5. Hydrogen Adsorption on Cobalt, Iron, and Nickel

TABLE 5 Demonstrated Volumetric and Static Methods for Measuring Monolayer Hydrogen Adsorption Capacities of Supported Nickel, Cobalt, and Iron Catalysts

Catalysts	Pretreatment	Equilibration and isotherm conditions	Limitations
Ni/alumina, Ni/silica	Reduce 15 h at 673-723 K, evacuate 1 h at -20 K of redn. T[b]	Equilibrate 45 min at 300 K, 350 torr; desorption isotherm[a] from 350 to 100 torr; extrapolate isotherm to P = 0[c]	Metal loadings must be > than 3 wt %; if adsorption is slow, equil. at 323 K; measure *total* uptake[d]
Co/alumina, Co/silica	Reduce 15 h at 673-723 K, evacuate 1 h at -20 K of redn. T[b]	Equilibrate 45 min at 400 K, 350 torr; desorption isotherm[a] from 350 to 100 torr; extrapolate isotherm to P = 0[c]	Metal loadings of > 3% for Co/silica and > 10% for Co/alumina; measure *total* uptake[d]
Fe/silica	Reduce 36 h at 723 K, evacuate 1 h at -20 K of redn. T[b]	Cool in measured amount of hydrogen (350 torr) from -20 K of redn. T to 300 K; desorption isotherm[a] at 300 K from 350 to 100 torr; extrapolate isotherm to P = 0[c]	Metal loadings > 15%; measure *total* uptake[d]

[a]Desorption isotherm is more linear than adsorption isotherm and can be obtained with 30-60 min (5-10 min/point) compared to 4-6 h (45-60 min/point) for adsorption isotherm.
[b]Short evacuation at slightly lower than reduction temperature avoids contamination of metal with water from support, pump oil, etc.
[c]Extrapolation eliminates contribution of physical adsorption or erroneous dead volume.
[d]Total uptake is the extrapolated uptake, which includes reversible and irreversible adsorption.

The procedure for nickel was tested on a commercial 12% Ni/alumina catalyst by a task force of the ASTM Committee D-32 in a round-robin involving eight laboratories with a standard deviation of only ±7% [69], which is well within the limits of error for conventional volumetric adsorption procedures.

Some of the assumptions underlying the choice of conditions or procedures as well as the limitations of the methods are also briefly summarized in Table 5. Some further clarification is appropriate here. Because of the strong interaction of base metal oxides with alumina [29,30] and of iron with silica [36], a 15-h reduction at 673 to 724 K is recommended for nickel and cobalt catalysts and a 36-h reduction for Fe/silica. These recommendations are based on measurements of the extent of reduction as a function of reduction time [29,36,47,49]. The relative short evacuation time of 1 h is chosen to avoid contamination of the metal surface by pump oil, water, or other such contaminants that might be present in the vacuum manifolds or traps and is based on experience from laboratories too numerous to mention. The purpose of evacuating at a slightly lower temperature than that used in the reduction is to avoid oxidation of the metal surface by water from the support, since it is well known that water can be removed incrementally by dehydroxylation of alumina and silica supports up to temperatures as high as 1200 K [70]. In fact, it was recently reported [71] that a lengthy evacuation of Pt/silica at a temperature higher than that of the reduction resulted in oxidation of a portion of the platinum, while there was no observable change in the state of reduction as measured by Mössbauer spectroscopy of a well-dispersed 1% Fe/carbon reduced at 673 K upon evacuation for 1 h at 653 K [72]. The equilibration at a relatively high pressure of 350 torr ensures that equilibrium will be attained within about 45 min [36,47,49,51]. By measuring the isotherm in increments of decreasing pressure (desorption method) it is generally possible to measure isotherm points every 5 to 10 min, while the adsorption method (increments of increasing pressure) may require waiting 45 min or longer for each point to reach equilibrium; moreover, the isotherm obtained by the desorption method is generally more linear than that obtained by the adsorption method [73]. Besides the enormous savings in time the desorption method avoids complications due to spillover that generally occur over a period of hours [17,18,53]. Because physical adsorption of hydrogen can occur on typical supports even at 300 to 400 K and because errors in the determination of the dead volume (cell and manifold volume) can cause significant variations in the isotherm slope, it is necessary to extrapolate the linear portions of the isotherm to zero pressure.

The methods described in Table 5 are generally limited to alumina- and silica-supported cobalt, iron, and nickel catalysts of moderate or high loading (>3% for Ni/silica and Co/silica and >10% for Ni/alumina, Co/alumina, and Fe/silica) since at lower loadings metal contamination by reduced support moieties, spreading of the crystallite on the support, or other forms of metal-support interaction limit accessibility of hydrogen to the metal surface. Thus these methods cannot be applied with any confidence to catalysts of low metal content nor to catalysts involving highly reducible supports such as titania or contaminating supports such as carbon without first establishing experimentally the conditions (if any) under which monolayer adsorption occurs.

While the methods described in Table 5 can be performed conveniently with conventional vacuum adsorption apparatus within 2 to 3 h following the completion of the pretreatment, reduction, and evacuation, there is clearly interest in developing faster, more convenient flow techniques for measuring hydrogen adsorption on supported metals. Unfortunately, previously proposed pulse-flow [63,74] or flow-desorption [68] techniques involve a purge at 273 to 300 K that removes reversibly chemisorbed hydrogen; thus the resultant estimate of hydrogen adsorption capacity is low by 30 to 50%, since typically 30 to 50% of the adsorbed hydrogen is easily desorbed under these conditions (see Sections II and III.A). A new method presently under development [75] involves cooling the reduced catalyst in hydrogen to about 200 K, purging with argon at 200 K, and continuing the purge in argon while the temperature is raised to 673 K; the amount of hydrogen originally adsorbed on the catalyst is determined from the area under the desorption peak. In principle, this technique should provide an accurate measure of monolayer capacity, since the initial purge is conducted below the temperature at which atomic states of hydrogen typically desorb.

IV. SUMMARY

In this chapter some of the fundamentals of the kinetics, energetics, reversibilities, and stoichiometries of hydrogen adsorption on supported cobalt, iron and nickel catalysts were presented and discussed in the light of presently available data. Several important issues relating to (a) the relationship of catalyst and adsorption properties and (b) the application of hydrogen chemisorption to the measurement of metal surface areas were discussed. The present evidence supports the following conclusions:

1. The kinetics and energetics of hydrogen adsorption on cobalt, iron, and nickel are apparently little affected by metal surface structure but are significantly changed by metal-support interactions, particularly in catalysts of low metal content or involving highly reducible supports such as titania. Contamination of the metal surface by reduced support moieties, spreading of the metal onto the support, or other forms of metal-support interactions cause (a) the appearance of new adsorption states of hydrogen at higher binding energies and (b) an increase in the adsorption activation energy for hydrogen which can lead to severe kinetic limitations in the adsorption process.

2. Precalcination treatments and promoters such as potassium also cause the appearance of new high-temperature adsorption states and significantly increase the adsorption activation energy for hydrogen. The appearance of new strongly bound states of hydrogen due to pretreatment, support, and promoter effects explains at least in part the origin of "hydrogen effects" discussed in earlier and later chapters of this book.

3. Hydrogen adsorption on cobalt, iron, and nickel is generally highly reversible at room temperature because of the existence of an atomic adsorption state which is easily desorbed at 200 to 300 K. The reversibility of hydrogen adsorption on supported metals varies from 15 to 90% and increases with increasing temperature, crystallite size, and metal content, and with decreasing extent of reduction. Hydrogen adsorption is more reversible on alumina-supported metals than on silica-supported metals.

4. Hydrogen chemisorbs on Ni/alumina and Ni/silica catalyst of moderate loadings at 300 K and 100 to 350 torr with a stoichiometry of one hydrogen atom per nickel surface atom if the adsorption method accounts for the total of reversible and irreversible chemisorption. A well-defined monolayer ($H/M_s = 1$) is also observed on cobalt and iron catalysts of similar loadings and supports if the total uptake is measured at 400 K or by cooling in hydrogen from about 673 K to 300 K in order to circumvent the limitations of highly activated adsorption. Under these conditions there is no evidence for "limited accessibility" of hydrogen to the metal surface for these nickel, cobalt, or iron catalysts.

5. Less than monolayer adsorption is observed on cobalt, iron, and nickel catalysts of low metal content or supported on highly reducible supports such as titania. These results are consistent with "limited accessibility" of the surface due to contamination by reduced support species.

6. Static volumetric methods for measuring monolayer hydrogen adsorption capacity have been demonstrated for alumina- and silica-supported cobalt, iron, and nickel catalysts of moderate loadings. Previously proposed pulse-flow or flow techniques measure only irreversibly chemisorbed hydrogen and are therefore *not* recommended.

REFERENCES

1. Toyoshima and G. A. Somorjai, *Catal. Rev.-Sci. Eng.*, 19:105 (1979).
2. O. Beeck, *Adv. Catal.*, 2:151 (1951).
3. M. Wahba and C. Kemball, *Trans. Faraday Soc.*, 49:1351 (1953).
4. D. F. Klemperer and F. S. Stone, *Proc. R. Soc. London*, A243:375 (1957).
5. D. Brennan and F. H. Hayes, *Trans. Faraday. Soc.*, 60:589 (1964).
6. F. J. Broecker and G. Wedler, *Discuss. Faraday Soc.*, 47:87 (1966).
7. J. Lapujoulade and K. S. Neil, *Surf. Sci.*, 35:288 (1972).

8. J. Lapujoulade and K. S. Neil, *J. Chem. Phys.*, *70*:798 (1973).
9. J. McCarty, J. Falconer, and R. J. Madix, *J. Catal.*, *30*:235 (1973).
10. K. Christmann, O. Schober, G. Ertl, and M. Neumann, *J. Chem. Phys.*, *60*:4528 (1974).
11. (a) K. Christmann, F. Chelab, V. Penka, and G. Ertl, *Surf. Sci.*, *152/153*:356 (1985); (b) K. Christmann, *Z. Naturforsch Teil*, *A 34*:22 (1979).
12. A. Winkler and K. D. Rendulic, *Surf. Sci.*, *118*:19 (1982).
13. N. M. Popora, L. V. Babenkova, and D. V. Sokol'skii, *Kinet. Katal.*, *10*:1171 (1969).
14. J. A. Konvalinka, P. H. Van Oeffelt, and J. J. F. Scholten, *Appl. Catal.*, *1*:141 (1981).
15. P. I. Lee and J. A. Schwarz, *J. Catal.*, *73*:272 (1982).
16. G. D. Weatherbee and C. H. Bartholomew, *J. Catal.*, *87*:55 (1984).
17. G. B. Raupp and J. A. Dumesic, *J. Catal.*, *95*:587 (1985).
18. G. B. Raupp and J. A. Dumesic, *J. Catal.*, *97*:85 (1986).
19. E. E. Ibok and D. F. Ollis, *J. Catal.*, *66*:391 (1980).
20. R. K. Herz, J. B. Kiela, and S. P. Marin, *J. Catal.*, *73*:66 (1982).
21. R. J. Gorte, *J. Catal.*, *75*:164 (1982).
22. J. L. Falconer and J. A. Schwarz, *Catal. Rev.-Sci. Eng.*, *25*:141 (1983).
23. J. S. Rieck and A. T. Bell, *J. Catal.*, *85*:143 (1984).
24. G. D. Weatherbee, J. L. Rankin, and C. H. Bartholomew, *Appl. Catal.*, *11*:73 (1984).
25. A. Clark, *The Theory of Adsorption and Catalysis*, Academic Press, New York (1970), p. 211.
26. J. K. Roberts, *Proc. R. Soc. London*, *A 152*:445 (1935).
27. K. B. Kester and J. L. Falconer, *J. Catal.*, *89*:380 (1984).
28. J.-G. Choi, H.-K. Rhee, and S. H. Moon, *Appl. Catal.*, *13*:269 (1985).
29. C. H. Bartholomew and R. J. Farrauto, *J. Catal.*, *45*:41 (1976).
30. M. Wu and D. M. Hercules, *J. Phys. Chem.*, *83*:2003 (1979).
31. R. J. Madix, *Catal. Rev.-Sci. Eng.*, *15*:293 (1977).
32. J. Dollimore and B. Harrison, *J. Catal.*, *28*:275 (1973).
33. J. M. Zowtiak and C. H. Bartholomew, *J. Catal.*, *82*:230 (1983).
34. J. M. Zowtiak and C. H. Bartholomew, *J. Catal.*, *83*:107 (1983).
35. Y. Amenomiya and G. Pleizier, *J. Catal.*, *28*:442 (1973).
36. J. L. Rankin and C. H. Bartholomew, *J. Catal.*, *100*:533 (1986).
37. R. Rudham and F. S. Stone, *Trans. Faraday Soc.*, *54*:421 (1958).
38. R. Dus and W. Lisowski, *Surf. Sci.*, *61*:635 (1976).
39. M. E. Bridge, C. M. Comrie, and R. M. Lambert, *J. Catal.*, *58*:28 (1979).
40. J. Benzinger and R. J. Madix, *Surf. Sci.*, *94*:119 (1980).
41. M. C. Hobson and A. D. Campbell, *J. Catal.*, *8*:294 (1967).
42. R. L. Garten and D. F. Ollis, *J. Catal.*, *35*:232 (1974).
43. M. Boudart, A. Delbouille, J. A. Dumesic, S. Khammouma and H. Topsoe, *J. Catal.*, *37*:486 (1975).
44. I. Sushumna and E. Ruckenstein, *J. Catal.*, *94*:239 (1985).
45. A. J. H. M. Kock, H. M. Fortuin, and J. W. Geus, *J. Catal.*, *96*:261 (1985).
46. A. A. Slinkin, A. V. Kucherov, and A. M. Rubinshtein, *Kinet. Catal.* *19*:415 (1978).
47. C. H. Bartholomew and R. B. Pannell, *J. Catal.*, *65*:390 (1980).
48. J. T. Richardson and T. S. Cale, *J. Catal.*, *102*:419 (1986).
49. R. C. Reuel and C. H. Bartholomew, *J. Catal.*, *85*:63 (1984).
50. J. R. Rostrup-Nielsen, *J. Catal.*, *11*:220 (1968).
51. C. H. Bartholomew, *Alloy Catalysts with Monolith Supports for Methanation of Coal-Derived Gases*, Final Technical Report to ERDA, FE-1790-9 (Sept. 6, 1977).

52. R. B. McLellan and C. G. Harkins, *Mater. Sci.*, *18*:5 (1975).
53. J. Xuan-Zhen, T. F. Hayden, and J. A. Dumesic, *J. Catal.*, *83*:168 (1983).
54. R. C. Reuel, M.S. thesis, Brigham Young University (1983).
55. P. W. Selwood, *J. Catal.*, *42*:148 (1976).
56. R. B. Pannell, K. S. Chung, and C. H. Bartholomew, *J. Catal.*, *46*:340 (1977).
57. D. G. Mustard and C. H. Bartholomew, *J. Catal.*, *67*:186 (1981).
58. C. H. Bartholomew, R. B. Pannell, and J. L. Butler, *J. Catal.*, *65*:335 (1980).
59. M. A. Vannice and R. L. Garten, *J. Catal.*, *56*:236 (1979).
60. J. S. Smith, P. A. Thrower, and M. A. Vannice, *J. Catal.*, *68*:270 (1981).
61. E. I. Ko, S. Winston, and C. Woo, *J. Chem. Soc. Chem. Commun.*, 741 (1982).
62. P. Turlier, J. A. Dalmon, and G. A. Martin, in *Metal-Support and Metal-Additive Effects in Catalysis* (B. Imelik et al., eds.), Elsevier, Amsterdam (1982), pp. 203-210.
63. P. Desai and J. T. Richardson, in *Catalyst Deactivation* (B. Delmon and G. F. Froment, eds.), Elsevier, Amsterdam (1980), pp. 149-158.
64. C. H. Bartholomew, R. B. Pannell, and R. W. Fowler, *J. Catal.*, *79*:34 (1983).
65. C. H. Bartholomew and W. L. Sorensen, *J. Catal.*, *81*:131 (1983).
66. V. K. Jones, L. R. Neubauer, and C. H. Bartholomew, *J. Phys. Chem.*, *90*:4832 (1986).
67. ASTM Committee D-32, in *1985 Annual Book of ASTM Standards*, *5*:940 (1985).
68. J. A. Amelse, L. H. Schwartz, and J. B. Butt, *J. Catal.*, *72*:95 (1981).
69. C. H. Bartholomew, *Progress Summary of Results for the First Round-Robin Testing of the Proposed ASTM Procedure for Measurement of Nickel Surface Area*, prepared for Metal Dispersion Task Force (November 9, 1976).
70. J. B. Peri, *J. Phys. Chem.*, *69*:211 (1965).
71. A. Frennet and P. B. Wells, *Appl. Catal.*, *18*:243 (1985).
72. L. R. Neubauer and C. H. Bartholomew (1987), paper in preparation.
73. C. H. Bartholomew, unpublished data.
74. E. A. Verma and D. M. Ruthven, *J. Catal.*, *19*:401 (1970).
75. C. H. Bartholomew and R. D. Jones (1987), paper submitted to *Appl. Catal.*

6 | Neutron Scattering Studies of Hydrogen in Catalysts

TERRENCE J. UDOVIC and RICHARD D. KELLEY

National Bureau of Standards, Gaithersburg, Maryland

I.	INTRODUCTION	167
II.	NEUTRON SCATTERING TECHNIQUES	168
	A. Inelastic Scattering: Vibrational Spectroscopy	168
	B. Quasielastic Scattering: Dynamics	169
	C. Diffraction: Structure	170
III.	NEUTRON SCATTERING STUDIES OF CATALYSTS: EXAMPLES	170
IV.	CONCLUSION	180
	REFERENCES	180

I. INTRODUCTION

Over the past two decades, the scattering of low-energy neutrons has become a well-established and very powerful tool for investigating the structure and dynamics of condensed matter [1-4]. Because of the neutron's penetrating power, neutron scattering is inherently a technique for investigating bulk phenomena. However, if the surface-to-bulk ratio of the material being studied is very high (the necessary condition for a useful heterogeneous catalyst), the scattering information can reflect predominantly surface phenomena. Moreover, the neutron possesses unique properties that render it a particularly useful surface probe: (a) the energies of thermal neutrons are comparable with the energies of most molecular motions; (b) the wavelengths of thermal neutrons are comparable with interatomic distances in condensed phases; and (c) the interaction time scales of low-energy neutrons with nuclei allow the measurement of residence times for molecular-scale reorientation and diffusion. Hence neutron scattering probes energies, distances, and times that are characteristic of surface processes.

Neutron scattering studies of catalysts have been summarized in a number of reviews [4-9], which illustrate the utility of this technique for characterizing the complex structures of catalytic materials as well as the vibrational, diffusive, chemical, and structural behavior of hydrogenous adsorbates on catalyst surfaces. Because hydrogen possesses such a relatively large incoherent neutron scattering cross section (ca. 80 barns, one to two orders

of magnitude greater than for most other nuclei [1]), it has been the focus
of adsorbate studies involving a variety of catalytic materials. This chapter
is designed to convey the range of neutron scattering research that has been
expended on studies directly related to hydrogen in catalysis.

II. NEUTRON SCATTERING TECHNIQUES

In a neutron scattering event, the incident neutron interacts directly with
the target nucleus. The neutron scattering cross section of a nucleus in
an ensemble of nuclei can be separated into coherent and incoherent terms.
The coherent cross section contains information about both the frequencies
of motion and the structure of the system. The incoherent cross section
contains no interference terms and therefore depends only on the average
motion of individual nuclei rather than on the correlations between the motions
of equivalent nuclei.

Most reported neutron scattering studies of catalysts have involved incoherent inelastic neutron scattering (IINS), incoherent quasi-elastic neutron scattering (QNS), or coherent elastic neutron scattering (i.e., powder neutron diffraction). In addition, instrumental improvements in small-angle neutron scattering (SANS) [10] have led to an enhanced capability for the study of particle size, shape, and pore structure of catalytic materials such as silicas and zeolites and undoubtedly will find considerably greater use in the future.

A. INELASTIC SCATTERING: VIBRATIONAL SPECTROSCOPY

Incoherent inelastic neutron scattering has been the predominant neutron
scattering technique used in catalyst studies. Because of its high sensitivity
to hydrogen, IINS has been used as a vibrational spectroscopic probe of
hydrogen and hydrogenous adsorbates on catalyst surfaces as well as organometallic complexes in the condensed phase [4].

General information on vibrational spectroscopy of surface hydrogen has
been given in Chapter 2 (with special emphasis on electron energy loss);
infrared spectroscopy is discussed in Chapter 7. Similar to photons or
electrons in infrared (IR) or electron energy loss spectroscopy (EELS),
neutrons in IINS can lose a portion of their energy to excite adsorbate vibrational modes. Yet, unlike photon or electron scattering, neutron scattering
(a) involves only well-understood nuclear interactions (in the absence of
neutron-electron spin magnetic interactions) independent of any dipole or
polarizability selection rules and (b) is capable of measuring all contributions
(in momentum space) to the vibrational density of states due to the similar
magnitudes of the scattering wave vectors Q and the reciprocal lattice vectors
[5]. For hydrogenous adsorbates, the incoherent inelastic neutron scattering
law [1] indicates that the neutron scattering intensity is roughly proportional
to the square of the hydrogen atom displacement vector (i.e., the mean-square hydrogen vibrational amplitude) for each normal mode. Hence an
IINS spectrum (i.e., neutron scattering intensity versus neutron energy
loss) represents the amplitude-weighted vibrational density of states of the
hydrogenous adsorbate averaged over all Q space. The observed neutron
scattering intensities are predictable and can be analyzed quantitatively in
conjunction with molecular-force-field models to test the validity of proposed
adsorbate-site geometries [9].

Experimentally, IINS requires the determination of the energies and wave
vectors of the incident and scattered neutrons. This is typically accomplished
using either a triple-axis or a time-of-flight spectrometer. The most fre-

quently used instrument for catalyst studies thus far has been a reactor-based, modified triple-axis spectrometer in which the incident neutron beam is monochromated by Bragg reflection. In a conventional triple-axis spectrometer, the scattered neutrons are energy analyzed by a second monochromator. To enhance the signal for catalyst studies, the modified spectrometer replaces the second monochromator with a low-energy bandpass filter (polycrystalline beryllium at 77 K) that transmits only neutrons with energies of less than 5 meV.[†] The IINS spectrum is developed by scanning the energy of the incident beam. The energy resolution of the modified triple-axis spectrometer is, in general, limited by the Be filter (ca. 5 meV) below 100 meV and by the monochromator crystal above 100 meV [$\Delta E/E \leq 5\%$ for a Cu(220) crystal monochromator]. Considerably better resolution can be achieved in a variety of ways at the expense of count rate [9].

For time-of-flight spectrometers, the monochromated incident neutron beam is chopped prior to the sample. This enables determination of the sample-to-detector flight times (and therefore, the energies) of the scattered neutrons. Use of multiple detectors covering a range of scattering angles permits the measurement of IINS spectra simultaneously over a wide range of energy and wave vector transfers.

In a research reactor, the energy range of an IINS spectrum (typically 5 to 250 meV) is determined by the nearly Maxwellian distribution of energies in the moderated neutron spectrum. This energy range makes IINS a complementary technique to IR spectroscopy. Both ends of the energy range can be extended through the use of a cold or a hot neutron source, which provides either a lower- or a higher-energy distribution for the incident neutron beam. In addition, fluxes of higher-energy neutrons (up to ca. 1 eV) can be produced using a pulsed-neutron source [2].

B. Quasielastic Scattering: Dynamics

Quasielastic neutron scattering is a low-energy scattering technique involving broadening of the incoherent, elastically scattered peak due to the translational or rotational motion of nuclei, much like the Doppler shift in optical spectroscopy. It has become a standard method for the study of diffusion in solids and liquids, particularly because diffusion models can be tested and distinguished on a microscopic scale. Analysis of the elastic peak width as a function of momentum transfer can produce detailed atomic-scale information on the diffusive motion of hydrogen in the bulk and on the surface of condensed-phase materials [8,11], including the residence times and distances involved in jumps from site to site and the mean-square vibrational amplitudes of the hydrogen atoms. Detailed theoretical treatments are presented elsewhere [1].

Because of the very small energy transfer involved, high-resolution spectrometers are required. Time-of-flight instruments are typically used in the time regime down to 10^{-10} s, together with a cold neutron source to improve resolution and signal intensity. Energy resolution better than 0.5 μeV (ca. 10^{-8} s) can currently be achieved using a backscattering spectrometer [2]. If the instrumental resolution and sensitivity are appropriate, QNS measurements can be used to elucidate the detailed reorientation mechanisms of molecular species, such as an adsorbate on the catalyst surface.

[†]In this chapter we use both of the common vibrational energy units: meV and cm^{-1}; 1 meV = 8.065 cm^{-1}.

C. Diffraction: Structure

Because the wavelengths of thermal neutrons are comparable to interatomic distances in condensed phases, the spatial distribution of coherent, elastically scattered neutrons reflects the spatial arrangement of atoms in the sample due to diffraction effects. Powder neutron diffraction is a well-established technique for determining the complex structures of condensed-phase materials. Experimental methods and recent instrumentation advances have been reviewed [12,13]. In contrast to x-ray or electron diffraction, where the scattering from heavy elements dominates the diffraction spectrum, the neutron coherent scattering cross section of hydrogen and other light elements is comparable to that of heavy elements. Hence neutron powder diffraction can be used to locate hydrogen (or deuterium) in catalyst structures such as zeolites and organometallic complexes, as well as hydrogenous and other low-mass adsorbates on high-surface-area materials [6,8].

III. NEUTRON SCATTERING STUDIES OF CATALYSTS: EXAMPLES

Neutron scattering has been used to characterize the interactions of hydrogen with metal surfaces [14]. For example, the adsorption of hydrogen on supported and unsupported nickel catalysts has been investigated by various groups [15-21]. Figure 1 illustrates the IINS spectra for H/Raney nickel at 80 K as a function of coverage and adsorption temperature. The uniform growth (in a 2:1 ratio) of the scattering features at 117 and 141 meV with increasing coverage above 150 K indicated hydrogen adsorption in predominantly one type of surface site. The constant 2:1 intensity ratio was suggestive of a threefold symmetric adsorption site as on Ni(111) planes with the degenerate asymmetric (E) stretching modes assigned to 117 meV and the symmetric (A1) stretching mode assigned to 141 meV. The weaker IINS feature near 73 meV was assigned to a smaller amount of hydrogen adsorption in fourfold symmetric sites in accord with EELS results for H/Ni(100) [22]. Solving the harmonic model for hydrogen bonded to three Ni atoms by Wilson FG matrix analysis yielded the results in Fig. 2. A relatively narrow range of force constants (k_{H-Ni} = 0.58 ± 0.03 mdyn Å^{-1}) and bond lengths (d_{H-Ni} = 1.88 ± 0.04 Å) were consistent with the observed scattering energies.

The IINS spectrum for H/D adsorbate mixtures on Raney nickel indicated additional scattering features at 86 and 104 meV corresponding to deuterium-nickel vibrations with the expected isotopic energy shifts. Moreover, isotopic dilution of the adsorbed hydrogen with deuterium caused a narrowing of the spectral line shapes and a slight shift in the energies of the hydrogen scattering features (e.g., 117 to 122 meV and 141 to 139 meV for 10% H/20% D). This was suggestive of significant repulsive H-H interactions for the asymmetric stretching modes (motions parallel to the surface) and a smaller interaction for the symmetric stretching mode (motion perpendicular to the surface) for undiluted H adsorbates. Hence dilution with deuterium caused a reduction of the dynamical coupling between hydrogen oscillators on the surface.

The IINS spectrum for H and CO coadsorption at 80 K on Raney nickel illustrated that the dominant hydrogen species adsorbed in threefold sites remain virtually unperturbed by coadsorbed CO [17,20]. Yet increasing the sample temperature to ca. 300 K (or catalytic reaction in flowing H_2/CO mixtures) resulted in significant redistribution of the scattering intensity from 117 and 141 meV to an increased scattering intensity over the entire spectral region, especially below 100 meV. This signaled a decrease in three-

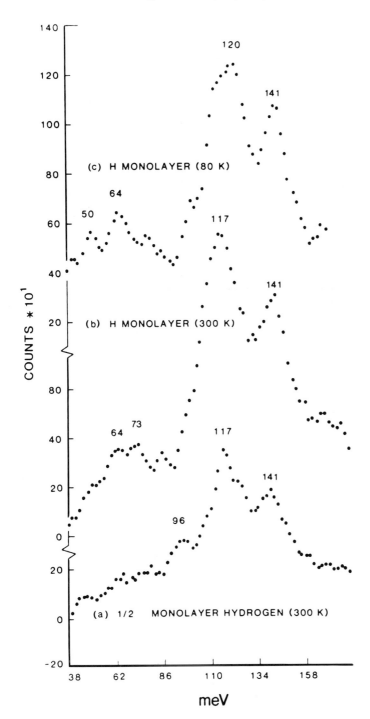

FIGURE 1 IINS spectra at 80 K of hydrogen adsorbed on Raney nickel: (a) half-saturation hydrogen coverage with sample annealed at 300 K prior to measuring spectrum; (b) saturation hydrogen coverage with sample annealed at 300 K prior to measuring spectrum; (c) saturation hydrogen coverage after adsorption at 80 K. (From Ref. 19.)

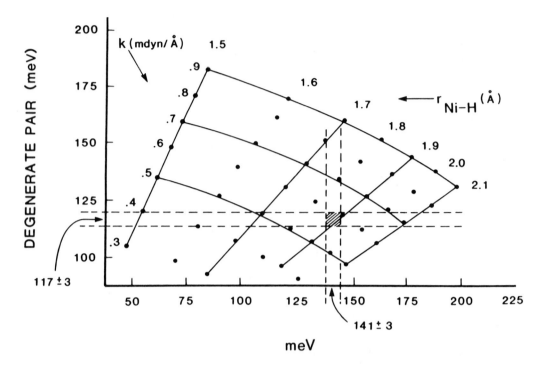

FIGURE 2 Calculated vibrational modes for hydrogen in a symmetric threefold adsorption site as a function of distance and force constant. The shaded region indicates the range of bond distances and force constants consistent with the observed frequencies and relative intensities. (From Ref. 19.)

fold hydrogen coverage concomitant with the appearance of vibrational features due to carbon-hydrogen adsorbate fragments [20].

It should be noted that the concept of delocalized hydrogen adsorption resulting in vibrational energy bands for hydrogen bound above Ni symmetry planes, as inferred by H/Ni wavefunction and energy-level calculations [23], is in disagreement with the IINS results for the H/Raney Ni system: (a) the intense ground-state to first-excited-state vibration bands expected at 40 to 60 meV are absent in the IINS spectra; (b) the neutron scattering peak widths are much lower than those predicted by the band theory and actually decrease (contrary to a predicted increase) with decreasing hydrogen coverage; and (c) harmonic behavior is observed, as evidenced by the appropriate isotopic shifts and overtones present in the neutron spectra [24]. However, it has been argued that H/Ni wavefunction calculations are based on the assumption of a perfect two-dimensional crystal and may not pertain to hydrogen adsorbed on the surfaces of small particles such as are found in Raney Ni [25]. Arguments for delocalized hydrogen adsorption on simple crystals of various metals can be found in Chapter 1.

Quasi-elastic neutron scattering has proved useful for quantifying the surface diffusion of hydrogen on Raney Ni at 423 K [26]. Analyzing the dependence of the quasi-elastic peak shape on neutron momentum transfer Q in terms of a scattering law for two-dimensional diffusion averaged over all directions of Q relative to the surface yielded a mean residence time for hydrogen adsorption [t_{H-Ni} = (2.7 ± 0.5) × 10^{-9} s], a mean hydrogen jump distance between adsorption sites (p_{H-Ni} = 3.0 ± 0.5 Å), and a surface diffusion coefficient [D_{H-Ni} = $(1/4)t^{-1}p^2$ = (0.8 ± 0.2) × 10^{-7} cm^2 s^{-1}]. A similar

QNS study was performed for hydrogen adsorbed on 12-Å platinum clusters encaged in Y-zeolite at 373 K [27]. The mean residence time for hydrogen adsorption was determined to be $t_{H-Pt} = (5.5 \pm 2.5) \times 10^{-9}$ s. Applying the Arrhenius equation with the reported activation energy for hydrogen diffusion on Ni (7 ± 1 kcal mol^{-1}) [28], the corrected residence time for hydrogen on Raney Ni at 373 K was estimated to be $t_{H-Ni} = (9 \pm 2) \times 10^{-10}$ s. Thus the tentative conclusion was that $D_{H-Pt} > D_{H-Ni}$ at the same temperature.

There exists good agreement between the hydrogen vibrational spectra for supported platinum catalysts [29, 30] and those for unsupported platinum black [31-35]. Figure 3a illustrates a typical vibrational spectrum for saturation hydrogen coverage at 80 K on platinum black with scattering features evident at 66, 115, and 166 meV. However, there has been some confusion concerning spectral peak assignments for the H/Pt adsorption system. Electron energy loss spectroscopic measurements for hydrogen adsorption at 90 K on Pt(111) [36] indicated two main vibrational features at 68 and 153 meV assigned to the symmetric (A1) and asymmetric (E) stretching modes, respectively, of hydrogen adsorbed in a threefold symmetric site. In addition, a weak feature at 107 meV was postulated as being due to hydrogen atoms in intermediate positions between bridge and threefold sites. On the assumption that Pt black exhibits predominantly (111) surface planes, Sayers [37] equated the three main scattering features in the IINS spectrum for H/Pt black from Howard et al. [33] to the three EELS features for H/Pt(111) [36]. Computing the amplitude-weighted vibrational density of states of hydrogen adsorbed at two- and threefold sites on Pt(111), it was suggested, contrary to the previous assignments, that the 68- and 153-meV features (corresponding to the 66- and 166-meV IINS features in Fig. 3a) are due to hydrogen in a twofold site.

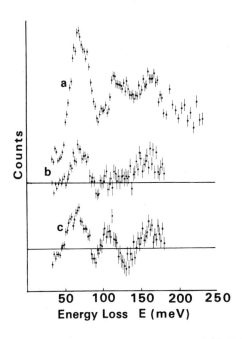

FIGURE 3 (a) IINS spectrum at 80 K for a monolayer of adsorbed hydrogen on platinum black; (b) difference spectrum after addition of ca. 1/3 monolayer of oxygen; (c) incremental difference spectrum after further addition of ca. 2/3 monolayer of oxygen (positive scattering intensities in difference spectra represent species removed by oxygen addition). (From Ref. 35.)

The asymmetric stretching mode along [2$\bar{1}\bar{1}$] was assigned to 68 meV and the two remaining normal modes, the asymmetric stretching mode along [0$\bar{1}$1] and the symmetric stretching mode along [111], being approximately degenerate, were assigned to 153 meV. This yielded reasonable values for the H-Pt bond length d_{H-Pt} = 1.93 Å and the H-Pt force constant k_{H-Pt} = 0.86 mdyn Å^{-1}. The weak EELS feature at 107 meV (close to the 115-meV IINS feature in Fig. 3a) was alternatively assigned to hydrogen in a threefold site with nearly degenerate symmetric and asymmetric normal modes assigned to 106 and 116 meV, respectively. This again yielded reasonable values for d_{H-Pt} = 1.90 Å and k_{H-Pt} = 0.49 mdyn Å^{-1}.

Figure 3b is the IINS difference spectrum following the addition of ca. 1/3 monolayer of oxygen to the hydrogen-saturated surface (i.e., O/H = 1/3) at 80 K [35]. The vibrational features at 66 and 166 meV diminish in the same ratio, unlike the feature at 115 meV. This was suggestive of at least two types of hydrogen adsorption sites, one associated with features at 66 and 166 meV (more susceptible to oxygen attack) and one associated with the feature at 115 meV (less susceptible to oxygen attack) consistent with the assignments by Sayers noted above. Figure 3c is the incremental difference spectrum following the further addition of ca. 2/3 monolayer of oxygen (i.e., total O/H = 1) at 80 K. Now the feature at 115 meV diminishes, along with further loss in intensity at 66 and 166 meV. Concomitant with this loss is the appearance of a scattering feature at 128 meV, associated with the bending mode of an adsorbed hydroxyl species now stabilized on the surface in agreement with EELS results for (OH)/Pt(111) [38].

The power of isotope dilution neutron spectroscopy is demonstrated clearly in experiments probing the dynamics of adsorbed hydrogen on platinum [39]. In particular, the substitution of chemically identical light (or heavy) mass defects, such as H for D, leaves the surface electronically and chemically unaltered, but can change the dynamic coupling between adsorbate oscillators. This results in changes in the vibrational density of states predictable by mass defect theory [40]. Figure 4 illustrates IINS difference spectra representing one of the dominant vibrational features of adsorbed hydrogen on platinum black at 80 K as a function of the H/D ratio at saturation (H + D) coverage. The spectrum for 100% H indicates a complex density of states centered at ca. 66 meV with a high-energy shoulder at ca. 76 meV. In comparison, the spectrum for 10% H distributed in a deuterated monolayer displays a significantly narrower Gaussian-shaped peak shifted from 66 meV to 73 meV. This spectral dependence on the H/D ratio was in agreement with calculations using simple mass defect theory and corroborated the existence of a dispersion of the surface optical vibrations for 100% H in a single type of adsorption site due to dynamic nonbonding interactions between H oscillators. The H-H force constant associated with these interactions was estimated to be ca. $0.2 k_{Pt-H}$.

In addition to the extensive neutron scattering studies concerned with characterizing bulk palladium hydrides [41-44], IINS investigations have also been undertaken to study the adsorption of hydrogen on unsupported palladium [45-48]. The IINS spectrum for hydrogen adsorbed on Pd black [45] indicated two scattering features, centered at 823 and 916 cm^{-1}, which were originally assigned to the symmetric and asymmetric stretching modes, respectively, of hydrogen adsorbed in a twofold (bridging) site. A broad band at ca. 1700 cm^{-1} was assigned to either the deformation (scissors) mode of coadsorbed water or a combination vibration of 823 cm^{-1} and 916 cm^{-1}. Later, in light of IINS studies of hydridocarbonyls, the three features were reassigned [46] to the symmetric stretching mode of H in a twofold site on Pd(100) (823 cm^{-1}) and Pd(111) (916 cm^{-1}), and the asymmetric stretches of hydrogen in twofold sites on both Pd(100) and Pd(111) (1700 cm^{-1}).

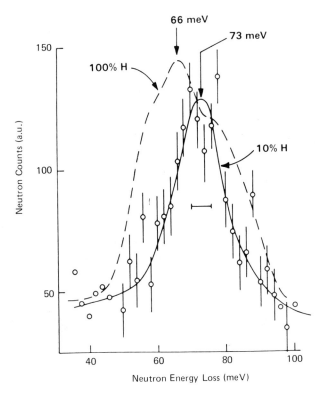

FIGURE 4 IINS spectra between 34 and 100 meV at 80 K for a saturation coverage of hydrogen isotopes on Pt black. Results for 100% H are shown by the dashed lines. The energy resolution (FWHM) is indicated by the horizontal bar. (From Ref. 39.)

These assignments have been questioned [49] since they imply an unreasonably short bond length d_{H-Pd} = 1.52 Å. Alternatively, it was suggested that hydrogen was adsorbed in a threefold site with the 823-cm^{-1} feature assigned to the symmetric stretching mode and the 916-cm^{-1} feature assigned to the degenerate asymmetric stretching modes. The harmonic model for a threefold site yielded a force constant k_{H-Pd} = 0.4617 mdyn Å^{-1} and a more reasonable bond length d_{H-Pd} = 1.881 Å. More recently, IINS results for hydrogen adsorbed on an oxidized Pd black surface [48] have indicated the presence of a scattering feature at 940 cm^{-1} assigned to the bending mode of adsorbed hydroxyl species (similar to Pt black [35]), in agreement with EELS results for (OH)/Pd(100) [50].

As for metal catalysts, IINS is useful for probing the vibrational modes of hydrogen adsorbed on nonmetallic materials. For example, the adsorption of hydrogen on ZnO was investigated utilizing IINS in conjunction with infrared and Raman spectroscopies [51]. Hydrogen adsorption at room temperature followed by cooling to 77 K resulted in the IINS difference spectrum in Fig. 5. Based on infrared results and neutron scattering intensity arguments, the spectral features at 829 and 1665 cm^{-1} were assigned to the bending and stretching modes, respectively, of Zn-H at type I sites formed by heterolytic dissociation of H_2 at Zn-O pair sites. The broad feature at 1125 cm^{-1} was assigned to the O-H bending mode at type I sites, and the spectral features below 600 cm^{-1} were assigned to H motions associated with lattice phonon modes.

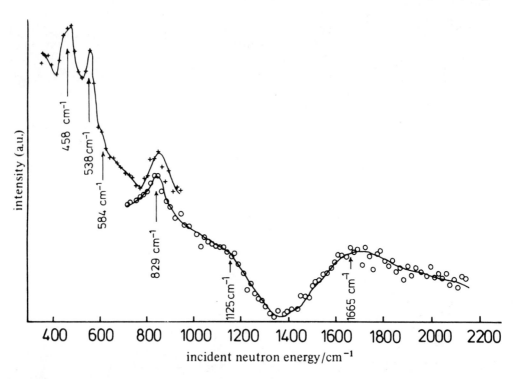

FIGURE 5 Difference spectrum: IINS spectrum of ZnO + H_2 minus IINS spectrum of ZnO. The symbols, + and o, represent data collected using the (200) and (220) planes of the Cu monochromator, respectively. (From Ref. 51.)

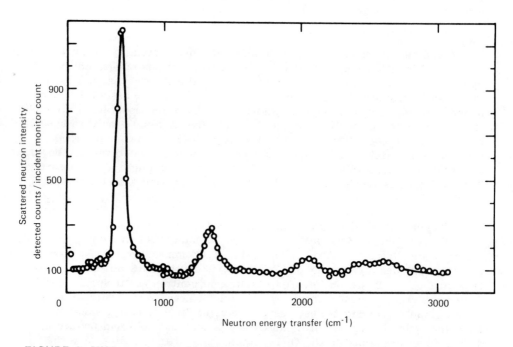

FIGURE 6 IINS spectrum of hydrogen adsorbed by WS_2. (From Ref. 52.)

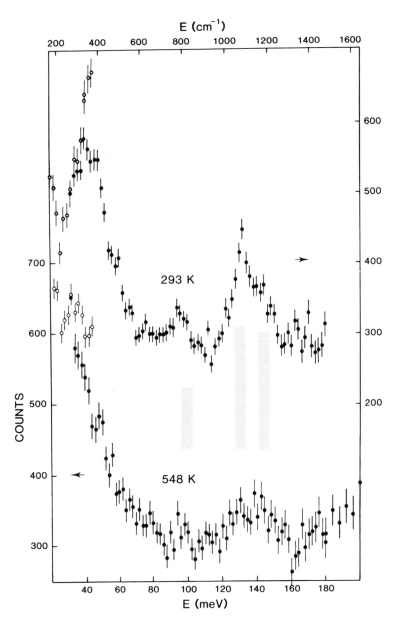

FIGURE 7 IINS spectra of fully dehydrated zeolite rho (acid form). Bars indicate calculated inelastic scattering intensities. The discontinuity at low energy results from the use of two monochromators. ●, Cu(220), 32 to 200 meV; ○, C(002), 20 to 44 meV. (From Ref. 57.)

Hydrogen adsorption studies have also been undertaken on WS_2 [52], MoS_2 [53,54], and RuS_2 [55] (see also Chapter 22). Figure 6 depicts the IINS spectrum following hydrogen adsorption on WS_2 at 673 K. Adsorption isotherms, infrared results, and models to predict neutron scattering intensity were used to suggest that H_2 dissociatively adsorbed at vacancies and diffused toward surface S atoms forming S-H bonds. The scattering feature at 694 cm^{-1} (and its first and second harmonics at 1380 and 2074 cm^{-1}) was assigned to the degenerate S-H bending mode, and the broad scattering feature at ca. 2500 cm^{-1} was assigned to the S-H stretching mode in agreement with the value found for thiols [56]. The similarity of the S-H bending mode feature for WS$_2$ to the scattering feature at 662 cm^{-1} for H/MoS$_2$ [54] and at 700 cm^{-1} for H/RuS$_2$ [55] is also suggestive of the formation of S-H species on these surfaces. Neutron scattering spectra for hydrogen adsorption on MoS$_2$ at higher H$_2$ pressures (up to 40 atm [54]) possessed an additional broad vibrational feature centered at ca. 400 cm^{-1}, indicating a second type of adsorbed hydrogen species whose coverage increased approximately proportional to pressure with a saturation coverage at ca. 50 atm. It was suggested that since the hydrodesulfurization reaction is first order in p_{H_2} up to 50 atm, this second type of adsorbed species may be actively involved in the HDS reaction mechanism [8].

Neutron scattering is equally well suited for investigating hydrogen interactions with zeolites. For example, Fig. 7 represents the IINS spectra as a function of temperature for the framework hydrogen in zeolite H-rho. The room-temperature vibrational spectrum was consistent with the existence of planar, symmetric AlO(H)Si units (i.e., hydroxyl groups symmetrically bridging Si and Al). Figure 8 displays the results of Wilson FG matrix analysis assigning the features at 1150 and 1060 cm^{-1} to the in-plane and out-of-plane T-O-H bending modes, respectively, and the feature at 750 cm^{-1} to the H motion associated with a symmetric T-O stretching mode. The lower intensity for the latter feature reflected the relatively smaller root-mean-square displacement of a zeolite framework mode. The intense feature at 360 cm^{-1} was suggestive of a large-amplitude vibration such as a T-OH torsional motion. Heating H-rho to 548 K caused a noncentrosymmetric-centrosymmetric transition accompanied by a shift of a significant fraction of the protons to a different type of bonding site of unknown nature. The high-temperature site

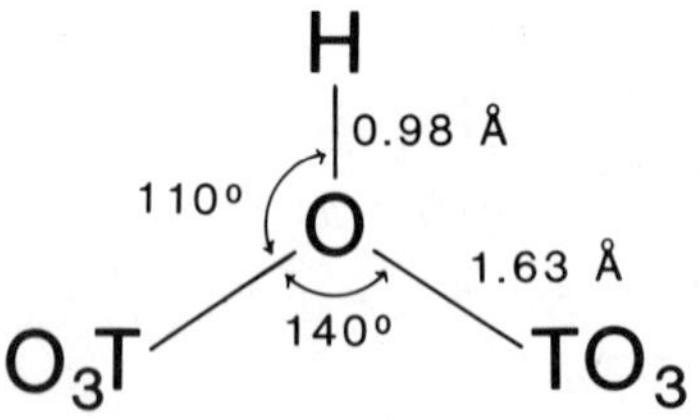

COORDINATE	FORCE CONSTANT
O-H	7.29
O-TO$_3$	3.9
H-O-TO$_3$ in plane	0.31
O$_3$T-O-TO$_3$	0.30
H-O-TO$_3$ out of plane	0.445

VIBRATION	FREQUENCY (cm^{-1})	INTENSITY
δ_{T-O-H}	1157	0.65
γ_{O-H}	1084	0.67
ν^{sym}_{T-O}	823	0.34

FIGURE 8 Model for the hydrogen-bonding site on zeolite rho. Force constants are given in mdyn Å^{-1}. Intensity refers to the calculated root-mean-square hydrogen displacement amplitude, which should be proportional to the inelastic scattering intensity. (From Ref. 57.)

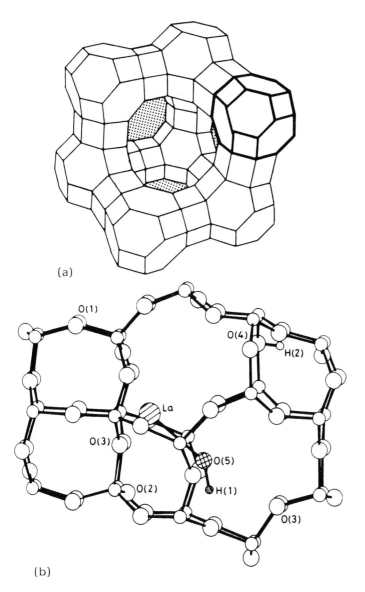

FIGURE 9 (a) View of the framework structure of zeolite-Y; the truncated octahedron is indicated by thicker lines. (b) Truncated octahedral element in La-Y, showing one of the hexagonal prismatic linkages to the adjacent element. Site SI', here shown occupied by La, is in the hexagonal window forming the outer end of this prism. (From Ref. 62.)

population is marked by the appearance of a large-amplitude-like intense feature centered at 260 cm$^-$, concomitant with a decrease in the room-temperature scattering features. This apparent site-to-site proton transfer at elevated temperatures provided the impetus to perform QNS experiments to study hydrogen diffusion on the zeolite framework. Quasielastic neutron scattering results for partially dehydroxylated H-rho (H/Al = 0.5) up to 673 K indicated no hydrogen diffusion on H-rho within the instrumental resolution. This put an upper limit on the two-dimensional diffusion coefficient of $D_{H\text{-rho}} = 4 \times 10^{-6}$ cm^2 s^{-1}.

Neutron diffraction has shown considerable utility for the structural characterization of catalytic materials [8,58]. In the present context, neutron diffraction has been useful in identifying the structural position of hydrogen in a variety of catalytic materials, ranging from zeolites [59-62] and high-surface-area materials [63] to transition-metal complexes [8,64]. For example, cation hydrolysis in lanthanum Y-zeolite has been directly observed by powder neutron diffraction [62]. Polycrystalline La-Y (Si/Al = 2.61) was prepared by ion exchange of the sodium form at 348 K. To introduce Brønsted acidity, protons were generated by cation hydrolysis of the rare-earth hydrate ($La(H_2O)_x^{3+} \rightarrow La(OH)_x^{(3-x)+} + xH^+$, $1 < x < 2$) at 873 K. The structural results are depicted in Fig. 9. Using the space group $Fd3m$, least-squares refinements of the powder diffraction data collected at 5 K indicated 67% ion exchange and identified the location of the La ions at site SI' and the unexchanged Na ions at SI' and SII. Difference Fourier calculations identified a single proton attached to nonframework La-bonded oxygen atoms O(5) with an OH bond length of 0.99 Å, and protons attached to framework oxygen atoms O(4) at a bond length of ca. 1.2 Å.

IV. CONCLUSION

It is clear from the examples presented that the special nature of neutron scattering interactions, coupled with the relatively large incoherent scattering cross section for hydrogen, renders neutron scattering a powerful surface-sensitive tool for probing the structure and adsorbate interactions of hydrogen as well as other hydrogenous molecules with real catalytic materials. Despite its potential, the use of neutron scattering in catalysis research has thus far been relatively infrequent compared to other surface probes, such as electrons or photons. Today, the continuing development of new, more-advanced instrumentation with higher resolution and sensitivity and the availability of higher-flux neutron sources are leading to an increased interest "as more chemical physicists are exposed to the unique power of neutron scattering, particularly at low energies (0.1 μeV-10 meV) where it has become an unparalleled probe of molecular dynamics in condensed systems" [65].

ACKNOWLEDGMENTS

We acknowledge, with thanks the Chemical Sciences Division, Office of Basic Energy Sciences, Office of Energy Research, U.S. Department of Energy for partial support of this work, and Drs. J. J. Rush and J. M. Nicol for careful reading and comments on the manuscript.

REFERENCES

1. S. W. Lovesey, *Theory of Neutron Scattering from Condensed Matter*, Vol. 1, Clarendon Press, Oxford (1984).
2. J. W. White and C. G. Windsor, *Rep. Prog. Phys.*, 47:707 (1984).
3. S. W. Lovesey and T. Springer (eds.), *Topics in Current Physics*, Vol. 3, *Dynamics of Solids and Liquids by Neutron Scattering*, Springer-Verlag, Berlin (1977).
4. J. Howard and T. C. Waddington, in *Advances in Infrared and Raman Spectroscopy*, Vol. 7 (R. J. H. Clark and R. E. Hester, eds.), Heyden, London (1978), p. 89.

5. P. G. Hall and C. J. Wright, in *Chemical Physics of Solids and Their Surfaces*, Vol. 7 (M. W. Roberts and J. W. Thomas, eds.), Billings, London (1978), p. 89.
6. R. K. Thomas, *Prog. Solid State Chem.*, *14*:1 (1982).
7. C. J. Wright and C. M. Sayers, *Rep. Prog. Phys.*, *46*:773 (1983).
8. C. J. Wright, *Catalysis*, 7:46 (1985).
9. R. R. Cavanagh, J. J. Rush, and R. D. Kelley, in *Vibrational Spectroscopy of Molecules on Surfaces* (J. T. Yates, Jr. and T. E. Madey, eds.), Vol. 4, *Methods of Surface Characterization*, Plenum Press, New York (1987).
10. C. J. Glinka, H. J. Prask, and C. S. Choi, in *Mechanics of Nondestructive Testing* (W. W. Stinchomb, ed.), Plenum Press, New York (1980), p. 143.
11. R. Hempelmann, *J. Less-Common Met.*, *101*:69 (1984).
12. R. Pynn, *Rev. Sci. Instrum.*, *55*:837 (1984).
13. G. E. Bacon, *Neutron Diffraction*, 3rd ed., Clarendon Press, Oxford (1975).
14. C. J. Wright, in *The Structure of Surfaces* (M. A. Van Hove and S. Y. Tong, eds), Springer-Verlag, Berlin (1985), p. 210.
15. R. Stockmeyer, H. M. Conrad, A. Renouprez, and P. Fouilloux, *Surf. Sci.*, *49*:549 (1975).
16. A. J. Renouprez, P. Fouilloux, G. Coudurier, D. Tocchetti, and R. Stockmeyer, *J. Chem. Soc. Faraday Trans. 1*, *73*:1 (1977).
17. R. D. Kelley, J. J. Rush, and T. E. Madey, *Chem. Phys. Lett.*, *66*:159 (1979).
18. R. Stockmeyer, H. M. Stortnik, I. Natkaniec, and J. Mayer, *Ber. Bunsenges. Phys. Chem.*, *84*:79 (1980).
19. R. R. Cavanagh, R. D. Kelley, and J. J. Rush, *J. Chem. Phys.*, 77:1540 (1982).
20. R. D. Kelley, R. R. Cavanagh, and J. J. Rush, *J. Catal.*, *83*:464 (1983).
21. H. Jobic and A. Renouprez, *J. Chem. Soc. Faraday Trans. 1*, *80*:1991 (1984).
22. S. Andersson, *Chem. Phys. Lett.*, *55*:185 (1978).
23. M. J. Puska, R. M. Nieminen, M. Manninen, B. Chakraborty, S. Holloway, and J. K. Norskov, *Phys. Rev. Lett.*, *51*:1081 (1983).
24. R. R. Cavanagh, J. J. Rush, and R. D. Kelley, *Phys. Rev. Lett.*, 52:2100 (1984).
25. M. J. Puska and R. M. Nieminen, *Surf. Sci.*, *157*:413 (1985).
26. A. Renouprez, P. Fouilloux, R. Stockmeyer, H. M. Conrad, and G. Goeltz, *Ber. Bunsenges. Phys. Chem.*, *81*:429 (1977).
27. A. J. Renouprez, R. Stockmeyer, and C. J. Wright, *J. Chem. Soc. Faraday Trans. 1*, *75*:2473 (1979).
28. R. Gomer, R. Wortman, and R. Lundy, *J. Chem. Phys.*, *26*:1147 (1957).
29. A. Renouprez, P. Fouilloux, and B. Moraweck, in *Growth and Properties of Metal Clusters* (J. Bourdon, ed.), Elsevier, Amsterdam (1980), p. 421.
30. A. J. Renouprez, J. M. Tejero, and J. P. Candy, *8th International Congress on Catalysis Proceedings*, Vol. 3, Berlin (1984), p. 47.
31. H. Asada, T. Toya, H. Motohashi, M. Sakamoto, and Y. Hamaguchi, *J. Chem. Phys.*, *63*:4078 (1975).
32. J. Howard, T. C. Waddington, and C. J. Wright, *J. Chem. Phys.*, *64*:3897 (1976).
33. J. Howard, T. C. Waddington, and C. J. Wright, in *Neutron Inelastic Scattering*, Vol. 2, IAEA, Vienna (1978), p. 499.
34. J. J. Rush, R. R. Cavanagh, and R. D. Kelley, *J. Vac. Sci. Technol.*, *A1*:1245 (1983).
35. T. J. Udovic, R. R. Cavanagh, J. J. Rush, and R. D. Kelley, private communication.

36. A. M. Baro, H. Ibach, and H. D. Bruchmann, *Surf. Sci.*, 88:384 (1979).
37. C. M. Sayers, *Surf. Sci.*, 143:411 (1984).
38. G. B. Fisher and B. A. Sexton, *Phys. Rev. Lett.*, 44:683 (1980).
39. J. J. Rush, R. R. Cavanagh, R. D. Kelley, and J. M. Rowe, *J. Chem. Phys.*, 83:5339 (1985).
40. R. J. Elliot and A. A. Maradudin, in *Inelastic Scattering of Neutrons*, Vol. 1, IAEA, Vienna (1965), p. 231.
41. W. Drexel, A. Murani, D. Toccheti, W. Kley, I. Sosnowska, and D. K. Ross, *J. Phys. Chem. Solids*, 37:1135 (1976).
42. D. G. Hunt and D. K. Ross, *J. Less-Common Met.*, 49:169 (1976).
43. J. J. Rush, J. M. Rowe, and D. Richter, *Z. Phys. Chem. Abt.*, B55:283 (1984).
44. J. J. Rush, J. M. Rowe, and D. Richter, *Phys. Rev.*, B31:6102 (1985).
45. J. Howard, T. C. Waddington, and C. J. Wright, *Chem. Phys. Lett.*, 56:258 (1978).
46. I. J. Braid, J. Howard, and J. Tomkinson, *J. Chem. Soc. Faraday Trans. 2*, 79:253 (1983).
47. H. Jobic, J. P. Candy, V. Perrichon, and A. Renouprez, *J. Chem. Soc. Faraday Trans. 1*, 81:1955 (1985).
48. J. M. Nicol, R. D. Kelley, and J. J. Rush, private communication.
49. C. M. Sayers and C. J. Wright, *J. Chem. Soc. Faraday Trans. 1*, 80:1217 (1984).
50. E. M. Stuve, S. W. Jorgensen, and R. J. Madix, *Surf. Sci.*, 146:179 (1984).
51. J. Howard, I. J. Braid, and J. Tomkinson, *J. Chem. Soc. Faraday Trans. 1*, 80:225 (1984).
52. C. J. Wright, D. Fraser, R. B. Moyes, and P. B. Wells, *Appl. Catal.*, 1:49 (1981).
53. C. J. Wright, C. Sampson, D. Fraser, R. B. Moyes, P. B. Wells, and C. Riekel, *J. Chem. Soc. Faraday Trans. 1*, 76:1585 (1980).
54. C. Sampson, J. M. Thomas, S. Vasudevan, and C. J. Wright, *Bull. Soc. Chim. Belg.*, 90:1215 (1981).
55. T. J. Udovic, J. J. Rush, W. H. Heise, K. Lu, Y.-J. Kuo, and B. J. Tatarchuk, private communication.
56. L. J. Bellamy, *The Infra-red Spectra of Complex Molecules*, Methuen, London (1958), p. 351.
57. M. J. Wax, R. R. Cavanagh, J. J. Rush, G. D. Stucky, L. Abrams, and D. R. Corbin, *J. Phys. Chem.*, 90:532 (1986).
58. J. M. Newsam, *Physica*, 136B:213 (1986).
59. Z. Jirak, S. Vratislav, J. Zajicek, and V. Bosacek, *J. Catal.*, 49:112 (1977).
60. Z. Jirak, S. Vratislav, and V. Bosacek, *J. Phys. Chem. Solids*, 41:1089 (1980).
61. V. Bosacek, S. Beran, and Z. Jirak, *J. Phys. Chem.*, 85:3856 (1981).
62. A. K. Cheetham, M. M. Eddy, and J. M. Thomas, *J. Chem. Soc. Chem. Commun.*, 1337 (1984).
63. J. P. Beaufils, Y. Barbaux, and B. Saubat, *J. Chem. Soc. Chem. Commun.*, 1212 (1982).
64. R. Bau, R. G. Teller, S. W. Kirtley, and T. F. Koetzle, *Acc. Chem. Res.*, 12:176 (1979).
65. J. J. Rush and J. M. Rowe, *Physica*, 137B:169 (1986).

7 | Infrared Spectroscopy of Adsorbed Hydrogen

TIBOR SZILÁGYI

Institute of Isotopes of the Hungarian Academy of Sciences, Budapest, Hungary

I.	INTRODUCTION	183
II.	SOME EXPERIMENTAL CONSIDERATIONS	184
III.	IR SPECTROSCOPY OF HYDROGEN ON DISPERSE CATALYSTS	186
	A. Adsorbed Forms of Hydrogen as Observed by IR Spectroscopy	186
	B. Properties of Weakly Bound Hydrogen Adsorbed on Supported Pt	188
	C. Interaction Between Hydrogen and CO on Pt Surface	191
IV.	CONCLUSIONS	192
	REFERENCES	192

I. INTRODUCTION

Infrared (IR) spectroscopy has been used for a long time to detect surface species at the solid-gas interface [1]. It became popular because its theory and measurement techniques are relatively simple and—as with all the vibrational spectroscopies (see, e.g., Chapter 2)—it usually supplies straightforward and ready-to-use information on the surface chemical bonds. The method became even more widespread with the advent of computer-controlled, rapid-scan interferometers. Fourier transform (FT) spectroscopy [2] helps to overcome the difficulties of severe optical conditions often encountered in the study of surface phenomena, and as computer facilities are always at hand for digital data storing and postprocessing, good-quality spectra can be produced on a reasonable time scale. In the last decades a number of excellent surface-probing methods have been developed, together constituting the experimental base of modern surface science. Many of them have acquired very high sensitivity; thus they surpass the efficiency of FTIR spectroscopy, especially when submonolayer coverages on low-area surfaces are investigated. Nevertheless, in certain respects, FTIR spectroscopy is still unique among all the experimental techniques applied for the study of surface processes.

In this chapter those properties of IR spectroscopy substantial from the viewpoint of a user specialist in catalysis are briefly discussed. Next, the reported infrared spectra of hydrogen adsorbed on different surfaces are reviewed.

II. SOME EXPERIMENTAL CONSIDERATIONS

Perhaps the most important property of IR spectroscopy is its high resolving power, which allows monitoring of very fine changes in the structure of surface species. In other words, since 4- to 5-cm^{-1} shifts in the frequency of a vibrational band are routinely measured, less than a 1% difference in bond energies during surface reactions is easily detected. Another point worth considering is the nature of the interacting agent (i.e., the infrared light): The maximum practical photon energy used in IR spectroscopy (4000 cm^{-1}, ca. 0.5 eV) falls far below the chemical bond energies (including the chemisorption energies). Other surface probing methods usually apply high-energy electrons, x-rays, ultraviolet photons, and so on. These can initiate several unwanted side reactions in the surface layer (e.g., ionization, dissociation, desorption, rearrangement, or reconstructions) which are difficult to take into account when the experimental results are interpreted. In IR spectroscopy, all such effects can be neglected.

The frequency range of FT spectroscopy seems to be extremely wide. Practical frequencies for *vibrational spectroscopy* extend from ca. 100 to 4000 cm^{-1}, as almost all the fundamental molecular vibrations occur in that range. However, modern FT spectrophotometers work essentially from the microwave up to UV frequencies (10 to 45,000 cm^{-1}). At the cost of changing some commercially available components (windows, beamsplitter, detector) the vibrational overtone region or the electronic spectrum can also be measured conveniently by the same instrument.

From a practical point of view, FTIR spectroscopy is suitable for the study of both metals and insulators of diverse geometrical shapes (single crystals, foils, films, powders or industrially preshaped catalyst pellets, etc.). There are no restrictions if high or low temperatures are necessary, and adsorbate pressure can range from UHV up to the MPa range. This makes IR spectroscopy an attractive tool for catalytic research. Practical catalysis means high adsorbate pressures and poorly defined surfaces, where the conditions (composition of surface layers, the kinetics and thermodynamics of surface reactions, etc.) are quite different from those observed at ultra-high vacuum on model surfaces. The science of catalysis is in severe need of structural information obtained under realistic conditions, and this can be supplied by IR spectroscopy.

The versatility of FTIR spectroscopy means that experiments can be performed by various optical setups best fitting the special requirements of the system under investigation; at least an advantageous compromise can be chosen. Thus spectra can be measured as (a) energy is absorbed from an incident light beam specularly or diffusely reflected by opaque samples, (b) light absorption occurs from a beam transmitted through transparent samples, or (c) energy is emitted by the sample itself as a result of its "perturbed" blackbody radiation. When vibrating molecules (approximated as oscillating dipoles) near highly polarizable media (i.e., adsorbed on metal surfaces) are studied, a special condition should be taken into account. As a result of the polarization effected by the electric field of the dipoles a so-called "image" dipole of equal magnitude is created inside the metal (Fig. 1). As a consequence of superposition of the electric fields of the surface dipole and its image, there is a net dipole moment change always perpendicular to the surface, and those vibrations having a dynamic dipole moment parallel to the surface are not observable in the IR spectrum. This is commonly

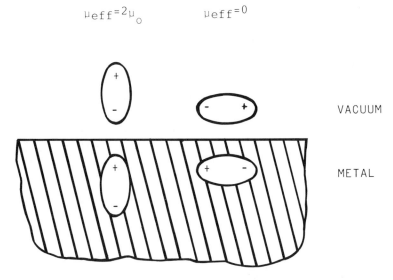

FIGURE 1 Surface dipole orientations. μ_0, Vibrational dipole moment; μ_{eff}, dipole moment effective in light absorption or emission.

referred to as the "metal surface selection rule" [3,4]. This is also valid in the case of dispersed metal particles with diameters small compared to the IR wavelengths [5-7], as long as these particles have the same electronic properties as in bulk metals [8,9].

Molecules adsorbed on low-area flat surfaces are investigated by reflection-absorption [10,11] or emission [12-14] techniques. In the former case a small change in a relatively high intensity beam (incident at high angles to the surface) should be detected; that is, a signal of very high dynamic range is produced which requires high-quality signal processing electronics (detector amplifiers, analog-to-digital converters, etc.). Sensitivity is considerably enhanced by special modulation techniques [15], and coverages down to 1% of a monolayer can be detected. In emission measurements very low intensities should be observed. This requires detectors of very low noise equivalent power and high-quality optical systems constructed to allow maximum energy throughput [12]. Moreover, light emitted at high angles to surface normal should be collected, since dipoles radiate essentially at high angles to their axis.

Dispersed metals (usually supported on high-area oxides) or other high-area powders can conveniently be examined by pressing them into a thin pellet that is studied by the conventional transmission method [1]. Obviously, in this case the advantage of having a well-characterized surface is lost, but this is compensated by the possibility of observing bands of low absorption coefficients since the effective surface of the samples (and consequently the number of light-absorbing molecules) is ca. 10^3 to 10^5 times larger than that of the low-area samples discussed above. However, the useful frequency range of the spectra is narrowed as oxide materials generally have strong absorption bands at lower frequencies. Thicker pellets absorb light totally, so that spectra in that range cannot be evaluated (Fig. 2).

Some other surface-probing IR methods, such as surface electromagnetic wave spectroscopy [16] and diffuse reflectance spectroscopy [2b,17], have been also reported. These usually apply special optical systems constructed of unusual components and have acquired limited catalytic applications up to now.

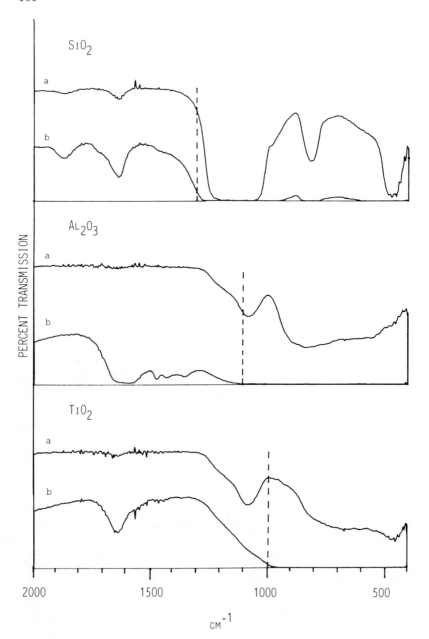

FIGURE 2 IR spectra of common support materials. a, standard spectrum (ca. 5 mg pressed in KBr pellet); b, spectrum of 0.1-mm-thick pellet. Dashed line shows the low-frequency limit of the transparent region.

III. IR SPECTROSCOPY OF HYDROGEN ON DISPERSE CATALYSTS

A. Adsorbed Forms of Hydrogen as Observed by IR Spectroscopy

Weak molecular adsorption of hydrogen (Table 1) occurs at low temperature on high-area nonmetallic surfaces such as evaporated alkali halides [18], zeolites [19], and porous silica (Vycor) glass [20]. (Recent HREELS studies also reported the existence of molecular H on metal surfaces at 10 K; see

TABLE 1 Vibrational Frequencies of Molecularly Adsorbed Hydrogen

Substrate	Frequency (cm^{-1})			Ref.
	HH	HD	DD	
—a	4160	3631	2993	44
NaCl (77 K)	4112			18
CsI (77 K)	4132	3602	2961	18
NaA, NaCaA zeolites (105-135 K)			2950-3000	19
Porous silica glass (90 K)	4131			20

aGas, observed in Raman spectrum.

Chapter 2.) As a result of perturbation by strong electrostatic forces of the surface, the otherwise forbidden H-H and D-D stretching band appears in the spectrum. The frequencies are slightly lowered compared to those measured in the gas phase (observable in the Raman spectrum [44], the heat of adsorption being small [18]. All these properties are characteristic of a physisorbed state (Table 1). Hydrogen is known to chemisorb dissociatively in different forms on metal surfaces [21]. One of these forms gives an infrared band at ca. 2000 cm^{-1} (the corresponding deuterium band is found at 1500 to 1400 cm^{-1}, Table 2). This frequency is rather close to that of the H_2^+ molecular ion (2297 cm^{-1}, [22]); thus it might be assigned to the bond stretching of a strongly electron-deficient molecular species on the surface. This supposition was elegantly disproved by Pliskin and Eischens [26]. Introducing $H_2 + D_2$ mixture to a supported Pt sample, they observed only the two bands mentioned above, whereas in the case of molecular adsorption, a third, intermediate band also should have been present, originating from the adsorbed HD molecules produced by the exchange reaction between H_2 and D_2. Hence, by analogy with the IR spectra of metal-hydride complex compounds [22-24], the bands are assigned to the M-H (M-D) stretching of hydrogen atoms coordinated to single metal atoms on the surface. IR spectra of such "on-top" hydrogen adsorbed on all the group VIII metals (at room temperature and above 10^{-4} Pa) except Os have been published [25]. However, detailed studies are available only for hydrogen adsorbed on Pt catalysts

TABLE 2 Frequency Ranges for Dissociatively Adsorbed Hydrogen as Observed by IR Spectroscopy

	Frequency (cm^{-1})		Refs.
	M-H	M-D	
On-top			
On group VIII metals	2120-1850	1520-1360	26-34
On Si	2060-2100		35
Multibonded	(below 1300)		
On Pt/MgO	950		32
On Pd film	880, 760		37
On W(100)	1070, 1270	767, 915	38

[26-33]. A band of similar frequency (2060 to 2100 cm^{-1}) was reported for atomic hydrogen adsorbed on a silicon surface [34]. Formation of SiH$_2$ groups on SiO$_2$ surfaces (bands at 2300 and 2227 cm^{-1}) was also observed [35].

Vibrational bands of "multibonded" hydrogen (i.e., H atoms coordinated to more than one Pt atom) occur at lower frequencies (Table 3). IR studies of these adsorbed forms are less abundant [31,36,37], since in reflection experiments the very weak bands are detected only with difficulty, and when investigated on disperse metals, are usually concealed by the strong absorption of the support. (However, excellent results were obtained by HREELS and neutron spectroscopy; see also Chapters 2 and 6). Multibonded H atoms have several vibrational degrees of freedom depending on the local symmetry of the adsorption site [25], but only one of the fundamentals (the symmetric stretch) is allowed by the metal-surface selection rule in the infrared. Certain overtones or combinations having the right symmetry are IR active as well [37].

In addition to the references cited above, IR studies dealing with the interaction of hydrogen and CO on single crystals [38], evaporated films [39], and supported metal surfaces [40] have also been reported. None of these mentions an IR band of adsorbed hydrogen. Instead, the effect of hydrogen on the shape and frequency of the CO vibrational band is discussed (see also Chapter 20).

The spillover of hydrogen on supported metal catalysts (see Chapter 12 for details) can be studied by IR spectroscopy [42]. If D$_2$ is adsorbed on a supported metal, the exchange of surface hydroxyl groups with D atoms spilled over the support is easily monitored by the appearance of and increase in the strong OD band in the spectra (Fig. 3). It was shown [41] that the rate of exchange diminished if CO was preadsorbed on Rh/Al$_2$O$_3$; that is, spillover was strongly retarded because CO blocked the metal sites for D$_2$ dissociation. By the use of D-labeled adsorbates, specific features of adsorption or surface reactions (e.g., dehydrogenation) can also be determined [42].

B. Properties of Weakly Bound Hydrogen Adsorbed on Supported Pt

Hydrogen is known to adsorb in several different forms on Pt surfaces [21]. The surface layer consists of strongly bound H atoms situated in trigonal or tetragonal holes (see Chapters 1 and 2). At room temperature (or below) saturation occurs below. ca. 10^{-6} Pa and a coverage value (expressed as H/Pt atomic ratio) of about unity is reached. If the pressure over the sample is increased further, a new form of adsorbed hydrogen appears above 10^{-4} Pa, as shown by the IR spectra of supported Pt catalysts [26-33]. This form may also be detected by temperature-programmed desorption (TPD) (see Chapter 4). The adsorption is reversible (i.e., hydrogen desorbs if

TABLE 3 Band Components of Hydrogen Adsorbed on Pt/SiO$_2$

	Frequency (cm^{-1})	Bandwidth (cm^{-1})
H1	2129	24
H2	2118	28
H3	2097	48

Source: Ref. 33.

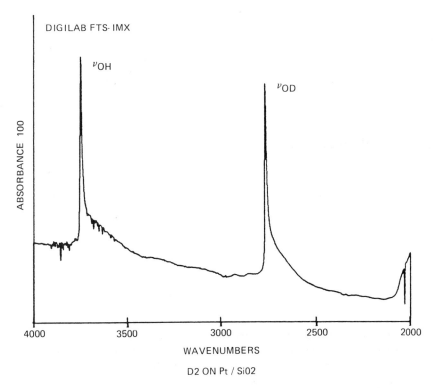

FIGURE 3 Spectrum of reduced Pt/SiO_2 after introducing 100-Pa D_2 to the sample.

the gas phase is removed). This weakly bound hydrogen gives an IR band at 2120 cm^{-1} which is assigned to the Pt-H stretching of H atoms coordinated to one Pt atom only. The corresponding Pt-D band is observed at 1520 to 1530 cm^{-1}. Usually, another band is found at 2080 to 2040 cm^{-1} as a shoulder or partly resolved from the high-frequency component. This is due to the adsorbed CO always present as a contamination of the samples [28,29,33]. A typical case is shown on Fig. 4. The Pt-H band is of asymmetrical shape and can be deconvoluted into (at least) three components. This refers to surface heterogeneity (i.e., the Pt-H vibrational frequency, and consequently the bond strength is influenced by the environment of the Pt atom to which the hydrogen is coordinated). A stronger bond is obtained if hydrogen interacts with Pt atoms of low coordination number (e.g., situated at the corners or edges of crystallites or at steps on high-index faces). Obviously, the adsorption site cannot be determined exactly from such measurements. To resolve this problem, reflection experiments on well-defined surfaces would be necessary.

The pressure dependence of the Pt-H band on Pt/SiO_2 is shown in Fig. 5. The band area (as a measure of the surface concentration) is a linear function of the logarithm of the H_2 pressure (i.e., surface coverage is described by a Temkin isotherm). The same behavior was found by chemisorption measurement on Pt black [43]. The relationship between the overall hydrogen coverage (θ_{tot}) and that of weakly bonded hydrogen (θ_w) is expressed as

$$\theta_{tot} = 1 + \theta_w$$

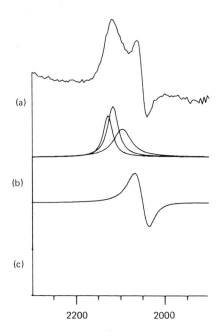

FIGURE 4 Spectrum of hydrogen adsorbed on Pt/SiO$_2$. p_{H_2} = 100 Pa. (a) Original difference spectrum (hydrogen-covered sample–hydrogen-free sample); (b) Lorentz components of the hydrogen band; (c) band shape of adsorbed CO (θ_{CO} < 0.05). (From Ref. 33.)

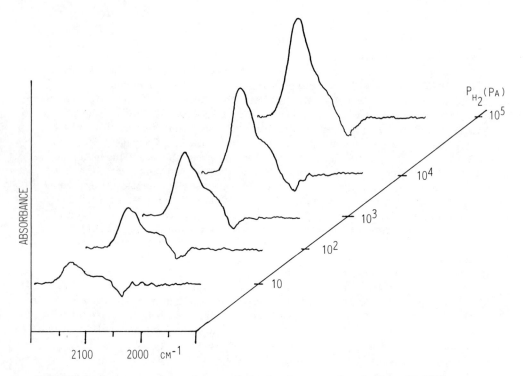

FIGURE 5 Pressure dependence of the hydrogen adsorbed on Pt/SiO$_2$. (From Ref. 33.)

7. Infrared Spectroscopy of Adsorbed Hydrogen

and $\theta_W = 0.35$ at $p_{H_2} = 10^5$ Pa at 300 K [43]. The slope of the isotherm was found not to change up to 10^5 Pa (i.e., no saturation occurred up to that pressure), but the coverage decreased drastically with increasing temperature and no Pt-H band was observed above 400 K up to atmospheric pressure [33]. Similar results, but with a dissociative Langmuir isotherm, were reported in another IR study on Pt/Al_2O_3 [30]. The differences in the coverage values and the shape of isotherm measured on Pt black, Pt/SiO_2, and Pt/Al_2O_3 may be due to support effects (e.g., surface composition changes caused by different crystallite size and shape, spillover of adsorbed H, etc.). On the other hand, the Pt-H band frequency did not change by increasing the coverage [30,33]. Further, the calculation of integrated band intensity from band areas measured on Pt/SiO_2 [33] with the use of coverage data on Pt black [43] gave results that were in good agreement with another report on Pt/Al_2O_3 [28]. This shows that the Pt-H band characteristics are not affected by the coverage (in contrast to, for example, the CO adsorption) and are essentially the same on all Pt surfaces.

The initial heat of adsorption for weakly bound H—as reported in the studies discussed above—is between 50 and 75 kJ mol^{-1} [30,33,43]. These values are comparable to those reported for strongly bound (multibonded) hydrogen at higher coverages (see Chapter 1). Hence neither form of adsorption is favored energetically; that is, it can be supposed that on-top adsorption also occurs before the multibonded monolayer is completed, but that on-top H atoms move immediately to multicoordinated positions (until such empty sites are available) in order to minimize surface entropy.

C. Interaction Between Hydrogen and CO on Pt Surface

The CO band in Fig. 4C has a derivative shape. This is a typical feature of difference spectra and is characteristic of a band shifted to higher frequencies. Hence the carbon-oxygen stretching frequency is increased by the influence of hydrogen adsorption [29,30]. However, the CO band is already shifted when the surface is saturated by multicoordinated hydrogen (i.e., prior to the appearance of the weakly bound hydrogen), and if the CO coverage is low, its frequency is not altered by a further increase in H_2 pressure [33]. Consequently, interaction occurs between strongly bound hydrogen and CO bonded to the same Pt atom (Fig. 6, interaction I). The relatively small upward shift of CO band (24 cm^{-1}, [34]) implies a weakly polar bond for the multibonded hydrogen. This is in accordance with other results (e.g., work function measurements; see Chapter 1).

H atoms of strong electron-acceptor character (i.e., hydride) would strongly decrease the back donation to the CO antibonding π orbital and it would cause a larger blue shift of the vibrational frequency. On the contrary,

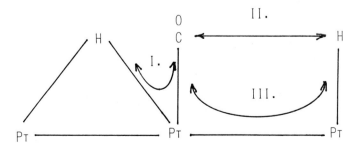

FIGURE 6 H-CO interactions in the adsorbed layer.

electron-donor H atoms would decrease the CO frequency as a result of the stronger back donation from the metal. However, at high CO coverages a red shift of the CO band was observed in the presence of weakly bound H [26], indicating that other effects became dominating. The decrease in CO frequency in such cases can be explained [11] by nonbonding interactions (dipole coupling or overlap of molecular orbitals between adjacent molecules) or by long-range forces through the metal (supposing a delocalized bond for chemisorbed molecules); see interactions II and III in Fig. 6, respectively.

IV. CONCLUSIONS

Study of surface hydrogen is a complex matter, so much so that some surface-probing experimental methods are unable to detect adsorbed hydrogen. Thus the contribution of IR spectroscopy to the determination of structural properties of surface hydrogen is very important, particularly in the case of the weakly bound on-top hydrogen which appears at pressures too high to be studied by electron spectroscopic methods.

While weak hydrogen chemisorption on Pt catalyst has been investigated on several occasions and conclusions seem to be well established, no detailed studies on other metals are available. Since the measurements are always performed on CO contaminated surfaces (and adsorbed CO has a very strong band near that reported for adsorbed hydrogen), one might venture to remark that even the existence of such an adsorbed form on other metals is doubtful. Obviously, more experimental work is necessary to clarify the situation. Measurements on well-defined surfaces would also be very useful.

REFERENCES

1. (a) L. H. Little, *Infrared Spectra of Adsorbed Species*, Academic Press, New York (1966); (b) M. L. Hair, *Infrared Spectroscopy in Surface Chemistry*, Marcel Dekker, New York (1967); (c) A. V. Kiselev and V. I. Lygin, *Infrared Spectra of Surface Compounds*, Wiley, New York (1975).
2. (a) R. J. Bell, *Introductory Fourier Transform Spectroscopy*, Academic Press, New York (1972); (b) P. R. Griffiths, in *Chemical Infrared Fourier Transform Spectroscopy* (P. R. Griffiths, ed.), Wiley-Interscience, New York (1975); (c) J. Chamberlain, *The Principles of Interferometric Spectroscopy*, Wiley, New York (1979).
3. S. A. Francis and A. H. Ellison, *J. Opt. Soc. Am.*, 49:131 (1959).
4. R. G. Greenler, *J. Chem. Phys.*, 44:310 (1966).
5. H. A. Pearce and N. Sheppard, *Surf. Sci.*, 59:205 (1976).
6. W. Jaisli, H. Kuhlman, and G. Schultz-Ekloff, *Surf. Sci.*, 112:L797 (1981).
7. R. G. Greenler, D. R. Snider, D. Witt, and R. S. Sorbello, *Surf. Sci.*, 118:415 (1982).
8. R. P. Messmer, S. K. Knudsen, K. H. Johnson, J. B. Diamond, and C. Y. Young, *Phys. Rev.*, B13:1396 (1976).
9. P. N. Ross, K. Kinoshita, and P. Stonehart, *J. Catal.*, 32:163 (1974).
10. F. M. Hoffmann, *Surf. Sci. Rep.*, 3:107 (1983).
11. R. F. Willis, A. A. Lucas, and G. D. Mahan, in *The Chemical Physics of Solid Surfaces and Heterogeneous Catalysis*, Vol. 2 (D. A. King and D. P. Woodruff, eds.), Elsevier, Amsterdam (1982), p. 59.
12. P. R. Griffiths, C. T. Foskett, and R. Curbelo, *Appl. Spectrosc. Rev.*, 6:31 (1972).
13. P. V. Huong, *Adv. Infrared Raman Spectrosc.*, 4:85 (1978).

14. J. B. Bates, *FTIR Spectrosc.*, 1:99 (1978).
15. M. J. Dignam, in *Vibrations at Surfaces* (R. Caudano, J. M. Gilles, and A. A. Lucas, eds.), Plenum Press, New York (1982), p. 265.
16. R. J. Bell, R. W. Alexander, and C. A. Ward, in *Vibrational Spectroscopies for Adsorbed Species* (A. T. Bell and M. L. Hair, eds.), *ACS Symposium Series 137*, American Chemical Society, Washington, D.C. (1980), p. 99.
17. I. M. Hammadeh, D. King, and P. R. Griffiths, *J. Catal.*, 88:264 (1984).
18. M. Folman and Y. Kozirovski, *J. Colloid Interface Sci.*, 38:51 (1972).
19. H. Förster and M. Schuldt, *J. Chem. Phys.*, 66:5237 (1977).
20. N. Sheppard and D. J. C. Yates, *Proc. R. Soc. London, A* 238:69 (1956).
21. Z. Paál and P. G. Menon, *Catal. Rev.-Sci. Eng.*, 25:229 (1983).
22. K. Nakamoto, *Infrared and Raman Spectra of Inorganic and Coordination Compounds*, Wiley, New York (1978).
23. D. M. Adams, *Metal-Ligand and Related Vibrations*, E. J. Arnold, London (1967).
24. H. D. Kaesz and R. B. Saillant, *Chem. Rev.*, 72:231 (1972).
25. U. A. Jayasooriya, M. A. Chesters, M. W. Howard, S. F. A. Kettle, D. B. Powell, and N. Sheppard, *Surf. Sci.*, 93:526 (1980) and references therein.
26. W. A. Pliskin and R. P. Eischens, *Z. Phys. Chem. (Neue Folge)*, 24:11 (1960).
27. D. D. Eley, D. M. Moran, and C. H. Rochester, *Trans. Faraday Soc.*, 64:2168 (1968).
28. D. J. Darensbourg and R. P. Eischens, *Proceedings of the 5th International Congress on Catalysis*, paper 21 (1972).
29. M. Primet, J. M. Basset, M. V. Mathieu, and M. Prettre, *J. Catal.*, 28:368 (1973).
30. M. Primet, J. M. Basset, and M. V. Mathieu, *J. Chem. Soc. Faraday Trans. 1*, 70:293 (1974).
31. J. P. Candy, P. Fouilloux, and M. Primet, *Surf. Sci.*, 72:167 (1978).
32. L. T. Dixon, R. Barth, and T. W. Gryder, *J. Catal.*, 37:368, 376 (1975).
33. T. Szilágyi, to be published.
34. G. E. Becker and G. W. Gobeli, *J. Chem. Phys.*, 38:2942 (1963).
35. C. Morterra and M. J. D. Low, *J. Phys. Chem.*, 73:321, 327 (1969); *Chem. Commun.*, 1491 (1968).
36. I. Ratajczikowa, *Surf. Sci.*, 48:549 (1975).
37. Y. J. Chabal, *Phys. Rev. Lett.*, 55:845 (1985).
38. P. Mahaffy and M. J. Dignam, *J. Phys. Chem.*, 84:2683 (1980).
39. J. F. Harrod, R. W. Roberts, and E. F. Rissmann, *J. Phys. Chem.*, 71:343 (1967).
40. (a) R. Bouwman and I. L. C. Freriks, *Appl. Surf. Sci.*, 4:21 (1980); (b) C. S. Kellner and A. T. Bell, *J. Catal.*, 75:251 (1982); (c) A. Palazov, G. Kadinov, Ch. Bonev, and S. Shopov, *J. Catal.*, 74:44 (1982); (d) M. M. Miura, M. L. McLaughlin, and R. D. Gonzalez, *J. Catal.*, 79:227 (1983).
41. R. R. Cavanagh and J. T. Yates, Jr., *J. Catal.*, 68:22 (1981).
42. T. Szilágyi, A. Sárkány, J. Mink, and P. Tétényi, *J. Catal.*, 66:191 (1980); *J. Mol. Struct.*, 60:437 (1980).
43. J. D. Clewley, J. F. Lynch, and T. B. Flanagan, *J. Catal.*, 36:291 (1975).
44. G. K. Teal and G. E. McWood, *J. Chem. Phys.*, 3:700 (1955).

8 | Effects of Adsorption of Hydrogen on the Magnetic and Electrical Properties of Metals

J. W. GEUS

University of Utrecht, Utrecht, The Netherlands

I.	INTRODUCTION	195
II.	MAGNETIC PROPERTIES OF BULK METAL HYDRIDES	198
III.	EFFECTS OF HYDROGEN ADSORPTION ON THE FERROMAGNETISM OF METALS	200
	A. Saturation Magnetization of Nickel	200
	B. Transport of Adsorbing Gas Molecules Through Porous Supported Catalysts	206
	C. Saturation Magnetization of Iron	207
	D. Ferromagnetic Anisotropy	210
	E. Conclusions	213
IV.	EFFECTS OF HYDROGEN ADSORPTION ON THE ELECTRICAL CONDUCTANCE OF METALS	213
V.	CONCLUSIONS	221
	REFERENCES	221

I. INTRODUCTION

When hydrogen is adsorbed onto metal surfaces, two extreme possibilities can be envisioned (Fig. 1): (i) the structure of the metal surface is not affected at all by the chemical bonding of the adsorbed hydrogen atoms, and (ii) the surface of the metal reacts to a structure analogous to that of the corresponding bulk hydride of the metal. The actual structure of a metal surface covered by adsorbed hydrogen atoms will be in between the extremes noted above. It is important to assess more precisely the position in between with different metals and different crystallographic surfaces of the same metal. When chemisorption of hydrogen calls for an appreciable modification of the structure of the metal surface, the adsorption is likely to be activated. Besides dissociation of the hydrogen molecule, modification of the structure of the metal surface is required, an example being the adsorption of hydrogen by alkali metals.

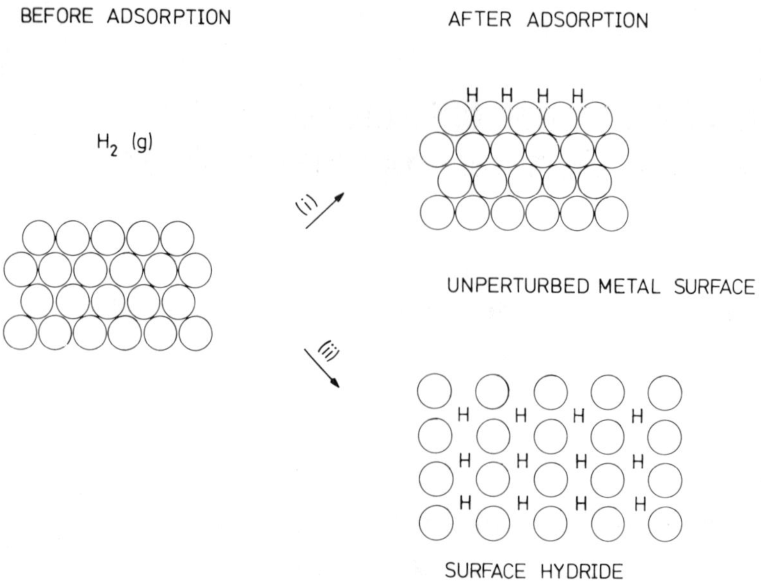

FIGURE 1 Adsorption of hydrogen onto a metal surface may lead to the extremes indicated: a metal structure completely unperturbed and a metal surface hydride of a structure analogous to that of a bulk metal hydride.

When hydrogen is dissociatively adsorbed without any effect on the metal surface structure, chemisorption can be expected to proceed more rapidly. Now it is possible that although hydrogen is dissociatively adsorbed onto the unperturbed metal surface, the hydrogen adatoms are bonded more strongly on or in a modified surface structure. In Chapter 1, Christmann discusses experimental evidence pointing to hydrogen being bonded more strongly at sites where the coordination with metal atoms is higher. It is likely that the transition from the state where the hydrogen adatoms are bonded to the unperturbed metal surface to that where they have been taken up into the modified metal surface proceeds smoothly. Then the rate of adsorption of hydrogen will still be high. However, the rate of desorption and the reactivity of the adsorbed hydrogen atoms now may be lower, since a (local) rearrangement of the metal surface may be necessary to release the hydrogen atoms taken up.

The possibility of accommodating hydrogen atoms into a modified metal surface is also important with competitive adsorption of, say, carbon monoxide and hydrogen. If adsorption of hydrogen cannot bring about a modification of the surface structure, exposure of the hydrogen-covered surface to gas molecules being more strongly adsorbed than hydrogen just leads to desorption of hydrogen. An example is platinum, where admission of carbon monoxide causes complete desorption of adsorbed hydrogen. If, however, hydrogen can penetrate the metal surface, exposure to carbon monoxide can be accompanied by penetration of hydrogen into the metal surface. There are indications that admission of carbon monoxide to a nickel surface covered by hydrogen leads to penetration of hydrogen into the nickel surface.

The usual techniques to study adsorption on monocrystalline metal surfaces pretreated and cleaned in ultrahigh vacuum are not very well suited to studying the interaction of metal surfaces with hydrogen. Auger electron spectroscopy cannot be used to assess the (extent of) adsorption of hydrogen, and LEED patterns different from that of the clean metal surface are often not exhibited by hydrogen-covered metal surfaces. Hence measurement of

8. Magnetic and Electrical Effects of Adsorbed Hydrogen

the effect of hydrogen adsorption on the work function and temperature-programmed desorption of hydrogen are used to investigate the adsorption of hydrogen. From the experimental results of these techniques, an effect on the metal surface structure is difficult to establish.

The effects of adsorption of hydrogen on the metallic properties can provide information about the surface structure of the metal being covered by adsorbed hydrogen. The metallic properties that can be studied most readily are the electrical conductance and the magnetic properties. Especially with supported metals, optical properties and microwave damping may also be used successfully.

When hydrogen adsorption leads to a modification of the surface structure of the metal, the hydrogen-covered surface is likely to display a structure resembling that of the corresponding bulk hydride. The effect on its structure of the reaction of a metal to a bulk hydride can be predicted from the electronegativities of hydrogen and the metal [1]. Metals whose electronegativity is appreciably lower than that of hydrogen react to form ionic solid hydrides. When an ionic solid is formed, the chemical bond between the metal atoms is completely changed. A smaller difference in electronegativity leads to the formation of metallic hydrides. The intermetallic bonds are much less affected in the formation of metallic hydrides than in that of ionic hydrides. The effects on the intermetallic bonds of reaction to bulk hydrides is evident from changes in physical properties. It is therefore important to review the differences in the physical properties brought about by reaction of the metal to the bulk hydride.

The magnetic properties of metal hydrides can provide information about the nature of the chemical bond in the hydrides. Especially with metallic hydrides, where the effects on the metal structure are relatively small, measurement of the magnetic properties can provide useful information. The effect on the paramagnetism of metals of the reaction to bulk hydrides has been interpreted along two lines. In most earlier papers a rigid band model was used, in which electrons from the hydrogen are assumed to fill d levels of the metal [2]. The resulting spin pairing leads to a drop in the paramagnetism. In more recent publications the difference between the magnetic properties of isolated metal atoms and a bulk metal is often used to rationalize the magnetic properties of bulk hydrides. Reaction to the hydride leads to a lower interaction between the metal atoms, and thus to the metal atoms approaching more closely the behavior of isolated metal atoms [3].

Ionic solid hydrides exhibit a low electrical conductivity. Examples include the alkali metals that react with hydrogen to saltlike hydrides. A smaller difference in electronegativity leads to the formation of metallic hydrides displaying a metallic conductivity, which is much higher and generally decreases at increasing temperatures. Although formation of an ionic or a metallic hydride can easily be assessed from measurement of the electrical conductance, the interpretation of the magnitude and temperature dependence of the electrical conductance of metallic hydrides is much more difficult.

Analogous to bulk hydrides the effect of hydrogen adsorption on the electrical conductance and magnetic properties of metals can provide information about the nature of the chemisorptive bond. To get effects that can be measured sufficiently accurately, the surface-to-volume ratio of the metal specimens must be appreciable. For the electrical conductance a thin metal film is generally required to assess reliably the effects of adsorption. The electrical conductance of small metal particles applied on a highly porous support is much more difficult to measure with the accuracy required. With magnetic measurements electrical contacts are not necessary, and thus the effects of hydrogen adsorption on the magnetic properties of supported metal particles can be measured relatively readily. Since the number of experimental methods providing information about the chemisorptive bond with small supported metal particles as utilized in technical catalysts is small, magnetic measurements on supported catalysts are especially interesting.

The effects of hydrogen adsorption on the electrical and magnetic properties of metals can also demonstrate the distribution of gas molecules over the surface of (porous) metal specimens. Especially with supported metal catalysts, the ability to provide information about the transport of hydrogen into the system can be important.

We will first review the evidence published on the effect of hydrogen adsorption on the magnetic properties of metals and subsequently that on the electrical conductance of vapor-deposited metal films. It will be evident that the results on the effect on the magnetic properties and the electrical conductance agree very well. The effect of hydrogen adsorption on the metal structure will turn out to depend strongly on the surface structure of the metal, which can for small supported metal particles be significantly different from that of vapor-deposited metal films.

To account for the effects of hydrogen adsorption on the magnetic properties of metals, the differences in magnetic properties of the metal and the bulk metal hydrides can be used as a guideline, as stated earlier. We therefore will first consider the properties of bulk hydrides and next deal with the effects of hydrogen adsorption on the magnetic properties of metals.

II. MAGNETIC PROPERTIES OF BULK METAL HYDRIDES

A system of noninteracting atoms having magnetic moments due to either spin or orbital moments exhibits paramagnetism [4-6], which implies that the overall magnetic moment of the system is determined by the strength of the magnetic field, on the one hand, and the thermal energy, on the other. Whereas the magnetic field seeks to orient the atomic magnetic moments parallel to the direction of the field, the thermal energy disorders the atomic magnetic moments. As a result the magnetic moment of the system is a function of $\mu H/kT$, where μ is the atomic magnetic moment, H the magnetic field strength, and kT the thermal energy.

With paramagnetic bulk metals, interaction between the individual atomic moments is usually considerable. Consequently, the resulting moment per atom is different from that of the isolated metal atoms. Examples are palladium, which is diamagnetic as individual noninteracting atoms and strongly paramagnetic as a metal; platinum, which is strongly paramagnetic as an atom and weakly paramagnetic in the metallic state; and copper and silver, which are paramagnetic as atoms and diamagnetic in the metallic state. The conduction electrons of metals exhibit a weak paramagnetism, which does not depend strongly on the temperature. This is a result of the Fermi-Dirac distribution, owing to which a small fraction of the conduction electrons, slowly rising with the temperature, can orient their magnetic moments along an external magnetic field (Pauli paramagnetism) [7,8].

Since the magnetic properties of metals are determined by both the atomic magnetic moments and the interaction between the metal atoms, an effect of reaction to the corresponding solid hydride can reflect a change in the atomic magnetic moments or in the interaction between the metal atoms. Reaction of palladium to diamagnetic palladium hydride was initially attributed to the filling of empty d levels of the palladium atoms by electrons of the hydrogen atoms [9]. Later, Gibb ascribed the decrease in the paramagnetism to a drop in the interaction between the metal atoms due to the insertion of hydrogen atoms [3]. He correlated the effect of the reaction to the bulk hydride on the paramagnetism for a number of metals, with the effect of expansion of the lattice of the pure metal at increasing temperatures and the magnetic properties of the individual metal atoms. Accordingly, the magnetic susceptibility of palladium strongly drops at both increasing temperature and increasing hydrogen content, and that of vanadium more weakly, whereas that of chromium increases in both cases. As further evidence for the decrease in interaction between metal atoms, Gibb mentions the observation of Michel and Gallisot [10] that chemically removing the hydrogen from palladium hydride at room temperature leads to a still-diamagnetic metal. Later he confirmed this observation experimentally [11].

With ferromagnetic metals an electrostatic interaction between the electrons on neighboring metal atoms causes the atomic moments to be oriented in parallel. Formation of Weiss domains [12] separated by Bloch walls leads to a decrease in the magnetostatic energy of large ferromagnetic metal specimens. A sufficiently large drop in the interaction can destroy the ferromagnetic coupling and lead to formation of a nonferromagnetic hydride exhibiting much lower magnetization.

The solubility of hydrogen is higher for face-centered cubic (fcc) metals than for body-centered cubic (bcc) metals. Nickel thus dissolves more hydrogen than iron, while the solubility of hydrogen in iron abruptly increases when iron is kept at higher temperatures, where it assumes the fcc structure [13]. As a result, nickel hydride can be prepared and measured much more easily than can iron hydride. Generally, nickel hydride is produced by electrochemically generating hydrogen atoms on a nickel cathode the surface of which is poisoned for the recombination of hydrogen atoms to molecular hydrogen atoms by sulfur (e.g., from thiourea) [14,15]. Baranowski and co-workers also produced nickel hydride by raising the hydrogen pressure up to about 3700 bar [16,17].

A considerable number of authors have published work on the magnetic properties of nickel hydride. The stoichiometry of nickel hydride is given as $NiH_{0.5}$ to $NiH_{0.7}$ [16,18]. Reaction to nickel hydride strongly decreases the ferromagnetism of the metal [18-20]; complete conversion to the hydride leads to a nonferromagnetic compound [21]. Baranowski and Bochenska determined the thermodynamic stability of nickel hydride [16], and Bauer et al. compared the formation and decomposition of NiH and NiD [22]. Ebisuzaki et al. investigated isotopic effects on the diffusion and solubility [23]. Cable et al. established by neutron diffraction the position of the hydrogen atom in nickel hydride [24]. In agreement with x-ray diffraction results they found a lattice constant of 0.372 nm, which is substantially higher than that of pure nickel, which is 0.352 nm. Accordingly, the lattice expands by 5.5% in the formation of nickel hydride. The hydrogen atoms take up octahedral positions leading to a NaCl structure.

It has been observed that the drop in magnetization does not vary when the temperature is decreased to 77 K [21]. This demonstrates that the loss of ferromagnetism is not due to the ferromagnetic coupling merely being destroyed. If the nickel atoms retain their magnetic moment, losing only the ferromagnetic coupling, the drop in magnetization must be smaller at lower temperatures. Faessler and Schmidt [25] measuring the K x-ray absorption edge of nickel and nickel hydride, interpreted their results to indicate that electrons of the absorbed hydrogen fill the 3d and 4s levels of the nickel atoms. However, theoretical calculations have shown that the energy differences of the $3d^84s^2$, $3d^94s$, and $3d^{10}$ levels do not differ appreciably. Accommodation of hydrogen in between the nickel atoms therefore may slightly shift the stability of the different configurations and may cause the diamagnetic $3d^{10}$ configuration to be the most stable. Hence Gibb's interpretation [3] of the decrease in interaction between the individual metal atoms by the takeup of hydrogen can still be maintained.

As mentioned above, the stability of iron hydride is much smaller than that of nickel hydride. Consequently, the magnetic properties of bulk iron hydride cannot be assessed. However, Mössbauer experiments on Fe-Ni alloys have provided information about the effects on the magnetic properties of iron to be expected. Wertheim and Buchanan [26], investigating the Mössbauer spectrum of a Ni-Fe alloy before and after saturation with hydrogen, observed that the nickel atoms lose their ferromagnetic moment, whereas the magnetic moments of the iron atoms are raised by the takeup of hydrogen. A slight increase in the magnetic moment of iron might thus be expected on reaction to iron hydride. Jech and Abeledo [27] carried out Mössbauer measurements on Fe-Pd alloys before and after absorption of hydrogen. Although dissolution of hydrogen decreased the Curie temperature, the hyperfine field around the iron atoms was not affected.

III. EFFECTS OF HYDROGEN ADSORPTION ON THE FERROMAGNETISM OF METALS

A. Saturation Magnetization of Nickel

The experiments on bulk hydrides have shown that for the adsorption of hydrogen on nickel a decrease in ferromagnetism can be expected, provided that the hydrogen is adsorbed into the nickel surface analogously to the absorption of hydrogen to form bulk nickel hydride. With iron no effect, or a small increase, is predicted from the results noted above. Besides changing the saturation magnetization, adsorption of hydrogen can also affect the ferromagnetic anisotropy energy. This energy causes the energy of a ferromagnetic particle to vary with the orientation of the magnetization with respect to the particle.

We will first consider the effect of hydrogen adsorption on the saturation magnetization of nickel. Measurements have been performed on small supported nickel particles and on vapor-deposited metal films. The effect of hydrogen adsorption on the saturation magnetization of small supported metal particles has been thoroughly investigated by different groups. As a result an extensive body of experimental evidence is available; generally, the data agree very well. Vapor-deposited metal films have been investigated much less thoroughly. Only three sets of measurements have been published.

Neugebauer [28-32] determined the saturation magnetization of nickel films deposited in ultrahigh vacuum by measuring the torque exerted by an external magnetic field. He established very thin (mean thickness 1.4 to 1.8 nm) nickel films to have a granular structure and to be superparamagnetic. In these films the nickel atoms thrown onto glass substrates have coalesced to small isolated nickel particles. The ferromagnetic anisotropy energy of these small particles is low compared to the thermal energy at, for example, 300 K. As a result, the magnetic moment μ_p of the nickel particles performs a Brownian-like rotation. The mean component of the magnetic moment in the direction of an external magnetic field is governed by the orientation energy $\mu_p H$ and the thermal energy kT. In contrast to paramagnetic systems, however, the magnetic moment μ_p is not the magnetic moment of one atom, but the combined moments of all the ferromagnetic atoms present in a nickel particle. Since the number of nickel atoms in a nickel particle can easily be of the order of 10^4, the apparent magnetic moments are very large, which leads to the designation "superparamagnetic" [33]. Nickel films of a higher mean thickness are continuous and display a magnetization analogous to that of bulk nickel.

Neugebauer measured the saturation magnetization of the nickel films after exposure to oxygen and hydrogen. Interaction at room temperature with oxygen (pressure 10^{-3} torr) led to a decrease in saturation magnetization, corresponding to a nickel layer of 0.8 to 1.5 nm becoming nonferromagnetic. Thicker nickel films, the roughness factor of which can be expected to be higher, display the loss of the ferromagnetism of a thicker layer. Exposures of nickel films of a mean thickness above about 3.0 nm at room temperature to high-purity hydrogen did not affect the saturation magnetization. Although the effect of oxygen on the saturation magnetic moment had shown that the accuracy of the measurement was sufficiently high to establish the loss of the ferromagnetism of even a fraction of a monolayer of nickel atoms, no drop in the saturation magnetic moment was observed on interaction with hydrogen. However, the saturation magnetic moment of films of a mean thickness below 3.0 nm did decrease on exposure to hydrogen. Evidently, the effect of the adsorption of hydrogen on the saturation magnetization of nickel depends on the particle size of the nickel. In the surface of thicker nickel films not being annealed for prolonged periods at high temperatures, atomically flat surfaces can be expected to be present, as will also be evident from the electrical conductance of vapor-deposited nickel films. The magnetic moments of the metal atoms in these atomically flat surfaces will not be affected by the adsorption of hydrogen. On the other hand, the ferromagnetic moment of the nickel atoms in the atomically more rough planes present in the surface of small nickel particles disappears through the adsorption of hydrogen atoms.

Geus and Koks [34] measured the ferromagnetic thickness of thin nickel and iron films by means of the Faraday effect, the rotation of the plane of polarization of light passing through a magnetized medium. These films had also been thrown on glass substrates in ultrahigh vacuum. Measuring films of a mean thickness of 10 nm or more, these authors found that the saturation magnetization was not affected by hydrogen adsorption. Also in this work, oxygen adsorption brought about a clearly measurable drop in the ferromagnetic thickness. If even 0.1 of a monolayer of nickel or iron atoms had become nonferromagnetic, the effect must have been measurable.

Blum and Göpel [35, 36] and Göpel [37] investigated extensively the ferromagnetism of vapor-deposited nickel and iron films by ferromagnetic resonance. They also studied the effect of exposure to hydrogen, carbon monoxide, and oxygen. The most extensive results are described by Göpel [37], while the experimental details are reported by Blum and Göpel [36]. With oxygen, the expected decrease in the ferromagnetism was observed. Reaction with oxygen leads to a surface layer of nickel oxide or iron oxide. The magnetic ions in this layer are no longer ferromagnetically coupled to the main magnetic moment of the nickel or iron. The thermal disordering causes the average magnetic moment to be very small at all but extremely low temperatures. The effect of exposure to hydrogen and carbon monoxide was measured for nickel films of 1.3, 4.0, and 10.0 nm thickness. The coverage caused by different exposures to hydrogen and carbon monoxide was determined by measuring the rise in pressure during flash desorption of the film at temperatures up to 500 K. To avoid changes in the film properties, the films were previously annealed at an elevated temperature. As reported by Neugebauer for nickel films, annealing vapor-deposited metal films at high temperatures leads to formation of a granular film structure. Grain boundary grooving leads to break up of the originally continuous film into small particles, as has been shown, for example, by Sundquist [38]. This author observed the formation of metal particles with rounded edges and corners after annealing. Hydrogen atoms can also penetrate the closely packed metal surfaces, since the small dimensions of the smooth planes allow the atoms present at the edges to shift sideward (Fig. 2).

The metal films investigated by Göpel are thus granular and contain metal particles that can take up hydrogen atoms in their surface. The number of

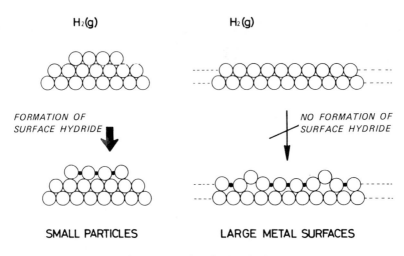

FIGURE 2 Since the metal atoms in a small plane present in the surface of a metal particle can readily shift sideward, hydrogen atoms can penetrate between the metal surface atoms. The resulting structure is that of a surface hydride. With more extended planes a sideward shift of the metal atoms leads to some metal atoms being lifted from the surface over a small distance, which is energetically very unfavorable.

nickel atoms rendered not ferromagnetic by adsorption of one carbon monoxide molecule is about 2.2, which agrees reasonably with the value of 1.85 measured for small supported nickel particles by Primet et al. [39]. The number of nickel atoms losing their ferromagnetism per hydrogen atom taken up is remarkably high at low coverages, dropping from 2.6 to 1.1 when the coverage is raised from 0.02 to 0.18. At coverages of 0.18 to 0.34, a number of 1.1 nickel atoms per adsorbed hydrogen atom lost their ferromagnetism.

These results show that adsorption of hydrogen onto closely packed atomic planes of nickel and iron does not cause a decrease in the saturation magnetization. In view of the stability of bulk nickel hydride, this is not unexpected. The lattice expansion of about 5.5% renders formation of the bulk hydride energetically rather unfavorable. Consequently, bulk nickel hydride is formed only at very high hydrogen pressures, which can be applied directly or indirectly by electrolysis of water in the presence of poisoning compounds. Although lattice expansion with closely packed surfaces can be expected to be less unfavorable, formation of a surface hydride is still not possible at low to moderate hydrogen pressures.

Penetration into less closely packed surfaces evidently proceeds more readily. Hence exposure to hydrogen leads to the formation of a surface hydride in which the nickel atoms are no longer ferromagnetic, as is evident from bulk nickel hydride not being ferromagnetic. The stoichiometry of bulk nickel hydride is $NiH_{0.5}$ to $NiH_{0.7}$. It is interesting to investigate the stoichiometry of nickel surface hydride, which can be formed at much lower hydrogen pressures than can bulk nickel hydride. The stoichiometry of nickel surface hydride can be inferred from the extensive experimental evidence on the decrease in ferromagnetism brought about by the adsorption of hydrogen onto small supported nickel particles.

According to the interpretation above, the effect of hydrogen on the saturation magnetization of supported nickel particles can be utilized to assess the surface structure of the particles. The presence of atomically rough surfaces is evident from the formation of a surface hydride that is not ferromagnetic on interaction with hydrogen. That penetration of hydrogen atoms in between metal surface atoms can proceed provided that the structure of the metal surface is sufficiently open can be inferred from LEED patterns of nickel surfaces. The effect of hydrogen adsorption on the LEED pattern of the atomically rough nickel (110) surface indicates an effect on the position of the metal surface atoms [40,41]. Also, the effect of hydrogen on the field-emission pattern of tungsten tips can be interpreted as the penetration of hydrogen in between metal atoms present at edges [42].

Dietz and Selwood [43,44] accurately measured the effect of hydrogen adsorption on the saturation magnetization of silica-supported nickel particles. They observed a decrease of $0.7\mu_B$ per adsorbed hydrogen atom (μ_B is a Bohr magneton). The effect of hydrogen adsorption on the saturation magnetization does not depend on the temperature at temperatures from about 4 to 300 K. Later, Reinen and Selwood [45] investigated both nickel-on-silica and nickel-on-alumina catalysts. For a coprecipitated nickel-on-silica catalyst they also found a decrease of $0.7\mu_B$ per adsorbed hydrogen atom, while the decrease measured with an impregnated nickel-on-silica catalyst was slightly lower, 0.5 to $0.6\mu_B$ per adsorbed hydrogen atom.

Martin et al. [46] also investigated the effect of hydrogen adsorption on the saturation magnetization of silica-supported nickel catalysts. A nickel catalyst prepared by reduction of nickel antigorite, a basic nickel silicate $Ni_3(OH)_4Si_2O_5$, exhibited a decrease in ferromagnetism, rising from $0.7\mu_B$ at 4 K to $0.86\mu_B$ at 300 K. An impregnated nickel-on-silica catalyst displayed slightly lower values, $0.64\mu_B$ at 4 K to $0.7\mu_B$ at 300 K. Later for the same impregnated catalyst, Martin and Imelik [47] reported a decrease of $0.7\mu_B$ being the same from about 178 K to about 733 K. Such consistent values of 0.6 to $0.7\mu_B$ per adsorbed hydrogen atom lead to stoichiometries of $NiH_{1.0}$ to $NiH_{0.87}$ of the surface hydride.

More important, however, the surface of silica-supported nickel particles evidently is able to react completely to a surface nickel hydride, which is by no means obvious. That the ability of the nickel surface to react to a surface hydride can also vary with supported metal particles is evident from results obtained with other supports. In addition to nickel-on-silica catalysts, Reinen and Selwood [45] investigated nickel-on-alumina catalysts prepared by impregnation. After treatment with hydrogen at 873 K for 5 h they observed the usual decrease in ferromagnetism of 0.6 to $0.7\mu_B$ per adsorbed hydrogen atom. The impregnated nickel-on-alumina catalyst reduced at 693 K exhibited a low degree of reduction and a decrease in ferromagnetism of 0.3 to $0.4\mu_B$ per adsorbed hydrogen atom. Martin et al. [48] investigated a range of supports: silica, silica-alumina, titania, and magnesia. They established the fact that titania-supported catalysts are relatively easy to reduce, as are nickel-on-silica catalysts. Nickel-on-alumina and nickel-on-magnesia catalysts were found to be more difficult to reduce. The degree of reduction appeared to be related to the effect of hydrogen adsorption on the saturation magnetization. At low degrees of reduction the drop in saturation magnetization amounted to $0.1\mu_B$ per adsorbed atom, and at high degrees of reduction to 0.6 to $0.7\mu_B$ per adsorbed hydrogen atom.

Martin et al. also observed that not completely reduced nickel catalysts, exhibiting a low decrease in ferromagnetism per adsorbed hydrogen atom, additionally showed an irreversible behavior on desorption of hydrogen. Remarkably, the saturation magnetization passed through a maximum during desorption of the hydrogen to display finally the saturation magnetization of the original catalyst. Figure 3 shows an example of the magnetic plots measured by Martin et al. [48]. The authors ascribe the maximum to the

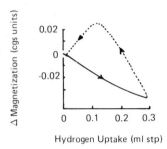

FIGURE 3 Results on nickel-on-alumina catalysts: catalyst reduced at 673 K, hydrogen adsorption measured at 298 K, hydrogen desorbed at 473 K. Whereas adsorption of hydrogen leads to a decrease in magnetization, desorption of hydrogen causes the magnetization to rise above the original value. The final magnetization is about the same as that of the freshly reduced catalyst. (From Ref. 48.)

presence of a second adsorbed state of hydrogen, raising the saturation magnetization, which is displayed only when not completely reduced nickel is present in the catalysts. The heat of adsorption of hydrogen increasing the magnetization is higher than that decreasing the magnetization. As a result, the former adsorbed hydrogen desorbs later than the latter, which causes the saturation magnetization to pass through a maximum.

That the degree of reduction also determines the effect on the saturation magnetization and not the chemical composition of the support alone can be concluded from the results of Knappwost and Schwarz [49]. These authors investigated nickel-on-magnesia catalysts prepared by decomposition of mixed magnesium-nickel oxalates. The nickel present in these catalysts is much more readily reduced than the nickel in impregnated nickel-on-magnesia catalysts. With nickel particles of a diameter of 4.0 to 7.0 nm, Knappwost observed a decrease in saturation magnetization of 0.5 to 0.6μ_B per adsorbed hydrogen atom, which dropped to 0.4 to 0.5μ_B when the mean size of the nickel particles decreased to 2.0 nm. Martin et al. [48] measured a decrease in saturation magnetization of only 0.2μ_B per adsorbed hydrogen atom for their nickel-on-magnesia catalyst.

To explain the curious maximum in the saturation magnetization, Martin et al. proposed adsorption of hydrogen into two states differing in adsorption energy and in the effect on the saturation magnetization. The experimental evidence also calls for the presence of unreduced nickel to be a prerequisite for the more strongly adsorbed state of hydrogen. The two different states of adsorbed hydrogen can easily explain the experimental results. However, the fact that the closely packed surfaces of vapor-deposited metal films have been shown unambiguously to adsorb hydrogen without any effect on the saturation magnetization leads us to propose an alternative explanation.

With all the supports noted above, nickel ions can react to compounds relatively stable with respect to reduction. In experiments on single crystals we have shown that reduction by hydrogen atoms procceeds much more rapidly than that by molecular hydrogen, which indicates that the dissociation of molecular hydrogen is the rate-determining step in the reduction. The excess of oxygen of bulk nickel oxide provides a sufficient number of sites where the dissociation of molecular hydrogen can proceed. With alumina nickel ions can react to nickel aluminate, and with magnesia to a magnesium nickel mixed oxide. These compounds do not exhibit a significant deviation from stoichiometry at low temperatures, thus do not readily dissociate hydrogen and therefore

8. Magnetic and Electrical Effects of Adsorbed Hydrogen   205

are difficult to reduce. Nickel ions can react with silica to nickel hydrosilicate, provided that water vapor is present. Consequently, nickel ions react at relatively low temperatures with silica to nickel hydrosilicate, which is more difficult to reduce than bulk nickel oxide. At high temperatures the nickel hydrosilicate decomposes and produces active nickel oxide, which is rapidly reduced before the nickel can react with the anhydrous silica. Consequently, the reactivity of nickel ions with silica to a compound difficult to reduce passes through a maximum at increasing temperatures, whereas the activity of nickel ions with alumina and magnesia to form compounds more difficult to reduce rises continuously with the temperature. The nickel-on-silica catalysts are therefore slowly but continuously reduced and nickel particles are produced that owing to the presence of an interlayer of hydrosilicate, are strongly bonded to the support. The fact that the nickel is wetting the support leads to metal particles that exhibit a rounded shape and almost exclusively atomically rough surfaces that can react with hydrogen to a surface nickel hydride.

With alumina and magnesia supports, only the most readily reducible nickel species is reduced at low temperatures. Presumably this involves exclusively nickel ions that do not interact strongly with the surface of the support. Probing the surface of silica-supported copper particles by the infrared spectra of adsorbed carbon monoxide, de Jong et al. [50] found that readily reducible copper species initially lead to copper metal crystallites exhibiting, almost exclusively, closely packed atomic planes in their surfaces. Copper metal atoms being rapidly supplied to the growing nuclei cannot be accommodated at sites leading to the equilibrium shape of the crystallite to be exhibited. The metal atoms generated on the support are taken up into sites on the metal particles leading to the growth of faceted particles (Fig. 4).

The argument runs parallel to that for the presence of faceted particles during the vapor deposition of metal atoms onto a nonmetallic substrate. Faceted particles were observed during deposition of gold atoms onto a molybdenite substrate in the electron microscope [51-55]. It has also been demonstrated that the relatively small driving force to the thermodynamic

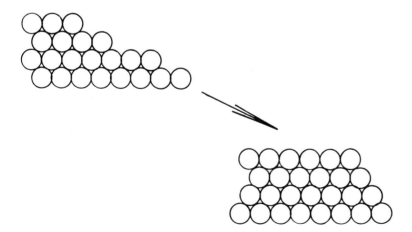

FIGURE 4 With a rapid supply of metal atoms migrating over the support, the atomically rough surfaces at the edges and corners of metal particles are rapidly filled. As illustrated, this leads to faceted particles. Nucleation of new planes, which is required to establish the equilibrium shape, usually cannot keep up with the supply of metal atoms.

equilibrium shape causes the equilibrium shape to be taken up relatively slowly. The readily reducible species on the support therefore leads to faceted metal crystallites. A small fraction of the surface of these crystallites only can react to a surface nickel hydride, the nickel atoms of which are not ferromagnetic. After reducing at more elevated temperatures for prolonged periods both the metal particles assume their equilibrium shape containing atomically rough surfaces, and the nickel ions, having reacted with the support, are being reduced. Models of nickel particles indicated that by shifting metal surface atoms parallel to the surface, penetration of hydrogen atoms in between the nickel surface atoms can proceed without expenditure of much energy (Fig. 2). Nickel-on-silica particles adhere strongly to the support. As a result, the particles have a hemispherical shape with mostly atomically rough planes in their surface. The drop in magnetization per adsorbed hydrogen atom thus does not depend strongly on the size of the particles. Hahn and Löckenhoff [56], investigating particles of 6 nm diameter, observed a decrease in magnetization per hydrogen atom of $0.72\mu_B$, which agrees with the value obtained by Dietz and Selwood for nickel particles of 12 nm diameter.

The foregoing explanation does not account for the peculiar maximum in the saturation magnetization exhibited during desorption of hydrogen. However, oxidation of nickel particles by water desorbing from the support during evacuation can proceed easily. Since low hydrogen pressure shifts the equilibrium to the other side, the surface can be reduced again during exposure to hydrogen at the higher temperatures which are required for the desorption of the hydrogen taken up. In the final stage of desorption the hydrogen pressures are again so low that reoxidation of the surface during evacuation at 723 K can proceed. Catalysts exhibiting a low degree of reduction have been treated under relatively mild conditions, where the surface of the support is likely to contain much adsorbed water. A higher degree of reduction, calling for a more severe thermal treatment, also removes the water from the surface of the support and thus lowers the extent of reoxidation at high temperatures in the absence of hydrogen. Experiments by Dutartre et al. [57] with supported iron catalysts provided unambiguous evidence for the reoxidation of metallic iron by water desorbing from the support.

Dalmon et al. [58,59] also investigated the effect of hydrogen adsorption on the saturation magnetization of nickel-copper alloys. This is interesting, since with alloys two effects may contribute to a decrease in magnetization: rendering ferromagnetic nickel atoms adsorbing hydrogen atoms nonferromagnetic, and decreasing the magnetic moments of nickel atoms not engaged in the adsorption of hydrogen by removal of neighboring ferromagnetic nickel atoms. Dalmon and coworkers measured a decrease in saturation magnetization corresponding to one nickel atom becoming nonferromagnetic per adsorbed hydrogen atom. Since the decrease in magnetization is not higher, it must be concluded that the magnetic moment of nickel atoms not adsorbing hydrogen is not affected by one or more neighboring nickel atoms becoming nonferromagnetic.

B. Transport of Adsorbing Gas Molecules Through Porous Supported Catalysts

The effect of adsorption on the magnetization of nickel catalysts measured at low magnetic field strengths has been used to establish the distribution of adsorbing gases over the nickel particles present in porous catalysts. Lee et al. [60] obtained evidence that at 77 K a substantial fraction of a dose of hydrogen admitted to a silica-supported nickel catalyst is physically adsorbed on the surface of the carrier, not on the surface of the nickel particles.

Geus et al. [61,62] found that molecular oxygen being admitted to cylindrical pellets of nickel-on-silica catalyst being filled with helium at about 0.5 torr virtually completely oxidizes the nickel particles present at the outermost layer of the cylindrical particles before penetrating the pores. In the plug flow exhibited in the pores, the nickel particles present in a small zone were virtually completely oxidized. Hydrogen was found to be adsorbed much more uniformly by the nickel particles in the catalyst. The decomposition of nitrous oxide was found to lead to a much more uniform distribution of oxygen throughout the catalyst than that resulting from exposure to doses of molecular oxygen.

In experimental work to be published, the effect of pelletizing a nickel-loaded support on the transport of oxygen and carbon monoxide has been investigated. Whereas the distribution of hydrogen was hardly affected by pelletizing the catalyst, that of oxygen and carbon monoxide became nonuniform to an extent that increased with pelletizing pressure.

Martin et al. [63] established that in a supported nickel catalyst the small nickel particles are preferentially oxidized by admission of molecular oxygen, whereas hydrogen is distributed much more evenly over the nickel particles present in the catalyst.

C. Saturation Magnetization of Iron

Although the adsorption of hydrogen on iron has been studied especially by Mössbauer spectroscopy, the effect on saturation magnetization has been investigated as well. Geus and Koks [34] and Göpel [37] found that the saturation magnetization of vapor-deposited iron films is not significantly affected by exposure to hydrogen.

Deportes et al. [64] investigated the effect of hydrogen interaction at 373 K with unsupported iron on the saturation magnetization measured at 4.2 K. They observed a strong increase in magnetization on exposure to hydrogen. Per adsorbed hydrogen molecule the increase was no less than $3.7\mu_B$, which is higher than the magnetic moment of an iron atom, which amounts to $2.2\mu_B$. Reduction by exposure to hydrogen of iron ions either remaining after the reduction or generated by oxidation through desorbing water could account for the increase in saturation magnetization, although the increase per hydrogen molecule taken up is too large. However, since the hydrogen taken up could be quantitatively recovered by thermal desorption, the authors rejected this explanation. They postulated the surface of their iron specimens to be covered by iron atoms the moments of which are not ferromagnetically coupled to the moment of the bulk. The adsorption of hydrogen brings about ferromagnetic coupling of the moments of the surface atoms. Accordingly, the magnetization increases on hydrogen adsorption. The authors assume that two iron atoms per hydrogen molecule taken up have their moments ferromagnetic coupled. They consider the experimental value of $3.7\mu_B$ to be near the expected value of $4.4\mu_B$.

Artyuk et al. [65] investigated the effects of exposure to hydrogen on the magnetization of silica-supported iron at low field strengths. At 77 K the magnetization decreased by 4.5%, and slightly above room temperature did not change at all, whereas at 573 K the magnetization rose by 6.3% on exposure to hydrogen. The authors assume that hydrogen is adsorbed into two different states, one increasing and another decreasing the magnetization. Since Artyuk et al. did not measure the saturation magnetization, an anisotropy increasing with the adsorption of hydrogen may also account for the experimental results. Later it will be argued that Mössbauer results indeed indicate that the anisotropy of silica-supported iron particles rises on adsorption of hydrogen. A more likely explanation of the results of Artyuk et al. thus is that the saturation magnetization which is measured at room temperature, where the system is superparamagnetic, is not affected by interaction with hydrogen. At 77 K the increase in anisotropy due to the interaction with hydrogen leads to a drop in the apparent magnetization. At 373 to 573 K reduction of the iron previously being oxidized by water desorbing from the support by interaction with hydrogen causes the magnetization to increase.

Boudart et al. [66,67] thoroughly investigated a 1% iron-on-magnesia catalyst measuring the magnetization at temperatures from 77 to 300 K at static magnetic fields of up to 300 G. The authors admitted the hydrogen at 710 K and subsequently cooled the sample under hydrogen to the measuring temperature. At room temperature the magnetization was not affected, while at 77 K the magnetization increased slightly. The authors established that the iron particles were not superparamagnetic at 77 K. From the effect of exposure to hydrogen on the Mössbauer spectra they could also show that interaction with hydrogen reversibly decreased the anisotropy. Hence the decrease in anisotropy can easily explain the slight increase in magnetization observed at 77 K. At the magnetic field strengths used by the authors the energy barrier between the favorable directions of the magnetic moments of many iron particles is too large to be surmounted at the low thermal energy at 77 K. When the anisotropy energy is decreased due to the interaction with hydrogen, the energy barrier is lower and more iron particles can line up their magnetic moments with the external magnetic field. As a result the magnetization increases. Analogously, an increase in the anisotropy as caused by the adsorption of hydrogen on silica-supported iron particles brings about a drop in the magnetization measured at low field strengths. The drop in magnetization at 77 K, where the magnetization of many particles cannot take up the mean orientation corresponding to thermodynamic equilibrium, observed by Artyuk et al. must be ascribed to an increase in the anisotropy energy by adsorption of hydrogen. After hydrogen adsorption the magnetization corresponding to thermodynamic equilibrium is approached to a smaller extent.

Pecora and Ficalora [68] investigated the magnetization of iron-on-alumina catalysts at low field strengths. After admission of hydrogen at 498 K they observed an increase in the saturation magnetization at this temperature. At room temperature the saturation magnetization did not change significantly, whereas at lower temperatures an appreciable decrease in magnetization was measured. The increase in magnetization at high temperatures can again be attributed to reduction of iron oxidized by water during the previous evacuation procedure. The decrease in magnetization at low temperatures must be ascribed to an increase in anisotropy; the experimental results point to an increase in the anisotropy of silica- and alumina-supported iron particles and a decrease with magnesia-supported iron particles.

Finally, Dutartre et al. [57] extensively studied coprecipitated iron-on-magnesia catalysts by magnetic and Mössbauer measurements. They showed that the effect of hydrogen on the saturation magnetization was very small below about 470 K. At higher temperatures an increase in magnetization was recorded on exposure to hydrogen. Dutartre et al. showed that evacuation at temperatures above about 470 K leads to evolution of hydrogen. Evidently, water desorbing from the magnesia support is oxidizing the metallic iron. The evolution of hydrogen is accompanied by a decrease in the magnetization, which varies from 1.5 to $2.2\mu_B$ per hydrogen molecule set free. The authors convincingly demonstrated that the rise in magnetization on interaction of iron at elevated temperatures with hydrogen is due to reduction of the surface. During the previous evacuation they could relate the evolved hydrogen with the decrease in magnetization. About one iron atom is oxidized per hydrogen molecule set free and hence stops contributing to the ferromagnetic moment at temperatures below about 670 K. At higher temperatures, presumably higher iron oxides than FeO will be formed and the drop in magnetization per hydrogen molecule set free will be smaller.

Above we ascribed the decrease in the saturation magnetization of nickel on adsorption of hydrogen to the formation of a surface nickel hydride on atomically rough surfaces. With iron, formation of a surface hydride is much less likely. The evidence as to the effect of hydrogen adsorption on the magnetization of iron particles indeed indicates that no iron surface hydride is formed. At room temperature, exposure to hydrogen does not significantly affect the saturation magnetization, whereas at lower temperatures an effect on the anisotropy may lead to an effect on the magnetization at low field strengths. The sign of the effect on the anisotropy depends on the support. Iron-on-magnesia exhibits a decrease and iron on silica or alumina an increase in anisotropy if the effect of hydrogen adsorption on the magnetization measured at low temperature is due to a change in anisotropy. Evacuation at high temperatures leads to oxidation of the iron surface by water desorbing from the support. Admission of hydrogen at high temperatures reduces the iron surface and the magnetization rises accordingly.

Although the explanation above seems to be consistent with the experimental evidence, investigations on iron single-crystal surfaces indicate that atomically rough iron surfaces cannot be freed from the last monolayer of oxygen by exposure to hydrogen at high temperatures. Since the atomically rough planes possibly capable of reaction to surface iron hydrides of a lower saturation magnetic moment hence have contaminated surfaces, formation of surface iron hydrides cannot proceed. As with vapor-deposited iron films that have noncontaminated surfaces, an effect of hydrogen adsorption on the saturation magnetization was not found; formation of a surface iron hydride is nevertheless unlikely.

Reaction of iron surface atoms to a different state can also be assessed very well by Mössbauer spectroscopy. A large number of investigations has been performed to investigate the effect of interaction with hydrogen on the Mössbauer parameters: the hyperfine field and the isomer shift [69]. It has been consistently found that exposure to hydrogen does not affect the amount of metallic iron in the catalysts. Although the ferromagnetic anisotropy energy is influenced by the adsorption of hydrogen, no change in the number of ferromagnetic iron atoms has been detected. The only exception is in the work by Clausen et al. [70] on an iron-on-silica catalyst. After reduction for 24 h at 720 K they measured both a bulk component and a surface component in the Mössbauer spectrum. The surface component was found to be affected by exposure to hydrogen and to carbon monoxide. The reduction procedure for this catalyst is, however, delicate. Morup et al. [71], studying the same catalyst, found after reduction for 94 h at 725 K no special surface component and no effect of hydrogen adsorption on the Mössbauer parameters. Consequently, it can be concluded that the Mössbauer results agree very well with the measurements of the saturation magnetization, in that no surface hydride is formed with iron.

The effect of adsorption of hydrogen on the saturation magnetization of cobalt has been studied less than for nickel and iron. Abeledo and Selwood [72] observed a drop in ferromagnetism of $0.54\mu_B$ per adsorbed hydrogen atom, which is much lower than the magnetic moment of a cobalt atom, which is $1.7\mu_B$. Dalmon et al. [73] also investigated cobalt-on-silica catalysts. These authors found a decrease in magnetization of about $0.17\mu_B$ per hydrogen atom on adsorption at 293 K and $0.35\mu_B$ at 473 K. The fact that the decrease in magnetization per adsorbed hydrogen atom is a fraction of the magnetic moment of one cobalt atom indicates that formation of a surface hydride that is not ferromagnetic can again proceed to a limited extent. Most of the hydrogen is adsorbed onto cobalt surfaces not capable of formation of a surface hydride. The fact that at higher temperatures a larger effect per adsorbed hydrogen atom is observed may be related to the solubility of hydrogen in cobalt increasing with temperature.

D. Ferromagnetic Anisotropy

A variation in the energy of a ferromagnetic single domain particle with the orientation of the magnetic moment with respect to the particle [4,6,74] can be due to (a) an anisotropic shape, (b) magnetocrystalline anisotropy, and (c) an anisotropic stress. Shape anisotropy is most obvious. When a single domain particle has an acicular morphology, the magnetostatic energy is lower when the magnetic moment is oriented along the axis of the needlelike particle. Any other orientation of the magnetic moment leads to a much higher number of magnetic poles at the surface and thus to a much greater magnetic field involving more energy.

The magnetocrystalline anisotropy energy is due to spin-orbit coupling. The orbital moments in solids are generally quenched, which implies that they can take up only a very restricted number of orientations. The magnetic moment connected to the spin of an electron, on the other hand, can take up any orientation without its energy being affected. However, as the direction of the spin and the orbital magnetic moment are coupled, the coupling must be broken if the spin moment takes up an orientation that is not allowed to the orbital magnetic moment. The spin-orbit coupling energy is usually relatively small. However, with ferromagnetic solids the spin-orbit coupling energy of the large number of atoms present in a single domain particle has to be added. As a result, the total spin-orbit coupling energy can be substantial. Since the orbital magnetic moment is bound to some crystallographic directions, the magnetocrystalline anisotropy energy depends on the crystallographic direction. For nickel the (111) directions are most favorable and for iron the (100) directions.

Finally, magnetostriction is connected with a nonuniform stress in a ferromagnetic particle. The deformation of the crystal lattice brought about by the stress can lead to a favorable orientation of the magnetic moment (e.g., perpendicular to the direction of the stress).

When ferromagnetic particles are in contact with another magnetic solid an additional source of anisotropy can be present, anisotropic exchange [75, 76]. The magnetic ions present in the contact layer with the other magnetic phase are coupled by exchange interaction with the neighboring magnetic moments of the particle and with the magnetic moments in the other magnetic material. Supported metals are generally produced by reduction of the corresponding oxides obtained by calcination of the hydrated precursors. The oxides are usually antiferromagnetic or ferrimagnetic [77]. With incompletely reduced catalysts an interfacial layer of oxide between the metal and the support is likely. It is also possible that some metal ions dissolving into the surface of the support are reacting to an antiferromagnetic or ferrimagnetic phase. Anisotropic exchange couples the moment of the ferromagnetic particles to those of the oxidic interfacial layers. Often the anisotropy of a ferrimagnetic or antiferromagnetic solid is larger, which causes anisotropic exchange to lead to an enhancement of the ferromagnetic anisotropy. With supported iron or nickel particles an exchange anisotropy is likely. As mentioned above, a reaction of nickel or iron ions with the support can easily proceed. The resulting ionic solids are often ferrimagnetic or antiferromagnetic. Examples include $FeAl_2O_4$, NiOMgO mixed oxides, and $FeTiO_3$. Interaction of ferromagnetic iron or nickel particles with the magnetic boundary layer of the ionic support may substantially increase the anisotropy of the ferromagnetic particles.

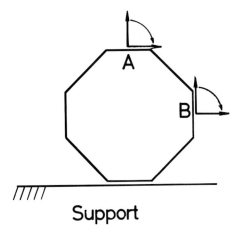

FIGURE 5 Symmetrical particles do not exhibit a contribution of surface anisotropy to the magnetic anisotropy. As can be seen, a contribution to the anisotropy (e.g., because the orientation of the magnetization perpendicular to the surface is energetically unfavorable) is prohibited. The drop in anisotropy at plane A is compensated by that of plane B.

With small ferromagnetic particles or thin films of ferromagnetic solids, surface anisotropy has often been assumed [75,76]. Surface anisotropy results from a more asymmetrical surrounding of the magnetic moments in the surface of a ferromagnetic. Two sources of surface anisotropy have been envisioned: magnetic interaction with neighboring moments, which can be more anisotropic for magnetic moments present in the surface of a particle than for moments present in the bulk, and an electrostatic environment more anisotropic for surface atoms than for bulk atoms. Néel [78,79] has developed a theory that accounts quantitatively for an anisotropic magnetic interaction. This theory has been used by a number of authors to explain the effects of adsorption on ferromagnetic anisotropy [67]. For surface anisotropy to affect the anisotropy of a ferromagnetic particle, the symmetry of the shape of the particle must deviate markedly from that of the symmetry of the crystal lattice of the ferromagnetic material [76]. As indicated in Fig. 5, the contributions of the surface anisotropy of equivalent faces are opposite and equal.

For nickel-on-silica particles it has been observed that adsorption of hydrogen can appreciably decrease the magnetic anisotropy. Geus and Nobel [62] could establish the effect of hydrogen on the ferromagnetic anisotropy from the apparent increase in magnetization measured at alternating fields. Whereas before adsorption the anisotropy of the nickel particles was too high to follow the alternating external magnetic field, after adsorption the anisotropy had decreased sufficiently for the particles to be magnetized in the alternating field. Geus et al. [80] more accurately measured the effect on the anisotropy by varying the frequency of the magnetizing field. From their high-field measurements at different temperatures, Reinen and Selwood [45] also concluded that the anisotropy of nickel-on-silica particles was decreased by the adsorption of hydrogen. The effect on the anisotropy of the alumina-supported nickel particles, on the other hand, was much smaller, as indicated by the results obtained by Reinen and Selwood.

The fact that at an equal decrease in magnetization, measured at a temperature where the system behaves superparamagnetically, hydrogen affects the magnetic anisotropy more strongly than oxygen shows that the decrease in anisotropy energy cannot be attributed simply to a drop in the volume of the ferromagnetic phase. An effect of hydrogen adsorption on the shape of the nickel crystallites, on the stress within the crystallites, or on the surface anisotropy can cause the change in anisotropy. With continuous nickel films the decrease in anisotropy observed by Neugebauer [29-32] is probably due to a decrease in stress. Adsorption of hydrogen decreases the surface energy of the nickel and therefore the tendency to take up a granular structure and to partially denude the substrate. The relief of the stress and thus the change in the anisotropy are established slowly, since a relatively large number of nickel atoms are involved in this process.

With small nickel crystallites a lowering of the surface energy of the metal may lead the particles to take up another shape. However, the experimentally observed drop in the anisotropy would call for a more symmetrical shape, whereas a smaller surface energy would lead to the particles raising their contact area with the support and thus becoming more asymmetrical. Also, a change in the stress within the crystallites can be considered. A compressive stress parallel to the interface with the support will raise the anisotropy of the magnetization. If the stress becomes smaller, the magnetic moment can more easily take up the direction of an applied magnetic field. A decrease in stress may therefore be the cause of the drop in anisotropy observed on hydrogen adsorption.

Since the nickel crystallites have a significant contact area with the support, they must have the asymmetric shape required for the surface anisotropy to be operative. Adsorption of hydrogen can lead to a change in both the magnetostatic coupling and the electrostatic environment of the nickel surface atoms, bringing about a decrease in the surface anisotropy. The effect on the saturation magnetization of nickel shows that the moments of the nickel surface atoms are annihilated by interaction with hydrogen. The effect on the saturation magnetization at 4 K shows that not only is the ferromagnetic coupling lost on the adsorption of hydrogen, but no remaining paramagnetism is exhibited. The surrounding by magnetic ions of the subsurface atoms that follow conversion of the topmost layer into nickel hydride at the surface of the ferromagnetic phase is not much different from that of the surface nickel atoms before adsorption. Although the number of atoms in the surface of the magnetic phase is decreased by conversion of the topmost layer to nickel hydride, an appreciable decrease in the surface anisotropy would not be expected if the coupling between magnetic moments would be the origin of the surface anisotropy decreasing on adsorption of hydrogen. If surface anisotropy is therefore determining the anisotropy of the ferromagnetic phase, the electrostatic interaction of the magnetic atoms situated in the surface of the ferromagnetic phase will dominate the anisotropy. Since the electrostatic surrounding of the magnetic subsurface atoms after reaction of the surface to nickel hydride is much more symmetrical than that of the nickel surface atoms before reaction with hydrogen, the decrease in anisotropy is obvious. Instead of vacuum the atoms in the surface of the ferromagnetic phase are now covered by a layer of nickel hydride.

The smaller effect of hydrogen adsorption on the anisotropy of alumina-supported particles may be due to the fact that the nickel particles are more symmetrical and have a relatively small contact area with the support. As a result, the effect of a difference in surface anisotropy on the total anisotropy of the nickel particle is smaller.

With nickel the drop in anisotropy was connected with a decrease in saturation magnetization pointing to at least a fraction of the nickel surface atoms losing their magnetic moment. With iron the situation is different. The magnetic and Mössbauer results demonstrate that the iron surface atoms are not changing their magnetic moment and the ferromagnetic coupling. Nevertheless, adsorption of hydrogen leads to a decrease in anisotropy with a magnesia support and an increase with a silica and presumably with an alumina support. Since the ferromagnetic coupling is very susceptible to the distance between neighboring magnetic atoms, a significant displacement of the surface iron atoms on adsorption of hydrogen is unlikely. With iron, therefore, Néel's theory, based on the interaction between atomic magnetic moments, is difficult to maintain. Adsorption of hydrogen causing the position of the atomic moments in the surface layer to be changed without an effect on the ferromagnetic coupling is highly improbable. Thus an effect on the surface anisotropy here must also be due to a different electrostatic environment for the surface atoms. The change in the electrostatic environment of the magnetic atoms in the surface of the ferromagnetic phase must therefore be caused by the hydrogen atoms adsorbed. Adsorption of hydrogen atoms apparently leads to a more symmetrical electrostatic field around the iron atoms in the surface and hence to a lower magnetic anisotropy.

The fact that with silica and alumina as a support the anisotropy of iron particles is raised by hydrogen adsorption and decreased with a magnesia support indicates that a simple change in the electrostatic surrounding of the iron surface atoms by adsorption of hydrogen alone cannot be the cause of the effect on the anisotropy. The nature of the support strongly affects the magnetic anisotropy of iron particles. Whereas the anisotropy of magnesia-supported iron particles is high, that of silica-supported particles is intermediate and that of carbon-supported particles is low. An effect of adsorption on the anisotropy having the same sign, but differing in magnitude, could easily be ascribed to a particle shape of a different extent of symmetry. However, the effect of a different sign cannot be explained along these lines. It is likely that exchange anisotropy determines the anisotropy of iron-on-magnesia particles, but it is difficult to explain the effects of hydrogen adsorption on the anisotropy in detail.

E. Conclusions

Ferromagnetism and ferromagnetic anisotropy energy are complicated physical phenomena. Beforehand it could therefore be doubted whether the characteristics of the chemisorptive bond of hydrogen could be elucidated from the effects on the ferromagnetic properties involved. The discussion above has shown, nevertheless, that unambiguous conclusions can be drawn from the experimental data available.

IV. EFFECTS OF HYDROGEN ADSORPTION ON THE ELECTRICAL CONDUCTANCE OF METALS

For an extensive discussion of the effects of chemisorption on the electrical conductance of metals [81], we refer to an earlier publication [34]. As mentioned above, thin metal specimens with very clean surfaces are required to establish reliably the effect of adsorption of hydrogen on the electrical conductance. Consequently, the measurements are performed on thin metal films thrown in ultrahigh vacuum onto nonconducting substrates. Generally, the substrate is glass, which is often cooled during deposition of the film. Especially when the film is thrown onto glass cooled at 77 K, a porous film exhibiting a surface area much larger than that of the substrate is obtained. Often, the surface area of the film increases linearly with the weight of the metal deposited. Electron microscopic investigation of specially prepared replicas of vapor-deposited films have shown that the linear increase in surface area with the weight of the film is due to the film consisting of columnar metal crystallites. The long axis of the columns is oriented in the direction of the incident metal atoms during the deposition of the film. Usually, the columns are oriented perpendicular to the plane of the substrate.

Adsorption can influence the electrical conductance of metal films thus produced by changing the work function of the metal, the reflection of the conduction electron against the metal-gas or metal-vacuum interface, and by changing the electrical conductivity of the surface layer of the metal. When the metal films are very thin, they have an islandlike structure. With these films the effect of adsorption on the work function dominates. With thicker continuous film, conduction by electrons tunneling across pores in the films does not contribute appreciably to the total conductance. The effects of adsorption on the conductance of continuous films is due to the other two effects mentioned above: the nature of the reflection of the conduction electrons and the conductivity of the surface layer.

Only when the metal surface is flat on an atomic scale, specular reflection of the conduction electrons can be expected. The wavelength associated with the motion of the conduction electrons in the metal is so small that a very small roughness is sufficient to bring about diffuse reflection of the conduction electrons.

Reflection of conduction electrons against the metal-gas or metal-vacuum interface does not depend on the temperature. Therefore, the additional resistivity due to nonspecular reflection of the conduction electrons is raising the residual resistivity, that is, the resistivity not dependent on the temperature. With thin metal films the collisions of conduction electrons with the film boundaries constitute a considerable fraction of the total number of collisions. The residual resistivity of vapor-deposited metal films must therefore indicate the nature of the reflection of the conduction electrons against the boundary planes of the metal film. However, vapor-deposited metal films thrown onto cooled substrates contain relatively small crystallites and thus have a relatively large number of grain boundaries. Since scattering of conduction electrons at the grain boundaries does not depend on the temperature, the high density of the grain boundaries adds to the residual resistivity. Thus from a high residual resistivity it cannot unambiguously be concluded that the reflection of conduction electrons against metal-vacuum interfaces is to a large extent diffuse.

Lucas [82] investigated the effects of deposition of gold atoms on the conductance of nonporous gold films. Without an effect on the reflection of the conduction electrons, an increase in the conductance with the number of gold atoms deposited would be expected. With films not annealed at 623 K, deposition of gold atoms just increased the film thickness and thus brought about an increase in conductance. However, films annealed at 623 K exhibited specular reflection of the conduction electrons, as was apparent from the low resistivity. Deposition of gold atoms onto specularly reflecting gold films caused the conductance to pass through a minimum. The gold atoms initially deposited onto the gold films are forming many two-dimensional nuclei, causing the gold surface to become rough on an atomic scale. Accordingly, the reflection of the conduction electrons becomes diffuse and the conductance drops. When the two-dimensional nuclei laterally grow and become interconnected, their size increases and the roughness decreases. As a result, the conductance rises again when the deposition of gold continues and a limited number of growing steps is established that rapidly accommodate the incident gold atoms without developing surface roughness on an atomic scale. Eventually, the increase in the thickness of the gold film caused by the deposition of gold leads to an increase in the conductance.

It is remarkable that the adsorption of hydrogen can affect the conductance of porous metal films in the same way as the deposition of gold onto specularly reflecting gold films. The mobility of nickel atoms over nickel surfaces is relatively high. As a result, the surface area of nickel films deposited onto glass kept at 77 K is rather low. Moreover, the residual resistivity of nickel films is relatively low. A relatively high fraction of specularly reflecting surfaces can therefore be expected with nickel films thus produced. Figure 6 shows the effect of hydrogen adsorption on the conductance of nickel films. It can be seen that the conductance passes through a minimum. At 77 K the scattering of the conduction electrons by thermal phonons is smaller and thus the effect of the change in the reflection of the conduction electron at the metal-gas interfaces is more important. Accordingly, the effect on the conductance at 77 K is larger than that measured at 273 K. Analogous behavior as to the conductance of nickel films on the adsorption of hydrogen has been observed by many authors [83-88].

The surface area of nickel films, the residual resistivity, and the influence of adsorption of hydrogen on the electrical conductance all show that continuous vapor-deposited nickel films not being annealed at high temperatures have atomically flat planes in their surface. Consequently, it cannot be expected that hydrogen atoms can penetrate the surface and thus affect the ferromagnetism.

The interpretation is that the first doses of hydrogen are adsorbed in an ordered structure at relatively large mutual distances. The large distances between the adsorbed hydrogen atoms bring about a diffuse reflection of the conduction electrons, and hence a decrease in the conductance. When more hydrogen is admitted to the film, rearrangement to a closely packed adlayer of hydrogen atoms proceeds and the reflection of the conduction electrons again becomes specular. It is interesting that after presentation of the foregoing explanation, LEED patterns have been taken at low temperatures of nickel surfaces covered by increasing amounts of hydrogen. As discussed by Christmann in Chapter 1, a nickel 111 surface covered by 0.5 monolayer of hydrogen atoms is relatively stable and exhibits a LEED pattern corresponding to a c2 × 2 structure.

As can be seen in Fig. 6, the coverages at which minimum conductance is exhibited and at which the conductance returns almost to its original value are much too small to agree with the required coverages of 0.5 and about 0.8 (Christmann argues that the maximum coverage of hydrogen is usually 0.8). The number of metal surface atoms of nickel can be taken to be about 1.7×10^{15} cm^{-2}, which corresponds with a number of hydrogen molecules of 0.85×10^{15} cm^{-2} at full coverage. The minimum conductance is exhibited at a coverage of about 0.1×10^{15} cm^{-2} and the conductance is almost at its original value at a coverage of 0.3×10^{15} cm^{-2}. The low values of the coverage at minimum conductance have also been observed by other authors.

The reason for the low hydrogen coverages is the fact that a considerable fraction of the surface area of vapor-deposited nickel films is present at narrow pores. The surfaces of narrow pores call for rather high hydrogen pressures to be covered. At these high pressures the parts of the nickel surface more easily accessible for molecules from the gas phase have already been covered by a full monolayer and are again exhibiting specular reflection of the conduction electrons. After covering the more easily accessible part of the nickel surface the adlayer moves with a sharp boundary into the narrow pores. At the boundary the coverage drops steeply from a monolayer to a negligible value. Since in this stage of the adsorption process every dose of hydrogen is covering part of the nickel by almost a monolayer, the conductance is not affected significantly.

The mobility of iron atoms over iron surfaces is lower than that of nickel atoms over nickel surfaces. An important reason can be that the fcc lattice planes are more smooth than the bcc lattice planes. As a result, iron films thrown onto glass kept at 273 K display surfaces of the same order of magni-

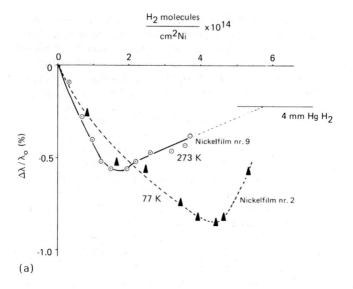

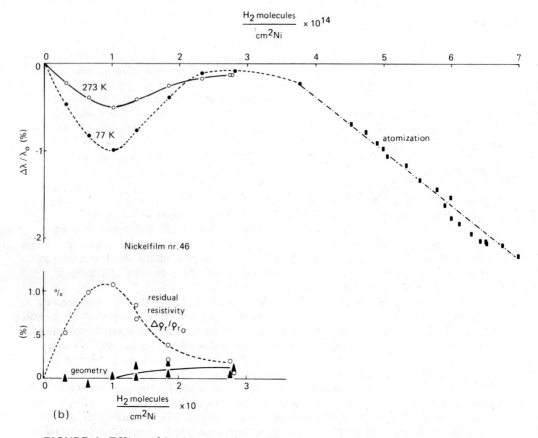

FIGURE 6 Effect of hydrogen adsorption on the electrical conductance of nickel films. (a) Prepared by vapor deposition onto glass kept at 77 K. Conductance measured at 77 and 273 K. (b) Prepared by vapor deposition onto glass kept at 273 K. Conductance measured at 77 and 273 K. Metal surface of film deposited onto glass kept at 273 K much more smooth.

8. Magnetic and Electrical Effects of Adsorbed Hydrogen

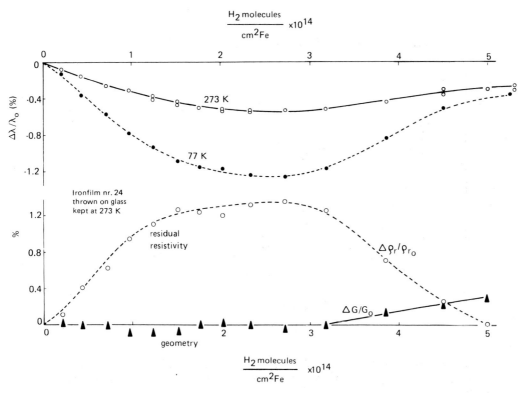

FIGURE 7 Effect of adsorption of hydrogen on the conductance of an iron film deposited on glass kept at 273 K. Hydrogen admitted at 273 K; effect on the conductance measured at 273 and 77 K. At the bottom the effect on the residual resistivity and the geometry of the conducting phase of the iron film is indicated.

tude as nickel films deposited on glass kept at 77 K. The fact that the iron and nickel films have about the same porosity indicates that the surfaces are equally smooth. The effect of adsorption of hydrogen on the conductance indeed has the same course as with nickel films, as shown in Fig. 7.

Again the minimum in the conductance is exhibited at a coverage that is low compared with the coverage corresponding to 0.5 monolayer. The minimum in conductance is displayed at a coverage of about 0.6×10^{15} H atoms cm^{-2} and the conductance has almost returned to the value of the clean film at a coverage of about 1.2×10^{15} cm^{-2}. The last coverage is lower than the density of iron atoms in the most closely packed (110) iron surface, which amounts to 1.7×10^{15} atoms cm^{-2}. The relatively low values of the hydrogen coverages can also be attributed to transport of hydrogen into narrow pores. A fraction of less closely packed iron surfaces may also be present. However, as is also apparent from the effect of adsorption of carbon monoxide on the conductance of iron and nickel films, transport into iron films proceeds more easily than into nickel films. This may be due to the fact that iron films deposited on glass at 273 K contain pores that are wider than those in nickel films deposited on glass kept at 77 K. Alternatively, the repulsive interaction between adsorbed atoms and molecules on iron surfaces may be smaller than that on nickel surfaces.

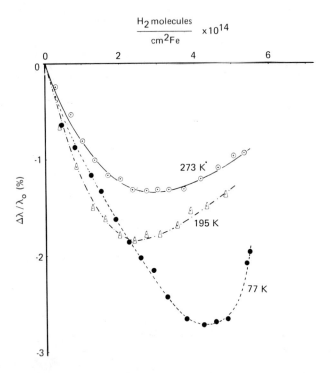

FIGURE 8 Effect of adsorption of hydrogen on the conductance of iron films thrown onto glass kept at 77 K. Films kept at the temperature indicated during measurement.

The effect of hydrogen adsorption on the residual resistivity of iron films clearly shows the effect on the reflection of the conduction electrons. Penetration of hydrogen atoms into the iron surface, which leads to a surface hydride of relatively low conductance, can barely proceed. It must be expected that iron films thrown on glass kept at 77 K contain a larger fraction of surfaces that hydrogen atoms can penetrate more easily. The effect of hydrogen adsorption on the conductance of iron films thrown on glass kept at 77 K reflects the larger fraction of more open crystallographic planes in the surface of the film. As can be seen in Fig. 8, the conductance remains appreciably lower than that displayed by the clean film. The residual resistivity shown at the bottom of Fig. 8 indicates that the effect on the reflection of the conduction electrons now is not almost completely reversible. This must be due to penetration of hydrogen atoms into more open surfaces, which renders the surfaces atomically more rough, and consequently the reflection of the conduction electrons, diffuse. The penetration of hydrogen atoms into the surface layer is also evident from the effect on the geometry of the conducting metal, which is also represented at the bottom of Fig. 8. Due to the penetration of hydrogen atoms, a surface layer of lower conductivity is established.

The high melting point of tungsten causes the mobility of tungsten atoms over tungsten surfaces to be very small. As a result, tungsten films thrown onto glass kept at 273 K can be expected to contain many atomically rough surfaces. Accordingly, adsorption of hydrogen does not affect the conductance, as with nickel and iron films. As can be seen in Fig. 9, the conductance continuously decreases on adsorption of hydrogen. Evidently, hydrogen penetrating the atomically rough surfaces affects mainly the electrical conductivity of the surface layer. Even before the adsorption of

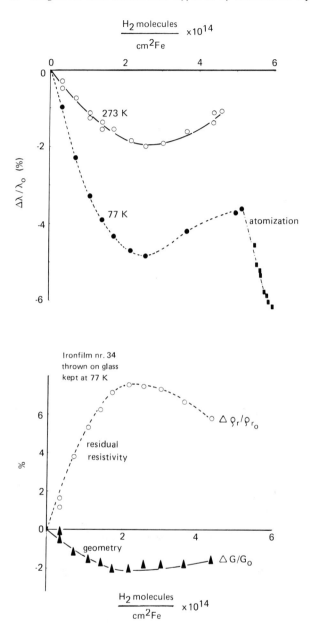

FIGURE 8 (continued)

hydrogen, the surface almost completely reflected the conduction electrons diffusely.

The discussion above has demonstrated that a change in both the reflection of the conduction electrons and the electrical conductivity of the surface layer indeed are affecting the conductance of metals with the adsorption of hydrogen. Nickel and iron films thrown on glass kept at 273 K mainly showed an effect on the reflection of the conduction electrons and tungsten films on the conductivity of the metal surface layer. With iron films thrown on glass kept at 77 K, both the reflectivity and conductivity of the surface layer were affected appreciably. However, the conductivity of the surface layer

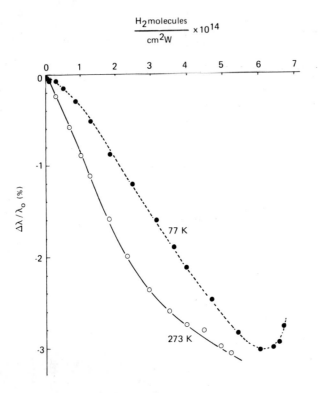

FIGURE 9 Effect of adsorption of hydrogen on the conductance of a tungsten film thrown onto glass kept at 273 K.

of nickel films can also be modified by penetration of hydrogen. Formation of a surface hydride can be brought about by exposure of the nickel film to hydrogen atoms produced at a tungsten filament kept at a high temperature. To prevent rapid recombination to hydrogen molecules that desorb, the nickel film has to be kept at 77 K during the bombardment with hydrogen atoms. The mobility of hydrogen adatoms below about 180 K is so low that recombination does not proceed rapidly. The result of bombardment with hydrogen atoms is also represented in Fig. 6. It can be seen that the conductance strongly decreases by reaction with the hydrogen atoms to a surface hydride.

We feel that the effects of hydrogen adsorption on the electrical conductance of metals can easily be accounted for. That the effect of adsorption of hydrogen on the electrical conductance depends on the surface structure of the adsorbing metal surface had to be expected from the magnetic results. The differences in the surface structure of nickel films deposited onto a substrate kept at 77 K and tungsten films thrown on a substrate kept at 273 K are obvious from transmission micrographs of the films. The tungsten films contain very small crystallites of about 5 nm diameter, which cannot have many atomically flat crystallographic planes in their surface. Consequently, specular reflection of the conduction electrons cannot be expected. On the other hand, the nickel and iron films exhibit the behavior to be predicted with penetration of hydrogen atoms into the surface layer when they are exposed to hydrogen atoms at low temperatures. Nickel and iron films thus exhibit the same behavior as tungsten films provided penetration into the surface can proceed.

It is remarkable that both magnetic and electrical conductance measurements consistently reflect the effect of the surface structure of the metal.

Less closely packed crystallographic surfaces or closely packed crystallographic surfaces of small dimensions display penetration of hydrogen atoms leading to a surface hydride. The discussion above has shown that not only with electrical conductance will the effects of hydrogen adsorption depend strongly on the metal surface structure, but also with magnetic properties.

Whereas with iron films deposited on glass kept at 77 K penetration into the surface layer proceeds, the ferromagnetism of iron is not affected by adsorption of hydrogen. The films of Geus and Koks presumably had mostly closely packed planes in their surface, but Göpel's films were annealed at high temperatures and are likely to consist of isolated iron particles. Especially with the thin films, an island structure has to be expected. Although hydrogen atoms must be considered able to penetrate the surface of the particles, no effect on the number of ferromagnetic atoms was observed. Presumably this is due to the fact that adsorption of hydrogen onto iron surfaces does not affect the ferromagnetic saturation magnetization. With the supported iron particles, the absence of an effect on the saturation magnetization may be due to the stable monolayer of oxygen, which is very difficult to remove. In view of the absence of an effect on the saturation magnetization of iron films, we prefer the explanation that penetration of hydrogen in the surface layer of iron does affect the conductivity, but not the saturation magnetization.

V. CONCLUSIONS

The above has shown that adsorption of hydrogen does not affect the metallic properties strongly unless an ionic (surface) hydride is formed. The structure of the metal surface is highly important in the effects on the metal structure. With closely packed metal surfaces the effect on the metallic properties is small. Adsorption of hydrogen onto metal surfaces of a more open structure can lead to the formation of surface hydrides that are stable under conditions where the bulk hydrides are very unstable. However, with a surface hydride the metallic properties of the surface layer are retained.

The gap between surface science and catalysis is often mentioned. A comparison of this chapter with Chapter 1 shows that the gap is appreciably narrower than is generally accepted. Measurements on supported catalysts and vapor-deposited metal films can effectively bridge the gap. However, merely measuring the extent of adsorption is clearly not sufficient. The effect of the structure of the adsorbing surfaces and the distribution of the admitted gas molecules over the usually large surface areas, a considerable fraction of which is often difficult to cover from the gas phase, must be assessed. Measurement of relevant physical properties of the adsorbing metal specimens can provide the required information. A difficulty is often the complicated structure and physical properties of the finely divided solids to be studied. Both the magnetic properties of small ferromagnetic particles and the electrical conductance of thin vapor-deposited metal films are complicated. Nevertheless, this work shows that very satisfactory results can be obtained and a large body of experimental evidence can be consistently explained.

REFERENCES

1. T. R. P. Gibb, Jr., in *Progress in Inorganic Chemistry*, Vol. 3 (F. A. Cotton, ed.), Interscience, New York (1962), p. 315.

2. J. M. Ziman, *Principles of the Theory of Solids*, 2nd ed., Cambridge University Press, Cambridge (1972), p. 133.
3. T. R. P. Gibb, Jr., in *Progress in Inorganic Chemistry*, Vol. 3 (F. A. Cotton, ed.), Interscience, New York (1962), p. 419.
4. L. F. Bates, *Modern Magnetism*, 4th ed., Cambridge University Press, Cambridge (1961), p. 14.
5. W. Klemm, *Magnetochemie*, Akademische Verlagsgesellschaft, Leipzig (1936), p. 21.
6. C. Kittel, *Introduction to Solid State Physics*, 3rd ed., Wiley, New York (1967), p. 432.
7. C. Kittel, *Introduction to Solid State Physics*, 3rd ed., Wiley, New York (1967), p. 446.
8. W. Hume-Rothery, *Electrons, Atoms, Metals and Alloys*, 3rd rev. ed., Dover, New York (1963), p. 246.
9. N. F. Mott and H. Jones, *The Theory of the Properties of Metals and Alloys*, Dover, New York (1958), p. 200.
10. A. Michel and M. Gallisot, *C. R. Acad. Sci. (Paris)*, 208:434 (1939).
11. T. R. P. Gibb, Jr., J. MacMillan, and R. J. Roy, *J. Phys. Chem.*, 70:3024 (1966). See, however, also J. C. Barton, F. A. Lewis, and I. Woodward, *Trans. Faraday Soc.*, 59:1201 (1963).
12. C. Kittel, *Rev. Mod. Phys.*, 21:541 (1949).
13. J. D. Fast, *Interaction of Metals and Gases*, Vol. 1, *Thermodynamics and Phase Relations*, Philips Technical Library, Eindhoven, The Netherlands (1965), p. 125.
14. B. Baranowski, in *Festkörperchemie* (V. Boldyrev and K. Meyer, eds.), VEB, Leipzig (1973), p. 364.
15. B. Baranowski and M. Smialowski, *J. Phys. Chem. Solids*, 12:206 (1959).
16. B. Baranowski and K. Bochenska, *Z. Phys. Chem.*, 45:140 (1965).
17. B. Baranowski and K. Bochenska, *Rocz. Chem.*, 38:1419 (1964).
18. H. J. Bauer, *Z. Phys.*, 164:367 (1961).
19. H. J. Bauer, *Z. Phys.*, 177:1 (1964).
20. S. von Aufschnaiter and H. J. Bauer, *Z. Angew. Phys.*, 18:209 (1964).
21. H. J. Bauer and O. Ruczka, *Z. Angew. Phys.*, 21:18 (1966).
22. H. J. Bauer, M. Becker, and J. Bofilias, *Naturwissenschaften*, 53:17 (1966).
23. Y. Ebisuzaki, W. J. Kass, and M. O'Keeffe, *J. Chem. Phys.*, 46:1373 (1967).
24. J. W. Cable, E. O. Wollan, and W. C. Koehler, *J. Phys. (Paris)*, 25:460 (1964).
25. A. Faessler and R. Schmidt, *Z. Phys.*, 190:10 (1966).
26. G. K. Wertheim and D. N. E. Buchanan, *Phys. Lett.*, 21:255 (1966).
27. A. E. Jech and C. R. Abeledo, *J. Phys. Chem. Solids*, 28:1371 (1967).
28. C. A. Neugebauer, in *Structure and Properties of Thin Films* (C. A. Neugebauer, J. B. Newkirk, and D. A. Vermilyea, eds.), Wiley, New York (1959), p. 358.
29. C. A. Neugebauer, *J. Appl. Phys. Suppl.*, 31:152S (1960).
30. C. A. Neugebauer, *Trans. Natl. Vac. Symp.*, 8:924 (1961).
31. C. A. Neugebauer, *Z. Angew. Phys.*, 14:182 (1962).
32. C. A. Neugebauer and M. B. Webb, *J. Appl. Phys.*, 33:74 (1963).
33. C. P. Bean and J. D. Livingston, *J. Appl. Phys. Suppl.*, 30:120S (1959).
34. J. W. Geus, in *Chemisorption and Reactions on Metallic Films*, Vol. 1 (J. R. Anderson, ed.), Academic Press, New York (1971), p. 472.
35. J. K. Blum and W. Göpel, *Ber. Bunsenges. Phys. Chem.*, 82:329 (1978).
36. J. K. Blum and W. Göpel, *J. Magn. Magn. Mater.*, 6:186 (1977).
37. W. Göpel, *Surf. Sci.*, 85:400 (1979).
38. B. E. Sundquist, *Acta Metall.*, 12:67 (1964).

39. M. Primet, J.-A. Dalmon, and G.-A. Martin, *J. Catal.*, 46:25 (1977).
40. L. H. Germer and A. U. MacRae, *J. Chem. Phys.*, 37:1382 (1962).
41. K. Christmann, F. Chehab, V. Penka, and G. Ertl, *Surf. Sci.*, 152/153: 356 (1985).
42. E. W. Müller, *Surf. Sci.*, 8:462 (1967).
43. R. E. Dietz and P. W. Selwood, *J. Chem. Phys.*, 35:270 (1961).
44. P. W. Selwood, *Chemisorption and Magnetization*, Academic Press, New York (1975).
45. D. Reinen and P. W. Selwood, *J. Catal.*, 2:109 (1963).
46. G.-A. Martin, G. Dalmai-Imelik, and B. Imelik, in *Proceedings of the 2nd International Conference on Absorption-Desorption Phenomena, Florence, 1971* (F. Ricca, ed.), Academic Press, New York (1972), p. 433.
47. G.-A. Martin, and B. Imelik, *Surf. Sci.*, 42:157 (1974).
48. G.-A. Martin, N. Ceaphalan, P. de Montgolfier, and B. Imelik, *J. Chim. Phys. (Fr.)*, 70:434 (1973).
49. A. Knappwost and W. H. Eugen Schwarz, *Z. Phys. Chem.*, 67:15 (1969).
50. K. P. de Jong, J. W. Geus, and J. Joziasse, *Appl. Surf. Sci.*, 6:273 (1980).
51. J. W. Geus, in *Chemisorption and Reactions on Metallic Films*, Vol. 1 (J. R. Anderson, ed.), Academic Press, New York (1971), p. 199.
52. D. W. Pashley, and M. J. Stowell, in *Proceedings of the 5th International Conference on Electron Microscopy* (S. S. Breese, ed.), Academic Press, New York (1962), p. GG-1.
53. D. W. Pashley and M. J. Stowell, *J. Vac. Sci. Technol.*, 3:156 (1966).
54. H. Poppa, *Z. Naturforsch.*, 19A:835 (1964).
55. J. F. Pócza, in *Proceedings of the 2nd Colloquium on Thin Films, Budapest, 1967* (E. Hahn, ed.), Van den Hoeck und Rupprecht, Göttingen (1967), p. 93.
56. A. Hahn and D. Löckenhoff, *Phys. Kondens. Mater.*, 2:284 (1964).
57. R. Dutartre, P. Bussière, J. A. Dalmon and G.-A. Martin, *J. Catal.*, 59:382 (1979).
58. J.-A. Dalmon, G.-A. Martin, and B. Imelik, *Proceedings of the 2nd International Conference on Solid Surfaces, 1974, Jpn. J. Appl. Phys. Suppl. 2*, Pt. 2, 261 (1974).
59. J.-A. Dalmon, G.-A Martin, and B. Imelik, *Surf. Sci.*, 41:587 (1974).
60. E. J. Lee, J. A. Sabatka, and P. W. Selwood, *J. Am. Chem. Soc.*, 79:5391 (1957).
61. J. W. Geus, A. P. P. Nobel, and P. Zwietering, *J. Catal.*, 1:8 (1962).
62. J. W. Geus and A. P. P. Nobel, *J. Catal.*, 6:108 (1966).
63. G.-A. Martin, P. de Montgolfier, and B. Imelik, *Surf. Sci.*, 3:675 (1973).
64. J. Déportes, J.-P. Rebouillat, R. Dutartre, J.-A. Dalmon, and G.-A. Martin, *C. R. Acad. Sci. (Paris)*, C276:1383 (1973).
65. Y. N. Artyuk, V. I. Yas'mo, N. K. Lunev, and M. T. Rusov, *Kinet. Katal.*, 16:167 (1975).
66. M. Boudart, J. A. Dumesic, and H. Topsoe, *Proc. Natl. Acad. Sci. USA*, 74:806 (1977).
67. M. Boudart, H. Topsoe, and J. A. Dumesic, in *The Physical Basis for Heterogeneous Catalysis* (R. Drauglis and R. I. Jaffee, eds.), Plenum Press, New York (1975), p. 337.
68. L. M. Pecora and P. J. Ficalora, *Metall. Trans. A.*, 8A:1841 (1977).
69. J. A. Dumesic and H. Topsoe, *Adv. Catal.*, 26:121 (1977).
70. B. S. Clausen, S. Morup, and H. Topsoe, *Surf. Sci.*, 106:438 (1981).
71. S. Morup, B. S. Clausen, and H. Topsoe, *J. Phys. (Paris)*, 41:C1-39 (1980).
72. C. R. Abeledo and P. W. Selwood, *J. Chem. Phys.*, 37:2709 (1962).

73. J.-A. Dalmon, G.-A. Martin, and B. Imelik, *Thermochimie*, Colloq. 201, Centre Nationale de la Recherche Scientifique, Marseille (1971), p. 593.
74. R. M. Bozorth, *Ferromagnetism*, Van Nostrand, New York (1951).
75. I. S. Jacobs and C. P. Bean, in *Magnetism*, Vol. 3 (G. T. Rado and H. Suhl, eds.), Academic Press, New York (1963), p. 271.
76. C. P. Bean, in *Structure and Properties of Thin Films* (C. A. Neugebauer, J. B. Newkirk, and D. A. Vermilyea, eds.), Wiley, New York (1959), p. 331.
77. J. Smit and H. P. J. Wijn, *Ferrite*, Philips Technical Library, Eindhoven, The Netherlands (1962), p. 35.
78. L. Néel, *C. R. Acad. Sci. (Paris)*, 237:1468 (1953).
79. L. Néel, *J. Phys. Radium*, 15:225 (1954).
80. J. W. Geus, A. P. P. Nobel, and H. Schutte, *J. Colloid Interface Sci.*, 26:266 (1968).
81. P. Wiszmann, *The Electrical Resistivity of Pure and Gas-Covered Metal Films*, Springer Tracts in Modern Physics (Ergebnisse de exakten Naturwissenschaften), Springer-Verlag, Berlin (1975).
82. M. S. P. Lucas, *Appl. Phys. Lett.*, 4:73 (1964).
83. W. M. H. Sachtler and G. J. H. Dorgelo, *Bull. Soc. Chim. Belg.*, 67:465 (1958).
84. R. Suhrmann, Y. Mizushima, A. Hermann, and G. Wedler, *Z. Phys. Chem.*, 20:332 (1959).
85. P. Zwietering, H. L. T. Koks, and C. van Heerden, *J. Phys. Chem. Solids*, 11:18 (1959).
86. V. Ponec and Z. Knor, *Collect. Czech. Chem. Commun.*, 25:2913 (1960).
87. Y. M. Mizushima, *J. Phys. Soc. Jpn.*, 15:1614 (1960).

9 | Electrochemical Investigation of Surface Hydrogen on Metal Catalysts

JÓZSEF PETRÓ, TAMÁS MALLÁT, AND ÉVA POLYANSZKY
Technical University of Budapest, Budapest, Hungary

TIBOR MÁTHÉ
Institute for Organic Chemical Technology, Hungarian Academy of Sciences, Budapest, Hungary

I.	INTRODUCTION	225
II.	METHODS IN THE LIQUID PHASE	226
	A. The Technique	226
	B. Experimental Examples	230
III.	MEASUREMENTS IN THE GAS PHASE	241
	A. The Measuring Cell	241
	B. Measurements in the Presence of Hydrogen	244
	C. Measurements During Hydrogenation of Benzene and Phenol	247
	REFERENCES	252

I. INTRODUCTION

The relationship between metal-catalyzed hydrogenation and the phenomena of electrochemistry was pointed out in a relatively early stage of electrochemistry by Tafel [1]. The principal advantages of electrochemical methods are their technical simplicity, cheapness, and short measurement times compared to several other methods. Among the restrictions, however, it should be mentioned that only good conductor metal or supported metal catalysts can be investigated by these methods.

The main areas of application are the investigation of the catalysts of liquid-phase catalytic hydrogenation and the investigation of the catalytic process. In addition, there is a possibility for the study of gas-phase, metal-catalyzed processes.

The basis of the methods is measurement of the electrochemical potential of metal catalysts. This is made possible by the fact that noble metals and transition metals important from the viewpoint of catalysis (e.g., Ni) behave as reversible hydrogen electrodes when saturated by hydrogen.

Consequently, electrochemical methods provide a unique, exceptional possibility for the study of metal-hydrogen systems, which enables not only quantitative analysis (i.e., amount of hydrogen sorbed on the metal) but also qualitative analysis (i.e., hydrogen bonded in different ways with different strengths) to be performed. In this respect, the results are similar to those obtained by the temperature-programmed desorption (TPD) method [2], with the differences that the measurement is carried out essentially at room temperature and the medium is liquid. A great advantage of electrochemical methods is that it is easy to provide near-equilibrium conditions, which is far from true in the case of, say, the TPD method.

It is a further advantage that since electrochemical phenomena take place on boundary layers, these methods give information about just the catalyst surface, which has a distinguished role in catalysis. The data obtained are, however, integrated values, which enable a more detailed analysis of surface conditions to be performed in certain cases only (e.g., polarization measurements).

Already in the classical book of Schwab [3], the catalytic application of electrochemistry was discussed in a separate chapter. In the analysis of the electrochemical aspects of heterogeneous catalytic processes taking place in aqueous solutions, the work of Müller and Schwabe [4] was of pioneering significance. It is pleasing that some areas of this subject are also discussed in modern work published recently (e.g., Refs. 2 and 5).

Electrochemical methods are applied in the following areas:

1. By means of galvanostatic and potentiodynamic polarization methods the surface area of metal catalysts can be determined on the basis of the measurement of their hydrogen sorption capacity [6], and the amount and bond strength of hydrogen sorbed on the surface [7-9]. By studying hydrogen sorption, the aging of catalysts can be monitored [10], and in the preparation of catalysts the composition of surface phase, and thereby processes taking place during the preparation of catalysts, can be studied.
2. The excess free surface energy of supported and unsupported metal catalysts can be measured.
3. Liquid- and gas-phase catalytic processes can also be studied by measuring the electrochemical potential of working catalysts. Thus the reaction conditions can be optimized, the progress of the process can be monitored, the selectivity and aging of catalysts, and the potential range of the reversible operation of transition metal catalysts can be investigated.

In view of all these advantages, it is an unfortunate situation that electrochemical methods are far less widespread for the characterization of catalysts than are the methods of surface chemistry, which often give less-direct data and involve the application of high-vacuum technology. For this reason, being aware of the importance of data provided by electrochemical methods, this subject is by all means worth discussing in a book dealing with the effects of hydrogen in catalysis.

Without aiming at completeness, some basic experimental techniques will be introduced, and the performance of electrochemical measuring methods will be illustrated by some practical examples.

II. METHODS IN THE LIQUID PHASE

A. The Technique

Of the polarization methods developed by electrochemists, galvanostatic and potentiodynamic polarizations have become widespread for the investigation of metal/hydrogen systems of catalysts.

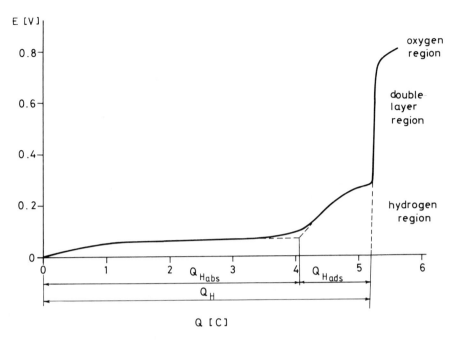

FIGURE 1 Galvanostatic curve of a disperse Pd catalyst in 0.5 M sulfuric acid. m = 10 mg, i = 2.5 mA. (From Ref. 11.)

In galvanostatic polarization, the catalyst (measuring electrode) is polarized at a constant current. By plotting electrode potential as a function of charge passing the system, the charging curve (galvanostatic curve), a name suggested by Frumkin and Sligin [6], can be obtained. The characteristic charging curve of disperse Pd can be seen in Fig. 1. Starting from the state saturated with hydrogen (0 to 0.05 V), the ionization of hydrogen dissolved in the β phase terminates at ca. 0.1 V, that of adsorbed hydrogen at ca. 0.3 V (the "hydrogen region"). Then the potential of the catalyst suddenly increases; in this region the charge is invested into recharging of the double layer only (the "double-layer region"), and then, at ca. 0.7 V, the adsorption of oxygen sets in (the "oxygen region"). If sufficiently low current density (generally below 1 μA cm^{-2}) is applied, the measurement can be performed under quasi-equilibrium conditions [12]. At present the potentiodynamic polarization method (common alternative names are "linear sweep" or "cyclic voltammetry") is much more widespread [13,14]. In such measurements the potential of the catalyst is changed with respect to the reference electrode at a constant rate, by means of a potentiostat, and the current passing the cell is simultaneously recorded [15-19]. The potentiodynamic curve is, in fact, a differential adsorption isotherm as a function of the potential.

Figure 2 shows the hydrogen region of the potentiodynamic curve of a Pd powder catalyst. On comparing Figs. 1 and 2, the higher sensitivity and "resolving power" of the latter method is obvious. The remark of McNicol [2] on the analogy between TPD and cyclic voltammetry is very instructive. In one of the methods, temperature, in the other, potential, is changed evenly in time, and the shapes of the curves are also similar.

Polarization measurements require three-electrode cells. The spaces of the reference electrode (usually platinized Pt/H$_2$), the measuring electrode,

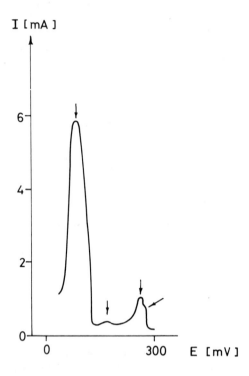

FIGURE 2 Potentiodynamic curve of a disperse Pd catalyst in 0.5 M sulfuric acid. m = 1 mg; sweep rate, 100 mV min^{-1}. (From Ref. 11.)

and the counter electrode (bright Pt or Au) are generally separated by diaphragms or ground taps. The fundamentals pertaining to the construction and practice of polarization measurements are given in several textbooks (e.g., Ref. 20).

When adapting the polarization measurements elaborated for bright or blackened sheet electrodes to catalysts, technical difficulties will be encountered: a proper electric contact must be ensured between the powder catalyst and the bright Pt sheet electrode (together, they form the measuring electrode). Three methods are commonly used for this purpose:

1. A sheet or rod is pressed from the powder by adding some filler, and this is the measuring electrode [18]. From the aspects of catalysis, it has disadvantages (e.g., the catalyst may be contaminated with the filler).
2. The catalyst is placed on a horizontal Pt sheet electrode [21]. This is technically the simplest solution, but the gravitational force is, for example, in the case of carbon-supported catalysts of low metal content, insufficient to ensure good contact. The scheme of a simple cell of this type is shown in Fig. 3.
3. Measurement in a slurry system, wherein a suspension of the catalyst is continuously directed to the immersed measuring electrode [22,23].

It is a further important requirement that the cell should make it possible to reduce the catalyst with hydrogen "in situ," before contacting it with the acidic electrolyte. Otherwise, the oxide layer on the surface may be dissolved, causing a very serious error in the case of catalysts of high dispersity, particularly in the case of alloys.

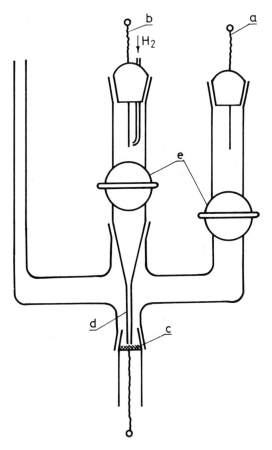

FIGURE 3 Design of an electrochemical cell for powder catalysts. a, Pt sheet polarizing electrode; b, platinized Pt/H$_2$ reference electrode; c, catalyst/Pt sheet working electrode; d, Luggin capillary; e, ground taps.

In general, aqueous acid solutions are applied as electrolyte. Electrochemists apply carefully purified electrolytes, electrodes, and gases; they use, for example, pyrodistilled or triple-distilled water. In the case of disperse catalysts, the situation is less demanding: The higher the specific surface area of the catalyst, the less the relative interference of "other" impurities (of noncatalyst origin), and thus the use of bidistilled water is sufficient.

From the charge (Q_H, see Fig. 1) invested into hydrogen sorption (the cathodic curve) or hydrogen ionization (the anodic curve), the specific amount of sorbed hydrogen (V_H) is readily obtained:

$$V_H = \frac{Q_H \times 22.4}{m \times 2 \times 96.5} \quad cm^3\ g^{-1} \tag{1}$$

where m is the mass of catalyst in grams. Q_H, in coulombs, can be evaluated from the potentiodynamic curve by integrating the area below the hydrogen maxima.

For calculation of the specific surface area, one should know the number n of metal atoms situated on a unit surface area of disperse catalyst and the stoichiometry of hydrogen sorption. If for Pt metals the approximation

$M_S:H_A = 1:1$ [16, 24, 25] is applicable (see also Chapter 4) the specific surface area S can be calculated as follows:

$$S = \frac{Q_H}{mn} \frac{6.03 \times 10^{23}}{9.65 \times 10^4} \quad m^2 \, g^{-1} \qquad (2)$$

For example, in the case of polycrystalline Pt, the generally accepted value for Q_H/S is 210 µC cm^{-2}. The surface area of disperse palladium can be calculated so that hydrogen ionizable below 0.1 V is regarded as dissolved in the β phase, whereas the strongly bonded one (ionized in the region 0.1 to 0.3 V is regarded as adsorbed hydrogen (see Fig. 1) [24]. The specific surface area of the catalyst can also be determined from oxygen sorption; this method has been used successfully for gold [16].

If the potential of the catalyst is set in the range where neither hydrogen nor oxygen is on its surface, and the electrolyte is substituted for by a solution containing the metal ions themselves, the excess surface free energy of the disperse system can be determined. The details of the experimental method and some applications of this interesting technique have been published elsewhere [26, 27].

The potential of the catalyst can also be measured in situ, in the course of a hydrogenation reaction, in a slurry system as mentioned above, and the figures obtained can be brought into correlation with the progesss of the chemical reaction. A great many examples of applications of this method are given in the works of Sokolskii [9, 28] and Sokolskii and Zakumbawa [29].

B. EXPERIMENTAL EXAMPLES

Metal-Hydrogen System of Group VIII and IB Metal Catalysts

On metal catalysts used for hydrogenation, bonds of various strength may be formed between the metal and hydrogen. Thus hydrogen in metals may be bonded by absorption, physical adsorption, and chemisorption. A detailed discussion is devoted to the amounts, ratios, and binding energies of the various types of hydrogen in Chapters 1, and 3, 4, and 5. These depend on the nature of the metal, but since adsorption is a surface phenomenon, also on the surface crystal structure [30], dispersity, support, and in the liquid phase, also on the organic impurities and the electrolyte.

If noble metal catalysts are investigated by potentiodynamic polarization, a characteristic curve is obtained for each metal [16]. Figure 4 shows the anodic and cathodic curves of Pt, Ir, Rh, and Au taken in 1 M H_2SO_4 at 25°C.

According to the figure, the hydrogen region of polycrystalline Pt has two characteristic peaks, at ca. 0.12 and 0.25 V and a shoulder in between. Other authors [31, 32] have found four to five types of peaks when carrying out polarization under extremely pure conditions.

The investigation of the adsorption isotherms of hydrogen has shown that there are two surface locations of different adsorption energies on Pt (weak and strong adsorption), and that significant differences could be detected between these two adsorption sites by infrared spectroscopy [33] (see also Chapter 7). Therefore, several authors attempted to correlate the two characteristic maxima of the potentiodynamic curve with two types of adsorption sites. Since the metal-H binding energy can also be calculated from the potentiodynamic curves, which is a lower value at high hydrogen coverage (low potential) and higher at low coverage, the 0.12-V peak was assigned to the weakly bonded hydrogen, and the 0.25-V peak, to the strongly bonded hydrogen [16].

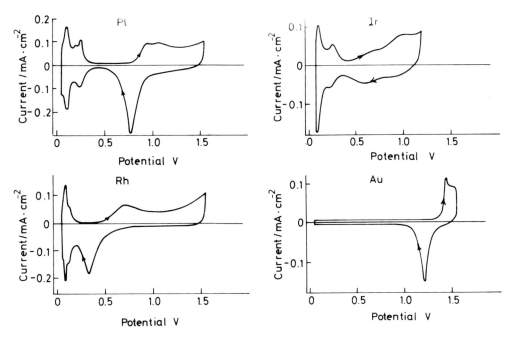

FIGURE 4 Potentiodynamic curves of noble metals in 1 M H_2SO_4 at 25°C; scanning rate v = 0.4 V s^{-1}. (From Ref. 16.)

There is not, however, unanimous opinion concerning the adsorption forms (atomic, molecular, ionic) and the site of adsorption. According to the single-crystal studies of Will [34], two hydrogen maxima were obtained on each of three crystallographic faces of Pt [(111), (100), and (110)] by potentiodynamic polarization, and the ratios of these hydrogen maxima were found to be different for the various faces. The two characteristic peaks obtained on the polycrystalline electrodes and the intermediate small local maximum were attributed to the adsorption on the above-mentioned three crystal faces. According to the surface optical spectroscopic measurements of McIntyre and Peck [35], the strongly bonded hydrogen has been assigned to atomic adsorption taking place at face (100), the weakly bonded one to molecules in covalent bonds, and the one with intermediate strength to species dissolved in the metal. The assumption of Stonehart [36] and Kinoshita et al. [18,37] is similar: in these works the strongly bonded hydrogen is assumed to be chemisorbed on the (100) plane, the weakly bonded hydrogen is attributed to a molecular ion (H_2^+), and the intermediate peak is attributed to adsorption taking place at edges and peaks. In contrast, Bagotzky et al. [38] interpret the appearance of the individual hydrogen maxima with the effect of ion adsorption from the electrolyte or by the effect of carrier, instead of sorption at different edges or peaks.

On the basis of infrared spectroscopic studies, Bewick [39] assumed that the weakly bonded hydrogen atom is in strong interaction with the water molecules on the (111) plane of Pt. This assumption is confirmed by the entropy measurements of Shibata and Sumino [40].

The energetic inhomogeneity of sorbed hydrogen is supported by other physicochemical methods. On platinum, generally three types of hydrogen are detected by TPD, and they are attributed to atomic or molecular hydrogen attached to one or two Pt atoms [41] (see also Chapter 4).

Despite these different behaviors, it is a widely accepted view in modern electrochemistry that the first characteristic maximum of the voltammogram of polycrystalline Pt can be attributed to adsorption at plane (111), and the second, to adsorption at plane (100) [18,42]. It is self-evident that if the surface pattern of Pt atoms and the proportion of various crystallographic faces is affected by any means, adsorption characteristics, and thus the voltammogram, will become different. This also explains the differences between the polarization curves published by different authors [43].

Returning to the curves of Fig. 4, it can be seen that of the other polycrystalline metals, a characteristic hydrogen peak and a shoulder appear in the curve of Rh, there is no sorbed hydrogen on Au, and on the curve of Ir, two characteristic hydrogen maxima appear, as on the curve of Pt. However, since the voltammograms obtained on single-crystal faces of low indexes [(111), (100), and (110)] of Ir are essentially different from those of the corresponding faces of Pt, it seems to be proved that the sorption of hydrogen is strongly influenced by the nature of metal in addition to geometric factors [43], as also evidenced by TPD data [44].

Of noble metals, the interpretation of the hydrogen sorption of Pd is the most complicated and most debated, since the metal may contain large amounts of dissolved hydrogen in addition to adsorbed hydrogen (a maximum of 0.69 H atoms/Pd), and their separation is not easy. We mentioned these problems and methods for their solution in Section II.A. Figure 1 illustrates the separation of the two types of hydrogen by galvanostatic polarization, and Fig. 2 shows the peaks obtained by potentiodynamic polarization. The first, larger maximum corresponds to dissolved hydrogen; the second, characteristic peak and smaller shoulders can be assigned to the adsorbed hydrogen species. Hydrogen dissolved in Pd and adsorbed in two sites may also be detected by TPD [44] (see also Chapter 1 for Pd single-crystal studies).

Hydrogen sorption on nickel catalysts, very frequently applied in catalysis, is similar to that of Pd. In the electrochemical investigation of sorption, some authors detected only one, others two hydrogen peaks, and significant differences can also be found among the potential values assigned to the maxima. According to Kinza [45], these controversies are due to the fact that the exchange current density of the Ni-H system is very low, and thus, when polarizing at too high a rate, an overpotential occurs which leads to false results. To eliminate this problem, Kinza recorded galvanostatic curves in 1 N KOH at a very low current density, between 20 and 200 mV, and by differentiating this curve, obtained a pseudo capacity curve representing the hydrogen region of Ni. According to the results, Ni can be characterized by a weakly bonded, adsorbed, and a strongly bonded, chemisorbed species: the former appears below 0.1 V, the latter at ca. 0.17 V (Fig. 5). The electron escape energy [46] and the results of TPD measurements (see Chapters 4 and 5) show that there are at least two types of sorbed hydrogen on Ni.

Effect of Organic Substances, Anions, and Prepolarization on Hydrogen Adsorption

Like other physicochemical measurements, electrochemical polarization poses a question as to whether the measurement itself or the preparation or pretreatment of the sample may change the surfaces to be studied and thus the hydrogen sorption ability. According to experience, organic contaminations, anion adsorption from the electrolyte, and electrochemical polarization may lead equally to changes in the surface.

Organic compounds such as alcohols adsorb irreversibly on noble metals, and can be removed from the surface only by oxidation. Being adsorbed

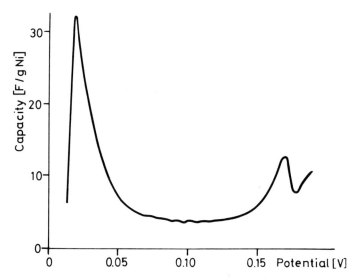

FIGURE 5 Differential pseudocapacity in 1 M KOH at 25°C of nickel catalyst prepared from nickel carbonyl. (From Ref. 45.)

more strongly than hydrogen, they displace hydrogen on the surface [16]. Similar behavior can also be observed for carbon monoxide.

The specific adsorption of anions on Ir single crystals was studied by Mooto and Furuya [47], and it was established that hydrogen sorption is strongly influenced by HSO_4^- ions. Comparing the results to those obtained on platinum, they found that anion adsorption is affected primarily by the physical structure of the metal (i.e., the nature of crystal faces), not by the chemical nature. It was established that this effect is strongest on the (111) plane of the highest density.

Clavilier [48] and Scherson and Kolb [49], investigating Au(111) and Pt(111) single-crystal surfaces, arrived at the conclusion that the peak found in the voltammogram of these metals in the vicinity of the double-layer region (i.e., 0.45 V) is caused by a phase transition due to the adsorption of anions.

The effects of electrochemical measurement and polarization pretreatment on the surface and adsorption properties are discussed by a great number of authors. In their investigations on Au(111) single-crystal surfaces, Kolb and Lehmpfuhl [50] have found changes on the surface accompanied by a decrease in surface energy. As found by Cervino et al. [51-53], fast ($\geqslant 10^4$ V s^{-1}) triangular polarization applied for a longer period (12 h) causes structural changes on the surface of Pt which are commensurate with the effect of thermal or chemical pretreatments. Upon treatment, the crystal structure of the surface layers of originally polycrystalline Pt undergoes such deep changes that the potentiodynamic curve of the sample, depending on the rate of the fast polarization and on the lower and upper limits of voltage, becomes similar to that of the (111) face (Fig. 6a) or the (100) face (Fig. 6b) of Pt single crystal. This structure is stable enough to remain unchanged upon a relatively slow polarization (0.5 V s^{-1}) below 0.9 V. Similar results were obtained by Canullo et al. [54], who, by applying fast, repetitive square-wave potential signals, obtained potentiodynamic curves characteristic of Pt(100) and Pt(111) planes, depending on polarization conditions, from polycrystalline platinum. It is pointed out by the authors that the purposive formation of planes oriented in desired ways may also be very important in electrocatalysis.

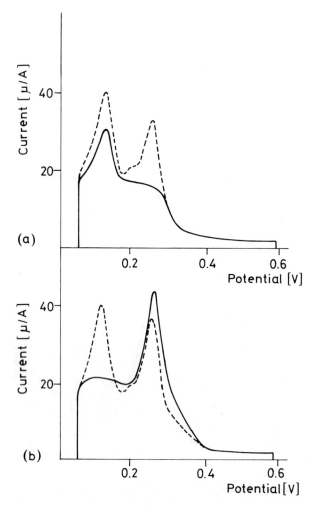

FIGURE 6 Potentiodynamic curves of platinum in 1 M H_2SO_4 at 0.1 V s^{-1} following a repetitive triangular potential sweep treatment for 12 h: (a) at 10^4 V s^{-1} between 0.42 and 1.1 V; (b) at 1.4×10^4 V s^{-1} between 0.06 and 1.5 V. Dashed curves, untreated polycrystalline electrode; solid curves, prepolarized electrode. (From Ref. 52.)

It has been found by Motoo and Furuya [55] that if prepolarization is carried out in a potential region that belongs to the oxygen adsorption region, one reason for the surface changes may be electrochemical oxygen adsorption and subsequent desorption (reduction), whereupon a new Pt atom is adsorbed into the hole left on the surface.

All these changes in surface and adsorption behavior on the effect of polarization can be eliminated by calculating the surface area and sorption capacity of the "living" polycrystalline catalyst from the first polarization curve.

Effect of Dispersity and Metal-Support Interaction on Hydrogen Sorption

If the dispersity of a metal is increased by either a special method of preparation or the application of a support, not only can a quantitative change be observed in its physical properties (i.e., surface area, sorption capacity), but qualitative changes occur as well. These changes are due to changes in geometry and electronic structure. It is one of the geometric changes that, with increasing dispersity, the orientation of the exposed faces of microcrystals change, and the amount of coordinatively unsaturated metal atoms increases. Detailed calculations on this effect can be found in the works of Van Hardeveld and Hartog [56,57].

The differences between bulk and disperse phases become even greater if the metal is brought onto a support and strong metal-support interaction (SMSI) takes place between them [58]. This phenomenon is discussed at length in Chapter 13.

Of course, changes in geometry and electronic structure also affect the sorption capacity of disperse metals. According to Nakamura and Yamada [59], the hydrogen sorption of unsupported, atomically dispersed Pt may reach a ratio of 2:1 instead of the generally assumed atomic ratio of 1:1. The number of hydrogen atoms related to one metal may be 3:1 or even higher on some supported catalysts, due to the spillover phenomenon [60] (see Chapter 12).

The effect of dispersity on hydrogen sorption can also be seen from the electrochemical polarization curve. According to the electrochemical investigations of Zakarina et al. [61] carried out on disperse, nonsupported Pd, there is only adsorbed hydrogen on the metal, with no dissolved hydrogen if the particle size decreases to below 2.5 nm.

Similarly, Polyánszky et al. [62] and Mallát et al. [63] found in studies on the hydrogen sorption of carbon-supported Pd catalysts (Table 1) that that when the amount of Pd in the catalyst decreases from 100% to 1% and the surface area of the metal increases simultaneously to more than three times the original area, the amount of strongly bonded (adsorbed) hydrogen increases due to the increased surface area, and thus the share of weakly bonded (dissolved) hydrogen decreases from 87% to 30% within the total amount of hydrogen. Simultaneously, the hydrogen content becomes more homogeneous in energy; only two hydrogen peaks appear instead of four in the potentiodynamic curve of the 1% Pd/C catalyst.

If supported on carbon, the sorption ability of platinum also undergoes a quantitative change [27], in terms of the ratio of the first and second characteristic hydrogen peaks of the potentiodynamic curves. Whereas on nonsupported platinum the ratio of weakly and strongly bonded hydrogen species is close to unity (Table 2), this value changes significantly with the effect of carbon support: on a 5 wt % Pt/C catalyst there is already five times as much weakly bonded hydrogen as strongly bonded hydrogen. In parallel, the bonding energy of weakly bonded species also increases (i.e., the hydrogen peaks are shifted toward higher potentials). The reason for

TABLE 1 Hydrogen Sorption on Carbon-Supported Pd Catalyst

Pd content (wt %)	Total amount of sorbed H (cm^3/g metal)	Ratio of weakly bonded and total amount of H (%)	Metal surface area (m^2/g Pd)
1	60	30	170
10	64	50	130
100	100	87	53

TABLE 2 Ratio of Weakly and Strongly Bonded
Hydrogen on Carbon-Supported Platinum

Metal content (wt % Pt)	Weakly bonded H / strongly bonded H
5	5.20
10	5.02
20	2.35
100	1.06

this phenomenon may be, in addition to structural changes accompanying higher dispersity, the spillover effect. This is further supported by the calculated H:Pt atomic ratio, from which the number of hydrogen atoms attached to surface metal atoms turns out to be 3.5 times higher on 5 wt % Pt/C catalysts than on nonsupported platinum.

From polarization studies on the hydrogen content of carbon-supported Pt catalysts, Bagotzky et al. also found [64] that the lower the metal content on the support (the more disperse the catalyst), the lower the ratio of strongly bonded hydrogen with respect to the total hydrogen content, and in the limiting case, on quasi-amorphous platinum, only weakly bonded hydrogen is adsorbed. This phenomenon is interpreted by the authors in terms of two factors. On the one hand, the energy of the crystal lattice and the coordination links between atoms in particles consisting of just a few atoms are significantly different from the corresponding features of large (nearly bulk) crystals, and on the other hand, the fraction of metal atoms directly interacting with the support is higher, and thus the probability of spillover increases.

Hydrogen Sorption on Bimetallic Catalysts

The preceding sections have shown the wealth of information present in electrochemical polarization curves: the characteristics of hydrogen and oxygen sorption can be determined from a single curve. With bimetallic catalysts there is a further possibility; from the reduction(cathodic curve) or dissolution (anodic curve) of a few atomic surface layers, the surface ratio and the "state" of the two metals (i.e., whether an alloy was formed, and whether this alloy is ordered) can be determined [16, 21]. In the foregoing, both cases will be illustrated by examples.

Zakumbaeva et al. [65] studied carbon-supported noble metal alloys (Pt-Rh, Pt-Ru, and Pd-Rh) and found that the amount of sorbed hydrogen and the energy distribution of metal-hydrogen bonds can be varied over a wide range with the metal content or the chemical composition. Figure 7 shows the potentiodynamic curves of some Pt-Rh/C catalysts. The energy distribution of metal-hydrogen bonds is similar to that of nonalloyed Pt in the case of an alloy with high Pt content, and similar to nonalloyed Rh in alloys of high Rh content.

If the measurements are carried out at different temperatures (e.g., 20 to 60°C) in addition to the amount of sorbed hydrogen, the heat of adsorption can be determined as a function of coverage (Table 3). Generally, the higher the Pt content of the catalyst, the higher the heat of adsorption obtained.

Table 3 illustrates the relationship between bulk composition and hydrogen sorption properties of catalysts. This is, however, only formal; it is widely

9. Electrochemistry of Surface Hydrogen

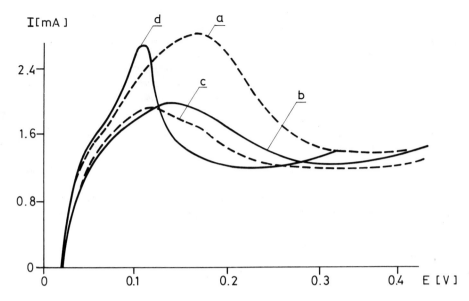

FIGURE 7 Potentiodynamic curves of Pt-Rh/C catalysts in 0.5 M H_2SO_4. Sweep rate = 0.1 mV s^{-1}. a, 10 wt % Pt/C; b, 7.5 wt % Pt + 2.5 wt % Rh/C; c, 2.5 wt % Pt + 7.5 wt % Rh/C; d, 10 wt % Rh/C. (From Ref. 65.)

TABLE 3 Hydrogen Sorption on Pt-Rh/C Catalysts

Pt loading (wt %)	Rh loading (wt %)	H_{ads} at 20°C (cm^3/g metal)	Q_{ads} (θ = 0.05-0.08) (kJ mol^{-1})
0	10	52.8	44.3-9.2
2.5	7.5	45.9	48.5-15.9
5	5	56.5	51.0-18.8
7.5	2.5	56.6	55.6-22.6
10	0	84.0	75.2-35.9
0	20	14.3	37.2-4.6
5	15	35.4	51.0-20.9
10	10	40.3	55.2-23.4
15	5	37.5	62.8-26.3
20	0	24.6	79.4-43.5

Source: Ref. 65.

known that surface and bulk compositions of alloys may be significantly different, and that the sorption and catalytic properties are determined by the structure and composition of a surface layer a few atoms thick [66-68]. It was shown by Woods et al. [16,69] for numerous noble metal alloy pairs including Pt-Rh alloys that the location of the cathodic oxide reduction maximum (E) for the alloy is between the potentials characteristic of the two pure metals (E_A and E_B), and its exact value is a linear function of surface composition:

$$E = C_A E_A + C_B E_B \qquad (3)$$

where C denotes atomic fractions (concentrations) and A and B refer to the components of the alloy.

Consequently, the amount of sorbed hydrogen, the energy distribution of metal-hydrogen bonds, and the surface composition of the alloy can be determined from a single cyclic voltammogram, and these parameters are already related by meaningful correlations.

It has been mentioned, but should be stressed again in connection with alloys, that upon repeating potential cycles, surface composition may be drastically changed even for noble metal alloys. Figure 8 shows the voltammogram of an alloy with surface composition 74 at% Rh-26 at% Pt [70]. In the first polarization cycle the cathodic oxide reduction peak is at 0.44 V, close

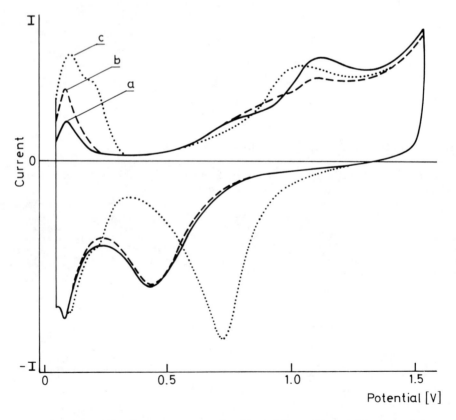

FIGURE 8 Cyclic voltammograms of a 26 at % Pt + 74 at % Rh electrode in 1 M H_2SO_4 at 25°C. Sweep rate = 40 mV s^{-1}. a, First cycle; b, second cycle; c, 100th cycle. (From Ref. 70.)

to the value characteristic of Rh (0.33 V). During the cycles, the two metals slowly dissolve from the surface; for example, after 200 cycles 2.4 µg cm^{-2} of Pt and 13.2 µg cm^{-2} of Rh were dissolved. Therefore, surface composition (and thus the location of the cathodic oxide reduction peak) has gradually shifted, and after 100 to 200 cycles practically Pt alone can be found in the surface layer (oxide reduction peak at 0.77 V). It is worth observing the significant change that takes place in hydrogen sorption as early as the second cycle. In the case of alloys, too, the correct method is to use the first cycle for the characterization of catalysts, and to suppress dissolution, the highest possible sweep rate and lowest anodic limit must be applied [16].

A contradiction has to be noted here. According to the so-called coherent potential approximation method (localized model) [66,71] developed in the 1960s for the description of electronic structures of disordered alloys, homogeneous noble metal alloys should produce two, energetically different cathodic oxygen peaks, at potentials different from those of pure components, owing to the interaction with neighboring atoms. In contrast, in the case of Pt-Rh (and some other) alloys, only one peak could be detected, which can be interpreted by assuming a delocalized electron orbital of chemisorption [16] (rigid band theory) [72].

The investigation of Pd-Cu alloys also proves the value of electrochemical polarization for studying bimetallic catalysts [73,74]. It can be established from the potentiodynamic curves of catalysts prepared by simultaneous reduction that when alloying palladium with increasing amounts of copper, the amounts of both weakly bonded (mostly dissolved) and strongly bonded (mostly adsorbed) hydrogen decrease, as well as the energy of the metal-hydrogen bond (i.e., the maxima are shifted gradually toward lower potentials).

Plotting the atomic ratio of hydrogen dissolved in the β phase to all metals as a function of copper concentration, the points fall on a single line to a good approximation. The line intercepts the abscissa at the copper content where the dissolution of hydrogen terminates (44 at%). From the quasi-equilibrium peak potentials (E) determined under relatively low sweep rates, the free enthalpy change of β → α phase transition can be determined by means of the equation $\Delta G = -2FE$, where F is the Faraday number. Plotting these values against copper content, again a straight line can be obtained, which intercepts the abscissa at 42 at%. The values obtained by the two methods for the limit of β-phase dissolution of hydrogen are in excellent agreement.

In the case of Pd-Cu alloys (and for some other noble metal-non-noble metal alloys, such as Pd-Hg [75] and Pd-Co [76]), the structure of the surface phase can be determined from the anodic branch of the polarization curve by the dissolution of the surface layer. Figure 9 shows the potentiodynamic curves of Pd-Cu/C catalysts obtained by consecutive reduction of Cu. The peak observed for the samples of low Cu content at ca. 0.5 V can be attributed to the dissolution of a $PdCu_3$ compound phase [77]. With increasing copper content, the amount of this phase increases, and according to the above, the change in hydrogen sorption indicates that Pd-rich (homogeneous) alloy is also formed. Starting from a copper content of 30 at%, unalloyed copper can also be found on the surface: the peak of the dissolution of copper starts below 0.3 V ($E°_{Cu/Cu^{2+}} = 0.34$ V). In parallel, hydrogen sorption decreases to a minimum. This method could be used, therefore, to show that in the consecutive reduction of copper onto palladium, three different phases containing copper were formed.

Metal adsorption, discovered by electrochemists and investigated recently with increasing intensity, can also be regarded as a special case of consecutive reduction. The phenomenon is due to the fact that if the potential of

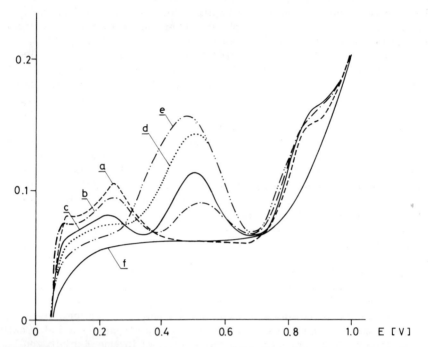

FIGURE 9 Potentiodynamic curves of Pd-Cu catalysts, supported on activated carbon. 0.5 M H_2SO_4, 10 mV min^{-1} sweep rate, m = 4 mg; 5 wt % Pd in all cases. a, Pd/C; b, 10 at % Cu-Pd/C; c, 20 at % Cu-Pd/C; d, 30 at % Cu-Pd/C; e, 40 at % Cu-Pd/C; f, activated carbon support. (From Ref. 74.)

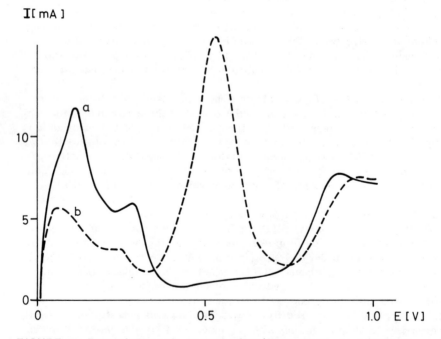

FIGURE 10 Potentiodynamic curves of Pt/Pt electrode in 1 M $HClO_4$. a, Clean surface; b, Re adsorption on Pt. (From Ref. 80.)

matrix metal (M_A) is slightly more positive than the metal/metal ion potential of the alloying metal (M_B) measured in the given solution during reduction, only adsorbed metal may be formed (M_A - M_B), so-called bulk metal formation (M_B - M_B) being impossible. Exhaustive discussions are devoted to metal-hydrogen systems in the reviews by Kolb [78] and Adzic [79]. In this work only one example is given, which is also relevant to catalysis (Fig. 10).

In studies on Re adsorbed on Pt, it was established [80] that Re adheres very strongly to the surface of Pt (it is ionized only in the region 0.3 to 0.7 V), while it causes the amount of adsorbed hydrogen to be decreased. It is to be noted that the adsorption of small metal atoms unable to sorb hydrogen (e.g., Cu) at low coverage decrease primarily the sorption of weakly bonded hydrogen, whereas for larger metal atoms (e.g., Pb) such a specific effect has not been found [78,79].

III. MEASUREMENTS IN THE GAS PHASE

A. The Measuring Cell

The great majority of electrochemical catalytic studies are carried out in the liquid phase, at around room temperature. However, a significant part of heterogeneous catalytic processes take place at a sufficient rate only above 100°C, and the reactants are in a gas or vapor state. Under such conditions the classical electrochemical cells can, of course, not be applied.

The liquid-phase electrochemical method was first extended to the investigation of catalytic gas-phase hydrogenation by Druz et al. [81-83]. The construction of the measuring cell was as follows. A platinum layer was deposited on the inner and outer surfaces of a glass tube sealed on one side. Air was led to the inner platinum layer; this was used as a reference electrode. Hydrogen and reactants were fed to the outer platinum layer: this was the measuring electrode. The voltage of this cell was measured by the compensation method. Investigations on the hydrogenation of benzene in this cell have shown that hydrogen replaces benzene on the surface of platinum.

In our laboratory, a measuring cell was constructed [84] which makes it possible to investigate the processes taking place on the surface of supported industrial metal catalysts in the presence of hydrogen. Measurements can be carried out both independently of the catalytic reaction and during hydrogenation reactions.

The scheme of the special galvanic cell is shown in sectional view in Fig. 11. The lower part of the glass tube (1) made of sodium glass is closed by the membrane (2), made of Dole-type glass of "high" conductivity, consisting of 22 wt % Na_2O, 6 wt % CaO, and 72 wt % SiO_2. This glass membrane serves as the solid electrolyte of the cell and also separates the reference and measuring electrodes. On the inner surface of the glass tube, silver film (3) is deposited. Against this layer a platinum ring (4) is pressed by a spring (5), ensuring electric contact. The electric cable (8) of the inner electrode is attached to the platinum ring. Air is fed into the inner space of the glass tube [i.e., the reference electrode of the cell is an $Ag/Ag_2O/O_2$(air) electrode]. This reference half-cell is in contact with a supported metal catalyst (6) placed in a basket (7) made of nickel or silver net, providing electric contact with the catalyst. The catalyst, contacted with hydrogen, is the measuring electrode of the cell, the terminal (9) of which is connected to an oscillating capacitor electrometer (10) of high input resistance.

The resistance of the measuring system is determined primarily by the electric resistance of the Dole-glass membrane closing the reference half-cell. The internal resistance of the entire electrode chain does not exceed 10^9 Ω; in the majority of cases it is between 10^6 and 10^8 Ω. The cell voltage is

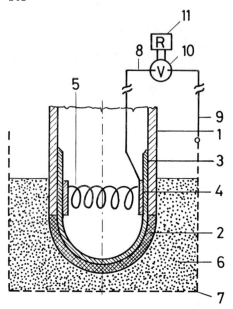

FIGURE 11 Cross section of the electrochemical cell for measurement of catalyst's potential in gas phase.

measured by an oscillating capacitor electrometer having an input resistance higher than 10^{15} Ω. With the use of screened cables [cables (8) and (9) in Fig. 11], external electric perturbations can be eliminated.

The Electrode Processes of the Measuring Cell

In the cell, one side of the glass electrode is contacted by a silver layer and oxygen, the other side with hydrogen and metal catalyst. For a description of the processes taking place on a catalyst/glass boundary, the electrode processes of the glass electrode [85, 86] and the hydrogen electrode [87] may be used as starting point.

1. Gas-phase hydrogen is adsorbed molecularly and then atomically on the surface of the catalyst:

$$H_2(g) \rightleftarrows H_2(ads) \rightleftarrows 2H(ads) \tag{4}$$

2. Protons cross the boundary between the catalyst and the glass, leaving electrons on the catalyst:

$$2H(ads) \rightleftarrows 2H^+ + 2e^- \tag{5}$$

3. Protons entering the surface layer of glass are attached to the so-called nonbridged oxygen ions (terminal oxygen atoms):

$$2H^+ + 2\ {-}\!\!\overset{|}{\underset{|}{Si}}\!\!{-}O^- \rightleftarrows 2\ {-}\!\!\overset{|}{\underset{|}{Si}}\!\!{-}OH \tag{6}$$

and start, simultaneously, a current through the glass.

9. *Electrochemistry of Surface Hydrogen*

By combining reactions 1 to 3 into a single equation, the overall electrode process of the measuring electrode can be obtained as

$$H_2(g) + 2\ |{-}Si{-}O^-\ \rightleftharpoons\ 2\ |{-}Si{-}OH + 2e^- \tag{7}$$

on the basis of Eq. (7), the potential of the catalyst with respect to the surface layer of the glass (the boundary potential) can be given by the equation

$$E_K = E_K^\circ + \frac{RT}{2F} \ln \frac{a_{pr}^2}{a_{sa}^2 p_{H_2}} \tag{8}$$

where E_K° is the normal potential of the electrode, a_{pr} the activity of the protonated groups, and a_{sa} the activity of the silicate anion groups.

On the basis of Eq. (8), at constant temperature the potential of the measuring electrode and the cell voltage measured by the measuring cell depend on the partial pressure of hydrogen only, since no changes take place in the reference electrode, and the mean activities and ratios of silicate anion groups and protonated groups presumably do not change.

For the reaction in Eq. (4) the equilibrium constant of the atomic adsorption of hydrogen (K_H) can be expressed as

$$K_H = \frac{a_H^2}{p_{H_2}} \tag{9}$$

where a_H is the activity of adsorbed hydrogen atoms, which is in conceptual relationship with surface concentration. Substituting Eq. (9) into Eq. (8), one obtains

$$E_K = E_K^\circ + \frac{RT}{F} \ln \frac{a_{pr}(K_H)^{1/2}}{a_{sa} a_H} \tag{10}$$

Similar relationships can be derived for the reference electrode. The inner surface of the glass membrane in the measuring cell is covered with a silver layer which is contacted with air [i.e., it is an $Ag/Ag_2O/O_2$(air) electrode]. This electrode in liquid phase was found by Hammer and Craig to be very stable [88]. According to our work [89], this electrode is extremely stable above 100°C. The overall electrode process of the reference electrode is as follows:

$$2\ |{-}Si{-}OH + 2e^- + (1/2)O_2(g) \rightleftharpoons 2\ |{-}Si{-}O^- + H_2O(g) \tag{11}$$

Reducing Eq. (7) and (11) into a single equation, one obtains the overall electrode process of the electrochemical cell:

$$H_2(g) + (1/2)O_2(g) \rightleftharpoons H_2O(g) \tag{12}$$

which shows that the cell is a hydrogen-oxygen cell. The electromotive force of this cell is 1240 mV at 25°C.

The voltage of the galvanic cell (E) is the difference of the potentials of the reference and measuring electrodes, which, after simplifications, can be written as

$$E = E_{S,T} + \frac{RT}{F} \ln \frac{a_H}{K_H^{1/2}} \tag{13}$$

where $E_{S,T}$ is the sum of all factors that are invariant at constant temperature.

Equation (13) shows the relationship between the voltage measured by an electrochemical cell and the activity or surface concentration of adsorbed hydrogen. Thus, on the basis of changes in cell voltage, conclusions can be drawn on the changes in the concentration of hydrogen sorbed on the surface of the catalyst.

B. Measurements in the Presence of Hydrogen

Temperature Dependence of the Cell Voltage

The measuring cell shown in Fig. 11 was placed in a tubular reactor in which the temperature and the feed rates of the reactants could be controlled precisely. The temperature dependence of cell voltage was studied with a great number of widely different catalysts, and a linear relationship was obtained in all cases [this corresponds to Eq. (13)]. As an example, Fig. 12 shows the results obtained with Ni/SiO$_2$, Pd/Klinosorb, and partially oxidized Pd/Klinosorb catalysts.

From thermodynamic data, the calculated temperature coefficient of the hydrogen/oxygen cell is -0.23 mV/°C. The measured temperature dependences were higher than this value for all three catalysts. This indicates that probably only the output voltage (not the electromotive force) of the cell could be measured.

Dependence of the Cell Voltage on the Partial Hydrogen Pressure

The dependence of cell voltage on the partial pressure of hydrogen was studied with the hydrogen fed into the tubular reactor diluted with high-

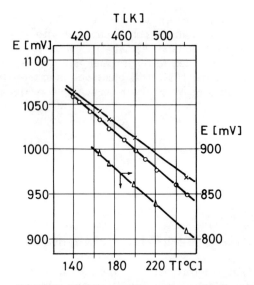

FIGURE 12 Temperature dependence of the cell voltage (E). Catalysts: +——+, Ni/SiO$_2$; o——o, Pd/Klinosorb; △——△, partially oxidized Pd/Klinosorb. Presence of hydrogen, 1 atm.

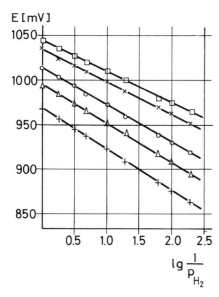

FIGURE 13 Dependence of the cell voltage on partial pressure of hydrogen. Catalyst, Ni/SiO$_2$. Temperature: □, 165°C; x, 175°C; ○, 200°C; △, 220°C; +, 250°C.

purity argon in the desired ratio. As an example, the cell voltages obtained with a Ni/SiO$_2$ catalyst were plotted against the logarithm of the reciprocal of partial hydrogen pressure in Fig. 13. In connection with the data measured it is noted that the cell voltages were constant within ± 0.5 mV, and these values could be reproduced to at least the same precision as a function of the partial pressure of hydrogen. The slopes of the straight lines shown in Fig. 13 and of those measured with the other two catalysts can be seen in Table 4, together with the slope calculated according to Eq. (13).

On the basis of the data in Table 4, it can be established that the experimental slope reaches the calculated one for none of the catalysts. Therefore, the relationship between the voltage measured with the cell and the partial pressure of hydrogen can be described by the following modified equation:

$$E = E_{S,T} + \left(\frac{RT}{2F} - w_{\underline{K},\underline{T}}\right) \ln p_{H_2} \tag{14}$$

Subscripts $\underline{K}$ and $\underline{T}$ of correction factor w in Eq. (14) refer to the fact that the value of this factor depends on the nature of the catalyst and on temperature. The value of correction factor decreases with increasing temperature for all catalysts studied.

Investigation of the Reduction and Activation of Catalysts

The variation of cell voltage as a function of time in the hydrogenic activation of supported nickel, palladium, and platinum catalysts can be seen in Fig. 14. The three curves are similar in character: the cell voltage is set at a value constant in time after an initially fast change.

In the case of Ni/SiO$_2$ catalyst, nickel oxide was in the catalyst at the beginning of measurement, and this was reduced to active metal on the effect of hydrogen, and the surface concentration of adsorbed hydrogen atoms increased. The progress and then the termination of this process were

TABLE 4 Dependence of the Cell Voltage on Partial Pressure of Hydrogen: Slopes of the Linearized Relationship

Temperature °C	$\dfrac{mV}{\ln(p_{H_2})^{-1}}$			
	Calc.	Ni/SiO$_2$	Pd/Klinosorb	Pd$_{ox}$/Klinosorb
165	42.6	35	35	25.5
175	43.6	36	37	28.5
200	46.0	41	37.5	32.5
220	47.9	43	46	34.5
250	50.8	46	47.5	40

shown by the change and subsequent stabilization of the cell voltage. By this time the catalyst, according to our investigations, has reached maximum activity in the hydrogenation of benzene.

Literature data on the reduction of supported nickel catalysts are rather contradictory. According to the measurements of Hill and Selwood [90], alumina-supported nickel oxide should be reduced at 450°C for 130 h in order to complete the formation of nickel metal. According to later investigations [91,92], shorter times (20 to 40 h) are sufficient. On the basis of studying the preparation of Ni/Al$_2$O$_3$ catalysts, Bartholomew and Farrauto [93] have shown that calcination before reduction increases the time required for reduction. This is the probable reason for the different reduction times obtained by different authors. (In our measurements the catalyst was not calcined.)

According to our magnetic measurements [94], during 4 h at 350°C, 80% of nickel was reduced into metal in a Ni/SiO$_2$ catalyst investigated in the electrochemical cell. Therefore (Fig. 14), the electrochemical measurement shows the reduction taking place in the surface layer of the catalyst. A comparison of the benzene hydrogenating activity of the catalyst and maximum cell voltage proves that the formation of surface metal atoms, which have a determining role in catalytic activity, can be followed satisfactorily by the electrochemical cell.

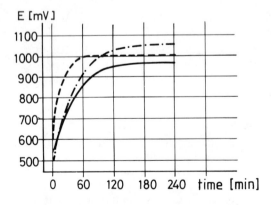

FIGURE 14 Cell voltage (E) as a function of time (min) during activation of catalysts. ———, Ni/SiO$_2$ at 250°C; ---, Pd/Klinosorb at 200°C; ·—·—, Pt/Al$_2$O$_3$ at 360°C.

The Pd/Klinosorb catalyst (Fig. 14, dashed line) was prepared by the reduction of palladium hydroxide. This catalyst was used to investigate the selective hydrogenation of phenol into cyclohexanone. With catalyst impregnated with palladium chloride (i.e., not converted into hydroxide), the cell voltage reaches the maximum value significantly more slowly. This maximum is lower by more than 150 mV than the value obtained with the catalyst prepared from palladium hydroxide, and the catalyst is less active and less selective.

In addition to following the progress of reduction, this experimental method may promote the development of catalysts with optimum properties for given purposes and the control of the reproducibility of catalyst preparation.

Effect of Air on the Cell Voltage

The experiment shown in Fig. 15 was performed in a way that in moments marked in the Figure by points 1 and 3, 5 vol % of air was admixed to the hydrogen feed for short periods at 250°C, the introduction of air was terminated at times 2 and 4. This experiment, too, proves that the electrochemical cell measures the surface concentration of adsorbed hydrogen. Upon the introduction of air, oxygen reacted with adsorbed hydrogen, decreasing its surface concentration. Consequently, the potential of the catalyst was shifted toward that of the reference electrode, and thus the measured cell voltage decreased. Upon terminating the introduction of air, an opposite process took place.

C. Measurements During Hydrogenation of Benzene and Phenol

The literature concerning the hydrogenation of benzene is rather contradictory. To illustrate the situation, let us quote some conclusions. According to van Meerten et al. [95, 96], benzene and hydrogen are adsorbed at different sites of Ni/SiO_2 catalysts (i.e., their adsorption is noncompetitive). In contrast, Franco and Phillips [97] consider the sorption of benzene and hydrogen competitive on the same catalyst. On the basis of the investigations of Nakano and Kusunoki [98], hydrogen is adsorbed on the metal, whereas benzene is adsorbed on the support, close to the metal-support boundary. Wang and Huang [99], in studies on TiO_2-ZrO_2-supported nickel catalysts, support the results of van Meerten et al. and attribute the contradictions between

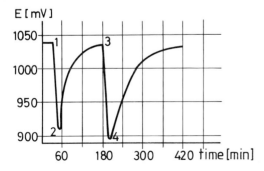

FIGURE 15 Changes of the cell voltage under the effect of oxygen. Catalyst, Ni/SiO_2. At times 1 and 3, a 2 vol % oxygen inlet is made into the pure hydrogen stream, and at times 2 and 4, oxygen is shut off, leaving pure hydrogen.

the results of various laboratories to the different metal contents of the catalysts investigated.

In our experiments the voltage of a special galvanic cell formed in the catalyst bed was recorded continuously during the operation of an integral tubular reactor. A characteristic recording obtained in the hydrogenation of benzene can be seen in Fig. 16. According to gas chromatographic measurements, a stable cell voltage corresponds to a constant conversion value (i.e., the reactor works in the steady state). Upon feeding benzene of 2.29 mol % tiophene content (Fig. 16, dashed line), the conversion has gradually decreased, and no constant cell voltage could be measured.
The stabilized cell voltage values are plotted in Fig. 17 as a function of contact time (the reciprocal of space velocity). This steady-state value is also influenced by the molar ratio of hydrogen to benzene.

The relationship between the partial pressures of reactants and cell voltage was also investigated under conditions where the extent of hydrogenation reaction was negligible. In these experiments the linear flow rate of reactants was significantly higher and the partial pressure of hydrogen was lower than during reaction studies. It can be established from Fig. 18 that the adsorption processes of benzene and hydrogen compete to a certain extent on the surface of the nickel catalyst. From the variation of the pressure of cyclohexane (Fig. 19) it can be concluded that hydrogen displaces cyclohexane on the surface of catalyst, and under conditions of hydrogenation the effect of the partial pressure of cyclohexane is less significant than that of benzene.

The results of benzene hydrogenation on Pd/Klinosorb catalyst are significantly different from those on Ni/SiO$_2$ catalyst. The steady-state cell voltages measured during the reaction hardly change with the variation of contact time. Hydrogen displaces benzene on the surface of palladium, and the reaction takes place on a palladium surface that is probably almost completely covered with hydrogen (Fig. 20).

Our results obtained with Ni/SiO$_2$ catalysts support the conclusions of Franco and Phillips, but contradict the results of van Meerten et al. The results on palladium catalysts support the work of Nakano and Kusunoki; it is hydrogen primarily that is sorbed on palladium, and the sorption of benzene on the metal surface is negligible. In contrast with the views of

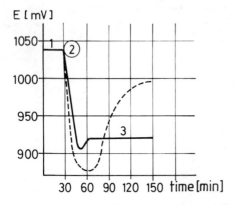

FIGURE 16 Changes of the cell voltage during hydrogenation of benzene on Ni/SiO$_2$ catalyst. The straight line, 1, indicates the E stabilized in hydrogen; benzene feed began at the time shown by 2; the stabilized section, 3, of $E(\pm 1$ mV) indicates the steady state of the reactor. The dashed line corresponds to hydrogenation of benzene containing 2.29 mol % thiophene.

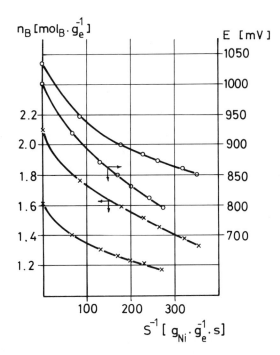

FIGURE 17 Cell voltage (○) and concentration of benzene (n_B, x) as a function of the contact time (s^{-1}) during hydrogenation of benzene in the steady state. Temperature, 165°C; catalyst, Ni/SiO_2; hydrogen-benzene mole ratio, $n_{H_2}^o/n_B^o$ = 6:1 and 10:1.

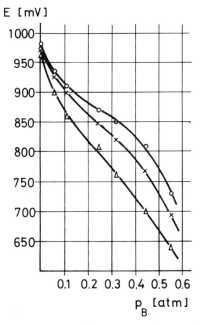

FIGURE 18 Dependence of the cell voltage on the partial pressure of benzene. Catalyst, Ni/SiO_2; temperature, 165°C. Partial pressure of hydrogen: ○, 0.015; x, 0.01; △, 0.005.

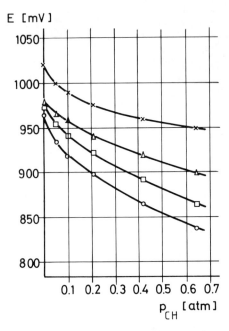

FIGURE 19 Dependence of the cell voltage on the partial pressure of cyclohexane. Catalyst, Ni/SiO$_2$; temperature, 165°C. Partial pressure of hydrogen: x, 0.21; △, 0.05; □, 0.01; ○, 0.005.

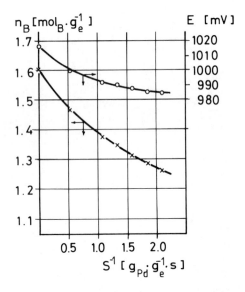

FIGURE 20 Cell voltage (○) and concentration of benzene (n_B, x) as a function of the contact time (s^{-1}) during hydrogenation of benzene in the steady state. Catalyst, Pd/Klinosorb; temperature, 165°C; mole ratio of hydrogen to benzene, 10:1.

Wang and Huang, we believe that it is mainly the quality of metal and the stability of the metal-hydrogen system, not the metal content of the catalyst, that determines sorption behavior. By means of the electrochemical method, the adsorption of hydrogen and benzene can be studied jointly, whereas by the older methods they must be measured separately.

The selective hydrogenation of phenol into cyclohexanone was also studied on Pd/Klinosorb catalyst [100]. The reaction can be characterized by the following kinetic equation:

$$w = k \frac{K_{H_2} p_F p_{H_2}}{[1 + (K_{H_2} p_{H_2})^{1/2}]^2} \tag{15}$$

In the case of the two mechanisms corresponding to this kinetic equation, selective hydrogenation takes place on a palladium surface almost completely covered with hydrogen [101]. This statement is supported by the cell voltages measured during selective hydrogenation (Fig. 21). Moreover, the relationship between the partial pressure of phenol and cell voltage proves that there is competitive adsorption between hydrogen and phenol (Fig. 22). With increasing partial pressure of hydrogen, the partial pressure of phenol influences cell voltage (i.e., the surface concentration of adsorbed hydrogen) to a decreasing extent. Extrapolating these data to the higher partial pressure of hydrogen used during hydrogenation reactions, it can be seen that in this case the adsorption of phenol is negligible. Comparing the results of electrochemical investigation of benzene with that of phenol, it can be assumed that the hydrogenation of these two compounds has a similar mechanism to that on the supported palladium catalyst.

The results described in this section illustrate that the electrochemical method can be applied to support reaction mechanisms assumed on the basis of kinetic data. A very significant advantage of the method is that measure-

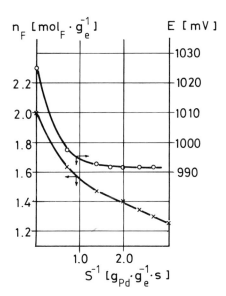

FIGURE 21 Cell voltage (○) and concentration of phenol (n_F, x) as a function of the contact time (s^{-1}) during selective hydrogenation of phenol to cyclohexanone. Catalysts, Pd/Klinosorb; temperature, 165°C; mole ratio of hydrogen to phenol, 6:1.

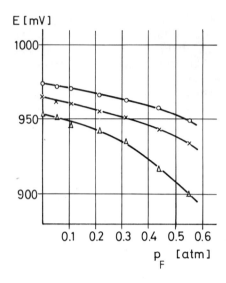

FIGURE 22 Dependence of the cell voltage on the partial pressure of phenol. Catalyst, Pd/Klinosorb; temperature, 165°C. Partial pressure of hydrogen: ○, 0.015; x, 0.01; △, 0.005.

ments can be carried out under conditions of catalytic reaction, using the acting catalyst as a measuring electrode. Therefore, it is not necessary to extrapolate experimental data obtained eventually under widely different conditions to the actual conditions of the catalytic reaction. By eliminating the great uncertainties arising from the extrapolation, electrochemical methods enable one to draw conclusions which to a certain extent, have stronger experimental foundations. This method may also have industrial relevance, since it can be used to check operation of the catalyst and the catalytic reactor [89,102].

REFERENCES

1. K. Müller, *J. Res. Inst. Catal. Hokkaido Univ.*, *17*(1): (1969).
2. B. D. McNicol, in *Characterization of Catalysts* (J. M. Thomas and R. M. Lambert, eds.), Wiley, Chichester, West Sussex, England (1980), p. 135.
3. G. M. Schwab (ed.), *Heterogene Katalyse*, Vol. 3, Springer-Verlag, Wier (1943), p. 133.
4. E. Müller and K. Schwabe, *Z. Elektrochem. Angew. Phys. Chem.*, *35*: 165 (1929); *Kolloid Z.*, *52*:163 (1930).
5. M. D. Birkett, A. T. Kuhn, and G. C. Bond, in *Catalysis, Specialist Periodical Reports*, Vol. 6 (G. C. Bond and G. Webb, eds.), Royal Society of Chemistry, London (1983), p. 61.
6. A. N. Frumkin and A. I. Sligin, *Dokl. Akad. Nauk SSSR*, *2*:173 (1934).
7. I. F. Tupitsin and L. P. Tverdovskii, *Zh. Fiz. Khim.*, *32*:345 (1958).
8. G. D. Zakumbaeva and D. V. Sokolskii, *Catalysis and Methods for Their Investigations* (in Russian), Nauka Kazakhskoi, Alma-Ata (1967).
9. D. V. Sokolskii, *Hydrogenations in Liquid Phase* (in Russian), Nauka Kazakhskoi, Alma-Ata (1979).
10. T. Biegler, *J. Electrochem. Soc.*, *114*:1261 (1967).

11. T. Mallát, É. Polyánszky, and J. Petró, *J. Catal.*, **44**:345 (1976).
12. I. Telcs and M. Jáky, *Acta Chim. Acad. Sci. Hung.*, **58**:275 (1968).
13. J. T. Maloy, *J. Chem. Educ.*, **60**:285 (1983).
14. D. E. Smith, *J. Chem. Educ.*, **60**:299 (1983).
15. F. G. Will and C. A. Knorr, *Z. Elektrochem.*, **64**:258, 270 (1960).
16. R. Woods, *J. Electroanal. Chem. Interfacial Electrochem.*, **9**:1 (1976).
17. P. Stonehart, H. A. Kozlowska, and B. E. Conway, *Proc. R. Soc. London,* **A310**:541 (1969).
18. K. Kinoshita, J. Lundquist, and P. Stonehart, *J. Catal.*, **31**:325 (1973).
19. S. Srinivasan and E. Gileadi, *Electrochim. Acta*, **11**:321 (1966).
20. A. J. Bard and L. R. Faulkner, *Electrochemical Methods Fundamentals and Applications*, Wiley-Interscience, New York (1980).
21. T. Mallát and J. Petró, *Acta Chim. Acad. Sci. Hung.*, **111**:477 (1982).
22. E. Keren and A. Soffer, *J. Catal.*, **50**:43 (1977).
23. G. Ja. Slaidin', Ja. E. Tchakste, and V. S. Bagotzkii, *Latv. PSR Izv. Akad. Nauk Latv. SSR Ser. Khim.*, 172 (1976).
24. D. A. Rand and R. Woods, *Anal. Chem.*, **47**:1481 (1975).
25. J. Bett, K. Knoshita, K. Routsis, and P. Stonehart, *J. Catal.*, **29**:160 (1973).
26. J. Petró, É. Polyánszky, and Z. Csürös, *J. Catal.*, **35**:289 (1974).
27. E. Polyanszky, J. Petro, and A. Sarkany, *Acta Chim. Acad. Sci. Hung.*, **104**:345 (1980).
28. D. V. Sokolskii, *Hydrogenation in Solutions* (in Russian), Nauka Kazakhskoi SSR, Alma-Ata (1962).
29. D. V. Sokolskii and G. D. Zakumbaeva, *Adsorption and Catalysis on Group VIII Metals in Solutions* (in Russian), Nauka Kazakhskoi SSR, Alma-Ata (1973).
30. A. T. Hubbard, J. L. Stickney, and M. P. Soriaga, *J. Electroanal. Chem. Interfacial Electrochem.*, **168**:43 (1984).
31. B. E. Conway, H. Angerstein-Kozlowska, and W. B. A. Sharp, *J. Chem. Soc. Faraday Trans. 1*, **74**:1373 (1978).
32. B. E. Conway and J. C. Currie, *J. Chem. Soc. Faraday Trans. 1*, **74**:1390 (1978).
33. A. Bewick and J. Russel, *J. Electroanal. Chem. Interfacial Electrochem.*, **132**:329 (1982).
34. F. G. Will, *J. Electrochem. Soc.*, **112**:451 (1965).
35. J. D. E. McIntyre and W. F. Peck, *Proceedings of the Symposium on Electrocatalysis* (1974), p. 212.
36. P. Stonehart, *Electrochim. Acta*, **15**:1853 (1970).
37. K. Kinoshita and P. Stonehart, *Electrochim. Acta*, **20**:101 (1975).
38. V. S. Bagotzky, J. B. Vasiliev, and J. J. Pisnograeva, *Electrochim. Acta*, **16**:2141 (1971).
39. A. Bewick, *J. Electroanal. Chem. Interfacial Electrochem.*, **150**:481 (1983).
40. Shigeo Shibata and Masae P. Sumino, *J. Electroanal. Chem. Interfacial Electrochem.*, **193**:135 (1985).
41. J. J. Stephan, V. Ponec, and W. M. H. Sachtler, *J. Catal.*, **37**:81 (1975).
42. A. T. Hubbard, *J. Electroanal. Chem. Interfacial Electrochem.*, **86**:271 (1978).
43. S. Motoo and N. Furuya, *J. Electroanal. Chem. Interfacial Electrochem.*, **167**:309 (1984).
44. N. M. Popova, L. V. Babenkova, and V. K. Solnyshkova, *Adsorption and Interaction of Simplest Gases with Group VIII Metals* (in Russian), Nauka Kazakhskoi SSR, Alma-Ata (1979).
45. H. Kinza, *Electrochim. Acta*, **24**:279 (1979).

46. A. N. Nigam and R. Rani, *Phys. Status Solidi*, *A41*:493 (1977).
47. S. Motoo and N. Furuya, *J. Electroanal. Chem. Interfacial Electrochem.*, *181*:301 (1984).
48. J. Clavilier, *J. Electroanal. Chem. Interfacial Electrochem.*, *107*:211 (1980).
49. D. A. Scherson and D. M. Kolb, *J. Electroanal. Chem. Interfacial Electrochem.*, *176*:353 (1984).
50. D. M. Kolb and G. Lehmpfuhl, *J. Electroanal. Chem. Interfacial Electrochem.*, *179*:289 (1984).
51. R. M. Cervino, W. E. Triaca, and A. J. Arvia, *J. Electroanal. Chem. Interfacial Electrochem.*, *182*:51 (1985).
52. R. M. Cervino, W. E. Triaca, and A. J. Arvia, *J. Electroanal. Chem. Interfacial Electrochem.*, *30*:1323 (1985).
53. R. M. Cervino, A. J. Arvia, and W. Vielstich, *Surf. Sci.*, *154*:623 (1985).
54. J. C. Canullo, W. E. Triaca, and H. J. Arvia, *J. Electroanal. Chem. Interfacial Electrochem.*, *175*:337 (1984).
55. S. Motoo and N. Furuya, *J. Electroanal. Chem. Interfacial Electrochem.*, *42*:339 (1984).
56. R. Van Hardeveld and F. Hartog, *Surf. Sci.*, *15*:189 (1969).
57. R. Van Hardeveld and F. Hartog, *Adv. Catal.*, *22*:75 (1972).
58. S. J. Tauster, S. C. Fung, and R. L. Garten, *J. Am. Chem. Soc.*, *100*:170 (1978).
59. Y. Nakamura and M. Yamada, *J. Catal.*, *39*:125 (1975).
60. M. Boudart, *Adv. Catal.*, *20*:153 (1969).
61. N. A. Zakarina, G. D. Zakumbaeva, N. F. Toktabaeva, A. Sh. Kuanyshev, and E. L. Litvyakova, *React. Kinet. Catal. Lett.*, *26*:441 (1984).
62. E. Polyánszky, T. Mallát, and J. Petró, *Acta Chim. Acad. Sci. Hung.*, *92*:147 (1977).
63. T. Mallát, J. Petró, and É. Polyánszky, *Magy. Kém. Lapja*, *36*:642 (1981).
64. V. S. Bagotzky, L. S. Kanevsky, and V. Sh. Palanker, *Electrochim. Acta*, *18*:473 (1973).
65. G. D. Zakumbaeva, L. A. Beketaeva, and R. M. Levit, *Elektrokhimija*, *15*:1138 (1979).
66. V. Ponec, *Catal. Rev.-Sci. Eng.*, *11*:1 (1975).
67. W. M. H. Sachtler and R. A. Van Santen, *Adv. Catal.*, *26*:69 (1977).
68. M. P. Seach, *Surf. Sci.*, *80*:8 (1979).
69. M. K. Aston, D. A. J. Rand, and R. Woods, *J. Electroanal. Chem. Interfacial Electrochem.*, *163*:199 (1984).
70. D. A. J. Rand and R. Woods, *J. Electroanal. Chem. Interfacial Electrochem.*, *36*:57 (1972).
71. N. D. Lang and H. Ehrenreich, *Phys. Rev.*, *168*:605 (1968).
72. F. Steitz, *The Modern Theory of Solids*, McGraw-Hill, New York (1940).
73. T. Mallát and J. Petró, *Acta Chim. Acad. Sci. Hung.*, *104*:381 (1981).
74. T. Mallát, J. Petró, and M. Schäffer, *Acta Chim. Acad. Sci. Hung.*, *98*:175 (1978).
75. J. Petró and T. Mallát, *Acta Chim. Acad. Sci. Hung.*, *95*:253 (1977).
76. T. Mallát, J. Petró, S. Szabó, and L. Marczis, *J. Electroanal. Chem. Interfacial Electrochem.*, *208*:169 (1986).
77. J. Petró, T. Mallát, S. Szabó, and F. Hange, *J. Electroanal. Chem. Interfacial Electrochem.*, *160*:289 (1984).
78. D. M. Kolb, *Adv. Electrochem. Electrochem. Eng.*, *11*:125 (1978).
79. R. R. Adzic, *Adv. Electrochem. Electrochem., Eng.*, *18*:159 (1985).
80. M. Ifandi, S. Szabó, and F. Nagy, *Elektrokhimija*, *18*:1205 (1982).
81. V. A. Druz and D. V. Sokolskii, *Tr. Inst. Khim. Nauk. Akad. Nauk. Kaz. SSR*, *8*:45 (1962).

82. V. A. Druz, D. V. Utugelov, and D. V. Sokolskii, *Dokl. Akad. Nauk SSSR*, *162*:373 (1965).
83. V. A. Druz, D. V. Utugelov, and D. V. Sokolskii, *Zh. Fiz. Khim.*, *40*: 1483 (1966).
84. Z. Csürös, J. Petró, T. Máthé, and A. Tungler, Hungarian Patent 158 593 (1971); U.S. Patent 3,645,111 (1972); German Patent 2,024,026 (1974).
85. M. Dole, *J. Chem. Phys.*, *2*:862 (1934).
86. D. J. G. Ives, *Reference Electrodes*, Academic Press, New York (1961), p. 150.
87. D. J. G. Ives, *Reference Electrodes*, Academic Press, New York (1961), p. 233.
88. W. S. Hammer and D. W. Craig, *J. Electrochem. Soc.*, *104*:206 (1957).
89. J. Petró, T. Máthé, and A. Tungler, *J. Catal.*, *42*:425 (1976).
90. F. N. Hill and P. W. Selwood, *J. Am. Chem. Soc.*, *71*:2522 (1949).
91. K. Hauffe and A. Rahmel, *Z. Phys. Chem.*, *1*:104 (1954).
92. D. J. C. Yates, W. F. Taylor, and J. H. Sinfelt, *J. Am. Chem. Soc.*, *86*:2996 (1964).
93. C. H. Bartholomew and R. J. Farrauto, *J. Catal.*, *45*:41 (1976).
94. A. Tungler, J. Petró, T. Máthé, and Z. Csürös, *Magy. Kém. Lapja*, *29*: 332 (1974).
95. R. Z. C. van Meerten, A. C. M. Verhaak, and J. W. E. Coenen, *J. Catal.*, *44*:217 (1976).
96. R. Z. C. van Meerten and J. W. E. Coenen, *J. Catal.*, *46*:13 (1977).
97. H. A. Franco and M. J. Phillips, *J. Catal.*, *63*:346 (1980).
98. K. Nakano and K. Kusunoki, *Int. Chem. Eng. Jpn.*, *23*:675 (1983).
99. I. Wang and W. Huang, *Appl. Catal.*, *18*:273 (1985).
100. T. Máthé, J. Petró, A. Tungler, Z. Csürös, and K. Lugosi, *Acta Chim. Acad. Sci. Hung.*, *103*:241 (1980).
101. T. Máthé, J. Petró and A. Tungler, *Heterogenous Catalysis Proceedings of the 4th International Symposium*, Varna, Part 2 (1979), p. 21.
102. J. Petró, T. Máthé, and A. Tungler, *Chem. Anlagen + Verfahren*, *74*: 64 (1974).

III

HYDROGEN AS A PARTNER IN PRODUCING ACTIVE CATALYSTS

10 | Hydrogen Effects in Sintering of Supported Metal Catalysts

ELI RUCKENSTEIN and IRUVANTI SUSHUMNA

State University of New York at Buffalo, Buffalo, New York

I.	INTRODUCTION	259
II.	MEASUREMENT TECHNIQUES	261
III.	MECHANISMS OF SINTERING	262
	A. Crystallite Migration Model	262
	B. Atomic Migration	264
IV.	EXPERIMENTAL OBSERVATIONS	267
	A. Effect of Time and Temperature	267
	B. Role of Atmosphere	267
	C. Effect of Coke	276
	D. Titania-Supported Catalysts	277
	E. Closing Remarks	280
V.	DISCUSSION	280
	A. Wetting	281
	B. Reactivity of Substrate and Stability Sequence	285
	C. Effect of Nonmetallic Elements on Wettability	286
	D. Interatomic Bond Strength and Stability Sequence of Metals Against Sintering	287
VI.	CONCLUSION	288
	REFERENCES	288

I. INTRODUCTION

Catalytic processes by metals are of great technological importance in a variety of fields such as petroleum refining, chemicals manufacturing, and automobile emission control. The activity of the solid catalyst is exclusively a surface characteristic. Therefore, to derive the maximum benefit from a given amount of a catalytic metal, the number of exposed atoms (or the surface area) has to be maximized. Metal powders, porous blocks, colloidal dispersions, and skeletal metals [1, 2] do not, however, make effective use of the metal and, in addition, they easily cluster together and lose their large surface area on treatment at high temperatures. The decrease in surface area and the corresponding decrease in the activity of these unsupported catalysts is the subject of Chapter 11.

To obtain thermally stable catalysts and, more important, to allow a more efficient utilization of the precious metals, the latter are dispersed on thermally stable (refractory), nonmetallic supports. A very large initial dispersion (the ratio of the surface to total atoms of the active component) is usually obtained by using highly porous supports such as alumina, silica, zeolites, etc., which provide a large specific surface area (300 m^2 g^{-1} or higher) for the dispersion of the active metal in the form of small particles [1,3]. In addition, the support influences the surface structure of the particles via their size and shape and consequently affects their activity and selectivity.

Such highly dispersed supported metal catalysts, however, deactivate with time due, among other things, to the agglomeration of the metal crystallites. Such a metal particle agglomeration, termed sintering, is a thermodynamically favored process, inherent in dispersed systems. Sintering, along with other forms of deactivation such as coking, causes concern because of the ensuing loss of activity and selectivity, and because of the consequent need to interrupt the reaction process to replace or to regenerate the deactivated catalyst (see also Chapter 25). The stability of supported metal catalysts is therefore very important and needs to be studied. In fact, in industrial applications, the high initial activity of a catalyst may be sacrificed in favor of its enhanced stability, since a relatively lower initial activity but a prolonged active catalyst life is in general preferable to a high initial activity that decays rapidly [4,5]. Apart from aiding in the design of stable catalysts, a knowledge of the mechanisms of, and the roles of the various factors in, the kinetics of sintering may also help subsequently in an effective redispersion of the sintered catalysts.

The transition metals and their oxides, which are usually the active catalysts, typically have high surface free energies (of the order of 10^3 erg cm^{-2} [6]). Sintering of these supported metal catalysts occurs in order to decrease the free energy of the system by decreasing the total surface area of these dispersed, high-surface free-energy, active components. The various interfacial free energies involved are, however, affected by the interactions between metal crystallites, support, and gaseous atmosphere. Further, irrespective of the mechanisms involved, agglomeration of highly dispersed supported particles must entail surmounting of energy barriers, and therefore the process of sintering is, in general, aided by higher temperatures. Finally, since the eventual and ultimate equilibrium state (although unattained) via sintering is a single large crystallite on the support, longer periods of heat treatment should in general lead to greater sintering. However, a decrease in the free energy of the system is not always achieved by a decrease in dispersion. In cases where sintering, especially via coalescence of particles, leads to accumulation of stresses and strains due to mismatched lattices of particles and support, a further decrease in the free energy of the system can be achieved by relaxing these stresses and strains, via a rupture of the particles, thus leading to an actual increase in dispersion. Similarly, when there is a change in the chemical atmosphere, from reducing to oxidizing, for example, leading to changes in the chemical composition of the crystallite or/and the support, the resulting changes in the surface and interfacial free energies may lead to extension- and spreading-generated rupture of the individual crystallites. Of course, the redispersed particles may subsequently sinter again in an attempt to decrease further the free energy of the system. However, for kinetic reasons the subsequent rate of sintering may be slow. Thus, during the transient stages, a decrease in the free energy of the system may be achieved by an actual redispersion of the catalysts.

10. *Hydrogen Effects in Sintering*

Sintering of supported metal catalysts occurs in all stages of the catalyst's life, although during the stages of activation (calcination and/or reduction) prior to use, and regeneration after use, it may be more serious. In addition, under severe conditions of usage, such as high temperatures and hydrothermal conditions, sintering may be very severe, especially if the support pore structure collapses and consequently aids the agglomeration of the metal crystallites. Except for the oxide catalysts used in oxidation reactions, activation of the conventional supported metal catalysts by reduction at moderate to high temperatures (depending on the reducibility, the desired degree of reduction of the active component, and the rate of loss of surface area) is routine in industry. In addition, hydrogen is present in some form in a vast majority of important chemical and petrochemical processes which utilize supported metal catalysts [2], as discussed in Chapters 4 and 5. Therefore, the role of hydrogen or hydrogen-containing atmospheres in affecting the sintering characteristics of supported metal catalysts is well worth reviewing.

However, water is present in any reaction environment, even if only adventitiously, either as trace contaminants in the reaction stream, or as a product of the reaction of hydrogen with adsorbed oxygen. Further, as just implied, trace amounts of oxygen as an adsorbed layer on metal surfaces are also present in most catalysts [7]. Thus it is to be recognized that even in a hydrogen atmosphere, the interactions between the crystallites and the support are affected by the trace impurities, as will be demonstrated subsequently with experimental results. In addition, even if these trace impurities could somehow be totally eliminated, some metals may in fact react, if they have a strong affinity for oxygen, with the oxygen ions of the oxide support, leading to modifications of the structure at the particle-support contact area [8].

Numerous papers [1, 3, 9-12] and a number of reviews [13-16] on sintering of supported metal catalysts are available in the literature. In what follows, a brief discussion of the mechanisms of sintering will be presented. (For a more complete presentation, the reader may consult Ref. 14.) Then, from the vast body of published literature on sintering of supported metal catalysts, a few of the investigations related to sintering in a hydrogen atmosphere will be reported as a representative sample. Finally, the factors, especially the metal-support interactions, which affect the behavior of supported metal catalysts will be discussed with a view to identifying the conditions that may lead to thermally more stable active catalysts.

II. MEASUREMENT TECHNIQUES

As noted in Section I, the primary characteristic of the sintering of supported metal catalysts is an increase in the average particle size and correspondingly, a decrease in the total exposed area of the crystallites. Therefore, to investigate the sintering behavior, changes in either the particle size or in the exposed surface area of the metal have to be followed. Measurement techniques useful in obtaining the particle size include, primarily, electron microscopy [1, 17-24] and x-ray diffraction peak broadening [1, 25, 26]. X-ray photoelectron spectroscopy (XPS), small-angle x-ray scattering (SAXS), extended x-ray absorption fine structure (EXAFS), and magnetic measurements are also useful [1, 4, 27-32]. Chemisorption of gases is, of course, used to follow changes in exposed surface area of the metal particles [1, 13-16, 27-30, 33-36] (see also Chapters 4 and 5). A number of articles [13-16, 27-29, 33] and monographs [1, 30] dealing with the techniques of measurement of catalyst dispersion are available in the literature and, recently, an excellent monograph on the characterization of heterogeneous catalysts has been published [4]. A detailed account of the widely used techniques for the measurement of dispersion, and also of the less common, though promising method of obtaining crystallite size distribution through XPS intensity measurement, has been provided by Lemaitre et al. [34]. Therefore, we shall not provide further discussion on the measurement techniques; for details, the reader is referred to the literature cited above.

III. MECHANISMS OF SINTERING

Two distinct mechanisms, which, however, are not mutually exclusive, have been proposed for the growth of particles dispersed on a support. The first model, proposed by Ruckenstein and Pulvermacher [9,10], envisages a random migration of crystallites, followed by collision with other stationary or mobile crystallites and their subsequent coalescence as the processes leading to the gradual decay of the total surface area. The other mechanism, proposed by Chakraverty [37] and discussed by Wynblatt and Gjostein [11], is based on Ostwald ripening and involves emission of single atoms by small fixed crystallites, their transport via the substrate surface or the vapor phase, and finally, the capture and incorporation of single atoms by other large fixed crystallites. The latter mechanism is termed *atomic migration*, and the former, *crystallite migration*. The atomic migration mechanism can, however, be further differentiated as follows. In the traditional atomic migration model, a large number of crystallites are involved, and the small crystallites lose atoms to a two-dimensional surface phase of single atoms, uniformly dispersed over the substrate, and thus decrease in size, while the larger crystallites capture atoms from this phase and increase in size. This process occurs when the substrate surface phase of single atoms is undersaturated with respect to the crystallites smaller than a critical size and supersaturated with respect to the larger crystallites.

The other possibility for growth by atomic migration, called direct ripening [38,39], involves the transfer of atoms released by a small particle directly to a neighboring larger particle without the involvement of a two-dimensional surface phase of single atoms on the entire substrate. Such ripening may occur even when the two-dimensional surface phase is, on the average, not supersaturated with respect to the larger crystallites. Whereas the first kind of ripening involves a large number of crystallites and is global, direct ripening is local and involves only two, or at best a handful of, neighboring crystallites.

A. Crystallite Migration Model

Ruckenstein and Pulvermacher established equations for the time dependence of the distribution of crystallite sizes, and for the decay of the exposed surface area of metal, for both homogeneous and nonhomogeneous surfaces with strongly interacting or trapping sites [9,10]. Two limiting cases, corresponding to diffusion control and coalescence control, have been considered. The analysis involves two different time scales. The small scale time is used to solve the diffusion equation in cylindrical coordinates to obtain the concentration profiles of different particles around a given-size particle, based on their relative diffusivities. These concentration profiles then yield the rates of collision and coalescence of different particles with the given size particle. The rate constants thus obtained are then used to determine the changes in particle size distribution over a large time scale (i.e., the time scale of the

10. Hydrogen Effects in Sintering

process). Considering only binary collisions, the surface number density of k-unit particles, N_k, increases by collisions of i-unit particles with (k - i)-unit particles and decreases by collisions of k-unit particles with any other particle. For a homogeneous surface, the rate of change of N_k over the process time scale is given by

$$\frac{dN_k}{dt} = \frac{1}{2} \sum_{i+j=k} K_{ij} N_i N_j - N_k \sum_{i=1}^{\infty} K_{ik} N_i \tag{1}$$

where the factor 1/2 that appears on the right-hand side ensures that the collisions are not counted twice. The second-order rate constants for the binary collisions, K_{ij}, depend on the mobility of the particles on the support and on the nature of the interactions between the particles, and are different for each of the rate-controlling processes. For coalescence control,

$$K_{ij} = 2\pi R_{ij} \alpha_{ij} \quad (R_{ij} = R_i + R_j) \tag{2}$$

where R_i is the radius of the particle-support interface and α_{ij} is a rate constant for the coalescence process. The rate constant for diffusion control can be approximated, after taking into account its weak dependence on the small scale time, by [10]

$$K_{ij} = \frac{4\pi D_{ij}}{\ln 4 T}$$

where

$$T = \frac{D_{ij} \theta'}{R_{ij}^2} \tag{3}$$

Here $D_{ij} = D_i + D_j$ is the relative diffusivity of particle i with respect to particle j, and θ' is the small scale time. Ruckenstein and Pulvermacher [9,10] solved Eq. (1) both numerically and by similarity techniques for an initial distribution of monodispersed particles and also for other initial distributions. Various size dependencies were considered for the diffusion coefficient and for the coalescence rate constant.

In the case of nonhomogeneous surfaces which give rise to immobilized particles because of interaction with the support (or chemical traps), in addition to N_k, the surface number density F_k of immobile k-unit particles has to be considered. As for the homogeneous surface case, equations have been set up for dN_k/dt and dF_k/dt, and together with the new rate constants K_{ij} (because the relative diffusivities are different when immobile particles are involved), the changes in particle size distributions have been calculated.

Based on their calculations for different initial size distributions, Ruckenstein and Pulvermacher noted that for the case of homogeneous surfaces and for diffusion control, the various size distributions eventually approach, after short times, that obtained for an initially unisized distribution. Thus the particle-size distributions result in this case in a universal distribution, independent of the initial distribution. For a nonhomogeneous surface, on the other hand, a steady-state size distribution, which is strongly dependent on the initial distribution, is reached after a sufficiently long time, and the metal surface area also reaches a steady-state value.

Ruckenstein and Pulvermacher [9,10] further showed that the exposed surface area of the particles (per unit area of the homogeneous support) decays with respect to time according to the equation

$$\frac{dS}{dt} = -KS^n \tag{4}$$

where K is a function of n. The exponent n varies with the size dependence of the diffusion coefficient when the process is diffusion controlled, and with the size dependence of the coalescence rate constant when the process is coalescence controlled. For example, if

$$K_{ij} \propto (R_i^{3m} + R_j^{3m}) \tag{5}$$

then $n = 4 - 3m$. When the process is diffusion controlled, m is expected to be negative and $n > 4$. For example, if the diffusion coefficient is inversely proportional to the base area of the crystallite, $m = -2/3$ and $n = 6$. When $m = -3$, $n = 13$, and so on. For coalescence control, m is considered to be positive and when $m = 2/3$, $n = 2$, when $m = 1/3$, $n = 3$, and so on. If the interactions between the crystallite and substrate are strong, then, as shown in Section V.A, the interfacial free energy between them will be small, the crystallite will have a large base in contact with the substrate, and as a result, its diffusion coefficient will be decreased.

Ruckenstein and Pulvermacher considered a number of existing experimental results and found that the results fit the above power-law equation with n values ranging from 2 to 8, the smaller values (less than 3) corresponding to coalescence control and the larger values (≥ 4) to diffusion control. Values of n larger than 8 are also possible, as sometimes observed experimentally, because of a very slow diffusion. For example, when $m = -3$, n is as large as 13. Wynblatt and Gjostein [11] noted, based on the work of Nichols [40] for the rate of coalescence by surface self-diffusion of two non-wetting contacting spheres, that the coalescence times for particles less than 50 nm in diameter are negligibly small (a few seconds at most) compared to the diffusion times (of hours) and concluded, therefore, that coalescence could not be a rate-limiting step. However, they emphasized that crystallite migration could be an important effect for crystallites below about 10 nm in diameter. With fresh catalysts most of the sintering occurs for particles below this size. The frequent occurrence of dumbbell-shaped particles in the electron micrographs of model supported metal catalysts clearly indicates that coalescence can be a much slower process than predicted from the Nichols equation (which was derived for clean surfaces). Therefore, migration and coalescence comprise an important mechanism to be reckoned with.

B. Atomic Migration

This mechanism involves the interparticle transport of atomic, or molecular, species from small to larger particles, thus leading to a gradual increase in the average particle size. Such a ripening mechanism was initially suggested by Ostwald for the growth of mercuric oxide particles in solution [41]. Lifshitz and Slyozov [42], Wagner [43], and Dadyburjor and Ruckenstein [39], theoretically analyzed the precipitation of solutes from supersaturated solutions by ripening. Chakraverty [37] extended the analysis to the two-dimensional case and obtained the size distribution of particles dispersed on a planar surface. The basis for the transfer of atoms from a particle to the surface phase (the bulk of the liquid in the three-dimensional case), or vice versa, is the Gibbs-Thompson relation, which determines the equilibrium concentration of single atoms at the interface of a curved surface. It is expressed as

$$n_j^S = n_\infty^S \exp\left(\frac{\delta}{r_j}\right) \qquad (6)$$

where n_j^S is the equilibrium concentration of single atoms or molecules (monomers) at the perimeter of a j-atom particle of radius r_j, n_∞^S is the corresponding quantity for a particle of infinite size, and

$$\delta = \frac{2\sigma V_m}{RT}$$

σ being the surface free energy of the metal particle, R the universal gas constant, T the absolute temperature, and V_m the molar volume of the monomer. It can be seen from Eq. (6) that the equilibrium concentration of single atoms on a small particle is larger than that on a larger particle. If the concentration of single atoms on the substrate away from the particles n_1 is such that it corresponds to the equilibrium interface concentration on a particle of radius r_c, the particles smaller than r_c will have a higher equilibrium concentration of single atoms than n_1 and therefore will lose atoms to the substrate, while conversely, particles larger than r_c will capture atoms from the substrate. This process in effect leads to the growth of larger particles at the expense of smaller ones. The particles of critical size, r_c, will not be affected. The critical particle size is obtained from Eq. (6) by replacing n_j^S by n_1:

$$n_1 = n_\infty^S \exp\left(\frac{\delta}{r_c}\right) \qquad (7)$$

Consequently,

$$r_c = \frac{\delta}{\ln(n_1/n_\infty^S)} = \frac{\delta}{\ln[1 + (n_1 - n_\infty^S)/n_\infty^S]} \qquad (8)$$

When $(n_1 - n_\infty^S)/n_\infty^S$ is small compared to unity

$$r_c \simeq \frac{\delta n_\infty^S}{n_1 - n_\infty^S} \qquad (9)$$

As shown by Eq. (9), r_c has significance only when $n_1 > n_\infty^S$.

Either the interfacial process (atom emission and incorporation) or diffusion on the substrate could be rate limiting for sintering. Chakraverty's analyses [37], for the cases of surface diffusion controlling and for interfacial process controlling, lead to different size distributions which, when recast in the form of Eq. (4), yield the rate expressions

$$\frac{dS}{dt} = -C_s S^5 \qquad (10)$$

for surface diffusion control and

$$\frac{dS}{dt} = -C_s' S^3 \qquad (11)$$

for interfacial process control.

In his analysis, Chakraverty [37] used the linear approximation (9). The approximation implies small values for $\delta/r_c = 0.1$, but for parameter values corresponding to typical sintering conditions (where δ/r_c could be as high as 0.75), the error is large. Lee extended the analysis by relaxing the assumption of small supersaturation and arrived at an equation of the form

$$\frac{dS}{dt} = -KS^n \exp(m'S) \qquad (12)$$

In Eq. (12), $m' = $ (constant) $(2\sigma V_m/RT)$, which can be assumed to be a constant if the temperature range is small. The exponent n has a value of 3 for interfacial control and 5 for diffusion control.

The large activation energies necessary for metal-atom emission (e.g., the heat of sublimation of platinum is 132 kcal gmol^{-1} [45]) has been the main argument against the atomic migration mechanism. Nevertheless, when conditions are favorable (e.g., high temperatures and/or the presence of a thermodynamic driving force such as that caused by strong metal-support interactions in an appropriate chemical atmosphere), the energy barriers could be overcome and sintering by atomic migration may take place. However, experiments indicate [46] that, in general, it is the diffusion that controls sintering, not the interfacial processes. The strong interactions, which facilitate atom emission as mentioned above, however, decrease the diffusion coefficient of the atoms on the substrate, and therefore, diffusion becomes rate limiting.

Flynn and Wanke [12] assumed single-atom migration and proposed a model for sintering which, however, ignores Ostwald ripening. They considered unrealistically that the rate of emission of single atoms from a crystallite to the support is independent of the crystallite size. Wanke later included the size dependence for the emission rate [47] and developed a treatment which is similar to that of Chakraverty for interfacial control.

While commenting on the work of Ruckenstein and Pulvermacher [9, 10] and Chakraverty [37], Wynblatt and Gjostein [11] suggested that large values of n (> 13) in Eq. (4), sometimes observed experimentally, may be due to the faceting of the particles. They postulated that if the particles tend to facet, the emission of atoms from such particles is not difficult, while incorporation is, since the growth of such faceted particles involves the difficult stage of formation of a nucleus on the facets of interest. This leads to inhibited particle growth and to large values of n.

Combining the crystallite migration and atomic migration mechanisms, Ruckenstein and Dadyburjor [48], developed a comprehensive model of sintering and redispersion based on the emission of single atoms and multiatom particles and migration and coalescence of particles of every size, including single atoms. Depending on the relative rates of emission and diffusion, metal loading, and the two-dimensional saturation concentration, this model can predict for the surface area: (a) an initial increase followed by a plateau or a continuous increase, (b) an initial plateau or an increase followed by a continuous decrease, or (c) a continuous decrease or an initial decrease followed by an increase.

The models considered so far have all involved the assumption of a planar substrate. However, when the substrate is curved and includes surface irregularities, as is often the case with practical supported catalysts, the sintering rates would be affected. Ahn et al. [49] quantitatively analyzed the effect of substrate curvature on sintering by particle migration as well as by atomic transport and substantiated the earlier conclusions [11, 29], which follow. The concavities of the substrate surface trap particles when the radii of curvature of the particles and those of the concavities are comparable. However, particles migrating toward the concave sites may collide and lead to enhanced initial particle growth. On the other hand, growth by

10. Hydrogen Effects in Sintering

interparticle transport would be retarded, since the particles in concave regions tend toward a stable limiting radius of curvature at the pore mouth, and once they attain the same radius of curvature as that of the concavity, they neither emit nor capture atoms.

IV. EXPERIMENTAL OBSERVATIONS

A number of investigations concerning sintering of supported metal catalysts have been reported in the literature [1, 10-15, 21-24, 28]. Here we shall, however, consider only some of the results, to indicate the diversity of the observations and to point out the complexity of the phenomena involved.

A. Effect of Time and Temperature

Chen and Ruckenstein [50] carried out a detailed investigation of the sintering of model Pd/Al_2O_3 catalysts as a function of time and temperature in a hydrogen atmosphere. A large number of events, which indicated the involvement of different mechanisms of sintering, were detected. The disappearance of small particles, crystallite migration, migration followed by coalescence, appearance of new crystallites, direct ripening, and faceting of crystallites were all observed. Schematic representations of the various events are provided in Fig. 1. At 650°C, during the early stages of heating (< 30 min), local events such as the disappearance of small crystallites, the appearance of particles in locations unoccupied before, coalescence of contacting particles, a decrease in size of a particle near a larger particle, etc., were detected. However, during the next stage of heating (between 30 min and 1 h), severe sintering leading to a new distribution of the crystallites resulted, suggesting that migration and coalescence played a major role. During subsequent heating for 3 h, sintering was relatively slow, but significant nonetheless, with crystallite migration and coalescence being the predominant event. Further heating for up to a total of 20 h led to a slow sintering via the coalescence of nearby particles. At the higher temperature of 725 °C, the sintering was so rapid and severe during the initial 30 min of heating that the individual particles could not be followed to detect the mechanism involved. At 725°C, faceted crystallites were observed following 20 h of heating, while at the higher temperature of 800°C, the crystallites were already large and started to facet after only 2 h of heating. Fiedorow et al. [51] also observed, in the case of Pt/alumina, a monotonic decrease in dispersion with increasing temperature and also a decrease in dispersion following longer heating times. Recently, Smith et al. [52] and White et al. [53], while investigating the behavior of Pt/alumina, noted that most of the sintering occurred during the first hour of heating, and they therefore investigated the sintering behavior by heating the specimens for only 1 or 2 h. These results indicate that sintering is faster and severe at higher temperatures, and at any particular temperature, it is very rapid and excessive during the initial stages (≤ 1 h) and subsequently is slow but still significant.

B. Role of Atmosphere

Effect of Pore Collapse and Role of Steam

As already mentioned, for a given metal-support combination, the atmosphere can be the major factor that affects sintering. Williams et al. [54], while investigating the loss in activity of Ni/alumina catalysts for steam methanation of hydrocarbons, observed that the nature of the atmosphere in which the

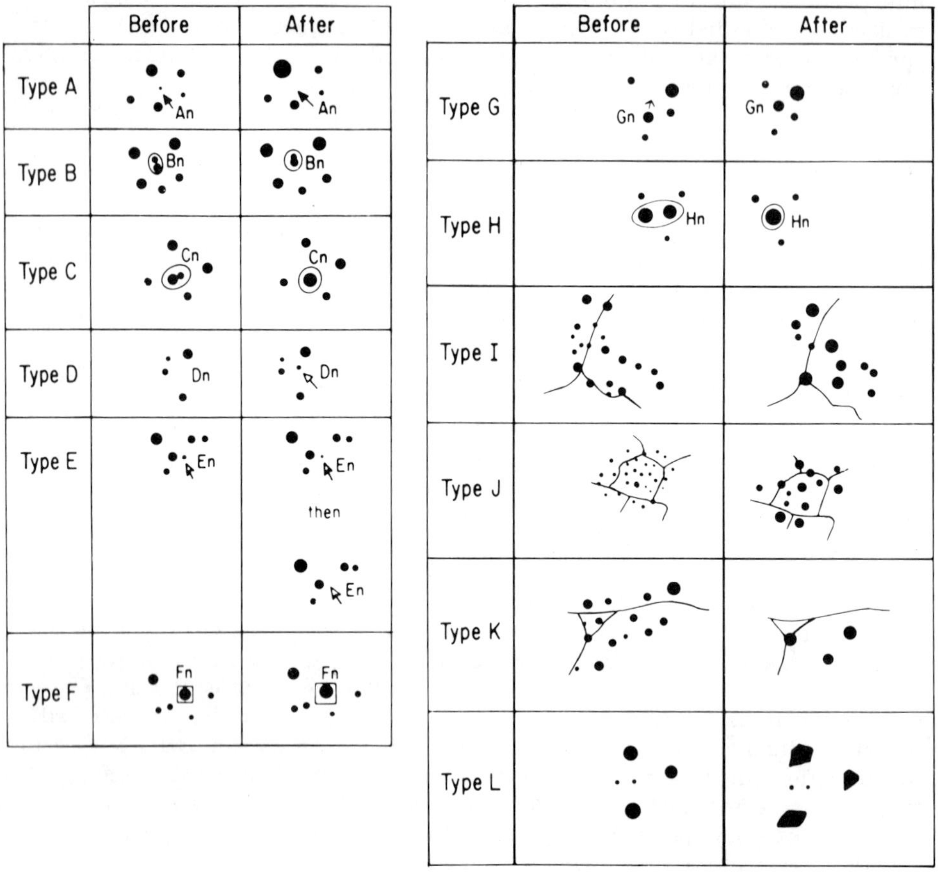

FIGURE 1 Schematic illustration of various events observed during the sintering process. Type A: Small crystallite disappears. Surrounding larger crystallites are stationary and either do not change or grow in size. Type B: Two neighboring crystallites come in contact with each other and either do not change or grow in their sizes. Type C: Two touching crystallites coalesce into a larger one. Type D: A new crystallite appears at a place where no crystallite was located before. The previous location of this new crystallite is unknown. Type E: A small crystallite very close to a larger one decreases in size and then disappears. Type F: Crystallite is stationary and grows in size. Type G: A crystallite migrates to a new position and grows or remains unchanged in size. Type H: Two or more neighboring crystallites coalesce into a new crystallite somewhere between their original positions. (Schemes I, J, and K contain grain boundaries. These events can, however, occur everywhere, not just near grain boundaries.) Type I: This is a global event that involves many crystallites. The former situation occurs after 30 min of heating at 650 °C. The latter situation occurs after an additional 30 min of heating at the same temperature. The process leads to a new distribution of crystallites with larger diameters. Type J: This event occurs during the first 10 min of heating of fresh catalyst at 725 and 800°C. The process is similar to a Type I event. Type K: This event occurs during the additional 20 min of heating at 725°C after the first 10 min. The process is similar to that of a type I event. Type L: The shape of large crystallites, with diameter of about 125 Å or beyond, changes from spherical to faceted while those of the small ones remain unchanged.

catalyst is used has a much greater effect on sintering than do its pressure and temperature. For instance, the area loss of a coprecipitated sample of Ni/alumina when heated in a hydrogen and steam mixture at only 400°C was as great as that on heating in hydrogen alone, but at 800°C. Based on the observation of roughly parallel decreases in the metal particle area and in the total (metal plus support) surface area, they concluded that the loss in nickel surface area was caused by the coalescence of the nickel particles pulled closer together by the collapse of the fine pore structure of the support. The pore collapse is a result of the accelerated transformation of γ-alumina to α-alumina under hydrothermal conditions, at temperatures substantially lower than the normal transformation temperature of about 1100°C. More recent results with Ni/silica [55, 56] support similar conclusions. Investigating a 15 wt % Ni on γ-alumina and Ni on $NiAl_2O_4$ specimens, Bartholomew et al. [57] concluded that at $T \leqslant 650°C$, the loss of nickel surface area on heating in hydrogen is due to the combined effect of crystallite growth and support collapse, predominantly to the latter, while at higher temperatures the decrease in surface area is due, predominantly, to the metal crystallite growth. The presence of water vapor in hydrogen accelerates the rate of sintering at $T \geqslant 530 °C$, because of enhanced collapse of the support. Below 750°C and in a reducing atmosphere, Ni on $\gamma\text{-}Al_2O_3$ is significantly more stable than on α or δ forms of alumina, and Ni on SiO_2 is less stable than on any of the forms of alumina noted above. They found that under the same conditions, Ni on $NiAl_2O_4$ is the most stable, perhaps, because the loss of Ni surface area due to sintering is compensated by the creation of new nickel sites via the reduction of the $NiAl_2O_4$ support, as also suggested earlier by Morikawa et al. [58] and Ross et al. [59]. The results above demonstrate the indirect role of the atmosphere (steam in these cases) in the sintering of supported metal crystallites via the collapse of the pore structure, and also the role of the crystalline or chemical structure of the support. Concerning the latter, severe sintering of Pt on alumina was reported by Chu and Ruckenstein [60] as a result of the transformation of the support from an amorphous to a crystalline form.

Effect of Traces of H_2O and O_2 and of Alternate Heating in H_2 and O_2

Changes in particle morphology and mobility, and consequently changes in the sintering behavior of supported metal catalysts as a result of direct interactions between support and crystallites in various atmospheres (as opposed to the indirect effect via support pore collapse and transformation discussed above), have also been reported [1, 3, 10, 60-69]. Detailed accounts of the phenomena observed with Fe/alumina model catalysts on heating in hydrogen and oxygen have been reported by Sushumna and Ruckenstein [67]. Interactions between the crystallites and support leading to unusual toroidal shapes, an alternation in the shapes of the crystallites, an initial fast growth followed by a very slow growth due almost entirely to coalescence of contacting adjacent particles, rupture of crystallites, etc., have been observed on heating Fe/alumina model catalysts in as-supplied, ultrahigh-purity hydrogen (99.999%, but containing traces of O_2 and moisture). Figure 2 shows one cycle of the alternation in the shapes of the crystallites between a torus and a core-and-ring structure, on heating a 0.6-nm-initial-film-thickness sample of evaporated Fe on model alumina catalyst at 400°C. Electron diffraction indicated the presence of iron aluminates and/or of solid solutions of Fe_xO in Al_2O_3, in addition to the stoichiometric oxide γ - Fe_2O_3 (Fe_3O_4). (γ - Fe_2O_3 and Fe_3O_4 have the same lattice parameter and therefore it is difficult to discriminate between them using electron diffraction.) Further, it was inferred from electron diffraction that the torus shape is

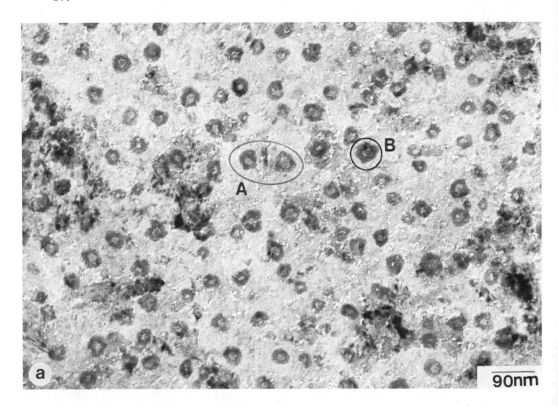

FIGURE 2 Alternation in crystallite shape in the same region of a 6-Å-initial-film-thickness iron/alumina specimen heated at 400°C in as-supplied hydrogen for (a) 6 h and (b) 12 h.

associated with a more oxidized state and the core-and-ring structure with a less oxidized state. As iron is very reactive, it is oxidized even by traces of oxygen and moisture. The oxide thus formed reacts with the substrate. Following this oxidation, the high excess of hydrogen reduces the metal oxides. The alternations in shape are a result of such repeated oxidation and reduction, as discussed in Ref. 68.

As expected, sintering was found to be more pronounced at higher treatment temperatures and in the early stages of heating. Sintering was faster and more severe even at the relatively low temperature of 500°C when the metal loading was greater, about 1.25 nm initial film thickness. When hydrogen which was purified further to eliminate the traces of oxygen and/or moisture was used, sintering was greater than in the as-supplied ultrahigh-purity hydrogen mentioned above. In the latter case, a relatively strong interaction between the partially oxidized crystallites and the support, due to the presence of the impurity oxidants in the hydrogen, hindered sintering. Both the impurities, and consequently, the interactions, were absent in the former case.

During the initial stages of heating a number of phenomena took place, often concurrently. For example, in the case of a 0.75-nm initial-film-thickness sample of Fe/Al_2O_3 heated in as-supplied hydrogen, neighboring crystallites extended, made contact, and coalesced; crystallites migrated and/or sintered by coalescence (Fig. 3); large particles disappeared, while small particles nearby remained unaffected (contrary to atomic migration mechanism) and sometimes both small and large particles disappeared; and

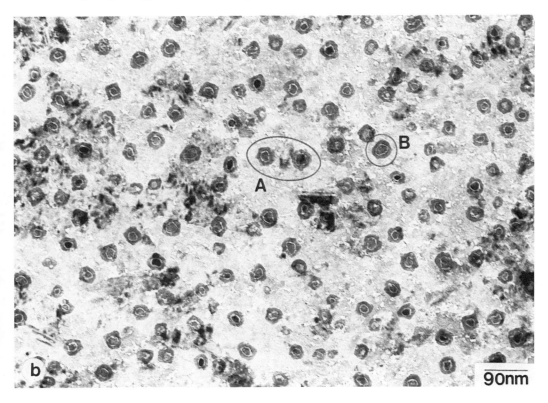

FIGURE 2 (continued)

small particles decreased in size and eventually disappeared (indicating ripening, direct or global). Apart from these observations, which suggest the occurrence of various mechanisms of sintering, events such as particle extension, a tendency for a particle to split into two and recontract to form a single particle again, particle rotation and rearrangement, disappearance of large particles and appearance in the region of a large number of smaller particles suggesting particle fracture, clearly detectable splitting of particles, and sometimes the subsequent migration of the resulting units were all observed following heating in as-supplied hydrogen [67]. In addition, repeated alternate changes in the shape of the particles from a torus to a core-and-ring structure (Fig. 2) led, finally, to their rupture, due perhaps to mechanical fatigue. Considerable breakup and redispersion of particles following repeated alternate heating in flowing O_2 and H_2 were also detected. Following heating in oxygen a number of phenomena, such as extension, cavity formation, and "tearing" of crystallites, were observed also by Chen and Ruckenstein [69] with model Pd/alumina catalysts.

The presence of moisture does not always promote stability. In the case of metals such as iron, nickel, cobalt, etc., which are easily oxidized by moisture, the particles will be stabilized because of the interaction at the interface between the metal oxide formed and the support. However, in the case of metals such as platinum, silver, and gold, which are not easily oxidized, moisture can lead to severe sintering because it cleans the support surface of impurities and makes the surface smooth. In addition, moisture, when adsorbed on the support, decreases the physical interactions between metal and support, leading to increased mobility of the particles [3]. Indeed, Chu and Ruckenstein [60] observed that in the case of Pt/alumina, significant

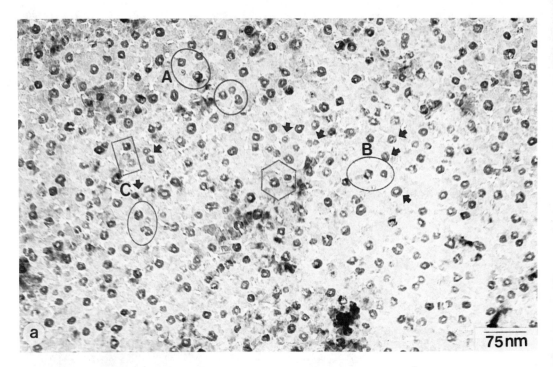

FIGURE 3 Sequence of changes in the same region of a 7.5-Å-initial-film-thickness iron/alumina specimen heated at 500°C in as-supplied hydrogen for (a) 1 h and (b) 5 h.

sintering occurred during heating in wet H_2. Smith et al. [52] recently reported that in the case of Pt/alumina, sintering in pure hydrogen is marginal, whereas it is serious in hydrogen containing only 1 vpm moisture and oxygen. Chen and Schmidt [70], however, reported that in the case of Pt/SiO_2, water in O_2 or in N_2 significantly reduced the rate of sintering. The difference in behavior is, perhaps, due to the differences in the chemical reactivity of the SiO_2 and Al_2O_3 support surfaces. The latter authors surmised that the inhibition of particle growth was probably caused by a reduction in the Pt surface diffusion coefficient on SiO_2, possibly associated with adsorbed H_2O and OH groups on the SiO_2 surface. Kuo et al. [71] reported that in the case of silica-supported nickel catalysts at above 500°C, heat treatment in hydrogen always resulted in greater sintering than on heating in N_2. The relative stability in nitrogen is thought to be due to the interaction between the unreduced, residual NiO and the support, and the greater sintering in hydrogen is thought to be due to the water formed by the reduction of NiO during heating in H_2, which aids sintering. It should be pointed out that the metal loading, which is an additional factor in sintering, was very large, about 58 wt % in their system. Glassl et al. [72] also reported that when a sample of Pt/Al_2O_3 was heated in oxygen following heating in hydrogen, sintering was more severe, probably due to the locally evolved heat of reaction between H_2 and O_2 and to the formation of water. Below 500°C, sintering was more pronounced in hydrogen than in oxygen, and at higher temperatures, sintering was stronger in an oxygen atmosphere. Ruckenstein and Chu [61] also reported that the migration of the crystallites is enhanced by periodic oxidation and reduction.

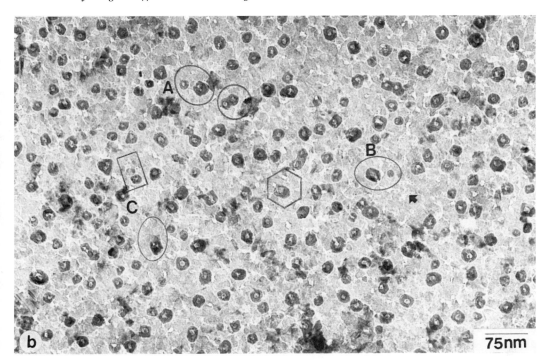

FIGURE 3 (continued)

Recent results from this laboratory with Pt/alumina model catalysts [73] indicate that in hydrogen, sintering occurs relatively slowly at temperatures below about 600°C. At 600°C and above, there is significant sintering and the disappearance of a number of small crystallites, particle migration, and coalescence have been observed to take place. More interesting results, however, have been observed during alternate heating in H_2 and O_2 at 500°C, as shown in Fig. 4. The particles appear larger and extended on heating in hydrogen and smaller on heating in oxygen, a behavior apparently in contrast to the normally observed extension of particles on oxidation on "nonreducible" substrates. It appears as if a part of the crystallite material is lost from the substrate during heating in oxygen. However, it is recovered following heating in hydrogen, when the particles regain their previous size. It is likely that on heating in oxygen (at 500°C), a type of film (probably of PtO, since the temperature is below the decomposition temperature of PtO of about 570°C) extends out from around each particle and remains in the immediate vicinity of and in contact with the parent particle. The residual particle is, therefore, smaller in the micrograph. The film is probably too thin to be detected in electron micrographs, as suggested previously for a similar phenomenon with Fe/alumina catalysts [67,74]. It is unlikely that the particles emit single atoms which remain as a two-dimensional phase on the support (as is often suggested for particle redispersion [13,75]), since if this were to happen, the likelihood of all the particles, small and large, regaining their original sizes on subsequent heating in hydrogen is very small. If a two-dimensional phase of single atoms is present on the surface of the substrate, at least some redistribution of particle sizes is expected to occur when the atoms are recaptured. Also, the films around the neighboring particles are not expected to be in contact with one another and form a

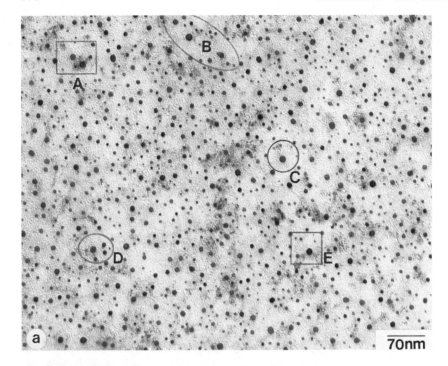
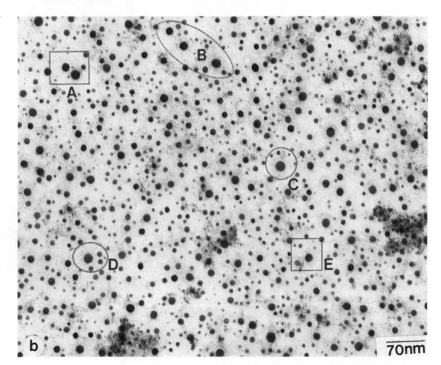

FIGURE 4. Alternation in the size of crystallites in the same region of a 15-Å-initial-film-thickness Pt/alumina specimen heated in as-supplied hydrogen and oxygen at 500°C (two cycles are shown here): (a) 4 h, O_2; (b) 2 h, H_2; (c) 2 h, O_2; (d) 1 h, H_2.

10. Hydrogen Effects in Sintering 275

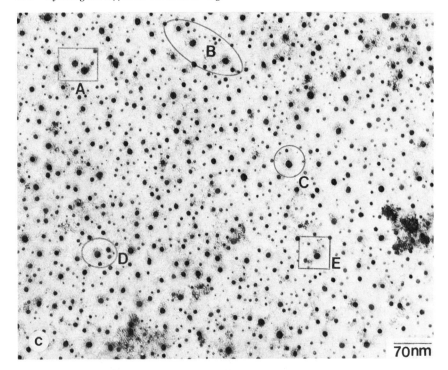

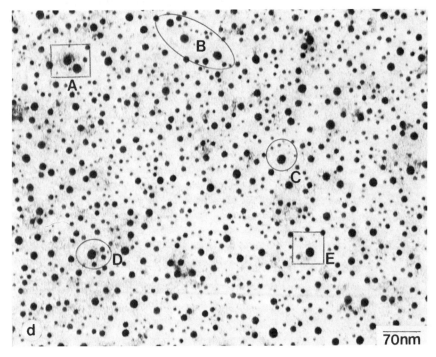

contiguous patch of surface film, unlike with Fe/alumina [74] at high temperatures, since in such a case again sintering by coalescence is likely to occur when the film contracts on subsequent heating in hydrogen and pulls the particles closer together. Since all the particles more or less regain their original sizes and since even the small particles continue to exist, it appears that the particle remains immobilized on the substrate and that, in an oxygen atmosphere, the film also remains only around the leading edge of, and in contact with, the particle. In a reducing atmosphere, the film contracts completely and merges with the respective particle. Another alternative is the partial reduction of the alumina substrate in a H_2 atmosphere, and the extension of the crystallites on the reactive, high-surface-free-energy compound formed, as was suggested for the case of reducible oxide supports such as TiO_2 [64].

C. Effect of Coke

Carbon deposition on catalysts during use in hydrocarbon environments, and the use of carbon as a substrate, also affect sintering, mostly aiding it but sometimes inhibiting it, as will be shown below. Channeling on a carbon substrate has been reported by Baker and co-workers in the case of Pd on graphite [24,76] and Ni on graphite [24,77], among other metals, during heating in hydrogen. Chu and Ruckenstein [78] observed channeling in the case of Pt on carbon support as well as on carbon-overlaid alumina. Ruckenstein and Lee [79] also observed such channeling in the case of Ni on carbon-overlaid alumina. During channeling, the gases resulting from the catalyzed gasification of carbon by hydrogen facilitate the migration of particles over the support and therefore lead to severe sintering.

The combined effect of hydrogen and coking on the deactivation of supported metal catalysts was investigated recently by Hu and Ruckenstein [80] using electron microscopy and model catalysts, by simulating an actual reaction environment: steam reformation of methane with Co, Fe, and Ni on alumina catalysts. Depending on the nature of the catalyst and the composition of the mixture, sintering and/or coking can occur much more rapidly in gas mixtures than in single-component atmospheres (hydrogen, steam, methane, or CO). For example, at 700°C, in a mixture of methane and steam, corresponding to the inlet composition of a reformer furnace, extension of particles on the cobalt catalyst and severe coking on the iron catalyst occurred, while sintering and coking took place on the nickel catalyst. The result of heating in hydrogen following heating in methane or in $CH_4 + H_2O$ was sintering in general with all three metals, and especially serious sintering with cobalt and nickel. As mentioned earlier, the sintering is due to enhanced mobility of the particles, caused by the gasification of the deposited coke.

The ternary mixture $CH_4 + H_2O + H_2$ had a drastic effect on dispersion. In the case of Ni, heating a short time at 700°C led to such severe sintering that the individual particles could not be followed to identify the operating mechanism. It is likely that in the absence of H_2, the amount of carbon that covers the region near the leading edge is relatively large and that the migration of the particles is thus impeded, even if a part of the carbon is gasified by steam. In addition, the likelihood of oxidation of the metal is high in the presence of steam, and the interactions between the oxide and the substrate also restrict particle mobility. In contrast, the presence of hydrogen, in addition to that of steam, inhibits carbon deposition, gasifies a part of the deposited carbon, and decreases the interaction of the reduced particle with the substrate. Consequently, the presence of less carbon and of weaker interactions lead to enhanced particle mobility. However, the behaviors of cobalt/alumina and iron/alumina in the mixture were different.

There was coalescence of the nearby particles, increased gasification, and hence decreased overall carbon deposition in the case of Co/alumina and a number of particles disappeared from the micrograph, perhaps by spreading undetectably (by electron microscopy) over the surface of the support. In the case of Fe/alumina, the particle extended to some degree on the substrate and, in addition, a part of the metal was lost to the gas stream because of its disintegration.

The addition of CO to the mixture had more or less the same effect on all the metals—a rapid and significant loss of material from the surface of the support, probably due to the formation of volatile carbonyls.

The foregoing results indicate that the severe deactivation observed in the case of steam reforming catalysts is due to the combined effect of sintering, coking, and transformations of the support in the presence of steam. In other words, the rapid deactivation reflects the composite effect of the components of the gas mixture on the supported metal catalysts.

D. Titania-Supported Catalysts

On alumina, silica, and magnesia supports, the particles in general assume contracted and globular shapes in reducing atmospheres, and extended shapes in oxidizing atmospheres. In these systems, the interactions between particle and support are stronger in an oxidizing environment and hence result in smaller contact angles and extended shapes, as will be shown in the next section. However, in the case of TiO_2 supports, it is reported that in H_2 (because of the catalyzed reduction of TiO_2 to a lower oxide and its subsequent strong interaction with the crystallites) Pt particles have an extended shape in H_2 compared to that in O_2 [64], suggesting that the sintering behavior on TiO_2 is also different. The associated phenomenon of the suppression of room-temperature H_2 chemisorption, following high-temperature ($\geqslant 500°C$) reduction, in the case of group VIII metal/reducible oxide support catalysts, called "strong metal-support interaction" (SMSI) [81], is discussed in detail in Chapter 13. Experimental observations of extended, thin, "pillbox" morphology for crystallites in a hydrogen atmosphere were reported by Baker et al. for Pt/TiO_2 [63,82] and by Tatarchuk and Dumesic for Fe/TiO_2 [83]. As regards the particle sintering, Tatarchuk and Dumesic reported that on reduction between 400 and 430°C, a growth in size to about 20 nm was observed at the expense of crystallites of $\leqslant 5$ nm in diameter. However, in the same specimen, at temperatures greater than 500°C, and especially at 600 and 650°C, they reported that a large fraction of the crystallites were less than 10 nm in size. They suggested that this decrease in size was due to the loss of observable material as a result of diffuse spreading of iron over the support, or more likely, to the diffusion of iron into the support.

Baker et al. [63,82] also observed an increase in the average particle size on heating Pt/TiO_2 catalysts in hydrogen at temperatures up to 600°C. The particles were reported to have assumed thin, flat, pillbox morphology above 500°C. A further increase in temperature to above 600°C, up to 750°C, did not cause any change in the average particle size or in the morphology, while on alumina, silica, and carbon supports, the growth in size continued up to 750°C.

Ni/TiO_2 behaved similar to Pt/TiO_2 on heating in hydrogen up to 550°C [84]. There was marginal sintering on heating a specimen from 150°C to 550°C, but no change in the particle morphology, which remained a hexagonal, thin, pillbox structure. However, on heating at 700 and 800°C, there was considerable sintering in the case of Ni/TiO_2, with the particles acquiring dense, dark, globular morphologies, in contrast to Pt/TiO_2, in which sintering as well as a change in the morphology were inhibited at the higher temperatures [84]. Nevertheless, as in the case of Pt/TiO$_2$, the support was reported to have been reduced to Ti$_4$O$_7$ in the case of Ni/TiO$_2$ also, as detected by electron diffraction.

Pd/TiO$_2$ behaved even more differently [84]. On heating in hydrogen, the particles sintered gradually at all temperatures and always had dense, compact structures. Above 500°C, the sintering was extensive and the particles acquired a more globular morphology. Pd/Al$_2$O$_3$, on the other hand, was reported to be relatively resistant to sintering, the behaviors of both Pd/TiO$_2$ as well as Pd/Al$_2$O$_3$ being different from that of the corresponding Pt/TiO$_2$ and Pt/Al$_2$O$_3$ systems. In contrast to Pd/TiO$_2$, there was very little change with Pd/Al$_2$O$_3$ at temperatures up to 500°C, while following heating at 700°C, there was a decrease of 50 to 60% in hydrogen uptake, corresponding to an approximately twofold increase in the average particle size.

There are differences in sintering and morphological behavior of Pt, Pd, and Ni on titania which have not been explained. It has, however, been suggested that all three cause a reduction of the support to Ti$_4$O$_7$ (and, consequently, are expected to react strongly with the support in two ways [86]: (a) via the interactions between the particle and the support beneath it and (b) via the interactions between the particle and the mono or submono layer of the reduced substrate species migrated onto it; in the literature, only the latter interactions are considered, but the former are equally if not more important and have to be considered). A possible explanation for the above differences in behavior could be sought in the differences in the interaction of hydrogen with the metals and the resulting extent of reduction of the support in the three cases. Nickel is known to be a good hydrogenation catalyst, implying that it dissociates hydrogen very effectively. Therefore, Ni reduces titania easily. At low temperatures, this reduction is perhaps incomplete and gives rise to the intermediate oxide Ti$_4$O$_7$. The interaction between Ni and this oxide (which, being nonstoichiometric, has high reactivity and high surface free energy compared to TiO$_2$) leads to the extended shapes for the following reasons (see Section V for notations): (a) σ_{sg} in this case is large, since the nonstoichiometric oxides in general have higher surface free energies than the stoichiometric oxides; (b) because of the higher reactivity of the reduced oxide, the interaction energy U_{cs} between the particle and the substrate beneath it is large and, consequently, σ_{cs} is expected to be small, as discussed in Section V; and (c) σ_{cg} is decreased because of the interaction between the particle and the mono or submono layer of the reduced Ti$_4$O$_7$ species migrated onto it. Since σ_{sg} is now larger and σ_{cs} and σ_{cg} are smaller, the contact angle decreases and the particle extends (see Section V).

However, at higher temperatures, the reduction of the support may lead to an even lower oxide, such as Ti$_3$O$_5$ or Ti$_2$O$_3$. These oxides, especially the latter (being stoichiometric), are expected to have lower surface free energies and reactivities than those of Ti$_4$O$_7$. Therefore, the metal-support interactions will be decreased and the particle is likely to assume a relatively contracted shape, as will become more evident from the discussion in Section V. The decreased metal-support interactions may result in the intensification of the dissociative chemisorption of H$_2$ by the metal crystallites, which, in turn, would enhance further reduction of the support toward the more stable Ti$_2$O$_3$. Therefore, at higher temperatures, the sintering and morphological behaviors are expected to be different from those at the lower temperatures. The sintering is expected to be more severe at high temperatures, as observed, because of the decreased metal-support interactions and perhaps also because of the phase transformations (of TiO$_2$ itself from one form to another [87] or of TiO$_2$ to lower oxides). It is likely that under such conditions, the lowest form of reduced oxide is present, although it has not been reported so far. The presence of Ti^{3+} (probably Ti$_2$O$_3$) was detected by ESR by Huizinga and

Prins [88] at low temperatures (up to 500°C), although they did not report having detected this species at higher temperatures. Both Ti_2O_3 and Ti_3O_5 are, however, stable over the entire temperature range of interest [89]. Ti_2O_3 may also be present at the higher temperatures, although Huizinga and Prins postulated that at higher temperatures Ti_2O_3 is transformed by dehydration into Ti_4O_7.

In the case of Pt, if the strong interactions which have been reported to exist between Pt and Ti ions at the low temperatures ($\simeq 500°C$) remain at higher temperatures, the continued presence of the mono or submono layer of TiO_x species on the crystallites would retard the dissociative chemisorption of hydrogen and consequently inhibit further reduction of the support beyond the Ti_4O_7 stage. In such a case, a change in morphology or in the sintering behavior with increasing temperatures is not likely to occur, as apparently observed experimentally. A strongly adsorbed layer of H_2 at 500°C, as suggested for the case of Pt/Al_2O_3 [90], could also inhibit further adsorption of H_2 at higher temperatures and hence explain the above observations.

In the case of Pd, because of the large amounts of hydrogen absorbed by the metal, especially at low temperatures, the titania support is perhaps not reduced at all, and the behavior observed is that of Pd on TiO_2 and not of Pd on Ti_4O_7. The particles are therefore globular in shape, as a result of the wetting characteristics of the Pd/TiO_2 system. On subsequent heating at higher temperatures, because the absorbed hydrogen is released and because there is no strong interaction between Pd and Ti ions established during prior heating at 500°C which may hinder reduction (unlike in the case of Pt/TiO_2), the support may be reduced to the lower oxide Ti_2O_3, the intermediate stages of nonstoichiometric oxides being transitory. The excessive sintering observed may be a result of the phase transformation involved and also a result of the wetting characteristics of the Pd/Ti_2O_3 system. In addition, as shown in Section V, the rate and extent of sintering of these three metals are in the order Pd > Ni > Pt. The sintering observed experimentally is in agreement with this order.

The interpretations above for the differences in sintering, shape changes, and the ability to chemisorb H_2 have relied significantly on the extent of reduction of the support in the three systems and the associated extent of the strong metal support interactions. It is to be recognized that there is at least one other plausible explanation for the suppression of hydrogen chemisorption (traditionally referred to as SMSI), even though it was originally proposed for a non-SMSI system, Pt/alumina, by Menon and Froment [90]. The room-temperature suppression of H_2 chemisorption following high-temperature ($\geqslant 500°C$) reduction in the case of Pt/alumina, reported originally by den Otter and Dautzenberg [90] and Dautzenberg and Wolters [92], was considered by Menon and Froment to be due to the activated, strong chemisorption of hydrogen at high temperatures (>500°C) [90]. This strongly chemisorbed hydrogen is not easily desorbed by evacuation at low temperatures (where dispersion measurements are made) and therefore prevents additional chemisorption of hydrogen at these temperatures [90,93]. Dispersion measurements are therefore not possible by hydrogen chemisorption. However, oxygen titration, or oxidation at low temperatures followed by hydrogen chemisorption at $T \leqslant 450°C$, can be used to measure dispersion. Similar observations of activated chemisorption were reported with titania (a SMSI support)-supported metals [94,95], although the suppression of chemisorption in these systems was still attributed only to the reduction of the support and the subsequent strong interaction between the crystallites and the reduced support species (TiO_x) migrated onto them (see also Chapter 13).

It is likely that a partial reduction of the support in the immediate vicinity of the metal particle does contribute to some extent to the suppression of chemisorption observed. However, it appears that this process is neither the sole cause of the suppression of chemisorption, nor is the suppression restricted to reducible oxide supports, such as TiO_2, Ta_2O_5, and Nb_2O_5.

E. Closing Remarks

There are cases of apparent or actual differences in observations or interpretations of sintering of supported metal catalysts, because of the instrumental limitations in obtaining accurate data on the atomic scale processes, as well as the difficulties involved in duplicating the exact conditions. For example, Graham and Wanke [96] in the case of Ir/Al_2O_3, and Wang and Schmidt [97] in the case of Ir/SiO_2, reported that under reducing conditions, these catalysts were more resistant to sintering than was Pt (or Rh) on the corresponding supports. On the other hand, Foger and Anderson [98] reported that treatment of a 0.9 wt % Ir/SiO_2 catalyst in hydrogen for 3 h at 700°C resulted in an 18-fold decrease in Ir dispersion. Graham and Wanke noted that they did not observe such large decreases in Ir dispersion on heat treatment in H_2 even at 800°C for 16 h, while Wang and Schmidt reported to have observed a four- to fivefold increase in the average particle size on heating in hydrogen at 750°C. As noted before in the context of the effect of moisture on sintering of Pt/alumina and Pt/silica, the reported differences in sintering rates are probably a result of the differences in supports. In general, sintering is faster on silica than on alumina, as will be noted again in Section V.

Further, there are a number of investigations that appear to provide evidence in favor of either of the two traditional mechanisms of sintering: crystallite migration and atomic migration. Frequently, unduly restricted claims have been laid to one of the mechanisms as being the sole mechanism of sintering. Suffice it to say that both, and maybe other mechanisms are involved. A review of only a small fraction of the published literature on sintering of supported metal catalysts would amply indicate that it is a complex phenomenon. It involves an intricate interplay between such factors as the interactions between metal, support, and atmosphere and their effect on the wetting characteristics of the metal-support system, the shape of the particles, and the mobility of the migrating species. In addition, the metal loading and initial size distribution, treatment of the support prior to and modifications during use, method of preparation and treatment prior to use of supported metal catalysts, presence and amount of reactive contaminants in the reaction environment, temperature and duration of heating, etc., also play a role.

However, a few general observations regarding sintering in a hydrogen atmosphere can be summarized as follows. Sintering is greater with higher metal loading and at higher temperatures, and at any particular temperature, it is more pronounced during the initial stages of heating and continues with time, although slowly. On supports such as alumina, silica, and magnesia, and for metals, such as Fe, Co, Ni, and Cu, which are oxidized easily, impurity moisture or oxygen, even if only in traces, hinders sintering to some extent, while for noble metals such as platinum, silver, and gold, which are not easily oxidized, moisture, in general, enhances sintering. The sintering characteristics and the shape of the particles on titania and similar reducible oxide supports are different from those on the usual refractory oxide supports. In general, the stability of the supports against sintering in a hydrogen atmosphere is in the order titania > alumina > silica $\simeq$ carbon.

V. DISCUSSION

It has been stated repeatedly in earlier sections that the interactions between crystallites, atmosphere, and support are responsible for the sintering and

10. Hydrogen Effects in Sintering

redispersion behavior of supported metal particles. These govern the wetting characteristics, the mobility of the crystallites, the ease of emission of atomic or molecular species, and their subsequent diffusion on the surface of the support. The atmosphere, in general, determines the chemical species in the crystallites and, in some cases, also those at the surface of the supports, and consequently, affects the interactions mentioned above. The temperature and, to a lesser extent, the pressure (or the partial pressure) of the gas affect these interactions via the energetics involved. Let us now consider some of the physicochemical factors.

A. Wetting

Changes in size and shape of the crystallites are determined by the wetting characteristics of the metal-support system. Young's equation, which relates the equilibrium contact angle of a particle on a support to the various interfacial free energies involved, provides the basic framework for discussion:

$$\sigma_{sg} - \sigma_{cs} = \sigma_{cg} \cos \theta \tag{13}$$

where θ is the equilibrium contact angle, and σ_{sg}, σ_{cs}, and σ_{cg} are the substrate-gas, crystallite-substrate, and crystallite-gas interfacial free energies, respectively. They are represented schematically in Fig. 5. Equation (13) shows that $\cos \theta > 1$ if

$$\sigma_{sg} - \sigma_{cs} > \sigma_{cg} \tag{14}$$

Under such conditions, the crystallite will spread out, since no equilibrium contact angle is compatible with $\cos \theta > 1$. On the other hand, when inequality (14) is reversed, θ assumes a value between 0 and 180°, depending on the magnitudes of the three interfacial free energies.

It is worth noting that the crystallite-substrate interfacial free energy, σ_{cs}, is a direct reflection of the interactions between the phases in contact (crystallite and support) and is given by the expression

$$\begin{aligned}\sigma_{cs} &= \sigma_{cg} + \sigma_{sg} - (U_{int} - U_{str}) \\ &\equiv \sigma_{cg} + \sigma_{sg} - U_{cs}\end{aligned} \tag{15}$$

where U_{int} is the interaction energy per unit contact area and U_{str} is the strain energy per unit area due to the mismatch of the overlying lattices. Adsorption decreases the surface free energy. When a crystallite merely

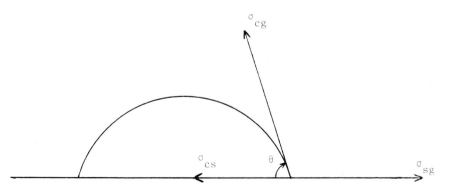

FIGURE 5 Schematic of a crystallite on a substrate at equilibrium.

adsorbs the atoms or the molecules of an ambient gas but does not undergo a reaction, as in the case of a hydrogen atmosphere in general, its surface free energy is decreased modestly. However, the surface free energies, σ_{cg}, of the metal oxides formed by the reaction between the metal and oxygen are much smaller than those of the corresponding metals. As a result of the chemical and structural similarities, in general, the interactions between two oxides are greater (U_{cs} is larger) and the strain energy due to the mismatch of their lattices smaller than the corresponding quantities for a metal/metal-oxide interface. Therefore, as Eq. (15) shows, σ_{cs} is also smaller in an oxidizing environment than in a reducing one. From the foregoing arguments one can conclude that since both σ_{cg} and σ_{cs} are smaller, the contact angle would be smaller and hence the particle would assume a more extended shape in an oxidizing atmosphere than in a reducing atmosphere. On similar lines, in a metal-refractory oxide system (i.e., in a reducing environment), due to the large differences in electronic and crystal structures of the two phases, σ_{cs} will be large and, since σ_{cg} of a metal is also large, the contact angle will be large and the particles would assume a contracted, globular morphology. Thus an extended particle shape indicates a stronger interaction with the support and provides a larger surface area and a higher dispersion. In addition, if sintering by crystallite migration were to occur, the catalyst would be more stable against sintering, since the strong interactions between crystallite and support would inhibit particle motion. On the other hand, a compact, globular morphology indicates a weaker interaction with the support and provides a smaller surface area and a lower dispersion. Also, because of the decreased interactions, the particles would migrate more easily and lead to a decrease in dispersion. It should be pointed out, however, that there is a possibility of easier emission of atoms from the particles to the substrate surface when the interactions are stronger. This may lead to sintering by a ripening mechanism. However, experimental observations indicate that when there is an extension of the crystallites over the surface of the substrate (be it with Fe, Co, Ni on Al_2O_3 in oxidizing atmospheres, or with Pt, Fe, Ni on TiO_2 in reducing atmospheres), there is an increase in dispersion. In the case of Pt/alumina and in an oxygen atmosphere, the interactions between crystallites and support are not as strong as in the above-mentioned cases, and the relatively higher sintering in O_2 than in H_2 reported in the literature (for details, refer to Refs. 1, 3, 10 to 14 and 73) is caused not only by ripening but also by crystallite migration.

Substituting for σ_{cs} from Eq. (15) in inequality (14), it results that the crystallite would spread out completely when

$$U_{cs} \geq 2\sigma_{cg} \tag{16}$$

Under equilibrium conditions, σ_{cs} has to be positive, but under nonequilibrium and dynamic conditions, it can be negative. Even if the inequality (16) is not satisfied, the quantities U_{cs} and $2\sigma_{cg}$ are closer in value (the ratio $U_{cs}/2\sigma_{cg}$ is closer to unity) in an oxidizing atmosphere, because of both a smaller σ_{cg} and a larger U_{cs}, than in a reducing atmosphere. If TiO_2 (or any other SMSI oxide) is the substrate, inequality (16) can be satisfied, or the ratio above will be closer to unity, in a hydrogen atmosphere, because in these systems σ_{cg} is smaller and U_{cs} larger in a reducing than in an oxidizing atmosphere. U_{cs} is large because of the strong interactions between the crystallite and the reduced substrate species, TiO_x. σ_{cg} is small, because of the migration of the reduced species onto the surface of the particle and the subsequent strong interactions with the latter, as explained before. The pillbox morphology of the Pt crystallites on TiO_2 supports in a reducing atmosphere is a result of a similar inequality [86]. The fact that the crystallite does not spread completely over the substrate but assumes a planar shape instead, with an abrupt variation of angle near the leading edge, was anticipated theoretically in Ref. 99 and confirmed later experimentally [63, 64,83].

Tables 1 and 2 list the surface free energies of a few metals, carbides, and oxides. If the interaction between the crystallite and support is very strong, of the order of 60 kcal/mol, then assuming a monolayer density of 10^{15} species cm^{-2}, U_{cs} is calculated to be 4300 erg cm^{-2}. Note, however, that with the conventional substrates, U_{cs} is much smaller than this value in a reducing atmosphere. Table 1 shows that even if U_{cs} is taken to be as large as 2500 erg cm^{-2}, the inequality (16) cannot be satisfied in a reducing atmosphere. Therefore, in the case of traditional supports, one can conclude that in a reducing atmosphere, the particles have a weak interaction with the substrate and a large contact angle. Consequently, they can have a higher mobility and exhibit low stability against sintering.

TABLE 1 Surface Free Energies of Metals

Metal	Temperature of measurement (melting point, K)[a]	Surface free energy, σ_{cg} (erg cm^{-2})	σ_{cg} at 773 K (erg cm^{-2})[b]
Aluminum	933	910	926
Barium	998	224	247
Cobalt	1768	1890	1990
Copper	1356	1350	1408
Gold	1337	1130	1186
Iridium	2683	2250	2441
Iron	1808	1850	1954
Molybdenum	2890	2250	2462
Nickel	1726	1800	1895
Osmium	3318	2500	2754
Palladium	1827	1500	1605
Platinum	2045	1800	1927
Potassium	336	101	—
Rhodium	2239	2000	2147
Rubidium	311	78	—
Ruthenium	2583	2250	2431
Silver	1235	930	976
Sodium	371	200	—
Titanium	1933	1600	1716

[a]From Ref. 101.
[b]Calculated assuming that $d\sigma/dT = -0.1$ erg cm^{-2} K^{-1}.
Source: Ref. 1.

TABLE 2 Surface Free Energies of Oxides and Carbides

Compound	Temperature of measurement (K)	Experimentally measured σ_{sg} (erg cm^{-2})	σ_{sg} at 773 K (erg cm^{-2})[a]
Al_2O_3	2143	925	1062
SiO_2	298	605	557
MgO	1870	1100	1210
ZrO_2	≤ 1423	1130	(1130) or 1195
TiO_2	2100	380	513
V_2O_5	963	90	109
Ta_2O_5	2100	280	413
WO_2	1773	100	200
FeO	1693	585	677
Fe_3O_4	1867	400	509
TiC	1373	1190 ± 350	1250 ± 410
WC	1423	2820 ± 30	2885 ± 95
TaC	1373	1290 ± 390	1350 ± 450
	1423	2690	2755
VC	1423	1677	1742
NbC	1423	2440	2505

[a] Calculated assuming $d\sigma/dt = -0.1$ erg cm^{-2} K^{-1}.
Source: Ref. 6.

Since it appears that a higher value of σ_{sg} leads to a better extension of the crystallites, one may be tempted to use the magnitude of σ_{sg} as a criterion for choosing a substrate that would lead to better stability toward dispersion. Let us examine this point by recasting Eq. (13) in the form

$$\cos \theta = \frac{\sigma_{sg} - \sigma_{cs}}{\sigma_{cg}} \tag{13'}$$

Assuming σ_{cg} and σ_{cs} to remain the same, the contact angle θ decreases with increasing values of σ_{sg} and there is, therefore, a tendency to increase the interfacial area of contact and consequently to increase the stability toward dispersion. From the values of the surface free energy, σ_{sg} of the oxides (Table 2), the stability against sintering of the crystallites should decrease in the order (TiC > TaC > WC) > ZrO_2 > MgO > Al_2O_3 > SiO_2 > (TiO_2 > Ta_2O_5 > V_2O_5). This suggests that TiC, TaC, WC, ZrO_2, and MgO with high surface free energies could be better supports than the conventional oxide supports, alumina and silica. In fact, Boudart et al. [100] reported that reduced iron with a high dispersion could be obtained on a MgO substrate, while reduced iron on alumina and silica supports could be achieved only with concomitant severe sintering.

In the consideration above, σ_{cs} has been assumed to remain the same when σ_{sg} is changed. However, σ_{cs} is not an independent quantity. Equation (15) shows that the value of σ_{cs} can remain constant with increasing σ_{sg} only when this increase is compensated by an increase in U_{cs}. Of course, if the increase in U_{cs} is greater (σ_{cs} becoming smaller), the dispersion will be additionally increased. As shown later, this can happen with ZrO_2 or MgO or other substrates that have a large σ_{sg} and are in addition sufficiently reactive. The traditional substrates with very high surface free energies, in general, have low reactivity and hence smaller U_{cs}, or larger σ_{cs}.

To summarize, let us emphasize that the classification of supports on the basis of the values of σ_{sg} alone is incomplete. To be complete, the analysis should take into account the values of both σ_{sg} and σ_{cs} or U_{cs}. There are cases in which σ_{sg} is very large, but U_{cs} is perhaps very small, so that the contact angle is still large; in other cases, σ_{sg} may be relatively small, but the interaction (U_{cs}) may be considerably larger, so that, over all, θ is smaller than in the previous case. TiO_2 has a much smaller σ_{sg} value than Al_2O_3. Under reducing conditions, when TiO_2 is reduced, the σ_{sg} of the lower oxide TiO_x may increase, but it may still be smaller than that of alumina. However, because of the defective, oxygen deficient, and hence reactive nature of the reduced support, the interaction energy, U_{cs}, between the particle and the support beneath may be tremendously increased, and consequently σ_{cs} is reduced. In addition, as a result of the interaction between the particle and the mono or submono layer of the reduced substrate species migrated onto it, σ_{cg} is also reduced. Therefore, the particle will wet this support better than alumina, even though σ_{sg} of TiO_x may be smaller than that of alumina. We shall now consider some of the factors that influence U_{cs}, such as the reactivity of the supports and the role of nonmetallic elements in the wettability of the supports by the crystallites.

B. Reactivity of Substrate and Stability Sequence

The wettability of the oxide supports by crystallites (especially in oxidizing atmospheres) depends on the relative strength of the interactions of the atoms or molecules of the particle among themselves and with those of the substrate. When the bonds in the substrate oxide are weaker, the substrate is more reactive than when the bonds are stronger. The bond strength of a compound is related to its heat of formation [101]. The greater the heat and the free energies of formation of the oxide substrate, the greater is its stability and the less likely it is to undergo a chemical modification. From their values [101], one can see that the bulk reactivities of the traditional supports should decrease in the order $MgO > SiO_2 > TiO_2 > ZrO_2 > Al_2O_3$. The difference in the heats of formation of MgO and Al_2O_3 is large and the former is expected to be more reactive. Of course, the reactivities at the surfaces of the oxides may be otherwise, depending on the type of metal that creates the interface. The surface free energy of magnesia is also greater than that of alumina, although the difference is not large. The greater reactivity (i.e., a smaller σ_{cs}) and the larger σ_{sg} of MgO compared to Al_2O_3 are, therefore, expected to lead to higher and more stable dispersions on MgO than on alumina. For similar reasons, ZrO_2 could also yield more stable dispersions than alumina.

It should be noted, however, that the stability against sintering depends not only on the reactivity of the substrate, but especially on the specificity of the reaction between crystallite and support. It is known that the surface of an oxide substrate exposes oxygen ions [8,102]. This is a result of the fact that the crystal field generated by the bulk ions displaces more easily the cations than the anions of the surface toward the interior. This is because the cations are smaller (hence more mobile) and less polarizable than the anions. Metals such as Fe and Ni which form oxides easily will react more strongly with the exposed oxygen ions of the substrate and therefore will have a smaller wetting angle. On the other hand, metals such as Pt which do not easily form an oxide will react less strongly with the substrate and will therefore have a larger wetting angle.

It has been shown above that when U_{cs} is large, there is extension and that the ratio $U_{cs}/2\sigma_{cg}$ is a good measure of the wettability of the support by the particles and hence a measure of the stability of the system toward higher dispersion. It is difficult to obtain the values of U_{cs}. However, the sequence of reactivities in terms of free energy of formation provides a partial guideline.

C. Effect of Nonmetallic Elements on Wettability

Nonmetallic solutes such as oxygen and sulfur, when present in very small amounts (smaller than needed to form oxides), are preferentially adsorbed both at the free surface of the crystallite as well as at the interface between the crystallite and the support. This is likely to happen if the interaction between two metal atoms, U_{22}, is stronger than the interaction between the solute and the metal atom, U_{12} [8]. As a result, the solute species will be preferentially rejected from the bulk and preferentially adsorbed at the surface where it decreases the surface free energy. In the case of Pt, which does not easily form an oxide, the Pt-Pt bond energy is considerably greater than the Pt-O bond energy and therefore the oxygen atoms would be preferentially adsorbed at the interface. In the case of metals such as Fe, Co, Ni, and Cu, which easily form an oxide, the solute forms a complex, $M^{x+}O^{x-}$, with the metal atoms. The metal atom that has, more or less, "donated" the electron to the oxygen atom would now be deficient in electrons and would therefore interact less strongly with the other atoms of the metal. In other words, the interaction between two metal atoms would be weaker when the nonmetallic solute is present (i.e., $U_{M^{x+}O^{x-}-M}$ would be weaker than U_{M-M}). As a result, $M^{x+}O^{x-}$ will accumulate at the interfaces. The oxide substrates expose the oxygen ions at the surface. Since the cation of $M^{x+}O^{x-}$ complex can interact with the oxygen ions of the substrate and thereby decrease the free energy, the enrichment of the complex would, in general, be greater at the particle-substrate interface and, because of the large interaction energy U_{cs}, the relative decrease in the interfacial free energy σ_{cs} will be greater than that at the free surface (σ_{cg}). These decreases lead to an extension of the particles and an increase in dispersion. In the absence of the metal cations of the crystallites, as is the case in a reducing atmosphere, a decrease in σ_{cs} is unlikely to occur, since a neutral metal atom interacts less strongly than a cation, with the negative (oxygen) ions exposed at the substrate surface.

The oxygen atoms in the $M^{x+}O^{x-}$ act as electron acceptor sites. Oxygen, of course, could be replaced by other electron acceptors, such as sulfur, halogens, complex anions of the SO_4^{2-} type, etc. [8]. Sulfur has the negative characteristic of poisoning platinum and other catalysts [103], although its presence may be necessary in catalytic reforming (see, e.g., Chapter 25). Halogens could be used to increase the wettability of the supports by the particles (when there is no loss of material through volatile compound formation), via the interfacial enrichment explained above. A neutral Pt atom may not interact with the alumina substrate. However, a Pt-halogen complex would, via the interactions of the Pt cation with the oxygen anion of the support, leading to extension of the particles and to an associated increase in the dispersion. This can explain the use of chlorine in regeneration of the sintered catalysts [14,104].

On similar lines, one can explain the role of alkali promoters. Since the alkali metals are more electropositive than any of the group VIII metals traditionally used as catalysts, they would be preferentially adsorbed at the interface, because of their stronger affinity for the oxygen ions of the support surface. They may therefore bind the particles to the support and increase the thermal stability of the particles against sintering. In addition, they act as chemical promoters, as they can facilitate electron exchanges.

D. Interatomic Bond Strength and Stability Sequence of Metals Against Sintering

As noted in Section III.A, sintering by atomic migration is likely to occur only when either the interactions between the metal atoms and those of the substrate are strong enough, or when the temperature is high enough, to cause the rupture of the metallic bonds in the particles and to dissociate single atoms from them. When there is no chemical interaction at the interface, the physical van der Waals interactions between the neutral metal atoms and the support are too weak to effect the dissociation of single atoms, because of the strong metallic bonds between the atoms of the metal. However, in such cases, once the single atoms are emitted to the substrate surface, they migrate very rapidly. (This is because the activation energy for surface diffusion is 0.1 to 0.2 times the energy of physical adsorption and the latter is of the order of only 1 to 5 kcal mol^{-1} [2,3].) The rate of dissociation of atoms is likely to be increased at high temperatures, because then the metallic bonds can be broken with higher probability. Also, since at high temperatures the conductivity and the reactivity of the support is increased (due to an increase in the number of electron-hole pairs with increasing temperatures) [8,102], the crystallite atoms may dissociate relatively easily, due to the added incentive to interact with the now more reactive support. If the crystallite migration mechanism is operative, sintering would be inhibited in reactive atmospheres, due to the same strong interactions with the support. Therefore, sintering by atomic migration is expected to occur at high temperatures and in reactive atmospheres, while at lower temperatures and in a hydrogen or inert atmosphere, sintering is expected to occur by crystallite migration. Of course, these are only guidelines; the actual process of sintering may be more complex. For example, ripening has been observed at low temperatures and particle migration has been observed at high temperatures [67,73].

From the arguments above it is clear that the greater the bonding strength between the metal atoms, the less likely they will sinter by ripening. It is thus expected that particles with higher melting points and sublimation energies should be more resistant to sintering. The resistance to sintering is in the following order [1], in parallel with their melting points (with the exception of Fe and Pd): Re > Os > Ir > Ru > Rh > Pt > Co > Ni > Fe > Pd > Au > Cu > Ag.

One of the mechanisms of crystallite migration is thought to be the advancement of the particle by the surface self-diffusion of atoms from one end of the particle to the other, perhaps, due to the variations in curvature or to nonuniform interactions with the substrate along the periphery. Thus, irrespective of whether the overall sintering mechanism is atomic migration or crystallite migration, the order of stability given above should, in general, hold in a hydrogen atmosphere. The stability order observed experimentally by Wang and Schmidt [97] and Fiedorow et al. [51], for a few of the metals noted above, namely Ir > Rh > Pt > Pd, is in agreement with the order given above.

VI. CONCLUSION

Hydrogen, because of its role in catalyst pretreatment or of its presence in reaction atmospheres, affects sintering of supported metal catalysts. In addition, other reactive gases, such as moisture, oxygen, chlorine, hydrocarbons, etc., affect, either concurrently or sequentially, the behavior of supported metal crystallites in various ways, depending on how they influence the metal-support interactions. The interactions between metal and support and the related wetting characteristics of the system explain, at least qualitatively, the thermal stability or sintering of the supported metal catalysts. Experimental observations are in general varied and, sometimes, even conflicting. Even though traditionally, either atomic migration or crystallite migration is claimed to be responsible for sintering, and experimental observations are provided to support one or the other mechanism, recent investigations show that sintering is coupled with a number of other phenomena, such as changes in shape, dissolution into, and/or chemical interactions with, the support, extension of particles, the presence of detectable, or possible presence of undetectable (by electron microscopy) multilayer surface films that coexist with crystallites, splitting of particles, etc. These constitute interconnected parts of complex surface phenomena that occur with supported metal catalysts.

REFERENCES

1. J. R. Anderson, *Structure of Metallic Catalysts*, Academic Press, New York (1975).
2. G. C. Bond, *Catalysis by Metals*, Academic Press, New York (1962).
3. J. W. Geus, in *Materials Science Research*, Vol. 10, *Sintering and Catalysis* (G. C. Kuczynski, ed.), Plenum Press, New York (1975), p. 29.
4. F. Delannay, in *Characterization of Heterogeneous Catalysts* (F. Delannay, ed.), Marcel Dekker, New York (1984).
5. G. J. K. Acres, A. J. Bird, J. W. Jenkins, and F. King, in *Catalysis, a Specialist Periodical Report*, Vol. 4 (C. Kemball and D. A. Dowden, eds.), Royal Society of Chemistry, London (1981).
6. S. H. Overbury, P. A. Bertrand, and G. A. Somorjai, *Chem. Rev.*, 75(5):547 (1975).
7. M. W. Roberts and C. S. McKee, *Chemistry of the Metal-Gas Interface*, Clarendon Press, Oxford (1978).
8. Ju. V. Naidich, in *Progress in Surface and Membrane Science*, Vol. 14 (D. A. Cadenhead and J. F. Danielli, eds.), Academic Press, New York (1981), p. 354.
9. E. Ruckenstein and B. Pulvermacher, *AIChE J.*, 19:356 (1973).
10. E. Ruckenstein and B. Pulvermacher, *J. Catal.*, 29:224 (1973).
11. P. Wynblatt and N. A. Gjostein, *Prog. Solid State Chem.*, 9:21 (1975).
12. P. C. Flynn and S. E. Wanke, *J. Catal.*, 34:390 (1974); 34:400 (1974).
13. P. C. Flynn and S. E. Wanke, *Catal. Rev.-Sci. Eng.*, 12:93 (1975).
14. E. Ruckenstein and D. B. Dadyburjor, *Rev. Chem. Eng.*, 1:251 (1983).
15. H. H. Lee and E. Ruckenstein, *Catal. Rev.-Sci. Eng.*, 25:475 (1983).
16. R. Hughes, in *Deactivation of Catalysts*, Academic Press, London (1984).
17. P. C. Flynn, S. E. Wanke, and P. S. Turner, *J. Catal.*, 33:233 (1974).
18. M. M. J. Treacy and A. Howie, *J. Catal.*, 63:265 (1980).
19. M. J. Yacaman, *Appl. Catal.*, 13:1 (1984).
20. F. Delannay, in *Characterization of Heterogeneous Catalysts* (F. Delannay, ed.), Marcel Dekker, New York (1984), p. 71.
21. T. Baird, in *Catalysis, a Specialist Periodical Report*, Vol. 5 (G. C. Bond and G. Webb, eds.), Royal Society of Chemistry, London (1982), p. 172.

22. E. Ruckenstein and M. L. Malhotra, *J. Catal.*, 41:303 (1976).
23. H. Glassl, R. Kramer, and K. Hayek, *J. Catal.*, 63:167 (1980).
24. R. T. K. Baker, *Catal. Rev.-Sci. Eng.*, 19:161 (1979).
25. H. P. Klug and L. E. Alexander, *X-Ray Diffraction Procedures*, Wiley, New York (1954).
26. P. Ganesan, H. K. Kuo, A. Saavedra, and R. J. De Angelis, *J. Catal.*, 52:310 (1978).
27. T. E. Whyte, *Catal. Rev.-Sci. Eng.*, 8:117 (1973).
28. R. L. Moss, in *Catalysis, a Specialist Periodical Report*, Vol. 4 (C. Kemball and D. A. Dowden, eds.), Royal Society of Chemistry, London (1981), p. 31.
29. B. Pulvermacher and E. Ruckenstein, *J. Catal.*, 35:115 (1974).
30. R. B. Anderson and P. T. Davison (eds.), *Experimental Methods in Catalytic Research*, Academic Press, New York (1976).
31. J. T. Richardson and P. Desai, *J. Catal.*, 42:294 (1976).
32. J. T. Richardson, J. R. Crump, and R. U. Osterwalder, in *Materials Science Research*, Vol. 13, *Sintering Processes* (G. C. Kuczynski, ed.), Plenum Press, New York (1980).
33. R. J. Ferrauto, *AIChE Symp. Series*, 70(143):9 (1974).
34. J. L. Lemaitre, P. G. Menon, and F. Delannay, in *Characterization of Heterogeneous Catalysts* (F. Delannay, ed.), Marcel Dekker, New York (1984), p. 299.
35. G. R. Wilson and W. K. Hall, *J. Catal.*, 17:190 (1970).
36. R. A. Dalla Betta and M. Boudart, *Proceedings of the 5th International Congress on Catalysis*, Vol. 2 (1973), paper no. 96, p. 1329.
37. B. K. Chakraverty, *J. Phys. Chem. Solids*, 28:2401, 2413 (1967).
38. E. Ruckenstein and D. B. Dadyburjor, *Thin Solid Films*, 55:89 (1978).
39. D. B. Dadyburjor and E. Ruckenstein, *J. Cryst. Growth*, 38:285 (1977).
40. F. A. Nichols, *J. Appl. Phys.*, 37:2805 (1966).
41. W. Ostwald, *Z. Phys. Chem.*, 34:495 (1900).
42. J. M. Lifshitz and V. V. Slyozov, *J. Phys. Chem. Solids*, 19:35 (1961).
43. C. Wagner, *Z. Electrochem.*, 65:581 (1961).
44. H. H. Lee, *J. Catal.*, 63:129 (1980).
45. C. Kittels, *Introduction to Solid State Physics*, 3rd ed., Wiley, New York (1966).
46. F. H. Huang and C. Y. Li, *Scr. Metall.*, 7:1239 (1973).
47. S. E. Wanke, in *Materials Science Research*, Vol. 10, *Sintering and Catalysis* (G. C. Kuczynski, ed.), Plenum Press, New York (1975), p. 107.
48. E. Ruckenstein and D. B. Dadyburjor, *J. Catal.*, 48:73 (1977).
49. T. M. Ahn, J. K. Tien, and P. Wynblatt, *J. Catal.*, 66:335 (1980).
50. J. J. Chen and E. Ruckenstein, *J. Catal.*, 69:254 (1981).
51. R. M. J. Fiedorow, B. S. Chahar, and S. E. Wanke, *J. Catal.*, 51:193 (1978).
52. D. J. Smith, D. White, T. Baird, and J. R. Fryer, *J. Catal.*, 81:107 (1983).
53. D. White, T. Baird, J. R. Fryer, L. A. Freeman, D. J. Smith, and M. Day, *J. Catal.*, 81:119 (1983).
54. A. Williams, G. A. Buttler, and J. Hammonds, *J. Catal.*, 24:352 (1972).
55. J. T. Richardson and J. G. Crump, *J. Catal.*, 57:417 (1979).
56. P. Desai and J. T. Richardson, in *Catalyst Deactivation* (B. Delmon and G. F. Froment, eds.), Elsevier, Amsterdam (1980), p. 149.
57. C. H. Bartholomew, R. B. Pannell, and R. W. Fowler, *J. Catal.*, 79:34 (1983).
58. K. Morikawa, T. Shirasaki, and M. Okada, *Adv. Catal.*, 20:98 (1969).
59. J. R. H. Ross, M. C. F. Steel, and A. Zeini-Isfahani, *J. Catal.*, 52:280 (1978).

60. Y. F. Chu and E. Ruckenstein, *J. Catal.*, 55:281 (1978).
61. E. Ruckenstein and Y. F. Chu, *J. Catal.*, 59:109 (1979).
62. E. Ruckenstein and J. J. Chen, *J. Colloid Interface Sci.*, 86(1):1 (1982).
63. R. T. K. Baker, E. G. Prestridge, and R. L. Garten, *J. Catal.*, 56:390 (1979).
64. R. T. K. Baker, *J. Catal.*, 63:523 (1980).
65. T. Wang and L. D. Schmidt, *J. Catal.*, 70:187 (1981).
66. E. G. Derouane, R. T. K. Baker, J. A. Dumesic, and R. D. Sherwood, *J. Catal.*, 69:101 (1981).
67. I. Sushumna and E. Ruckenstein, *J. Catal.*, 94:239 (1985).
68. I. Sushumna and E. Ruckenstein, *J. Catal.*, 90:241 (1984).
69. J. J. Chen and E. Ruckenstein, *J. Phys. Chem.*, 85:1696 (1981).
70. M. Chen and L. D. Schmidt, *J. Catal.*, 55:348 (1978).
71. H. K. Kuo, P. Ganesan, and R. J. De Angelis, *J. Catal.*, 64:303 (1980).
72. H. Glassl, R. Kramer, and K. Hayek, *J. Catal.*, 68:388 (1981).
73. I. Sushumna and E. Ruckenstein, submitted for publication.
74. I. Sushumna and E. Ruckenstein, *J. Catal.*, 97:1 (1986).
75. H. C. Yao, M. Sieg, and H. K. Plummer, Jr., *J. Catal.*, 59:365 (1979).
76. R. T. K. Baker and J. A. France, *J. Catal.*, 39:481 (1975).
77. R. T. K. Baker, R. D. Sherwood, A. J. Simoens, and E. G. Derouane, in *Metal-Support and Metal-Additive Effects in Catalysis* (B. Imelik et al., eds.), Elsevier, New York (1982).
78. Y. F. Chu and E. Ruckenstein, *Surf. Sci.*, 67:517 (1977).
79. E. Ruckenstein and S. H. Lee, *J. Catal.*, 86:457 (1984).
80. X. D. Hu, M.S. thesis, State University of New York at Buffalo, 1985.
81. S. J. Tauster, S. C. Fung, and R. L. Garten, *J. Am. Chem. Soc.*, 100(1):170 (1978).
82. R. T. K. Baker, E. B. Prestridge, and R. L. Garten, *J. Catal.*, 59:293 (1979).
83. B. J. Tatarchuk and J. A. Dumesic, *J. Catal.*, 70:308 (1981).
84. A. J. Simoens, R. T. K. Baker, D. J. Dwyer, C. R. F. Lund, and R. J. Madon, *J. Catal.*, 86:359 (1984).
85. R. T. K. Baker, E. B. Prestridge, and G. B. McVicker, *J. Catal.*, 89:422 (1984).
86. E. Ruckenstein, in *Strong Metal-Support Interactions* (R. T. K. Baker, S. J. Tauster, and J. A. Dumesic, eds.), *ACS Symposium Series 298*, American Chemical Society, Washington, D.C. (1986).
87. G. Skinner, H. L. Johnston, and C. Beckett, *Titanium and Its Compounds*, Herrick L. Johnston Enterprises, Ohio (1954).
88. T. Huizinga and R. Prins, *J. Phys. Chem.*, 85:2156 (1981).
89. G. V. Samsonov (ed.), *The Oxide Handbook*, IFI/Plenum, New York (1973).
90. P. G. Menon and G. F. Froment, *J. Catal.*, 59:138 (1979).
91. G. J. den Otter and F. M. Dautzenberg, *J. Catal.*, 53:116 (1978).
92. F. M. Dautzenberg and H. B. M. Wolters, *J. Catal.*, 51:26 (1978).
93. Z. Paal and P. G. Menon, *Catal. Rev.-Sci. Eng.*, 25(2):229 (1983).
94. X. Jiang, T. F. Hayden, and J. A. Dumesic, *J. Catal.*, 83:168 (1983).
95. G. B. Raupp and J. A. Dumesic, *J. Phys. Chem.*, 88:660 (1984).
96. A. Graham and S. E. Wanke, *J. Catal.*, 68:1 (1981).
97. T. Wang and L. D. Schmidt, *J. Catal.*, 66:301 (1980).
98. K. Foger and J. R. Anderson, *J. Catal.*, 59:325 (1979).
99. E. Ruckenstein and P. S. Lee, *Surf. Sci.*, 52:298 (1975); *J. Colloid Interface Sci.*, 86:573 (1982).
100. M. Boudart, A. Delbouille, J. A. Dumesic, S. Khammamouma, and H. Topsoe, *J. Catal.*, 37:486 (1975).

101. R. C. Weast (ed.), *Handbook of Physics and Chemistry*, 62nd ed., CRC Press, Boca Raton, Fla. (1981), p. 102.
102. D. Beruto, L. Barco, and A. Passerone, in *Oxides and Oxide Films*, Vol. 6 (K. Vijh Ashok, ed.), Marcel Dekker, New York (1981).
103. T. Halachev and E. Ruckenstein, *Surf. Sci.*, *108*:292 (1981).
104. D. J. C. Yates, U.S. Patent 1,433,864 (1976).

11 | Hydrogen-Induced Sintering of Unsupported Metal Catalysts

ZOLTÁN PAÁL

Institute of Isotopes of the Hungarian Academy of Sciences, Budapest, Hungary

I.	INTRODUCTION: ON THE EQUILIBRIUM CRYSTAL SHAPE AND HOW TO REACH IT	293
II.	HYDROGEN-INDUCED RECONSTRUCTIONS	295
	A. Single Crystals	296
	B. Disperse Catalysts	297
III.	CONCLUSIONS	308
	REFERENCES	308

I. INTRODUCTION: ON THE EQUILIBRIUM CRYSTAL SHAPE AND HOW TO REACH IT

Metal catalysts exist as crystals of various shapes and sizes, the surfaces of which are usually covered by impurities of unknown amount and composition. The outer shape of these metal crystals rarely corresponds to an "equilibrium shape" when the free energy of the total surface is minimal:

$$\sum_i S_i \gamma_i = \text{minimum} \tag{1}$$

where S_i is the area and γ_i the surface free energy of face i and the summation has to be carried out for all crystal faces. Instead, various "kinetic shapes" may occur in practice which represent some frozen stages the surface reconstruction of which can occur under appropriate conditions (although the rate may be negligible).

Wulff [1] derived an expression concerning the shape of the crystal as determined by the surface free energies of various crystal planes:

$$\frac{\gamma_1}{h_1} = \frac{\gamma_2}{h_2} \cdots \frac{\gamma_i}{h_i} = \text{const.}$$

where h is the distance of the plane from the geometric center of the crystal (Wulff point). If one plots the values of γ in polar coordinates around the Wulff point, a self-contained curve is obtained with "cusps" in various

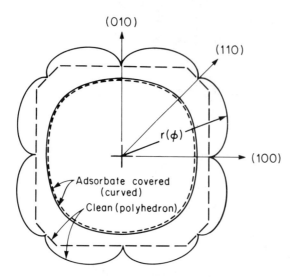

FIGURE 1 Polar plot of surface free energy γ as a function of orientation (ϕ) in the (100) - (110) direction. Cubes with (110) edges will be obtained if the anisotropy is as shown in the outer curve. Adsorption should reduce its anisotropy to produce spherical particles as shown in the inner curve. (From Ref. 2; used with permission.)

directions. The cusps correspond to stable crystal planes with local minima of the surface energy. Such a Wulff plot is shown in Fig. 1.

Any adsorbed layer on the surface will change the surface free energy and consequently, the equilibrium crystal shape. An extreme case of this situation is depicted in Fig. 1, where a "rounded" equilibrium shape is proposed under the effect of a strong adsorbate which brings about an almost isotropic surface free energy distribution [2].

The change from a nonequilibrium shape toward equilibrium must involve mass transport which shifts atoms from one position of the crystal to another [3]. Surface reconstruction of larger single crystals occurs on their exposed two-dimensional faces [4]. It can be enhanced largely by the presence of adsorbates. Here interatomic forces may determine the type and extent of restructuring. With field-emission tips, usually of dimension ca. 100 nm, the reconstruction may affect the entire three-dimensional crystal. Here surface diffusion is believed to play the main role in reconstruction (e.g., Ref. 5). Also, the penetration of adsorbates below the surface may influence the surface structure and energy; with hydrogen atoms being small, this process is more pronounced than with other adsorbates (see Chapter 1, Section III.A; see also Section II.B of the present chapter).

In the case of supported small crystals, the interface free energy between the crystal and the support may be decisive in determining the particle shape [6]. In the case of "poor wetting," three-dimensional small crystallites may exist; with "good wetting," however, two-dimensional spreading of the particle may occur [6]. The support itself may enhance the mobility of grains as well as that of individual metal atoms as discussed in detail in Chapter 10. Some relevant information will, however, also be used here.

Unsupported materials cannot float in air; they must be held up by some other material. However, the contact area and the intimacy between a disperse metal and its support (e.g., between grains and an electron microscopy carbon foil) are much lower here. As opposed to small separate crystals in a thin film on a support, the grains of a polycrystalline material always have much contact with each other. Their surface and bulk reconstruction are therefore easier. In principle, there is no difference between a polycrystalline massive metal (e.g., a ribbon) and a loose stack of individual crystallites; however, it is easy to see that the importance of individual processes can be enormously different in these cases. Here we shall concentrate on loose, powderlike materials. The phenomenological description as well as the theory of sintering are of paramount importance in several areas, such as metallurgy and the ceramic industry (see, e.g., Refs. 3 and 7 to 12). However, they would be far beyond the scope of the present book.

Briefly, when two small particles come in contact, the loss of surface area between them is thermodynamically favorable. Therefore, a "neck" is formed between them. Its high curvature represents a driving force for atom transport to this zone [10,12]. This occurs mainly by surface diffusion. In this initial stage grain growth does not occur. This neck growth stage is followed by an "intermediate stage" where material transport is ensured either by bulk diffusion or by grain boundary diffusion. The situation is somewhat analogous to film growth on support layers during metal evaporation, described in detail by Geus [13]. The neck formation is followed here by migration of the grain boundary sweeping through one of the crystals. Here the driving force is that this migration will gradually decrease the boundary area. The energy liberated during the combination of two crystallites into one may even cause melting of the entire particle if they are small enough; with sizes somewhat above this, a phenomenon such as zone melting may accompany grain boundary migration [14]. There is one decisive difference between film and powder growth: In the former case evaporation ensures a constant supply of adatoms which are much more mobile than those already located in lattice positions [13]. Alternative mechanisms are also possible [11] (see also Chapter 10).

The kinetics of grain growth in the intermediate stage can be described by the following equation:

$$D^m - D_0^m = Kt \exp\left(\frac{-E}{kT}\right) \qquad (3)$$

where D_0 and D are the most likely initial and instantaneous grain diameters, t is the time, E the activation energy of the rate-determining process, and K is a coefficient interpreted by Feltham [7]. The actual values of the exponents M vary: It is claimed that $m = 2$ when grain boundary diffusion is rate determining [7], and that $m = 3$ when the process is controlled by bulk diffusion [8], and that $m = 2$ [9] or 3 [15] with coalescence control in a network of densely packed grains.

Grain growth will slow down or stop after reaching a size such that diffusion cannot supply enough materials for reconstruction. This may be the situation in films when several particles get into contact with each other [13] and the grain boundaries can no longer diffuse out of the crystals. In the presence of adsorbates, the mobility of surface atoms decreases and the resulting grain sizes also decrease [13]. Obviously, an adsorbate layer on the contacting surfaces between grains may also represent a considerable barrier for diffusion [5].

II. HYDROGEN-INDUCED RECONSTRUCTIONS

Wulff plots may help one predict the expected crystal shapes and faces exposed in small crystallites. Unfortunately, the situation today is full of contradictory explanations and even speculations, since reliable data are not available for the surface energies of various crystal faces except for the simplest low-index ones. The situation is even worse for crystal planes in the presence of various adsorbates.

A. Single Crystals

Wang et al. [2] based their conclusions on the equilibrium shapes of clean, adsorbate-covered crystals (Fig. 1) on electron microscopic observations of small particles (see Section II.B). Drechsler [16], on the other hand, argued that field-emission tips had their equilibrium shapes in rounded form with small, flat, low-index facets exposed. He attributed flat planes observed in polyhedral crystallites to large stepped surfaces stabilized by strong adsorbates. By saying so, the stability of large low-index flat planes was implicitly disclaimed. This is certainly not true. Blakely and Somorjai [17] studied 22 various—flat, stepped, and kinked—planes of Pt. They found all three basic low-index planes—(111), (110), and (100)—stable in vacuum and in the presence of oxygen and carbonaceous adsorbates up to ca. 1500 K. Some more open surfaces, such as (211), (311), (210), (221), and (331), were also stable up to ca. 1200 K. Other faces were liable to reconstruction, the temperature of which decreased markedly in the presence of the adlayers mentioned. Three basic types of reconstructed surfaces were observed. Polyhedral structures postulated by Drechsler [16] correspond to the "hill and valley" structure postulated by Blakely and Somorjai [17].

Hydrogen causes strong surface reconstruction of some metals, most conspicuously on Ni(110) (see, e.g., Ref. 18 and Chapter 1).

Kumar and Grenga [19-21] investigated the surface reconstruction of various field-emission tips under annealing in vacuum and in hydrogen. They found a definite increasing of facets (100), (110), and (111), as well as (210) and (311) of Ir, upon annealing between 890 and 1330 K [19]. (Note that the latter two are among the stable single-crystal surfaces reported by Blakely and Somorjai [17].) The growth rate of facets (110) and (311) was increased more by hydrogen than was the growth rate of the other facets. The latter two facets correspond to the so-called B_5 sites; that is, they represent five-coordinated adsorption sites [22].

Steps with a (110) symmetry have been claimed to represent peculiar hydrogen adsorption sites on Pt having face-centered cubic (fcc) crystals [23]. Hydrogen increased the growth rate of the body-centered cubic (bcc) W(310) region when annealing between 650 and 1330 K; facets were obtained that were larger by 10 to 20% [20]. Facets more than twice as large (110) as those obtained in vacuum [21] were obtained when annealing a Fe tip (also having a bcc structure) in hydrogen at 710 K. Thus, of all three metals (Ir, W, Fe) studied, hydrogen exerted the greatest influence on the surface energy and/or surface diffusion of iron.

Nielsen and Adams [24] reported a stable disordered superstructure when annealing Pt(100) up to 1150 K. After Ar^+ ion bombardment, this reconstruction could be reproduced by subsequent heating. If, however, about 1/3 monolayer of surface hydrogen was present, this reconstruction only needed heating up to 400 K (i.e., the superstructure was produced at a temperature 750 K lower than that for pure surfaces).

Tung and Graham [25] studied the intrinsic self-diffusion coefficients of various Ni planes. Heating the surface in hydrogen up to 350 K caused an increase in the self-diffusion coefficient on Ni(110) in both perpendicular directions by about 10^7. Hydrogen-etched Ni surfaces showed as high diffusivities at 80 to 90 K as observed with vacuum-annealed tips at 150 to 180 K. This effect can be related to the reported reconstruction of this particular single-crystal face (Chapter 1) and it must be emphasized that this reconstruction must also have caused long-range surface effects. Similar but less pronounced enhancement was also found on (311) and (331) planes.

Hydrogen adsorption occurred on stepped surfaces under conditions when low-index flat surfaces did not adsorb appreciable amounts [26]. Steps did not serve just as sites for hydrogen accumulation; rather, they represented "gates" through which hydrogen may have diffused to positions where it was "retained" by the metal. After surface cleaning, H may have reappeared from these sites. Maire et al. [27] studied a Pt single crystal with monoatomic steps--Pt(S)-[6(111) × (100)]--in a hydrogen atmosphere. A hydrogen pressure of 10^{-6} torr applied for several minutes between 470 and 770 K resulted in doubling both the step height and the terrace width. This array was stable under heating in vacuum up to 870 to 970 K; above this temperature the original structure reappeared. The authors proposed the appearance of (311) intermediate steps, in agreement with the changed catalytic properties of reconstructed stepped surfaces.

Considering the observations listed, it seems that hydrogen effects in surface reconstruction and/or diffusion may be more numerous than believed and the lack of more experimental facts may be due to the difficulty of their observation rather than to their absence. A much lower number of surface reconstructions have been observed so far under the effect of hydrogen than upon adsorption of other gases, and they occur mostly on stepped surfaces, as summarized by Somorjai [28].

B. Disperse Catalysts

In this section we deal mostly with polycrystalline metal blacks, and in a few cases, information obtained with ribbons, wire, and so on, is included. It is unavoidable here to refer to analogies obtained with supported systems, but this will be done sparsely and with caution.

Equilibrium Shapes of Supported and Unsupported Metals

Wulff constructions have been used to predict equilibrium shapes and exposed faces of small crystals as a function of their size [29]. Recently, a few interesting studies have been published concerning electron microscopy (EM) of metal particles deposited on oxide supports. Wang et al. [2] vapor deposited Pt and Rh particles onto SiO_2 and Al_2O_3 support. A subsequent annealing (e.g., 24 h at 870 K) at "high-purity" (not specified) flowing H_2 of atmospheric pressure resulted in rectangular metal particles. The same treatment in N_2 gave hexagonal particles. Rh gave hexagonal outlines in all cases. On heating the "cubes" in nitrogen or air at 870 K, rounded shapes were obtained; heating in hydrogen reconstructed most (but not all) crystallites back to rectangular outlines. Internal twin boundaries were unaffected during these treatments.

One has to remember that these experiments concern *supported* particles. Here the interface free energy also has an important role in determining particle shapes [6]. Noble metals on SiO_2 and Al_2O_3 have contact angles near 90°, and this may be one of the reasons why rectangular particles appear. Their rounding was attributed by Wang et al. [2] to adsorption of impurities; hydrogen was claimed to remove these while it itself would not stay on. The latter statement is at least uncertain in view of H incorporation at even higher temperatures [26]. Instead, we believe in the alternative explanation that the "particles are contaminated in all situations and that H_2 merely causes a different contamination in which polyhedra are stable" [2]. These surfaces may also contain hydrogen in subsurface regions.

Harris [30] attributed sharp polyhedra exposing (111) faces in a Pt/Al_2O_3 film after heating at 1170 K in vacuum to carbon deposition on the surface, whereas heating in air produced rapid growth and rounded "kinetic" shapes. Fast growth—partly giving triangular large plates—occurred upon heating in 1-bar air at 970 K. Many of the particles were twins [30].

A recent high-resolution EM study [31] found rounded Pt particles after vapor deposition of Pt onto Al_2O_3 (1.8 ± 0.6 nm). They had perhaps some hexagonal symmetry. Heating this in "pure" H_2 (770 K, 1 h) increased grain size to 2.1 ± 0.7 nm; as little as 1 volume per million (vpm) water, O_2, and hydrocarbons each in H_2 enhanced growth to 3.8 ± 1.5 nm. Oxidation and oxychlorination gave much larger particles (11 and 23 nm, respectively). Shapes closer to polyhedral were obtained with another catalyst, prepared by wet impregnation of alumina with H_2PtCl_6. Twins were frequent here, too. Mostly triangular Pd particles (half cubo-octahedra or truncated tetrahedra, size 4 to 6 nm) were seen by EM in a fresh Pd on mica film [32]. These were rounded when heating in O_2; no H_2 treatment is reported. More information on supported systems is supplied in Chapter 10 and the references therein.

The reduction of H_2PtCl_6 in situ (dried solution on a carbon support film used for EM) by the electron beam itself resulted in mostly hexagonal clear-cut crystals (Fig. 2). Here traces of HCl and Cl_2 given off during reduction may have been adsorbed; chlorination, however, did not result in polyhedra in Pt/Al_2O_3 [32]. An appreciable amount of hydrocarbon from the rest gas

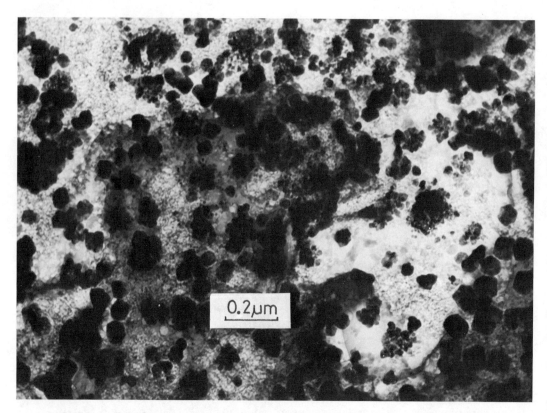

FIGURE 2 H_2PtCl_6 decomposing under the beam of an electron microscope giving Pt particles (dark areas). Note the hexagonal outline of most metal particles. (From Ref. 33; reproduced by permission.)

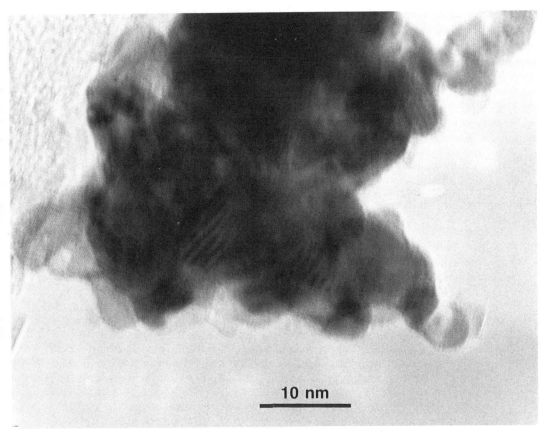

FIGURE 3 Lattice-resolution EM picture of a freshly reduced Pt black. Lattice fringes can be seen; particle size 5 to 10 nm. (Courtesy of J. R. Günter, results to be published.)

could not have adsorbed during the 1 or 2 s of formation. Therefore, we suggest that this shape must be characteristic of a clean Pt surface.

The size of crystallites produced under a beam is between 20 and 60 nm, larger than that of Pt grains produced via reduction of H_2PtCl_6 in aqueous medium by alkaline formaldehyde [33-35]. Here ample amounts of C, O, and/or CO can be adsorbed which originates from the decomposition of organics. Indeed, their presence on the Pt surface was observed by x-ray photoelectron spectroscopy (XPS) [36]. The agglomerates of these crystallites consisted of rounded crystals of 5 to 12 nm, as shown by lattice-resolution electron microscopy (Fig. 3). This is in agreement with the suggestion of Wang et al. [2] that clean surfaces would produce polyhedral shapes and contaminated ones, rounded shapes. Support for this theory is given by extensive sintering of Pt black under the beam of an electron microscope. This produces large (1 to 2 μm), rounded crystals (see Fig. 8a in Ref. 33), as opposed to clear-cut polyhedra of sintered Au [37]. It is well known from adsorption studies that Au is much more reluctant than Pt to adsorb impurities such as CO or hydrocarbons; the Au surface can thus be regarded as much cleaner both before and after sintering than that of Pt.

The shape itself—rounded or polyhedral—cannot be regarded as an indicator of surface purity. Some crystal planes may have lower free energy in the presence of an adsorbate. For example, adsorption of sulfur on Pt(100) lowers its surface energy to an extent that a recrystallization of other crystal planes is induced into (100) faces [65]. This is clearly observed on a Pt-Rh wire gauze after its exposure to 100 ppm H_2S for 110 h at 1400 K during HCN production from NH_3 and CH_4 in air [66]. The appearance of rectangular particles could also be seen in a disperse Pt-on-alumina film prepared on an electron microscopic grid [30] after its treatment in hydrogen containing 100 vpm H_2S at 773 K for 16 h [67]. Air-heated particles seemed to be rounded octahedra. The (100) faceting was almost perfect and was observed in the case of twinned particles too.

Morphology Changes of Metal Blacks Promoted by Hydrogen

Of the individual steps inducing particle restructuring, hydrogen was shown to facilitate surface reconstruction as well as surface self-diffusion. Here a few results concerning Pd and Pt metals will be presented showing that hydrogen can promote their sintering, accompanied by their major recrystallization.

Severe sintering of Pd black was reported when exposed to hydrogen at temperatures as low as 330 K [38]. "Meniscus formation" was observed between contacting crystallites after hydrogen treatment, corresponding to the first ("neck growth") stage of sintering, as discussed in Section I.

Freshly reduced Pt black (Fig. 3) represents a system with mostly point-like contacts between largely irregular-shaped, rounded grains. Heating this sample in air, helium, or vacuum up to 630 K resulted in crystallite growth of low extent [39]. Slightly larger crystallites were obtained after heating in ethylene ambient. Heating up in hydrogen, however, caused a marked crystallite growth, and even more pronounced sintering was observed when the sample was heated in air, then flushed with helium, and subsequently contacted with hydrogen at the final isothermal temperature of the thermal cycle. Maximum crystallite sizes were obtained with final temperatures between 570 and 670 K [40]. Figure 4 reveals that again smaller crystallites were found in samples sintered above 700 K.

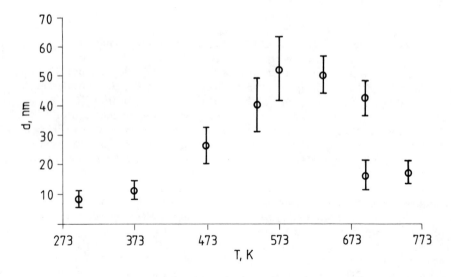

FIGURE 4 Average crystallite sizes determined by EM as a function of the final temperature of a standard thermal cycle of 3 h (heating up in air, He flushing, 1 h H_2, then He, cooled in He). (After Ref. 40; used with permission.)

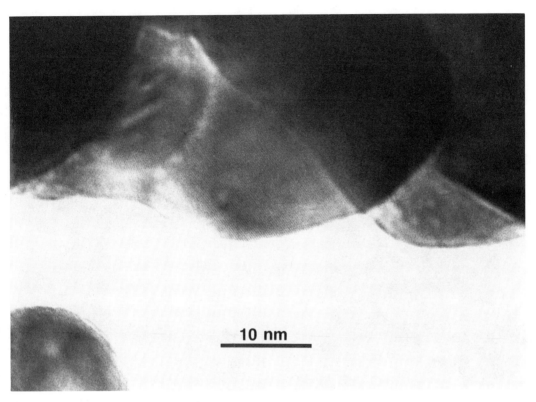

FIGURE 5 Lattice-resolution EM picture of a Pt-black sample after a thermal cycle at 470 K. Crystallite between 20 and 30 nm, on the limits of transparency in EM. Samples sintered at 630 K are no longer transparent. (Courtesy of J. R. Günter, results to be published.)

The catalyst particles sintered at 470 K were still at the limit of transparency by electron beams and exhibited well-ordered lattice fringes in crystallites of about 20 to 30 nm (Fig. 5). More regular, less rounded crystal faces were seen after heating at 573 K [41]. Here also cavities were observed, due, perhaps, to incomplete coalescence. Catalysts sintered at 630 K were no longer transparent (crystallite size 50 to 60 nm).

EM crystallite size determinations are usually based on measuring particle diameters at the edges of the agglomerate. One has to be cautious, however, since a photograph of a model supported catalyst (Fig. 6) shows that single crystals may exhibit very irregular shapes [42].

In situ x-ray diffraction (XRD) monitoring of fresh Pt-black lines reveals line broadening. This arises from both small crystallite size ($\bar{D}$) and "lattice strain" ($\bar{e}$) expressed as $\Delta \bar{d}/\bar{d}$, where $\bar{d}$ is the interplanar spacing [43]. Plotting line broadening during in situ heating as a function of time reveals that sintering in hydrogen is completed within 15 to 40 min after a temperature jump (100 K min^{-1}) to 470, 570, or 670 K. Again, maximum grain size has been found at 570 K. The values of $\bar{e}$ and $\bar{D}$ can be separated from XRD data [44]. Crystallite size measurements by EM and XRD are in fair agreement (Table 1). However, surface measurement gives too-large particle sizes, indicating two-dimensional contact between crystallites in sintered Pt, which, consequently, does not expose all faces to adsorption.

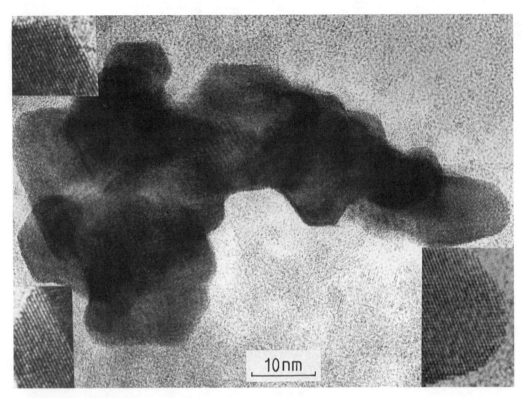

FIGURE 6 Lattice-resolution image of a platinum particle supported in Al_2O_3 (model catalyst). Three lattice images have been magnified showing that the entire particle is a single crystal. (After Ref. 42; used with permission.)

Fresh Pt black has a rather high lattice strain (ca. 0.4%), showing that its structure is far from being in equilibrium and is full of various defects. Hydrogen treatment even at 300 K caused a dramatic drop in the lattice strain [43]; this may correspond to the neck growth stage without crystallite increase (Fig. 7). Indeed, the EM of a system subjected to hydrogen treatment near 300 K reveals small crystallites which apparently underwent the first, neck growth stage of sintering and represent a network with two-dimensional contacts between crystallites [39]. The strain decreased further during heating without detectable crystallite growth. Heating in air did not cause any change in lattice strain up to 500 to 550 K; then a sudden collapse of the crystals is seen, together with rapid crystallite growth. (The fact that particle growth started at the same temperature in air and hydrogen in Fig. 7 may be due to different experimental conditions. Slow heating was applied from 375 K upward, as distinct from isothermal hydrogen introduction in Refs. 39 and 40; also, the sample pressed to a glass filter represented a much more intimate contact between individual grains than in the case of a loose stack of crystallites used in EM studies.)

The facts mentioned are in agreement with a mechanism in which adsorbed oxygen is removed from the metal surface by hydrogen. Although this exchange may not be complete (cf. oxygen detected by XPS after heating in hydrogen [36]), it is sufficient to bring about spontaneous welding between particles followed by true coalescence. Indeed, the exponents of Eq. (3) depended on the nature of gas adsorbed on the metal. Crystallite growth in hydrogen showed an exponent m = 2 (energy of activation E = 40 kJ mol^{-1}).

TABLE 1 Crystallite Sizes of Pt-Black Samples Subjected to Various Treatments

Treatment (ambient)	Temperature	Crystallite size (nm) as determined by:		
		EM[a]	XRD	BET surface[b]
None	—	8 ± 3	11.5 ± 8 [39]	19.5 (14 m² g⁻¹) [34]
			12 ± 2 [43]	13 (21 m² g⁻¹) [35]
			10 [34]	
Hydrogen[c]	300	9 ± 2		
STC[d]				
Air, H₂	473	26 ± 6		
		30 [33]	14; 17 [34]	72 (3.9 m² g⁻¹) [34]
	633	39 ± 9		
		70 [33]	30; 27 [34]	156 (1.9 m² g⁻¹) [34]
Helium	633	12.5 ± 3.5		
Air	633	12.5 ± 3.5		
Ethylene	633	18 ± 5		

[a]From Ref. [40] if not stated otherwise.
[b]Assuming spherical particles.
[c]Neck growth; see Ref. 40.
[d]STC, standard thermal cycle: heating up in air to T, kept there for another 30 min, flushing by He, H₂ for 1 h, He to complete cycle to 3 h, cooling in He. In the latter three cases, H₂ was replaced by the corresponding gas.

Oxygen-covered particles showed an exponent m = 3 with E = 128 kJ mol⁻¹ [45]. One may risk the conclusion that different growth mechanisms prevail in these cases. Rapid neck growth at 300 K is followed by grain-boundary diffusion between "welded" particles, whereas bulk diffusion should be the route of growth where initial welding is hindered by adsorbed oxygen. Khassan et al. [46] reported m = 2 and E = 77 kJ mol⁻¹ for Pt-black sintering in a rather poor "vacuum"; apparently, rest gas adsorption contaminated the surface of their sample.

The very rapid completion of sintering and the great differences observed in hydrogen and other atmospheres may suggest that hydrogen has a more important role than merely to clean the surface for welding. We may interpret these results in terms of hydrogen penetration into the crystal lattice. Gibb [47] postulates that hydrogen may get into the bulk of a metal through discrete surface "gates"; these may be either surface steps, corners [26], or defect sites. By penetrating subsurface layers, hydrogen can loosen the structure of the solid, thus facilitating the transport of metal atoms into thermodynamically more favorable positions (i.e., decrease lattice strain). We have already seen this possibility in Chapter 1 and in previous sections of this chapter. Adsorption, absorption, and hydride formation involve a number of elementary steps of hydrogen penetration, each of which can be identified with various positions of hydrogen. Eley and Pearson [48] combined their observations with previous data [49,50] and proposed that on palladium, the precursor of dissolved hydrogen is "a hydrogen molecule adsorbed with its axis perpendicular to the surface with one atom embedded in an octahedral interstitial site in the surface (α state)." They assigned the two different β states to oppositely charged H atoms (demonstrated by work function measurements [51]; Fig. 8a). They also constructed a potential curve for various types of hydrogen [Fig. 8b; this is very similar to that shown in Fig. 24 in Chapter 1 (note the different definition of "metal surface")]. Under these conditions no β*-PdH could be present, although

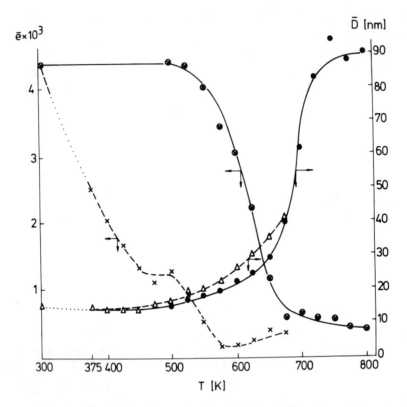

FIGURE 7 Lattice strain and average crystallite size in Pt black during heating. ⊗ and ●, Heating in air, heating rate: 100 K min^{-1} 300 to 473 K, 2 K min^{-1} 473 to 823 K. x and △, Heating in hydrogen, heating rate: 100 K min^{-1} 300 to 375 K, 2 K min^{-1} 375 to 700 K. (Adapted from Ref. 43.)

the existence of two extra "monolayers" of dissolved hydrogen might suggest "something analogous to a surface $\alpha^* \to \beta^*$ phase transition . . . in the top two or three layers of metal atoms."† Kiskinova et al. [52] found that the equilibration between adsorbed and absorbed hydrogen was complete within about 12 s at 300 K on a Pd ribbon after removing adsorbed hydrogen at 450 K.

Platinum is a metal known not to dissolve considerable amounts of hydrogen in its bulk [53]. Subsurface hydrogen was nevertheless detected by low-energy electron diffraction (LEED)-Auger studies [26], radio-tracers [54], and catalytic activity [55]. Due to the less facile penetration of hydrogen, equilibration between surface and subsurface layers required several minutes at 630 K [54]. Unstable Pt-hydride was suggested to promote recrystallization of small Pt particles on zeolite [68] and also Pt/C [69].

Electrochemical measurements on hydrogen diffusion through a thin platinum electrode indicated that 90 to 95% of the hydrogen inside the metal is located near cracks, faults, and other defect sites [56]. No evidence has been reported until recently that hydrogen may penetrate the lattice. Such a movement should cause lattice expansion as measured by XRD in the case of

---

†The literature uses α and β for both surface hydrogen states and phases of PdH. Here the latter meaning is denoted by an asterisk.

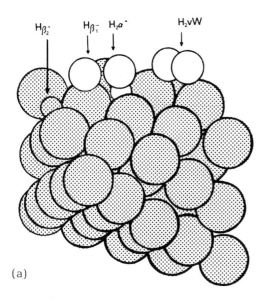

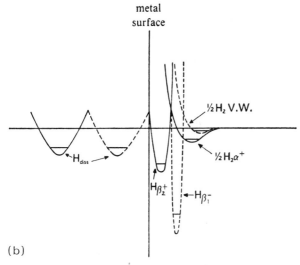

FIGURE 8 (a) Proposed positions of various chemisorbed hydrogen species (white spheres) on Pd(100) (shaded spheres). The partial electric charges are noted; H_2 vW is van der Waals adsorbed hydrogen over two Pd atoms. (b) Potential energy diagram (to scale) for surface and subsurface hydrogen on Pd(100). Solid lines denote sections through surface octahedral holes, dashed lines section through neighboring Pd atoms. (After Ref. 48; used with permission.)

metals such as Nb [57]. A Chinese group recently reported a well-observable lattice expansion up to 2% when a Pt film was heated in hydrogen between 573 and 973 K [58]. The change was largest at 773 K. It could be envisaged that diffusion of hydrogen occurs, indeed, mainly along crystal defects at lower temperatures and that its major penetration into the lattice starts at higher temperatures. The slight increase in lattice strain above 570 K seen in Fig. 7 can be attributed to this penetration. Heating in Ar or in O_2 could

TABLE 2 Pt Lattice Spacings: d Value of Pure Platinum Film Subjected to Various Treatments

Treatment conditions[b]	Lattice spacing, d_{hkl} (Å)[a]						
	(111)	(200)	(220)	(311)	(331)	(420)	(422)
Ref. value[c]	2.265	1.962	1.387	1.183	0.9000	0.8773	0.8008
573 K Ar	2.26	1.96	1.38	1.18	0.893	0.867	
573 K H_2	2.29	1.99	1.40	1.19	0.893		
773 K H_2	2.33	2.03	1.42	1.22			
973 K Ar	2.22	1.92	1.35	1.15	0.875	0.850	0.774
973 K H_2	2.31	2.00	1.41	1.21	0.913	0.889	0.813
773 K O_2	2.26	1.96	1.39	1.18	0.897	0.871	0.794

[a]Measured by a Hitachi H-600 electron microscopy by electron diffraction at 100 kV energy. Material: Pt film, thickness about 30 to 50 Å, evaporated onto a carbon film on an EM grid.
[b]Gas flow at atmospheric pressure, typically for 1 h.
[c]Ref. value from *Powder Diffraction File*, Inorganic Vol. PDIS-5iRB, 04-802, Vol. PDIS-18iRB, 17-64, published by the Joint Committee on Powder Diffraction Standards Pennsylvania.
Source: Adapted from Ref. 58.

reverse the process completely. The same sample subjected to various treatments (Table 2) could produce the changes noted, but samples brought directly to various stages of treatment (last rows of the table) gave the same results. This is the first direct proof of hydrogen penetration into Pt lattice at high temperatures. The same group also reported changes in the lattice spacing of Pt supported on TiO_2. No differences in the Pt-Pt distance were found with Pt/SiO_2 of different metal percentage and dispersion even after reduction in hydrogen. The amplitude of vibration of Pt atoms, which is higher at the surface than in the bulk [59], increased with the temperature of H_2 reduction [59,60]. The least disperse Pt sample (particle size 8 to 12 nm) also showed lattice strain, perhaps because its dimensions were comparable to the pore diameter. Hydrogen treatment did not change this lattice strain but "cut off" from these particles very small Pt grains that remained on the surface of the support [60].

The sintering process can be hindered by carbonaceous residues. Such species are formed when sintering is carried out above 700 K. Carbon segregation is a process competing with hydrogen penetration and having the opposite effect: its completion protects the particles from sintering. The segregation of a partly graphitized overlayer has been detected by EM [33] and by soft x-ray emission (SXES) [34]. The result is a smaller-crystallite-size Pt black which has rather poor catalytic activity [61]. Carbon segregation was apparently completed during the first 30 min of sintering in air at 750 K; subsequent hydrogen treatment at 630 K did not result in larger crystallites [40]. Heating Pt/C in helium at 773 K for 1 h prevented particle-size growth during subsequent hydrogen treatment at 673 K [69]; we suggest this was also due to carbonization of Pt surface. Sintering in an ethylene atmosphere at 573 K produced small crystallites [40] and the lattice strain decreased considerably, to 0.26% (Fig. 7). Apparently, two effects took place here simultaneously.

Ethylene broke up partly on the catalyst surface and its hydrogen could diffuse into subsurface layers like hydrogen from H_2. At the same time, its carbonaceous residues could effectively block crystallite coalescence. Still, since they do not form a continuous overlayer, the activity of this catalyst in cyclohexane dehydrogenation was rather good [61].

Changes of Chemical Composition and Electronic Structure During Sintering of Pt Black in Hydrogen

The electronic structure of freshly reduced high-surface-area Pt blacks is far from equilibrium. This is reflected by their soft x-ray emission spectra. Here only the largest transition (conduction band to $4f_{7/2}$) is observed; the other two characteristic transitions ($5p_{1/2}$ and $4f_{5/2}$) disappear. The latter transitions both appeared in sintered samples and also when Pt was heated in situ under an electron beam, also causing crystal size growth (Fig. 9). Sintering at 473 K gave curves closest to those of the clean metal, in agreement with XPS and secondary ion mass spectroscopy (SIMS) data [36,62].

One of the reasons why the electronic structure is somewhat distorted may be that the surface of freshly reduced Pt black is covered by carbon and oxygen and to a lower extent, by silicon [33,36]. Hydrogen treatment between 470 and 750 K removed these impurities only partially [36,62]. Sintering must also have caused their encapsulation into subsurface layers.

The chemical state of these impurities can also be followed. Soft x-ray emission spectra [34] show that oxygen adsorbed on the surface at fresh Pt black transforms upon heating into chemically bonded species and penetrates subsurface layers, in accordance with the findings of Akhtar and Tompkins [63]. The chemical state of carbon changes from isolated atoms at 470 K to an overlayer at 630 K which becomes considerably graphitized at 750 K [34]. XPS shows that hydrogen treatment at 570 K removes, almost entirely, oxygen bound to Pt atoms [36].

Hydrogen had a dramatic effect on the Si content of Pt black. The original Pt salt being contaminated, the fresh catalyst had a Si content of up to 4.5%.

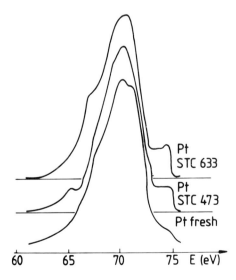

FIGURE 9 Soft x-ray emission spectra of Pt black subjected to various treatments, compared to the theoretical spectra. (Adapted from Ref. 34.)

Heating in hydrogen decreased this value below 0.5%. The Si content was proportional to reciprocal grain size, $1/\bar{D}$ (i.e., to the specific surface, $\bar{D}^2/\bar{D}^3$). This indicates the segregation of Si to the surface, where at elevated temperatures it can react with H (activated and dissociated by Pt) to form gaseous SiH, which then leaves the solid [33]. Of course, in-beam sintering in the evacuated EM chamber did not cause any changes in the silicon content.

III. CONCLUSIONS

Both single-crystal and disperse catalyst studies reveal that hydrogen interacting with catalyst surfaces may cause considerable changes in the metal morphology. Surface reconstruction of low-index planes may also involve an enormous increase in surface diffusivity. High-index stepped planes may also undergo reconstruction in hydrogen. Heating in hydrogen usually does not cause dramatic crystallite growth on supported metals [64].

Hydrogen adsorption may also enhance surface diffusion on more imperfect, highly dispersed unsupported metals, causing their sintering to involve grain growth and a decrease in lattice strain, as well. It is highly probable that especially the latter process is caused by the penetration of hydrogen into subsurface regions. This process is probably a diffusion of hydrogen to defect sites, which also increases their mobility. The metal structure therefore gets closer to equilibrium. This relaxation can be followed at higher temperatures by hydrogen diffusing into the lattice, causing its expansion and also a slight increase in the lattice strain. These all also influence the electronic structure of freshly reduced high-surface-area metal blacks.

It seems likely that hydrogen-induced sintering does not bring about a profound cleaning of the surface concerned (as often believed), but that it does involve the removal of impurities from critical sites where hydrogen penetration into subsurface layers may occur.

REFERENCES

1. G. Wulff, *Z. Kristallogr.*, 34:449 (1901).
2. T. Wang, C. Lee, and L. D. Schmidt, *Surf. Sci.*, 163:181 (1985).
3. Vu Thien Binh (ed.), *Surface Mobilities on Solid Materials: Fundamental Concepts and Applications*, NATO ASI Series, Vol. 86, Plenum Press, New York (1981).
4. M. A. Van Hove, R. J. Koestner, P. C. Stair, J. P. Bibérian, L. L. Kesmodel, I. Bartos, and G. A. Somorjai, *Surf. Sci.*, 103:218 (1981).
5. H. P. Bonzel, in *Surface Mobilities on Solid Materials: Fundamental Concepts and Applications* (Vu Thien Binh, ed.), Plenum Press, New York (1981).
6. W. L. Winterbottom, *Acta Metall.*, 15:303 (1967).
7. P. Feltham, *Acta Metall.*, 5:97 (1957).
8. R. L. Coble, *J. Appl. Phys.*, 32:787, 793 (1961).
9. M. Hillert, *Acta Metall.*, 13:227 (1965).
10. J. G. R. Rockland, *Acta Metall.*, 15:227 (1965).
11. E. Ruckenstein and D. B. Dadyburjor, *Rev. Chem. Eng.*, 1:251 (1983).
12. W. Schatt, W. Hermel, E. Friedrich, and P. Lanyi, *Sci. Sintering*, 15:5 (1983).
13. J. W. Geus, in *Chemisorption and Reactions on Metallic Films*, Vol. 1 (J. R. Anderson, ed.), Academic Press, New York (1971), p. 129.

14. J. F. Pócza, Á. Barna, and P. B. Barna, *J. Vac. Sci. Technol.*, 6:472 (1969).
15. I. M. Lifshitz and V. V. Slyozov, *Zh. Eksp. Teor. Fiz.*, 35:479 (1958).
16. M. Drechsler, in *Surface Mobilities on Solid Materials: Fundamental Concepts and Applications* (Vu Thien Binh, ed.), Plenum Press, New York (1981), p. 405.
17. D. W. Blakely and G. A. Somorjai, *Surf. Sci.*, 65:419 (1977).
18. K. H. Rieder and W. Stocker, *Surf. Sci.*, 164:55 (1985).
19. R. Kumar and H. E. Grenga, *Surf. Sci.*, 50:399 (1975).
20. R. Kumar and H. E. Grenga, *Surf. Sci.*, 59:612 (1976).
21. H. E. Grenga and R. Kumar, *Surf. Sci.*, 61:283 (1976).
22. R. Van Hardeveld and F. Hartog, *Adv. Catal.*, 22:75 (1972).
23. A. M. Baro and H. Ibach, *Surf. Sci.*, 92:237 (1980).
24. H. B. Nielsen and D. L. Adams, *Surf. Sci.*, 97:L351 (1980).
25. R. T. Tung and W. R. Graham, *Surf. Sci.*, 97:73 (1980).
26. B. Lang, R. W. Joyner, and G. A. Somorjai, *Surf. Sci.*, 30:454 (1972).
27. G. Maire, P. Bernhardt, P. Legare, and G. Lindauer, in *Proceedings of the 7th International Vacuum Congress and 3rd International Conference on Solid Surfaces, Vienna*, Vol. 1 (1977), p. 861.
28. G. A. Somorjai, *Chemistry in Two Dimensions: Surfaces*, Cornell University Press, Ithaca, N.Y. (1981).
29. L. D. Marks, *Surf. Sci.*, 150:358 (1985).
30. P. J. F. Harris, in *Electron Microscopy and Analysis, 1985, Institute of Physics Conference Series 78*, Newcastle (1985), p. 489; *Appl. Catal.*, 16:439 (1985); *J. Catal.*, 97:527 (1986).
31. D. J. Smith, D. White, T. Baird, and J. R. Fryer, *J. Catal.*, 81:107 (1983); D. White, T. Baird, J. R. Fryer, L. A. Freeman, D. J. Smith, and M. Day, *J. Catal.*, 81:119 (1983).
32. M. F. Gillet and S. Channakhone, *J. Catal.*, 97:427 (1986).
33. A. Barna, P. B. Barna, L. Tóth, Z. Paál, and P. Tétényi, *Appl. Surf. Sci.*, 14:87 (1982-83).
34. Z. Paál, P. Tétényi, L. Kertész, A. Szász, and J. Kojnok, *Appl. Surf. Sci.*, 14:101 (1982-83).
35. Z. Paál and S. J. Thomson, *Kém. Közl.*, 43:463 (1975).
36. Z. Paál, P. Tétényi, D. Prigge, X. Zh. Wang, and G. Ertl, *Appl. Surf. Sci.*, 14:307 (1982-83).
37. I. Hansson and A. Thölén, *Sci. Sintering*, 10:3 (1978).
38. P. Sermon, *J. Catal.*, 24:460, 467 (1972).
39. T. Baird, Z. Paál, and S. J. Thomson, *J. Chem. Soc. Faraday Trans. 1*, 69:50 (1973).
40. T. Baird, Z. Paál, and S. J. Thomson, *J. Chem. Soc. Faraday Trans. 1*, 69:1237 (1973).
41. Z. Paál, Á. Barna, P. B. Barna, and L. Tóth, *Mater. Chem.* 6:95 (1981).
42. M. Spanner, T. Baird, J. R. Fryer, L. A. Freeman, and D. J. Smith, in *EUREM '84, Proceedings of the 8th European Congress on Electron Microscopy, Budapest*, Vol. 2 (Á. Csanády, P. Röhlich, and D. Szabó, eds.) (1984), p. 1153.
43. I. Manninger, Z. Paál, and P. Tétényi, *Z. Phys. Chem. (Neue Folge)*, 143:247 (1985).
44. I. Manninger, Z. Paál, and P. Tétényi, *Z. Phys. Chem. (Neue Folge)*, 132:193 (1982).
45. I. Manninger, *J. Catal.*, 89:164 (1984).
46. S. A. Khassan, S. G. Fedorkina, G. I. Emelyanova, and V. P. Lebedev, *Zh. Fiz. Khim.*, 42:2507 (1968).
47. T. P. R. Gibb, in *Progress in Inorganic Chemistry*, Vol. 3 (F. A. Cotton, ed.), Interscience, New York (1962), p. 315.

48. D. D. Eley and E. J. Pearson, *J. Chem. Soc. Faraday Trans. 1*, *74*:223 (1978).
49. W. Auer and H. J. Grabke, *Ber. Bunsenges. Phys. Chem.*, *78*:58 (1974).
50. A. W. Aldag and L. D. Schmidt, *J. Catal.*, *22*:260 (1971).
51. R. Dus, *Surf. Sci.*, *42*:325 (1973).
52. M. Kiskinova, G. Bliznakov, and L. Surnev, *Surf. Sci.*, *94*:169 (1980).
53. E. Raub, in *Gase und Kohlenstoff in Metallen* (E. Fromm and E. Gebhardt, eds.), Springer, Berlin (1976), p. 648.
54. Z. Paál and S. J. Thomson, *J. Catal.*, *30*:96 (1973).
55. P. B. Wells, *J. Catal.*, *52*:498 (1978).
56. E. Gileadi, M. A. Fullenwider, and J. O'M. Bockris, *J. Electrochem. Soc.*, *113*:926 (1966).
57. H. Peisl, in *Hydrogen in Metals*, Vol. 1 (F. Alefeld and J. Völkl, eds.), Springer, Berlin (1978), p. 53.
58. Tang Sheng, Xiong Guoxing, and Wang Hongli, *J. Catal. (China)*, *8*:225 (1987).
59. S. R. Sashital, J. B. Cohen, R. L. Burwell, Jr., and J. B. Butt, *J. Catal.*, *50*:479 (1977).
60. R. K. Nandi, F. Molinaro, C. Tang, J. B. Cohen, J. B. Butt, and R. L. Burwell, Jr., *J. Catal.*, *78*:289 (1982).
61. Z. Paál, M. Dobrovolszky, I. Manninger, and P. Tétényi, *Z. Phys. Chem (Neue Folge)*, *124*:75 (1981).
62. Z. Paál and D. Marton, *Appl. Surf. Sci.*, *26*:161 (1986).
63. M. Akhtar and F. C. Tompkins, *Trans. Faraday Soc.*, *67*:2454 (1971); *67*:2461 (1971).
64. R. Burch, in *Catalysis*, Vol. 7 (G. C. Bond and G. Webb, eds.), Royal Society of Chemistry, London (1985), p. 149.
65. G. A. Somorjai, *J. Catal.*, *27*:453 (1972).
66. L. D. Schmidt and D. Luss, *J. Catal.*, *22*:269 (1971).
67. P. J. F. Harris, *Nature*, *323*:792 (1986).
68. B. A. Dalla Betta and M. Boudart, *Proceedings of the 5th International Congress on Catalysis, Palm Beach, 1972*, Vol. 2 (J. Hightower, ed.), North-Holland, Amsterdam (1973), p. 1329.
69. F. Rodriguez-Reinoso, I. Rodriguez-Ramos, C. Moreno-Castilla, A. Guerrero-Ruiz, and J. D. López-González, *J. Catal.*, *99*:171 (1986).

12 | Spillover of Hydrogen

WILLIAM CURTIS CONNER, JR.

University of Massachusetts, Amherst, Massachusetts

I.	INTRODUCTION	311
II.	BACKGROUND	312
III.	ENHANCED ADSORPTION	314
IV.	WEAK AND STRONG INTERACTIONS WITH OXIDE SURFACES	317
	A. Reduction of Oxides	317
	B. Bronze Formation	318
	C. Metal-Support Interactions	318
V.	CATALYTIC ACTIVITY INDUCED ON REFRACTORY OXIDES	319
	A. Hydrogenation on Activated Silica	321
	B. Reactions of Benzene and Cyclohexene on Silica	321
	C. Reactions over Activated Alumina	322
	D. Reactions on Magnesia	323
VI.	RETENTION OF CATALYTIC ACTIVITY	323
VII.	KINETICS AND MECHANISM OF SPILLOVER	325
	A. Mechanism of Spillover	325
	B. Solid-State Reaction Mechanisms	327
	C. Amount of Spillover and the Density of Active Sites	329
	D. Kinetic Control and the Rate of Spillover	331
VIII.	OVERVIEW	335
	A. The Phenomena	335
	B. The Mechanisms	340
	C. Concluding Remarks	341
	REFERENCES	342

I. INTRODUCTION

As noted throughout the present book, hydrogen has a significant place in catalysis by metals. In addition to reactions involving hydrogen, it is used to activate most metal systems, and to retain or to rejuvenate the activity of metal-containing "catalyst systems." Until recently, hydrogen exposure to metal-containing catalysts was envisioned to occur by a stoichiometric reaction of the hydrogen with the metal (i.e., $H_2 + 2M \rightarrow 2HM$, where M is a sorbing metal). The transport of the atomic hydrogen from the metal onto other phases of the system was not imagined.

Limited mobility of sorbed species has long been understood. The exchange of species from one position to another, either on the surface or through the bulk, has been well established [1]. More unique is the mobility of a sorbed species from one phase onto another phase where it does not directly adsorb. This has been defined as "spillover."

Indeed, the transport of hydrogen across the surface from one phase to another is a phenomenon that may have a profound influence on heterogeneous catalysis. As a concept, spillover affords a new perspective on the interactions that may occur among the phases in catalytic systems.

Several aspects of hydrogen spillover involving metal-containing catalyst systems are described below. In the first we describe the background of our understanding of hydrogen spillover. Next, the resultant phenomena associated with spillover are discussed. The initially measurable influence of spillover is an enhanced sorption, described in Section III.

In Section IV we focus on hydrogen that may react with the phase onto which it is spilled over. The reaction may be weak, involving a mild reduction of the accepting surface, or it may involve the incorporation of the spiltover hydrogen into the bulk of the accepting phase. The specific example of strong metal-support interactions (SMSI) has drawn considerable attention and is discussed in detail in Chapter 13. The role of spillover in this phenomenon is then described briefly.

Associated with the possible reduction of even refractory oxide surfaces (e.g., SiO_2 or Al_2O_3), there is overwhelming evidence that these surfaces become catalytically active. This activity may be significant and may differ from catalysis taking place on the metal and will depend on the specific system. Hydrogen spillover has also been implied in the retention of catalytic activity by the removal of coke or its precursors from the surface and this is discussed in Section V.

To assess the potential influence and significance of hydrogen spillover, it is necessary to understand the rates and mechanism of spillover based on the evidence discussed in the preceding sections. In Section VII we offer a perspective of hydrogen spillover and the areas where the phenomena are still not well understood.

Spillover, as any new complex phenomenon or concept, has not been accepted universally. Initially, the phenomena were not well defined and spillover was invoked to explain a broad range of experimental observations. Hopefully, this chapter will provide a perspective on the potential implications of these phenomena as they relate to hydrogen spillover to and from metals.

II. BACKGROUND

The first direct experimental evidence for spillover was presented in 1964 by Khoobiar, who documented that the formation of tungsten bronzes (H_xWO_3) was possible *at room temperature* for a mechanical mixture of WO_3 with Pt/Al_2O_3 [2]. Sinfelt and Lucchesi postulated that reaction intermediates (presumably H) had migrated from Pt/SiO_2 onto Al_2O_3 to effect ethylene hydrogenation [3]. During the ensuing score of years over 500 papers involving spillover have been published. The First International Symposium on the Spillover of Adsorbed Species was held in 1983 in Lyons [4].

Review of the publications have appeared in 1973 [5], 1980 [6], 1983 [7], and most recently in *Advances in Catalysis* [8]. The focus of this chapter is to discuss the current understanding of the phenomena. During the recent International Symposium a definition was proposed and affirmed by the congress [9]. "Spillover involves the transport of an active species sorbed or formed on a first phase onto another phase that does not under the same condition sorb or form the species. And offered as a comment: The result may be the reaction of this species on the second phase with other sorbing gases and/or reaction with, and/or activation of the second phase."

Mechanistically, this can generally be visualized as a sequence of steps:

$H_2 + M \rightleftarrows H_2M$	sorption	(1)
$H_2M + M \rightleftarrows 2H_aM$	activation of sorbing species	(2)
$H_aM + M' \rightleftarrows M + H_aM'$	exchange	(3)
$H_aM' + \theta \rightleftarrows M' + H_{sp}\theta$	spillover to second phase	(4)
$H_{sp}\theta + \theta' \rightleftarrows \theta + H_{sp}\theta'$	exchange or surface diffusion	(5)
$H_{sp}\theta \rightleftarrows H_{sp}\theta^*$	creation of active surface site on second phase	(6)
$H_{sp}\theta^* \rightleftarrows {}^* + P_{gas}$	desorption	(7)

where

M, M'	= sites able to adsorb and dissociate H_2
H_a	= activated sorbed species
H_{sp}	= spilt-over species (usually $\equiv H_a$)
θ, θ'	= phase unable to sorb H_2 directly
$H_{sp}\theta^*, {}^*$	= possible active site or new phase resulting from θ
P_{gas}	= possible product of the creation of the new active site

These seven steps initially simplify the discussion of the phenomena related to hydrogen spillover. Hydrogen is adsorbed from the gas phase onto a metal (Pt, Pd, Ni, etc.), where it dissociates to atomic hydrogen. Spillover occurs if the atomic hydrogen is then able to be transported onto the support ($PtH + \theta \rightarrow Pt + H_{sp}\theta \equiv$ step (4)].

Spillover may result in a spectrum of changes in the phase onto which it occurs. In the weakest sense, the spiltover species is transported across the surface of this phase as a two-dimensional gas. It may exchange with similar surface species (e.g., $OH + D_{sp} \leftrightarrow OD + H_{sp}$). The spiltover hydrogen may react with the surface (e.g., $H_{sp} + M-O-M' \rightarrow M-O-H + M'$). This can result in the creation of surface defects and/or active sites. Further, the bulk of the solid may be transformed into a different structure (as in the transformation of oxides to hydrogen bronzes). In each of these cases, the second phase is no longer an inert. It is not serving to promote the inherent activity on the first phase. The second phase is participating directly in the transport, exchange, and reaction with the spiltover hydrogen. In some examples the second phase is able to become catalytically active on its own and thereby to participate directly in subsequent catalysis.

The most important consequences of spillover are discussed and analyzed in the following six sections. It should be noted that the reaction of adsorbed spiltover hydrogen with organic adsorbents is discussed in detail in Chapter 18, where hydrogenation by spiltover hydrogen is also mentioned. Hydrogenations are discussed only briefly in Section V.A.

III. ENHANCED ADSORPTION

Most studies have involved hydrogen adsorption on groups VIII, VI, and I− metals supported on Al_2O_3, SiO_2, zeolites, or carbons. The studies involving SMSI have focused on TiO_2 as a support. Enhanced adsorption of H_2 is generally reported to occur at temperatures about 200°C. However, many cases are described for adsorption at ambient temperature (and below).

The initial studies [5,6] of the effects of spillover results were expressed as H/M_t ratios, where H stands for the number of H atomic species adsorbed per total number of metallic atoms, M_t, in supported catalysts. Typical results are shown in Table 1 from the review published by Sermon and Bond in 1973 [5]. In general, the ratio H/M_s, where M_s is the total number of *surface* metal atoms of the catalyst, rarely exceeds 2 [10]. Large values of H/M_t indicate enhanced adsorption attributed to a spillover of adsorbed hydrogen. Values of H/Pt_t ratios as high as 64 and 6.8 for Pt/Al_2O_3 and Pt/SiO_2 catalysts, respectively, have been measured by Altham and Webb [11], while H/Pt_t ratios between 3 and 277 have recently been calculated by Sermon and Bond [12] for platinum-containing bronzes.

In addition to the amount of hydrogen adsorbed, the rate of adsorption is found to change due to spillover from the metal that originally adsorbs the hydrogen. Fujimoto and Toyoshi [13] found increased rates of H_2 adsorption with active carbon as a support for or added as a diluent to cobalt. The rate increase depended on the contact between the metallic component and the active carbon. Moreover, they found that the amount of adsorption did not vary with the source of spillover hydrogen (Ni, Cu, Fe), the metal loading, or the use of pure or sulfided metals. The amount of the hydrogen uptake at 400°C at equilibrium was the same. The extent depended only on the number of accepting sites on the carbon. Similar results were reported for 10% Pd/C [14] and for Cs, Mo, Ni, or Cu on carbon catalysts [15]. Since active carbon adsorbs hydrogen by itself at high temperatures (T > 300°C [14]), the effect of spillover is to increase the rate of uptake but not the extent. With the inorganic oxides, spillover most often results in an increase in both the rate and the amount of adsorption.

Berzins et al. [16] recently studied isothermal chemisorption on oxide-supported platinum and demonstrated that in most cases the gradient dH_a/dp (H_a ≡ the amount of adsorbed gas: H_a or H_{sp} above; p ≡ the partial pressure) never falls to zero with increasing p. They hypothesized that the inclined plateau (where $dH_a/dp \neq 0$) may be due to spillover. Indeed, such a case has been clearly reported by Anderson et al. [17] in their study of H_2 chemisorption on several Pt-Au-on-aerosil catalysts at 20°C.

In a series of two papers, Apple et al. [18,19] studied the spillover of H_2 on Rh/TiO_2 samples using NMR and TPD and came to the conclusion that hydrogen spillover gives rise to an extra upfield NMR resonance line due to an H species interacting with paramagnetic Ti^{3+} centers. They also demonstrated that the additional hydrogen adsorption was suppressed by CO pre-adsorption [19] as well as by coadsorption. A recent investigation of hydrogen adsorption made by Jiang et al. [20] with Ni, NiFe, or Pt on the same support, TiO_2, indicated very slow equilibrium of hydrogen during adsorption-desorption. This slow uptake was attributed to hydrogen spillover from the metal to either Ti^{3+} cations (TiO_x) or to OH groups on TiO_2.

Bimetallic catalyst systems exhibit all the characteristics noted for monometallic-supported catalysts. Hydrogen spillover was reported to occur on many Pt-Re-on-alumina systems by Dowden et al. [21] and more recently by Barbier et al. [22]. A maximum in the amount of spillover occurred with a metal-metal ratio of Pt 40%-Re 60%. In both studies the catalysts were treated by hydrogen at high temperatures, 480°C and 500°C, respectively.

TABLE 1 Maximum Ratios of Hydrogen Adsorbed: Total Metal Atoms (H/M_t)

Metal	Support	H/M_t	Temp. (°C)	H_2 pressure (mmHg)
Pd	SiO_2	4.0	-196	400
	Al_2O_3	5.0		
		3.2	25	
Pt	SiO_2	1.3	-196	0.1
		1.5	-196	
		1.6	-196	$10^{-2}-10^{-5}$
	Al_2O_3	2.4	250	120
		2.2	250	120
		2	300	
		1.5	200	9
		>1	100	9
		10		
	C	75	250	600
	Zeolite	1.4	21	
		2	100	
		2	200	50-200

Source: After Ref. 5.

Spillover depends on two conditions: a source for the spilling species (e.g., a group VIII metal) and an acceptor (e.g., an oxide or an active carbon). Anderson et al. present evidence for hydrogen spillover occurring on group VIII-group Ib metal alloys: Pt-Au on aerosil [17]. The contribution of spillover was deduced by comparing the adsorption measurements to the surface compositions of the alloys. It was assumed that the group IB metals were able to chemisorb hydrogen atoms but not dihydrogen.

Spillover also occurs in systems where there is not a support. Crucq et al. [23] measured the adsorption of hydrogen as a function of composition for Ni-Cu alloys and attributed the additional adsorption to spillover. Goodman and Peden have studied hydrogen chemisorption on Ru(0001) single crystals covered with up to a monolayer of copper [24,25]. Individually, Ru is known to chemisorb hydrogen by dissociation to form a 1:1 ratio of hydrogen to exposed Ru atom. If the surface is first totally covered with Cu ($\theta_{Cu} = 1$), hydrogen chemisorption is inhibited as found in prior studies [26]. However, if less than a monolayer (e.g., $\theta_{Cu} = 0.7$) of copper is evaporated onto the surface, a condition confirmed by Auger spectroscopy, the TPD results for preadsorption of hydrogen at -173°C and -43°C are different, as shown in Fig. 1. Curve c on the figure is the difference between the two TPD spectra and is attributed to hydrogen first adsorbed and dissociated on Ru and then spilled over onto Cu. Grenter and Plummer [27] had demonstrated that the adsorption of predissociated, atomic hydrogen on Cu and the TPD traces are essentially the same as represented in curve c. Although the spiltover hydrogen did not effect a total coverage of the copper (i.e., $H_{sp}/Cu < 1$), the amount spilled over did not vary with the time or temperature (above -43°C) of exposure to H_2 and was therefore kinetic and not diffusion controlled. From these results the authors conclude that spillover occurs below -43°C but above -173°C and can influence the calculation of exposed metal in bimetallic systems (e.g., Cu/Ni, Cu/Ru, etc.) characterized by H_2 adsorption at or above -40°C.

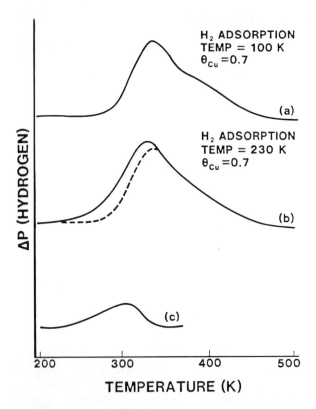

FIGURE 1 Temperature-programmed desorption following hydrogen saturation at 100 K (a) and at 230 K (b) on a Ru(0001) surface that is 70% covered by Cu (i.e., $\theta = 0.7$). The dashed line in part (b) indicates the superimposed trace of part (a). The difference between parts (a) and (b) is shown in part (c).

Several factors influence the enhanced adsorption (extent and rate) associated with hydrogen spillover of hydrogen. Following is a list of these factors.

1. Range of the temperatures of hydrogen adsorption (below room temperature or significantly above)
2. Possible necessity of a cocatalyst such as water or other proton acceptors [28]
3. Amount and the percent dispersion of the source of hydrogen spillover [9,11]
4. Nature and area of contact between the source and the acceptor [29,30]
5. Partial pressures of H_2
6. Chemical nature of the acceptor and source [9]
7. Presence of chlorine and sulfur ions [13,15,31-33]
8. Duration of chemisorption [20]
9. Effects of coadsorbents, O_2 [34] and CO [19]
10. Nature of the spiltover hydrogen [9] and its bond with the acceptor sites

Among these numerous factors some are detrimental to hydrogen spillover, especially O_2 and CO, or limited contact between the source and the acceptor.

12. Spillover of Hydrogen

Added water is either an accelerating promoter or has no influence, although it may be formed from O_2 in the system [35].

Reverse spillover or back-spillover is observed to proceed by surface migration of the spiltover species from the accepting sites to the metal, where it desorbs as H_2 molecules or reacts with another hydrogen acceptor, such as O_2, pentene, ethylene, and so on.

IV. WEAK AND STRONG INTERACTIONS WITH OXIDE SURFACES

A. Reduction of Oxides

The temperature of reduction by H_2 of numerous metal oxides is lowered by addition of a transition metal and has been attributed to spillover [5, 6, 36]. Besides the catalyzed reduction of metal oxides either to metals or suboxides, the result may be the formation of new and specific reduced oxides (e.g., the hydrogen bronzes of W, Mo, and V) [5, 37-39]. In many cases the reduction of the corresponding oxides by spillover of H_2 led to reduced compounds not otherwise obtainable [40].

Two groups of metal oxides (that do not necessarily form bronzes) have been reduced by spillover from Pt or Pd on silica:

1. Co_3O_4, V_2O_5, UO_3, Fe_2O_3, MoO_3, WO_3, Re_2O_7, and CrO_4 [40-42]
2. Ni_3O_4, MnO_2, NiO, CuO, Cu_2O, ZnO, CdO, and SnO_2 [40, 43]

Group 1 is easily reduced catalytically and stepwise by small amounts of Pt metals. Each step leads to a decrease in the oxidation state of a metal one at a time, as in the hydrogen bronze formation. All metals in this groups were in their heighest oxidation valency state. In general, these catalyzed reductions proceeded without evidence of an induction period. However, for H_2 reduction, the process may be autocatalytic. Spillover has been invoked to explain the reduction. Spiltover H atoms are known to react with nonrefractory metal oxides very rapidly even at low temperatures [44].

Group 2 of the oxides was found to be reduced in the presence of Pt metals without either a large increase in the rate or a decrease in reduction temperatures. These compounds contained relatively stable, low-valency cations. These oxides are more or less readily reduced to the metallic state in a one-step process. Further, the metal may then be able to activate H_2. Therefore, the reduction would also be autocatalytic, which is often found for nonspillover reductions of these oxides at higher temperatures.

For both series of metal oxides, the apparent activation energies of the catalyzed (spillover) reduction reactions were found to be similar to those of the noncatalytic reductions. The reduction and not the spillover would seem to be rate controlling. Generally, the effect of the Pt was to increase the available reactive hydrogen and/or to increase the rate of the nucleation (preexponential factors). Thus this "catalysis" increases the availability of H but does not ("classically") decrease the activation energy.

The reduction of one supported metal by another has been noted. Mieville [45], used TPR to demonstrate that rhenium oxide cosupported with Pt on Al_2O_3 can be reduced by H_2 spillover from Pt. This process is autocatalytic, as the Re forms metal centers which then catalyze the further reduction of Re oxides. However, Isaacs and Petersen [46] have calculated that there may be too few spiltover species to account for these results.

The formation of alloys between the metal source of spillover and the supported oxide being reduced has been suggested. Praliaud and Martin [47] proposed the formation of Ni–Si and Ni–Cr alloys on silica and chromia supports, respectively, under H_2 at sufficiently high temperatures. They suggested that hydrogen spilled over from Ni to the Cr_2O_3 carrier and partially reduced it to $Cr^°$, which was then alloyed with Ni as indicated by magnetic measurements.

Dalmon et al. used TPD in conjunction with IR spectroscopy and volumetric adsorption of H_2 to study the reduction of Ni on alumina or on zeolite catalysts [48]. These supported Ni systems contained Ni^0 and Ni^+. H_2 was found to be activated only when the couple Ni^0/Ni^+ was present:

$$H_2 \rightarrow 2H \quad \text{on } Ni^0 \tag{8}$$

and

$$2H + 2Ni^+ \rightarrow 2Ni^0 + 2H^+ \tag{9}$$

Hydrogen spillover across intermetallics such as $SmMg_3$ treated with aromatics was detected by Imamura and Tsuchiya [49]. Hydrogen adsorption and sorption on $SmMg_3$ was enhanced (and observed also on Mg_2Ni [50]). The reaction of the intermetallic with spiltover hydrogen to produce samarium hydride (SmH_2) and metallic magnesium was detected.

B. Bronze Formation

Special consideration must be given to the numerous studies of hydrogen bronze formation induced by spillover. In particular, tungsten and molybdenum oxide hydrogen bronzes have been thoroughly studied since Khoobiar's initial experiments [2].

The kinetics and mechanism of the formation of molybdenum bronzes was studied in detail and published in a series of three papers by Erre et al. [51-53], who studied reaction on $MoO_3(100)$. Pt was added and the sample was exposed to H_2 at 10^{-8} to 10^{-5} torr. Monoclinic $H_{1.6}MoO_3$ was formed. The kinetics unexpectedly seemed to occur by three steps:

1. Activation of H_2 by Pt with an increasing rate with time on stream
2. A period where the rate remained constant
3. A decrease in the rate of bronze formation.

The process is autocatalytic, as once step 1 starts, Pt is not required for the dissociation and spillover. The hydrogen is now able to adsorb and dissociate molecular hydrogen directly from the gas phase. A similar situation was found for δ or amorphous aluminas at much higher temperature (430°C). Erre et al. were able to duplicate all their findings on MoO_3 using H atoms provided by a hot W wire [53].

C. Metal-Support Interactions

Metal-support interactions have recently been reviewed by Bond [54], who focused attention on catalyst systems that gave evidence for strong metal-support interaction (SMSI). This condition was first observed in 1978 by Tauster et al. [55] for Pt on titania catalysts..

In each of the mechanisms proposed by the various authors to account for SMSI (chiefly on M/TiO_2 systems), hydrogen spillover plays an important and often major role. Hydrogen spillover can be considered as a necessary condition for creation of the SMSI state. Common to each proposed SMSI mechanism is the reduction of the support by hydrogen. The reductions occur near 500°C in the presence of a metal able to chemisorb and dissociate molecular hydrogen.

12. Spillover of Hydrogen

Baker et al. [56] used high-resolution electron microscopy studies to study the morphological changes that are involved in hydrogen spillover onto TiO_2. These authors developed a model for the SMSI state (reduction at $T > 500°C$) [57]. The metal dissociates H_2 into atoms that spill over onto the TiO_2 and reduce it to Ti_4O_7. The Pt particles assume a "pillbox" microstructure on the Ti_4O_7. To determine if the microstructure was unique to the platinum or was due to the reduction of the support spillover, Ag/TiO_2 systems were also studied, as Ag does not dissociate H_2. After the reduction by H_2 at 550°C of this system, the metal particles appeared to be globular in outline, which is normal, and the TiO_2 remains as an unreduced rutile. Pt was then added and the sample was reexposed to H_2 at 550°C. The Ag now assumes a pillbox structure and the TiO_2 is reduced to Ti_4O_7. The presence of Pt and exposure to H_2 causes the silver to take on the morphological characteristics associated with the SMSI state of Pt on Ti_4O_7. It appears that the function of Pt is to provide a source of H atoms which spill over to titania and are responsible for its conversion to Ti_4O_7. The reduction of TiO_2 is also observed after low-temperature (~300°C) treatment, but the inhibition of H_2 adsorption is not found (as in the SMSI state).

The ESR studies by Huizinga and Prins [58] of Pt/TiO_2 reduced at 300°C gave evidence for the formation of Ti^{3+} with hydrogen present. If the sample is evacuated, the signal disappears; however, it reappears if H_2 is reintroduced at room temperature. If TiO_2 is reduced by H_2 at 875°C before deposition of Pt, the SMSI state is also observed [59]. If this reduced TiO_2 is reoxidized before the deposition of Pt, the catalyst irreversibly adsorbs H_2 at room temperature, like any Pt/TiO_2 reduced at low temperatures, and the SMSI state is not achieved. Further, Pt deposited on Ti_2O_3 or on TiO does not adsorb H_2 (as in the SMSI state). Migration (spillover?) of a titanium suboxide (with TiO_x due to reduction of TiO_2 by hydrogen spillover) onto Pt particles has also been advanced by several authors [20,60,61] to account for SMSI states. SMSI phenomena are discussed in detail in Chapter 13.

At least on titania, transition metals promote the spillover of hydrogen to the support; this is a necessary step in the reduction of the support (and hence modification of the global solid's catalytic properties). Hydrogen spillover merely facilitates the reduction of more or less easily reducible metal oxides, as mentioned in the preceding subsection.

V. CATALYTIC ACTIVITY INDUCED ON REFRACTORY OXIDES

After the studies of Khoobiar et al. [2] showing that hydrogen spilled over from Pt can convert WO_3 into a hydrogen bronze at room temperature, a trend developed to consider this hydrogen only as a reactant for various hydrogenations. In particular, it has been presumed only to increase the amount of available hydrogen due to spillover from a supported metal admixed or in contact with a support.

It has been shown in previous sections that the addition of small amounts of a transition metal to various metal oxides lowers the temperature required for their reduction by H_2. This phenomenon has been attributed to hydrogen spillover. It follows that a partial reduction of the host oxide can induce or modify the catalytic activity of the host material.

Hydrogen atoms (or ions) can migrate from a metal such as Pt or Pd, or even from a nonmetal [62,63], to another substance in contact with the first. In the case of a metal-supported catalyst, a carrier such as alumina is the first substance that accepts the spillover hydrogen (primary spillover). However, this migration may extend further, to a second hydrogen acceptor in mechanical contact with the catalyst, such as tungsten trioxide [2] or molybdenum trioxide [64,65], which forms bronzes.

The contribution of activity on the support to the activity on the metal is difficult to determine precisely in systems where both are still present. To yield unambiguous results a reactor was developed by Gardes et al. [66] in which the catalyst (e.g., Pt/Al_2O_3) was first in contact with the carrier (acceptor) and exposed to H_2. The catalyst may be withdrawn and isolated from the acceptor after hydrogen spillover. The interaction between the hydrocarbon to be hydrogenated and hydrogen (molecular and spillover) is observed in the absence of the supported metal, with only the carrier present in the reactor. The first results obtained with this "reactor with an elevator" involved Ni/Al_2O_3 activator and various aluminas as acceptors of the spiltover hydrogen [66-70]. This system (catalyst and acceptors) is rather a complicated one; the behavior of Pt/Al_2O_3 catalyst and SiO_2 acceptor is easier to understand. Hence the Pt/Al_2O_3 + SiO_2 system will be described first. The reactions of ethylene, benzene, and n-heptane are catalyzed by silica activated by hydrogen spillover after the Pt/Al_2O_3 has been removed.

The glass reactor used is shown schematically in Fig. 2 and is similar to that described by Gardes et al. [66,67]. More recently, a system using greaseless valves and a magnetic lifting mechanism has been used [71], but the results were exactly the same. A porous glass bucket which is filled with the supported metal catalyst (spillover source) is lowered by a winch

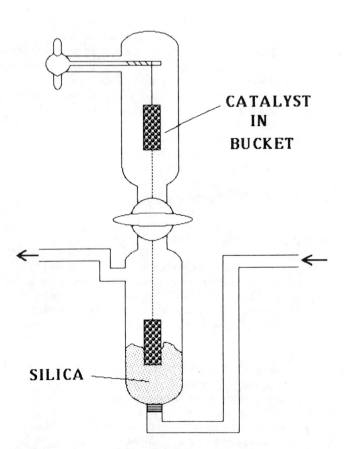

FIGURE 2 Schematic of the "reactor with an elevator" used to study the activity induced by hydrogen spillover. The source of spillover in the bucket is raised by the windlass from the lower position to the upper position and is then isolated by the stopcock.

mechanism from the upper compartment into the lower compartment. The accepting oxide is present in the lower compartment. After activation, when the bucket is lifted and stopcock A is closed, the activated oxide is effectively isolated from the supported metal catalyst. The reactor and inlet and outlet arms are surrounded by an insulated heating mantle capable of temperatures to 800 K. For both silica and alumina the activation procedure consists of evacuation, heating to 430°C, and exposure of the oxide in contact with the Pt/Al_2O_3 to hydrogen for over 12 h at 430°C. The temperature is then lowered to the reaction conditions and the metal is removed and isolated. The reactants can then be added to the circulating reactor system. Various tests, including neutron activation analysis, showed that traces of Pt from the catalyst did not migrate onto SiO_2 during activation.

A. Hydrogenation on Activated Silica

Silica may be converted by hydrogen spillover activation into a very unusual hydrogenation catalyst, active at about 150°C (for ethylene hydrogenation, E_a = 12 kcal mol^{-1}) and not poisoned by O_2 (or even H_2O). The spillover hydrogen that is required for the creation of active sites on SiO_2 inhibits hydrogenation for the first run (i.e., there is an induction period). Note that no catalytic activity is recorded if there is no contact, even indirectly, between the silica and the Pt/Al_2O_3 in the activation procedure or if He is used instead of H_2 with contact. Other silicas (e.g., Degussa fumed silica) behave in the same manner.

The kinetic studies of the hydrogenation of ethylene indicate [72] that the interaction of both reactants can be represented by Langmuir-Hinshelwood kinetics. They are strongly adsorbed on distinct sites (order zero with respect to each reactant for a stoichiometric feed). However, with hydrogen in large excess, a competition in adsorption is observed, and ethylene is displaced from its sites by hydrogen (positive order with respect to C_2H_4 and negative with respect to hydrogen).

The reaction is initially slow due to the presence of spiltover hydrogen. This is not true for the second dose or for hydrogenation performed after the evacuation and desorption of the spiltover hydrogen. This behavior shows again that the spiltover hydrogen is strongly adsorbed on silica and is less reactive than the gaseous, molecular hydrogen. The conversion of acetylene into ethylene and ethane is sequential and a maximum in the conversion into ethylene is found. The hydrogenation of either ethylene or acetylene is insensitive to ammonia or to an oxygen pretreatment of silica. Although 1,4-cyclohexadiene does not affect the hydrogenation of ethylene into ethane, it completely inhibits the conversion of acetylene into ethylene. The 1,3-cyclohexadiene isomer has no effect on either reaction. This behavior shows that silica activated by hydrogen spillover acquires at least two types of hydrogenating sites, those poisoned by 1,4-cyclohexadiene in the conversion of acetylene into ethylene and those unpoisoned by this cyclo-olefin in the conversion of ethylene into ethane.

B. Reactions of Benzene and Cyclohexene on Silica

The spiltover hydrogen can simply be added to benzene (forming cyclohexane and cyclohexene) in a noncatalytic reaction which entirely exhausts this hydrogen species. If the reaction of benzene is carried out after evacuation of silica that has been activated, the spiltover hydrogen desorbs and no hydrogenation is found. However, there is evidence that ethane and initially acetylene are produced at 200°C [73]. The first gas-phase product of the reaction of benzene with H_2 is acetylene, which is then normally hydrogenated into ethane. The net result is therefore a catalytic hydrogenolysis of benzene into ethane at 170°C on silica activated by hydrogen spillover. The reoxidation of silica inhibits the reaction of benzene, whereas there is no effect on the subsequent hydrogenation of ethylene. This cracking of benzene is poisoned by ammonia and inhibited by oxygen pretreatment. Therefore, a third type of site (for cracking) seems to be created on silica activated by hydrogen spillover and may be of an acidic nature. The energetic balance for cracking of benzene at 170°C is quite unfavorable and some other concerted reactions are probably involved. This very unusual reaction requires further detailed study. It is of interest to point out that silica activated by hydrogen atoms from a plasma, instead of spiltover hydrogen species from Pt/Al_2O_3, behaves in exactly the same way in reactions with ethylene or benzene [74].

These results help discriminate between the types of spiltover hydrogen or the types of sites adsorbing the spiltover hydrogen. Indeed, the amount of spiltover hydrogen adsorbed on all sites is of the order of 1.5 cm^3 per gram of SiO_2. The amount sorbed on sites poisoned by NH_3 is of the order of 0.6 cm^3 per gram of SiO_2. Only this type of spiltover hydrogen may be added to cyclohexene to form cyclohexane. But spiltover hydrogen fixed on other sites or of a different nature (0.9 cm^3 per gram of SiO_2) can react with benzene (even in the presence of NH_3) to convert it into cyclohexene. NH_3 inhibits any further reaction. It may be speculated that a protonic form of the spiltover hydrogen is inhibited (as NH_4^+) by NH_3, whereas the nonprotonic forms are insensitive to this poison. Alternatively, acid sites (also active in the cracking of benzene or cyclohexadienes) are poisoned by NH_3, whereas other sites (also active in hydrogenation) are insensitive to NH_3.

C. Reactions over Activated Alumina

There are differences among the supports and even between different sources of spillover (Pt/alumina or Ni/alumina). These results will be compared to those for silica discussed above. The behavior of alumina aerogel [71] or δ-alumina in the hydrogenation of ethylene after hydrogen spillover activation was very similar to that of silica. The corresponding reaction progress on δ-alumina at 145°C [75] is identical to that on silica at 200°C. The spiltover hydrogen was exhausted by the first dose of ethylene or by the evacuation after activation. Again ammonia or oxygen at 430°C did not influence the catalytic activity. The main difference with silica was that the spiltover hydrogen was reformed by heating a previously activated, evacuated alumina in H_2 at 430°C without the Pt/Al_2O_3 catalyst. The subsequent hydrogenation of ethylene (at 145°C) also exhibited an induction period.

There is a noticeable difference between alumina activated by Pt or by Ni. A very small amount of NO (0.5 cm^3) introduced during the reaction did not affect the kinetics of ethylene hydrogenation on δ-Al_2O_3 activated by Pt/Al_2O_3 at 430°C. If the Pt/Al_2O_3 catalyst was used at 300°C in the activation of δ-Al_2O_3, no catalytic activity was recorded in the hydrogenation of ethylene. Conversely, if Ni/Al_2O_3 activator was used at 430°C, no activity was recorded [76]; however, activation by Ni/Al_2O_3 at 300°C produces an active catalyst.

Another significant difference was found for the lower-temperature Ni/Al_2O_3 activation. After activation of δ-Al_2O_3 or amorphous alumina aerogel by Ni/Al_2O_3 at 300°C, these aluminas were active for ethylene hydrogenation only while H_{sp} was present [67]. If the spiltover hydrogen left after the activation (1.3 cm^3 per gram of Al_2O_3 aerogel at 110°C) was evacuated, only minimal activity with molecular H_2 was found. Since NO, which may be a radical scavenger, inhibits the reaction for Ni/Al_2O_3-activated alumina, a scheme involving H· radicals was proposed for this system.

12. Spillover of Hydrogen

Another interesting reaction on amorphous alumina activated by Pt/Al_2O_3 activator (at 430°C) was the isomerization of many doses of methylcyclopropane at 25°C, in H_2 or in He. The products were exclusively cis-2-butene and 1-butene; no trans-2-butene was detected [76]. These catalytic properties were permanent and did not require the spillover hydrogen.

Activated δ-alumina behaved differently from silica in the reaction with n-heptane. In the absence of molecular and spiltover hydrogen (in He) alumina exhibited (at 270°C) only a very limited catalytic activity for dehydrocyclization and hydrocracking [77]. In the presence of molecular and spiltover hydrogen, in contrast to silica, the reaction proceeded easily, giving heptene and heptadienes. The reaction was catalytic, as many doses of n-heptane and could be converted. Now, the spillover type of hydrogen can be reformed on activated alumina (at 270°C) from molecular hydrogen in the absence of the Pt/Al_2O_3 activator [71,75]. The difference in the behavior toward n-heptane of activated silica and activated alumina may be summed up by proposing that hydrogen spillover activation of silica creates a bifunctional catalyst with hydrogenating and acidic centers, whereas the activation of alumina gives mainly a monofunctional catalyst where acidic sites are weak or limited in number.

Finally, δ-alumina activated by Pt/Al_2O_3 (at 430°C) exhibited a completely different pattern from silica, activated in the same manner, for the conversion of benzene, cyclohexadienes (1,3- and 1,4-), and cyclohexene [72]. The reaction occurred only in the presence of molecular hydrogen (at 160°C), giving cyclohexane in all four cases. No cracking was observed in the presence of He (contrary to silica). Ammonia and oxygen at 430°C did not affect the activity. This behavior is therefore similar to the activity for ethylene hydrogenation, which was also insensitive to NH_3 or O_2. The (acid) sites that were developed by the activation of silica do not seem to be induced on alumina. The activation by hydrogen spillover is, therefore, specific to the nature of the oxide to be activated. This conclusion is reinforced by the behavior of magnesia activated by hydrogen spillover, described briefly below.

D. Reactions on Magnesia

Magnesia may also be activated by molecular H_2 at 430°C, without the need of a Pt/Al_2O_3 activator [75,78]. However, the sites active for hydrogenation of ethylene, unlike those on alumina or silica, were destroyed by oxidation at 430°C. Spiltover hydrogen (but not molecular H_2) activates MgO at 200°C, and the active sites are insensitive to O_2 treatment but are blocked by NH_3. Therefore, they are not the same as those created on SiO_2 and Al_2O_3. The poisoning effects of NH_3 and O_2 on reactions with cyclohexadienes are also different on activated MgO from those observed on SiO_2 or Al_2O_3.

VI. RETENTION OF CATALYTIC ACTIVITY

Inherent in the process of spillover is the concept that a metal is providing a "gate" by which hydrogen can gain easy access onto and off surfaces. As hydrogen is involved both in the formation of coke by dehydrogenation of sorbed hydrocarbon species and the removal of coke by gasification, spillover has a special place in the retention of activity for supported metal catalysts. It is known that coke forms on all portions of the exposed surface, both near and at a distance from the "active" metal centers. Thus, on single-crystal surfaces, where active sites may be present only at "steps and kinks," coke is found to form on the low-activity planes on the surface. For a metal supported on an oxide, for example, the oxide becomes covered with coke.

It has long been known that most hydrocarbon reactions (e.g., catalytic reforming) require an excess of hydrogen present to inhibit catalyst deactivation by coke formation. It is easy to envision that the active metal (or any other part of the system that can activate hydrogen) provides the gate for hydrogen sorption, spillover, and eventual reaction with hydrocarbon species on other parts of the surface.

The spillover of hydrogen from Pt onto carbon was proposed early in the studies of spillover by Boudart et al. [79]. Indeed, studies by Neikam and Vannice [80] and Bond and Tripathi [33] have concluded that a carbon interface may provide a bridge for hydrogen spillover onto other surfaces. The spiltover hydrogen is then able to react with solid carbon to form methane [81]. Even in the presence of water, Pt is found to be the most efficient in promoting the gasification of carbon supports to produce carbon monoxide and hydrogen (presumably involving the spillover of H, O, and OH species) [82]. In the absence of water, only methane is produced.

Parera et al. studied the reaction of naphtha and methylcyclohexane over alumina and alumina admixed with Pt on alumina [83]. The coke that is formed on the alumina is only partially removed from pure alumina if it is exposed to hydrogen after coking by the hydrocarbon species. It Pt/Al_2O_3 is added (a physical mixture), all of the coke is readily eliminated by exposure to hydrogen at elevated temperatures. This observation led these investigators to suggest that spillover is a primary step in the elimination and/or control of coke for metal-containing catalyst systems. Gates et al. have also suggested this significant implication of spillover [84]. The spillover of oxygen from a supported metal has also been implicated in the removal of coke and gasification of solid-crabon phases.

Conversely, spillover may take an active part in the formation of coke and its precursors. Fujimoto and Toyoshi [13] have shown that the presence of a transition metal accelerates the rate of dehydrogenation of isopentane and cyclohexane at around 400°C. Just as the presence of a metal accelerates the rate but not necessarily the amount of adsorbed hydrogen on carbon, the rate of recombination and desorption would be increased. Via "reverse spillover" the metal provides a center for recombination and a gate for the facile desorption of hydrogen originally associated with hydrocarbons sorbed on the surface. TPD studies were used to demonstrate the enhanced rate of hydrogen desorption and the involvement of the support in reverse spillover.

In a related process, the spillover of hydrogen can provide an active reductant that not only creates activity on refractory supports but may be involved in the retention of activity. Many reactions involve redox couples in which reduction may be promoted by spillover from another phase as well as by direct adsorption from the fluid phase. Atomic hydrogen may also be involved in promoting specific mixed-metal or defected phases on which catalytic activity is found.

Spillover may also be involved in the reaction and removal of other poisons to the catalytic activity that may form or adsorb on the surface. As an example, sulfur, a notorious poison for transition metals, has been shown to be removed from the surface as H_2S via the spillover of hydrogen [85,86].

In summary, spillover has been implied in both the formation and removal of hydrogen-deficient hydrocarbons. As it provides a gate for a potentially active hydrogen across heterogeneous surfaces, it may lead to the formation or removal of a variety of undesired sorbents.

VII. KINETICS AND MECHANISM OF SPILLOVER

Studies of spillover have concentrated on one aspect or another of the phenomenon. Many have involved characterizing the species adsorbing and spilling over. Another focus has involved transformation occurring on the surface or throughout the solid. The specific nature of the active site created on the "support" surface and of the catalysis proceeding on these sites has also been investigated. Underlying these mechanistic studies is the rate by which the processes occur and the extent of their influence.

A. Mechanism of Spillover

The nature of spiltover hydrogen is the subject of considerable controversy. Common to each proposed mechanism is the dissociative sorption of hydrogen onto a metal or other sorbing surface (or other oxides capable of sorbing H_2: ZnO, Cr_2O_3, Co_3O_4). The controversy involves the interpretation of the mechanism of the initial adsorption (or of the subsequent spillover) as homolytic or heterolytic. This can result in four possible species being created and subsequently transported on the sorbing surface: H^+ or H^-, $H\cdot$ and $H-$. The atomic species may be an ion, a radical, or a bonded species.

Sinfelt and Lucchesi first proposed that activated hydrogen migrated from Pt onto the SiO_2 or Al_2O_3 support, where it reacted with ethylene from the gas phase [3]. The discussion implies that the migrating species is atomic hydrogen (not necessarily ionic). No proposals were offered to differentiate between a radical or bound hydrogen. Khoobiar utilized electron conductivity measurements to suggest that protonic hydrogen is formed on Pt and diffuses across Al_2O_3 onto the separate WO_3 particles [2].

Neikam and Vannice used ESR to measure the enhanced decrease in anthracene and perylene radicals by the presence of Pt supported on base-exchanged zeolites [80]. The hydrogen free radicals, $H\cdot$, formed and spilt over from the Pt are able to react with the organic radicals sorbed onto the zeolite. In a subsequent [87] study, Vannice and Neikam found that the rate and extent of hydrogen uptake corresponded to the decrease in the ESR-measured perylene radicals. The perylene was suggested as a quantitative sink for the radicals, $H\cdot$, created and diffusing from Pt. They also suggested that the hydrogen radicals spilling over eventually react with oxygen sorbed onto the zeolite surface.

Studies of the effect of other sorbed species give considerable insight into the nature of spillover. Boudart et al. studied the influence of sorbed water on the rate of spillover [79]. The rate of WO_3 conversion at room temperature to the hydrogen bronzes and the sorption of hydrogen from the gas phase were dramatically increased by the presence of water. Levy and Boudart used a spectrum of alcohols and acids as coadsorbents and found that the increase in the rate of reduction of WO_3 was related to the proton affinity of the coadsorbent [28,88]. The proton, produced on the Pt, is proposed to be solvated by the coadsorbent and to diffuse via the layer of coadsorbent to the WO_3 surface, where the tungsten bronze is readily formed.

Jiehan et al. interpreted the "blue shift" in the IR band of CO with H_2 sorption as indicative of proton (not atomic H) spillover from Pt onto TiO_2 (containing CO) [89]. The studies were reinforced with parallel studies of the conductivity of the samples.

Gadgil and Gonzalez created atomic hydrogen species in the gas phase from a tungsten filament [90]. These species were then sorbed onto a variety of oxide surfaces at low temperatures. The rate of adsorption was different depending on the nature of the "support" surface, with a constant rate of hydrogen atom production from the gas phase. This implies that the sorption process was activated. The rate of adsorption of atomic hydrogen and the amount adsorbed were small compared to the rate of hydrogenation found to occur on the surfaces. This underscores the conclusion that the dominant effect is activation of the surface, not reaction of the hydrogen being spilled over.

Che has confirmed the latter interpretation with similar experiments [44, 91]. H atoms produced in a microwave discharge were found to activate SiO_2 to catalyze ethylene hydrogenation. The proposal is that hydrogen radicals create F center defects that are catalytically active in hydrogenation. Nogier et al. used a hydrogen plasma to induce hydrogenation activity on commercial (Davison and Degussa) silicas [74]. The hydrogenation activity was found to be an order of magnitude less than for silica activated by hydrogen spilt over from Pt; without the plasma treatment there was no activity. In similar studies Minachev et al. concluded that only the external exposed surface is able to interact with gaseous atomic species [92].

The foregoing studies are not necessarily in conflict with more conventional methods of spillover activation of an oxide in the presence of a metal. Each of these studies involved attempts to activate oxides at low temperatures. However, the results of these studies and more conventional spillover experiments confirm that the process is activated and may be enhanced by treatment at higher temperatures. Yet each study suggests that atomic hydrogen (as a radical or an ion) is involved in the creation of active sites on various oxides.

Studies by Dimitriev et al. [93] and Minachev et al. [94] involved a variety of surfaces between the source of spillover and a reacting surface (OH-OD exchange). They found that n- and p- semiconductors and insulators equally promoted the transport from the source to the reacting surface, although the rate of transport was much less on stainless steel. They concluded that the species spilling over was uncharged and that its transport did not depend on the semiconductor properties of the oxide. However, the oxide (or hydroxide) surface was involved.

Duprez and Miloudi studied the activation of Eh/TiO_2 by magnetic measurements. They proposed that hydrogen radicals created on the metal were able to spillover as hydrogen protons, while electrons diffused in parallel within the conduction band of the TiO_2 support [95].

The studies seem to contradict each other in specific proposals concerning the nature of spillover species. This conflict is resolved, however, by suggesting that hydrogen spillover does not occur via a single entity. Indeed, atomic hydrogen may spillover in a charged, bound, or radical state. If a cocatalyst is present (such as alcohol, acid, or water), protonic spillover can dominate the transport. The subsequent activation of the support to create catalytic activity involves a slower process with a finite activation energy. Even above 400°C considerable time is required to activate SiO_2 or Al_2O_3 (or other refractory oxides) fully. Lower-temperature adsorption processes may involve similar species of atomic hydrogen, although the activation of the support may not be complete. For processes with smaller activation energies, such as sorption or bronze formation, kinetic control is shifted to the formation, spillover, or transport of the activated atomic hydrogen.

To complicate the picture further, there is strong evidence that more than a single form of hydrogen can exist as a spiltover species under similar conditions. The evidence is that the possibility of multiple states of spiltover hydrogen species varies depending on the support; similarly, the various species can be involved in the reaction on and activation of the surface in different ways. This multiplicity will therefore be discussed in more detail in Sections VI.B and VI.C, even though the evidence implies that studies focusing on proving or disproving a single form of hydrogen on the support surface may not negate the existence of spillover.

B. Solid-State Reaction Mechanisms

There is a broad spectrum of solid changes that are induced by spillover. In the extreme, spillover facilitates a bulk crystallographic change in the solid. This transformation may involve the incorporation of the spiltover species into the solid lattice or the reduction of the lattice by reaction with the spiltover species. The effect may be confined to the surface. Other adsorbed species may facilitate the transport or be removed by the spiltover species. In the mildest sense, the surface may only exchange with the spilling-over species.

As we have seen, isotopic exchange of the hydroxyls on oxide surface by spilt-over deuterium has been known since 1965 [96]. For exchange to occur over a distance from the source of atomic hydrogen (e.g., Pt), a mobile hydrogen must be present on the surface. The picture is, however, somewhat clouded. If either water or oxygen is present, the exchange may occur via H_2O–HDO–D_2O intermediates [97].

More profound transformations can sometimes occur, also. The incorporation of spiltover hydrogen into the lattice of bronze-forming oxides was recognized early in studies of spillover [2]. A variety of studies focused on the nature of the changes that occur in the solid, and both molybdenum and tungsten bronzes have been used as indicators for spillover. The extent of hydrogen incorporation into MoO_3 was measured as $H_{1.6}MoO_3$ and $H_{0.4}WO_3$ for tungsten trioxide. It has been suggested that reduction occurs directly to the four-valent metallic species [6]. X-ray analysis shows that the starting monoclinic tungsten oxide converts to a simple cubic structure, while infrared gives no evidence of hydroxyl formation. For an initial hexagonal tungsten trioxide structure the maximum extent of hydrogen incorporation is $H_{0.4}WO_3$. The minor shifts in x-ray structure and the stoichiometry led Gerand and Figlarz to propose that the atomic spiltover hydrogen resides in the hexagonal holes of the host lattice (Fig. 3).

The results for bronzes indicate that spiltover hydrogen can become an integral part of oxide lattices. It can change the crystallographic structure and influence the oxidation state of the metal without lattice oxygen removal or hydroxide formation. The hydrogen can come from metal impregnated onto the trioxide or from a mechanical mixture, although the ease of hydrogen incorporation depends on the nature of the bronze that is formed. Indeed, a variety of other oxides and mixed metal oxides are able to incorporate the atomic hydrogen within the lattice [98].

The host oxide lattice, moreover, is able to be reduced by spiltover hydrogen, producing water. Spillover induces lower temperatures of reduction for vanadium, uranium, chromium, cobalt, cadmium, and tin oxides [99], among others. The reduction may involve bulk transformations or it may be confined to the surface. The most studied example of this phenomena involves TiO_2 and the resultant changes in sorption capabilities of the surface, SMSI, as discussed above. SMSI seems to be an extreme example of the change in chemisorptive properties with reduction and subsequent occultation of the supported metal.

A variety of surfaces has been shown to react with spiltover hydrogen. The discussion above focused on oxide surfaces where polar association is possible. Spillover occurs from Pt to carbon at modest temperatures [79]. Indeed, it has been suggested that a hydrocarbon bridge assists in spillover from Pt onto the supports [35,80]. The spiltover hydrogen is able to react

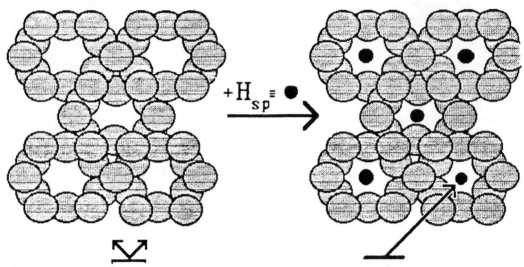

Hexagonal WO$_3$ Reduction by Spiltover Hydrogen

FIGURE 3 Schematic representation of the proposed transformation of hexagonal WO$_3$ into H$_x$WO$_3$ (x = 0.3 to 0.4) induced by hydrogen spillover from platinum at room temperature. (After Ref. 38.)

with the carbon lattice to form methane [81]. By enhancing the spillover there is the potential of increasing the lifetime or decreasing the regeneration requirements for reforming catalysts. It has been suggested that interparticle transport of spiltover hydrogen is inhibited between carbon surfaces [29].

Spillover has also been shown to induce the reduction of sulfides, specifically silver sulfide [85,86]. In the studies by Fleisch and Abermann [86], electron microscopy was used to follow the decrease of the area of a silver sulfide island as a function of time. Baker et al. investigated these phenomena in relation to the spillover-induced reduction of titania (giving rise to SMSI) [56]. The phenomena surrounding spillover are not distinct but reflect diverse extents of interaction.

Section V concerned the observations that spillover can induce catalytic activity on the support. The nature of the active site created on the support may result from the surface reduction, or the adsorbed hydrogen may be a center and site for reaction [100]. On the other extreme, spiltover hydrogen has been shown to inhibit ortho-para conversion over sapphire and ruby surfaces [101]. Similarly, it is found to inhibit catalytic reaction on spillover-activated surfaces; this gives rise to an induction period for ethylene hydrogenation [102].

This brings the discussion of the changes in the solid full circle. Spiltover hydrogen can exchange with the surface. It may react with and replace methoxyls with hydroxyls. It may be incorporated into the bulk with a change in the bulk crystal structure. Bulk reduction may occur. The species spilling over may react only with the surface, with coke, or with other sorbed species. In addition, spillover may promote or inhibit reaction on the surface.

Additional evidence of the ability of active sites to sorb hydrogen directly comes from recent studies of H$_2$-D$_2$ exchange [103]. A silica surface that has been activated by H$_2$ spillover is able to promote H$_2$-D$_2$ exchange without the metal present.

12. Spillover of Hydrogen

Further evidence of the multiple nature of spiltover hydrogen comes from the pioneering studies of Beck and White [33,104,105]. Hydrogen and deuterium were sorbed separately on Pt/TiO_2 at 227 and 27°C, respectively; the sorption and evacuation between the sorptions were performed in rapid sequences and the sample was cooled to -133°C and subsequently programmed with an increasing temperature to over 480°C. Separate D_2 (at 77°C) and H_2 (at 327°C) peaks were observed (Fig. 4). An insignificant amount of HD was produced and when Pt was not present, little H_2 adsorption-desorption was found. When both species were sequentially sorbed at 27°C, considerable HD was found and no peak at 327°C was evident.

The authors interpreted the separate peaks and lack of HD as due to the spatial separation of H_{sp} and D_{sp} on the surface. One of the states was located in the vicinity of Pt and assumed to be H on reduced titania TiO_x where $1 < x < 2$ (Ti^{3+} ions), the other one being situated in the bulk.

This explanation is not totally satisfactory. Unless subsequent spillover displaces previous spillover, diffusion will involve a monotonic gradient from the source. Subsequent spillover should not displace but rather, intermix with previously sorbed species. Since it is generally accepted that the spillover species is atomic, it is difficult to accept that little HD is formed and the desorption peaks occur with a 250°C difference in temperature. These studies seem to give credence to the hypothesis that multiple states of spiltover hydrogen (or deuterium) exist on the surface. Further studies are needed to clarify the nature of the sorbed states, their energetics, and the concentration on the support surface.

C. Amount of Spillover and the Density of Active Sites

It follows from the discussion above that a sequence of steps is involved in the process of spillover, although few studies have isolated them. A variety of assumptions has been made to simplify the analysis. Further, some of the phenomena associated with spillover, particularly the induction of support activity, have not been universally accepted. This is in part due to the few quantitative studies present in the literature.

First, it is desirable to estimate the concentration of spiltover hydrogen on the surface. Bianchi et al. [67] measured by volumetric adsorption a concentration of ca. 10^{12} sites cm^{-2} for alumina aerogel at temperatures ranging between 300°C and the ambient. Kramer and Andre [106], using TPD of hydrogen on Pt/Al_2O_3 samples, reported values of ca. 2×10^{12} sites cm^{-2} at 400°C and 710 torr of H_2. The same site density ca. 10^{12} sites cm^{-2}, was also found on silica by Bianchi et al. [107] using volumetric H_2 chemisorption, and confirmed by Lacroix et al. [108], who titrated the spillover hydrogen with ethylene as the quantity corresponding to the end of the induction period. Sermon and Bond [30] measured the same value by titration of the hydrogen spilled over onto silica by pentene-1. A similar value [i.e., 1.5×10^{12} sites (or H) cm^{-2} of SiO_2] was deduced from nuclear magnetic resonance (NMR) experiments carried out on a $Pt-SiO_2$ catalyst at room temperature by Sheng and Gay [109].

For alumina, the values of the site density of the accepting centers, 10^{12} cm^{-2}, were fully confirmed in experiments where a physical rather than a chemical means of supplying atomic hydrogen species was employed (e.g., Pt on alumina). Gadgil and Gonzalez [90] generated the atoms in the gas phase on a hot tungsten filament and let them adsorb on pure alumina powders at 0°C. They calculated a value of 10^{11} cm^{-2} H atoms based on desorption data at 250°C for the same alumina aerogel described by Gardes et al. [68]. Kramer et al. [106] reported a value of the same order at 400°C in their

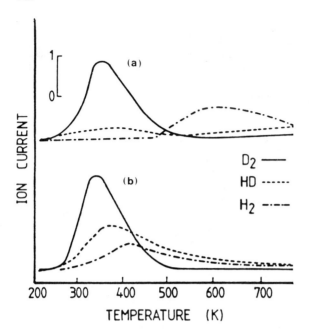

FIGURE 4 Temperature-programmed desorption from TiO_2 + Pt following sequential doses of hydrogen followed by deuterium (for 100 s at 0.077 torr): (a) H_2 is dosed at 500 K and D_2 at 300 K; (b) both are dosed at 300 K.

work with partially dissociated H_2 obtained through a high-frequency discharge.

Conner and Lenz were able to estimate the number of active sites as 10^{12} cm^{-2} from measurements of the adsorption of ethylene on an activated silica surface, noting that the hydrogenation of ethylene is effectively zero order in ethylene, which suggests rapid ethylene saturation of the surface [110]. On this bases they were able to calculate an effective turnover frequency (TOF) for the ethylene hydrogenation and H_2-D_2 exchange reactions on activated silica. For the hydrogenation, the TOF varied between 9×10^{-3} and 1.1×10^{-1} s^{-1}, increasing with temperature (135 to 200°C) and hydrogen partial pressure. The TOF for H_2-D_2 exchange was ca. 10^{-1} s^{-1} without ethylene and decreased to ca 10^{-2} s^{-1} in the presence of ethylene. This means that the reaction rates for these reactions on the sites created by spillover are more than an order of magnitude less than the rates over the metals (e.g., Ni, Pt, or Pd).

On alumina and silica, it seems that the density of sites accepting hydrogen spillover is in the range of 10^{12} cm^{-2}, assuming that one accepting site corresponds to one monoatomic hydrogen-spiltover species. The capacity for spiltover hydrogen varies from surface to surface, although there is less discrepancy on similar supports. For oxides, in general, the coverage is less than 1% of the surface and is, therefore, far less than the number of surface hydroxyls [20].

The situation is strikingly different on carbon, where concentrations of 10^{13} to 10^{16} sites cm^{-2} have recently been reported by many authors [13,35]. These are in good agreement with results published earlier by Robell et al. [111] and Boudart et al. [29]. It appears that carbon is a better acceptor than alumina or silica by several orders of magnitude.

D. Kinetic Control and the Rate of Spillover

Quantification of rate constants for this multistep process hinges on the assumed rate-controlling step. Depending on the steps that are assumed to dictate the rate, reaction rates or diffusion constants are calculated from the net kinetics of reaction or sorption. Various studies have assumed that either of two steps are rate controlling: either the surface diffusion or the actual spillover from the source. All analyses have assumed a first-order dependence on the concentrations of atomic hydrogen for each step in the sequence.

During the early studies of spillover, the diffusion of hydrogen away from the source was assumed to be rate controlling. This was based on the pioneering studies of Kramer and Andre [106] for Pt/Al_2O_3 as well as those of Fleisch and Abermann [86], and Schwabe and Bechtold [85], involving the reduction of Ag_2S by spiltover hydrogen. A dependence of the initial rate of reduction on the square of the distance from the islands of Pt to the receding Ag_2S surface led the authors to conclude that diffusion was rate controlling. The agreement is expressed in terms of classical diffusion theory, where the concentration is dependent on the square of the distance from the source. However, the production of H_2S would involve two atomic hydrogens. Is the rate therefore first order in hydrogen? For a very dilute (generally, <1% of the surface) two-dimensional system, are homogeneous analogs appropriate?

On other surfaces both the concentration and strength of bonding of the spiltover species will vary. The rate-controlling step may shift, and since the activation energies of the individual steps are not equal, the rate control can shift with temperature. For example, the diffusion coefficient may behave as $T^{3/2} \to T^{1/2}$ for two-dimensional diffusion as a surface gas. If a "jump-like" diffusion occurs from point to point, the dependence may contain an appropriate activation energy.

Nevertheless, based on the assumption that diffusion was the rate-controlling step in the process, the effective diffusion coefficients and activation energies listed in Table 2 were found.

There is considerable variation between the estimated diffusion coefficients and activation energies for similar systems. The techniques used to measure the rate were TPD (temperature-programmed desorption) [106], sorption studies [37,80], and electron microscopy [30,85,86]. NMR has also been used to attempt to estimate hydrogen surface diffusion. In 1972, Mestdagh et al. [112 estimated the diffusion coefficient as 10^{-15} cm^2 s^{-1} at 25°C on Y zeolites. More recently, pulse NMR has been used involving both the temperature dependence of the longitudinal relaxation time and pulsed field gradients [113,114]. For tungsten bronzes at room temperature, diffusion coefficients of hydrogen have been estimated as 7×10^{-6} and 7.9×10^{-6} cm^2 s^{-1}, respectively.

All the earlier investigators have deduced or assumed that surface migration is rate controlling. Indeed, diffusion over a distance from the source of spillover was clearly shown by infrared (IR) and Fourier-transform IR (FTIR) techniques on $Pt-Al_2O_3/SiO_2$ at 90°C and Pt/SiO_2 wafers at 240°C using OH-OD exchange [71,115]. In each series of experiments, a concentration gradient for the OD band appearance was found: a Pt spot was either on the edge or on the center of the pellet.

Recent studies by Cevallos-Candau and Conner provide additional detail regarding the analysis of surface diffusion rate and mechanism [116]. A 1-mm "spot" of Pt/Al_2O_3 was pressed into the center of a 1-in. silica wafer. An infrared cell was developed for these studies that is able to pretreat the

TABLE 2 Estimated Surface Diffusion Parameters

Surface	Temp. (K)	D_{eff} (cm^2 s^{-1})	Distance (Å)	E_a (kcal m^{-1})	Refs.
Al$_2$O$_3$	673	10^{-15}	2000	28.5	106
WO$_3$	323	10^{-16}			37
MO$_3$	323	10^{-13}			37
C or SiO$_2$	373	10^{-14}	2000	15.5	86
C	473	10^{-12}	1000	21	85
C	392	10^{-17}		39.2	30
Ce/Y-zeolite	293	10^{-10}			80
Y-zeolite	298	10^{-15}			112
TiO$_2$	500	10^{-16}		7.6	105
ZnO	343	10^{-7}		nrc	117
WO$_3$(H)	300	10^{-5}		nrc	113

sample at up to 500°C and then measure the infrared spectra at different positions on the disk with ca. 1-mm resolution. Deuterium was introduced over the sample. The deuterium does not directly exchange with the surface hydroxyls; however, after adsorption on the Pt spot, it can spill over as atoms and readily exchange with the surface hydroxyls. By recording the OD and OH bands as a function of time and position, the rate and extent of spillover can be measured. A schematic for the experiment is shown in Fig. 5

Figure 6 shows the changes in the infrared spectra at 1 mm from the Pt spot as a function of time. The difference spectra show that the OD bands at 2757 cm^{-1} increase with time as the OH bands at 3740 cm^{-1} decrease (i.e., become negative by subtraction of the initial spectra). However, note that the OD infrared bands do not increase uniformly. The broad band at 2300 to 2740 cm^{-1} (which represents hydroxyls, as OD, that are interacting with neighboring hydroxyls) increases rapidly, but the rate of increase slows down after a period of time. The narrow band at 2740 cm^{-1} represents isolated hydroxyls, as OD, and does not increase as rapidly initially but does so after a period of time. A possible explanation for this is that the associated hydroxyls (broader band) provide the primary mode of transport, as a bucket brigade. In a secondary process, the isolated hydroxyls exchange with the spilt-over deuterium. This hypothesis is supported by the observation that the rate of transport is decreased if the sample is pretreated by degassing at temperatures above 480°C, where the associated hydroxyls are beginning to be removed.

The increase in the OD bands as a function of time at three positions, 1, 5, and 9 mm from the Pt spot, is shown in Fig. 7. The gradient as a function of the distance from the source of spillover, the spot, is evident. The effective diffusion coefficients calculated from these studies vary between 10^{-3} and 10^{-5} cm^2 s^{-1}, depending on the temperatures of the experiment and the sample pretreatment. These results agree with the latter studies shown in Table 3 and demonstrate that spillover, not surface diffusion, is the rate-controlling step in the sequential processes.

12. Spillover of Hydrogen 333

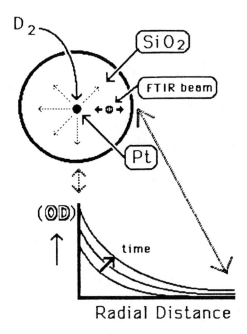

FIGURE 5 Schematic representation of the measurement of spillover and surface diffusion from a platinum spot across a silica wafer. The expected radial concentration of deuterium measured as OD (by FTIR) is shown below.

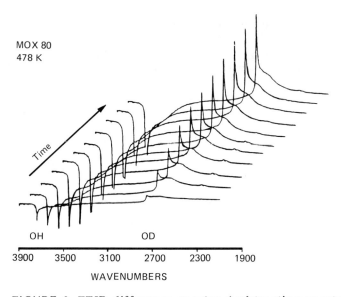

FIGURE 6 FTIR difference spectra (subtracting spectra at time = 0) at a distance of 1 mm from the platinum spot for a sequence of times after deuterium introduction.

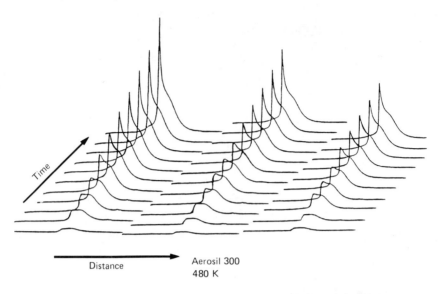

FIGURE 7 Change in time for the integrated OD infrared bands measured 1, 5, and 9 mm from the source of spillover.

By measuring the changes of the electroconductivity of a film of ZnO exposed to hydrogen spiltover from Pt/SiO_2 or Pt/Al_2O_3, Lobashina et al. [117] produced evidence for the surface migration of spilling hydrogen atoms across a SiO_2 surface. At 100°C the rate of diffusion was 2×10^7 particles s^{-1} with an estimated diffusion coefficient on quartz of 1.3×10^{-7} $cm^2\ s^{-1}$. These agree fairly well with corresponding values published elsewhere for comparable metal oxides (as discussed above).

Several recent studies have involved the kinetics of exchange [118] and demethylation of a silica aerogel surface [100] with a variation of contact between the source of spillover and the reacting surface. Each concluded that the surface migration is the easy step. In agreement with the latter studies (cited above), Smitriev et al. estimated the diffusion coefficients at room temperature as $10^{-10}\ cm^2\ s^{-1}$ and at 200°C as $10^{-6}\ cm^2\ s^{-1}$ [118]. Separately, they estimated the OH group diffusion coefficients as 10^{-16} $cm^2\ s^{-1}$ (at 20°C) to 10^{-11} (at 200°C), which agrees with the estimates of the earlier studies. Depending on the dominant mechanism of transport and the temperature, the diffusion may vary by more than 10 orders of magnitude.

TABLE 3 Estimated Diffusion Coefficients ($cm^2\ s^{-1}$)

Temp. (K)	A-300	A-200	MOX-80
458	10^{-5}		
473		5×10^{-4}	10^{-4}
480	5×10^{-4}		
495	$5 \times 10^{-4} - 5 \times 10^{-5}$		
508	10^{-4}		
523		$10^{-3} - 10^{-4}$	

The variation in rate-determining step with coadsorbed species was noted by Bond in his review [7]. Bond cited the extensive evidence that spillover (passage from the metal to the support) can be rate determining. For systems where H_2O (or other sorbed species) promotes spillover, it may do so by increasing the interfacial transport, a potentially slow step in the process.

It seems from these diverse studies that for most systems the spillover from metal to support is the rate-controlling step, although this may not always be the case. It seems reasonable that spiltover hydrogen has only a weak bond with the support, compared to its bond with the metal. The spillover step must be either endothermic or close to energetically neutral— at least it should not be very exothermic. Unless the activation energy from position to position on the support is high, the spillover will be rate determining. If the activation energy from point to point were great, desorption would be favored during the transition. In the experiments involving atomic hydrogen produced by discharge, the ability of this atomic hydrogen to induce activity was reduced compared to hydrogen spillover at higher temperatures on dispersed metallic systems [20, 24, 90, 91].

The process of activation is not simple. Whereas transport of hydrogen may occur at temperatures well below 400°C, the induction of catalytic activity on the support by spiltover hydrogen is an activated process and requires considerable time (up to 12 h of treatment at 430°C in hydrogen). The initial studies by Lenz and Conner [110] indicate that there is an optimum in the temperature of activation, at least on silica. Exposure of silica to spiltover hydrogen above 500°C or below 300°C does not induce activity independent of the time of exposure. Comparison of catalytically active surfaces must be done with similar temperatures and times of spillover pretreatment. To complicate the analysis further, there is evidence that an activated support (e.g., Al_2O_3 or SiO_2) may be able to dissociate hydrogen. The process may therefore be autocatalytic; that is, the support first activated by spillover may be able to adsorb, dissociate, spillover, and consequently activate more support surface [103].

VIII. OVERVIEW

There are several areas where the consequences of spillover may be significant. Initially, it is necessary to underscore that these effects will depend on the conditions and nature of the surface: the phenomena are not the same for different surfaces or under different conditions.

In this section we consider the myriad studies discussed above in order to address the following questions: What may be involved? What is the contribution to reaction mechanisms? What is the effect on catalysis?

A. The Phenomena

The foregoing discussion demonstrates the multifaceted nature of spillover. The interphase transport of atomic hydrogen onto a surface (and sometimes into the bulk) where it is unable to be formed without the activator can induce a variety of changes on, and reactions with, the surface. All the reactions of atomic hydrogen are found to be induced by spillover: exchange, bronze formation, reduction, demethoxylation, and catalytic activation. An activated species is able to gain indirect access to the nonsorbing surface.

Three figures will be discussed that help us visualize some of the phenomena associated with spillover. Figure 8 uses three schematics to depict the general phenomena associated with spillover onto an oxide surface. A surface is shown that represents a metal deposited on an oxide. Hydroxyls (represented by the

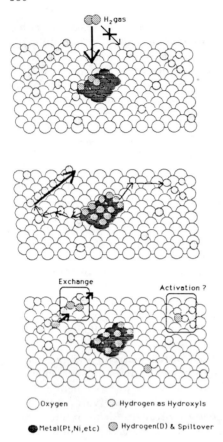

FIGURE 8 Representation of the processes associated with hydrogen spillover from a metal onto a partially hydroxylated oxide surface. Three sequential schematics are depicted.

lightly spotted hydrogen) are found in groups (upper left) or as isolated species. The metal (dark spheres) is shown as a cluster in the center of each representation.

As represented at the top of Fig. 8, molecular hydrogen adsorbs onto the metal, where it dissociates. In the center schematic the adsorbed hydrogen may spill over and has several paths for surface diffusion. In general, the paths tend to involve association with surface hydroxyls, although random surface diffusion may play a part. Any associated hydroxyls (the string in the upper left and the group in the upper right) may greatly facilitate the transport (as a bucket brigade). The spiltover hydrogen can then react with the surface, as shown in the lower schematic, leading to exchange (as $OH + D_{sp} \rightarrow OD + H_{sp}$) or the creation of active sites. The grouped or associated hydroxyls seem to enhance these processes.

The specific example of the role of spiltover hydrogen in the creation of the SMSI state is depicted in Fig. 9. In the top schematic, hydrogen is shown to adsorb and dissociate on the metal. Atomic hydrogen then spills over and reduces the supported oxide. In the center schematic the now-reduced oxide is then able to "spill over" onto the metal, where in the lower schematic, it is seen able to block the sorption of hydrogen.

The last of these series of figures represents the roles of spillover in the formation of surface coke. In the top schematic of Fig. 10 the metal is shown to provide a gate by which hydrogen from an adsorbed hydrocarbon is able

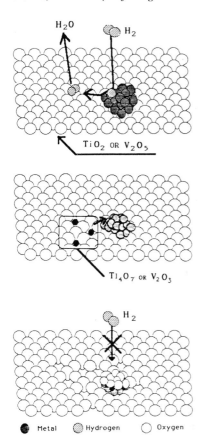

FIGURE 9 Representation of the creation of a state of strong metal-support interaction (SMSI). Three sequential schematics show the hydrogen spillover, reduction of the host oxide (TiO_2), and the migration of the reduced oxide onto the metal.

to recombine and desorb via "reverse spillover." This may occur if there is insufficient hydrogen in the gas phase. In an excess of hydrogen, shown in the lower schematic, the hydrogen may adsorb and dissociate on the metal and spill over across the oxide. The spillover hydrogen (or oxygen) can then remove the surface, hydrogen-deficient carbon as methane. Water may play a role in facilitating the spillover. The surface near the metal is thereby effectively kept clean in the process.

The effects of spillover are dependent on the receiving surface and the temperature of the system. Most oxides are able to exchange their hydroxyls via spillover of deuterium. The temperature must be sufficient for the deuterium to spill over onto the oxide surface. Coadsorbents (such as water) can facilitate the transport [Eq. (4)] or the transport can be facilitated by an intermediate solid phase (such as a carbonlike or hydrocarbon phase). This implies that the method of catalyst preparation can influence the rate of spillover and the subsequent exchanges. Therefore, for example, not all types of Pt/SiO_2 will behave the same. This also means that the phenomena associated with Pt/SiO_2 may differ from Pt/Al_2O_3, Pd/SiO_2, Ni/SiO_2-Al_2O_3, and so on.

Various phenomena associated with spillover occur after the spillover process; spillover is a necessary but not sufficient process to induce the effects that have been found. These effects can be independent of each

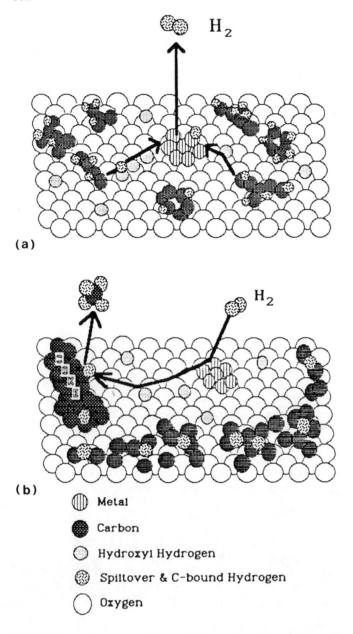

FIGURE 10 Representation of the roles of hydrogen reverse spillover in the creation of surface coke (a) and of hydrogen spillover in the removal of the coke (b).

other. As an example, the exchange of hydroxyls with deuterium on silica occurs readily at ca. 100°C, but the surface is not catalytically activated at this temperature (T > 400°C is required). As above, this relative contribution depends on the systems being investigated. Atomic hydrogen generated by a microwave discharge induces the same catalytic properties in silica as the spiltover hydrogen [74].

Even for a single system, the conditions may be such that the history of the sample can have an effect. Spillover involves a sequence of steps, and each of these steps may be influenced by the pretreatment of the sample. As an example, surface transport and surface activation may depend on hydroxyl content (via $H_{sp} + OH' \rightarrow OH + H'_{sp}$ and $H_{sp} + MOH \rightarrow M + H_2O$ mechanisms, respectively). Dehydroxylation of an oxide surface may, therefore, reduce the rate and extent of H_{sp} transport or subsequent surface activation [103].

The influence of spillover species on an acceptor phase can be in the extreme either subtle or profound. Many of the phenomena associated with hydrogen spillover are as subtle as the influences of type 2 hydrogen on the activity of ZnO [119] or as significant as bulk reduction, bronze formation, or catalytic activation. The effects may be similar to the exposure of a surface to a hydrogen plasma.

The effects of temperature on the phenomena associated with spillover differ from study to study. We can, however, generally depict an energy coordinate for the various states. The relative energies will depend on the specific system. The hypothetical relationships are depicted in Fig. 11 for

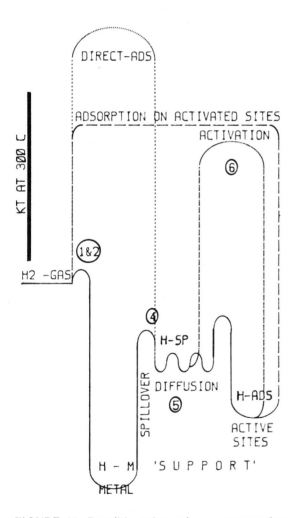

FIGURE 11 Possible schematic representation of the energetic coordinate for the phenomena associated with hydrogen adsorption, spillover, surface diffusion, and the creation of active sites. The numbers reflect the mechanistic steps from the equations in Section II.

a metal, M, and a support (such as an oxide). The circled numbers refer to the mechanistic steps proposed in Eqs. (1) to (6). The dashed lines indicate mechanistic steps involving the sites formed on the support, which does not directly sorb H_2 (the high activation energy indicated by the dashed line). H_2 is readily adsorbed and activated on the metal [steps (1) and (2)]. Spillover [step (4)] is activated but may be facilitated if a coadsorbent (such as H_2O) is present and provides an alternative path. Diffusion across the support [step (5)] is relatively easy. It is shown as a uniform sequence of steps although there is evidence that more than one path may exist (i.e., with or without association with hydroxyls, or intra- and interparticle paths). The support can become activated if the thermal energy is sufficient [step (6)]. This creates sites that *then* accept H_{sp} from the support or directly from the gas phase (e.g., for amorphous or δ-alumina), but the direct adsorption may be difficult (a higher-energy barrier, as an example, for silica). The relative energies of the states and the transitions will dictate the stability of the various species. As an example, it is evident that the spiltover hydrogen adsorbed on the sites created on δ-alumina differs in energy from the spiltover hydrogen on the corresponding sites of amorphous alumina. This gives rise to differences in behavior during activation or exposure to hydrogen gas at high temperatures.

B. The Mechanisms

The species being spilled over is not always unique. It can be ionic or radical-like; again, the mechanism depends on the surface of the acceptor, on the source of spillover, and on the interface. Further, parallel mechanistic paths may exist that give rise to the same net effect. As an example, bronze formation can occur by hydrogen spillover directly to the oxide; it can occur by spillover assisted by coadsorbents; it can (autocatalytically) occur by spillover from sites created on the metal oxide. Further, the relative contribution of the parallel paths may change in the course of the process.

Each of the intermediate steps in the mechanistic sequence involved in spillover can differ. As an example, surface transport may occur as a two-dimensional gas, as species associated with an activated site, or as two-dimensional exchange (via hydroxyls). The literature has suggested a variety of mechanisms for spillover and mechanisms induced by spillover. The studies have assumed that only one of the mechanisms was possible and have not focused on discriminating between the alternative possibilities. This more open (albeit more disconcerting) approach is suggested for an understanding of the phenomena of spillover.

For reactions that are the same on metal and other catalytic sites (e.g., hydrogenation or total oxidation), the reaction may seem to proceed in a similar fashion on the metallic source of spillover and on the "diluent" support. For reaction that occurs with different selectivities or kinetics on the induced sites, the implications are far more complex. Teichner has found that the selectivities for the isomerization of methylcyclopropane and for cracking of benzene and cyclohexadienes reaction may be substantially different on "activated" silica or alumina than on the "normal" oxides (see above). In these cases, a separate rate expression may need to be included. Depending on the rate constants and relative kinetics, this can substantially change the reaction rate expression. Further, differences in the activation energies may affect the contribution at different temperatures. If the reaction temperature is sufficient, the activation of the support may be able to occur during the course of the experiment.

The crucial question is the relative activity on the metal surface and the activity induced on the "support." The number of sites potentially induced on the support is low per unit area for most refractory supports (SiO_2 and Al_2O_3). The dispersion of the support and ratio of solid diluent to supported catalyst will dictate the potential effect of activity on the support. A 1% Pt catalyst on a 200-m^2/g support will have a maximum (100% dispersion) of 1.5×10^{13} Pt/cm^2. As discussed above, the estimated maximum number of spiltover hydrogen atoms on silica or alumina is 10^{12} cm^{-2}. If we assume that one site is produced for each spiltover hydrogen atom and that the specific activity on the support and on the Pt is the same, the ratio of activity on the metal to that on the support would be 15:1. Obviously, the relative activity for a supported metal catalyst will depent on the percentage and dispersion of metal, the number of active sites of all types, and their relative activity. For the example cited, under these constraints the relative activity of the support would probably be minimal (<1%). For systems where the concentration of induced active sites is larger or systems where a solid diluent is present, the contribution of the support can be significant. Studies that focus on the relative activity (concentration of active sites and their specific activity) are needed to *quantify* the potential contribution of the induced activity compared to the inherent activity.

Hydrogen spillover by itself is not tremendously significant; however, as the first crucial step in a sequence, it may be very important. The indirect contribution of spillover to catalysis seems to be the more important. The availability of potential reactant species is changed by spillover from one site, where reaction does not occur, to another site, where reaction may occur. This involves the induction of activity on seemingly inert supports or other otherwise inactive metals and the retention of activity on sites that would deactivate (by coke formation, by oxidation and/or reduction, etc.) without spillover. Further, spillover can promote the indirect activation of one metal by another in bimetallic catalyst systems. Conversely, the spillover may be detrimental by contributing to undesirable side reactions such as hydrogen transfer in catalytic hydrotreating. Also, the behavior of an activated support toward poisons is of interest, as the catalytic activity of the support may be modified by the poison in a manner that differs from the effect of the poison on the metal.

C. Concluding Remarks

In this chapter we have attempted to put hydrogen spillover in perspective. At this time, there seem to be more questions than definitive conclusions. As with any newly discovered phenomenon, "spillover" has been used to explain a variety of phenomena on heterogeneous solids. Even if limited to interphase phenomena the picture is clouded. The influence is well documented, but the extent of the *relative* influence has been studied in less quantitative terms. Further, at this time it is necessary to focus the discussion of spillover to gain understanding but not to limit the approaches.

ACKNOWLEDGMENTS

The author thanks Professors S. J. Teichner and G. M. Pajonk for their assistance in assembling this information and their invaluable discussions of spillover, and NSF (CPE 81-21800) and PRF (13703-AC7) for their generous support of the research that contributed to these studies.

REFERENCES

1. A. Ozaki, *Isotopic Studies of Heterogeneous Catalysis*, Academic Press, New York (1977).
2. S. Khoobiar, *J. Phys. Chem.*, 68:411 (1964); and S. Khoobiar, R. E. Peck, and B. J. Reitzer, *Proceedings of the 3rd International Congress on Catalysis, Amsterdam* (1965), p. 338.
3. J. M. Sinfelt and P. J. Lucchesi, *J. Am. Chem. Soc.*, 85:3365 (1963).
4. G. M. Pajonk, S. J. Teichner, and J. E. Germain (eds.), *Proceedings of the First International Symposium on the Spillover of Adsorbed Species*, Elsevier, Amsterdam (1983); a volume of *Discussions*, University of Claude Bernard-Lyon 1, Villeurbanne (1984).
5. P. A. Sermon and G. C. Bond, *Catal. Rev.*, 8:211 (1973).
6. D. A. Dowden, *Catalysis*, Vol. 8, Chemical Society, London (1980), Chap. 6.
7. G. C. Bond, in *Proceedings of the First International Symposium on the Spillover of Adsorbed Species* (G. M. Pajonk, S. J. Teichner, and J. E. Germain, eds.), Elsevier, Amsterdam (1983), p. 1.
8. W. C. Conner, G. M. Pajonk, and S. J. Teichner, *Advances in Catalysis*, Vol. 34, *Spillover of Sorbed Species* (1986), p. 1.
9. W. C. Conner, in *Discussions*, University of Claude Bernard-Lyon 1, Villeurbanne (1984), p. 71.
10. G. M. Pajonk and S. J. Teichner, *Adsorption at the Gas-Solid and Liquid-Solid Interface* (J. Rouquerol and K. S. W. Sing, eds.), Elsevier, Amsterdam (1982).
11. J. A. Altham and G. Webb, *J. Catal.*, 18:133 (1970).
12. P. A. Sermon and G. C. Bond, *J. Chem. Soc. Faraday Trans. 1*, 76:889 (1980).
13. K. Fujimoto and S. Toyoshi, *Proceedings of the 7th International Congress on Catalysis*, Kodansha, Tokyo (1981), p. 235.
14. E. Keren and A. Soffer, *J. Catal.*, 50:43 (1977).
15. K. Fujimoto, A. Ohno, and T. Kunugi, in *Proceedings of the First International Symposium on the Spillover of Adsorbed Species* (G. M. Pajonk, S. J. Teichner, and J. E. Germain, eds.), Elsevier, Amsterdam (1983), p. 241.
16. A. R. Berzins, M. S. W. Lau Vong, P. A. Sermon, and A. T. Wurie, *Adv. Sci. Technol.*, 1:51 (1984).
17. J. R. Anderson, K. Foger, and R. J. Breakspere, *J. Catal.*, 57:458 (1979).
18. T. M. Apple, P. Gajardo, and C. Dybowski, *J. Catal.*, 68:103 (1981).
19. T. M. Apple and C. Dybowski, *J. Catal.*, 71:316 (1981).
20. X. Z. Jiang, T. F. Hayden, and J. A. Dumesic, *J. Catal.*, 83:168 (1983).
21. D. A. Dowden, I. H. B. Haining, J. D. N. Irving, and D. A. Whan, *J. Chem. Soc. Chem. Commun.*, 631 (1977).
22. J. Barbier, H. Charcosset, G. Periera, and J. Riviere, *Appl. Catal.*, 1:71 (1981).
23. A. Crucq, L. Degols, G. Lienard, and A. Frennet, in *Proceedings of the First International Symposium on the Spillover of Adsorbed Species* (G. M. Pajonk, S. J. Teichner, and J. E. Germain, eds.), Elsevier, Amsterdam (1983), p. 137.
24. D. W. Goodman and C. F. Peden, *J. Catal.*, 94:576; 95:321 (1985).
25. D. W. Goodman and C. F. Peden, *Ind. Eng. Chem. Fundam.*, 25:58 (1986).
26. K. Christmann, G. Ertl, and J. C. Vickerman, *J. Catal.*, 71:175 (1981).
27. F. Grenter and E. W. Plummer, *Solid State Commun.*, 48:37 (1983).
28. R. B. Levy and M. Boudart, *J. Catal.*, 32:304 (1974).

29. M. Boudart, A. W. Aldag, and M. A. Vannice, *J. Catal.*, *18*:46 (1970).
30. P. A. Sermon and G. C. Bond, *J. Chem. Soc. Faraday Trans. 1*, *72*:745 (1976).
31. J. M. Parera, N. S. Figoli, E. L. Jablonski, M. R. Sad, and J. N. Beltramini, *Studies in Surface Science and Catalysis*, Vol. 6, *Catalyst Deactivation*, Elsevier, Amsterdam (1980).
32. N. S. Figoli, M. R. Sad, J. N. Beltramini, E. L. Jablonski, and J. M. Parera, *Ind. Eng. Chem. Prod. Res. Dev.*, *19*:545 (1980).
33. D. D. Beck, A. O. Bawagan, and J. M. White, *J. Phys. Chem.*, *88*:2771 (1984).
34. G. C. Bond and J. B. P. Tripathi, *J. Less-Common Met.*, *36*:31 (1974).
35. G. C. Bond and T. Mallat, *J. Chem. Soc. Faraday Trans. 1*, *77*:1743 (1981).
36. N. I. Il'chenko, *Russ. Chem. Rev.*, *41*:47 (1972).
37. P. A. Sermon and G. C. Bond, *J. Chem. Soc. Faraday Trans. 1*, *72*:730 (1976).
38. B. Gerand and M. Figlarz, in *Catalysis*, Vol. 8, Chemical Society, London (1980), p. 275.
39. D. Tinet, H. Estrade-Szwarckopf, and J. J. Fripiat, in *Metal-Hydrogen Systems* (T. N. Veziroglu, ed.), Pergamon Press, Oxford (1982).
40. G. C. Bond and J. B. P. Tripathi, *J. Chem. Soc. Faraday Trans. 1*, *72*:933 (1976).
41. A. Ekstrom, G. E. Batley, and D. A. Johnson, *J. Catal.*, *34*:106 (1974).
42. G. A. L'Homme, M. Boudart, and L. D'Or, *Bull. Ch. Sci. Acad. R. Belg.*, *52*:1206, 1249 (1966).
43. G. E. Batley, A. Ekstrom, and D. A. Johnson, *J. Catal.*, *34*:368 (1974).
44. M. Che, B. Canosa, and A. R. Gonzales-Elipe, *J. Chem. Soc. Faraday Trans. 1*, *78*:1043 (1982).
45. R. L. Mieville, *J. Catal.*, *87*:437 (1984).
46. B. H. Isaacs and E. E. Petersen, *J. Catal.*, *77*:43 (1982).
47. H. Praliaud and G. A. Martin, in *Proceedings of the First International Symposium on the Spillover of Adsorbed Species* (G. M. Pajonk, S. J. Teichner, and J. E. Germain, eds.), Elsevier, Amsterdam (1983), p. 191.
48. J. A. Dalmon, C. Mirodatos, P. Turlier, and G. A. Martin, in *Proceedings of the First International Symposium on the Spillover of Adsorbed Species* (G. M. Pajonk, S. J. Teichner, and J. E. Germain, eds.), Elsevier, Amsterdam (1983), p. 169.
49. H. Imamura and S. Tsuchiya, *J. Chem. Soc. Faraday Trans. 1.*, *79*:1461 (1983).
50. H. Imamura, T. Takahashi, and S. Tsuchiya, *J. Catal.*, *77*:289 (1982).
51. R. Erre and J. J. Fripiat, in *Proceedings of the First International Symposium on the Spillover of Adsorbed Species* (G. M. Pajonk, S. J. Teichner, and J. E. Germain, eds.), Elsevier, Amsterdam (1983), p. 285.
52. R. Erre, H. Van Damme, and J. J. Fripiat, *Surf. Sci.*, *127*:69 (1983).
53. R. Erre, M. H. Legay, and J. J. Fripiat, *Surf. Sci.*, *127*:69 (1983).
54. G. C. Bond, in *Metal-Support and Metal-Additive Effects in Catalysis*, Vol. 11 (B. Imelik et al., eds.), Elsevier, Amsterdam (1982), p. 1.
55. S. J. Tauster, S. C. Fung, and R. L. Garten, *J. Am. Chem. Soc.*, *100*:170 (1978).
56. R. T. K. Baker, E. B. Prestridge, and L. L. Murrell, *J. Catal.*, *80*:348 (1983); R. T. K. Baker, E. B. Prestridge, R. L. Garten, *J. Catal.*, *56*:390 (1979).
57. R. T. K. Baker, E. B. Prestridge, and R. L. Garten, *J. Catal.*, *59*:293 (1979).
58. T. Huizinga and R. Prins, *J. Phys. Chem.*, *85*:2156 (1981).

59. B.-H Chen and J. M. White, *J. Phys. Chem.*, *86*:3534 (1982).
60. D. E. Resasco and G. L. Haller, *J. Catal.*, *82*:279 (1983).
61. H. R. Sadeghi and V. E. Henrich, *J. Catal.*, *87*:279 (1984).
62. P. Delmon, *React. Kinet. Catal. Lett.*, *13*:203 (1980).
63. D. Pirote, P. Grange, and P. Delmon, *Proceedings of the 7th International Congress on Catalysis*, Kodansha, Tokyo (1981), p. 1422.
64. J. P. Marcq, X. Wispenninckx, G. Poncelet, D. Keravis, and J. J. Fripiat, *J. Catal.*, *73*:309 (1982).
65. J. P. Marcq, G. Poncelet, and J. J. Fripiat, *J. Catal.*, *87*:339 (1984).
66. G. E. E. Gardes, G. M. Pajonk, and S. J. Teichner, *C.R. Acad. Sci.*, *C227*:191 (1973).
67. D. Bianchi, G. E. E. Gardes, G. M. Pajonk, and S. J. Teichner, *J. Catal.*, *38*:135 (1975).
68. G. E. E. Gardes, G. M. Pajonk, and S. J. Teichner, *J. Catal.*, *33*:145 (1974).
69. G. E. E. Gardes, G. M. Pajonk, and S. J. Teichner, *C.R. Acad. Sci.*, *C278*:659 (1974).
70. S. J. Teichner, A. R. Mazabrard, G. M. Pajonk, G. E. E. Gardes, and C. Joang-Van, *J. Colloid Interface Sci.*, *58*:88 (1977).
71. D. Maret, G. M. Pajonk, and S. J. Teichner, in *Proceedings of the First International Symposium on the Spillover of Adsorbed Species* (G. M. Pajoni, S. J. Teichner, and J. E. Germain, eds.), Elsevier, Amsterdam (1983), p. 215.
72. M. Lacroix, G. M. Pajonk, and S. J. Teichner, *Bull. Soc. Chim. Fr.*, 94 (1981).
73. M. Lacroix, G. M. Pajonk, and S. J. Teichner, *C.R. Acad. Sci.*, *C287*:499 (1978); *Bull. Soc. Chim. Fr.*, 265 (1981).
74. J. P. Nogier, J. Bonardet, and J. Fraissard, in *Proceedings of the First International Symposium on the Spillover of Adsorbed Species* (G. M. Pajonk, S. J. Teichner, and J. E. Germain, eds.), Elsevier, Amsterdam (1983), p. 233.
75. M. Lacroix, G. M. Pajonk, and S. J. Teichner, *Bull. Soc. Chim. Fr.*, 101 (1980).
76. C. Hoang-Van, A. M. Mazabrard, C. Michel, G. M. Pajonk, and S. J. Teichner, *C.R. Acad. Sci.*, *C281*:211 (1975).
77. M. Lacroix, G. M. Pajonk, and S. J. Teichner, *Ind. Eng. Chem. Fundam.*, in press.
78. M. Lacroix, G. M. Pajonk, and S. J. Teichner, *React. Kinet. Catal. Lett.*, *12*:369 (1979).
79. M. Boudart, J. Benson, and H. Kohn, *J. Catal.*, *5*:307 (1966).
80. W. C. Neikam and M. A. Vannice, *J. Catal.*, *27*:207 (1972).
81. D. Tomita and Y. Tamai, *J. Catal.*, *27*:293 (1973).
82. R. T. Ravick, P. R. Wentrcek, and H. Wise, *Fuel*, *53*:274 (1974).
83. J. Parera, E. Traffano, J. Masso, and C. Pieck, in *Proceedings of the First International Symposium on the Spillover of Adsorbed Species* (G. M. Pajonk, S. J. Teichner, and J. E. Germain, eds.), Elsevier, Amsterdam (1983), p. 101
84. B. Gates, J. Katzer, and G. Schuit, *Chemistry of Catalytic Processes*, McGraw-Hill, New York (1979), p. 289.
85. U. Schwabe and E. Bechtold, *J. Catal.*, *26*:427 (1972).
86. T. Fleisch and R. Abermann, *J. Catal.*, *50*:268 (1977).
87. M. A. Vannice and W. C. Neikam, *J. Catal.*, *20*:260 (1971).
88. R. Levy, Ph.D. thesis, Stanford University (1974).
89. Hu Jiehan, Hong Zupei, Song Yongze, and Wang Hongli, in *Proceedings of the First International Symposium on the Spillover of Adsorbed Species* (G. M. Pajonk, S. J. Teichner, and J. E. Germain, eds.), Elsevier, Amsterdam (1983).

(G. M. Pajonk, S. J. Teichner, and J. E. Germain, eds.), Elsevier, Amsterdam (1983).
90. K. Gadgil and R. D. Gonzalez, *J. Catal.*, 40:190 (1975).
91. M. Che, *Proceedings of the 7th International Congress on Catalysis*, Kodansha, Tokyo (1981), Discussion, p. 287.
92. C. Minachev, R. Dmitriev, K. Steinberg, A. Detjuk, and H. Bremer, *Iz. Akad. Nauk SSSR Ser. Khim.*, 2670 (1975).
93. R. V. Dmitriev, K. H. Steinberg, A. N. Detjuk, F. Hofmann, H. Bremer, and C. M. Minachev, *J. Catal.*, 65:105 (1980).
94. C. Minachev, R. Dimitriev, K. Steinberg, and H. Bremer, *Iz. Akad. Nauk SSSR Ser. Khim.*, 2682 (1978).
95. D. Duprez and A. Miloudi, in *Proceedings of the First International Symposium on the Spillover of Adsorbed Species* (G. M. Pajonk, S. J. Teichner, and J. E. Germain, eds.), Elsevier, Amsterdam (1983), p. 163.
96. J. L. Carter, P. J. Lucchesi, P. Corneil, D. J. C. Yates, and J. H. Sinfelt, *J. Phys. Chem.*, 69:3070 (1965).
97. D. Bianchi, D. Maret, G. M. Pajonk, and S. J. Teichner, in *Proceedings of the First International Symposium on the Spillover of Adsorbed Species* (G. M. Pajonk, S. J. Teichner, and J. E. Germain, eds.), Elsevier, Amsterdam (1983), p. 45.
98. M. Daage, and J. Bonnelle, in *Proceedings of the First International Symposium on the Spillover of Adsorbed Species* (G. M. Pajonk, S. J. Teichner, and J. E. Germain, eds.), Elsevier, Amsterdam (1983), p. 261.
99. G. Bond, P. Sermon, and J. Tripathi, *Ind. Chim. Belg.*, 38:506 (1973).
100. D. Bianchi, M. Lacroix, G. M. Pajonk, and S. J. Teichner, *J. Catal.*, 68:411 (1981).
101. P. Selwood, *Proceedings of the 4th International Congress on Catalysis*, Moscow (1968), p. 248.
102. M. Lacroix, G. M. Pajonk, and S. J. Teichner, *Bull. Soc. Chim.*, 84 (1981).
103. W. C. Conner, J. Cevallos-Candau, and D. Lenz, unpublished results.
104. D. D. Beck and J. M. White, *J. Phys. Chem.*, 88:174 (1984).
105. D. D. Beck and J. M. White, *J. Phys. Chem.*, 88:2764 (1984).
106. R. Kramer and M. Andre, *J. Catal.*, 58:287 (1979).
107. D. Bianchi, M. Lacroix, G. M. Pajonk, and S. J. Teichner, *J. Catal.*, 59:467 (1979).
108. M. Lacroix, G. M. Pajonk, and S. J. Teichner, *Bull. Soc. Chim.*, 258 (1981).
109. T. C. Sheng and I. D. Gay, *J. Catal.*, 71:119 (1981).
110. D. H. Lenz and W. C. Conner, *J. Catal.* (1986), in press.
111. A. J. Robell, E. V. Ballou, and M. Boudart, *J. Phys. Chem.*, 68:2748 (1964).
112. M. Mestdagh, W. Stone, and J. Fripiat, *J. Phys. Chem.*, 76:1220 (1972).
113. R. E. Taylor, M. M. Silva-Crawford, and B. C. Gerstein, *J. Catal.*, 62:401 (1980).
114. A. Cirillo, Ph.D. dissertation, University of Wisconsin-Milwaukee (1979).
115. W. C. Conner, Jr., J. F. Cevallos-Candau, N. Shah, and V. Haensel, in *Proceedings of the First International Symposium on the Spillover of Adsorbed Species* (G. M. Pajonk, S. J. Teichner, and J. E. Germain, eds.), Elsevier, Amsterdam (1983), p. 31.
116. J. F. Cevallos-Candau and W. C. Conner, *J. Catal.* (1986), in press.
117. N. E. Lobashina, N. N. Savvin, and I. A. Myasnikov, *Dokl. Akad. Nauk SSSR*, 268:1434 (1983); *Kinet. Katal.*, 24:747 (1983).
118. R. V. Dmitriev, A. N. Detjuk, C. M. Minachev, and K. H. Steinberg, in *Proceedings of the First International Symposium on the Spillover of*

Adsorbed Species (G. M. Pajonk, S. J. Teichner, and J. E. Germain, eds.), Elsevier, Amsterdam (1983), p. 17.
119. R. J. Kokes and A. L. Dent, *Adv. in Catal.*, 22:1 (1972).

13 | Strong Metal-Support Interactions

ROBBIE BURCH

University of Reading, Reading, Berkshire, England

I.	METAL-SUPPORT INTERACTIONS: HISTORICAL PERSPECTIVE	348
II.	STRONG METAL-SUPPORT INTERACTIONS	349
III.	CHEMISORPTION OF H_2 AND CO: EFFECT OF HIGH-TEMPERATURE REDUCTION	349
	A. Oxide Supports Other Than TiO_2	350
	B. Heats of Adsorption of H_2 and CO	351
	C. Recovery of CO and H_2 Adsorption Capacity	355
IV.	CATALYTIC PROPERTIES OF SMSI METALS	357
	A. Structure-Insensitive Reactions	357
	B. Structure-Sensitive Reactions; Hydrogenolysis and Isomerization of Hydrocarbons	359
	C. Selectivity Changes in Hydrocarbon Reactions	359
	D. Selectivity in Hydrogenolysis Reactions	360
	E. Reactions of Methylcyclopentane and of Neohexane	360
	F. Reactions of Butane	360
	G. CO/H_2 Reaction	363
V.	ROLE OF TITANIA IN THE CO/H_2 REACTION: SMSI OR SYNERGY?	363
	A. Indirect Action	363
	B. Direct Action	364
VI.	ORIGIN AND NATURE OF SMSI	364
	A. Electronic Factor in SMSI	365
	B. Localized Electronic Factors	366
	C. Structural Effects in SMSI	366
	D. Encapsulation, Decoration, and Support Migration	366
VII.	EFFECT OF SUPPORT MIGRATION ON CHEMISORPTION AND CATALYTIC PROPERTIES	367
	A. Suppression of Adsorption	367
	B. Changes in Catalytic Activity	367
VIII.	CONCLUDING REMARKS	368
	REFERENCES	368

I. METAL-SUPPORT INTERACTIONS: HISTORICAL PERSPECTIVE

Supported metals are used extensively as heterogeneous catalysts, the traditional role of the support being to disperse and stabilize the small metal particles. However, it has been known for some years, and suspected for much longer, that interaction between a metal and its support is a frequent occurrence [1,2]. Boudart and Djega-Mariadassan [3] have identified 10 known types of metal-support interaction (see Table 1). Some of these interactions (Table 1, items 3 to 6) involve the support as an active participant in a catalytic reaction. Others result in a modification of the properties of the metal or the support, or in the creation of a unique type of active site at the metal-support interface.

It is almost 60 years since Schwab and Pietsch [4] identified the interfacial region as a center of anomalous catalytic activity, and the concept is still invoked to explain unusual effects in supported catalysts [5]. After the original Schwab work there followed a succession of isolated observations in which the importance of metal-support interactions was recognized. The most detailed early investigations of these effects were made independently by Schwab [6] and Solymosi [7], who came to the conclusion that support effects were a consequence of electronic interactions. Although the current value of some of this early research is questionable because of the absence then of techniques for reliable characterization of the systems, the suggestion that some metal-support effects are electronic in nature has survived and is still being advanced [8-16].

Geometric and structural factors have also been invoked to account for the occurrence of metal-support interactions. In addition to obvious structural influences of the support in creating or maintaining a particular distribution of small metal particles, the support may alter the morphology of the metal particles. There is evidence that this may affect the catalytic activity and selectivity of the metal [17,18].

There is a need to differentiate between real and apparent metal-support effects [1]. The latter category includes specific particle-size effects, bifunctional, and spillover catalysis. To this must be added poisoning of the metal by impurities, or encapsulation by the support. These interactions can produce a change in the observed catalytic properties. However, they are not genuine metal-support interactions.

TABLE 1 Known Examples of Metal-Support Interactions

1. Genesis and preservation of high metal dispersion
2. Metal particle shape
3. Dual function—the reaction starts on the metal
4. Dual function—the reaction starts on the support
5. Spillover from metal to support
6. Spillover from support to metal
7. Contamination of the metal by the support
8. Interfacial effects—adlineation
9. Modification by the metal of the catalytic activity of the support
10. Modification by the support of the catalytic activity of the metal

Source: Ref. 3.

13. Strong Metal-Support Interactions

An unambiguous definition of a metal-support interaction is required. Foger [8] has suggested the following: "a *direct* influence of the support on the chemisorption and catalytic properties of the metal phase either by stabilising unusual metal particle structures, by changing the electronic properties due to electron transfer processes between the metal particles and the support, or chemical bonding—compound formation—between metal and support."

Bond [19] has pointed out the diversity of metal-support effects and suggested the following classification:

1. Weak metal-support interactions (WMSI), found with refractory, nonreducible oxides, and involving van der Waals interactions, and producing only minor modification in the metal (e.g., a change in particle structure)
2. Medium metal-support interactions (MMSI), found with metal particles in zeolite channels [20]
3. Strong metal-support interactions (SMSI), found for many metals supported on reducible transition metal oxides after reduction at moderately high temperatures

II. STRONG METAL-SUPPORT INTERACTIONS

Interest in strong metal-support interactions originated with the observation [21, 22] that reduction of supported metal compounds by hydrogen at high temperatures could, in certain instances, result in unusual and unexpected effects. The whole concept of SMSI has evolved from studies of the effects of high-temperature hydrogen reduction. However, it is now recognized that SMSI is much more than just a simple hydrogen effect, so it is appropriate here to provide a more comprehensive discussion of this phenomenon. Evidence for a change in catalytic properties with increasing temperature of reduction was first observed for alumina-supported Pt [21, 22]. However, much larger effects, produced at lower temperatures, were reported by the Exxon group [9-11, 23] for transition metals supported on transition metal oxides (TiO_2, V_2O_3, Nb_2O_5, Ta_2O_5). The term "strong metal-support interaction" (SMSI) was coined by Tauster et al. [9] to describe the interaction "consisting of a covalent bond between the metal atoms of the supported phase and . . . cations . . . of the support." The term has become established in the vocabulary of catalysis even though there is now a school of thought which believes SMSI to be an artifact that should be given a proper burial. At a more general level, SMSI is taken to be the state into which a metal supported on a reducible oxide enters after heating in hydrogen at elevated temperatures. We shall discuss later the nature of SMSI, but first let us accept the existence of such a state and consider the consequences of this so-called SMSI on the chemisorptive and catalytic properties of metals.

III. CHEMISORPTION OF H_2 AND CO: EFFECT OF HIGH-TEMPERATURE REDUCTION

Metals supported on TiO_2 have a greatly reduced capacity to adsorb H_2 or CO after reduction at temperatures above about 750 K. A detailed survey of chemisorption data for TiO_2-supported metals has been published previously [1]. All the TiO_2-supported metals that have been studied behave similarly, although there are some differences in detail and more recent work [24] has shown some metals to have different sensitivities to the reduction treatment.

TABLE 2 Adsorption of H_2 and CO on Metals Supported on "SMSI Oxides"

Metal	Support [surface area $(m^2 g^{-1})$]	T_{red} (K) [time (h)]	H/M	CO/M	Ref.
10% Ni	Nb_2O_5 [10]	573 [1]	0.16		25
		773 [1]	0.02		25
2% Ni	Nb_2O_5 [10]	573 [1]	0.03		25
		773 [1]	0.002		25
0.77% Pd	La_2O_3 [12.6]	448	0.056	0.083	16
		773	0.044	0.053	16
0.74% Pd	CeO_2 [15.9]	448	0.078	0.072	16
		773	0.046	0.014	16
0.70% Pd	Pr_6O_{11} [10.1]	448	0.085	0.064	16
		773	0.021	0.014	16
0.81% Pd	Nd_2O_3 [23.9]	448	0.071	0.033	16
		773	0.039	0.022	16
0.77% Pd	Sm_2O_3 [8.4]	448	0.102	0.079	16
		773	0.099	0.035	16
0.61% Pd	Eu_2O_3 [37.1]	448	0.087	0.148	16
		773	0.091	0.065	16
1.3% Rh	ZrO_2	525 [2]	0.75		26
		673 [2]	0.36		26
		873 [2]	0.05		26
1.0% Rh	ZrO_2	525 [2]	0.35		26
		673 [2]	0.04		26
		873 [2]	0.01		26

A. Oxide Supports Other Than TiO_2

The early Exxon papers [9,11] showed that a suppression of H_2 and CO adsorption could occur with other binary oxides, such as V_2O_3, Ta_2O_5, Nb_2O_5, and with ternary oxides, such as $BaTiO_3$ and $ZrTiO_4$. Table 2 gives an updated survey of chemisorption data on these "SMSI-oxide" supports.

The results in Table 2 show that hydrogen adsorption is suppressed after reduction at high temperatures, there being a large decrease for Nb_2O_5 and and ZrO_2, and a smaller, but nevertheless clear, effect for La_2O_3, CeO_2, Pr_6O_{11}, and Nd_2O_3. The adsorption of CO has not been investigated to the same extent. For the rare earth oxides Mitchell and Vannice [16] have reported a suppression of CO adsorption over all the oxides investigated (see Table 2). It is unfortunate that more data on CO adsorption are not available since the CO/H_2 reaction is the only one where it would appear that the rate is enhanced by SMSI (see later).

Table 3 summarizes the data reported for the CO/H adsorption ratio on metals supported on "SMSI oxides" after low- and high-temperature reduction. There is a problem when analyzing these data due to the variability in the results. Thus, for Pd/TiO_2 the CO/H ratio sometimes increases and sometimes decreases as the reduction temperature is increased. Even under equivalent conditions the CO/H ratio varies quite significantly; for example,

TABLE 3 CO/H Ratio for Metals After Low- and High-Temperature Reduction

Metal (wt %)	Support	CO/H LTR	CO/H HTR	Refs.
2% Ir	TiO_2	—	1.0	23
		0.74	—	9, 11
6.8% Pt	TiO_2	1.08	0.94	27
2% Ru	TiO_2	2.78	1.83	9, 11
		1.62	2.00	9, 11
		0.55	0.70	28
0.95% Rh	TiO_2	2.21	7.13	29
2% Pd	TiO_2	0.57	0.40	9, 11
1.86% Pd	TiO_2	1.28	2.33	30, 31
2% Pd	TiO_2	1.07	2.00	32
		1.33	1.00	33
1.9% Pt	TiO_2	0.57	0.50	34
TiO_x/Pt	TiO_2	0.32	0.89	34
1.9% Pt	TiO_2	0.69	0.83	35
10% Fe	TiO_2	0.58	0.88	36
0.77% Pd	La_2O_3	1.48	1.20	16
0.74% Pd	CeO_2	0.92	0.37	16
0.70% Pd	Pr_6O_{11}	0.75	0.67	16
0.81% Pd	Nd_2O_3	0.46	0.56	16
0.77% Pd	Sm_2O_3	0.77	0.35	16
0.61% Pd	Eu_2O_3	1.70	0.71	16

after a low-temperature reduction (LTR) of Pd/TiO_2, values from 0.57 to 1.33 have been reported. Caution is required, therefore, if changes in the adsorption of CO and H_2 are taken as evidence for parallel changes in surface properties, which, in turn, are assumed to account for changes in the activity of an SMSI catalyst for the CO/H_2 reaction.

For most Pd/rare earth oxides the CO/H ratio decreases as T_{red} is increased. The few data available for other metals show increases in the CO/H ratio for Ir, Rh, Fe, and Pt and a decrease for Ru.

B. Heats of Adsorption of H_2 and CO

The decrease in the quantity of H_2 or CO adsorbed on metals in the SMSI state is well established. It is also clear from Table 3 that in the SMSI state the adsorption of CO is generally enhanced as compared with H_2 adsorption (i.e., the CO/H ratio usually increases). It is important, therefore, in terms of catalytic activity to know whether the heat of adsorption of the residual H_2 or CO is altered since this could be one reason for the enhanced activity of TiO_2-supported catalysts for the CO/H_2 reaction (see later). It is important to note, however, that when a catalyst in the SMSI state is exposed to a CO/H_2

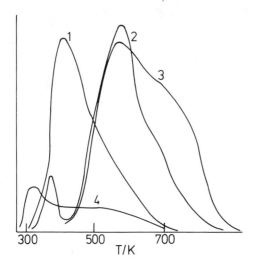

FIGURE 1 Temperature-programmed desorption of H_2 from Ni catalysts. 1, Ni/SiO_2; 2, Ni/TiO_2 after low-temperature reduction; 3, Ni/Al_2O_3; 4, Ni/TiO_2 after high-temperature reduction. (After Ref. 37.)

mixture under reaction conditions the SMSI state may be at least partially destroyed.

Temperature-programmed desorption (TPD) spectra of H_2 from Ni/TiO_2 [37] and of H_2 and CO from TiO_x/Ni [38] have been reported but give conflicting information. Figure 1 shows spectra reported by Weatherbee and Bartholomew [37] for the TPD of H_2 from Ni on different supports. These results show that the TiO_2 is intermediate between SiO_2 and Al_2O_3 after reduction at 673 K (R673). For details, see Table 1 in Chapter 5.

When the Ni/TiO_2 is put into the SMSI state (see Fig. 1) the size of the high-temperature peak declines and the low-temperature peak moves to about 323 K. This suggests that when Ni is fully in the SMSI state the heat of adsorption of H_2 is greatly reduced. However, this is probably not very relevant to the CO/H_2 reaction since the SMSI state is unstable under reaction conditions (see Section IV). Consideration of the R673 data is more appropriate and this shows that compared with SiO_2 both the TiO_2- and Al_2O_3-supported Ni adsorb some hydrogen with a large heat of adsorption (see Table 1 of Chapter 5). The hydrogen in this state may be important in the CO/H_2 reaction (see Chapter 20).

Raupp and Dumesic [38] have studied Ni foil partially covered with TiO_x. They observe a complex behavior with three adsorption states for H_2, one less strongly and two more strongly held than on a clean Ni surface. Adsorption into the most strongly held state was activated and it is suggested that dissociative adsorption of H_2 on the metal surface is followed by diffusion of atomic hydrogen to adjacent Ti^{3+} sites, where it may be associated with hydroxyl groups. If this is correct, the reduced TiO_x could act as a reservoir for H atoms.

Raupp and Dumesic [38] also studied the TPD of CO. Figure 2 shows that in contrast to the H_2 results, the presence of a TiO_x surface species weakens significantly the CO adsorption, the T_{max} moving down from about 400 K to 300 K. This corresponds to a decrease in the binding energy of about 50 kJ mol^{-1}. A significant conclusion is that all CO states on TiO_x/Ni are weaker than those on clean Ni, whereas there are H binding states on TiO_x/Ni which are stronger than on clean Ni. This should result in a decrease in the CO/H

13. Strong Metal-Support Interactions

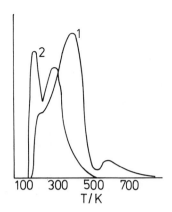

FIGURE 2 Temperature-programmed desorption of CO from Ni catalysts. 1, Clean Ni; 2, Ni partially covered with TiO_x. (After Ref. 38.)

ratio, which is contrary to the general trend seen in Table 3. However, it is possible that the higher CO/H ratios found in adsorption experiments may be observed because the adsorption of H is slow, so the amount of H adsorbed may be underestimated.

Menon and Froment [22a] have investigated the TPD of H_2 from Pt/TiO_2 catalysts and observed that as the reduction temperature was increased there was a change in the shape of the profile with more H_2 being desorbed at high temperatures (see Fig. 5 in Chapter 4, p. 128). Belton et al. [39] have studied the TPD of H_2 and CO from Pt and Rh films on TiO_2. For Rh/TiO_2 encapsulation decreases the CO peak area by 68%, but there is no change in the temperature of the peak maximum (see Table 4). The same trend is found for H_2. For Pt/TiO_2, however, there is a shift downward of T_{max} of 45 K for CO and 30 K for H, which is attributed partly to electronic interactions. In contrast, however, Ko and Gorte [40] in a study of TPD from Pt foil partially covered with TiO_x found no shifts in the positions of the desorption peaks for either CO or H_2. They concluded that the uncovered parts of the Pt surface have normal adsorption properties. Vannice and co-workers [35,41] have studied Pt/TiO_2 catalysts

TABLE 4 Temperature-Programmed Desorption Data for Rh and Pt Films on TiO_2

Sample	Metal thickness (nm)	Adsorbate	Annealing temp. (K)	T_{max}	Peak area
Rh/TiO_2	3.0	CO	525	490	1
			760	490	0.32
		H_2	525	270	1
			760	270	0.20
Pt/TiO_2	3.0	CO	525	400	1
			760	355	0.33
		H_2	525	270	1
			760	240	0.36
	0.24	H_2	130	263	1
			370	230	0.30

Source: After Ref. 39.

by infrared (IR) spectroscopy and concluded that in the SMSI state adsorption sites are altered to allow H_2 to compete more favorably with CO. This is indicated by the fact that CO is displaced from the surface by H_2 at 300 K.

Foger [42] has studied the TPD of H_2 from Ir/TiO_2. Although the quantity of H_2 adsorbed decreases as the reduction temperature is increased, the position of T_{max} does not change. Tau and Bennett [36] have studied the TPD of H_2 from Fe/TiO_2 and observed that the T_{max} moves down about 30 K after reduction at 773 K.

Vannice and Chou [32] have made direct measurements of the heats of adsorption of H_2 and CO on Pd/TiO_2 by differential scanning calorimetry. Their results, summarized in Table 5, show that the heat of adsorption of H_2 on Pt/TiO_2 after low-temperature reduction (LTR) is initially larger than for Pd on other supports. After high-temperature reduction (HTR) the heat of adsorption decreases. However, even when the Pd is in the SMSI state, the heat of adsorption is higher than for Pd/SiO_2 or Pd/Al_2O_3. The heat of adsorption of CO on Pd/TiO_2 after LTR is comparable to that over conventional Pt/SiO_2 or Al_2O_3 catalysts, but then decreases as the reduction temperature is increased. However, once again, even in the SMSI state, the CO that does adsorb is bound as strongly as on a $Pd/SiO_2-Al_2O_3$ catalyst.

Vannice et al. [43-45] have used calorimetry to measure the heat of adsorption of H_2 and CO on Pt catalysts. Table 5 includes some of their results. When Pt/TiO_2 is compared with other Pt catalysts it is found that even after LTR the heat of adsorption of H_2 is decreased significantly and the heat of adsorption of CO is also slightly lower. After HTR the heat of adsorption of H_2 is very small, but the heat of adsorption of CO is comparable to the value obtained for Pt/Al_2O_3.

These results conflict with those of Belton et al. [39], who observed a decrease in the binding energy for CO, and with those of Ko and Gorte [40], who observed no change in the binding energy for H_2. It is unclear whether these differences originate in the experimental procedure (calorimetric versus temperature-programmed desorption), or because of the different types of catalyst used (foil, film, powder).

To summarize, there are some metals (e.g., Ir, Rh) where SMSI does not seem to have much effect on the heat of adsorption of H_2 or CO, but in most cases (e.g., Ni, Fe, Pt, Pd) changes are observed. In the case of Pd the heat of adsorption of H_2 is increased and of CO is unchanged; with Pt the calorimetric data indicate that the heat of adsorption of H_2 is greatly decreased but CO is unaffected; and with Ni H_2 binds more strongly and CO more weakly in the SMSI catalyst. How these changes in binding energies will affect the activity of a metal for the CO/H_2 reaction will depend on which rate-determining step operates for a particular metal. Thus, over Ni where CO adsorption is normally very strong, weakening of the Ni-CO bond could be beneficial, whereas over Pd or Pt a strengthening of the M-CO bond, or at least an adjustment in the relative binding energies of CO and H_2 may be advantageous.

There are, of course, difficulties in relating these thermodynamic data to catalytic processes and it needs to be emphasized that adsorption data are dominated by the most abundant surface species present. Wang et al. [30] and Vannice et al. [35, 41] have suggested that in the CO/H_2 reaction over metals perhaps as little as 1% of the metal surface contains active sites. This is a view that is gaining in popularity [46]. If it is correct, then clearly the activity of a metal could change dramatically without the overall adsorption characteristics appearing to change very much. The converse is also true. Furthermore, it is probably also the case that strongly bound adsorbed species, which are those most easily detected in calorimetric experiments, are the least important intermediates in catalytic reactions. It is worth recall-

13. *Strong Metal-Support Interactions*

TABLE 5 Heats of Adsorption of H_2 and CO on Pd/TiO_2 and Pt/TiO_2 Catalysts

Support	Reduction temp. (K)	H/M	CO/M	Heat of adsorption (kJ mol^{-1}) H_2	CO
		Pd catalysts [32]			
SiO_2	673	0.12	0.12	88	96
		0.12	0.12	113	100
SiO_2-Al_2O_3	673	0.31	0.29	88	71
		0.31	0.29	92	67
Al_2O_3	673	0.28	0.27	105	71
TiO_2	448	0.32	0.34	138	100
	773	0.004	0.017	109	71
		Pt catalysts [43-45]			
SiO_2	723	0.17	0.15	144	133
SiO_2-Al_2O_3	723	0.39	0.35	114	96
Al_2O_3	723	0.74	0.46	115	100
TiO_2	473	1.00	0.68	87	83
	773	0.04	0.06	24	97

ing the well-known "volcano curve" relating catalytic activity to heats of adsorption. The most "active" surface species are adsorbed neither too strongly nor too weakly.

C. Recovery of CO and H_2 Adsorption Capacity

Reoxidation Treatment

In the early Exxon work it was recognized [9] that the SMSI state could be reversed by reoxidation at 673 K followed by reduction at a low temperature. Whether reversibility is possible under less severe reoxidation conditions has been disputed, and seems to depend on the metal, the metal loading, and the precise details of the reoxidation treatment. The effect of H_2-O_2 cycles on the reproducibility of hydrogen chemisorption or titration on supported Pt and Pt-Re catalysts has been known for some time [22]. With regard to Pt/TiO_2 catalysts, Menon and Froment [22a] have reported that a few air-H_2 cycles at 293 K restored part of the H_2 chemisorption capacity.

Ellestad and Naccache [13] and Meriaudeau et al. [15] observed that when Pt, Ir, and Rh in the SMSI state are contacted with O_2 at 293 K (i.e., oxygen titration experiment), they recover their capacity to adsorb hydrogen. Kunimori et al. [47] obtained 40% recovery for Pt after exposure to O_2 at room temperature. Foger [42] has reported that exposure of Ir to 1% O_2 in He at 473 K gives total recovery. Herrmann and Pichat [48,49] obtained essentially complete recovery for Pt after H_2/O_2 titration at 295 K. Jiang et al. [50] have observed significantly higher O/M ratios than H/M ratios for Ni and Ni-Fe/TiO_2 catalysts, although it is unclear whether or not part of the adsorbed oxygen is taken up by the partially reduced titania support.

Although the reversibility of SMSI even under relatively mild conditions has been observed there are still some interesting differences observed depending on the precise experimental procedure adopted. Thus Anderson et al. [24] have found that SMSI Pt or Rh adsorb O_2 at 298 K, suggesting a reversal of the SMSI state. Indeed, this O_2 may be titrated with H_2, and after outgassing at 673 K normal H_2 adsorption is observed. However, if the adsorbed O_2 is removed by re-reducing at 573 K (i.e., at a temperature well below the temperature required to create SMSI in a fresh catalyst), the SMSI state is partially recreated. The model proposed to account for these different effects is reproduced in Fig. 3.

Anderson et al. [24] have also noted that Rh/TiO_2 and Pt/TiO_2 behave differently. Pt enters the SMSI state more readily and more completely than Rh, and it is more difficult to destroy the SMSI state with Pt. For example, whereas 55% of the adsorption capacity of Rh is recovered after exposure to O_2 at 298 K, there is only a 13% recovery with Pt. Furthermore, after a single O_2/H_2 titration, Rh recovers almost completely its capacity to adsorb H_2, whereas the recovery for Pt is less than 25%. It was concluded that the stability of the SMSI state is controlled by the strength of the $M-TiO_x$ bond.

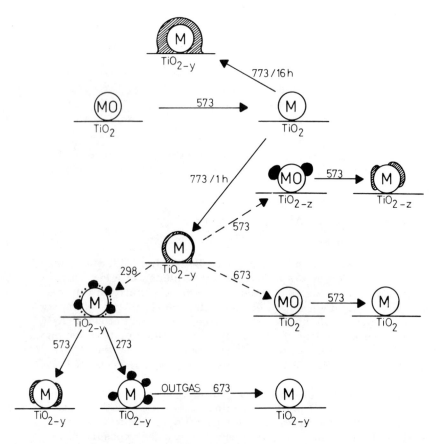

FIGURE 3 Changes in the physical state of titania-supported catalysts after various reduction and oxidation treatments. M, metal particles; MO, metal oxide particles; ▨, surface TiO_x phase; ●, surface TiO_2 phase; •, adsorbed oxygen atoms. Solid arrows indicate reduction, dashed arrows indicate oxidation, numbers indicate temperatures in K. (After Ref. 55; reproduced by permission of Elsevier Science Publishers.)

CO/H_2 Treatment

Particularly significant is the reversibility of the SMSI state under the reaction conditions used for the CO/H_2 reaction. This possibility has been recognized for some time [47, 51-53] and has been partially confirmed by different workers. Thus, Duprez and Miloudi [53] have demonstrated partial reversibility of the SMSI state for Rh, Pt, and Ni by pulsing H_2O over titania-supported catalysts at 773 K. Moreover, even at ambient temperature some water was retained and hydrogen released. Kunimori et al. [47] have shown that the activity of a Pt/TiO_2 catalyst for the CO/H_2 reaction actually increases from an initially low value [turnover number (TON), 2.7×10^{-4} s^{-1}] when the Pt is fully in the SMSI state to a much higher value (TON, 9.7×10^{-4} s^{-1}) after about 10 h of exposure to a CO/H_2 mixture at 503 K.

Anderson et al. [54] have shown that exposing an SMSI Ir/TiO_2 catalyst to a CO/H_2 mixture at 548 K for 20 min produces almost complete recovery of the H_2 adsorption capacity. Further work [55] has confirmed that recovery may be very rapid in a CO/H_2 mixture, and that even in a H_2O/N_2 mixture partial recovery is possible at these relatively low temperatures. However, in general, the recovery is not instantaneous and the rate varies from one metal to another, being much slower and less complete for Pt than for Rh. The degree of recovery depends on the reducing power of the H_2/H_2O couple generated by the reaction. Consequently, when "SMSI" catalysts are used in the CO/H_2 reaction, whether the catalyst remains in the SMSI state or is converted partially or totally into the normal metallic state will depend on the reaction conditions (i.e., temperature, pressure, percentage conversion).

IV. CATALYTIC PROPERTIES OF SMSI METALS

The general trends in catalytic activities for SMSI metals have been reviewed recently [1]. The main features may be summarized as follows:

1. For structure-insensitive reactions, such as hydrogenation, the activity decreases by less than about an order of magnitude and the selectivity for partial hydrogenation increases [56, 57].
2. For structure-sensitive reactions, such as hydrogenolysis, the activity decreases by several orders of magnitude.
3. For the CO/H_2 reaction the activity increases by about one order of magnitude and the selectivity toward higher hydrocarbons is enhanced.

It should be cautioned that whereas for reaction types 1 and 2 the SMSI state is stable under reaction conditions, this is not the case for the CO/H_2 reaction.

A. Structure-Insensitive Reactions

Table 6 summarizes the data available on hydrogenation and dehydrogenation reactions. With Pt and Ir the decrease in activity after HTR is about an order of magnitude for all the reactions studied. Pd similarly shows a large decrease in activity for the hydrogenation of benzene. Rh, however, behaves differently and shows only a small decrease in activity for hydrogenation and dehydrogenation reactions. The contrast between Pt and Rh parallels the different sensitivities of these two metals to SMSI effects (see earlier). Pt is more stable than Rh in the SMSI state. The high activity of SMSI Rh may indicate that particles in a partial SMSI state remain active for hydrogenation/dehydrogenation reactions.

TABLE 6 Relative Activities of Titania-Supported Metals for Structure-Insensitive Reactions After Reduction at Low or High Temperatures

Metal	LTR/HTR[a]	Reactant	Ref.
Pt	12.5	Benzene	58
	15.4	Benzene	27
	11.4	Benzene	15
Rh	1.3	Benzene	13
	13.4	Benzene	15
Pd	51.4	Benzene	33
Ir	∞	Benzene	15
	12.5	Benzene	13
Rh	1.3	Cyclohexane	28
	2.1	Cyclohexane	15
Ir	4.6	Cyclohexane	15
Pt	8.0	Cyclohexane	15
	8.1	Ethylene	15
Ir	9.4	Ethylene	15
Pt	5.0	Styrene[b]	15
Ir	14.1	Styrene[b]	15
Rh	3.4	Styrene[b]	15

[a] Ratio of activities after reduction at low (LTR) or high (HTR) temperatures.
[b] Product was ethyl cyclohexane.
Source: Ref. 33.

The data given in Table 6 for benzene hydrogenation over Pd/TiO_2 disguise the fact that Pd/TiO_2, even in the SMSI state, has a turnover number which is comparable to that of Pd/Al_2O_3 or Pd/SiO_2, and an order of magnitude higher than that of Pd powder. The fact is that non-SMSI Pd/TiO_2 has an enhanced activity (compare the CO/H_2 reaction). Vannice and Chou [33] have attributed this to the presence of acidic sites at the metal-support interface. Evidence that Cl^- ions may contribute to the formation of acidic sites was obtained by adding HCl to a chloride-free catalyst. An increase in turnover number by a factor of 4 was observed. These authors propose that the active site is a Ti^{3+} cation adjacent to a Pd particle. Benzene, being a basic molecule, adsorbs at the acidic site and hydrogenation occurs via spillover of hydrogen.

A sequence of this model is that any particle of a metal which is in a partial SMSI state may have a high activity. Of course, this means that the observed decrease in activity for hydrogenation reactions may be misleading and a metal fully in the SMSI state may have a very low activity (note the different data for Ir in Table 6). In the case of Rh, mentioned above, a 90% suppression of H_2 adsorption would still leave 10% of the Rh accessible to H_2. If this residual Rh is in a partial SMSI state, then an enhanced hydrogenation activity at the metal-support interface could account for the fact that the measured activity is comparable to that of all the original Rh.

B. Structure-Sensitive Reactions; Hydrogenolysis and Isomerization of Hydrocarbons

Den Otter and Dautzenberg [21] have reported that after reduction at 973 K Pt/Al$_2$O$_3$ catalysts show an increased selectivity toward aromatization and a decreased selectivity toward hydrogenolysis. These changes in selectivity with increasing reduction temperature are accompanied by a decrease in hydrogen coverage. Paál [59] has interpreted these trends in terms of competitive adsorption; that is, the reaction which requires less hydrogen on the surface (aromatization in this case) becomes predominant. It is reasonable to expect similar trends to be observed with metals in the SMSI state. There now seems little doubt that metals which are fully in the SMSI state have exceedingly low activities for hydrocarbon hydrogenolysis reactions (Ref. 1 and references therein). The residual activity sometimes reported for SMSI metals probably reflects as much as anything the activity of a small fraction of the metal which remains in a non-SMSI state. Justification for this view comes from the fact that compared with the changes in activity, the effect of SMSI on the activation energies and product distributions is insignificant [24,60,61]. It is interesting nevertheless to examine the selectivity of metals in a partial SMSI state for hydrogenolysis reactions.

C. Selectivity Changes in Hydrocarbon Reactions

Because hydrogenolysis reactions are more severely curtailed than hydrogenation/dehydrogenation reactions, selectivity differences are observed for SMSI metals. Tauster et al. [23] have reported that Ru and Ir have greatly enhanced selectivities for heptane reforming at high temperatures (about 750 K) and pressures (20 bar). Table 7 shows that hydrogenolysis is greatly suppressed and that the selectivity to isomerization and aromatization products is much enhanced.

Haller et al. [28] have reported a modest increase in selectivity for dehydrogenation versus hydrogenolysis of cyclohexane with Rh/TiO$_2$ catalysts. In contrast, with Pt/TiO$_2$ Meriaudeau et al. [58] find only a small decrease in selectivity for hydrogenolysis (from 22.6% to 18.7%). The major changes they observe are a decrease in isomerization (47.3% to 20.0%), an increase in ring closure to methylcyclopentane (26.0% to 43.3%), and an increase in the formation of C$_6$-ring products (4.1% to 18.10%). Dauscher et al. [62] have studied the isomerization of various labeled methylpentanes on Pt/TiO$_2$ catalysts. Since reduction was limited to 663 K their results should correspond to catalysts in a partial or non-SMSI state. Even so, they observe drastically reduced turnover numbers (lower by a factor of 50) as compared with Pt/Al$_2$O$_3$ and 10 times lower than for Pt black. Only a slight increase

TABLE 7 Reforming of n-Heptane over Ir and Ru Catalysts

Catalyst	Selectivity		
	< C$_7$	iso-C$_7$	Aromatics
Ir/Al$_2$O$_3$	98.4	0.5	1.1
Ir/TiO$_2$	87.2	5.0	7.8
Ru/Al$_2$O$_3$	100	—	—
Ru/TiO$_2$	38.7	51.6	9.7

Source: Ref. 23.

in selectivity for isomerization versus hydrogenolysis was observed. Evidence was obtained that under the same experimental conditions the mechanism for the isomerization over Pt/TiO$_2$ changes to a bond shift, in contrast to Pt/Al$_2$O$_3$, where the C$_5$-cyclic mechanism is predominant. This change in mechanism was attributed to an electronic factor. However, interfacial effects, such as those reported by Kramer and Zuegg [63,64], may also contribute to changes in product distribution.

D. Selectivity in Hydrogenolysis Reactions

As mentioned earlier the product distribution obtained in hydrogenolysis reactions over SMSI catalysts often differs only slightly from that found with normal metals. This seems to be true for the hydrogenation of n-hexane irrespective of which metal is chosen; similar results have been obtained with Ru [60], Rh [65,66], and Pt [65]. Alternative explanations for this behavior are: residual activity is due to a small fraction of metal in a non-SMSI state; or, the blockage of the surface of the metal with TiO$_x$ species may deactivate an SMSI catalyst in the same way that self-poisoning deactivates a normal metal. The latter explanation is preferred since, as we shall now see, small but definite changes in selectivity are observed when other reactants are used.

E. Reactions of Methylcyclopentane and of Neohexane

Changes in the product distribution for the ring opening of methylcyclopentane (MCP) over SMSI catalysts depends on the metal. With Pt, the SMSI catalyst is totally inactive for the MCP reaction even though the same catalyst hydrogenolyzes n-hexane [65]. Rh shows a small increase in the selectivity for the formation of n-hexane [65,66] and the 2MP/3MP ratio decreases. Ru in the SMSI state shows the largest change in product distribution [60]. No n-hexane is detected for either the normal or the SMSI Ru catalyst. However, the 2 MP/3MP ratio changes from 1.3 for a non-SMSI catalyst to 2.4 for an SMSI catalyst. Furthermore, a Ru catalyst in a partial SMSI state, and which is believed to be further poisoned by Cl ions gives the same product distribution as the SMSI catalyst.

The most pronounced changes in selectivity for Rh catalysts are found for the neohexane reaction [66]. Three different types of two-point adsorption are possible ($\alpha\beta$, $\alpha\gamma$, and $\alpha\gamma'$). In the hydrogenolysis reaction these produce neopentane, 2-methylbutane, and 2-methylbutane + 2-methylpropane, respectively. The results obtained by Schepers et al. [66] are summarized in Table 8. When the test is performed at about 560 K there is little change in the product distribution. However, at 620 K, the $\alpha\beta$ adsorption is suppressed and the $\alpha\gamma$ mode is enhanced in the SMSI catalyst. As yet there is no clear understanding of these various changes in selectivity, although the results are consistent with a blocking model for SMSI.

F. Reactions of Butane

Table 9 summarizes some data for the reaction of butane over various titania-supported metals. Activities have not been included because, as indicated earlier, it is probably only metal particles in a partial SMSI state which retain any activity. The selectivity obtained should reflect this state of affairs and the possibility that interfacial effects make a major contribution to the selectivity must not be overlooked.

Table 9 shows for Pt/TiO$_2$ that as the metal is transformed into the SMSI state the selectivity for isomerization decreases sharply. Among the hydro-

TABLE 8 Selectivity in the Neohexane Reaction over Rh/TiO$_2$ Catalysts

Reduction	Selectivity (%)			
	$\alpha\beta_{cr}$	$\alpha\gamma_{cr}$	iso	C3 + C4
LTR[a]	77.2	17.2	1.0	4.6
HTR[a]	79.8	17.6	0.7	1.9
HTR[a,b]	72.0	21.5	2.3	4.3
LTR[c]	59.7	28.9	0.8	10.6
HTR[c]	53.6	36.5	10.8	9.2
HTR[b,c]	43.8	40.6	3.4	8.0

[a]Tested at about 560 K.
[b]Reduced for 28 h.
[c]Tested at about 620 K.
Source: Ref. 66.

genolysis products there is little change in selectivity, with ethane remaining the major product. In the case of Rh/TiO$_2$ the HTR catalyst produces more C$_2$ and C$_3$ and less C$_1$. Foger [42] has studied impregnated Ir/TiO$_2$ catalysts and observed a significant decrease in selectivity for the formation of ethane with the SMSI catalyst. He also reports that the activity of the SMSI catalyst increases by 100-fold after 30 h on-stream. The partial reversibility of the SMSI state due to traces of O$_2$ or H$_2$O in the feed was excluded. Instead, it was proposed that the reaction itself is able to break the metal-support interaction, the active species being unsaturated fragments spilled over from the metal to the support. To test this possibility, pulses of butene were injected over the catalyst. The activity was measured and found to be about three times higher than before.

TABLE 9 Selectivity of TiO$_2$-Supported Metals for the Reaction of Butane

Metal	Reduction	Selectivity[a]				Ref.
		C1	C2	C3	iC4	
Pt	LTR	18	24	13	45	15
	HTR	27	45	28	0	15
Rh	LTR	51	46	3	0	15
	HTR	30	54	16	0	15
Ir	LTR	9	72	19	0	42
	HTR	b	56	b	0	42
	LTR	b	71	b	0	68
	HTR	b	33	b	62[c]	68
Ru	LTR	47	24	29	0	67
	HTR	43	38	19	0	67

[a]See original references for definitions of selectivities.
[b]No details given in reference.
[c]Includes 56% butenes.

Resasco and Haller [68] have studied ion-exchanged Ir and Rh catalysts. For Ir after LTR only hydrogenolysis was detected and the selectivity to ethane was 71% (see Table 9). After HTR the activity was very much lower (six orders of magnitude) and as a consequence isomerization and dehydrogenation products were detected. The selectivity to ethane was much lower at 33%. When Rh/TiO_2 catalysts were prepared by impregnation, although there was a large decrease in activity after HTR (almost five orders of magnitude), no isomerization or dehydrogenation products were detected. However, when the catalysts were prepared by ion exchange, some isomerization and a significant amount of dehydrogenation was observed. Clearly, the hydrogenolysis activity of a metal such as Rh or Ir is so large that isomerization and dehydrogenation are important only when all the Rh has been almost totally deactivated by SMSI.

Bond and Xu Yide [67,69] have studied Ru/TiO_2 catalysts over a range of temperatures. The product distribution is temperature dependent, but after LTR the approximate order is $S_1 > S_2 = S_3$, whereas after HTR it is essentially $S_1 = S_2 \gg S_3$. In this work Bond and Xu Yide also observe that it is difficult initially to put Ru into the SMSI state. However, after a reduction/oxidation/reduction cycle the SMSI state is obtained and the activity declines. It was also observed that after the intermediate oxidation a very high activity was obtained. It was suggested that chloride ions originally present on the catalyst after impregnation with Ru chloride solution affect the dispersion of the Ru by inhibiting the spreading of Ru oxide. It was proposed that Cl^- controls the proportions of particulate and dispersed forms of Ru. These two forms of Ru are thought to provide alternative mechanisms for the chemisorption and reaction of butane. It was suggested that particulate Ru produces comparable amounts of methane and ethane, whereas the dispersed Ru produces predominately methane and only a small amount of ethane.

Different reaction mechanisms have also been postulated to occur over SMSI Rh and Ir [42,68]. The change in selectivity for the hydrogenolysis of butane from ethane to methane has been interpreted [42,68] as being due to a change in mechanism from the C_2-unit mode to the iso-unit mode [70]. The essential difference between these two mechanisms is that the latter may occur on isolated metal atoms, whereas the former, which is claimed to be the mechanism over clean metal surfaces, may require a larger ensemble of metal atoms.

Resasco and Haller [71] have drawn attention to the similarity between the effects of SMSI and of alloy formation (between a group VIII metal and a group Ib metal), the implication being that both the creation of the SMSI state and the formation of such an alloy change the distribution of ensemble sizes. However, since SMSI seems to be created by migration of TiO_x moieties *onto* the surface of the metal rather than substitution *into* the surface, a more appropriate analogy would be with poisoning by strongly adsorbed species, such as S and As. The proposed changeover from the C_2-unit mode to the iso-unit mode would be consistent with this type of surface contamination. Of course, we cannot exclude the possibility that TiO_x species on the metal surface create localized electronic perturbations of the metal atoms and that it is this which is responsible for the observed changes in selectivity. However, this should produce changes in the activation energy, and this appears not to happen.

The appearance of dehydrogenation products over very inactive SMSI Rh and Ir catalysts prepared by ion exchange has led Resasco and Haller [68] to propose that the active site for dehydrogenation may be isolated Rh^+ ions. The fact that some metal oxides are known to be capable of catalyzing dehydrogenation reactions would support this contention. However, it is

13. *Strong Metal-Support Interactions*

difficult to exclude the alternative possibility that a heavily contaminated Rh metal surface is capable of removing hydrogen from a hydrocarbon but not of breaking C-C bonds. The formation of olefinic products in the gas phase reflects a change in the balance between the rate of dehydrogenation and the rate of hydrogenation. Presumably only a small amount of H_2 is adsorbed on an SMSI catalyst, so hydrogenation would be expected to be slow.

G. CO/H_2 Reaction

The CO/H_2 reaction has been studied in detail over titania-supported Fe [36,72], Co [72], Ni [5,73-81], Ru [72,82,83], Rh [29,84-86], Pd [30,31,54, 84,87,88], Ir [72], and Pt [34,35,41,47,58,72]. In most cases there is a substantial increase in the turnover number and this is often accompanied by an increase in the selectivity for the formation of higher hydrocarbons. The observation that metals which were unable to adsorb CO or H_2 could catalyze a reaction between these molecules presented a paradox at first. However, it is now known that metals in the SMSI state do not catalyze the CO/H_2 reaction for the very simple reason that under reaction conditions the SMSI state is unstable. As soon as the reaction mixture is introduced, the SMSI state is at least partially reversed [65]. The enhanced activity for the CO/H_2 reaction of titania-supported metals in the absence of SMSI is now well documented. It is more appropriate to consider the titania as providing a promoter action rather than a strong metal-support interaction.

V. ROLE OF TITANIA IN THE CO/H_2 REACTION: SMSI OR SYNERGY?

There is a consensus [30,35,80,85,86,89-94] that the rate-determining step in the formation of methane from CO/H_2 is the dissociation of CO. This seems very possible over metals such as Pd and Pt which do not dissociate CO readily. Vannice and co-workers have proposed a hydrogen-assisted dissociation of CO for Pd and Pt [30,35]. Metals such as Ni or Ru can dissociate CO, but there is evidence [80,94] that the rate of hydrogenation of surface carbon is much faster than the rate of hydrogenation of adsorbed CO. This is particularly true in the case of Ni/TiO_2 catalysts [80].

Assuming that the dissociation of CO is the rate-determining step, what is the role of titania? Two alternatives should be considered:

1. *Indirect action*. The titania modifies the metal and the reaction proceeds solely on metal sites—SMSI effect.
2. *Direct action*. The titania is an active constituent of the active site—synergy effect.

Indirect action can, in turn, create two different effects:

1a. A change in the electronic properties of a metal particle
1b. A change in the structure of the metal particle and hence in the distribution of surface sites with different coordination numbers

A. Indirect Action

Bulk Electronic Changes

There has been much discussion and speculation on the electronic factor in SMSI (see Section VI.A). In the context of the CO/H_2 reaction Vannice and

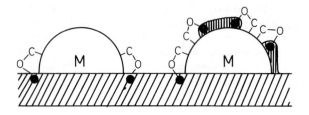

FIGURE 4 Model of active sites for the CO/H_2 reaction over Ni/TiO_2 catalysts. M, metal particle; ●, Ti^{3+} ion; ▥, TiO_x; ▨, TiO_{2-y}. (After Ref. 54.)

Sudhakar [34] have performed a very neat model experiment in which titania was deposited onto 1-μm Pt particles. The addition of as little as 1 monolayer of TiO_2 was sufficient to increase the turnover number by a factor of 4. This experiment, in which there is a massive excess of Pt to titania, would seem to exclude, as an explanation of the role of titania, bulk electronic changes in the Pt due to alignment of the Fermi levels at the metal-semiconductor junction. However, this does not rule out very localized perturbations at the interface (see later).

Structural Changes

Vannice and Sudhakar [34] have also argued that these same results exclude changes in the structure of the Pt as an explanation. This is also consistent with the fact that irrespective of which type of support is used, the activity of metals for the CO/H_2 reaction is relatively insensitive to particle-size effects.

B. Direct Action

When it was realized that titania-supported catalysts were very active for the CO/H_2 reaction in the absence of SMSI, Burch and co-workers [5, 54, 87] proposed that the role of the titania was to create new active sites at the metal-support interface which were unique for this reaction. The uniqueness of the CO/H_2 reaction was illustrated by the fact that a variety of other hydrogenation and hydrogenolysis reactions proceeded as normal. The active site originally proposed for the CO/H_2 reaction involved a metal atom and a Ti^{3+} ion exposed at an anion vacancy at the metal-support interface. Since migration of TiO_x onto a metal is now known to occur after high-temperature reduction the model has been refined to include interfacial sites on top of the metal (Fig. 4). Vannice and Sudakhar [34] have proposed independently on almost identical model. In both cases the preferred action involves binding a CO molecule via the C to a metal atom and via the O to a Ti^{3+} cation. Of course, a localized electronic interaction cannot be fully discounted [5].
The crucial point, however, is the creation of new sites at the metal-support interface. Rooney [46] has pointed out the parallels with this model to be found in homogeneous systems.

VI. ORIGIN AND NATURE OF SMSI

SMSI has been attributed to several causes and we consider these under the general headings: (a) electronic; (b) structural; and (c) encapsulation, decoration, and support migration. SMSI was originally attributed to a direct electron transfer between a reduced support and a metal particle [9, 15, 86, 95]. The change in adsorption and catalytic properties was believed to occur as a consequence of such an electron transfer. This model, which closely parallels the very much earlier electronic models of Schwab [6] and Solymosi [7], has been widely criticized on the grounds that charging effects will restrict electron transfer to such an extent that no sensible change in bulk properties would be anticipated [96]. However, a localized electronic interaction is possible in principle, so we now consider the evidence relating to an electronic factor in SMSI.

A. Electronic Factor in SMSI

X-ray photoelectron spectroscopy (XPS) and Auger spectroscopy have been used to study the electronic state of supported metals. Data have been reported which it is claimed provide evidence for a substantial amount of electron transfer between a metal and its support [29,74,97-102]. Supporting evidence has also been obtained from electron spin resonance (ESR) [103,104] and electrical conductivity measurements [12,14,48,49,100,105]. Equally, however, a large body of evidence has been accumulated which indicates no significant amount of electron transfer [106-109]. Furthermore, it has been argued that earlier XPS data which it was claimed showed evidence for electron transfer were incorrect because of particle size and relaxation effects [96].

The basis of the theoretical model [95] in which electron transfer from Ti^{3+} to Pt was first proposed has also been challenged recently [110]. Additional arguments against a bulk electronic effect have been put forward. For example, SMSI is still observed with large metal particles [5,102,111] and very similar effects are observed when a small amount of support is deposited on bulk metal [34,40].

The bulk electronic model is derived from the Schottky theory of metal-semiconductor junctions [112,113]. In essence this states that at thermodynamic equilibrium at the interface the Fermi levels of the metal and the semiconductor must be equal. If the work function of the electrons in the semiconductor is smaller than in the metal, electrons will flow from the semiconductor to the metal. Several authors have used this simple model to account for SMSI [15,48,49,86,105].

The direction of electron transfer apparently observed by XPS has been correlated with variations in the work function of different metals [15,36,58]. Chen and White [12] have correlated the occurrence of SMSI with the bulk electronic properties of the supports. They propose that supports which produce SMSI effects have high electrical conductivity (n-type or semimetals) and a low work function.

Resasco and Haller [71] have conducted an elegant test of the delocalized electronic model by reducing a Rh/TiO_2 catalyst with H atoms. Although reduction of the TiO_2 occurs under these conditions and causes an increase in the n-type conductivity no change in the capacity to adsorb H_2 was observed. Resasco and Haller conclude that the suppression of H_2 adsorption is not due to a change in the collective properties of the support.

There are two serious problems with a bulk electronic model. First, only a very small amount of electron transfer is required to compensate for the difference in work functions. Second, whereas in a semiconductor the low electron density means that an electron or positive hole is only screened out over quite large distances, the screening distance in a metal is extremely short. Calculations indicate that the electronic perturbation is screened out within two atomic distances [114,115]. It is difficult to see, therefore, how electronic effects could be transmitted through anything other than a very small and very thin metal raft. Although raft formation was originally believed to be important in SMSI, it is now known that three-dimensional crystallites also have SMSI characteristics. Recently, the electronic factor in SMSI has been interpreted more cautiously and more correctly as a very localized interaction.

B. Localized Electronic Factors

Joyner et al. [114] have presented a theoretical analysis of metal-support interactions. The main conclusion of their study is that the effect of the support on catalytic activity is very localized. For example, for spherical particles (dispersion D) the influence of the support will be proportional to D^5. They further show that the poisoning effect of an atom adsorbed on a metal surface is localized to nearest-neighbor metal atoms.

The proposition underlying the localized electronic factor is that at the metal-support interface metal atoms nearest neighbor to support ions may undergo charge transfer [36,39,50,71,74,107,108,116,117]. Provided that allowance is made for the fact that moieties from the reduced support may migrate onto the metal surface, thus creating new "interfacial regions," this model could explain some of the consequences of SMSI. On the other hand, it seems impossibly difficult to exclude nonelectronic localized effects at the interface since even if the support cation is directly bonded to surface metal atoms, the associated oxide ions and anion vacancies may also participate in particular catalytic reactions. Simoens et al. [107] have suggested that the interaction could be of the ligand type, giving a chemical-electronic interaction between the support moiety on the metal surface and adjacent surface metal atoms.

C. Structural Effects in SMSI

The evidence relating to structural changes in SMSI catalysts is conflicting. TEM evidence has been reported which, it is claimed, shows that SMSI metals adopt a two-dimensional "pillbox" morphology. However, even though there is evidence of structural effects in some instances, these seem not to be responsible for SMSI since in many cases the SMSI effect is observed when no changes are observed in the structure or morphology of the metal particles [15,86,97,107,108,118,119].

D. Encapsulation, Decoration, and Support Migration

Encapsulation was considered in the original Exxon work but rejected on the grounds that no loss of support surface area was observed after HTR and also because the effects of HTR were completely reversible after oxidation. Although no experimental evidence was available, Meriaudeau et al. [58] proposed for Pt/TiO_2 that high-temperature reduction in the presence of Pt would lead to the formation of TiO suboxide, which could cover the Pt surface.

Direct experimental evidence for support migration has since been obtained for several different metal-support systems using Rutherford backscattering [120], Auger electron spectroscopy [39,40,107,121-124], Mössbauer spectroscopy [107,125], TEM [126], HREELS [121], UPS and XPS [122,123,127], EXAFS [128], and SIMS [116,124]. Figure 5 shows the Auger sputter profiles obtained by Sadeghi and Henrich [122,123] for a model Rh/TiO_2 catalyst. They conclude that the decrease in Ti and O signals that accompanies the increase in the Rh signal during the first few minutes of sputtering must indicate that after a high-temperature treatment some Ti oxide is physically covering the Rh. Indirect support for an encapsulation or decoration model has been obtained from catalytic and chemisorption studies [34,36,38,39,50, 81,108,117,129].

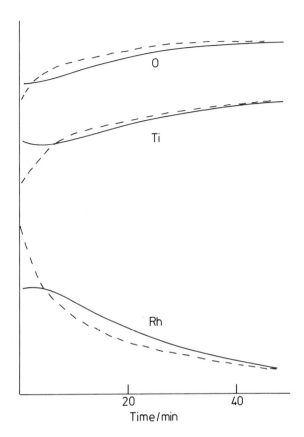

FIGURE 5 Auger sputter profiles (peak amplitude versus sputtering time) for unreduced (dashed lines) and reduced (solid lines) model Rh/TiO_2 catalysts. (After Refs. 122 and 123.)

VII. EFFECT OF SUPPORT MIGRATION ON CHEMISORPTION AND CATALYTIC PROPERTIES

A. Suppression of Adsorption

The loss of chemisorption capacity for H_2 or CO after HTR is readily explained by the encapsulation or decoration model. Since the surface of the metal is inaccessible to gaseous molecules, the absence of adsorption is easily understood and no electronic interactions localized or otherwise are required. This does not exclude an electronic interaction, since it is clear that a metal atom which is bonded to a TiO_x species on the metal surface may have different chemical bonding characteristics than a metal atom in a clean surface. This could contribute to a change in the heat of adsorption, as seems to be indicated by the (small) shifts in the positions of the desorption peaks for CO or H_2 in SMSI systems.

B. Changes in Catalytic Activity

Suppression of Activity

In most cases the effect of HTR is to reduce the activity of the metal for catalytic reactions. The extent of this suppression of activity is related to the type of reaction—hydrogenolysis reactions are almost totally eliminated

[21,22] and hydrogenation, dehydrogenation or isomerization reactions are suppressed by about an order of magnitude. This pattern of changes in activity is exactly what would be expected whatever the exact nature of SMSI, since demanding reactions (i.e., hydrogenolysis) are much more likely to be seriously affected than facile reactions (i.e., hydrogenation, etc.). These effects would be observed if SMSI arises because of an encapsulation or decoration of the metal particle by the support. Equally, however, a short-range electronic interaction, such as has been proposed for bimetallic catalysts [130], could reduce the apparent ensemble size with essentially the same results.

Enhancement of Activity

The observation that the activity for the CO/H_2 reaction is increased for titania-supported metals has attracted most attention. However, it is now known that this enhancement cannot be attributed to SMSI in the conventional sense because under reaction conditions the SMSI state is at least partially destroyed. The role of the support in the CO/H_2 reaction is still a matter of debate.

VIII. CONCLUDING REMARKS

The effect of a high-temperature reduction on the catalytic and chemisorptive properties of metals supported on reducible oxides is well established. As yet these effects are incompletely understood. The original bulk electronic model has been rejected and a migration of the reduced support onto the surface of the metal is now thought to be responsible for the suppression of adsorption capacity and of catalytic activity for hydrocarbon reactions. The analogy with strong poisoning of a metal surface is apparent.

The enhanced activity observed in the CO/H_2 reaction remains to be explained. The support is clearly important in this reaction, but its precise role is unknown. The fact that the CO/H_2 reaction proceeds at a high rate even in the absence of SMSI adds a further complication. Two alternative models have been proposed in which the support is held to play either a direct or an indirect part in the reaction. In the former it is suggested that CO is activated at a new site created at the metal-support interface. In the latter the role of the support is to modify the properties of the metal atoms adjacent to the metal-support interface. It is clearly almost impossible to differentiate between these two alternatives. However, in both cases the support may be considered to have a synergistic effect on the metal, analogous in some respects to the promoter effects observed in sulfide catalysts.

In conclusion, it appears that under strong reducing conditions SMSI is essentially a type of poisoning, while under mild reducing conditions SMSI may have a promoting or synergistic effect.

REFERENCES

1. G. C. Bond and R. Burch, in *Catalysis*, Vol. 6 (G. C. Bond and G. Webb, eds.), Royal Society of Chemistry, London (1983), p. 27.
2. B. Imelik, C. Naccache, G. Coudurier, H. Praliaud, P. Meriaudeau, P. Gallezot, G. A. Martin, and J. Vedrine (eds.), *Metal-Additive Effects in Catalysis*, Elsevier, Amsterdam (1982).
3. M. Boudart and G. Djega-Mariadassou, in *Kinetics of Heterogeneous Catalytic Reactions*, Princeton University Press, Princeton, N.J. (1984), p. 201.

4. G. M. Schwab and E. Pietsch, *Z. Phys. Chem. Abt.*, *B1*:385 (1928).
5. R. Burch and A. R. Flambard, *J. Catal.*, *78*:389 (1982).
6. G. M. Schwab, *Adv. Catal.*, *27*:1 (1978).
7. F. Solymosi, *Cat. Rev.*, *1*:233 (1967).
8. K. Foger, in *Catalysis: Science and Technology*, Vol. 6 (J. R. Anderson and M. Boudart, eds.), Springer-Verlag, Berlin (1984), p. 227.
9. S. J. Tauster, S. C. Fung, and R. L. Garten, *J. Am. Chem. Soc.*, *100*:170 (1978).
10. S. J. Tauster and S. C. Fung, *J. Catal.*, *55*:29 (1978).
11. S. J. Tauster, S. C. Fung, R. T. K. Baker, and J. A. Horsley, *Science*, *211*:1121 (1981).
12. B. H. Chen and J. M. White, *J. Phys. Chem.*, *86*:3534 (1982).
13. O. H. Ellestad and C. Naccache, in *Perspectives in Catalysis* (R. Larsson, ed.), Gleerup, Lund, Sweden (1981), p. 95.
14. J. M. Herrmann, J. Disdier, and P. Pichat, in *Metal-Support and Metal-Additive Effects in Catalysis* (B. Imelik et al., eds.), Elsevier, Amsterdam (1982), p. 27.
15. P. Meriaudeau, O. H. Ellestad, M. Dufaux, and C. Naccache, *J. Catal.*, *75*:243 (1982).
16. M. D. Mitchell and M. A. Vannice, *Ind. Eng. Chem. Fundam.*, *23*:88 (1984).
17. R. Burch, in *Catalysis*, Vol. 7 (G. C. Bond and G. Webb, eds.), Royal Society of Chemistry, London (1985), p. 149.
18. G. Maire and F. C. Garin, in *Catalysis Science and Technology*, Vol. 6 (J. R. Anderson and M. Boudart, eds.), Springer-Verlag, Berlin (1984), p. 161.
19. G. C. Bond, in *Metal-Support and Metal-Additive Effects in Catalysis* (B. Imelik et al., eds.), Elsevier, Amsterdam (1982), p. 1.
20. P. Gallezot, *Cat. Rev.-Sci. Eng.*, *20*:121 (1979).
21. G. den Otter and F. Dautzenberg, *J. Catal.*, *53*:116 (1979).
22. P. G. Menon and G. F. Froment, *J. Catal.*, *59*:138 (1979).
22a. P. G. Menon and G. F. Froment, *Appl. Catal.*, *1*:31 (1981).
23. S. J. Tauster, L. L. Murrell, and S. C. Fung, U. S. Patent 1,576,848 (1976).
24. J. B. F. Anderson, R. Burch, and J. A. Cairns, *Appl. Catal.*, *25*:173 (1986).
25. E. I. Ko, J. M. Hupp, F. H. Rogan, and N. J. Wagner, *J. Catal.*, *84*:85 (1983).
26. T. Beringhelli, A. Gervasini, F. Morazzoni, D. Strumolo, S. Martinengo, L. Zanderrighi, F. Pinna, and G. Strukul, *8th International Congress on Catalysis, West Berlin*, Vol. 5 (1984), p. 63.
27. P. Meriaudeau, B. Pommier, and S. J. Teichner, *C. R. Acad. Sci.*, *289*:395 (1979).
28. G. L. Haller, D. E. Resasco, and A. J. Rouco, *Faraday Discuss. Chem. Soc.*, *72*:109 (1981).
29. J. R. Katzer, A. W. Sleight, P. Gajardo, J. B. Michel, E. F. Gleason, and S. McMillan, *Faraday Discuss. Chem. Soc.*, *72*:121 (1981).
30. S. Y. Wang, S. H. Moon, and M. A. Vannice, *J. Catal.*, *71*:167 (1981).
31. M. A. Vannice, S. Y. Wang, and S. H. Moon, *J. Catal.*, *71*:152 (1981).
32. M. A. Vannice and P. Chou, *Chem. Commun.*, 1590 (1984).
33. M. A. Vannice and P. Chou, *8th International Congress on Catalysis, West Berlin*, Vol. 5 (1984), p. 99.
34. M. A. Vannice and C. Sudhakar, *J. Phys. Chem.*, *88*:2429 (1984).
35. M. A. Vannice and C. C. Twu, *J. Catal.*, *82*:213 (1983).
36. L. M. Tau and C. O. Bennett, *J. Catal.*, *89*:285, 327 (1984).
37. G. D. Weatherbee and C. H. Bartholomew, *J. Catal.*, *87*:55 (1984).
38. G. B. Raupp and J. A. Dumesic, *J. Phys. Chem.*, *88*:660 (1984).

39. D. N. Belton, Y. M. Sun, and J. M. White, *J. Phys. Chem.*, 88:5172 (1984).
40. E. I. Ko and R. J. Gorte, *J. Catal.*, 90:59 (1984).
41. M. A. Vannice, C. C. Twu, and S. H. Moon, *J. Catal.*, 79:70 (1983).
42. K. Foger, *J. Catal.*, 76:225 (1982).
43. M. A. Vannice, L. C. Hasselbring, and B. Sec, *J. Phys. Chem.*, 89:2972 (1985).
44. M. A. Vannice, L. C. Hasselbring, and B. Sec, *J. Catal.*, 95:57 (1985).
45. M. A. Vannice, L. C. Hasselbring, and B. Sec, *J. Catal.*, 97:66 (1986).
46. J. J. Rooney, *J. Mol. Catal.*, 31:147 (1985).
47. K. Kunimori, S. Matsui, and T. Uchijima, *Chem. Lett.*, 359 (1985).
48. J. M. Hermann, *J. Catal.*, 89:404 (1984).
49. J. M. Hermann and P. Pichat, *J. Catal.*, 78:425 (1982).
50. X. Z. Jiang, T. F. Hayden, and J. A. Dumesic, *J. Catal.*, 83:168 (1983).
51. S. R. Morris, R. B. Moyes, P. B. Wells, and R. Whyman, in *Metal-Support and Metal-Additive Effects in Catalysis* (B. Imelik et al., eds.), Elsevier, Amsterdam (1982), p. 247.
52. G. L. Haller, V. E. Henrich, M. McMillan, D. E. Resasco, H. R. Sadeghi, and S. Sakellson, *8th International Congress on Catalysis, West Berlin*, Vol. 5 (1984), p. 135.
53. D. Duprez and A. Miloudi, in *Metal-Support and Metal-Additive Effects in Catalysis* (B. Imelik et al., eds.), Elsevier, Amsterdam (1982), p. 179.
54. J. B. F. Anderson, J. D. Bracey, R. Burch, and A. R. Flambard, *8th International Congress on Catalysis, West Berlin*, Vol. 5 (1984), p. 111.
55. J. B. F. Anderson, R. Burch, and J. A. Cairns, *Appl. Catal.*, 21:179 (1986).
56. C. U. I. Odenbrand and S. L. T. Andersson, *J. Chem. Technol. Biotechnol.*, A33:150 (1983).
57. K. H. Stadler, M. Schneider, and K. Kochloefl, *8th International Congress on Catalysis, West Berlin*, Vol. 5 (1984), p. 229.
58. P. Meriaudeau, J. F. Dutel, M. Dufaux, and C. Naccache, in *Metal-Support and Metal-Additive Effects in Catalysis* (B. Imelik et al., eds.), Elsevier, Amsterdam (1982), p. 95.
59. Z. Paál, *Adv. Catal.*, 29:273 (1980).
60. G. C. Bond, R. Burch, and R. Rajaram, *J. Chem. Soc. Faraday Trans. 1*, 82:1985 (1986).
61. D. E. Resasco and G. L. Haller, in *Metal-Support and Metal-Additive Effects in Catalysis* (B. Imelik et al., eds.), Elsevier, Amsterdam (1982), p. 105.
62. A. Dauscher, F. Garin, F. Luck, and G. Maire, in *Metal-Support and Metal-Additive Effects in Catalysis* (B. Imelik et al., eds.), Elsevier, Amsterdam (1982), p. 113.
63. R. Kramer and H. Zuegg, *J. Catal.*, 80:446 (1983).
64. R. Kramer and H. Zuegg, *8th International Congress on Catalysis, West Berlin*, Vol. 5 (1984), p. 275.
65. J. B. F. Anderson, R. Burch, and J. A. Cairns, *J. Catal.*, submitted.
66. V. J. Schepers, J. G. van Senden, E. H. van Broekhoven, and V. Ponec, *J. Catal.*, 94:400 (1985).
67. G. C. Bond and Xu Yide, *J. Chem. Soc. Faraday Trans. 1*, 80:3103 (1984).
68. D. E. Resasco and G. L. Haller, *J. Phys. Chem.*, 88:4552 (1984).
69. G. C. Bond and Xu Yide, *Chem. Commun.*, 1284 (1983).
70. K. Foger and J. R. Anderson, *J. Catal.*, 59:325 (1979).
71. D. E. Resasco and G. L. Haller, *J. Catal.*, 82:279 (1983).
72. M. A. Vannice, *J. Catal.*, 74:199 (1982).

73. M. A. Vannice and R. L. Garten, *J. Catal.*, *66*:242 (1980).
74. C. C. Kao, S. C. Tsai, and Y. W. Chung, *J. Catal.*, *73*:136 (1982).
75. M. A. Vannice and R. L. Garten, *J. Catal.*, *56*:236 (1979).
76. C. H. Bartholomew, R. B. Pannell, and J. L. Butler, *J. Catal.*, *65*: 335 (1980).
77. C. H. Bartholomew, R. B. Pannell, J. L. Butler, and D. G. Mustard, *Ind. Eng. Chem. Prod. Res. Dev.*, *20*:296 (1981).
78. C. K. Vance and C. H. Bartholomew, *Appl. Catal.*, *7*:169 (1983).
79. Y. I. Yermakov, Y. A. Ryndin, O. S. Alekseev, and M. N. Vassileva, *Chem. Commun.*, 1480 (1984).
80. S. Z. Ozdogan, P. D. Gochis, and J. L. Falconer, *J. Catal.*, *83*:257 (1983).
81. Y. W. Chung, G. Xiang, and C. C. Kao, *J. Catal.*, *85*:237 (1984).
82. M. A. Vannice and R. L. Garten, *J. Catal.*, *63*:255 (1980).
83. C. H. Yang and J. G. Goodwin, *8th International Congress on Catalysis, West Berlin*, Vol. 5 (1984), p. 263.
84. K. Kunimori, H. Abe, E. Yamaguchi, S. Matsui, and T. Uchijima, *8th International Congress on Catalysis, West Berlin*, Vol. 5 (1984), p. 251.
85. A. Erdöhelyi and F. Solymosi, *J. Catal.*, *84*:446 (1983).
86. F. Solymosi, I. Tombácz, and A. Erdöhelyi, *J. Catal.*, *75*:78 (1982).
87. J. D. Bracey and R. Burch, *J. Catal.*, *86*:384 (1984).
88. Y. A. Ryndin, R. F. Hicks, and A. T. Bell, *J. Catal.*, *70*:287 (1981).
89. J. A. Rabo, A. P. Risch, and M. L. Poutsma, *J. Catal.*, *53*:295 (1978).
90. J. G. McCarty and H. Wise, *J. Catal.*, *57*:406 (1979).
91. E. Zagli and J. L. Falconer, *Appl. Catal.*, *4*:135 (1982).
92. M. Araki and V. Ponec, *J. Catal.*, *58*:170 (1979).
93. F. Solymosi, A. Erdöhelyi, and I. Tombácz, *Appl. Catal.*, *14*:65 (1985).
94. F. Solymosi and A. Erdöhelyi, *Surf. Sci.*, *110*:L630 (1981).
95. J. A. Horsley, *J. Am. Chem. Soc.*, *101*:2870 (1979).
96. V. Ponec, in *Metal-Support and Metal-Additive Effects in Catalysis* (B. Imelik et al., eds.), Elsevier, Amsterdam (1982), p. 63.
97. S. C. Fung, *J. Catal.*, *76*:225 (1982).
98. S. H. Chien, N. Shelimor, D. E. Resasco, E. H. Lee, and G. L. Haller, *J. Catal.*, *77*:301 (1982).
99. M. K. Bahl, S. C. Tsai, and Y. W. Chung, *Phys. Rev.*, *B21*:1344 (1981).
100. Y. W. Chung and W. B. Weissbard, *Phys. Rev.*, *B20*:3456 (1979).
101. C. C. Kao, S. C. Tsai, M. K. Bahl, Y. W. Chung, and W. J. Lo, *Surf. Sci.*, *95*:1 (1980).
102. B. A. Sexton, A. E. Hughes, and K. Foger, *J. Catal.*, *77*:85 (1982).
103. B. H. Chen and J. M. White, *J. Phys. Chem.*, *87*:1327 (1983).
104. A. R. Gonzalez-Elipe, J. Soria, and G. Munuera, *J. Catal.*, *76*:254 (1982).
105. J. Disdier, J. M. Herrmann, and P. Pichat, *J. Chem. Soc. Faraday Trans. 1*, *79*:651 (1983).
106. D. R. Short, A. N. Mansour, J. W. Cook, D. E. Sayers, and J. R. Katzer, *J. Catal.*, *82*:299 (1983).
107. A. J. Simoens, R. T. K. Baker, D. J. Dwyer, C. R. F. Lund, and R. J. Madon, *J. Catal.*, *86*:359 (1984).
108. J. Santos, J. Phillips, and J. A. Dumesic, *J. Catal.*, *81*:147 (1983).
109. A. Crucq, L. Degols, G. Lienard, and A. Frennet, *Acta Chim. Hung.*, *124*:109 (1987).
110. V. E. Henrich, *J. Catal.*, submitted (private communication).
111. S. C. Chung, *J. Catal.*, *76*:225 (1983).
112. W. Schottky, *Z. Phys.*, *113*:367 (1939).

113. L. J. Brillson, *Appl. Surf. Sci.*, *11/12*:249 (1982).
114. R. W. Joyner, J. B. Pendry, D. K. Saldin, and S. R. Tennison, *Surf. Sci.*, *95*:1 (1984).
115. J. R. Smith, F. J. Arlinghaus, and J. G. Gay, *Phys. Rev.*, *B26*:1072 (1982).
116. D. N. Belton, Y. M. Sun, and J. M. White, *J. Phys. Chem.*, *88*:1690 (1984).
117. T. H. Fleisch, R. F. Hicks, and A. T. Bell, *J. Catal.*, *87*:398 (1984).
118. L. L. Murrell and D. J. C. Yates, *Stud. Surf. Sci. Catal.*, *7*:1470 (1981).
119. J. S. Smith, P. A. Thrower, and M. A. Vannice, *J. Catal.*, *68*:270 (1981).
120. J. A. Cairns, J. E. E. Baglin, G. J. Clark, and J. F. Ziegler, *J. Catal.*, *83*:301 (1983).
121. S. Takatani and Y. W. Chung, *J. Catal.*, *90*:75 (1984).
122. H. R. Sadeghi and V. E. Henrich, *J. Catal.*, *87*:279 (1984).
123. H. R. Sadeghi and V. E. Henrich, *Appl. Surf. Sci.*, in press (private communication).
124. D. N. Belton, Y. M. Sun, and J. M. White, *J. Am. Chem. Soc.*, *106*:3059 (1984).
125. B. J. Tatarchuk and J. A. Dumesic, *J. Catal.*, *70*:308, 323, 325 (1981).
126. B. R. Powell and S. E. Wittington, *J. Catal.*, *81*:382 (1983).
127. T. Huizinga, J. C. Vis, H. F. J. van't Blik, and R. Prins, *Recl. Trav. Chim. Pay-Bas*, *102*:496 (1983).
128. S. Sakellson, M. McMillan, and G. L. Haller, *J. Phys. Chem.*, *90*:1733 (1986).
129. R. T. K. Baker, E. B. Prestridge, and G. B. McVicker, *J. Catal.*, *89*:422 (1984).
130. R. Burch, *Acc. Chem. Res.*, *15*:24 (1982).

14 | Catalytic Properties of Metal Hydrides

WACŁAWA PALCZEWSKA

Institute of Physical Chemistry, Polish Academy of Sciences, Warsaw, Poland

I.	INTRODUCTION	373
II.	FORMATION, STRUCTURE, AND PROPERTIES OF METAL HYDRIDES	374
	A. Phase Diagram for Metal-Hydrogen Solutions	375
	B. Electronic Structure of Metallic Hydrides	376
	C. Metal Dispersion and Hydride Formation	377
	D. Activation and Poisoning of a Metal Hydriding Process	378
III.	PALLADIUM HYDRIDE AS HYDROGENATION CATALYST	378
	A. Reactions of Hydrogen: o-p Conversion, H_2-D_2 Exchange, H Atom Heterogeneous Recombination	381
	B. Hydrogenation of Acetylene	381
	C. Hydrogenation of Olefinic and Aromatic Compounds	384
IV.	HYDRIDES OF INTERMETALLIC COMPOUNDS AS CATALYSTS IN CHEMICAL REACTIONS	386
	A. Catalytic Properties of Hydrides of Intermetallics: RE M_5	386
	B. Catalytic Properties of Hydrides of Intermetallics Formed by Ti, Zr, Hf, or Th with Other Metals	388
V.	CATALYTIC MEMBRANE SYSTEMS	390
IV.	CONCLUDING REMARKS	391
	REFERENCES	392

I. INTRODUCTION

The field of research and application of metal-hydrogen-metal hydride systems is large. The long list of the subjects studied include hydrogen permeation through metallic membranes, commonly used for hydrogen purification and its isotope segregation, as well as complex phenomena involved in large-scale potential technologies of the "hydrogen economy" [1,2]. Hydrogen generation and storage have attracted particular attention.

Metal hydrides are the most efficient materials for hydrogen storage, owing to their high density of hydrogen atoms. At present, many transition, rare-earth, and actinide metal hydrides are known; their properties and structure have been described. The respective intermetallic compounds, with still higher hydrogen retentive capacity, were prepared. However, despite the large number of metal hydride systems investigated and the considerable amount of theoretical and experimental information assembled, a metal hydride meeting all requirements has still not been acquired.

The progress of knowledge on metal hydride systems has imposed new problems on heterogeneous catalysis. At first, for the activation of the process of hydrogen entry into metal, catalytic promoters are often to be present and poisons are to be avoided. Second, the hydrides proposed and investigated as hydrogen storage materials became, in turn, subjects of research as new catalysts in the hydrogenation of unsaturated hydrocarbons, nitrobenzene-to-aniline reduction, synthesis of ammonia, CO methanation, or alkane conversion reactions.

In this chapter we present a survey of information on those properties of metal-hydrogen-hydride systems which may influence their catalytic behavior. Knowledge of phase equilibria, rates of hydrogen uptake and discharge, and geometric and electronic structures is of fundamental significance for an understanding of the initial behavior of hydrides in catalytic reactions. Moreover, the given metal-hydrogen-metal hydride system under complex dynamic conditions in the particular reactant mixture is to be studied in order to describe properly the role of the hydride catalyst and to interpret the mechanism of its catalytic activity.

II. FORMATION, STRUCTURE, AND PROPERTIES OF METAL HYDRIDES

A large number of metals form hydrides. They have been classified according to the character of their metal-hydrogen bond. The alkali metals and most of the alkaline earth metals form ionic (i.e., saltlike) hydrides, with hydrogen anions and metal cations in their crystal lattice. The representatives of covalent hydrides are beryllium, germanium, and probably, magnesium hydrides. Most of the d-transition metals, all the lanthanides, and the actinides form metallic hydrides (Table 1) [3-8]. Alkali and alkaline earth metals, aluminum, d-transition metals, lanthanides, and actinides, able to be hydrided, form an immense number of bi- or tricomponent alloy systems, disordered or

TABLE 1 d-Transition Elements, Lanthanides and Actinides, Which Form Metallic Hydrides

IIIA	IVA	VA	VIA	VIIA		VIIIA						
Sc	Ti	V	Cr	Mn	Fe	Co	Ni					
Y	Zr	Nb	Mo			Rh	Pd					
La	Hf	Ta										
Ac												

La	Ce	Pr	Nd	Pm	Sm	Eu	Gd	Tb	Dy	Ho	Er	Tm	Yb	Lu
Ac	Th	Pa	U		Np	Pu				Am	Cm			

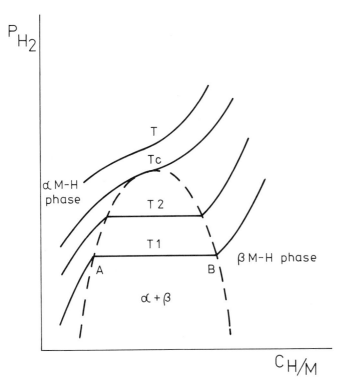

FIGURE 1 Representative p_{H_2}-$C_{H/M}$ isotherms at T_1, T_2, and so on. T_c is the critical isotherm of a metal hydride formation.

ordered solutions, which in turn react with hydrogen to give metallic hydrides. In a number of cases those ternary, $M_I M_{II} H_x$, (or higher) hydrides can accommodate more hydrogen in their unit cell and release it more easily than can the respective parent metal components. A short survey of basic information on hydrides might be useful to provide an insight into methods of identification of a metal hydride phase in a catalyst and for following its behavior in a catalytic system.

A. Phase Diagram for Metal-Hydrogen Solutions

Absorption of dissociatively adsorbed hydrogen, H, by a metal, M, results at first in the formation of a disordered solution M-H, called the α phase, with low hydrogen concentration. For treatment of adsorption and absorption of hydrogen on pure metals, see Chapter 1. Hydrogen pressure-temperature-concentration of the M-H solution (H/M atomic ratio), the P-T-C relation, is often represented by means of a series of P-C isotherms, represented in their general and simplest case in Fig. 1. When at T = const. the α solution becomes saturated and the β phase, an ordered interstitial solid solution called β metal hydride, begins to form (point A in Fig. 1). Further introducing of hydrogen has, as the only effect, the progress of α-to-β conversion at invariant pressure. After complete formation of the β hydride (point B in Fig. 1), the concentration of hydrogen increases again and, eventually, successive hydride phases of higher stoichiometry form: ζ, ε, γ, λ, σ, and so on, phases.

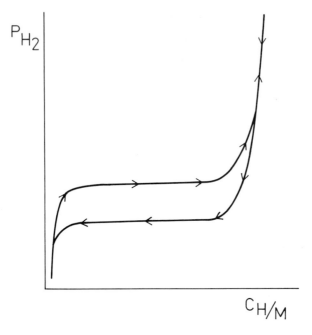

FIGURE 2 Representative hysteresis loop of a hydrogen sorption isotherm for a metal-hydrogen-metal hydride system.

The isotherm T_c represents a critical temperature above which the two-phase region disappears; the isotherm is analogous to a critical isotherm for a one-component liquid-vapor system.

The sorption isotherms are not strictly reversible; this is represented schematically in Fig. 2. Qualitatively, the increase and decrease of hydrogen pressure reproduce the α-β phase transformation, but a higher p_{H_2} is necessary to form the β phase than to convert it back into the α phase. The shape of that hysteresis loop varies with the type of metal specimen concerned and has been interpreted qualitatively in terms of its compressive strain, effects of contaminations, and metal crystallite size. However, quantitative descriptions and respective predictions of the shape and position of a hysteresis loop are still lacking.

To be of value in potential applications in hydrogen storage or in catalysis, metallic hydrides have to release their hydrogen by a change of conditions. When exhausted, those hydrides have to absorb hydrogen anew as easily as they had desorbed it before.

In all known metal-hydrogen systems the expansion of the crystal lattice of the parent metal takes place under the influence of the dissolved hydrogen. That change of lattice parameter provides information on the hydrogen in metal content and on the α-β phase transformation. The phenomenon has an important bearing on the catalytic behavior of a metal that is able to form a hydride. A sequence of sorption-desorption processes may bring about cracking and disintegration of metal crystallites, their restructuring, enhanced segregation of a component, or a contamination.

B. Electronic Structure of Metallic Hydrides

The electronic structure of a d-transition metal changes on its conversion into a hydride. Thus immediate consequences are observed for physical properties of a metal (e.g., its magnetic susceptibility, electrical resistance, and thermoelectric power). The simplistic concept of a metallic hydride electronic structure has treated "hydride" hydrogen as a quasi-metal, and a metal hydride, as a whole, as a quasi-intermetallic compound. The host transition metal was accommodating s electrons of hydrogen in the empty states of the d band at the Fermi energy. The model of a metallic hydride with "screened protons" in its lattice has been very fruitful. However, the pioneer research by Switendick [8] led to the conclusion that filling of the d band and shifting the Fermi level of a parent metal were not the only changes induced by the conversion of a metal into hydride. The simultaneous appearance of a group of electron energy levels beneath the metal d band was stated. Thus Switendick formulated his final conclusion: "Aspects of both simplistic pictures— (1) proton model: electrons added at the Fermi level, and (2) anion model: low lying states associated with electronic charge in the vicinity of the hydrogen—are common to all hydride systems" [8]. It is evident that the conversion of a given metal into its hydride phase, which is so different in its electronic structure, must result in a profound alteration of the metal catalytic properties. Moreover, metal catalysts possess a high degree of dispersion. The change in the electronic properties when passing from the bulk metal to clusters and surfaces must also be taken into account.

Molecular orbital calculations (EH and CNDO) for small metal clusters of different shape and size have shown significant changes in the electronic structure compared with the respective bulk metal. The width and position of conduction band states revealed their close relation to the coordination and number of metal atoms in the cluster investigated. However, the results of theoretical research prove that in the case of palladium, for example, even extremely small clusters are qualitatively similar to the bulk metal, because of the high density of states at the Fermi level and the incomplete configuration of d electrons in the cluster [9-11]. The results have been confirmed by the rare experimental evidence given by photoelectron spectra and by the fact that magnetic susceptibility values decrease with the increase of metal (Pd on SiO_2) dispersion [12].

C. Metal Dispersion and Hydride Formation

The question: "At what size does a cluster of metal atoms behave as a bulk metal?" [11] is of particular importance for the phenomenon of metal hydride formation. The hydride is a tridimensional bulk phase and it seems justifiable to expect, intuitively, that its formation would be improbable below a certain size of parent metal crystallite. The experimental results support this hypothesis.

Aben [13] found that the hydrogen uptake in the palladium hydride decreased with the increase of the palladium dispersion. Boudart and Hwang [14] also proved that with a series of systematic measurements of hydrogen solubility in Pd crystallites of different dispersion, supported on SiO_2. Dispersion, defined as the ratio of surface atoms to bulk atoms in a crystallite (or cluster), was related to respective crystallite sizes (73 to 12% dispersions and 0.8- to 7.6-nm diameters). The authors conclude that "dispersed palladium dissolves less hydrogen in the β-phase region than bulk palladium." The results by Nandi et al. [15] go still further. Using extended x-ray absorption fine structure (EXAFS) and wide-angle x-ray scattering methods they found that palladium on silica, with metal dispersion equal to or greater than 49.8% (corresponding to the Pd particle-size diameter of 2.6 nm), does not form the β-hydride phase at p_{H_2} = 1 atm and T = 298 K.

Summarizing the related survey of equilibrium properties of the system metal-hydrogen-metal hydride, one must point out that any attempt to predict the conversion of a dispersed metal catalyst into its hydride phase may be hazardous when based on the thermodynamics of the bulk metal-hydrogen system. The usefulness of knowledge on the thermodynamic properties of bulk metal-hydrogen-metal hydride systems is still limited by slowness of the equilibration of the system, particularly when the solid phase has not been dispersed sufficiently.

D. Activation and Poisoning of a Metal Hydriding Process

As already mentioned, the interconversion metal-metallic hydride, applied in both hydrogen storage and catalysis, should be an easily occurring process. Therefore, in most cases, samples are to be activated. The processes of activation are of various characters. Repetitive hydriding and dehydriding of a metal alloy (or an intermetallic) at a sufficiently high temperature successfully enhances penetration of hydrogen into the metal lattice. The cycling in hydrogen removes and disrupts oxide films used to coat almost all hydride-forming metals and to protect them against hydrogen attack. Simultaneously, large crystals break down into finely divided metal, which is advantageous in the case of catalytic applications [16-21].

There are, however, processes of activation of typical catalytic character. Wicke and co-workers used the following substances to catalyze hydrogen entry into massive metals: Pd black or Ti, Th, Ce, or U hydrides into palladium; U and Ti hydrides into tantalum; and U hydride into titanium. Palladium black or Pd salts, as precursors, have been commonly employed for hydriding of V, Nb, and Ta. Raney Ni, Co, Cu, and Cu fine powder itself accelerate hydrogen sorption in Pd. Wicke even introduced a special term, "transfer catalysts" [22-25], for promoters of hydrogen penetration across the hydrogen-metal interface.

The presence of certain chemical compounds or their precursors in a gaseous or liquid hydrogenating environment has proved their utility for enhancement of the kinetics of hydrogen penetration into metals. These include compounds of S, Se, Te, As, Sb, and Bi with hydrogen. When adsorbed on the metal sample surface, they activate co-adsorbed hydrogen to cross the barrier of the metal-gas or metal-liquid interface and to become incorporated into the lattice [26-28].

On the other hand, much experimental evidence has been collected on the deactivating effect of certain compounds, which consists of poisoning of hydrogen sorption in some metals. Moisture, oxygen, and, most effective, carbon monoxide protect metals against hydrogen attack. Sulfur dioxide prevents absorption of hydrogen and slows down its desorption, keeping hydrogen in the bulk metal [18,29-32].

There are many such observations, and the utility of the activators or poisons is unquestionable. However, much less is known regarding their kinetic role and the mechanism of their action, particularly in the case of polymetallic systems. TiFe and $LaNi_5$ have been relatively well recognized because of their potential role in hydrogen storage, but knowledge even of those systems hardly permits us to formulate generalizations on the nature of catalytic phenomena, which influence in a positive or negative way the formation-decomposition of the respective hydrides under a wide range of conditions [29,33,34].

III. PALLADIUM HYDRIDE AS HYDROGENATION CATALYST

Among metal-hydrogen-metallic hydride systems the palladium-hydrogen-β-palladium hydride system has attracted most interest and has been studied systematically in all its aspects for years (since the 1930s). The monograph by Lewis [35] and the review by Wicke and Brodowsky [36] represent contemporary fundamental knowledge on the system (including Pd alloys) and its properties.

On the other hand, palladium has been investigated extensively and usually applied as a catalyst in hydrogenation reactions, particularly when remarkable bond and stereo selectivities were necessary. In those reactions the probability that the Pd-H system will undergo conversion into the β hydride must be taken into consideration, because the T and p_{H_2} conditions of the majority of those reactions do not exclude such transformation of the catalyst. Of course, the metal hydride, eventually formed during a catalyst preparation, will retain its state under those conditions.

The role of palladium, nickel, and some other binary hydrides in catalysis was presented in an earlier review by the author [37] in which were analyzed changes of catalytic properties of a metal after it was transformed into β hydride, stable under specific reaction conditions. The hydride, not the initial parent metal, became in fact the catalyst of a given reaction. During dynamic equilibrium in such "stable" metal-hydrogen systems, the flux of hydrogen exchanged between the metal and the reactant medium is very low. In general, experimental evidence showed [37] a pronounced decrease in catalytic activity, induced by the transformation of the metal into its hydride. In terms of kinetic equations (in the rare cases when they were available) it meant an increase in activation energy and a decrease of preexponential factor. The poisoning effect of a metal catalyst hydriding was interpreted in terms of the change of electronic structure of a parent metal.

However, the question arises as to whether that effect will still be valid when molecules to be hydrogenated will be so "eager" for hydrogen that they could compete successfully with the metal itself for surface hydrogen adatoms. The intuitive answer will be that, in order to keep the hydride phase, the hydrogen partial pressure should be increased sufficiently to satisfy both needs—the hydrogenation reaction and the metal hydriding rates. Moreover, simultaneously, under the conditions of catalytic hydrogenation (e.g., at a relatively elevated temperature, or at a high degree of metal dispersion), the flux of hydrogen exchanged across the metal interface might be large enough to keep the β-hydride content in the metal catalyst and to ensure an increased supply of active hydrogen for the reactant admolecules. In such a case an increase in rate of hydrogenation could be anticipated, although the metal catalyst would still continue to be present in its hydride form. Thus an enhanced catalytic activity of metal hydrides in hydrogen reaction may be observed as well. One can speculate about the mechanism of that catalytic effect, but one is tempted to interpret a part played there by the metallic hydride, mainly as that of a reversible storage medium of active hydrogen.

The model research on ethylene and acetylene adsorption and hydrogenation on palladium and β-palladium hydride films deposited under UHV conditions, performed by Duś [38] and Borodziński et al. [39], illustrates that apparently ambiguous role of the hydride. Values of the change of work function, $\Delta \phi$, were monitoring adsorption and surface reactivity in the Pd-H-C_2H_4-C_2H_2 system and the Pd-to-β-PdH phase transformation as well (Fig. 3). At 195 K and still at 298 K, ethylene is unable to adsorb or to react with hydrogen on the β-PdH (see also Chapter 18). The hydrided palladium became "poisoned" for that reaction. However, acetylene introduced onto the β-PdH surface under the same conditions undergoes hydrogenation at the cost of the "hydride" hydrogen: during the reaction the evident positive shift of $\Delta \phi$ value from that characteristic of the β-PdH is observed. However, at 348 K ethylene can also be hydrogenated by the "hydride" hydrogen coming from the β-Pd hydride film [40].

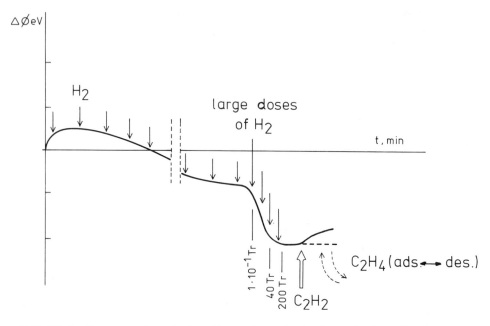

FIGURE 3 Changes of work function values ($\Delta\phi$) of a palladium film induced by hydrogen sorption, C_2H_2 introduction onto β Pd-hydride film (solid line), and alternative C_2H_4 introduction onto β Pd-hydride film (dashed line). T = 298 K.

For a chosen system—palladium-hydrogenation reactants—the entire body of thermodynamic and kinetic conditions accounts for a metal or metal hydride state in a catalyst, as well as for consequences of the catalyst state with regard to its effectiveness in a given hydrogenation reaction. The same concerns all the metal catalysts able to form hydrides. Since the complete information necessary is still lacking, any interpretation of experimental data on kinetics and mechanism of hydrogenation reactions of those metal catalysts, in terms of the role of a metal hydride phase, requires not only routine research on the reaction rate, but also an examination in situ of the metal catalyst phase composition. The progress of the process

Me + H_2 → Me - H (hydride phase)

should be followed by means of monitoring lattice parameter [by means of the x-ray diffraction (XRD) method], magnetic susceptibility, or electrical resistivity changes.

Additionally, the results are to be verified, eventually, by a process of hydride decomposition. The recent publication by Stachurski and Frackiewicz provides the necessary argument [41]. The authors found the formation of a Pd-C solution during the exposure of palladium to acetylene or its mixture with hydrogen. The observed increase of lattice parameter amounts to 3.992 Å for the highest concentration of carbon in palladium, corresponding to $PdC_{0.13}$. That is a potential source of misinterpretation of any positive shift of the palladium lattice parameter, observed in a reactant mixture of hydrogen and hydrocarbons, as due to the β Pd hydride formation.

A. Reactions of Hydrogen: o-p Conversion, H_2-D_2 Exchange, H Atom Heterogeneous Recombination

The research by Scholten and Konvalinka [42] has been a classical example of a study on (a) the o-p conversion of H_2, and D_2 as well, and (b) the H_2-D_2 exchange, performed on palladium and β palladium hydride catalysts. In those reactions hydrogen (or deuterium) and metal (Pd in this case) are sole partners in the entire process (i.e., in chemical reaction stages at a metal catalyst surface and in the α-β M-H phase transformation of a catalyst). That is why those reactions suit ideally for performance of the aimed comparative catalytic study.

The authors were aware of the consequences due to the α-β M-H phase interconversions, particularly of cracking and disintegration of metal crystallites. This effect itself may change the surface reactivity of a metal. The experiments concern catalytic activity of both phases of the Pd-H system, and the kinetics and mechanism of the reacting system. The decrease of preexponential factor and increase in activation energy that accompanied the α-β Pd-H phase transformation was studied within a wide temperature range.

In a recent study on the H_2-D_2 exchange, Nishiyama et al. [43] compared two metal catalysts: Pt and Pd (powders; supported metals). The enhancement of catalytic activity was observed after a pretreatment in ultrahigh vacuum (UHV) at 673 to 1373 K. The authors related the results with the surface rearrangement of the metal crystallites, induced by the form of pretreatment chosen. However, the authors did not take into account an eventual transformation of palladium (powder or Pd on silica) into the β-hydride phase when an equimolar mixture, H_2 + D_2, at 2.6 kPa, reacted on the catalyst at 273 K (Pd powder) or 159 K (Pd/SiO_2). The H_2 and D_2 pressures employed were high enough for the phase transformation. In such a case the observed restructuring of Pd crystallites would be reinforced by the effect accompanying the α-β transformation of the Pd-H system. A lack of direct information on XRD data taken under reaction conditions does not permit more than these speculations.

A decrease in the coefficient of recombination of hydrogen atoms at the surface of palladium transformed into its hydride phase was reported by Dickens et al. [44]. The loss of catalytic activity of palladium was quantitatively similar to that induced by the alloying of gold with palladium. An interpretation based on the respective change of the electronic structure of palladium seemed obvious to the authors.

B. Hydrogenation of Acetylene

Palladium is known as the best metal catalyst for the highly selective semi-hydrogenation of acetylenic bonds in a great variety of reactions carried out in industry and in the laboratory. They involve (a) a large-scale industrial operation of acetylenics removal from ethylene and propylene streams obtained in the hydrocarbon cracking process; and (b) small-scale syntheses in the pharmaceutical and fine chemical industries, where bond- and stereoselective hydrogenation is to be performed.

The question of the true character of a palladium catalyst is not only of a fundamental significance. The change of state of a Pd-H catalyst system due to the α-β Pd-H phase interconversion may be brought about by a slight change of p_{H_2}-T-P_i (i-reactants) conditions. In consequence, a sudden change of reaction rate, and its selectivity and mechanism, may occur.

The work of Borodziński et al. [39] and Frak [45] as well as the recent study of Pielaszek et al. [46] concentrate on the reaction of acetylene hy-

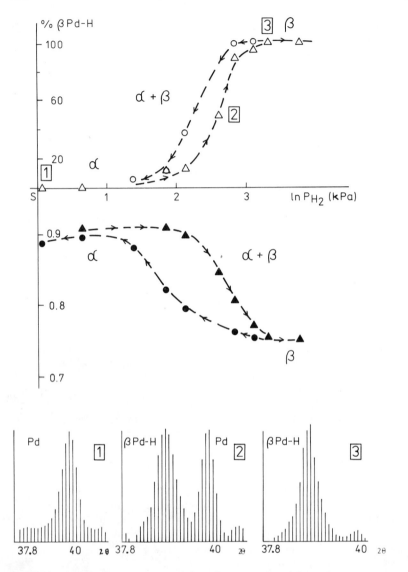

FIGURE 4 Relations between the phase composition of Pd-H system, expressed in percent β Pd-H, and the selectivity of catalytic semihydrogenation of acetylene ($C_2H_2 \rightarrow C_2H_4$); (△,▲) p_{H_2} increasing, (○,●) p_{H_2} decreasing. The x-ray diffraction patterns 1, 2, 3 were recorded using a step-by-step counting method; the figures were obtained after proper subtraction of background introduced by the support. (Data from Refs. 45 and 46.)

drogenation on supported Pd catalysts, conducted in a flow differential reactor-XRD chamber [47, 48]. The acetylene conversion and the phase composition of a palladium catalyst were followed simultaneously. The purpose was to find relationships between the initial composition of a reaction mixture, the state of a Pd-H catalyst system, and the distribution of the reaction products. An example of the results obtained [45, 46] is presented in Fig. 4. The $p_{C_2H_2} = 0.9$ kPa was constant, while p_{H_2} was variable; T = 303 K. The catalyst, 3 wt % Pd on γ-Al_2O_3, with a mean particle size diameter $\bar{d}$ = 18 nm, was metallic palladium or the α Pd-H phase (point 1;

the XRD pattern); it was selectively hydrogenating C_2H_2 to C_2H_4. When p_{H_2} reached about 7 kPa the phase transition of palladium into the β Pd-H phase began. The degree of the α-β Pd-H phase transformation was proceeding (point 2; XRD) and simultaneously, the selectivity, s, dropped until about s = 70% for the 100% β Pd-H phase (point 3; XRD). Along a decreasing branch of the hysteresis loop, with p_{H_2} diminishing, the β Pd-H phase decomposed and, simultaneously, the selectivity of C_2H_2 to C_2H_4 hydrogenation improved.

Summarizing the conclusions: For the given Pd catalyst with the mean particle size of the diameter $\bar{d}$ = 18 nm, at 303 K and $p_{C_2H_2}$ = 0.9 kPa:

1. Palladium transforms into the β hydride at a much higher p_{H_2} than predicted by the Pd-H solubility diagram for the α-β (dispersed) Pd-hydrogen phase transition. Acetylene is obviously competing for consumption of surface hydrogen atoms with the palladium bulk metal.
2. However, the dynamic equilibrium of the hydrogen absorption-desorption process on palladium may supply enough active hydrogen for the effective and nonselective hydrogenation of acetylene. The hypothesis was put forward that the desorbing "hydride" hydrogen transformed the postacetylenic admolecules (or ethylenic admolecules before their desorption) directly into ethane. The analogous mechanism was postulated by Moses et al. [49], based on the ethylidyne admolecules being hydrogenated to ethane. Bond and Wells [50] reported much earlier that acetylene hydrogenation on a Pd/Al_2O_3 catalyst became nonselective when the hydrogen pressure employed was sufficiently high. The authors explained a high activity of palladium (observed also in Ref. 39 on the β Pd-H phase catalyst) in terms of a high availability of surface hydrogen under conditions of its dynamic equilibrium with the bulk "hydride" hydrogen in the β Pd-H phase.
3. The highly dispersed palladium in a Pd 0.3 wt % (or less) on γ-Al_2O_3 ($\bar{d}$ = 1.7 nm) did not give any experimental evidence of undergoing the α-β Pd-H phase transformation; the selectivity of C_2H_2 to C_2H_4 hydrogenation was 90% or more up to a hydrogen partial pressure of about 30 kPa [46].

Carturan et al [51] carried out the phenylacetylene hydrogenation in a liquid solution at 298 K and constant p_{H_2}. Palladium catalysts of different metal dispersion manifested their changing ability for hydrogen absorption: The smallest Pd particles (diameter 1.3 to 1.9 nm) formed only the α Pd-H phase, the largest ones (diameter 5.4 to 10.6 nm) underwent the α-β Pd-H phase transformation. The best selectivity of the triple-bond semihydrogenation was characteristic of the highly dispersed Pd. As in Refs. 39 and 46, the Pd catalysts able to be hydrided showed the poorest selectivity, which was attributed by the authors to the β hydride, which furnished styrene admolecules with enough hydrogen to transform them into ethylbenzene. Unfortunately, XRD patterns were not taken in situ and the conditions inside the reactive medium could not be described in terms of the hydrogen availability for the Pd hydride formation.

An interplay of the variables p_{H_2}, $p_{C_2H_2}$, their mutual ratio, T, size of Pd crystallites and their real structure, and the presence of contaminants— in short, all the complex static and dynamic factors that influence the adsorption-desorption of H_2-C_2H_2-C_2H_4-C_2H_6, their reactions, and absorption-desorption in the Pd-H system—should be taken into account when predicting the best conditions for a desirable high yield of an olefin.

When analyzing the papers by various authors dealing with acetylene hydrogenation on palladium catalysts, one is generally unable to answer the question as to what was really acting as a catalyst: palladium or palladium hydride. The interpretation often raises doubts. The most convincing argument supporting the lack of the β Pd-H phase, within a very large range of conditions, is the application of a highly diluted, supported palladium catalyst, whose particles should be too small to form the β hydride. However, in many industrial (and laboratory) Pd catalysts, prepared for the acetylene semihydrogenation, palladium is deposited only near the outside of the pellets, because of a special peripheral impregnation of Al_2O_3 grains with palladium compound [49,52,53]. That outer layer of a catalyst grain, where the entire metal loading is concentrated, should be separately examined in order to determine the metal dispersion.

Another indirect argument supporting an eventual statement on the absence of the β Pd-H phase in the palladium catalyst employed is the similarity of adsorptive and catalytic behavior of palladium and platinum in a reactive hydrogenating medium.

In practice, an experimentally chosen set of conditions and correct composition of a catalyst ensure, within certain limits, good effectiveness of the process. When dealing with the problem of metallic hydrides in catalysis, one must mention their role, either potential or real, in the procedures involved in palladium catalyst preparation. Repeated hydride formation-dissociation, under the chosen conditions for each, results in the formation of well-reduced, finely dispersed, and sometimes, favorably restructured palladium crystallites [43,54-56].

C. Hydrogenation of Olefinic and Aromatic Compounds

When studying the hydrogenation of acetylene on palladium, Bond et al. [57] observed that prolonged exposure of the catalyst to hydrogen poisoned the hydrogenation of ethylene. The effect was interpreted in terms of the electronic theory of catalysis applied to the β palladium hydride. Systematic studies were carried out by Duś [38] and Frąckiewicz et al. [40]. The β palladium hydride, which was stable at sufficiently low temperature, was catalytically inactive in ethylene hydrogenation, but at higher temperatures (T = 348 K) the flux of hydrogen exchanged between palladium hydride and the reactant mixture $p_{C_2H_2}$ = 2 kPa and p_{H_2} = 27 to 40 kPa) supplied enough "hydride" hydrogen for the ethylene hydrogenation and simultaneously regenerated the β Pd-H phase by the hydrogen absorption process. The partially hydrided palladium sample studied by Rennard and Kokes [58] was an effective catalyst, even at low temperature, in propylene and ethylene hydrogenation via the desorbing hydrogen. One might expect, as appears likely, that active hydrogen spilled over to α Pd-H crystallite faces hydrogenated there by ethylene admolecules. In numerous papers published recently, the ability of palladium to absorb/adsorb hydrogen has not been exploited purposely in order to study relationships with the rate and mechanism of hydrogenation reactions.

The possibility of palladium hydride formation revealed its utility for interpretation of the dependence of the hydrogenation rate on palladium dispersion in the range of crystallite size, which could not justify structure sensitivity in its commonly applied meaning. Gubitosa et al. [59] found that 1-octene hydrogenation (1 atm H_2, 303 K, solvent: n-octane) on Pd/SiO_2 catalysts decreased with increased palladium dispersion, even within the relatively low dispersion range 0.1 to 0.2. The analogous character of the hydrogen solubility in Pd dispersion entitled the authors to explain the apparent "structure sensitivity" of 1-octene hydrogenation through the influence of the presence of the β Pd-H phase in palladium crystallites, vanishing with increased dispersion.

One might be tempted to propose a similar interpretation for the effect of 1-butyne, 1,3-butadiene, and isoprene hydrogenation (2 MPa H_2, 298 K, liquid phase) sensitive to palladium dispersion observed by Boitiaux et al. [60]. At a dispersion of 0.1 to 0.2 the activity of the hydrogenation of the unsaturated hydrocarbons is high, but it diminishes dramatically, by a factor of 12 to 15, within dispersion values of 0.2 to 0.4. The authors' interpretation based on the change of electronic character of the metal as its dispersion increases might also be valid as an indirect argument. That change would influence the hydriding of palladium bulk as well as the adsorption and reactivity of hydrocarbon molecules at its surface. The hydrogenation of propylene on Pd/SiO_2 and Pt/SiO_2 catalysts was studied by Rorris et al. [61]. The sensitivity of the rate of reaction (TOF) on catalyst pretreatment conditions and metal dispersion was compared. The relationships observed for palladium were distinct from those for platinum. The TOF values of propylene hydrogenation ($C_3H_6:H_2$ = 1:17; 201 K) passed through a maximum with increasing Pd dispersion (dispersion at the maximum TOF rate, 65.5%).

During the hydrogenolysis of methylcyclopentane investigated by Nandi et al. [62] on palladium catalysts of different metal dispersion and pretreatment, the α-β Pd-H phase transformation was followed in situ by the XRD method. The larger palladium hydride crystallites were less active than the smaller ones.

In the propylene hydrogenation [61] the authors could not detect the palladium hydride phase. However, they admitted its occurrence for lower values of metal dispersion. With decreasing crystallite size the propylene admolecules "extract" hydrogen more easily from the dispersed bulk β Pd-H phase which itself, moreover, decomposes more easily, being thermodynamically less stable. Finally, at a dispersion of about 60% the palladium particles become too small to be transformed into the hydride phase and the rate or reaction diminishes.

Following that reasoning, one might suppose that for a particular hydrogenation reaction, under chosen p_i and T conditions, a certain optimum dispersion of palladium could be found. Palladium supported on TiO_2 was chosen for a search of the additional effect in SMSI, characteristic of a metal able to be transformed into metallic hydride under reaction conditions. Vannice et al. [63] and Baker et al. [64] observed, in contrast to Pd/Al_2O_3 (or Pd/SiO_2, $Pd/Al_2O_3/SiO_2$), that Pd/TiO_2 exhibited diminishing hydrogen uptake with increasing reduction temperature. The total quantity of adsorbed and absorbed hydrogen fell to zero after reduction of the catalyst at 773 to 873 K [64]. A concomitant decrease in activity in benzene hydrogenation was also observed [63]. The authors of both studies considered various possible explanations in terms of SMSI (see in detail in Chapter 13). However, another attempt at interpretation might be presented as well. The palladium particles supported on TiO_2 and reduced at 773 K were small enough (d = 3 nm) [63] to undergo the SMSI of electronic character. In consequence, palladium could lose its ability to be transformed into the β hydride under reaction conditions (T = 417 K; p_{H_2} = 91 kPa; p_{O_2} = 6.6 kPa). If the availability of "hydride" hydrogen was responsible for the high values of the TOF of benzene hydrogenation, the observed loss of the catalyst activity after its reduction at high temperature could be adequately explained. The same reaction was studied by Orozco and Webb [65] over platinum and palladium supported on alumina or silica. The interesting observation concerned the lack of retention of benzene admolecules (or rather their carbonaceous deposits) on palladium crystallites during benzene hydrogenation. The authors interpreted that result by the limited dissociation of benzene admolecules due to the high surface coverage of palladium with hydrogen, provided by the bulk hydride.

Therefore, it seems justified to expect that the hydrided palladium catalyst should be more resistant to coke formation than most transition metal catalysts are.

IV. HYDRIDES OF INTERMETALLIC COMPOUNDS AS CATALYSTS IN CHEMICAL REACTIONS

The binary, MH_x, saltlike hydrides of alkali and alkaline earth metals, metallic hydrides of most d-transition metals, all rare earth metals, and actinides are very stable up to high temperatures. Their dissociation pressures are low under ambient conditions and reach 1 atm in the temperature range which extends from about 600 K (Mg) to 1600 K (Y), compared to about 400 K for Pd. A large hysteresis effect and, in practical hydriding, a necessity to apply much higher pressures put those hydrides out of the question in a search for materials suitable for reversible hydrogen storage [4]. However, with regard to catalytic applications, particularly for reactions at higher temperatures and hydrogen pressures, the potential utility of certain M-H binary systems must be considered. Lisichkin et al. [66] showed a stable catalytic activity of titanium and zirconium hydrides dispersed on supports in model hydrogenations of hexene-1 and cyclohexene in the temperature range 370 to 420 K. The large crystalline unsupported samples were much less effective in those reactions even at 570 to 720 K, although they catalyzed isomerization and disproportionation of those hydrocarbons.

Numerous alloys and intermetallic compounds of the foregoing metals absorb hydrogen and form ternary hydrides $M_I M_{II} H_x$. Systems that had gained a good "reputation" in hydrogen storage also became the subject of catalytic studies because of the similarity of some properties necessary for materials relevant to both fields. Moreover, it was an advantage for catalytic studies that those metallic and intermetallic hydride systems had been recognized relatively well and much earlier and their properties had been studied extensively because of their probable utility for "hydrogen economy."

Among those ternary metallic hydrides, described now as "probably the best yet found for stationary applications" [4], are $LaNi_5H_6$ and $TiFeH_2$. Moreover, they may be taken as prototypes for series of other alloys and intermetallics.

The intermetallic compound $LaNi_5$, and derived alloys such as $LaNi_{5-x}Cu_x$ or $MmNi_5$ (Mm stands for "Mischmetal," a commercial mixture of rare earths), are known by the ability to form hydride. In the intermetallic TiFe, a part of the iron has been replaced by Cr, Mn, Co, Ni, or Cu, and the titanium by zirconium. All are capable of being transformed into the respective hydride phase(s). Among the remaining ternary systems, those based on thorium [i.e., $ThNi_5$ (where nickel may be replaced by Co, Fe)] deserve particular attention. It is noteworthy that the parent metal in all the alloys and intermetallics mentioned (i.e., La, Ti, Zr, Th) becomes oxidized easily and then the bimetallic system breaks down into a respective metal oxide and a transition (or group IB) metal.

A. Catalytic Properties of Hydrides of Intermetallics: RE M_5

The hydrides of intermetallics RE M_5 (RE stands for rare earth metal, and M for Ni, Co, Fe, Cr, Pd, Cu, or Ag) are alike with regard to their dissociation pressure being much lower than that of the host rare earth metal itself. Therefore, they are much more suitable for hydrogen storage and heterogeneous catalysis.

The effectiveness of intermetallic compounds as catalysts was anticipated in 1977: "The rapidity with which the rare earth intermetallics absorb hydrogen suggests that they are unusually effective in breaking the hydrogen bond. This in turn suggests that these materials . . . might be effective as hydrogenation catalysts" [67].

Takeshita et al. [68] studied ammonia synthesis over RE-Fe and RE-Co and RE-Ru intermetallics, under 5 to 7 MPa of the $N_2:H_2 = 1:3$ reaction mixture at 673 K (RE-Ce, Pr, Er). They compared the yield of NH_3 with that obtained over a conventional, industrial iron catalyst. The RE-M catalysts studied were superior to it, but an interesting transformation of the intermetallics was observed: They converted into finely dispersed transition metal supported on a respective RE nitride. The authors note the absorption of a large quantity of hydrogen by the catalysts.

Coon et al. [69], Luengo et al. [70], Atkinson et al. [71], and Elattar et al. [72] investigated the (CO, H_2) reaction in the temperature range 520 to 700 K and pressure range 0.1 to 1 MPa, over the $LaNi_5$ and the other RE-M (here M stands for Fe, or Co). The activity of those catalysts exceeded that of a typical industrial nickel catalyst by about an order of magnitude. The main product was methane, although some tendency to higher hydrocarbon formation was noticed. The intermetallics were found to decompose during their exposure to reaction conditions. The active catalysts were identified by XRD and EDAX as heterogeneous mixtures consisting of the dispersed nickel or iron supported on the RE oxide. The authors stressed that all of the intermetallics readily absorbed hydrogen. Luengo et al. [70] observed an increased resistance of those catalysts on carbonization or carbide formation and a subsequent loss of activity and reproducibility of the catalytic system studied.

Recently, Barrault et al. [73] have studied $LaNi_5$ and $LaNi_4Mn$ catalysts in methanation, hydrodealkylation of toluene, and ethane hydrogenolysis. In the last case only, the intermetallics were catalytically less active than an industrial Ni/Al_2O_3 catalysts. The observed decomposition of the intermetallic compound leading to a dispersed metal-on-oxide support system was described as a new, unconventional method of preparation of highly active supported metal catalysts. The research carried out by Schaper et al. [74] confirmed that even the addition of 5 mol % of La_2O_3 to Ni/Al_2O_3 catalysts increased their methanation activity.

Although the authors interpret the high catalytic effectiveness of the intermetallics studied in terms of their ability to absorb hydrogen, there is insufficient experimental evidence to claim that hydride phases are responsible for the phenomenon. For instance, the pressure versus composition isotherms for the system $LaNi_5-H_2$ predict that within the relatively low range of pressures and high temperatures applied, the $\alpha-LaNi_5H_x$ solution was forming with a much lower hydrogen content than in the hydride phase. Nevertheless, the dynamics of its sorption-desorption process could influence reactions involving hydrogen on the catalyst surface.

Soga et al. [75] studied catalytic hydrogenation of ethylene over partially hydrided intermetallics: $LaNi_5H_n$ (n = 0.5 to 2.4) as prototype, also with La substituted by Pr, Ce, Sm, and Ni by Co. In most experiments the hydride content was low (e.g., n_{max} equals 7 for the $LaNi_5$ compound) and the $LaNi_5H_n$ should be treated as being far from homogeneity, similar to the earlier study [58]. In a temperature range 188 to 227 K, ethylene hydrogenation occurred solely at the expense of the "hydride" hydrogen, when only ethylene had been introduced over the hydride surface. When a mixture $C_2H_4 + H_2$ was used, the reaction still proceeded by using mainly the lattice hydrogen; the gaseous H_2 reacted comparatively slowly. The ethylenic admolecules retarded the hydrogen absorption and the hydride restoration.

$LaNi_5H_6$ has also been tested as an electrocatalytic material of an electrode in the electrooxidation of hydrogen dissolved in the hydride $LaNi_5$ [76]. The discharge curves on $LaNi_5H_6$ electrodes in 7 N KOH, at 293 K, after an initial jump on the curve: potential versus current density showed a long plateau of potential in the large range of current density values. In the stationary state, the rate of the "hydride" hydrogen release from the bulk of the electrode was equal to the rate of consumption of the "surface" hydrogen in the electrochemical discharge step

$$H_a + OH^- \to H_2O + e$$

The electrocatalytic activity of the $LaNi_5H_6$ in the hydrogen oxidation reaction has been expressed by the value of the exchange current; it was one order of magnitude higher than that of metallic nickel.

The $LaNi_5$ with its nickel partially substituted by copper (i.e., $LaNi_{5-x}Cu_x$) and the respective α and β $M_IM_{II}M_{III}$-H phases have been studied by Konenko et al. in catalytic hydrogenation of propene [77]. The system had been investigated earlier, and the authors had stated that the replacement of Ni by Cu in $LaNi_5$ reduced the dissociation pressure of the β hydride, thus indicating its higher stability. However, the increase of copper content lowered the hydrogen concentration in the alloy. The studies also concerned the electronic structure of those trimetallic alloys [78,79].

Konenko et al. [77] found that the catalytic activity of the intermetallics $LaNi_3Cu_2$ and $LaNi_2Cu_3$ was one order of magnitude lower than that of the respective hydrides. The mechanism of the reaction studied was different. In the case of intermetallics the rate equation was of zero order with respect to propene concentration and first order with respect to hydrogen. With the hydrides the reverse was found; the zero order in respect to hydrogen partial pressure permits one to anticipate that the dynamic equilibrium between the "hydride" and the gaseous hydrogen was not disturbed by the catalytic reaction [45,46]. Therefore, the propylene hydrogenation occurred on the metallic hydride catalyst.

B. Catalytic Properties of Hydrides of Intermetallics Formed by Ti, Zr, Hf, or Th with Other Metals

The intermetallic compound FeTi reacts with hydrogen and undergoes the transformation into hydrides: FeTiH and then $FeTiH_2$. Their dissociation pressures are over 1 atm at 273 K, unlike the TiH_2 itself, which is very stable. The intermetallic compounds and alloys, which can be regarded as derivatives of the Ti-Fe hydrides (e.g., $TiFe_{1-x}Ni_xH_y$), dissociate at a hydrogen pressure one order lower than their prototype [4,80].

Numerous studies have been performed by Lunin et al. (Refs. 1 to 85 and references cited therein) aiming to compare the catalytic activity of a long series of intermetallics with that of their hydrides. The intermetallic hydrides studied were $TiCuH_x$, $TiMoH_x$; $ZrFeH_x$, $ZrCoH_x$, $ZrNiH_x$; $ZrCr_2H_x$, $ZrMo_2H_x$, ZrW_2, $ZrNiCuH_x$, $ZrCoCuH_x$, $HfNiH_x$, and $HfCoH_x$. The experimental evidence was generally similar and may be best represented by the behavior of $ZrNiH_{2.8}$, $HfNiH_{2.8}$, and the respective hydrides substituted by Co. Their catalytically active surface has been forming under the influence of the reaction environment during a long induction period (8 to 10 h). Breakdown of the intermetallic structure, subsequent oxidation of its components in the layer of the hydride and segregation of the finely dispersed metallic nickel (or cobalt) are represented in Fig. 5 [81]. The similar behavior of the lanthanum-based intermetallic hydrides has been described earlier.

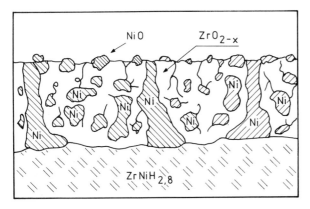

FIGURE 5 Spectacular presentation of $ZrNiH_{2.8}$ hydride after its exposure to an oxidizing medium [81].

However, the $ZrNiH_{2.8}$ phase remains inside catalyst grains and determines the activity and stability of the catalyst. The respective conventional model systems (e.g., Ni supported on ZrO_2) prepared by impregnation and precipitation, are much less active and stable, for example in toluene conversion at 453 to 573 K (hydrogenation, hydrogenolysis, hydrodemethylation).

Physical properties of the complex catalyst system formed have been studied and the model represented in Fig. 5 is based on those results. The final conclusions are as follows:

1. The superficial oxidation of the intermetallic hydride is to be considered as the process of active catalyst formation.
2. The hydride phase plays an important role in catalytic activity, based on indirect exchange of "hydride" hydrogen with the reaction medium.
3. The complex metal-metal oxide, M_I-$M_{II}O_x$, layer exerts a highly specific influence on the direction of catalytic reaction, thus determining its selectivity.

The intermetallic hydrides were studied in many hydrocarbon conversions: alkane hydrogenolysis, hydroisomerization, dehydrogenation, dehydrocyclization, unsaturated hydrocarbon hydrogenation, aromatic hydrocarbon transformation, and so on. The intermetallic hydrides have always demonstrated their superiority over conventional catalysts of a similar composition.

$ZrNiH_3$ has recently been investigated by Frąk [86] in the catalytic hydrogenation of benzene. As in previous studies, the hydride was a precursor of a complex M_I-$M_{II}O_x$-$M_IM_{II}H_x$ catalyst. The catalyst precursor transformations could be followed during the catalytic reaction in situ in a special reactor-x-ray camera. However, the author claims that under the reaction conditions, dispersed metallic nickel and gaseous hydrogen, not the "hydride" hydrogen, plays the principal role in the catalytic reaction. The kinetic equation of the nonzero order in respect to hydrogen supports the proposed mechanism.

The contradictions between the observations and interpretations of different authors might be explained in terms of different behavior of the same bulk hydride systems. The easy or difficult accessibility to "hydride" hydrogen, its ability to be exchanged with the necessary rate across the interface, determines the "hydride" hydrogen reactivity. The "art" of an adequate choice of pretreatment would influence the subsequent catalytic reaction mechanism.

An outstanding example of the hydride phase utility in catalysis has been presented by Fanelli et al. [87]. The hydride-forming metal powders Ti, Ti-V-Fe, and Ti-V-Fe-Pt were used as reactants in alkane dehydrogenation, for example, at 641 K,

$$i\text{-}C_4H_{10}(g) + Ti(s) = i\text{-}C_4H_8(g) + TiH_2(s)$$

Titanium "extracts" hydrogen from the alkane molecule and then absorbs it. The sufficiently high negative value of the standard free enthalpy of TiH_2 formation, $\Delta G^\circ_{TiH_2}$, overcompensates the positive value of the respective dehydrogenation reaction itself. However, the additional introduction of a conventional catalyst, Pt/Al_2O_3, is necessary for acceleration of the reaction. Titanium, when transformed into TiH_2, is to be regenerated (873 K, 0.1 Pa) or used in a reaction of hydrogenation. The regenerated titanium shows a higher activity in the absorption of hydrogen; that is understandable as the effect of the hydriding-dehydriding process which develops the surface area of the metal-sorbent. The intermetallic compounds or alloys mentioned above, Ti-V-Fe or Ti-V-Fe-Pt, were applied by the authors in order to avoid the additional use of a separate catalyst. Unlike titanium, those alloys exhibited a certain, but rather low catalytic efficiency. The mechanism proposed by the authors interprets elementary acts of the hydrogen transfer from hydrocarbon molecule to titanium crystallites. However, the rate-determining step of the process is the hydrogen uptake by titanium.

V. CATALYTIC MEMBRANE SYSTEMS

Metals and alloys of outstanding solubility and diffusivity of hydrogen have been used frequently for the purification of hydrogen. Their transformation into hydride phase may be disadvantageous because of irreversible changes of their real structure; the formation of cracks and blisters; and so on.

Gryaznov et al. (Refs. 88-90 and references cited herein) initiated and have developed reactors, based on the idea of a controlled permeation of hydrogen across metallic membranes (e.g., Pd, PdRh, PdRu, PdNi, etc.) into a reaction chamber filled with hydrocarbon to be hydrogenated. In dehydrogenation, dehydrocyclization, and other reactions studied, the membrane, permeable for hydrogen, is like the titanium in Fanelli's [87] study, but its superiority resides in the continuous character of the process. The area of membrane catalysts is covered by numerous papers published by Gryaznov and co-workers. They concern reactors with membranes designed in different ways; the permeability of palladium alloys and palladium metal used as the membrane materials; and saturated, unsaturated, aromatic, cyclic, and so on, hydrocarbons undergoing conversion to desired products—with high selectivity, regulated by the change in the hydrogen flux across the membrane, and with large effectiveness, due to the chosen material of the membrane and to the active hydrogen desorbing from the membrane onto the reactive surface (see also Chapter 26).

Soga et al. [91] also employed a membrane for hydrogen transfer into a reactive medium. In their study, however, the $LaNi_5H_X$ plate was used as membrane. As in earlier studies [58,75] ethylene hydrogenation was carried out also by the "hydride" hydrogen in the case of that hydride membrane. The source of hydrogen supplied to ethylene molecules had a constant concentration and was not reduced gradually during the reaction. The reported study represents one more case of a hydride as a medium just providing active hydrogen on the catalytic surface.

VI. CONCLUDING REMARKS

When metals, alloys, or intermetallic compounds able to absorb a large quantity of hydrogen and to form hydride phases are employed as catalysts in hydrogen reactions, the conversion of metallic phase into its hydride, below the critical temperature of hydride formation, must be taken into account. The changed geometry of the crystal lattice as well as the different electronic structure of the hydride might be equivalent to the change of a catalyst itself. One then observes changes of catalytic activity and selectivity, resistance on deactivating influence of carbonaceous deposits, and so on.

The knowledge of a phase diagram of the particular metal-hydrogen-metal-hydride system itself is of prime importance. Nevertheless, that is not sufficient. The shape of the hysteresis loop, the state of surface activation or deactivation, the metal dispersion—indeed, the entire body of properties influencing dynamics of hydrogen sorption-desorption process—plays an important role, but unfortunately, it cannot be predicted on the ground of existing experimental evidence and theoretical research.

The high dispersion of a metal (alloy) catalyst, typical for those systems, creates further problems for anticipation of a metal-hydrogen behavior. Highly dispersed hydride metal shows lower stability, its dissociation pressure has a large value, and hydrogen concentration is small.

When under reaction conditions the competition for surface hydrogen between the metal and the reactant to be hydrogenated becomes significant, an increase in hydrogen partial pressure is necessary if the hydride phase has to be dynamically preserved. The value of the optimum p_{H_2} under particular reaction conditions is not predictable: Experimental evidence is necessary, based on studying the metal-hydrogen-hydride transformation in situ during adequately changed conditions (p_i, T_i, p_i/p_{H_2}) of a given reaction.

The stable d-transition metal β hydride, with a characteristic low dissociation pressure, owing to its electronic structure, may be inactive in hydrogenation reactions. There are, however, reactants whose affinity to hydrogen is high enough to enable them to extract "hydride" hydrogen from the metal lattice.

The metallic β hydride and α solution, with high hydrogen content unstable under given reaction conditions but exhibiting a high rate of exchange of hydrogen across the metal-gas interface, are sources of the active hydrogen for a reactant medium. The rate of catalytic reaction becomes higher than in the case of a respective metal depleted of hydrogen. Suitable selection of the given reaction conditions may secure dynamic stability and continuous activity in the hydride catalyst. The reaction studied is decisive for an adequate choice of conditions under which it is to proceed, in order to avoid, or to form, a metal catalyst of hydride character.

Apart from their direct participation in the reaction as a peculiar catalyst, metal hydrides may be useful in a catalyst-activating pretreatment. A cycling hydriding-hydrogen desorption process changes the catalyst precursor morphology, disintegrates it, and in some cases favors a certain texture dominating. Moreover, intermetallic hydrides ($RE-MH_x$; M_iMH_x; where M_i stands for Ti, Zr, Hf, Th, and M for Fe, Co, Ni, Cu, etc.) react easily with oxygen (even of low partial pressure) and break down, transforming into highly dispersed Fe, Ni, Co, etc., on RE or Ti, Zr, Hf, Th, etc., oxides. The nucleus of the intermetallic grain eventually retains the initial hydride phase and may become a steady-state source of hydrogen. The complex catalytic system formed is highly effective in many catalytic reactions, much more so than the respective intermetallic compound.

Summing up, one can express an optimistic opinion that rate and selectivity of many hydrogen reactions might be directed better by the presence of hydride phase or by a controlled supply of "hydride" hydrogen. Further progress as to knowledge on metal hydrides and their catalytic effects is necessary to achieve that goal.

REFERENCES

1. G. Alefeld, in *Proceedings of the Japanese Institute of Metals International Symposium, Minakami*, Vol. 2, Japanese Institute of Metals, Sendai, Japan (1979), p. 25.
2. G. G. Libowitz, in *Proceedings of the International Symposium, Geilo, Norway, 1977* (A. F. Andersen and A. J. Maeland, eds.), Pergamon Press, Oxford (1978), p. 1.
3. A. J. Maeland, in *Proceedings of the International Symposium, Geilo, Norway, 1977* (A. F. Andersen and A. J. Maeland, eds.), Pergamon Press, Oxford (1978), p. 19.
4. W. E. Wallace and S. K. Malik, in *Proceedings of the International Symposium, Geilo, Norway, 1977* (A. F. Andersen and A. J. Maeland, eds.), Pergamon Press, Oxford (1978), p. 33.
5. F. E. Wagner and G. Wortmann, in *Hydrogen in Metals*, Vol. 1 (G. Alefeld and J. Völkl, eds.), Springer-Verlag, Berlin (1978), p. 131.
6. R. Wiswall, in *Hydrogen in Metals*, Vol. 1 (G. Alefeld and J. Völkl, eds.), Springer-Verlag, Berlin (1978), p. 201.
7. K. H. J. Buschow and H. H. van Mal, in *Intercalation Chemistry* (M. S. Whittingham and A. J. Jocobson, eds.), Academic Press, London (1982), p. 405.
8. A. C. Switendick, *Ber. Bunsenges. Phys. Chem.*, 76:535 (1972); in *Hydrogen in Metals*, Vol. 1 (G. Alefeld and J. Völkl, eds.), Springer-Verlag, Berlin (1978), p. 101.
9. R. C. Baetzold and R. E. Mack, *J. Chem. Phys.*, 62:1513 (1975).
10. R. C. Baetzold, in *Catalysis in Chemistry and Biochemistry: Theory and Experiment* (B. Pullman, ed.), D. Reidel, Dordrecht, The Netherlands (1979), p. 191.
11. F. Cyrot-Lackmann, in *Catalysis in Chemistry and Biochemistry: Theory and Experiment* (B. Pullman, ed.), D. Reidel, Dordrecht, The Netherlands (1979), p. 217.
12. S. Ladas, R. A. Dalla Betta, and M. Boudart, *J. Catal.*, 53:356 (1978).
13. P. C. Aben, *J. Catal.*, 10:224 (1968).
14. M. Boudart and H. S. Hwang, *J. Catal.*, 39:44 (1978).
15. R. K. Nandi, P. Georgopoulos, J. B. Cohen, J. B. Butt, R. L. Burwell, and D. H. Bilderback, *J. Catal.*, 77:421 (1982).
16. P. D. Goodell, *J. Less-Common Met.*, 89:45 (1983).
17. H. Uchida, Y. Uchida, and Y. C. Huang, *J. Less-Common Met.*, 101:459 (1984).
18. D. Khatamian, G. C. Weatherly, F. D. Manchester, and C. B. Alcock, *J. Less-Common Met.*, 89:71 (1983).
19. J. J. Reilly, R. H. Wiswall, *Inorg. Chem.*, 13:218 (1974).
20. M. M. Antonova, S. N. Endreevskaia, V. S. Puk'anchikov, A. G. Shablina, and O. T. Khorp'akov, *Neorg. Mater.*, 15:1939 (1979).
21. E. Wicke and G. H. Nernst, *Ber. Bunsenges. Phys. Chem.*, 68:224 (1964).
22. A. Küssner and E. Wicke, *Z. Phys. Chem. (Neue Folge)*, 24E:152 (1960).
23. A. Küssner, *Z. Phys. Chem. (Neue Folge)*, 64:225 (1969).
24. E. Wicke and K. Meyer, *Z. Elektrochem.*, 66:675 (1962).

25. N. Boes and N. Zücher, *Z. Naturforsch. Teil, A31*:760 (1976).
26. M. Śmiałowski and H. Jarmołowicz, *J. Catal.*, *1*:165 (1962).
27. M. Hashimoto, in *Proceedings of the Japanese Institute of Metals International Symposium, Minakami*, Vol. 2, Japanese Institute of Metals, Sendai, Japan (1979), p. 209.
28. H. Murayama, in *Proceedings of the Japanese Institute of Metals International Symposium, Minimaka*, Vol. 2, Japanese Institute of Metals, Sendai, Japan (1979), p. 297.
29. T. Schober, *J. Less-Common Met.*, *89*:63 (1983).
30. F. R. Block and H. J. Baks, *J. Less-Common Met.*, *89*:77 (1983).
31. D. M. Gualtieri, K. S. V. L. Narasimnan, and T. Takeshita, *J. Appl. Phys.*, *47*:3432 (1976).
32. R. Wiswall, *Hydrogen in Metals*, Vol. 2 (G. Alefeld and J. Völkl, eds.), Springer-Verlag, Berlin (1978), p. 201.
33. B. Kasemo and E. Törngvist, *Appl. Surf. Sci.*, *3*:307 (1979).
34. G. Sicking and B. Jungblut, *Surf. Sci.*, *127*:255 (1983).
35. F. A. Lewis, *The Palladium Hydrogen System*, Academic Press, New York (1976).
36. E. Wicke and H. Brodowsky, *Hydrogen in Metals*, Vol. 2 (G. Alefeld and J. Völkl, eds.), Springer-Verlag, Berlin (1978), p. 73.
37. W. Palczewska, *Adv. Catal.*, *24*:245 (1975).
38. R. Duś, *Surf. Sci.*, *50*:241 (1975).
39. A. Borodziński, R. Duś, R. Frąk, A. Janko, and W. Palczewska, *Proceedings of the 6th International Congress on Catalysis, London, 1976* (G. C. Bond, P. B. Wells, and F. C. Tompkins, eds.), Chemical Society, London (1977), p. 150.
40. A. Frąckiewicz, R. Frąk, and A. Janko, *Rocz. Chem.*, *51*:2395 (1977).
41. J. Stachurski and A. Frackiewicz, *J. Less-Common Met.*, *108*:249 (1985).
42. J. J. S. Scholten and J. A. Konvalinka, *J. Catal.*, *5*:1 (1966).
43. S. Nishiyama, S. Matsumura, H. Morita, S. Tsuruya, and M. Masai, *Appl. Catal.*, *15*:185 (1985).
44. P. G. Dickens, J. W. Linnett, and W. Palczewska, *J. Catal.*, *4*:140 (1965).
45. R. Frąk, Ph.D. thesis, Institute of Physical Chemistry, Polish Academy of Science (1979).
46. J. Pielaszek, J. Sobczak, A. Borodziński, and W. Palczewska, unpublished results.
47. A. Borodziński and A. Janko, *React. Kinet. Catal. Lett.*, *7*:163 (1977).
48. J. Zieliński and A. Borodziński, *Appl. Catal.*, *13*:305 (1985).
49. J. M. Moses, A. H. Weiss, K. Matusek, and L. Guczi, *J. Catal.*, *86*:417 (1984).
50. G. C. Bond and P. B. Wells, *J. Catal.*, *5*:65 (1965).
51. G. Carturan, G. Facchin, G. Cocco, S. Enzo, and G. Navazio, *J. Catal.*, *76*:405 (1982).
52. W. T. McGown, Ch. Kemball, D. A. Whan, and M. S. Scurrell, *J. Chem. Soc. Faraday Trans. 1*, *73*:632 (1977); W. T. McGown, Ch. Kemball, and D. A. Whan, *J. Catal.*, *51*:173 (1978).
53. L. Guczi, Z. Schay, A. H. Weiss, V. Nair, and S. LeViness, *React. Kinet. Catal. Lett.*, *27*:147 (1985).
54. O. Leonte, M. Birjega, N. Popescu-Pogrion, C. Sarbu, D. Macorei, P. Pausescu, and M. Georgescu, *Proceedings of the 8th International Congress on Catalysis, West Berlin*, Vol. 2 (1984), p. 683.
55. A. Janko, W. Palczewska, and I. Szymerska, *J. Catal.*, *61*:264 (1980).
56. M. J. Ledoux and F. G. Gault, *J. Catal.*, *60*:15 (1979).
57. G. C. Bond, D. A. Dowden, and N. Mackenzie, *Trans. Faraday Soc.*, *54*:1537 (1958).

58. R. J. Rennard and R. J. Kokes, *Catal. Rev.*, *6*:1 (1972).
59. G. Gubitosa, A. Berton, M. Camia, and N. Pernicone, in *Preparation of Catalyst*, Vol. 3 (G. Poncelet, P. Grange, and P. A. Jacobs, eds.), Elsevier, Amsterdam (1983), p. 431.
60. J. P. Boitiaux, J. Cosyns, and S. Vasudevan, in *Preparation of Catalyst*, Vol. 3 (G. Poncelet, P. Grange, and P. A. Jacobs, eds.), Elsevier, Amsterdam (1983), p. 123; *Appl. Catal.*, *6*:41 (1983); *Appl. Catal.*, *15*: 317 (1985).
61. E. Rorris, J. B. Butt, Jr., and J. B. Cohen, *Proceedings of the 8th International Congress on Catalysis, West Berlin*, Vol. 4 (1984), p. 321.
62. R. K. Nandi, F. Molinaro, C. Tang, J. B. Cohen, J. B. Butt, and R. L. Burwell, Jr., *J. Catal.*, *78*:289 (1982).
63. M. A. Vannice and P. Chou, *Proceedings of the 8th International Congress on Catalysis, West Berlin*, Vol. 5 (1984), p. 99.
64. R. T. K. Baker, E. B. Prestridge, and G. B. McVicker, *J. Catal.*, *89*: 422 (1984).
65. J. M. Orozco and G. Webb, *Appl. Catal.*, *6*:67 (1983).
66. G. V. Lisichkin, A. V. Guriev, and E. A. Viktorova, *Neorg. Mater.*, *14*: 1710 (1978).
67. W. E. Wallace in *Proceedings of the International Symposium, Geilo, Norway, 1977* (A. F. Andersen and A. J. Maeland, eds.), Pergamon Press, Oxford (1978), p. 501.
68. T. Takeshita, W. E. Wallace, and R. S. Craig, *J. Catal.*, *44*:236 (1976).
69. V. T. Coon, T. Takeshita, W. E. Wallace, and R. S. Craig, *J. Phys. Chem.*, *80*:1878 (1976).
70. C. A. Luengo, A. L. Cabrera, H. B. Mackay, and M. B. Maple, *J. Catal.*, *47*:1 (1977).
71. G. B. Atkinson and L. J. Nicks, *J. Catal.*, *46*:417 (1977).
72. A. Elattar, W. E. Wallace, and R. S. Craig, *Hydrocarbon Synthesis from Carbon Monoxide and Hydrogen, Advanced Chemistry Series 178* (E. L. Kugler and F. W. Steffgen, eds.), American Chemical Society, Washington, D.C. (1979), p. 15.
73. J. Barrault, D. Duprez, A. Percheron-Guégan, and J. C. Achard, *J. Less-Common Met.*, *89*:537 (1983).
74. H. Schaper, E. B. M. Dvesburg, P. H. M. Korte, and L. L. Van Reijen, *Appl. Catal.*, *14*:371 (1985).
75. K. Soga, H. Imamura, and S. Ikeda, *J. Phys. Chem.*, *81*:1762 (1977); *J. Catal.*, *56*:119 (1979).
76. A. G. Pshenichnikov, *J. Res. Inst. Catal., Hokkaido Univ.*, *30*:137 (1982).
77. I. R. Konenko, E. V. Starodubtseva, P. Yu. Stepanov, E. A. Fedorovskaya, A. A. Slinkin, E. I. Klabunovskii, E. M. Savitskii, V. P. Mordovin, and T. P. Savostiyanova, *Kinet. Katal.*, *26*:340 (1985).
78. W. E. Wallace, *J. Less-Common Met.*, *88*:141 (1982).
79. J. Shimar, I. Jacob, D. Davidov, and D. Shaltiel, in *Proceedings of the International Symposium, Geilo, Norway, 1977* (A. F. Andersen and A. J. Maeland, eds.), Pergamon Press, Oxford (1978), p. 61.
80. J. J. Reilly and R. H. Wiswall, *Inorg. Chem.*, *13*:218 (1974).
81. V. V. Lunin and A. Z. Khan, *J. Mol. Catal.*, *25*:317 (1984).
82. V. V. Lunin, A. Z. Khan, L. A. Erivanskaia, O. V. Chetina, G. V. Antoshin, E. I. Shpiro, and Kh. M. Minachev, *Proceedings of the 8th International Congress on Catalysis, West Berlin*, Vol. 4 (1984), p. 859.
83. V. V. Lunin, O. V. Kryukov, V. N. Verbetskii, and A. L. Lapidus, *Vest. Mosk. Univ. Khim. 2*, *26*:94 (1985).
84. E. N. Anisochkina, A. R. Akbasova, and V. V. Lunin, *Vest. Mosk. Univ. Khim. 2*, *26*:99 (1985).

85. V. V. Lunin and Yu. I. Solovetskii, *Kinet. Katal.*, 26:694 (1985).
86. R. M. Frąk, *J. Less-Common Met.*, 109:279 (1985).
87. A. J. Fanelli, A. J. Maeland, A. M. Rosan, and R. K. Crissey, *J. Chem. Soc. Chem. Commun.*, 8 (1985).
88. V. M. Gryaznov, V. S. Smirnov, and M. G. Slinko, *Catalysis*, Vol. 2, p. 39, *Proceedings of the 5th International Congress on Catalysis, Miami Beach, Fla., 1972* (J. W. Hightower, ed.), North-Holland, Amsterdam (1973); *Proceedings of the 6th International Congress on Catalysis, London, 1976*, Vol. 1 (G. C. Bond, P. B. Wells, and F. C. Tompkins, eds.), Chemical Society, London (1977), p. 894; *Proceedings of the 7th International Congress on Catalysis, Tokyo*, Part A (1980), p. 224.
89. V. M. Gryaznov and M. G. Slinko, *Discuss. Faraday Soc.*, 72:73 (1981).
90. V. M. Gryaznov and V. E. J. Klabunowskii (eds.), *Metals and Alloys as Membrane Catalysts* (in Russian), Izdatelstvo "Nauka," Moscow (1981).
91. K. Soga, K. Otsuka, M. Sato, T. Sano, and S. Ikeda, *J. Less-Common Met.*, 71:259 (1980).

IV

HYDROGEN EFFECTS IN CATALYTIC REACTIONS

15 | General Kinetics of Hydrogen Effects: Hydrocarbon Transformations over Metals as Model Reactions

ALFRED FRENNET

Université Libre de Bruxelles, Brussels, Belgium

I.	INTRODUCTION; HISTORY OF THE PROBLEM	399
II.	CORRELATION BETWEEN HYDROGEN PRESSURE DEPENDENCY AND HYDROGEN COVERAGE	404
	A. Uniform Surface	406
	B. Nonuniform Surface: Temkin Isotherm	406
	C. Nonuniform Surface: Freundlich Isotherm	408
III.	HYDROGEN PRESSURE EFFECT ASSOCIATED WITH DISSOCIATIVE OR ASSOCIATIVE ADSORPTION MECHANISM OF THE ALKANE	409
IV.	MULTISITE EFFECT	411
V.	LANDING SITE	412
	A. Coverage Function Associated with Landing Site	413
	B. Hydrogen Coverage Function	413
	C. Combined Hydrogen and Hydrocarbon Coverage	415
VI.	MULTISITE AND MULTIBONDING	418
VII.	GENERAL CONCLUSIONS	418
	REFERENCES	421

I. INTRODUCTION; HISTORY OF THE PROBLEM

Contrary to homogeneous gas-phase reactions, reactions between gases catalyzed by metals often exhibit a negative-pressure dependency on the partial pressure of one of the reactants, even under initial conditions. This is frequently the case for hydrogen in catalytic reactions involving hydrocarbons such as hydrogenolysis, isomerization, and deuterium exchange. The first concepts making it possible to formulate rate equations able to account for such inhibiting effects were introduced in 1911 by Langmuir [1, 2] and have been applied by Hinshelwood [3]. The active surface, able to chemisorb, often dissociatively, the reactants, is composed of localized sites at the surface of the catalyst in a limited number, and all sites are potentially available to both reactants. This important concept involves competition for chemisorption and appears in the rate equation in the form of an inhibiting term. In the presence of many metal surfaces, hydrogen adsorption and desorption are fast phenomena and adsorption-desorption equilibrium of that reactant is generally assumed.

It soon appeared that simple competition for adsorption was not able to account for the observed inhibiting effects in the frame of these concepts, as concluded by Kemball in 1966 [4]. In fact, the fraction of free sites (θ_S) is related to p_{H_2} through the hydrogen coverage (θ_H):

$$\theta_S = 1 - \theta_H \tag{1}$$

Using a Langmuir-type adsorption isotherm

$$\theta_H = \frac{(Kp_{H_2})^{1/2}}{1 + (Kp_{H_2})^{1/2}} \tag{2}$$

$$\theta_S = \frac{1}{1 + (Kp_{H_2})^{1/2}} \tag{3}$$

At high hydrogen coverages this equation reduces to

$$\theta_S = (Kp_{H_2})^{-1/2} \tag{4}$$

It is thus clear that the maximum negative-pressure dependency to be expected in that case is -1/2. On the other hand, it is well known that many catalytic reactions [5] exhibit a much more important inhibiting effect in H_2 pressure.

Throughout this chapter the general kinetics of hydrogen effects will be analyzed using the wide range of information obtained in the numerous works published on ethane hydrogenolysis, taken as a model reaction. In the case of that reaction, the order in H_2 pressure in the formal rate equation

$$R_H = k_H (p_{H_2})^\alpha (p_{HC})^\beta \tag{5}$$

is generally negative and may be as large as -2.5 [5-8,10]. This is, in a general way and even more recently, interpreted in terms of degree of dehydrogenation of the active intermediate on the surface. Owing to the complexity of most reaction schemes, a complete steady-state treatment is often not manageable. Therefore, the classical procedure may be summarized as follows. One step is considered to be the rate-determining step (RDS), all steps preceding the RDS are at equilibrium, and all steps following the RDS are kinetically not significant. Following that procedure, if the RDS follows a series of dehydrogenation steps, each of these steps contributes to the inhibiting effect in hydrogen pressure.

Let us briefly summarize the successive interpretations published to account for the H_2 partial pressure effect on the rate of ethane hydrogenolysis (R_H) catalyzed by metals. The first reaction scheme and rate equation were proposed in 1954 by Cimino et al. [9]. In that scheme, the RDS is the breaking on the surface of the C-C bond of a more-or-less dehydrogenated C_2 radical (C_2H_x) involving an H_2 molecule:

$$(C_2H_x)_a + H_2 \rightarrow CH_y + CH_z \tag{6}$$

15. Kinetics of Hydrogen Effects

This radical C_2H_x is in equilibrium with the gas phase through successive dehydrogenation steps, summarized as

$$C_2H_6 + S \rightleftarrows (C_2H_x)_a + \left(\frac{6-x}{2}\right) H_2 \tag{7}$$

where S represents a free site. Considering that the surface coverage by components other than the MASI (most abundant surface intermediates) (C_2H_x) is negligible, this leads to

$$\alpha_1 p_{C_2H_6}(1 - \theta_{C_2H_x}) = \alpha_2 \theta_{C_2H_x}(p_{H_2})^{(6-x)/2} \tag{8}$$

and

$$\theta_{C_2H_x} = \frac{\alpha p_{C_2H_6}/(p_{H_2})^{(6-x)/2}}{1 + \alpha p_{C_2H_6}/(p_{H_2})^{(6-x)/2}} \quad \text{with } \alpha = \frac{\alpha_1}{\alpha_2} \tag{9}$$

In a restricted range of pressure, this equation may be approximated as

$$\theta_{C_2H_x} = \left[\frac{\alpha p_{C_2H_6}}{(p_{H_2})^{(6-x)/2}}\right]^n \tag{10}$$

Thus the rate of hydrogenolysis (R_H) is written as equal to the rate of the RDS taking into account Eqs. (6) and (10):

$$R_H = k(p_{C_2H_6})^n (p_{H_2})^{1-n(6-x)/2} \tag{11}$$

This mechanism has been used by several authors, sometimes with slight modifications. For example, Sinfelt has considered two modifications. One of these [7] was to split the equilibrium (7) into

$$C_2H_6 \underset{k'_1}{\overset{k_1}{\rightleftarrows}} (C_2H_5)_a + (H)_a \tag{12}$$

followed by

$$(C_2H_5)_a + (H)_a \overset{K_2}{\rightleftarrows} (C_2H_x)_a + \left(\frac{6-x}{2}\right) H_2 \tag{13}$$

In that scheme the breaking of the C-C bond on the surface is still the RDS, but the rate of the step (12) in the reverse direction is considered as of the same order as that of the desorption of the C_1 species, thus fast.

The coverage in C_2H_5 radicals ($\theta_{C_2H_5}$) can only be determined by the use of steady-state treatment:

$$\frac{d}{dt}\theta_{C_2H_5} = 0 = k_1 p_{C_2H_6} - k'_1 \theta_{C_2H_5} \theta_H - k \theta_{C_2H_x} p_{H_2} \tag{14}$$

The derived rate equation for hydrogenolysis is

$$R_H = \frac{k_1 p_{C_2H_6}}{1 + b(p_{H_2})^{(6-x)/2-1}} \quad \text{with } b = \frac{k'_1}{kK_2} \tag{15}$$

Another modification made by the same author [11] was to leave out the molecular hydrogen from the RDS and the rate equation becomes

$$R_H = k(p_{C_2H_6})^n (p_{H_2})^{-n(6-x)/2} \tag{16}$$

An improvement has been introduced in the analysis by Leclercq et al. [12,13]. In fact, they did not consider the approximation as in Eq. (10) and they used the rate equation as

$$R_H = k \frac{\alpha p_{H_2} p_{C_2H_6}}{(p_{H_2})^{(6-x)/2} + \alpha p_{C_2H_6}} \tag{17}$$

where α has the same meaning as in Eq. (9). Equation (17) may also be written in the form

$$\frac{p_{C_2H_6} p_{H_2}}{R_H} = \frac{(p_{H_2})^{(6-x)/2}}{k\alpha} + \frac{p_{C_2H_6}}{k} \tag{18}$$

The values of $p_{C_2H_6} p_{H_2}/R_H$ are plotted as a function of $(p_{H_2})^{(6-x)/2}$ for an optimized value of x to give a straight line. One thus obtains the value of x, k, and α. Let us notice that α contains information on the thermodynamics of the formation of the MASI. This analysis thus provides information less formal than the exponent n in Eqs. (10), (11), and (16).

In 1972, Boudart [14] introduced the concept that the site on which the hydrocarbon radical is chemisorbed is not the same as the one on which hydrogen atoms are chemisorbed. The scheme is thus written

$$2S + H_2 \rightleftarrows 2H_a \tag{19}$$

$$* + yS + C_2H_6 \rightarrow C_2H_x^* + yH_a + \left(\frac{6-x-y}{2}\right) H_2 \tag{20}$$

$$C_2H_x^* + H_2 \rightarrow \tag{21}$$

Furthermore, the adsorption of ethane is considered as irreversible. This mechanism results in an equation where the rate is a function of $(1 - \theta_H)^Y$, θ_H being the hydrogen coverage in equilibrium with the gaseous hydrogen and Y being the number of H atoms lost by the hydrocarbon that are held on Y neighboring S sites. In order to account for an important inhibiting effect in H_2 pressure, one needs to consider large values of θ_H. In fact, it is assumed that the coverage by hydrogen is near saturation, so that the hydrogen coverage function reduces to $(Kp_{H_2})^{-Y/2}$. In this analysis, the inhibiting effect in H_2 pressure is still related to the degree of dehydrogenation of the chemisorbed hydrocarbon radical that constitutes the MASI.

As a final example of an interesting improvement, let us mention the analysis made by Gudkov et al. [15] in 1982. In this analysis, the new feature considered is that the rate-determining step is not independent of the experimental conditions: the degree of dehydrogenation of the MASI is dependent on the amount of H_2. A fitting with experimental data is made considering that with excess hydrogen on the surface the MASI is a C_2H_5 adsorbed radical. The scheme considered is

$$H_2 + 2S \rightleftarrows 2H_a \tag{22}$$

$$C_2H_6 + 2S \rightleftarrows (C_2H_5)_a + H_a \tag{23}$$

$$(C_2H_5)_a + S \rightarrow (CH_3)_a + (CH_2)_a \tag{24}$$

from which the following rate equation is derived:

$$R_H = k \frac{p_{C_2H_6}(p_{H_2})^{1/2}}{(p_{C_2H_6} + k'p_{H_2})^2} \tag{25}$$

If there is not enough hydrogen on the surface, the C_2H_5 radical dehydrogenates through steps of the type of Eq. (23), and the MASI becomes a C_2H_2 adsorbed radical. The corresponding rate equation is written

$$R_H = k \frac{p_{C_2H_6}(p_{H_2})^2}{[p_{C_2H_6} + k'(p_{H_2})^{2.5}]^2} \tag{26}$$

These interesting considerations are unfortunately associated with the use of simple Langmuir isotherm and the inhibiting effect in H_2 pressure is still associated with the degree of dehydrogenation of the MASI.

Treatments of kinetic equations considering the surface as uniform for H_2 chemisorption, and considering one chemisorbed radical per site, are often applied to another widely studied reaction—the methanation of CO [16-22]. The kinetic treatment is of the classical type. As an example [20], the mechanism of CO hydrogenation on nickel is written as follows:

$$(CO)_g + 2S \overset{0}{\rightleftarrows} (CO)_a$$

$$(CO)_a \overset{1}{\rightleftarrows} C_a + O_a$$

$$(H_2)_g + 2S \overset{2}{\rightleftarrows} 2H_a$$

$$O_a + H_a \overset{3}{\rightleftarrows} (OH)_a + S$$

$$(OH)_a + H_a \overset{4}{\rightarrow} (H_2O)_g$$

$$C_a + H_a \overset{5}{\rightleftarrows} (CH)_a + S$$

$$(CH)_a + H_a \overset{6}{\rightleftarrows} (CH_2)_a + S$$

$$(CH_2)_a + H_a \overset{7}{\rightleftarrows} (CH_3)_a + S$$

$$(CH_3)_a + H_a \overset{8}{\rightarrow} (CH_4)_g$$

As mentioned by the authors, the experimental data fit surprisingly well the only rate equation derived considering step 4 in the consumption of the surface oxygen atom O_a, and step 6 in the consumption of C_a as the RDSs. This equation is rather complex:

$$R = \frac{Ap_{H_2}(p_{CO})^{1/2}}{[1 + B(p_{H_2})^{1/2} + C(p_{CO})^{1/2} + D(p_{H_2})^{1/2}(p_{CO})^{1/2}]^2} \quad (27)$$

but when studying the influence of modifications of the partial pressure of one reactant it is shown that for both CO and H_2 pressure, Eq. (27) can be linearized as follows:

At constant H_2 pressure:

$$\left[\frac{(p_{CO})^{1/2}}{R}\right]^{1/2} = a + b(p_{CO})^{1/2}$$

At constant CO pressure:

$$(R)^{-1/2} = e + f(p_{H_2})^{-1/2}$$

The data fit such equations for R values spanning five orders of magnitude, in a temperature range from 560 to 840 K and variations of CO partial pressure by a factor of 650 and of H_2 partial pressure by a factor of 14. This analysis is then taken as a verification of the validity of the proposed mechanism.

Let us finally mention that such a kinetic analysis for hydrogenation of hydrocarbons [23], namely in the frame of selective hydrogenation of acetylenic compounds and olefins, based on the Langmuir isotherm for the reactants in adsorption-desorption equilibrium and also based on all steps preceding the RDS considered as at equilibrium is still presently made. On the other hand, it is well known, and discussed in other chapters of this book, that the surface of a metal catalyst does not behave as uniform toward H_2 chemisorption. We have already criticized procedures considering the equilibrium constant of the chemisorption reaction of hydrogen as independent of coverage in kinetic treatments [24]. This is the subject of Section II.

In all the interpretations of the negative order characteristic of the H_2 pressure, in reactions of alkanes that we have summarized in this introduction, it is striking that in a general way, the inhibiting effect in H_2 pressure is related to the degree of dehydrogenation of the MASI. On the other hand, when analyzing the experimental data concerning catalytic reactions of alkanes, and namely ethane hydrogenolysis on metals, it appears that the value of the inhibiting effect in H_2 pressure is not independent of the experimental conditions [25]: It varies continuously with both temperature and pressure conditions. In fact, that inhibiting effect is increasing both when the H_2 pressure increases and when the temperature is lowered. This of course leads to the difficulty that the degree of dehydrogenation should increase when the H_2 pressure is increased or when the temperature is lowered! There is thus a need to search for other interpretations leading to more logical conclusions. In the next section we discuss how to analyze the variations of the thermodynamics of hydrogen adsorption with coverage.

II. CORRELATION BETWEEN HYDROGEN PRESSURE DEPENDENCY AND HYDROGEN COVERAGE

As we have seen in the introduction, a very general way to analyze the kinetics of a catalytic reaction is to eliminate the coverages θ_i in the surface species

15. Kinetics of Hydrogen Effects

appearing in the chemical equations of the steps preceding the RDS by combining the equations expressing the equilibrium of these steps with the equation expressing the finite character of the surface, that is,

$$\theta_s = 1 - \Sigma \theta_i \tag{28}$$

where θ_s is the fraction of unoccupied sites. As a result, the rate equation is composed of a function of the partial pressure of the reactants (p_i), the rate constant of the RDS and the constants of equilibrium of the steps preceding the RDS, some of the partial pressures, and of the equilibrium constants at a power 0.5, 1, or 2, depending on the stoichiometry of some steps.

In the formal experimental rate equation

$$R = k_e (P_{H_2})^{\alpha_e} (p_{HC})^{\beta_e} \tag{29}$$

the experimental hydrogen pressure dependence α_e is then compared to the one derived from a kinetic scheme. The way α_e is derived from the kinetic measurements assumes that locally the experimental rate constant k_e is really constant and does not include any pressure dependence. Such an analysis does not take into account the dependence of the constants of equilibrium of the steps preceding the RDS on the partial pressures, namely the H_2 pressure. It is well known that the thermodynamics of hydrogen chemisorption is strongly dependent on the hydrogen coverage [26]. This dependence appears in a very general way as a decrease in the heat of adsorption with coverage, leading to a dependence of coverage toward the pressure that is much less important than that predicted by the Langmuir isotherm.

Let us hereafter analyze briefly what the result of that variation of thermodynamics with coverage can induce on the local apparent dependence of θ_H on p_{H_2}, with α_t defined as

$$\alpha_t = \frac{d \ln \theta}{d \ln p} \tag{30}$$

In a small interval of θ and p the adsorption isotherm may be written

$$\theta = K_t p^{\alpha_t} \tag{31}$$

where $\alpha_t = \alpha_t(\theta)$ but where K_t is really constant, that is, independent of θ. In the general adsorption isotherm

$$\frac{\theta}{1 - \theta} = K^a p^a \tag{32}$$

the value of a is 1/2 for dissociative chemisorption.

Let us derive an expression of α_t [Eq. (30)] from Eq. (32): The derivative of Eq. (32) divided by $K^a p^a \theta$ gives

$$\frac{d\theta}{\theta} = a(1 - \theta) \frac{dp}{p} + a(1 - \theta) \frac{dK}{K} \tag{33}$$

which may be written

$$d \ln \theta = a(1 - \theta) d \ln p + a(1 - \theta) d \ln K \frac{d \ln \theta}{d \ln \theta} \tag{34}$$

One thus finally obtains

$$\alpha_t = \frac{d \ln \theta}{d \ln p} = \frac{a(1 - \theta)}{1 - a(1 - \theta)(d \ln K/d \ln \theta)} \qquad (35)$$

A. Uniform Surface

Of course, in the case of a uniform surface $d \ln K/d \ln \theta = 0$ and Eq. (35) reduces to

$$\alpha_{tL} = a(1 - \theta) \qquad (36)$$

where α_{tL} is the value of α_t in the case of a Langmuir isotherm. Thus in Eq. (35), the term α_C (corrective term)

$$\alpha_C = \left[1 - a(1 - \theta) \frac{d \ln K}{d \ln \theta}\right]^{-1} \qquad (37)$$

represents in the case of a nonuniform surface the correction to apply to the dependence of θ on p as derived from a Langmuir isotherm.

We have seen in the introduction of Section II that use is made of Eq. (32) in the more general form

$$\frac{\theta_H}{\theta_S} = (K_H)^{1/2}(p_{H_2})^{1/2} \qquad (38)$$

without making explicit the pressure dependence of K on p through its dependence on θ. On the other hand, the function α_C [Eq. (37)] expresses the deviation of the H_2 pressure dependence from the equivalent pressure dependence α_{tL} for a uniform surface, taking into account the variation of K_H with θ_H. It thus represents the pressure dependence included in the part of the constant that is dependent on θ. So we may replace $K(\theta)$ by $\alpha(\theta)$ in Eq. (38), where K_t is really constant.

$$\frac{\theta_H}{\theta_S} = [K_H(\theta)]^{1/2}(p_{H_2})^{1/2} \to (K_t)^{1/2}(p_{H_2})^{\alpha_C/2} \qquad (39)$$

Thus, when comparing the parameters k_e, α_e, . . . [Eq. (29)] of the formal experimental rate equation with those appearing in a rate equation derived from a given reaction scheme, one may consider the whole H_2 pressure dependence associated with one dehydrogenation step as expressed in the equation of the hydrogen adsorption equilibrium [Eq. (39)] as $\alpha_C/2$. To evaluate quantitatively the relative importance of the corrective term α_C we need to know the function of the dependency of K on θ.

B. Nonuniform Surface: Temkin Isotherm

A quasi-linear decrease of the heat of adsorption is often observed in a wide range of coverage [24, 27]. In that case, the adsorption constant as a function of θ is written

$$K = K_0 \exp\left[\frac{Q_0(1 - m\theta)}{RT}\right] = K_0^* \exp(-u\theta) \qquad (40)$$

where Q_0 is the heat of adsorption near zero coverage and where

$$u = Q_0 \frac{m}{RT} \qquad (41)$$

15. Kinetics of Hydrogen Effects

The corresponding hydrogen adsorption isotherm is the Temkin isotherm,

$$\ln \frac{\theta}{1 - \theta} = a \ln p + a \ln K_0^* - au\theta \tag{42}$$

In a wide range of coverage around $\theta = 0.5$, the term $\ln[\theta/(1 - \theta)]$ is negligible and it follows that the quasi entire dependence of θ on p is included in the last term expressing the variation of heat of adsorption with coverage. Using Eq. (40), one calculates

$$\left(\frac{d \ln K}{d \ln \theta}\right)_T = \frac{d \ln K}{d\theta/\theta} = \theta \frac{d \ln K}{d\theta} = -u\theta \tag{43}$$

Thus in the case of a Temkin-like isotherm, Eq. (35) is written

$$(\alpha_t)_T = \frac{d \ln \theta}{d \ln p} = \frac{a(1 - \theta)}{1 + a(1 - \theta)u\theta} \tag{44}$$

Values of u to consider are in the range 5 to 50, as in Eq. (41) values of m mentioned in the literature range around 0.5 ± 0.2 and the heat of adsorption of hydrogen at zero coverage has a value of 20 kcal mol^{-1} within a factor of 2 on most metals used as catalysts.

As shown in Fig. 1, the result is that as soon as the hydrogen coverage is not small ($\theta_H > 0.2$), the local pressure dependence α_t drops to very small values. For a value of $u = 15$ (corresponding to frequent values of m and Q_0 under working catalytic conditions), the pressure dependence of θ_H takes such small values ($\alpha_t \approx 0.05$ at $\theta_H = 0.5$) that in first approximation, when varying the H$_2$ pressure by factors 2 to 5, as usually done in kinetic studies, the hydrogen coverage remains quite constant. Of course, this is already

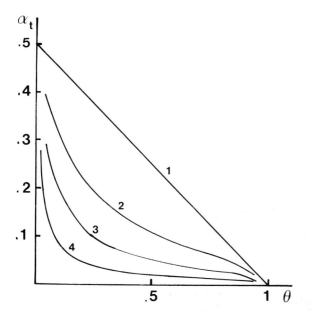

FIGURE 1 Values of α_t versus θ as calculated from Eq. (44) for $a = 1/2$ and $u = 0$ (curve 1, $\alpha_t = \alpha_L$), $u = 5$ (curve 2), $u = 15$ (curve 3), and $u = 50$ (curve 4).

contained in a Temkin-like isotherm. In this case, the corrective term α_{CT} to the Langmuir-type dependence α_{tL} is

$$\alpha_{CT} = \frac{1}{1 + u\theta(1 - \theta)/2} \tag{45}$$

C. Nonuniform Surface: Freundlich Isotherm

In spite of reported quasi-linear variations of the heat of adsorption Q with the coverage θ_H (as derived, for example, from an isoteric analysis of a series of isotherms [24, 27, 28]), isotherms measured in a not-too-large range of pressure (up to three orders of magnitude) may be accounted for as well by using the Temkin formalism as by using the Freundlich one. Some authors [8] thus also use the Freundlich isotherm. This isotherm is based on a linear decrease of the heat of adsorption with the logarithm of the coverage. Thus in this case:

1. The function $d \ln K/d \ln \theta$, needed to calculate α_t [Eq. (35)] and α_C [Eq. (37)] reduces to $-Q_0/RT$.
2. The local apparent dependence on the hydrogen pressure is, in the case of dissociative chemisorption,

$$(\alpha_t)_F = \frac{(1 - \theta)/2}{1 + (1 - \theta)(Q_0/2RT)} \tag{46}$$

3. The corrective term to apply to the dependence of θ on p, as derived from a Langmuir isotherm, is

$$(\alpha_C)_F = \frac{1}{1 + (1 - \theta)(Q_0/2RT)} \tag{47}$$

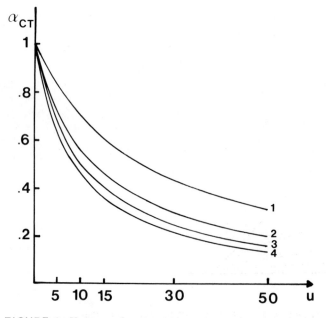

FIGURE 2 Values of α_{CT} versus u, as calculated from Eq. (45) for some values of θ. 1, $\theta = 0.1$ and 0.9; 2, $\theta = 0.2$ and 0.8; 3, $\theta = 0.3$ and 0.7; 4, $\theta = 0.5$.

15. Kinetics of Hydrogen Effects

The values of these parameters are to be compared to the equivalent ones derived for a Temkin isotherm [Eqs. (44) and (45)]. If the term Q_0/RT is sufficiently large as compared to 1, which is mostly the case if θ_H is not too large, $(\alpha_t)_F$ reduces to a constant, equal to RT/Q_0, that corresponds effectively to the exponent of the pressure in the Freundlich isotherm.

Conclusions

Values of α_{CT} are shown in Fig. 2. The agreement between these calculated values of α_{CT} and the ones derived from experimental hydrogen adsorption isotherms is striking [24]. One sees that in a very wide range of coverage ($0.2 < \theta_H < 0.8$) and for usual values of u ($15 < u < 30$), the pressure dependence to consider is smaller by a factor of 2 to 5 than the one expressed in Eq. (38). Similar conclusions arise for values of $(\alpha_C)_F$. In this way, a dehydrogenation step leading to the formation of a chemisorbed hydrogen atom in fast equilibrium with gaseous hydrogen may only account for a smaller inhibiting effect (compared to the one of $\frac{1}{2}$ as usually considered). One thus needs to find other ways to interpret the important inhibiting effects in H_2 pressure as observed on hydrogenolysis.

III. HYDROGEN PRESSURE EFFECT ASSOCIATED WITH DISSOCIATIVE OR ASSOCIATIVE ADSORPTION MECHANISM OF THE ALKANE

In a very general way, the adsorption step of the hydrocarbon, in catalytic reactions involving alkanes on metals, is considered as being of dissociative type. In a review paper of his extensive study on isotopic exchange between alkanes and D_2 [29] Kemball showed that in general the exchange of hydrogen is easier and is observed at much lower temperature than any other reaction for the same hydrocarbon. This is associated with a dissociative chemisorption of the hydrocarbon corresponding to the breaking of a C-H bond. Even more precisely, the observation of single exchange (SE; that is, the substitution by a deuterium atom of only one hydrogen atom in the hydrocarbon during its residence time on the surface) is taken as evidence of dissociative chemisorption of the alkane, and the observation of multiple exchange (ME; that is, the substitution of more than one hydrogen atom in the hydrocarbon under initial conditions) is taken as evidence of further dehydrogenation of the chemisorbed hydrocarbon radical. We would like to show that there are a series of arguments in favor of another type of mechanism for the adsorption step of the alkane:

1. In the case of exchange between deuterium and methane, it is observed that the ratio of the rate of single exchange (R_{SE}) to the rate of multiple exchange (R_{ME}) has the largest value the more difficult it is to desorb the chemisorbed hydrocarbon radicals from the surface of a metal in the presence of gaseous hydrogen [30].
2. It is shown in the measurement of the reaction between D_2 and CH_4 [31] or C_2H_6 [32,33] under transient effects associated with modifications of the hydrocarbon partial pressure, that R_{SE}/R_a, where R_a is the adsorption rate of the hydrocarbon, is independent of the measured hydrocarbon coverage.
3. At constant temperature, the ratio R_{SE}/R_a measured under steady activity conditions is independent of the hydrogen pressure [34]. For ethane this is observed whatever the relative importance of hydrogenolysis to multiple exchange. On the other hand, the ratio R_{ME}/R_H,

where R_H is the hydrogenolysis rate, also measured at steady activity and constant temperature, varies nearly proportionally with the hydrogen pressure. In other words, the probability for a chemisorbed radical to desorb in the form of a multiply exchanged hydrocarbon molecule instead of undergoing breaking of the C-C bond is simply proportional to the hydrogen pressure.
4. In the framework of multisite effect on the adsorption rate of the hydrocarbon [35, 36] discussed in Section IV, the results are not in agreement with a dissociative adsorption step, but with an associative mechanism.

All these arguments are in favor of a mechanism of the adsorption step of the type

$$C_2H_6 + H_a \rightleftarrows (C_2H_7)_a \rightleftarrows (C_2H_5)_a + H_2 \tag{48}$$

The exchange between the pool of hydrogen radicals chemisorbed on the metal surface and gaseous hydrogen is a very fast phenomenon compared to the reactions involving the hydrocarbon. Thus, under initial conditions, H_a is a deuterium atom. In this scheme, single exchange of ethane is the result of backward decomposition of the associative complex $(C_2H_7)_a$ formed in between an ethane molecule and a chemisorbed hydrogen atom.

Let us remember that within the deuterated isomers formed by multiple exchange, the perdeuterated compound is the most abundant [32-34]. It has even been shown that in some cases the perdeuterated hydrocarbon is the only initial product at low hydrogen pressure ($p_{D_2} < 1$ torr) [32,33,37]. This means that the rate of exchange between the pool of H atoms directly bound to the metal surface and the H atoms belonging to a hydrocarbon radical bound to the surface by a carbon atom is very fast compared to the rate of desorption of the hydrocarbon. Furthermore, when working at much higher deuterium pressure (10 to 200 torr), all deuterated isomers are formed as initial products [29, 38]. This is in agreement with the following picture. The surface dehydrogenation steps, in contrast with the adsorption step, proceed via a mechanism in which the hydrogen atom lost by the chemisorbed hydrocarbon radical is chemisorbed on the metal

$$C_aH_x + S \rightleftarrows C_aH_{x-1} + H_a \tag{49}$$

where the site S might even be a site hindered for chemisorption from the gas phase [39]. In this case there is no competition between the H atom formed in this step and hydrogen from the gas phase for chemisorption on the site S. In any case, the H atoms so formed and bonded to the metal may be as mobile as the H atoms resulting from H_2 chemisorption. They just enter the pool of H atoms bonded to the metal. In such a mechanism, the rate of substitution of the hydrogen atoms of the hydrocarbon radicals by deuterium atoms is dependent on θ_H, whereas the desorption of the hydrocarbon is directly dependent on the H_2 pressure [Eq. (48)] [34]. Thus, increasing the H_2 pressure by several orders of magnitude increases the rate of desorption by an equivalent factor, but modifies θ_H by only a small factor (see Section II) and thus modifies very little the rate of exchange between the chemisorbed hydrocarbon radical and the pool of deuterium atoms bonded directly to the metal. Such a mechanism, which is rather different from a Langmuir-type dissociative adsorption-desorption mechanism, has received important experimental support and thus must be considered.

IV. MULTISITE EFFECT

During the last decade, some analyses have been made of the important inhibition of the rate of catalytic reaction of hydrocarbons by hydrogen pressure based on the concept of multisite effect. All these analyses start from observations according to which the residues of adsorption of a molecule occupy more than two sites. We made an attempt in 1963 [39] to account for the very large slowdown of the rate of methane adsorption with coverage during one experiment. This analysis is based on the observation according to which at saturation of the surface by the hydrocarbon [30], an important fraction of the surface sites are still free for hydrogen chemisorption. The idea is then that some chemisorbed radicals are larger than the surface site, the size of which is derived from the number of H atoms that may be chemisorbed per unit surface area. In that model [39], three types of sites may be distinguished on a surface covered with large radicals: (a) the sites to which the chemisorbed radicals are bonded, (b) the sites that are covered by the large chemisorbed radical and that are hindered for any adsorption, and (c) the sites that are hindered for chemisorption of other large radicals, but that are free for small ones such as hydrogen.

Later, Martin et al. [40] and Dalmon et al. [41], studying the interaction of ethane with nickel using magnetic measurements, concluded that, after having chemisorbed ethane at low temperature, molecule is progressively dissociated by steps when raising the temperature. It appeared that there is concordance between the temperature at which the hydrocarbon is decomposed on the nickel surface into two carbon atoms and six hydrogen atoms according to

$$2 \underset{Ni\ Ni\ Ni}{\diagdown\overset{C}{|}\diagup} + 6 \underset{Ni}{\overset{H}{|}} \tag{50}$$

thus involving 12 bonds with the surface, and the temperature at which hydrogenolysis starts to be measurable. These authors suggest that the active intermediate leading to hydrogenolysis corresponds to the highly dissociated species involving X sites, in competition with hydrogen for chemisorption. In the rate equation there is a requirement to find an ensemble of X free sites [42] expressed as $(1 - \theta_H)^X$. When the partial pressure of ethane is maintained constant, the resulting rate equation is written

$$R_H = k(1 - \theta_H)^X \tag{51}$$

The experimental dependence α_e of R_H on p_H [see Eq. (29)] may be written [8]

$$\alpha_e = \frac{\partial}{\partial \ln p_{H_2}} \ln R_H = \left(\frac{\partial}{\partial \ln \theta_H} \ln R_H\right)\left(\frac{\partial}{\partial \ln p_{H_2}} \ln \theta_H\right) \tag{52}$$

where $\partial (\ln \theta_H)/\partial \ln p_{H_2}$ is the slope a of the Freundlich isotherm for H_2 adsorption. Equation (51) may then be written

$$\ln R_H = \ln k + X \ln(1 - \theta_H) \tag{53}$$

which combined with Eq. (52) gives

$$\frac{\alpha_e}{a} = -X \frac{\theta_H}{1 - \theta_H} \tag{54}$$

If the rate equation is of a more general form,

$$R_H = kp_{HC}(\theta_H)^Y (1 - \theta_H)^X \tag{55}$$

The same treatment gives

$$\frac{\alpha_e}{a} = Y - X \frac{\theta_H}{1 - \theta_H} \tag{56}$$

From experimental determinations of a and of θ_H, the graph of α_e/a versus $\theta_H/(1 - \theta_H)$ is in agreement with a rather good straight line, thus giving the number X of surface sites involved in the definition of the required ensemble needed to form the active intermediate for hydrogenolysis. Values of 15 for X and of -1 for Y have been derived for ethane hydrogenolysis on Ni [42] and values of 9 to 20 for X in the case of propane hydrogenolysis [43], depending on the depth of hydrogenolysis.

In the framework of this analysis, the dilution of an active metal (nickel) with an inactive one (copper) plays a role equivalent to that of H_2 pressure by lowering the number of active sites for hydrogenolysis by a factor equal to $(1 - \theta_H)^X$ [8]. The old ideas of competition for adsorption on the same sites between H_2 and the hydrocarbon, together with the dissociation of the hydrocarbon (progressive dehydrogenation to form the MASI, followed by the breaking of the C-C bond), are still present in this model. The main feature that differentiates this model from the preceding ones is the requirement to find the number of sites necessary to accommodate the hydrocarbon as an ensemble, leading to the appearance in the rate equation of the function $(1 - \theta_H)^X$, with large values of X. Despite the interesting formalism, the model is based on some weak assumptions. Correlations are made between the values of X and the number of bonds formed by chemisorption of the fragments of the hydrocarbon with the metal as derived from magnetic measurements. These last measurements are conducted in the absence of gaseous hydrogen, whereas the hydrogenolysis measurements are conducted in the presence of a large excess of hydrogen. It would be surprising that the degree of dehydrogenation should be the same under such different conditions. Furthermore, it is very hard to accept a complete cracking at once of the hydrocarbon, cracking that should not be the result of a series of successive steps of dehydrogenation and of C-C bond breaking. This, associated with the high mobility of hydrogen on the surface, at the temperature where hydrogenolysis proceeds, makes it difficult to understand why all these X sites should be found as an ensemble in the framework of the reaction scheme described above.

V. LANDING SITE

Some years ago, it was noticed that when the isotopic exchange reaction between light hydrocarbons and D_2 is studied under conditions such that there is no other reaction, an inhibiting effect on H_2 pressure is also observed, which is as important as that for hydrogenolysis of the same hydrocarbon [35]. It is shown that the total exchange rate also measures the adsorption rate. Later, hydrogenolysis of ethane was conducted with deuterium, making it possible, in contrast to the work of Martin, to differentiate the adsorption rate of ethane (R_a), the rate of ethane single exchange (R_{SE}), the rate of multiple exchange (R_{ME}), and the rate of hydrogenolysis (R_H) [32-34]. It has thus been possible to provide evidence that *there is an important inhibiting effect of hydrogen on the only adsorption step whatever the reaction*

15. Kinetics of Hydrogen Effects

follows on the surface: exchange only or exchange and hydrogenolysis [34]. Thus the inhibiting term in H_2 pressure needs to be formulated on the only adsorption step of ethane. This adsorption step of alkane, mentioned in Eq. (48), must be written in detail as

$$C_2H_6 + H_a + ZS \overset{1}{\rightleftarrows} (C_2H_7)_a \overset{2}{\rightleftarrows} (C_2H_5)_a + WS + H_2 \qquad (57)$$

where S represents a free site [36]. In the reverse reaction of step 2, W is the eventual difference between the number of sites S occupied by the chemisorbed ethyl radical and the number of sites needed to form the complex $(C_2H_7)_a$. From chemisorption experiments it is clear that the C_2H_5 chemisorbed radical [23] covers more than one H chemisorption site, and thus that $Z > W$. The desorption step of the chemisorbed C_2 radicals is then written

$$R_{des} = k'_2 \theta_{C_2H_5} (\theta_S)^W p_{H_2} \qquad (58)$$

A. Coverage Function Associated with Landing Site

The surface structure of a metal crystallite defines positions where the bonding of a chemisorbed radical is most stable. These positions are here defined as the potential sites, thus defined a priori by the topography of the metal surface. In the use we have made of the model, this potential site is associated with the hydrogen chemisorption site [35]. The density of potential sites is thus measured by the amount of chemisorbed hydrogen atoms at saturation of the surface. In a general way, the fraction of free sites θ_S is equal to $\theta_S = (1 - \Sigma \theta_i)$ [see Eq. (38)], where the θ_i are the coverages by the different i species present on the surface. If these species are in equilibrium with the gas phase, the problem is then to define the dependence of θ_i to p_i. If there are no impurities on the surface, the $\Sigma \theta_i$ reduces to

$$\Sigma \theta_i = \theta_H + \theta_C \qquad (59)$$

in the case of catalytic reactions involving hydrogen and hydrocarbons. θ_H represents the coverage by the H atoms directly bound to the metal and θ_C the total coverage in hydrocarbon residues in equilibrium with the gas phase.

B. Hydrogen Coverage Function

We shall mainly concentrate here first on an analysis of the surface coverage function where the coverage by species other than hydrogen is negligible, that is, $\theta_S = 1 - \theta_H$. A series of works [30-36, 40-43] have shown that the number of potential sites involved in the required ensemble that defines the active site for reactions of alkanes with hydrogen catalyzed by metals is large, often in the range 8 ± 2, and on nickel, up to twice that value. It is clearly shown [33, 34, 36] by the use of deuterium that this ensemble is required on the only adsorption step of the alkane, whatever the following reaction may be. Even in the case of hydrogen, it is shown in the frame of studies of the sticking coefficient of that molecule on single-crystal faces of Ni and Ni-Cu alloys that the active site for chemisorption is composed of an ensemble of four nickel atoms [44, 45]. When the active site for adsorption (landing site) involves an ensemble of Z potential sites, the rate equation of that step contains the function

$$(\theta_S)^Z = (1 - \theta_H)^Z \tag{60}$$

If the mechanism is of dissociative type [36],

$$C_nH_{2n+2} + ZS \to \tag{61}$$

the adsorption rate equation is

$$R_a = k_a p_{HC} (\theta_S)^Z = k_a p_{HC} (1 - \theta_H)^Z \tag{62}$$

If the mechanism is of associative type,

$$C_aH_{2n+2} + H_a + ZS \to \tag{63}$$

the adsorption rate equation is

$$R_a = k_a p_{HC} \theta_H (\theta_S)^Z = k_a p_{HC} \theta_H (1 - \theta_H)^Z \tag{64}$$

Due to the large values of Z [35, 36], as soon as θ_H is not small ($\theta_H > 0.3$), which is generally the case under catalytic working conditions, the function $(\theta_S)^Z$ is predominant and it is difficult to differentiate which of these two mechanisms is more probable. When plotted versus θ_H, which is a function of the hydrogen pressure, the function $(1 - \theta_H)^Z$ decreases monotonously, whereas the function $\theta_H(1 - \theta_H)^Z$ goes through a maximum at $\theta_H = 1/(Z + 1)$ [35, Fig. 2].

As θ_H is dependent on p_{H_2} and on the thermodynamics of hydrogen chemisorption, the surface coverage function $GD = (\theta_S)^Z$ and $GR = \theta_H(\theta_S)^Z$ [35] contains a dependence on these two parameters. The coverage function G may thus be written

$$G = A \left(\frac{P}{P_0}\right)^{\alpha_G} \exp\left(\frac{-E_G}{RT}\right) \tag{65}$$

The pressure dependency is defined as

$$\alpha_G = \frac{\partial}{\partial \ln(p/p_0)} \ln G \tag{66}$$

and the temperature dependency

$$E_G = -R \frac{\partial}{\partial (1/T)} \ln G \tag{67}$$

It is shown [34] that the isosteric heat of hydrogen adsorption

$$(Q)_{iso} = -R \left[\frac{\partial}{\partial (1/T)} \ln \frac{p}{p_0}\right]_\theta \tag{68}$$

relates E_G to α_G through

$$E_G = -\alpha_G Q_{iso} \tag{69}$$

It is important to notice that this relation holds whatever the function $G(\theta)$ may be. This equation has received experimental support [8].

15. Kinetics of Hydrogen Effects

For values of Z as large as 8, the corresponding values of α_G may be of the same order as the one observed on the overall catalytic reaction [35], such as hydrogenolysis [5]. In fact [35], in the case of a dissociative mechanism, α_{GD} varies from 0 to $-Z/2$ when θ varies from 0 to 1, whereas in the case of an associative mechanism, α_{GR} varies from $+0.5$ to $-Z/2$.

In a similar way, the apparent activation energy E_G, associated with the coverage function varies with the coverage in the case of a dissociative mechanism from 0 up to $-(Z/2)(Q_{iso})_{\theta \to 1}$. Values of E_G may thus be of the same order of magnitude as the apparent activation energy observed for reactions such as hydrogenolysis.

The temperature dependence of G may also be written as

$$G = (G)_0 \exp\left(\frac{-E_G}{RT}\right) \tag{70}$$

It is interesting to notice that there exists an important compensation effect between the parameters G_0 and E_G [46]. Another interesting feature included in the function G is the demanding character as defined by Boudart et al. [47]. In fact, the function G expresses the fraction of active sites, which is dependent on the thermodynamics of adsorption of the surface species. If one compares values of G associated with hydrogen coverage (for values of Z in a range of experimentally determined values, around atmospheric pressure, and estimated at two temperatures, 400 K and 500 K, differing by about 20%), differences of up to five orders of magnitude in these values of G are obtained [48,49]. Thus two catalysts, at the same temperature and pressure but presenting a difference of about 20% in heat of adsorption, may have a fraction of active sites differing by a similar factor. This provides an explanation for the large demanding character of the isotopic exchange reaction of D_2 with alkanes [48]. On the other hand, these exchange reactions are very little structure sensitive and alloying sensitive. In a more general way, reactions involving breaking of the C-H bond are very much less sensitive to the structure and alloying effects than reactions involving breaking of the C-C bond [5,50].

C. Combined Hydrogen and Hydrocarbon Coverage

All that has been considered above concerns the effect of hydrogen contained in the only hydrogen coverage function. Due to the very large value of the exponent Z of θ_S, starting with very small values (only a few percent) of the coverage in hydrocarbon residues, it is necessary to take that coverage into account [35,55]. The function θ_S then becomes much more complex, because if θ_C is dependent on the hydrocarbon pressure, it is also an inverse function of $(p_{H_2})^\delta$ [35,53] with δ related to the degree of dehydrogenation of the hydrocarbon residues and to the H_2 pressure through the desorption step of the hydrocarbon [see Eqs. (48) and (57)]. The fraction of free sites θ_S, here equal to $(1 - \theta_H - \theta_C)$, thus contains an inhibiting term in p_{H_2} through θ_H and a positive dependency to p_{H_2} through θ_C divided by $p_{HC}/(p_{H_2})^\delta$. A dependence of G on the hydrocarbon pressure is thus defined as

$$\beta_G = \frac{\partial \ln G}{\partial \ln p_{HC}} \tag{71}$$

There is, unfortunately, a significant lack of information on the values of the hydrocarbon coverage really present on the surface under working catalytic conditions. There is even less information about the hydrogen pressure dependence of θ_C on the H_2 pressure.

This problem has received a quantitative analysis in the particular case of the isotopic exchange of D_2 with methane [35]. It is shown that as soon as θ_C is not negligible, G contains an inhibiting effect on p_{HC} and an important positive dependence on the hydrogen pressure. Considering that if $(\theta°)_H$ is the hydrogen coverage in the absence of chemisorbed hydrocarbon radicals, the hydrogen coverage in presence of chemisorbed hydrocarbon radicals is given by

$$\theta_H = (\theta°)_H (1 - \theta_C) \tag{72}$$

the fraction of free sites is then expressed by

$$\theta_S = [1 - (\theta°)_H](1 - \theta_C) \tag{73}$$

The adsorption rate equation, according to a reactive chemisorption mechanism, is then written

$$R_a = k_a p_{HC} (\theta°)_H [1 - (\theta°)_H]^Z (1 - \theta_C)^{Z+1} \tag{74}$$

One observes (Fig. 3) that the adsorption rate, as evaluated at constant values of p_{HC}/p_{H_2}, obeys a volcano-shaped curve when plotted versus p_{H_2}. The maximum of that curve becomes sharper when the ratio $p_{HC} p_{H_2}$ increases such that the hydrocarbon coverage θ_C plays a more important role the larger the value of the p_{HC}/p_{H_2} ratio, but the position of the maximum H_2 pressure

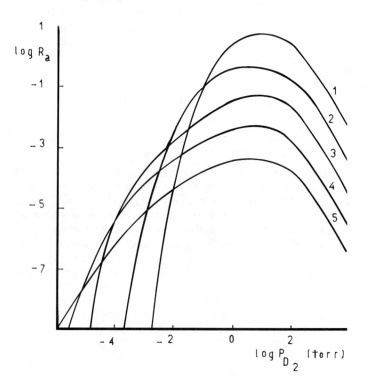

FIGURE 3 Influence of the hydrogen pressure on the rate of hydrocarbon adsorption R_a, according to Eq. (74), for Z = 6, at constant ratios of CH_4/D_2 pressures. 1, 10^2; 2, 10; 3, 1; 4, 10^{-1}; 5, 10^{-2}. The numerical values of coverages are obtained on rhodium at 400 K. (See Fig. 7 in Ref. 35.)

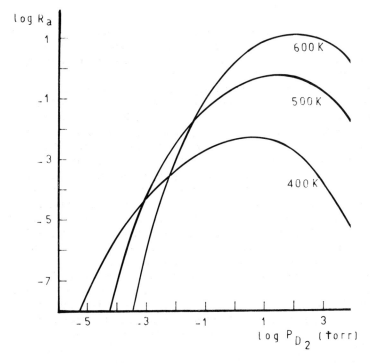

FIGURE 4 Variation of the hydrocarbon adsorption rate R_a [Eq. (74)] with hydrogen pressure. The values are computed for a constant pressure ratio of $CH_4/D_2 = 10^{-1}$, at different temperatures, with the same numerical values as in Fig. 3. (See Fig. 8 in Ref. 35.)

is hardly affected. The formalism includes the fact that the larger the value of Z, the steeper is this maximum. On the other hand, when R_a is plotted versus p_{H_2} but for one given value of the ratio p_{HC}/p_{H_2}, and at different temperatures (Fig. 4), it is observed that the position of the maximum of R_a moves toward the higher H_2 pressures when the temperature increases. Thus, in the adsorption rate equation

$$R_a = k_a (p_{H_2})^{\alpha_a} (p_{HC})^{\beta_a} \qquad (75)$$

the total hydrocarbon pressure dependence $\beta_a = \beta_G + 1$ drops from +1 to 0 and even negative values. Simultaneously, the absolute value of α_G decreases, goes to zero, and α_G becomes positive before β_a reaches zero. Furthermore, the maximum activity for adsorption is attained for conditions such that θ_C reaches the largest possible value before playing a significant inhibiting role in the function G. It is interesting to nitice that when β_a reaches negative values, the adsorption rate drops in a very important way.

These concepts are qualitatively in agreement with pressure and temperature effects as well as with effects of the size of the hydrocarbon (related to the value of Z) widely observed by the group of Paal [51-53]. Despite the fact that these authors used these concepts successfully in the discussion of their results, a quantitative treatment in the framework of the formalism above is unfortunately hopeless, because too many parameters are unknown [53,56].

VI. MULTISITE AND MULTIBONDING

Reactions like hydrogenolysis are characterized, on the one hand, by a hydrogen surface coverage function equivalent to that characterizing exchange reactions, but on the other hand are sensitive to the structure and alloying effects [48]. The first-order dependence of the hydrogenolysis rate on the hydrocarbon pressure observed by Martin suggests that θ_C is negligible in his experiments. In the interpretations of that author, no account is taken of the hydrocarbon coverage. These arguments are certainly in favor of the adsorption step being rate determining. In any case it would be difficult to answer that question because hydrogenolysis was not conducted with D_2.

On the other hand, when hydrogenolysis of ethane is studied under both transient and steady-state conditions using hydrogen in the form of deuterium [32-34] it is shown that (a) there is often no real rate-determining step in the reaction scheme of hydrogenolysis [33], (b) there is no discontinuity in the variations of the important inhibiting effect of hydrogen on the adsorption rate of ethane when going from conditions where only exchange is measurable to conditions where mainly hydrogenolysis is observed [34], (c) in the whole range of conditions studied, R_{ME}/R_H is about proportional to p_{H_2} [34], (d) in the H_2 pressure range where $R_{ME}/R_H < 1$, the hydrogen coverage function appearing in the hydrogenolysis rate equation is due mainly to the one characteristic of the adsorption step, which also means that the requirement for finding an ensemble of Z sites is not at all related to the decomposition of the hydrocarbon molecule into numerous fragments [34], and (e) in the pressure range where $R_{ME}/R_H > 1$, the hydrogenolysis rate is even more strongly inhibited by the H_2 pressure than the adsorption rate.

An important question, then, is: *Do demanding character and sensitivity to structure and alloying effect correspond to the same phenomena?* A strong demanding character is associated with the H_2 coverage function that is related to the only adsorption step of the alkane, leading to the chemisorption of the hydrocarbon with bonding to the metal by only one carbon atom [48]. An explanation for the structure sensitivity could be the requirement of a further surface step, namely the bonding of the hydrocarbon to the surface by a second or more carbon atoms, associated with surface dehydrogenation steps, leading finally to breaking of the C-C bond. This is related to the concept of ensemble as discussed by many authors (e.g., Sachtler and Van Santen [54], Davis and Somorjai [5], Gault [58], and Anderson [59], and recently reviewed by Van Broekhoven and Ponec [60]), where the important requirements for catalytic activity in breaking the C-C bond are associated with the possibility of multibonding of the hydrocarbon. On the other hand, in the framework of the ensemble associated with our model of landing site, the coverage function G is associated with the requirements for building up the associative complex of the hydrocarbon with a H chemisorbed atom, leading finally to breaking of the C-H bond.

VII. GENERAL CONCLUSIONS

Hydrogen plays a kinetic role in the two forms of a molecule (directly from the gas phase or molecularly adsorbed as a precursor state) or of a chemisorbed atom. There are a series of arguments in favor of the direct role of a hydrogen molecule in the desorption step of a chemisorbed hydrocarbon.

The contribution of dehydrogenation steps of the chemisorbed hydrocarbon to the effect of hydrogen pressure on the rate of the catalytic reaction seems to be rejected for three reasons:

1. It leads to an internal contradiction: that the dehydrogenation of the chemisorbed radical should increase when the hydrogen pressure increases or when the temperature decreases.
2. The contribution of a dehydrogenation step to the hydrogen pressure dependency of the catalytic reaction is too small by a large factor in a wide range of hydrogen coverage ($0.2 < \theta_H < 0.8$). In association with the widely observed Freundlich-like or Temkin-like adsorption isotherm, that coverage probably corresponds to most of the catalytic working conditions.
3. The rate of the only adsorption step of the alkane has a hydrogen pressure dependence of the same order as the overall catalytic reaction, whatever may be the reaction that follows the adsorption (exchange, hydrogenolysis, both, . . .).

On the other hand, the multisite effect on the adsorption step provides a formalism predicting the observed kinetic features experimentally observed:

1. A hydrogen pressure dependence α_G with a sign and value similar to that observed for overall catalytic reactions α_e. Furthermore, the variations of α_G with pressure and temperature correspond to the variations of α_e with these parameters.
2. There is an interdependence between the H_2 pressure dependence α_G and the apparent activation energy E_G associated with the coverage function G through the isosteric heat of adsorption of hydrogen [Eq. (69)].
3. There is also an interdependence between α_G and β_G similar to that observed for catalytic reactions: that is, when β_a is equal to +1, α_a has negative values, and as soon as α_a becomes less negative or even positive, β_a drops to 0 and may even become negative. The negative values of β_a correspond to a very important decrease of R_a.

The role of the chemisorbed hydrogen atom is more complex:

1. They play a role in the progressive dehydrogenation of the chemisorbed hydrocarbon radical through dissociative steps. These phenomena contribute rather little to the hydrogen pressure effect on the catalytic reaction.
2. They play the most complex but important role in the definition of the active site for hydrogen chemisorption (landing site) expressed in the kinetic equations by the coverage function G. In this model hydrogen has a dual function: an H atom plays the role of center for adsorption, whereas other ones simultaneously block the surface for hydrocarbon adsorption.

The model of landing site developed [35,36] differs from all the other models in that it is not related at all either to the degree of dehydrogenation or to the nature of the MASI. If it definitely contains some steric effect, it is not clear if there exists a contribution of some type of electronic effect [48]. Very small changes in the energetics of chemisorption bonds may have as real an effect on the kinetics of a process in which that bond is involved as it plays in an exponential dependence. In studies of alloying effects, measurements concerning the electronic structure by the methods of solid-state physics (XPS, UPS) provide information that is not characteristic of the sole surface monolayer. IR spectra are more significant for the surface monolayer and do not provide information in favor of an electronic effect. On the other hand, TPD spectra and isosteric analysis of series of chemisorption isotherms are also significant of the energetics of chemisorption [57]. In that case the question of the role of electronic structure is relevant. The study of the chemisorption (sticking probability, TPD) of H_2 and CO on well-characterized Cu-Ru surfaces also leads to the conclusion that there exists some type of ligand effect associated with the ensemble defining the chemisorption site [61-64]. We thus agree with van Broekhoven and Ponec [60] that this question concerning an eventual contribution of a ligand effect did not yet receive a general and definitive answer.

Another way to visualize an electronic effect in the landing site model comes from the polarization of the hydrogen-metal chemisorption bond. Having in mind the well-known solubility of alkanes in strong acids by proton capture (the proton capture by methane is exothermic by more than 10 kcal mol^{-1}), it is easier to understand the associative complex of the alkane with the $H^{\delta+}$ on the surface [30]. Furthermore, the presence of another dipole, adjacent to the active H atom, can play a role other than simple steric hindrance to the formation of the associative complex. This can explain why the replacement of active metal atoms by inactive ones does not modify the D_2 exchange rate very much, but simultaneously, does decrease the value of the hydrogen pressure dependence [48]. *All these features concern only the C-H bonds* and are associated with the hydrogen pressure dependency as expressed by the coverage function G and also contribute to the demanding character as defined by Boudart [47,48].

On the other hand, *the breaking of the C-C bond*, and probably many reactions concerning the carbon skeleton, must need a hydrocarbon radical bound to the metal by more than one carbon atom as an active intermediate. In the framework of our reaction scheme, the formation of that multibonded hydrocarbon radical is restricted by the population in hydrocarbon species bonded by one carbon atom, which may be dependent on $(p_{H_2})^{-1}$ [34] [see Eqs. (57) and (58)], and by some coverage function. Of course, the bonding by a second carbon atom certainly also needs to satisfy all the geometric requirements discussed widely for many years [5,54,58,60]. These requirements need to be associated with the sensitivity to dispersion (or to the structure of the metal surface) and to alloying effects. Let us remember that if in the analysis of the effect of dilution of nickel with copper on the rate of ethane hydrogenolysis [8], the surface concentration in Cu appears in a function similar to the H coverage function, the value of the exponent is systematically smaller than the value of the exponent of the H coverage function. *Thus, in the general frame of ensemble effects, the ensemble defining the landing site responsible for the adsorption step, concerns the C-H bond. It needs to be differentiated from the ensemble allowing multibonding (bonding of the hydrocarbon by more than one carbon atom) leading to reactions that concern the C-C bond.*

Finally, if one remembers that (a) when a method is used making the measurement of the rates of the successive steps of a reaction such as hydrogenolysis of ethane feasible, there is no real RDS, (b) the only adsorption step of ethane is characterized by kinetic parameters (orders, activation energy, and their variations) of the same order as the overall catalytic reaction, and (c) the surface steps involving the carbon skeleton are certainly also affected by the partial pressures and the topography of the metal surface, then correlations of the kinetic parameters as measured on the catalytic reaction under steady conditions and their variations with temperature, partial pressures, and with dispersion, alloying effects, and so on, contain very complex information. In any case, the orders as a function of the partial pressures of the reactants, namely hydrogen, as measured under steady conditions cannot be taken as evidence of the nature of the reaction scheme.

This does not mean that kinetic measurements are useless. But there is an absolute need to obtain information on the kinetic features of at least some steps and on coverages in the various chemisorbed species. Besides the identification of surface species, surface properties, and so on, by all the recently developed physical techniques, this can be achieved in kinetic studies by the use of labeled molecules, essentially with stable isotopes, together with the use of both steady-state and transient conditions [31-33].

ACKNOWLEDGMENTS

It is a pleasure to acknowledge my colleagues A. Crucq, L. Degols, and G. Liénard, with whom the research work of our laboratory summarized here has been realized and discussed. This work was conducted under the sponsorship of Fonds National de la Recherche Scientifique of Belgium.

REFERENCES

1. I. Langmuir, *J. Am. Chem. Soc.*, 38:2221 (1916).
2. I. Langmuir, *Trans. Faraday Soc.*, 17:621 (1921).
3. C. N. Hinshelwood, *Kinetics of Chemical Change*, Oxford University Press, New York (1926), p. 145.
4. C. Kemball, *Trans. Faraday Soc.*, 41:190 (1966).
5. S. M. Davis and G. A. Somorjai, in *The Chemical Physics of Solid Surfaces and Heterogeneous Catalysis*, Vol. 4 (D. A. King and D. P. Woodruff, eds.), Elsevier, Amsterdam (1982).
6. J. H. Sinfelt, *Catal. Rev.*, 3:175 (1969).
7. J. H. Sinfelt, *Adv. Catal.*, 23:91 (1973).
8. J. A. Dalmon, in *Fundamental and Industrial Aspects of Catalysis by Metals* (B. Imelik, G. A. Martin, and A. J. Renouprez, eds.), CNRS, Paris (1984), p. 253.
9. A. Cimino, M. Boudart, and H. Taylor, *J. Phys. Chem.*, 58:796 (1954).
10. J. H. Sinfelt and W. F. Taylor, *Trans. Faraday Soc.*, 64:3086 (1968).
11. J. H. Sinfelt, *J. Catal.*, 27:468 (1977).
12. G. Leclercq, L. Leclercq, and R. Maurel, *Bull. Soc. Chim. Fr.*, 11:2329 (1974).
13. G. Leclercq, L. Leclercq, and R. Maurel, *J. Catal.*, 44:68 (1976).
14. M. Boudart, *AIChE J.*, 18:465 (1972).
15. B. S. Gudkov, L. Guczi, and P. Tétényi, *J. Catal.*, 74:207 (1982).
16. D. F. Ollis and M. A. Vannice, *J. Catal.*, 38:514 (1975).
17. G. C. Bond and B. D. Turnham, *J. Catal.*, 45:128 (1976).
18. J. G. Ekerdt and A. T. Bell, *J. Catal.*, 58:170 (1979).
19. J. R. H. Ross, *J. Catal.*, 71:205 (1981).
20. R. Z. C. Van Meerten, J. G. Vollenbroek, M. H. J. M. de Croon, P. F. M. T. van Nisselrooy, and J. W. E. Coenen, *Appl. Catal.*, 3:29 (1982).
21. P. F. M. T. van Nisselrooy, J. A. M. Luttikholt, R. Z. C. van Meerten, M. H. J. M. de Croon, and J. W. E. Coenen, *Appl. Catal.*, 6:271 (1983).
22. J. Klose and M. Baerns, *J. Catal.*, 85:105 (1984).
23. J. Cosyns, in *Fundamental and Industrial Aspects of Catalysis by Metals* (B. Mielik, G. A. Martin, and A. J. Renouprez, eds.), CNRS, Paris (1984), p. 371.
24. A. Crucq, G. Liénard, L. Degols, and A. Frennet, *Appl. Surf. Sci.*, 17:79 (1983).

25. L. Guczi, A. Frennet, and V. Ponec, *Acta Chim. Hung.*, *112*:127 (1983).
26. Z. Paál and P. G. Menon, *Catal. Rev.-Sci. Eng.*, *25*:229 (1983).
27. A. Frennet and P. B. Wells, *Appl. Catal.*, *18*:243 (1985).
28. A. Crucq, L. Degols, G. Liénard, and A. Frennet, *Acta Chim. Acad. Sci. Hung.*, *111*:547 (1982).
29. C. Kemball, *Adv. Catal.*, *11*:223 (1959).
30. A. Frennet, *Catal. Rev.-Sci. Eng.*, *10*:37 (1974).
31. A. Frennet, G. Liénard, A. Crucq, and L. Degols, *Proceedings of the 7th International Congress on Catalysis, Tokyo, 1980*, Part B, Elsevier, Amsterdam (1981), p. 1482.
32. A. Frennet, A. Crucq, L. Degols, and G. Liénard, *Acta Chim. Acad. Sci. Hung.*, *111*:499 (1982).
33. A. Frennet, A. Crucq, L. Degols, and G. Liénard, *Proceedings of the 9th Ibero-American Symposium on Catalysis*, Lisboa (1984), p. 493.
34. A. Frennet, A. Crucq, L. Degols, and G. Liénard, *Acta Chim. Acad. Sci. Hung.*, *124*:1 (1987).
35. A. Frennet, G. Liénard, A. Crucq, and L. Degols, *J. Catal.*, *53*:150 (1978).
36. A. Frennet, G. Liénard, A. Crucq, and L. Degols, *Surf. Sci.*, *80*:412 (1979).
37. A. Frennet and G. Liénard, *Surf. Sci.*, *18*:80 (1969).
38. H. F. Leach, C. Mirodatos, and D. A. Whan, *J. Catal.*, *63*:138 (1980).
39. R. Coekelbergs, A. Frennet, G. Liénard, and P. Resibois, *J. Chem. Phys.*, *39*:604 (1963).
40. G. A. Martin, J. A. Dalmon, and C. Mirodatos, *Bull. Soc. Chim. Belg.*, *88*:559 (1979).
41. J. A. Dalmon, J. P. Candy, and G. A. Martin, *Proceedings of the 6th International Congress on Catalysis*, Chemical Society, London (1977), p. 903.
42. G. A. Martin, *J. Catal.*, *60*:345 (1979).
43. M. F. Guilleux, J. A. Dalmon, and G. A. Martin, *J. Catal.*, *62*:235 (1980).
44. K. Y. Yu, D. T. Ling, and W. E. Spicer, *J. Catal.*, *44*:373 (1976).
45. E. M. Silverman, R. J. Madix, and P. Delrue, *Surf. Sci.*, *109*:127 (1981).
46. L. Degols, A. Frenne´, A. Crucq, and G. Liénard, *Bull. Soc. Chim. Belg.*, *88*:631 (1979).
47. M. Boudart, A. W. Aldag, J. E. Benson, N. A. Dougharty, and C. G. Harkins, *J. Catal.*, *6*:92 (1966).
48. A. Frennet, G. Liénard, L. Degols, and A. Crucq, *Bull. Soc. Chim. Belg.*, *88*:621 (1979).
49. A. Frennet, in *Proceedings of the 12th Swedish Symposium on Catalysis*, Lund, Sweden (R. Larson, ed.) (1981), p. 49.
50. G. A. Somorjai, *Adv. Catal.*, *26*:2 (1977).
51. Z. Paál, *Adv. Catal.*, *29*:273 (1980).
52. H. Zimmer, Z. Paál, and P. Tétényi, *Acta Chim. Acad. Sci. Hung.*, *111*:513 (1982).
53. H. Zimmer and Z. Paál, *Proceedings of the 8th International Congress on Catalysis*, West Berlin, Vol. 3 (1984), p. 417.
54. M. W. H. Sachtler and R. Van Santen, *Adv. Catal.*, *26*:69 (1977).
55. A. Frennet, G. Liénard, A. Crucq, and L. Degols, *Proceedings of ECOSS III*, Vol. 1 (D. A. Degras and M. Costa, eds.) (1980), p. 419.
56. P. Parayre, V. Amlr-Ebrahimi, F. G. Gault, and A. Frennet, *J. Chem. Soc. Faraday Trans.*, *1*, *76*:1704 (1980).
57. A. Crucq, L. Degols, G. Liénard, and A. Frennet, *Acta Chim. Acad. Sci. Hung.*, *111*:487 (1982).
58. F. G. Gault, *Adv. Catal.*, *30*:1 (1981).

59. J. R. Anderson, *Adv. Catal.*, *23*:1 (1973).
60. E. H. van Broekhoven and V. Ponec, *Prog. Surf. Sci.*, *19*:351 (1985).
61. K. Christmann, G. Ertl, and H. Shimuzu, *J. Catal.*, *61*:397 (1980).
62. H. Shimuzu, K. Christmann and G. Ertl, *J. Catal.*, *61*:412 (1980).
63. J. C. Vickerman, K. Christman, and G. Ertl, *J. Catal.*, *71*:175 (1981).
64. J. C. Vickerman and K. Christmann, *Surf. Sci.*, *120*:1 (1982).

16 | Role of Hydrogen in Low- and High-Pressure Hydrocarbon Reactions

FRANCISCO ZAERA

University of California at Riverside, Riverside, California

GABOR A. SOMORJAI

University of California at Berkeley, Berkeley, California

I.	INTRODUCTION	425
II.	CHEMISORPTION STUDIES	426
	A. H_2 Chemisorption	426
	B. Hydrocarbon Chemisorption	427
III.	SURFACE REACTIONS AT LOW PRESSURES (10^{-8} to 10^{-5} torr)	429
	A. H_2-D_2 Exchange	430
	B. Ethylene Hydrogenation and H-D Exchange	430
	C. Cyclohexene Reactions	432
IV.	ATMOSPHERIC PRESSURE REACTIONS OVER SINGLE-CRYSTAL SURFACES	433
	A. Reactions Involving Ethylidyne	434
	B. Ethylene Hydrogenation	435
	C. Other Hydrogenation and Dehydrogenation Reactions	437
	D. Deuterium Exchange Reactions	439
	E. Hydrocarbon Skeletal Rearrangement	441
V.	CONCLUSIONS	444
	REFERENCES	445

I. INTRODUCTION

The chemisorption and surface reactions of hydrocarbons on metal surfaces have been investigated intensively from the viewpoint of heterogeneous catalysis. Hydrogenation, dehydrogenation, and skeletal rearrangement reactions are of particular importance for the catalytic reforming of petroleum feedstocks. These reactions are selectively catalyzed by only a small group of transition metals and alloys, most notably platinum and platinum-based alloys. The chemisorption of hydrocarbons over well-defined surfaces of these metals and under controlled environments, generally ultrahigh vacuum (UHV), has been widely studied since the development of new surface-sensitive analytical techniques over the last two decades. A better understanding of the organic-metallic interface at the atomic scale has emerged from these investigations. The next step toward the comprehension of catalytic reactions requires bridging the gap between surface science studies under UHV and catalytic kinetic investigations at atmospheric pressures. These types of investigations have been initiated in our laboratory over the past 10 years.

Most experiments have been performed in standard ultrahigh-vacuum chambers equipped with several surface-sensitive techniques, such as Auger electron spectroscopy (AES), low-energy electron diffraction (LEED), high-resolution electron energy loss spectroscopy (HREELS), and mass spectrometry for thermal desorption studies (TDS). Additionally, these systems include an environmental cell for catalytic reaction studies, as described in more detail later. The samples used consist mainly of metal single crystals cut at different angles in order to expose particular crystallographic phases. Most work has been done on close-packed surfaces, such as (111) and (100) faces of face-centered cubic metals. In a few instances reactions have been performed on surfaces with controlled amount of defects (steps and kinks) in order to determine the dependence of activity on surface structure.

In the present chapter we present a summary of the research that has been carried out in this field, with emphasis on the extrapolation of knowledge obtained under UHV to high-pressure conditions and on studies of well-characterized model single-crystal catalyst systems. The role of hydrogen as a unique reactant is also emphasized.

II. CHEMISORPTION STUDIES

A. H_2 Chemisorption

Detailed description of H_2 chemisorption over metal surfaces have already been given in previous chapters (see, in particular, Chapter 1), and therefore only its most relevant aspects will be mentioned here. H_2 chemisorption over metals is generally dissociative and is not activated. Several adsorption states have been observed even over smooth single-crystal surfaces. Activation energies for desorption range from 15 to 30 kcal mol^{-1}, with no clear trend across the periodic table [1,2]. A weakly chemisorbed state has also been observed in several cases, but the nature of such hydrogen and its importance to catalysis are not clear.

The results reported for hydrogen chemisorption on platinum single crystals are typical of those obtained for other metals and will be used here to illustrate the main features of the process. This system has been studied by many groups using a wide variety of techniques [3-10]. Thermal desorption spectra reveal the existence of more than one adsorption state even over flat (111) surfaces. Additional, stronger-bonded states appear when steps and kinks are present (Fig. 1). The values for the activation energies of desorption are estimated to range from 5 and 9 kcal mol^{-1} at low coverages over (111) terraces, to about 20 and 30 kcal mol^{-1} on the steps and kinks, respectively [3,10]. The heats of adsorption also change with coverage due to lateral interactions between adsorbed atoms. The main peak in the thermal desorption spectra (TDS) for H_2 on Pt(111) indicates a change in activation energy from 9.1 kcal mol^{-1} at low coverages to 6.6 kcal mol^{-1} at near saturation [3]. The area under the TDS can also be used to estimate the saturation coverage; Christmann and co-workers estimated $\theta_{sat} \sim 1$ using this procedure [3]. Unity saturation coverage was independently measured by Lee et al. [12] using He diffraction. They determined that hydrogen forms a (1 × 1) structure where the H atoms sit on top of the threefold hollow sites of the Pt(111) surface. Low-energy electron diffraction (LEED), high-resolution

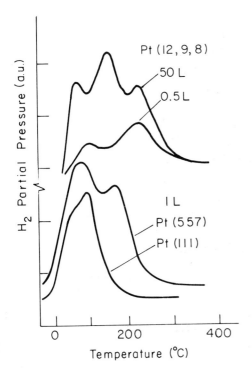

FIGURE 1 Thermal desorption spectra for hydrogen chemisorbed on the flat (111), stepped (557), and kinked (12,9,8) platinum single-crystal surfaces. (From Ref. 11.)

electron energy loss spectroscopy (HREELS), ultraviolet photoelectron spectroscopy (UPS), electron energy loss spectroscopy (ELS), and work function measurements have also been used to study this system [3,5,10]. Extensive theoretical work has also been published. More comprehensive reviews are given elsewhere [2,13,14].

B. Hydrocarbon Chemisorption

Studies on the chemisorption of hydrocarbons on metal single crystals are extensive, and a complete review of the subject is beyond the scope of this chapter. Here we only mention representative examples in order to present the relevant information needed for subsequent discussions.

The chemisorption of saturated low-molecular-weight alkanes (methane, ethane, propane) is activated, and therefore is generally not observed at room temperature. Physisorption of heavier alkanes can be attained at close to liquid-nitrogen temperature, but such adsorption is followed only by molecular desorption as the metal is warmed up [15]. Chemisorption under UHV may occur at higher temperatures, and it generally starts with an initial C-H bond-breaking step. Further heating of the system in such cases leads to full decomposition of the hydrocarbon fragments with simultaneous hydrogen desorption, while carbon is left behind on the surface [16].

Unsaturated hydrocarbons, on the other hand, can easily be adsorbed; they have sticking coefficients close to 1 at either liquid-nitrogen or room temperature. Chemisorption is followed by stepwise decomposition as the temperature of the metal is increased. Studies for olefins, alkynes, cyclic

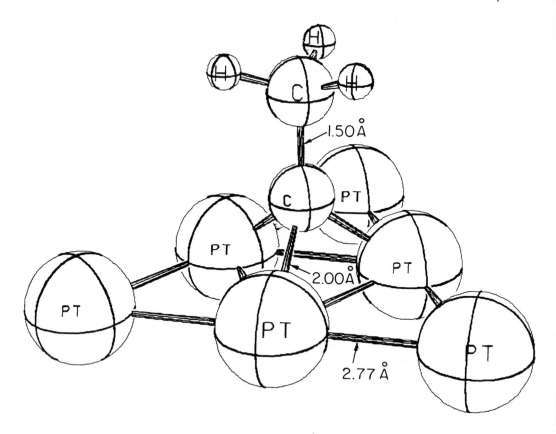

FIGURE 2 Atomic surface structure for ethylidyne species chemisorbed on Pt(111).

olefins, and aromatic compounds have been reported on many surfaces [17-24]. The chemisorption of ethylene and acetylene over Pt, Rh, and Pd (111) surfaces have received special attention, since they form an ordered structure at room temperature that has been identified as being composed of ethylidyne moieties, with a structure that is shown schematically in Fig. 2 [25-35]. Ethylidyne then further decomposes above 400 K to form C_2H and CH fragments, and the final dehydrogenation product is graphite, which is left on the surface at higher temperatures [32,36-38]. Chemisorption of ethylene over Ni(100), on the other hand, is followed by stepwise decomposition and the successive formation of a vinyl group, an acetylene-like fragment, and a C_2H moiety [86].

Adsorption studies of other, heavier alkenes have been performed as well [35, 36, 39, 40]. H_2 TDS for ethylene, propylene, and butene over Pt(111) are shown in Fig. 3. It can be seen that for all three olefins three decomposition regimes can be identified. At room temperature the first H_2 peak is observed, associated with the formation of ethylidyne or the corresponding alkylidyne analog [39,40]. Between 350 and 500 K further decomposition takes place, with formation of smaller hydrocarbon fragments such as CH and C_2H. Finally, total dehydrogenation above 600 K leads to the formation of a graphitic overlayer on the Pt(111) surface.

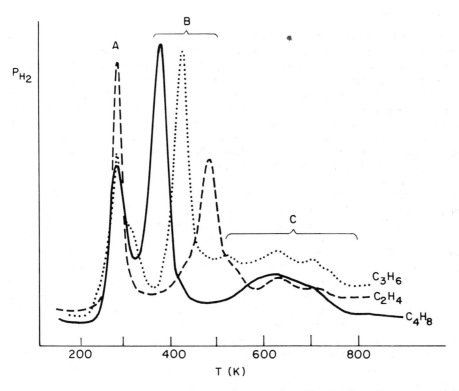

FIGURE 3 Hydrogen thermal desorption spectra illustrating the sequential dehydrogenation of 0.5 L ethylene (---), propylene (···), and *cis*-2-butene (——) chemisorbed on Pt(111) at about 120 K ($\beta = 12$ K s^{-1}).

Aromatic molecules follow a similar decomposition process as a function of temperature. HREELS and LEED studies of benzene adsorption over Rh(111) indicate that the molecule adsorbs molecularly at low temperatures with the ring parallel to the surface [22,41]. H_2 TDS show that the first major decomposition peak occurs at 470 K, with formation of surface species that are believed to consist of CCH and CH fragments from HREELS studies [42]. Further gradual decomposition occurs until graphite is formed at 800 K. Similar results have been reported for other surfaces [21].

III. SURFACE REACTIONS AT LOW PRESSURES
(10^{-8} to 10^{-5} torr)

In addition to thermal decomposition studies such as those reported in Section II, a few hydrocarbon reactions have been investigated under UHV as well. Two types of experiments can be differentiated: those where gas coadsorption is followed by thermal desorption of products, and those where reactions are carried out under steady-state conditions, generally by using molecular beams. In this section we report results published for H_2-D_2 exchange, ethylene hydrogenation, and cyclohexene reactions over platinum single-crystal surfaces.

A. H_2-D_2 Exchange

H_2-D_2 exchange is probably one of the simplest reactions that involves formation and breakage of chemical bonds, and therefore it has been studied extensively by several research groups [43-48]. Engel and Ertl have published a review on the subject [49]. The first step for the overall reaction involves the dissociative chemisorption of both H_2 and D_2. It appears that once atomic hydrogen is present on the surface, the reaction proceeds through a Langmuir-Hinshelwood mechanism where the limiting step is the recombination of H and D atoms, followed by molecular desorption [43]. The process is more complicated over platinum surfaces, where two different types of sites are believed to coexist, and where the limiting step could be the migration of H atoms from one site to the other [50].

The rate of exchange increases rapidly as the catalyst temperature is raised, until it reaches a maximum value above 500 to 600 K. This behavior has been attributed to a coverage dependence of the sticking coefficient, $s(\theta)$, for hydrogen. The rate of HD production is proportional to this sticking probability under the molecular beam experimental conditions [43], and since the steady-state coverage changes with temperature, so does $s(\theta)$.

H_2-D_2 exchange have also shown a strong dependence on surface topography. Several research groups have proven that steps are at least an order of magnitude more active than flat (111) terraces on platinum [44,45,50]. Somorjai et al. [50] have studied the angular dependence of HD production on Pt(s)-[6(111) × (111)] by using molecular beams, and proved that the reaction at the bottom of the steps is about seven times faster than that on the terraces (Fig. 4). These results also suggest that the existence of a precursor state is not important for the reaction mechanism.

Finally, the angular and velocity distributions of the outgoing HD from molecular beam experiments have cosine and Maxwell-Boltzmann distributions over Pt(111) and Pd(111), but deviation from such behavior have been reported for Cu(110) and Ni(111) [43,44,51,52].

B. Ethylene Hydrogenation and H-D Exchange

Ethylene self-hydrogenation has been observed over several supported metals [53]. The same phenomenon has been reported on Ni(111) [54], Ni(100) [86], and Pt(111) [55-57] under UHV. Thermal desorption experiments revealed the formation of ethane at around room temperature after ethylene saturation of Pt(111). The activation energy was estimated to be 18 kcal mol^{-1}, and a C-H bond breaking was proposed to be the limiting step [57]. If hydrogen is adsorbed prior to ethylene, the activation energy for ethane production drops to a value of 6 kcal mol^{-1}, and its yield increases by about an order of magnitude (Fig. 5). Based on the experimental results, the following mechanism was proposed:

$$H_2(g) \rightarrow 2H(a) \tag{1}$$

$$C_2H_4(g) \rightarrow C_2H_4(a) \tag{2}$$

$$C_2H_4(a) \rightarrow C_2H_3(a) + H(a) \tag{3}$$

$$C_2H_4(a) + H(a) \rightleftharpoons C_2H_5(a) \tag{4}$$

$$C_2H_5(a) + H(a) \rightarrow C_2H_6(g) \tag{5}$$

where (g) and (a) stand for gas and adsorbed, respectively. A step was included in this scheme to account for the formation of ethylidyne, CCH_3

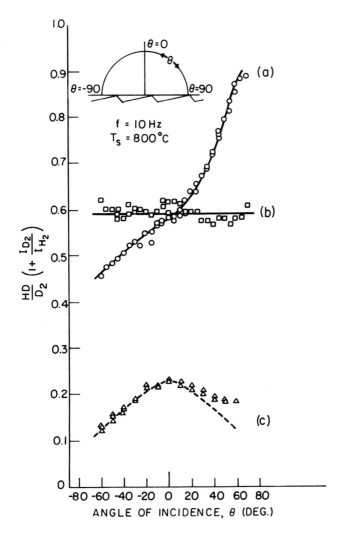

FIGURE 4 HD production as a function of angle of incidence, θ, of the molecular beam, normalized to the incident D_2 intensity: (a) Pt(332) surface with the step edges perpendicular to the incident beam ($\phi = 90°$); (b) Pt(332) where the projection of the beam on the surface is parallel to the step edges ($\phi = 9°$); (c) Pt(111).

(Fig. 2). A computer simulation using this model gave results that were in excellent agreement with the experiments for both self-hydrogenation and hydrogen preadsorbed cases.

H-D exchange is a related reaction that takes place simultaneously with hydrogenation. If deuterium is preadsorbed on Pt(111), deuterated ethane and ethylene are produced from ethylene TDS [57]. Ethylidyne can also exchange hydrogen atoms in the methyl group. Exchange has been observed by using TDS, HREELS, and secondary ion mass spectroscopy (SIMS) at submonolayer ethylidyne coverages over Pt(111) and Rh(111) [36, 58-60], but once saturation is reached, atmospheric pressures of deuterium are needed for the exchange to occur.

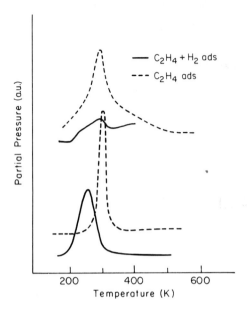

FIGURE 5 C_2H_4 (27 amu) and C_2H_6 (30 amu) TDS curves corresponding to the decomposition of C_2H_4 adsorbed over clean (dashed lines) and hydrogen predosed Pt(111) surfaces at 150 K. Exposures were 6 L for C_2H_4 and 30 L for hydrogen.

C. Cyclohexene Reactions

The reaction of cyclohexene with hydrogen over several platinum single-crystal surfaces was studied at total pressures of 10^{-8} to 10^{-5} torr [61-64]. Typical experimental results are shown in Fig. 6. The carbon buildup, as followed by Auger electron spectroscopy (AES), is also presented. The rate for benzene production shows a maximum after an induction period of 3 to 4 min, followed by a decay due to poisoning from the irreversibly adsorbed carbonaceous residues left on the surface. Little cyclohexane formation was also detected. These results clearly show that the reaction probability for cyclohexene conversion under vacuum is close to 1 over the clean platinum surface, but is reduced quickly by poisoning. Total turnover numbers of about 0.2 reacted molecule per platinum atom were obtained, so this reaction cannot be considered catalytic.

Benzene production from cyclohexene at 10^{-5} torr total pressures displayed little structure sensitivity over Pt(111), Pt(557), and Pt(10,8,7) [63]. The activation energy for the dehydrogenation was calculated to be less than 4 kcal mol^{-1} for all three surfaces, and the rates were first order in cyclohexene and positive fractional order in hydrogen. Some n-hexane production was detected as well, with rates comparable to benzene production over Pt(10,8,7) but an order of magnitude slower over Pt(111).

The effect of gold deposition over Pt(100) have also been investigated [62]. Surprisingly, the activity for benzene formation increases by a factor of 4 at one monolayer of gold. Further gold deposition leads to slow poisoning and reaction rate reduction. If platinum is deposited over Au(100), a broad maximum in rate is observed at one to five platinum layers, followed by an asymptotic decrease to the activity of pure platinum. A clear explanation for this activity enhancement has not yet been given. Hydrogen coadsorption has no effect in this reaction. Experiments carried out in excess

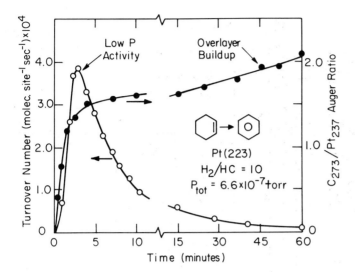

FIGURE 6 A comparison at 150°C of the cyclohexene dehydrogenation rate over Pt(223) at low pressures with the simultaneous buildup of the irreversibly chemisorbed carbonaceous overlayer. A C_{273}/Pt_{237} ratio of 2.8 corresponds to monolayer coverage.

hydrogen ($H_2:C_6H_{10}$ = 17) yielded the same activity and selectivity results as those performed in the total absence of H_2. No changes were observed when deuterium was used instead of hydrogen.

IV. ATMOSPHERIC PRESSURE REACTIONS OVER SINGLE-CRYSTAL SURFACES

Although UHV reaction studies are relevant to catalysis, their results cannot always be directly extended to high-pressure conditions such as those existing in most industrial processes. In fact, most surface reactions under vacuum are not catalytic in nature. Atmospheric pressures of reactants allow for the existence of steady-state concentrations of weakly bonded species that cannot easily be studied under vacuum. In some cases the presence of such new chemisorbed states open new pathways for reaction mechanisms. Two main approaches have been taken in order to link the two reaction regimes: (a) Surface-sensitive techniques have been developed that can be used in situ to characterize the catalytic system while the reactions are taking place; and (b) a low pressure-high pressure apparatus was constructed that permits the transfer of the catalytic sample from UHV to atmospheric pressures and back so that standard vacuum techniques can be used to study changes occurring in the surface before and after exposure to reaction conditions.

Most work that has been carried out on single crystals have been performed using the second approach, and in this section we review their results. Typical experimental apparatus is shown schematically in Fig. 7. It consists of a standard UHV chamber equipped with several surface analytical techniques, such as LEED, AES, TDS, and HREELS, and a retractable environmental cell that can be used to isolate the sample from vacuum and to insert it in a loop that can be pressurized up to 100 atm with reactant gases. The reaction kinetics can then be studied by circulating the gases and periodically

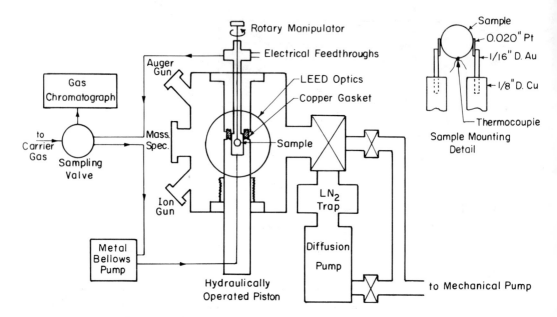

FIGURE 7 Schematic diagram of the low pressure-high pressure apparatus for combined surface analysis and catalysis studies. (From Ref. 65.)

analyzing small amounts of the gas mixture using either gas chromatography or mass spectrometry. After reactions the loop is pumped and the cell opened so that the catalyst is returned to UHV for further analysis. Several hydrocarbon reactions have been studied this way, including hydrogenation of olefins and aromatic compounds, H-D exchange, and reforming reactions of several model hydrocarbon molecules.

A. Reactions Involving Ethylidyne

It was mentioned in Section III that when ethylene is chemisorbed at room temperature under UHV over Pt(111), Pt(100), Rh(111), or Pd(111), ethylidyne is formed (Fig. 2) [25,66,67,33]. This moiety is stable upon hydrogen treatment even at atmospheric pressures, as revealed by the use of ^{14}C ethylene and a radiotracer technique [37], and it can only be rehydrogenated at temperatures above 350 K. On the other hand, CH fragments formed from ethylene decomposition at 470 K under UHV cannot be completely removed from the platinum surface even with high H_2 pressures. If ethylene adsorption is carried at above 600 K, irreversibly chemisorbed carbon with little or no hydrogen content is left on the surface. Hydrogen-to-carbon ratios for the different fragments formed during ethylene chemisorption and their ability to be rehydrogenated are shown as a function of adsorption temperature in Fig. 8.

If ethylidyne is exposed to atmospheric deuterium, not only hydrogenation reactions are seen, but H-D exchange is observed as well. The exchange is possible under UHV only with submonolayer coverages of ethylidyne, but at saturation high pressures of D_2 are required for this reaction to occur. HREELS and TDS have been used to determine that the exchange occurs in a stepwise fashion, one hydrogen atom exchanged at a time, and at comparable rates to ethylidyne hydrogenation [59,60,68,69]. An example of such results is presented in Fig. 9 for Rh(111). After a 5-min exposure of ethyli-

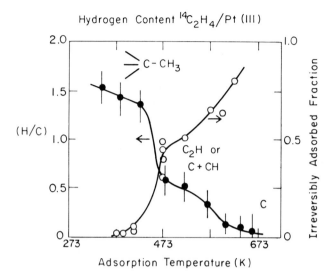

FIGURE 8 Composition and reactivity of ^{14}C-ethylene chemisorbed on Pt(111) at 320 to 670 K. The irreversibly adsorbed fraction determined by radiotracer analysis displays an excellent correlation with the average hydrogen content (H/C) of the strongly bound surface species.

dyne to 1-atm D_2, new peaks appear in the vibrational spectrum, corresponding to Pt_3C-CH_2D.

Hydrogen can also be coadsorbed with ethylidyne. However, after ethylidyne saturation, H_2 chemisorption is observed only after exposures at pressures above 10^{-5} torr at 150 K [57]. Furthermore, the exposure of ethylidyne to high H_2 pressures at 320 K results in the formation of a new species, probably ethylidene (=CH-CH$_3$), as identified by TDS (Fig. 10).

B. Ethylene Hydrogenation

Few studies of ethylene hydrogenation over single-crystal surfaces have been published to date [54,68-71]. The activity over single crystals has been found to be comparable to that of supported metal particles in all cases [68-70], suggesting that similar reaction mechanisms take place over both types of catalysts. Dalmai-Imelik and Massardier [70] have reported a structure dependence of the reaction rates as a function of crystallographic orientation of the exposed surface for nickel crystals. While the (100) surface is practically inactive for hydrogenation, activities for the (111) and (110) faces were much higher and differ by about a factor of 2 between themselves. These differences were explained in terms of the cracking of C_2H_4 observed over the (100) face. Self-hydrogenation at pressures of up to 0.1 torr has also been reported over nickel in the absence of H_2 [54].

The reaction over Pt(111) and Rh(111) crystals has been studied using several surface-sensitive techniques [68,69,72]. Ethylidyne was found to be present on the surfaces after high-pressure reactions, as confirmed by LEED, TDS, and HREELS results (Fig. 11). All three techniques yielded similar results for Pt(111) saturated with ethylidyne and after high-pressure ethylene hydrogenation reactions. Furthermore, reactions over ethylidyne-saturated surfaces had identical kinetic parameters to those performed over clean platinum. However, rates for hydrogenation of ethylidyne are several

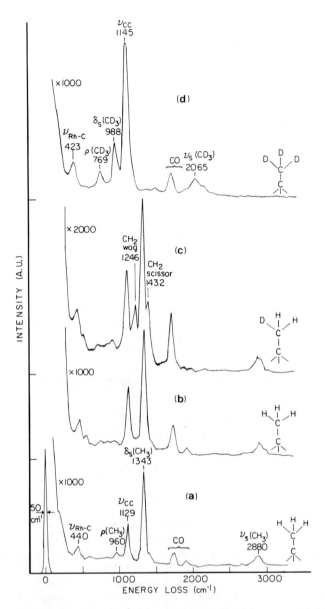

FIGURE 9 Vibrational spectra of ethylidyne on Rh(111) spectra at 310 K illustrating the deuterium exchange in the methyl group. Spectra (b) and (c) were obtained from ethylidyne after 5 min exposure to 1-atm H_2 and D_2, respectively. CCH_3 (a) and CCD_3 (d) spectra are shown for comparison.

orders of magnitude lower than for ethylene hydrogenation. Turnover frequencies for the relevant processes are presented in Table 1. Therefore, ethylene hydrogenation must take place in the presence of the strongly bonded hydrocarbon fragments. Furthermore, TDS and HREELS have shown that direct chemisorption of new ethylene molecules over the metal is sterically hindered, so the hydrogen incorporation must occur on a second layer of weakly chemisorbed ethylene molecules. A mechanism similar to that proposed by Thomson and Webb [88], involving hydrogen transfer from the

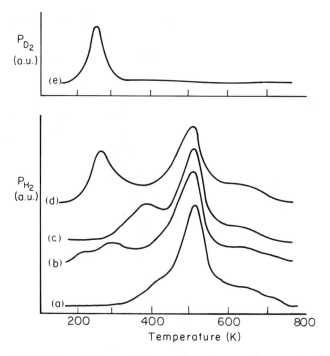

FIGURE 10 Thermal desorption (amu = 2) spectra from H_2-ethylidyne co-adsorption experiments. (a) 6 L C_2H_4 dosed at 320 K (saturated ethylidyne); followed by (b) 1200 L H_2, 10^{-5} torr at 150 K; (c) 1200 L H_2, 10^{-5} torr at 320 K, then cooled immediately to 150 K; (d) $\sim 10^{10}$ L H_2, 1 atm at 240 K; (e) $\sim 10^{10}$ L D_2, 1 atm at 240 K, 4 amu TDS.

surface through the ethylidyne fragments, has been proposed to explain these results [68]. It includes the reversible interconversion of ethylidyne into ethylidene, and vice versa, as illustrated in Fig. 12. Further proof of the model awaits the unequivocal identification of ethylidene moieties on the surface.

C. Other Hydrogenation and Dehydrogenation Reactions

The hydrogenation of benzene over nickel single-crystal surfaces has been studied by Dalmai-Imelik and Massardier [70]. They found no structure dependence for this reaction, contrary to what was observed for ethylene hydrogenation. The rates for dehydrogenation of cyclohexane over platinum, on the other hand, depend on the atomic structure of the catalytic surface [73]. Product accumulation curves for four different platinum surfaces are shown in Fig. 13. The stepped Pt(557) surface is about twice as active as Pt(111), and both kinked surfaces are appreciably more active than the stepped surface. The effect of gold deposited over Pt(111) or alloyed with it is to increase the turnover frequency for benzene formation, up to a maximum at a gold coverage of about half a monolayer, followed by a monotonic decrease at higher Au coverages [74]. The bimetallic results have been explained in terms of a competitive mechanism where the chemisorbed species either dehydrogenate and desorb as products or decompose and deactivate the catalyst. If this second pathway requires big ensembles of platinum

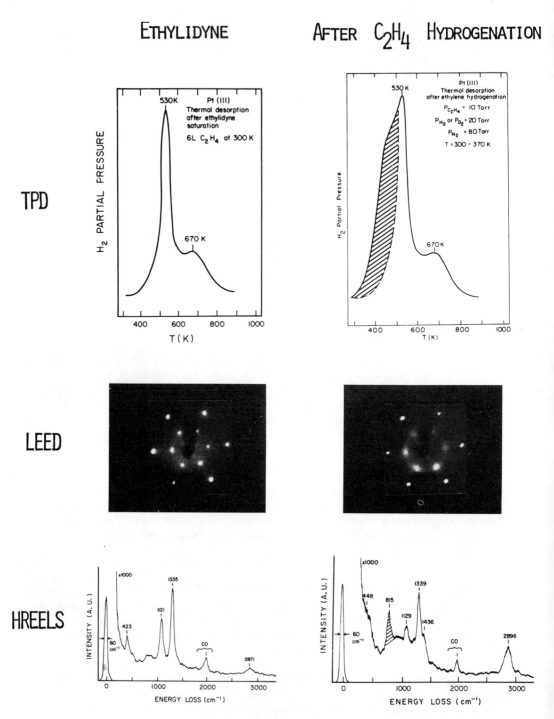

FIGURE 11 Evidence for the presence of ethylidyne on the Pt(111) surface after hydrogenation of gas-phase ethylene at atmospheric pressures over this surface. Temperature-programmed desorption, low-energy electron diffraction, and high-resolution electron energy loss spectroscopy data for ethylidyne are compared with the results of these techniques on Pt(111) after ethylene hydrogenation at atmospheric pressure and return of the crystal to vacuum.

16. Role of Hydrogen in Hydrocarbon Reactions

TABLE 1 Comparison of Hydrogenation and Exchange Rates over Pt(111) Single-Crystal Surfaces at Near Room Temperatures

Process	Estimated turnover rate (reactions/metal atom s^{-1})	Refs.
H-D exchange in saturation CCH$_3$ methyl group	10^{-5}	68, 69, 72
Hydrogenation and removal of saturation CCH$_3$ from surface	6×10^{-5}	37, 69
C$_2$H$_4$ hydrogenation $P_{C_2H_4}$ = 20 torr P_{H_2} = 100 torr	25	68, 69

atoms, it would then be inhibited by the addition of gold, which would favor the desorption of products. An alternative hypothesis involving new hydrogen chemisorption sites at the platinum-gold interface was not supported by CO TDS experiments, although Biloen et al. [87] have used related arguments to explain a change on hydrogen partial pressure dependence in the rate of propane dehydrogenation when alloying gold with platinum.

The reaction of cyclohexene with hydrogen has also been studied over platinum surfaces in the pressure range 10^{-8} to 10^2 torr [61]. The selectivity for hydrogenation over dehydrogenation changes drastically with changing pressures (Fig. 14). Benzene is produced predominantly at low pressures (ca. 10^{-7} torr), while cyclohexane is observed mainly at high pressures (ca. 10^2 torr). The low-pressure reactions are structure sensitive and proceed over the clean metal surface. At high pressures such reactions are structure insensitive, due to the continuous presence of close to a monolayer of carbonaceous species, similar to the case for ethylene. Widely differing coverages of reactive, weakly adsorbed hydrogen could be crucial in the product selectivity reversal between high and low pressures.

D. Deuterium Exchange Reactions

H-D exchange of ethane has been studied over Pt(111) surfaces [69, 75]. The activation energy and pressure dependences with respect to deuterium and

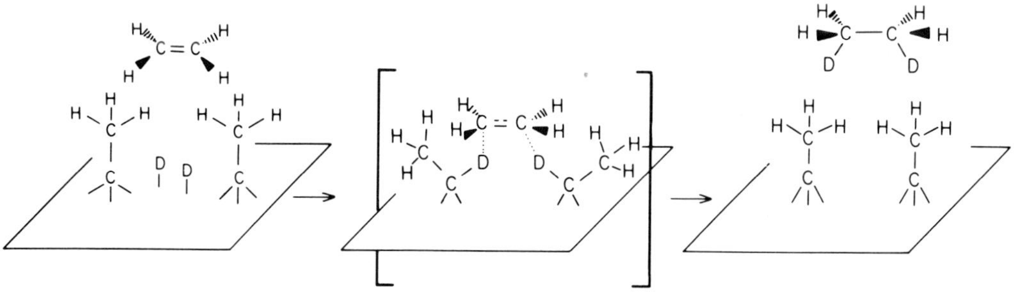

FIGURE 12 Schematic representation of the mechanism for ethylene hydrogenation over Pt and Rh(111) single-crystal surfaces.

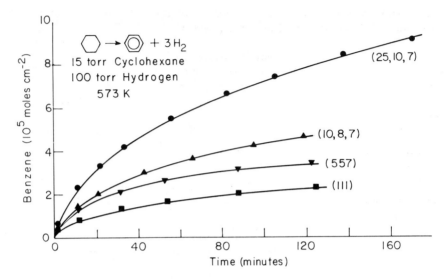

FIGURE 13 Benzene production accumulation curves over platinum single crystals as a function of surface structure.

hydrogen were similar to those reported for supported catalysts. The resulting ethane product distribution was U-shaped, peaking at one and six deuterium atoms per ethane molecule (Fig. 15). A competitive mechanism with two branches was proposed. It included multiple sequential hydrogenation-dehydrogenation equilibria on one of the ethene carbon atoms after adsorption. The presence of ethylidyne and ethylidene as intermediates was proposed.

Exchange reactions for isobutane, n-hexane, and n-heptane have been reported as well [76]. As for ethane, the exchange product distributions

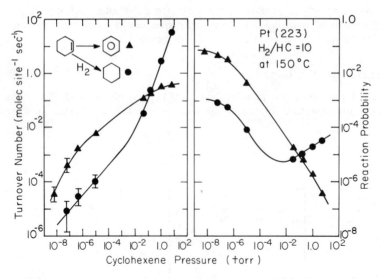

FIGURE 14 Correlation of cyclohexene reaction rates and reaction probabilities over a 10-order-of-magnitude pressure range. The reactions were performed at 425 K over the stepped Pt(223) crystal surface with $H_2/HC = 10$.

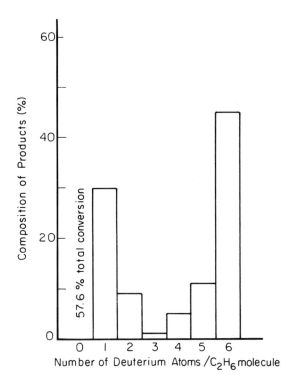

FIGURE 15 Deuterium atom distribution in the resulting ethane from the exchange with deuterium over Pt(111). $P_{C_2H_6}$ 10 torr, P_{D_2} = 100 torr, T = 550 K.

were U-shaped for n-hexane and n-heptane, but had local maxima at d_4 and d_6 for isobutane. This difference is easily explainable, since ethane, n-hexane, and n-heptane are all linear molecules, while isobutane has a tertiary carbon atom with three methyl groups attached. In addition, the exchange reactions were all shown to be structure insensitive and at least one order of magnitude faster than any other conversion process.

E. Hydrocarbon Skeletal Rearrangement

There is a marked structure sensitivity in the rates for hydrogenolysis for ethane, n-butane, and cyclopropane over nickel surfaces, as reported by Goodman [77,78]. Hydrogenolysis over (111) faces is about one order of magnitude slower than that over (100) crystals. This trend parallels that reported for ethylene hydrogenation [70], but in this case the differences in rate were established to be intrinsic to the surface topography and not due to selective poisoning by carbon.

Hydrogenolysis and isomerization reactions over several platinum crystals have been reported for many hydrocarbons, including ethane, isobutane, n-butane, neopentane, methylcyclopentane, n-hexane, and n-heptane [75, 79-83]. Changes in reaction rates with differences in surface structure have been observed for most cases. Examples of initial isomerization turnover frequencies for different hydrocarbons and platinum surfaces are shown in Fig. 16. For light alkanes there is an increase in activity when going from (111) to (100) terraces. Faster rates are also obtained when defects are

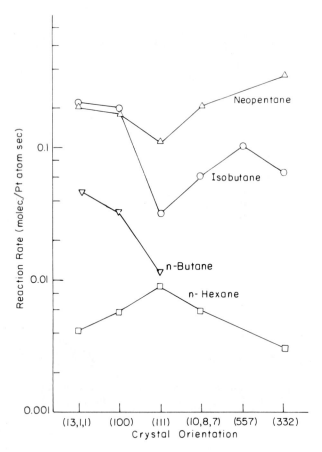

FIGURE 16 Structure sensitivities for alkane isomerization reactions over platinum single-crystal surfaces. Shown are the initial reaction rates as a function of crystallographic orientation.

present on the surface [as is the case for Pt(10,8,7), Pt(557), and Pt(332) as compared to Pt(111)]. These changes are not observed for methylcyclopentane or n-hexane. Total conversion rates also decrease with molecular weight of the reactant molecule.

For n-hexane and n-heptane, cyclization and aromatization products are also obtained. Benzene formation from n-hexane is about four times faster on (111) terraces than on (100) surfaces [82]. However, no significant differences in rates occur when steps and kinks are present, and the activities for Pt(111), Pt(557), Pt(332), and Pt(10,8,7) are all comparable. The same insensitivity to the presence of low-coordination platinum atoms is seen for toluene formation from n-heptane, where the rates are within a factor of 2 for the four platinum surfaces studied, all having (111) terraces but different defect concentrations [83]. The hydrogen pressure dependence of the benzene formation rate from n-hexane has been shown to depend on temperature [82]. Increasing positive orders were reported with increasing reaction temperatures, but at 573 K a maximum was seen at about 200 torr of H_2 (H_2: n-hexane = 10), followed by a decrease in reactivity at higher H_2 pressures. Only positive orders in P_{H_2} were seen for hydrogenolysis reactions, strongly suggesting that these reactions occur through different mechanisms.

16. *Role of Hydrogen in Hydrocarbon Reactions*   443

The formation of carbonaceous deposits over the catalyst surfaces is always detected after all hydrocarbon reactions. These deposits have been characterized through the use of several techniques, including AES, LEED, TDS, CO titration, and ^{14}C radiotracer detection [37,84]. It was found that such fragments are mainly irreversibly adsorbed during reactions, covering most of the surface area. There is, however, a small fraction of platinum atoms uncovered that are the responsible for catalytic activity. The amount and type of these metal atoms can be measured by CO titration experiments [84]. Steps and kinks can be differentiated because CO chemisorbs more strongly on those defects as compared to (111) terraces, a difference that becomes evident by the presence of a high-temperature peak in the thermal desorption spectra of CO. This is shown in Fig. 17 for Pt(557). It can be seen that the high-temperature peak is observed in titrations after cyclohexane, neopentane, and isobutane reactions, but not after methylcyclopentane or *n*-hexane reactions. These results correlate well with those from kinetic studies, where surfaces with defects displayed higher activity for light alkanes than (111) terraces, while no difference was noticed for methylcyclopentane or *n*-hexane conversion.

Quantitative hydrogen thermal desorption studies were carried out as a function of surface structure and reaction temperature. The most important

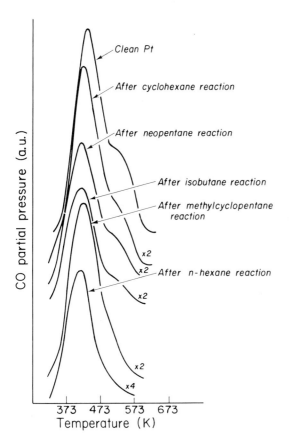

FIGURE 17 Comparison between CO thermal desorption from the clean (557) platinum surface and Pt(557) following hydrocarbon reactions. The adsorption temperature was 310 to 315 K, 80 K s^{-1}, and CO exposure = 36 L. (From Ref. 85.)

chemical properties of the carbonaceous deposits are their ability to store
and exchange hydrogen with reactant surface species and to provide desorption
sites for product molecules. A model was proposed where dehydrogenation
of hydrocarbons occurred over the deposits, followed by surface migration
to the bare platinum sites, skeletal rearrangement, and migration back to
the carbonaceous fragments for rehydrogenation and desorption [84]. The
hydrogenation-dehydrogenation role of the strongly bonded hydrocarbon
fragments is consistent with the model proposed previously for ethylene
hydrogenation.

V. CONCLUSIONS

The preceding are representative studies of hydrocarbon reactions over metal
single-crystal surfaces, performed under controlled environments. They show
that high-pressure experiments are recently being added to the vast knowledge
available on chemisorption under ultrahigh vacuum using well-characterized
single-crystal catalyst surfaces in order to bridge the gap between the two
pressure regimes. One of the conclusions to be reached from these new
results is that although the information obtained under UHV is useful for the
understanding of gas-solid interactions, they are in general noncatalytic
processes, and extrapolation to high-pressure catalytic reactions has to be
done with caution. While chemisorption of hydrocarbons under vacuum
generally leads to decomposition and hydrogen desorption, catalytic reactions
take place at atmospheric pressures, including those where hydrogen-carbon
bonds are formed. A good example of these differences was illustrated for
the cyclohexene reactions in Fig. 14: There is a marked inversion in selectivity from almost total dehydrogenation to benzene in vacuum to hydrogenation
and cyclohexane formation in the 10^2 torr range. High pressures of H_2 are
also indispensable for most skeletal rearrangement processes. A second main
difference between the two pressure regimes, also illustrated in Fig. 14, is
the change in reaction probabilities, from almost unity under UHV down to
less than 10^{-4} at 1 atm. This change is explained in part by the formation
of carbonaceous deposits over the catalyst surface under atmospheric reactant
pressures, leading to changes in the reaction mechanisms. For example,
while ethylene hydrogenation takes place over clean platinum in UHV with
an activation energy of 6 kcal mol^{-1}, the same reaction is catalytic and takes
place in the presence of an ethylidyne layer at atmospheric pressures with
an activation energy of 11 kcal mol^{-1}. Similar differences exist in other
hydrogenation reactions. Hydrogenolysis and isomerization processes take
place at higher temperatures and do occur over the metal atoms, but again
hydrogenation and dehydrogenation steps are necessary to make these
reactions catalytic, and hydrogen transfer through strongly bonded hydrocarbon residues makes the difference between high pressure and vacuum.

In summary, the combination of surface science techniques with transfer
devices that allows catalyst samples to be transported from UHV to high
pressures and back is helping to obtain a better overall picture of catalytic
reactions. It has become evident that hydrogen plays a vital role in hydrocarbon conversion. Hydrogen transfer mechanisms are also becoming better
understood. They may not always proceed directly from the metal surface
to the reactant molecule but may sometimes involve intermediate states where
the strongly bonded hydrocarbon fragments may be key components. In
situ spectroscopies together with studies over well-characterized surfaces
are still needed to clarify further the mechanistic details of these reactions
on the atomic scale.

ACKNOWLEDGMENT

This work was supported by the Director, Office of Energy Research, Office of Basic Energy Sciences, Materials Sciences Division of the U.S. Department of Energy under Contract DE-ACO3-76SF00098.

REFERENCES

1. G. A. Somorjai, *Chemistry in Two Dimensions: Surfaces*, Cornell University Press, Ithaca, N.Y. (1981).
2. Z. Knor, in *Catalysis Science and Technology*, Vol. 3, Springer-Verlag, Berlin (1982), pp. 231-280.
3. K. Christmann, G. Ertl, and T. Pignet, *Surf. Sci.*, 54:365 (1976).
4. P. N. Ross, Jr., *J. Electrochem. Soc.*, 126:67 (1979).
5. K. Christmann and G. Ertl, *Surf. Sci.*, 60:365 (1976).
6. M. Salmeron, R. J. Gale, and G. A. Somorjai, *J. Chem. Phys.*, 70:2807 (1979).
7. K. E. Lu and R. R. Rye, *Surf. Sci.*, 45:677 (1974).
8. D. M. Collins and W. E. Spices, *Surf. Sci.*, 69:85 (1977).
9. R. W. McCabe and L. D. Schmidt, *Proceedings of the 7th International Conference on Solid Surfaces*, Vienna (1977), p. 1201.
10. A. M. Baro and H. Ibach, *Surf. Sci.*, 92:237 (1980).
11. S. M. Davis and G. A. Somorjai, *Surf. Sci.*, 91:73 (1980).
12. J. Lee, J. P. Cowin, and Z. Wharton, *Surf. Sci.*, 130:1 (1983).
13. J. P. Muscat, *Surf. Sci.*, 110:389 (1981).
14. R. A. Van Santen, *Surf. Sci.*, 53:35 (1975).
15. M. Salmeron and G. A. Somorjai, *J. Phys. Chem.*, 85:3835 (1981).
16. T. S. Wittrig, P. D. Szuromi, and W. H. Weinberg, *J. Chem. Phys.*, 76:3305 (1982).
17. J. C. Hamilton, N. Swanson, B. J. Waclawski, and R. J. Celota, *J. Chem. Phys.*, 74:4156 (1981).
18. J. E. Demuth and D. E. Eastman, *Phys. Rev. Lett.*, 32:1123 (1974).
19. J. A. Stroscio, S. R. Bare, and W. Ho, *Surf. Sci.*, 148:499 (1984).
20. R. Ducros, M. Hausley, M. Alnot, and A. Cassuto, *Surf. Sci.*, 71:433 (1978).
21. R. B. Moyes and P. B. Wells, in *Advances in Catalysis*, Vol. 23, Academic Press, New York (1980), p. 121.
22. B. E. Koel and G. A. Somorjai, *J. Electron Spectrosc. Relat. Phenom.*, 29:187 (1983).
23. S. J. Thomson, in *Catalysis*, Vol. 3, Chemical Society, London (1980), p. 1.
24. S. M. Davis and G. A. Somorjai, in *The Chemical Physics of Solid Surfaces and Heterogeneous Catalysis*, Vol. 4 (D. A. King and D. P. Woodruff, eds.), Elsevier, Amsterdam (1982), p. 217.
25. L. L. Kesmodel, L. H. Dubois, and G. A. Somorjai, *Chem. Phys. Lett.*, 56:267 (1978).
26. L. L. Kesmodel, L. H. Dubois, and G. A. Somorjai, *J. Chem. Phys.*, 70:2180 (1979).
27. J. E. Demuth, *Surf. Sci.*, 93:L82 (1980).
28. P. Skinner, M. W. Howard, I. A. Oxton, S. F. A. Kettle, D. B. Powell, and N. Sheppard, *J. Chem. Soc. Faraday Trans. 2*, 77:1203 (1981).
29. H. Ibach and S. Lehwald, *J. Vac. Sci. Technol.*, 15:407 (1978).
30. N. Freyer, G. Pirug, and H. P. Bonzel, *Surf. Sci.*, 125:327 (1983).
31. R. J. Koestner, J. Stöhr, J. L. Gland, and J. A. Horsley, *Chem. Phys. Lett.*, 105:332 (1984).

32. J. A. Gates and L. L. Kesmodel, *Surf. Sci.*, 124:68 (1983).
33. L. L. Kesmodel and J. A. Gates, *Surf. Sci.*, 111:L747 (1981).
34. D. R. Lloyd and F. P. Netzer, *Surf. Sci.*, 129:L249 (1983).
35. R. J. Koestner, M. A. Van Hove, and G. A. Somorjai, *J. Phys. Chem.*, 87:203 (1983).
36. M. Salmeron and G. A. Somorjai, *J. Phys. Chem.*, 86:341 (1982).
37. S. M. Davis, F. Zaera, B. E. Gordon, and G. A. Somorjai, *J. Catal.*, 92:240 (1985).
38. A. M. Baro and H. Ibach, *J. Chem. Phys.*, 74:4194 (1981).
39. M. A. Van Hove, R. J. Koestner, and G. A. Somorjai, *J. Vac. Sci. Technol.*, 20:886 (1982).
40. R. J. Koestner, J. C. Frost, P. C. Stais, M. A. Van Hove, and G. A. Somorjai, *Surf. Sci.*, 116:85 (1982).
41. M. A. Van Hove, R. Lin, and G. A. Somorjai, *Phys. Rev. Lett.*, 51:778 (1983).
42. B. E. Koel, J. E. Crowell, B. E. Bent, C. M. Mate, and G. A. Somorjai, *J. Phys. Chem.*, 90:2949 (1986).
43. T. Engel and H. Kuipers, *Surf. Sci.*, 90:162 (1979).
44. S. L. Bernasek and G. A. Somorjai, *J. Chem. Phys.*, 62:3149 (1975).
45. I. E. Wachs and R. J. Madix, *Surf. Sci.*, 58:590 (1976).
46. K. E. Lee and R. R. Rye, *Surf. Sci.*, 45:647 (1974).
47. K. Christmann and G. Ertl, *Surf. Sci.*, 60:365 (1976).
48. M. Salmeron, R. J. Gale, and G. A. Somorjai, *J. Chem. Phys.*, 67:5324 (1977).
49. T. Engel and G. Ertl, in *The Chemical Physics of Solid Surfaces and Heterogeneous Catalysis*, Vol. 4 (D. A. King and D. P. Woodruff, eds.), Elsevier, Amsterdam (1982), p. 195.
50. M. Salmeron, R. J. Gale, and G. A. Somorjai, *J. Chem. Phys.*, 70:2807 (1979).
51. G. Comsa, R. David, and B. J. Schumacher, *Surf. Sci.*, 85:45 (1979).
52. M. J. Cardillo, M. Balooch, and R. E. Stickney, *Surf. Sci.*, 50:263 (1975).
53. J. Horiuti and K. Miyahara, *Hydrogenation of Ethylene on Metallic Catalysts*, NSRDS-NBS 13 (1968).
54. W. Hasse, H.-L. Günter, and M. Henzler, *Surf. Sci.*, 126:479 (1983).
55. H. Steininger, H. Ibach, and S. Lehwald, *Surf. Sci.*, 117:685 (1982).
56. P. Berlowitz, C. Megiris, J. B. Brett, and H. H. Kung, *Langmuir*, 1:206 (1985).
57. D. Godbey, F. Zaera, R. Yates, and G. A. Somorjai, *Surf. Sci.*, 167:150 (1986).
58. J. R. Creighton, K. M. Ogle, and J. M. White, *Surf. Sci.*, 138:L137 (1984).
59. B. E. Koel, B. E. Bent, and G. A. Somorjai, *Surf. Sci.*, 146:211 (1984).
60. A. Wieckowski, S. D. Rosasco, G. N. Salaita, A. Hubbard, B. E. Bent, F. Zaera, D. Godbey, and G. A. Somorjai, *J. Am. Chem. Soc.*, 107:5910 (1985).
61. S. M. Davis and G. A. Somorjai, *J. Catal.*, 64:60 (1980).
62. J. W. A. Sachtler, J. P. Biberian, and G. A. Somorjai, *Surf. Sci.*, 110:43 (1981).
63. C. E. Smith, J. P. Biberian, and G. A. Somorjai, *J. Catal.*, 57:426 (1979).
64. D. W. Blakely and G. A. Somorjai, *J. Catal.*, 42:181 (1976).
65. D. W. Blakely, E. Kozak, B. A. Sexton, and G. A. Somorjai, *J. Vac. Sci. Technol.*, 13:1901 (1976).
66. H. Ibach, in *Proceedings of the International Conference on Vibrations in Adsorbent Layers*, Jülich, West Germany (1978), pp. 64-75.

67. L. H. Dubois, D. G. Castner, and G. A. Somorjai, *J. Chem. Phys.*, 72: 5234 (1980).
68. F. Zaera and G. A. Somorjai, *J. Am. Chem. Soc.*, 106:2288 (1984).
69. F. Zaera, Ph.D. thesis, University of California, Berkeley (1984).
70. G. Dalmai-Imelik and J. Massardier, in *Proceedings of the 6th International Congress on Catalysis, London* (1976), pp. 90-100.
71. D. W. Goodman, *Acc. Chem. Res.*, 17:194 (1980).
72. B. E. Bent, F. Zaera, and G. A. Somorjai, to be published.
73. R. K. Herz, W. D. Gillespie, E. E. Petersen, and G. A. Somorjai, *J. Catal.*, 67:371 (1981).
74. J. W. A. Sachtler and G. A. Somorjai, *J. Catal.*, 81:77 (1983).
75. F. Zaera and G. A. Somorjai, *J. Phys. Chem.*, 89:3211 (1985).
76. S. M. Davis and G. A. Somorjai, *J. Phys. Chem.*, 84:1545 (1983).
77. D. W. Goodman, *Surf. Sci.*, 123:L679 (1982).
78. D. W. Goodman, in *Proceedings of the 8th International Congress on Catalysis, Berlin*, Vol. 4 (1984), p. 3.
79. S. M. Davis, F. Zaera, and G. A. Somorjai, *J. Am. Chem. Soc.*, 104: 7453 (1982).
80. F. Zaera, D. Godbey, and G. A. Somorjai, *J. Catal.*, 101:73 (1986).
81. F. Garin, S. Aeiyach, P. Legare, and G. Maire, *J. Catal.*, 77:323 (1982).
82. S. M. Davis, F. Zaera, and G. A. Somorjai, *J. Catal.*, 85:206 (1984).
83. W. D. Gillespie, R. K. Herz, E. E. Petersen, and G. A. Somorjai, *J. Catal.*, 70:147 (1981).
84. S. M. Davis, F. Zaera, and G. A. Somorjai, *J. Catal.*, 77:439 (1982).
85. F. Zaera and G. A. Somorjai, *Langmuir*, 2:686 (1986).
86. F. Zaera and R. Hall, *Surf. Sci.*, 180:1 (1987).
87. P. Biloen, F. M. Dautzenberg, and W. M. H. Sachtler, *J. Catal.*, 50:77 (1977).
88. S. J. Thomson and G. Webb, *J. Chem. Soc. Chem. Commun.*, 526 (1976).

17 | Hydrogen Effects in Skeletal Reactions of Hydrocarbons over Metal Catalysts

ZOLTÁN PAÁL

Institute of Isotopes of the Hungarian Academy of Sciences, Budapest, Hungary

I.	INTRODUCTION	449
	A. Metal Catalysts for Skeletal Reactions of Hydrocarbons	450
II.	EFFECT OF HYDROGEN PRESSURE ON HYDROCARBON REACTIONS	452
	A. Dehydrogenation	454
	B. Ring Closure	456
	C. Skeletal Rearrangement of Alkanes	466
	D. Hydrogenolysis	470
	E. Homologation	472
	F. Cyclic Hydrocarbons	474
	G. Catalyst Deactivation	477
III.	PERMANENT EFFECTS OF VARIOUS HYDROGEN TREATMENTS	480
	A. High-Temperature Hydrogen Treatment of Supported Metal Catalysts	480
	B. Structure Sensitivity on Supported Pt, Pd, and Rh Pretreated in Hydrogen	481
	C. Catalytic Properties of Unsupported Metals Pretreated at Different Temperatures	487
IV.	CONCLUSIONS	491
	REFERENCES	492

I. INTRODUCTION

Metal catalysts were used as hydrogenating-dehydrogenating catalysts during the first 30 years of this century. Their importance in such reactions is still remarkable (and their possibilities are far from being exhausted), as shown in Chapters 18, 19, and 24. Various metal-catalyzed "skeletal" reactions of hydrocarbons (i.e., those involving the rearrangement of the original carbon skeleton) were discovered in a rapid sequence in the 1930s: hydrogenolysis of alkanes [1], cyclopentanes [2], dehydrocyclization to give aromatics [3], and skeletal isomerization [4]. C_5 cyclization followed in 1954

[5]. After several early observations (e.g., Ref. 6), the buildup reaction was termed "homologation" as late as in 1979 [7].

Aromatization obtained large-scale industrial application in technologies developed in the late 1940s called, in general, "reforming." It applied Pt on acidic silica-alumina support and high hydrogen pressures, as discussed in Chapter 25. The dual-function mechanism developed in 1953 [8] confined the function of metal to dehydrogenation and attributed other reactions (isomerization, ring closure and enlargement, hydrocracking) to the acidic function. The study of other reactions over metals began again in the early 1960s; for example, metal-catalyzed isomerization was "rediscovered" in 1963 [9]. Since then the mechanisms of these processes have been studied extensively, as indicated by a large number of reviews and books (e.g., Refs. 10 to 22). Despite counterclaims, there are still many obscure areas in this field.

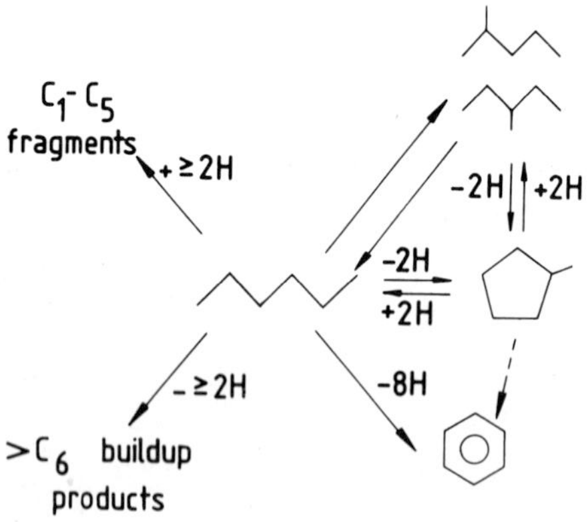

Scheme 1

Scheme 1 shows that the liberation or uptake of hydrogen may be quite different for these processes. Therefore, it is expected that hydrogen may exert a remarkable effect on their occurrence. In fact, a long, steady catalyst activity can be achieved only by applying a large excess (10-to-20-fold) of hydrogen together with the hydrocarbon reactant (see also Chapter 25). This is supported by industrial practice in reforming. Initial Soviet work usually applied pure hydrocarbon feed; here, however, hydrogen given off during aromatization in relatively large flow reactor vessels may have supplied some hydrogen.

That aromatization has maximum yield as a function of hydrogen pressure was first reported by Rohrer et al. [23]. Later studies revealed that most reactions have maximum yields as a function of the hydrogen pressure on laboratory samples [24, 25] as well as on industrial catalysts (see Chapter 25). Hydrogen effects of this type were summarized in our previous review [26].

A. Metal Catalysts for Skeletal Reactions of Hydrocarbons

Reactions included in Scheme 1 are usually catalyzed by group VIII metals. The general experience is that a large number of metals exhibit some catalytic effect but that good catalysts are few: Pt, Pd, Rh, Ir, and Ni. Aromatization occurs on almost every good dehydrogenating metal [14]. Hydrogenolysis has been observed on Re, Tc, Mn [11], and W [10], whereas groups VI and VII metals appear rather active in homologation [7]. Isomerization was observed over films of Ti, V, Ni, Mo, Ta, W, Re, and Ir [27]; also, Rh, Ir,

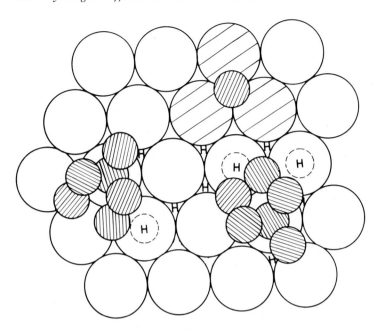

FIGURE 1 Schematic representation of hydrogen and carbon effects over Pt(111). Large open circles denote Pt atoms, small shaded ones C atoms (to scale). Hydrogen adsorbed interstitially and on top of atoms is denoted by "H." Interstitial hydrogen prevents the formation of deeply dehydrogenated C_5 species Pt_3C and opens the possibility for atom-top adsorption of C_5 species (shown with methylcyclopentane and 3-methylpentane). Interstitial carbon adsorption deactivates surrounding Pt atoms; here only reactions such as dehydrogenation is possible. (After Ref. 32; used by permission of The Royal Society of Chemistry.)

Ru, and Au were found to be active in addition to the best catalysts: Pd and Pt (for a review, see Ref. 18). C_5 cyclization and ring opening are, in turn, confined to four metals: Rh, Ir, Pd, and Pt, in order of increasing activity [28].

Earlier theories of catalysis (e.g., Balandin's multiplet theory [29]) attribute catalytic activity to the geometry of the catalyst surface. Interatomic distance can be one of the common important factors with the four metals catalyzing C_5-cyclic reactions [18,28]. Electronic theories [30] assigned catalytic activity to the (collective) electronic properties of the catalyst metal. This presents one possible interpretation of the alloying effect. The concept of Montarnal and Martino [31] relating catalytic properties to the "hard" or "soft" Pearson-acid character of the metal is also a type of electronic theory.

Hydrogen can be regarded as a peculiar alloying element adsorbed or absorbed by the metal (see Chapter 1). Its small atom may occupy certain (e.g., interstitial) positions on the surface, thus activating other positions. An example is shown in Fig. 1. Here hydrogen is compared with one C atom in a similar position, which, in turn, *deactivates* neighboring metal atoms, such as inactive alloying metals (e.g., Cu) [33]. This geometric interpretation may be satisfactory for explaining C_5-cyclization activity [18,28]. On the other hand, the electron of hydrogen may become a part of the electron gas of the metal [34]; by doing so, it can also influence electronic properties.

In principle, working metal catalysts should always be regarded as *metal-hydrogen systems*—unfortunately, of unknown composition. One may recall here the "hydrogen fog" on metal surfaces postulated in Chapter 2. In hydrocarbon reactions, the catalysts always contain carbon, too [13]. The correlation between surface carbon and hydrogen content is one of the big unsolved problems of present-day catalysis.

Two types of hydrogen effects are discussed in this chapter. First, the larger part of the chapter deals with the effect of varying hydrogen pressures. Its changes may promote or suppress individual reactions and can be attributed to the fact that there exists an *optimum hydrogen coverage* for each reaction. Second, pretreatment of catalysts with hydrogen may result in rather permanent changes in catalyst activity and selectivity. Here some catalytic aspects of phenomena are treated which are related to spillover (Chapter 12), SMSI (Chapter 13), or sintering (Chapters 10 and 11).

II. EFFECT OF HYDROGEN PRESSURE ON HYDROCARBON REACTIONS

Kiperman [35] discussed the classification of possible hydrogen effects on various hydrocarbon reactions. He suggested that hydrogen may have (a) a direct, kinetic effect when hydrogen is an active participant in the activated complex of the rate-determining slow step; or (b) an indirect, "thermodynamic" effect when it participates in fast, practically equilibrated steps or influences side products or the reverse reaction or when it "stabilizes the surface layer by interactions of other type (such as dissolution in the subsurface layers, interactions with impurities, etc.)." We think that the latter types of effects are those which have been least recognized so far and which will occupy a large part of what follows.

In Chapter 15 the above-mentioned direct kinetic effects were discussed in detail. We stress the observation reported there distinguishing between ensemble effects as defined by the size of "landing site" (i.e., how many Z hydrogen adsorbing sites participate in the adsorption of the reactive intermediate *bound by one of its C atoms*) and between the ensemble size required to adsorb the reactive entity by more than one of its C atoms. This picture will be used throughout our discussion, but since the reactions and reactants are rather complex moieties here, no quantitative treatment will be possible like that for simpler reactions (e.g., methane adsorption on clean surfaces) [36].

One very important feature of the concept outlined in Chapter 15 is that the *primary adsorption* may be an *associative* complex capable of H-D exchange. This may correspond to the "weak I" interaction proposed by Guczi et al. [37]. This complex may desorb or—perhaps with participation of surface hydrogen atoms—can be progressively dehydrogenated in steps such as Eqs. (61) and (63) in Chapter 15. This progressive dehydrogenation leads to various surface species, each being the intermediate of different reaction. Their abundance can be regulated by the amount of surface hydrogen, that is, the way hydrogen influences selectivity.

This idea was put forward first for the formation of aromatics versus C_6 saturated products (isomers plus methylcyclopentane) from *n*-hexane [38] (Scheme 2).

Comparing actual values with thermodynamical equilibrium data, it seems as if hydrogen concentration on the surface corresponded to higher hydrogen pressure than that in the gas phase. The idea was developed further by Margitfalvi et al. [39] for hydrocarbon reactions and by Menon et al. [40] for $CO + H_2$ synthesis (see also Chapter 23).

17. *Hydrogen Effects in Skeletal Reactions*

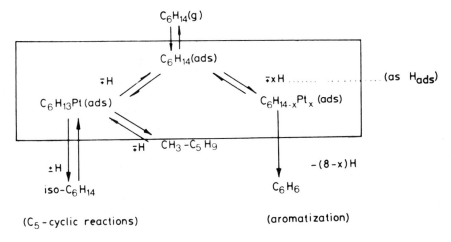

Scheme 2

The simplified order of hydrogen optima for hydrocarbon reactions given in [26] is as follows: aromatization < hydrogenolysis < production of saturated hydrocarbons (isomers, C_5 cyclics). The corresponding optimum hydrogen coverage for each reaction can be created by appropriate hydrogen pressure. However, three other factors should be considered here. At higher *temperatures*, the pressure necessary to maintain this optimum coverage will be higher and higher; thus maximum yield or selectivity will be shifted toward higher p_{H_2} values. Alternatively, at constant pressure, the increase of temperature is equivalent to decreasing hydrogen pressure. An interesting consequence was experienced in temperature-programmed reaction of n-hexane over Pt catalysts [41]. As temperature increased, first saturated C_6 products, then fragments, and finally, benzene predominated. Also, hydrogen effects may be regarded as hydrogen and hydrocarbon competing for surface sites. Here the *relative adsorption coefficient* of the reacting hydrocarbon may be important. With increasing molecular mass, this value also increases [42]. These effects explain why the same types of reactions (e.g., hydrogenolysis) of various hydrocarbons measured in the same pressure range give positive, zero, or negative hydrogen orders [42]. For example, C_5 cyclization of n-pentane showed a strong negative hydrogen order [43], that of 2,3-dimethylpentane [44], a positive hydrogen order in the near-atmospheric pressure range. Figure 2 illustrates that this is, indeed, the fact if the *same* pressures are compared; however, in a sufficiently wide pressure range, both reactions have maxima as a function of hydrogen pressure, but the position of maxima is at different p_{H_2} values. An analysis of the possible surface processes in terms of the "landing site" model predicts the bell-shaped curve for the rate as p_{H_2} increases; however, such curves were actually shown for methane H-D exchange only with adsorption being the rate-determining step. For larger molecules, the possibilities offered by the formally derived equations were not fully exploited [44].

Ruckenstein and Dadyburjor [47] proposed a similar multiple landing area model. Using the idea that the surface concentration of intermediates of various dehydrogenation is regulated by hydrogen, they derived theoretically that reaction rates of various n-pentane reactions (as reported by Garin and Gault [43]) should have maxima as a function of the hydrogen pressure. Maxima very similar to those were reported actually (e.g., in [25, 45, 46]).

The structure of the reactant represents a third factor to be considered. For example, aromatization is facilitated if the alkane has at least six C atoms

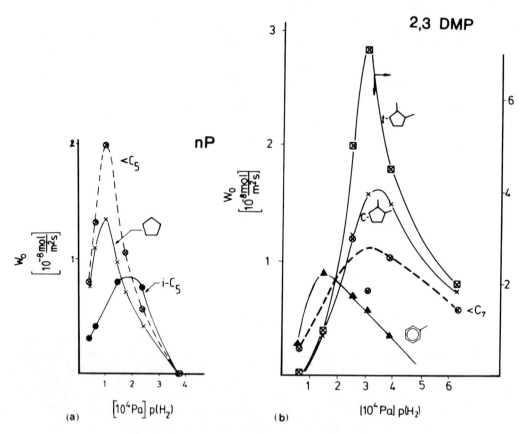

FIGURE 2 Rates of formation of various products form (a) n-pentane (nP) and (b) 2,3-dimethylpentane (2,3 DMP), as a function of the hydrogen pressure. T = 603 K, 0.1 g of Pt black (pretreatment 633 K), p_{HC} = 1.23 kPa. (From H. Zimmer, unpublished data (nP) and adapted from Ref. 45.)

in its carbon chain and C_5 cyclization if the main carbon chain contains five C atoms [45]. With quaternary C atoms in the molecule, olefin formation is necessarily suppressed, with the tertiary C atom enhanced. Actual selectivities at two p_{H_2} values are illustrated in Table 1.

The different character of surface intermediates has been supported by the transient kinetic experiments of Margitfalvi et al. [39,47]. Hexane aromatization and hydrogenolysis involve strongly bound surface intermediates and require large sites. Isomerization, C_5 cyclization, and n-hexene formation, on the other hand, should proceed via weakly bound surface intermediates with less dehydrogenation. The same reaction (e.g., ethane hydrogenolysis [48]) may also have involved two types of intermediates: a higher degree of dehydrogenation at lower hydrogen pressure and a lower degree of dehydrogenation at higher hydrogen pressure. The former is the only one giving maximum rate as a function of hydrogen. In the forthcoming sections, hydrogen effects on individual reactions of open-chain hydrocarbon are discussed. Cyclic compounds will be treated separately.

A. Dehydrogenation

Alkanes and cycloalkanes usually give some olefin products under the conditions of skeletal reactions. However, the olefin concentration permitted by

TABLE 1 Selectivity of Formation of Various Products from Alkanes[a]

Reactant	Selectivity (%) for:				
	Hydrogenolysis	Aromatization	C_5 cyclization	Isomerization	Olefin formation
	p_{H_2} = 120 torr				
n-Heptane	59	14	18	8	1
3-Methylhexane	50	22	19	2	7
2,4-Dimethylpentane	27	6	44	3	20
3,3-Dimethylpentane	49	5	37	5	4
	p_{H_2} = 320 torr				
n-Heptane	42	6	35	17	—
3-Methylhexane	46	16	27	5	6
2,4-Dimethylpentane	11	1	84	4	—
3,3-Dimethylpentane	28	—	64	8	—

[a]Static-circulation system, 0.1 g of Pt black, T = 603 K, p_{HC} = 10 torr.
Source: Data from Ref. 46.

thermodynamics is rather low at typical temperatures of these reactions (500 to 700 K) and olefins can also undergo reactions similar to those of alkanes. Thus it is sometimes difficult to state without other evidence whether olefins are by-products or intermediates of skeletal transformations.

One expects that increasing p_{H_2} values suppress olefin yields. In some cases, however, maxima are observed as a function of hydrogen pressure (Fig. 3). These maxima, as postulated in Section I, are shifted to higher p_{H_2} values as temperature increases. The optimum hydrogen coverage for dehydrogenation is rather low.

Dramatic hydrogen effects are observed as far as the structure of the olefin produced is concerned. Higher hydrogen pressures favor the formation of internal olefins; less hydrogen leads to terminal olefins [45,50]. The production of olefins with double bonds adjacent to a tertiary C atom is especially favored; these olefins are usually thermodynamically much more stable than others. Typical data have been collected in Table 2. Actual compositions are compared with equilibrium ones in one case. The 100% terminal olefin in helium is very far from equilibrium; however, the composition in hydrogen does not correspond to it either: the formation of 3-methyl-2-hexene is clearly preferential. Much more olefin was always formed in a pulse system than in static-circulation apparatus. Apparently, olefins are primary products formed under transient conditions and are consumed under long contact times.

Olefins were formed with selectivities between 0.01 and 0.06 over Pd, Re, Ir, and Pt; less active Ag and Cu showed selectivities of 0.16 and 0.56, respectively (pulse system, hydrogen carrier gas, 633 K) [51]. No olefins were detected with Co, Ni, Ru, Rh, and Os. Biloen et al. [52] suggested that the splitting of the second H atom from an alkane would be rate determining over Pt or Pt-Au catalysts, the active center being a single Pt atom also holding a H atom. Thus direct participation of hydrogen in the active site is likely. These experiments also pointed to the fact that gold decreases

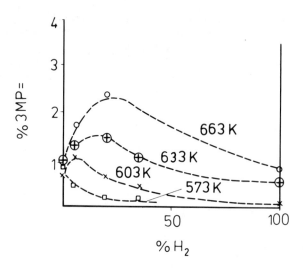

FIGURE 3 Rates of formation of 3-methylpentene isomers from 3-methylpentane as a function of the hydrogen content of the carrier gas. Pulse system, 0.4 g of Pt black, 3-μl Pulses into 60 ml min^{-1} carrier gas (He + H$_2$ mixtures), T = 633 K. (Adapted from Ref. 50.)

the ability of Pt to dissociate and/or to chemisorb hydrogen. Similar effects were also found on Pt-Sn [14, 53]. The presence of small amounts of carbon on Pt black can even slightly increase dehydrogenation activity [54]; at the same time, yields of fragments and saturated products decreased considerably.

Dehydrogenation of cyclohexane gives benzene; thus it is an aromatization reaction related somewhat to those treated in the next section. Hydrogen was reported to hinder this reaction on unsupported Ni [175] and Pt [176] catalysts below about 573 K. Above this temperature the rate of benzene formation was almost independent of hydrogen pressure. This must correspond to the descending branch of maximum curves of the type shown in Figure 2. The ascending branch was reported in the case of Ni on sepiolite between 598 and 648 K [177]. Here also the shift of maximum toward higher hydrogen pressures could be seen as the temperature increased, although in each case the maxima were not reached. The temperature effects mentioned in Section II.A also manifest themselves in these results: the negative hydrogen order below 573 K becomes zero order around this temperature [175,176] and positive order [177] above 600 K.

B. Ring Closure

Cyclic products with C$_6$ and C$_5$ rings can be formed from open-chain alkanes. C$_6$-cyclic products are almost always aromatics, whereas saturated ones predominate among C$_5$ cyclics. Both can be formed on *metallic functions* and, we believe, their production involves basically different routes. Their ratio can be controlled by hydrogen pressure. Also, single-crystal studies showed that the reaction intermediate leading to aromatization has a lower hydrogen content than that leading to alkylcyclopentane formation [55]. Increasing hydrogen pressure and decreasing temperature favor C$_5$ cyclization. Table 3 compares selected results obtained on Pt black with those reported for single-crystal faces in Ref. 55. The aromatic/C$_5$ cyclic ratio also increases, as expected, at constant hydrogen pressure but at increasing temperature (Table 3c, single crystals; Table 4, dispersed catalysts).

TABLE 2 Structure of Olefinic Products in Helium and Hydrogen[a]

Starting C_7H_{10}	Conversion to olefins (mol %) in:		Product olefin	Selectivity of individual olefins (%) in:		Equilibrium olefin concentration [b]
	He	H_2		He	H_2	
3-Ethylpentane	0.262	0.858	3-Ethyl-1-pentene	100	10.2	—
			3-Ethyl-2-pentene	—	89.8	—
2,3-Dimethylpentane	0.261	0.517	2,3-Dimethyl-1-pentene	56.3	31.2	—
			2,3-Dimethyl-2-pentene	—	57.0	—
			2,3-Dimethyl-4-pentene	18.3	3.0	—
			2-Isopropyl-1-butene	25.4	8.8	—
3-Methylhexane	0.101	0.513	3-Methyl-1-hexene	—	4.3	2.4
			3-Methyl-2-hexene	—	50.6	28.6
			3-Methyl-3-hexene	—	17.0	43.6
			3-Methyl-4-hexene	—	28.1	23.0
			3-Methyl-5-hexene	100	—	2.4
2,2-Dimethylpentane	0.303	0.148	2,2-dimethyl-3-pentene	35.6	78.8	—
			2,2-dimethyl-4-pentene	64.4	21.1	—

[a] Pt-black, pulse system, T = 633 K.
[b] From D. R. Stull, E. F. Westrum, and G. C. Sinke, *The Chemical Thermodynamics of Organic Compounds*, Wiley, New York, 1969. The sum of detectable olefins was normed to 100%. In fact, they only represent 91.0% of all possible olefins, the remaining 9.0% is 2-ethyl-1-pentene which, however, could not be separated from the parent alkane under the conditions of gas chromatography applied.
Source: After Ref. 45.

TABLE 3 Ratios of Aromatic to C_5-Cyclic Products from n-Hexane at Various Hydrogen Pressures

a. Pulse system, T = 633 K, 0.4 g of Pt black, 3-µL pulses into 60-ml min^{-1} carrier gas (after Ref. 25)

Ratio	Hydrogen content in the carrier gas (%)				
	0	5	20	50	100
Bz/MCP[a]	∼200	38	14.5	3.5	0.64

b. Static-circulation system, 0.1 g of Pt black, T = 603 K. (H. Zimmer, unpublished data)

Ratio	p_{H_2} (torr)				
	50	120	250	320	520
Bz/MCP[a]	7.6	4.8	0.92	0.31	0.55

c. Static system, Pt single-crystal faces (after Ref. 55)

Ratio	Crystal plane	T(K)	p_{H_2} (torr)	
			220	620
Bz/MCP[a]	(111)	573	0.6	0.27
		638	2.6	--
		693	--	2.8
	(100)	573	0.14	--

[a]Bz, benzene; MCP, methylcyclopentane.

TABLE 4 Ratios of Aromatic to C_5-Cyclic Products over Pt Catalysts at Various Temperatures

a. Pulse system, 0.32 g of Pt black, carrier gas: hydrogen, 60 ml min^{-1}, 3-µL n-hexane pulses (after Ref. 56)

Ratio	T(K)			
	633	663	678	693
Bz/MCP	<1	3.1	14.8	36

b. Pulse system, 30 mg of 0.6% Pt/Al$_2$O$_3$, carrier gas: H$_2$, 14 ml min^{-1}, 1-µL pulses of n-hexane and n-heptane, respectively (after Ref. 57)

Ratio	T(K)				
	723	743	763	793	813
Bz/MCP	0.41	0.49	0.79	1.07	1.31
T/C_5-cyclic[a]	0.88	1.08	1.52	2.73	3.63

[a]Toluene/Σ (dimethylcyclopentanes + ethylcyclopentane).

The hydrogen sensitivity of the two reactions confirms the suggestion that these two processes are essentially different. Figure 4a gives an example of the temperature effect. The main difference is that the rate of aromatization increases with temperature at every p_{H_2}, while inversion can be observed at low p_{H_2} values in the rates of C_5 cyclization. This means that moderate hydrogen pressures promote aromatization by, for example, removing precursors of carbonaceous deposits (see later), but hydrogen must be an *active participant* in the surface intermediate of C_5 cyclization. Below a certain coverage, this reaction does not take place. The possible intermediates are shown in Fig. 4b.

The formation of an aromatic ring requires the splitting of eight H atoms from an alkane—evidently favorable under low hydrogen pressures. This is the situation with Pt [14, 45, 57, 59] and Pd [60, 61] catalysts. Maximum yields are at relatively low pressures, compared to those for other skeletal reactions. The same optimum p_{H_2} values were found for aromatization of *n*-hexane, *n*-heptane, and methylhexanes [45]. Increasing the hydrogen content of the carrier gas accelerates the aromatization of unsaturated *n*-C_6 hydrocarbons (Table 5).

One of the possible explanations of hydrogen effects in aromatization is as follows. In the stepwise dehydrogenation hexane → hexene → hexadienes → hexatriene, several geometric isomers can be formed. Of these, only two (Scheme 3), the cisoid conformations of *cis*-1, 3, 5-hexatriene and *cis*-2, 4-

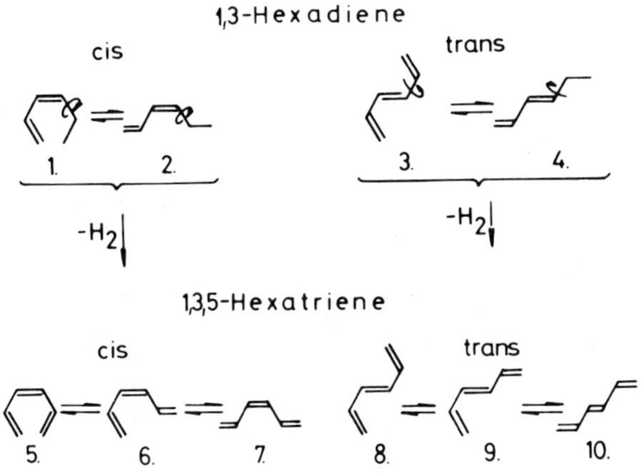

Scheme 3

hexadiene, are able to direct ring closure without geometric rearrangement [14]. The surface intermediates are liable to excessive dehydrogenation and form species where, for example, a C=C unit is bound to the surface by one π and two σ bonds [63]. If they have the right geometry, overdissociated benzene species (such as $C_6H_4M_2$ or $C_6H_2M_4$) can be formed and one of the rate-determining steps may be their hydrogenative desorption. This was the probable interpretation of benzene formation from *n*-hexane, hexenes, and hexadienes in temperature-programmed reaction (TPR) experiments [64] where all these starting hydrocarbons had a common rate-determining step, and was not in contradiction with transient kinetic experiments, either [47]. *cis*-1, 3, 5-Hexatriene not only gave more benzene, but the majority of it involved a more facile rate-determining step, with a lower T_{max}. This value was similar but not the same as those when cyclohexane, cyclohexene, and cyclohexadiene were the starting hydrocarbons [65]. If, however, the dehydrogenated products are produced in geometries not favorable for ring

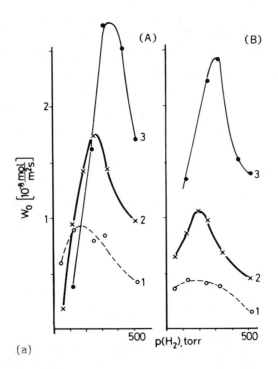
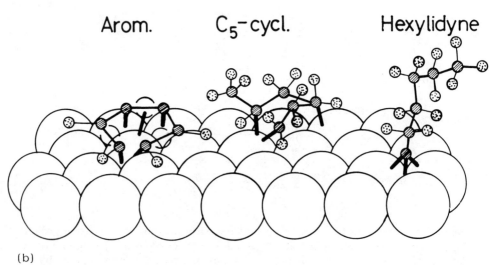

FIGURE 4 (a) Yields of ethylcyclopentane (A) and toluene (B) from n-heptane over Pt black (pretreatment: 473 K) as a function of hydrogen pressure. 1, 573 K; 2, 603 K; 3, 633 K. For conditions, see Fig. 2. (After Ref. 58; used by permission of Academic Press, Inc.) (b) Possible surface intermediates for aromatization and flat-lying "hydrogenative" C_5 cyclization. Triadsorbed "hexylidyne" is also shown; interstitial hydrogen (Fig. 1) can hinder the formation of the latter species. (After Ref. 46, copyright © 1986 by The American Chemical Society, used by permission.)

TABLE 5 Aromatizing of Various n-C_6 Hydrocarbons over Pt Black in the Presence of Carrier Gases of Different Composition[a]

Reactant	Benzene (mass %) in the product of the first hydrocarbon pulse if the carrier gas is:		
	He	5% H_2 + 95% He	20% H_2 + 80% He
n-Hexane	8.1	10.0	16.0
1-Hexene	13.1	–	11.0
1,5-Hexadiene	7.1	8.3	8.7
1,4-Hexadiene	5.7	10.3	10.5
1,3-Hexadiene	4.3	7.9	8.4
2,4-Hexadiene			
trans-trans	4.5	9.5	12.8
cis-trans	4.7	10.0	11.6
1,3,5-Hexatriene			
trans	13.4	19.4	17.7
cis	92.7	90.1	87.7

[a]Catalyst: 0.4 g of Pt; T = 633 K; carrier gas: 60 ml min^{-1}, 5-µL pulses.
Source: After Refs. 56 and 62.

closure, they need isomerization to form aromatics. Geometric isomerization of double bonds, in turn, involves half-hydrogenated surface intermediates [66], and therefore requires hydrogen. Without sufficient hydrogen, they stick to the surface as they form. That is why cis-1,3,5-hexatriene aromatizes on Pt black much more rapidly than does any other n-C_6 hydrocarbon (Table 5). Hydrogenation of cis-triene by hydrogen retained on the catalyst produces hydroaromatic products as opposed to open-chain hydrogenation products of trans-triene (Table 6). These are secondary products of cyclohexadiene hydrogenation rather than products of ring closure to cyclohexane, which step was disproved on Pt black by radiotracers [14,67,68]. Since benzene is an irreversible sink and the metal-catalyzed triene cyclization is very rapid, an exceptionally low steady-state triene concentration must be sufficient to ensure reasonable aromatization rates. This does not exclude "thermal" triene formation and subsequent cyclization involving no desorption of intermediates, which may be one possible explanation of "direct" C_6 ring closure, found experimentally by several researchers under various conditions [69-72].

That aromatization requires the presence of low amounts of hydrogen has also been shown by aromatization of n-hexane on a Pd membrane where hydrogen could be removed from the reaction space through the membrane (see also Chapter 26). The maximum rate was observed at 767 K, when about 85% of the hydrogen produced went through the membrane [73]. Dilution of Pt atoms to a small extent by alloying can result in a situation in which each atom of a carbon chain could not be attached to the catalyst by strong bonds. Thus the desorption (and subsequent isomerization) of di- and triene precursors is facilitated. Much more hexadiene appeared in the gas phase over alumina-supprted Pt-Sn and Pt-Pb than on pure Pt catalysts [74]. However, excessive alloying (10% Pt in Au) decreased aromatization activity to a higher extent than dehydrogenation activity [75, 76], indicating that ensembles of two or three metal atoms are necessary for benzene formation [41].

TABLE 6 Catalytic Reactions of cis- and trans-1,3,5-Hexatriene[a]

Number of pulse	Composition (mass %)						1,3,5-Hexatriene		Benzene
	< C_6	Hexane	Hexenes[b]	Hexa-dienes	Cyclo-hexane	Cyclo-hexene	1,3-Cyclo-hexadiene[c]	trans	cis	
				Starting hydrocarbon: trans-1,3,5-hexatriene						
1	2.2	0.1	1.7	1.3	—	—	18.8	62.6	—	13.4
2	—	—	—	0.6	—	—	30.6	65.9	—	3.0
5	—	—	—	0.4	—	—	29.5	69.2	—	0.8
				Starting hydrocarbon: cis-1,3,5-hexatriene						
1	6.2	0.7	0.4	—	—	—	—	—	—	92.7
2	2.5	—	1.2	2.3	0.4	2.5	12.1	1.0	—	78.3
5	0.2	—	0.1	3.5	1.8	2.6	55.8	3.7	0.4	32.0
8	—	—	—	1.9	1.1	0.9	71.8	7.7	1.3	15.4

[a]Catalyst: 0.4 g of Pt; T = 633 K, carrier gas: 60 ml min^{-1} helium, 5-µL hydrocarbon pulses each.
[b]With 1,5-hexadiene.
[c]With cis-trans- and cis-cis-2,4-hexadiene.
Source: After Ref. 62.

17. Hydrogen Effects in Skeletal Reactions

Single-crystal studies confirm that hexagonal symmetry of flat (111), stepped (332), and kinked (10,8,7) planes promotes aromatization [55].

The right-hand side of the bell-shaped curve at higher hydrogen pressures indicates conditions where increasing hydrogen pressures no longer favor multiply dehydrogenated intermediates of aromatization. On supported catalysts, however, this section can be shifted toward higher p_{H_2} values or even be absent. This may be connected with the better "hydrogen economy" [57] of these catalysts. Excess hydrogen can be removed here by spillover instead of competing with hydrocarbons for active sites (see Chapters 12 and 25). Alloying Pt with Pd, however, again caused a decrease in benzene yield at higher hydrogen pressures, even on SiO_2-supported catalysts (Fig. 5).

Platinum is by far the best catalyst for C_5 cyclization of alkanes. The next best metal, Pd, showed two to five times less activity in its various forms (Pd black [28], Pd/SiO_2 [77], Pd/Al_2O_3 [14]). Several less [14,19,78,79] or more [15,22,80] dissociated species were suggested as being surface intermediates. We proposed earlier [50] that a flat-lying half-hydrogenated intermediate (probably bound to the catalyst via the tertiary C atom) should be responsible for C_5 cyclization, as opposed to other, more dissociated intermediates (see, e.g., Ref. 15). For this intermediate, the distances between catalyst atoms must be important [18,28]. That is the reason C_5 cyclization is much less crystal-face sensitive than is aromatization [55]. The presence of hydrogen in interstitial positions (Fig. 1) may be important to block these sites that are able to form multiple C-M bonds for hydrogenolysis [32]. Another, atom-top hydrogen may participate in the surface complex of C_5 cyclization [58]. Kinetic analysis of cyclization of 2,2,4-trimethylpentane on Pt/C [81] and Pt/Al_2O_3 [82] also points to the active participation of hydrogen in the rate-determining step of ring closure. Even the shift of maximum rate with increasing temperature toward higher hydrogen pressures was similar on Pt/C to that observed on Pt black [81].

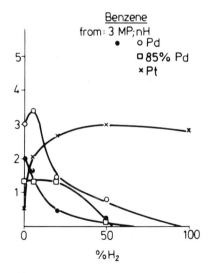

FIGURE 5 Yields of benzene from n-hexane and 3-methylpentane as a function of the hydrogen content of the carrier gas. Catalyst: 1% metal/SiO_2; 0.2 g of Pt/SiO_2, 0.5 g of Pd/SiO_2 and Pt-Pd/SiO_2; T = 633 K; pulse system, 3-μL pulses (see also Fig. 3). (After Ref. 77; used by permission of Academic Press, Inc.)

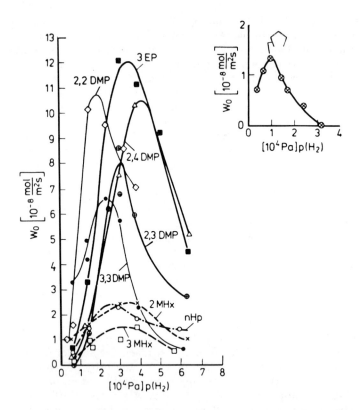

FIGURE 6 Rates of formation of C_5-cyclic products from various heptane isomers and n-pentane (inset), as a function of the hydrogen pressure. For conditions, see Fig. 2. (Data on the main figure from Ref. 46; copyright © by The American Chemical Society, used by permission.)

As opposed to aromatization, the maximum yields for C_5 cyclization are not at the same p_{H_2} values [45, 46, 83]. Approximately, the larger the patch of surface covered by the flatly adsorbed hydrocarbon, the higher the p_{H_2} for maximum cyclization yield (Fig. 6). With n-pentane, the interaction of the chain ends with one metal atom, as proposed by Finlayson et al. [79], may be likely. Flat-lying dual-site adsorption can be suggested for 2,3- and 2,4-dimethyl- and also 3-ethylpentane, while 1,4 interactions may be possible for *gem*-dimethylpentanes [46]. On Pt/C, n-hexane and methylpentane showed the same rates of cyclization [84], as opposed to Pt black [25, 56].

On Pt/SiO_2, the maximum yield of ethylcyclopentane was at lower hydrogen pressure than that of 1,2-dimethylcyclopentane [59]. Tentatively a more dissociated surface intermediate could be proposed for the former type of cyclization.

An interesting directing effect of hydrogen has been discussed by Csicsery [85] based on the experiments of Erivanskaya et al. [86, 87]. 1-Propylnaphthalene can give two types of tricyclic products by C_5 cyclization: peri dehydrocyclization gives 1-methylacenaphthene and 1-methylacenaphthylene, while β dehydrocyclization gives 4,5-benzindan and 4,5-benzindene (Scheme 4). In the presence of hydrogen on 0.5% Pt/Al_2O_3, peri cyclization was about four times faster than β cyclization. The latter reaction proceeds with the same rate in helium; however, the rate of peri cyclization decreased by a

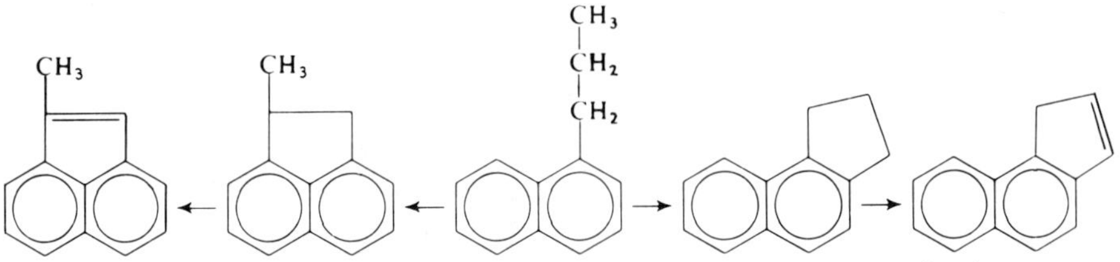

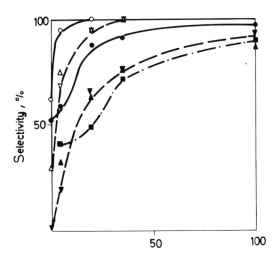

Scheme 4

factor of 7. This fact reveals that C_5 cyclization can proceed in the side chain of alkylaromatic compounds. In fact, the first dehydrogenative mechanism was developed by Shephard and Rooney on the basis of cyclization of n-propylbenzene [88]. Their "alkyl-alkene" insertion mechanism was then transferred to explain the cyclization of alkanes [89].

Some of the C_5-cyclic products appear as unsaturated. Their yields were so high with n-pentane cyclization that dehydrogenation to surface pentadienyl and subsequent cyclization to cyclopentene or cyclopentadiene was suggested [7]. In fact, with decreasing hydrogen pressure, higher and higher fractions of cyclization products of 3-methylpentane and 3-methylpentenes appear as unsaturated. Figure 7 also illustrates the temperature effect. The methylcyclopentane selectivity was always lower with unsaturated reactants. The fact that the original olefin geometry is not preserved during cyclization [50,84] can be explained either by the half-hydrogenated intermediates discussed previously or by the cyclohexadienyl complex. Methylcyclopentene and methylcyclopentadiene were observed in hydrogen among the products of reaction of various n-C_6 unsaturated hydrocarbons over Pt black [90]

FIGURE 7 Selectivity of methylcyclopentane formation within the sum of C_5-cyclic products, as a function of the hydrogen content of the carrier gas. For conditions, see Fig. 3. 100% selectivity means that no methylcyclopentenes are present. Symbols mean different reactants: ○, 3-methylpentane (3MP) at 573 K; ▽, trans-3-methyl-2-pentene (3M2P=), 573 K; △, cis-3M2P=, 573 K; ●, 3MP, 633 K; □, 3M1P=, 633 K; ▼, trans-3M2P=, 633 K; ▲, cis-3M2P=, 633 K. (Calculated from Ref. 50.)

TABLE 7 Ratios of Benzene/C_5 Cyclics from Various Open-Chain Reactants[a]

Bz/C_5-cyclic[b] from:	T(K)				Ref.
	573	673	723	773	
n-Hexane	–	0.17	0.52	0.97	91
n-1-Hexene	–	0.14	0.37	0.83	91
1,5-Hexadiene	–	0.12	0.32	0.89	91
1,3,5-Hexatriene[c]	1.33	8.9	8.4	14.8	92
1,3,5-Hexatriene[c,d]	30.7	22.7	40.4	Very high	92

[a] Pulse system, 0.35-μL pulses into 50 ml min^{-1} hydrogen. Catalyst: 0.6% Pt/Al_2O_3, 30 mg.
[b] Bz, benzene; C_5 cyclics, methylcyclopentane + methylcyclopentenes + methylcyclopentadiene.
[c] Isomeric composition not given in Ref. 92; probably a mixture of 70% trans- plus 30% cis-triene as formed in the given method of preparation.
[d] In 50 ml min^{-1} helium.

and Pt/Al_2O_3 [91,92]. Deactivation of Pt black increased not only their relative amount compared to benzene but also their absolute yield [90]. The benzene/C_5 cyclic ratio was lower here than with n-hexane (Table 7). Of all the reactants, 1,5-hexadiene gave the highest yields of 1-methylcyclopentene [90]. Scheme 5 suggests the possibility of dehydrogenative C_5 cyclization on single metal atoms of deactivated metal catalysts [93].

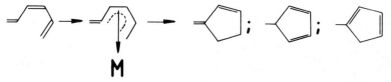

Scheme 5

C. Skeletal Rearrangement of Alkanes

The literature suggests two different alkane isomerization pathways on metals: one involves a C_5 cyclic, the other a C_3-cyclic intermediate on the surface [10,14,15,18,22,94]. The latter route is often called "bond shift." The C_5-cyclic mechanism involves the intermediate(s) already discussed in Section II.B. Two essentially different pathways have been postulated for bond shift: One has a less dehydrogenated intermediate attached to only one metal atom [95] (Scheme 6). A much deeper dehydrogenation is proposed for the other mechanism [96] (Scheme 7). Foger and Anderson [97] found a higher energy of activation for the former pathway. This route was related to hydrogenolysis (also shown in Scheme 6). Scheme 7 proposes the total separation of a CH_2 unit from the molecule. Thus it seems that the same surface species may lead to both isomerization and hydrogenolysis.

The regulating effect of hydrogen on skeletal isomerization is not generally recognized, although the literature provides numerous examples. The most obvious one deduced from Schemes 6 and 7 (i.e., the hydrogen regulation of isomerization to hydrogenolysis selectivity) has been recognized by Guo

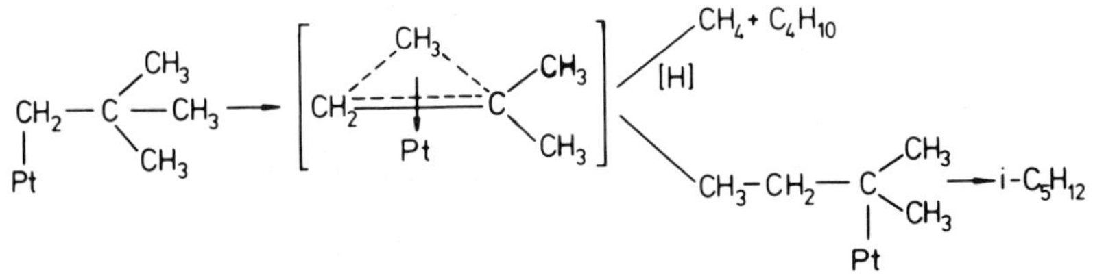

Scheme 6

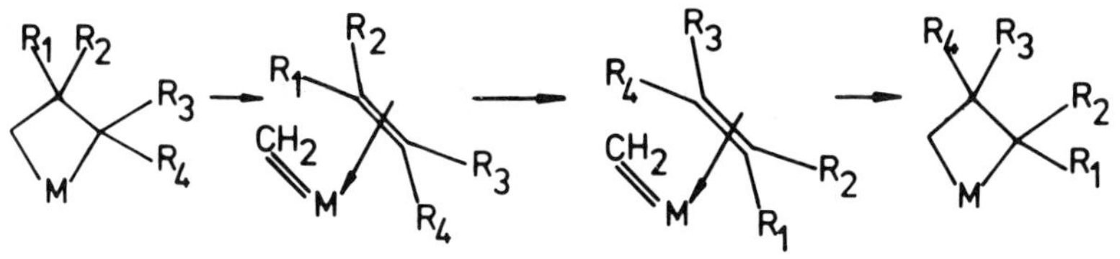

Scheme 7

et al. for the bond-shift process [98]; this is discussed in Section II.D. Hydrogen may also regulate (a) the amounts of isomers and C_5 cyclics in a reaction when only the C_5-cyclic pathway is important; (b) the ratio of the C_5- and C_3-cyclic pathways; and (c) the ratio of unsaturated and saturated skeletal rearrangement products. These effects are discussed below.

a. Hydrogen availability may be one of the factors determining whether the product desorbed from the surface intermediate will be cyclic or open-chain alkane (Scheme 8). More hydrogen usually favors isomerization. The

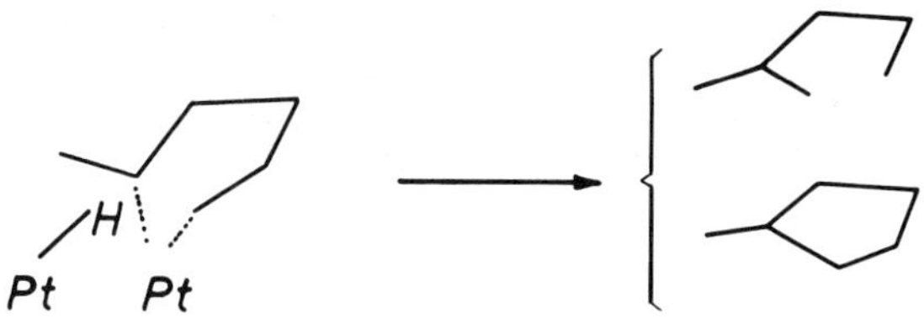

Scheme 8

isomer/cyclic ratio increases with increasing hydrogen pressure (Table 8); similar data are presented in Section III. The suggestion of Guo et al. [98], who attributed cyclization to fewer dehydrogenated intermediates, and C_5-cyclic isomerization to more dehydrogenated intermediates (their ratio being controlled by surface hydrogen), is in disagreement with these data.

b. Hydrogen pressure sensitivity of yields of various isomers permits us to distinguish between C_3- and C_5-cyclic isomerization [46]. Figure 8a shows isomers obtained over Pd black which could be formed via the C_5-cyclic route. They show a hydrogen dependence very similar to that observed with the corresponding cyclization reaction (see Fig. 6), and all yields show strong maxima. On the other hand, the yields of isomers that could only be produced by the C_3-cyclic route show a slight monotonic increase with increasing hydrogen pressure (Fig. 8b). This is in agreement with the results (Table 9; see later) that the importance of the C_3-cyclic route increases somewhat at higher pH_2 values, but this depends very much on the reactant structure.

The hydrogen dependencies shown in Fig. 8 are in agreement with Pt-Au alloy studies revealing that the C_3-cyclic route requires fewer (probably,

TABLE 8 Ratio Skeletal Isomers/C_5-Cyclic Products from 2,3-Dimethylpentane over 10% Pt/Al_2O_3 Catalyst[a]

p_{H_2} (torr)	Conversion (%)	Isomers / Cyclopentanes
200	8.4	0.42
400	11.6	0.98
760	8.7	1.64
1150	7.4	1.49

[a]Pulse system, 5 torr hydrocarbon admixed to H_2 and passed as a square pulse over the catalyst, T = 533 K.
Source: After Ref. 44.

one) metal atoms in its active site [75,76]. Alloying Pt with Pd decreased considerably the overall isomerizing activity on SiO_2-supported catalysts; at the same time, maximum yields were observed at lower p_{H_2} values [77].

Gault [15], Van Schaik et al. [75], Amir-Ebrahini and Gault [96], and Parayre et al. [99] studied the isomerization pathways of alkanes labeled in various positions by ^{13}C. Hydrogen pressure was not the main parameter in their work; nevertheless, p_{H_2} sensitivity of various routes can also be obtained from their work. For example, the fraction of bond-shift isomerization changed slightly as a function of the hydrogen pressure with 2-methylhexane [99] and 2,3-dimethylpentane [44] feed. Also, the pronounced effect of reactant structure is seen (Table 9). The C_5-cyclic route may be more important with a molecule more liable to C_5 cyclization (2,3-dimethylpentane; see Fig. 2).

c. Saturated skeletal isomers are not formed from isomer hexanes in the absence of hydrogen, but all these form aromatics over Pt [100] and Pd

TABLE 9 Percentage of Isomers Formed via C_3-Cyclic ("Bond Shift") and C_5-Cyclic Pathways as a Function of the Hydrogen Pressure

| Reactant | p_{H_2} (torr)[a] | Percent of isomers via: | | Ref. |
		C_3-cyclic pathway	C_5-cyclic pathway	
2-Methylhexane[b]	200	41.5-43.5	56.5-58.5	99
	760	57.5-59	41-42.5	
	1150	62-63	37-38	
2,3-Dimethylpentane	200	6.5	93.5	44
	760	11.4	98.6	
	1150	10.8	90.2	

[a]p_{HC} = 5 torr.
[b]Since the calculation was done on the basis of fragments in the mass spectrometric analysis, two different assumptions were applied to estimate the ^{13}C concentration in the C_5 fragment ion. The two values correspond to these two limits.

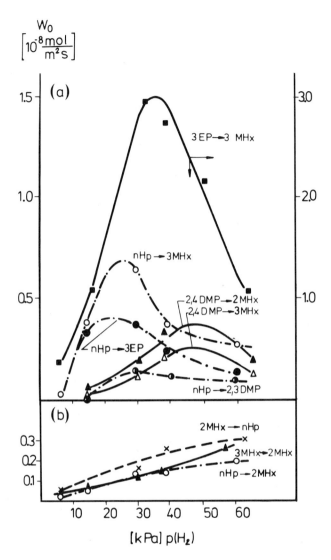

FIGURE 8 Rates of formation of various skeletal isomers from heptanes as a function of the hydrogen pressure. For conditions, see Fig. 2. (After Ref. 46; copyright © 1986 by The American Chemical Society, used by permission.)

blacks [60]. Figure 9 shows that aromatization of dimethylbutanes on Pd drops quite abruptly with increasing p_{H_2}; at the same time, isomerization yields increased slowly. The overall yield of skeletal rearrangement remained nearly constant from medium hydrogen pressures. Pt black behaved similarly [101]. We propose that the two types of end products correspond to different intermediates: The more dehydrogenated species (Scheme 7) ultimately gives aromatics (with no other desorbed intermediate) and the less dehydrogenated intermediate (Scheme 6) leads to saturated isomers. No doubt that in the case of dimethylbutanes no C_5-cyclic intermediates are possible.

Sárkány [102,103] proposed isomerization via carbene ("CH_x") addition and its subsequent abstraction under hydrogen-deficient conditions over Pd, Ni, Co, and Ni-Ag catalysts. These reactions often give aromatic end

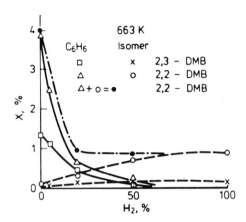

FIGURE 9 Yields of benzene and skeletal isomers from dimethylbutanes as a function of the hydrogen percentage in the carrier gas. Pulse system, catalyst: 0.31 g of Pd black. (After Ref. 60; used by permission of Elsevier Scientific Publ. Co.)

products. Low hydrogen pressures are favorable for them and they are related to homologation reactions (see Section II.E).

D. Hydrogenolysis

Foger and Anderson [97] report a maxiumum of the rate of neopentane transformation (hydrogenolysis selectivity up to 50%) over Pt-aerosil. In the same paper they postulated that hydrogenolysis and isomerization were related processes with different H-exponents. It is generally accepted that hydrogenolysis involves multibonded surface species which are rather deeply dehydrogenated [10,11,47,49,98]. One has to expect, therefore, that the isomerization/hydrogenolysis ratio (or, more correctly, the ratio of non-degradative saturated products to hydrogenolysis) should increase at higher hydrogen pressure. This is, in fact, the situation with several metal catalysts and hydrocarbon reactants: n-butane and isobutane on Pt [104] and Pd [105]; neopentane on Pd [106] (Table 10); n-butane on Ru, Rh, and Ir [107]; hexane isomers on Pt [49] and Pd [58] (Table 11). Very marked hydrocarbon pressure effects can also be seen in Table 10, indicating strong competition between hydrocarbon and hydrogen for the hydrogenolytic sites, as predicted by the transient kinetic model [47].

Guo et al. [98] suggests that less dehydrogenated 1,3-adsorbed species give isomers; multiply dehydrogenated ones lead to fragmentation. This may be valid for the hydrogen pressure range included in Tables 10 and 11.

The *depth* of hydrogenolysis is defined as the number of fragments formed when a molecule is broken up. Various characteristic numbers have been developed for its characterization. Montarnal and Martino [31] use the value $P_f = C_1/C_{n-1}$ for this purpose. This value is different for different metals. They mention after Ref. 108 that P_f on Ni decreases from 25 to 1 when p_{H_2} increases from 5 bar to 30 bar. Similar changes can be induced by hydrogen with other metals, too. Figure 10 also contains the ranges given by Montarnal and Martino. The important feature here is that the *same* P_f values can be produced by different metals—only the appropriate temperature range and hydrogen pressure are necessary.

Another ratio, $C_1/\Sigma C_2-C_5$, was used to characterize n-hexane conversion on Pt/Al$_2$O$_3$ at 673 K [109]. This value increased gradually from 0.3 to 0.9 as the hydrogen content of the carrier gas in the pulse reactor decreased

TABLE 10 Isomerization/Hydrogenolysis Ratio in the Reaction of Neopentane over Pd Films[a]

p_{H_2} (torr)	p_{HC} (torr)	r[b] (% min^{-1})	$\dfrac{\text{Isomerization}}{\text{Hydrogenolysis}}$
6.6	1.6	0.047	0.55
15	1.5	0.0053	4.99
20	1.8	0.0046	2.00
25	2.5	0.0061	0.49
28.5	1.8	0.0025	2.50

[a]Circulation apparatus, T = 563 K. Catalyst: Pd film on mica, annealed in H_2 (4 torr, 773 K, 2 h); (111) oriented.
[b]Overall conversion, percent neopentane consumed per minute.
Source: After Ref. 106.

from 100% to 0%. Switching over from He to H_2 *without regeneration* suddenly increased this value to 5.7, but it dropped gradually to 4.4 and 3.6 in the next two pulses. Perhaps surface carbon may also participate in fragment formation [39]. Partial carbonization of the surface also enhanced hydrogenolysis on Pt/SiO$_2$ [59] and Pt black [54].

Leclercq et al. [110] defined the factor ω to characterize the *pattern* of hydrogenolysis as follows:

$$\omega = \frac{\text{actual rate of rupture}}{\text{random rate of rupture}} \tag{1}$$

The reactivity of individual C-C bonds depends on the nature of the hydrogenolyzing metal [111,112] and the structure of the reactant [110,112], but

TABLE 11 Isomerization/ Hydrogenolysis Ratio as a Function of Hydrogen Pressure in Alkane Reactions over Metal Blacks[a]

H_2 % in carrier gas	$\dfrac{\text{Isomerization}}{\text{Hydrogenolysis}}$ over:			
	Rh[b]	Pd[c]		Pt[d]
	n-Hexane	n-Hexane	3-Methylpentane	3-Methylpentane
0	0.00093	0.14	0.083	0.045
5	0.00088	0.085	0.14	0.63
20	—	0.34	0.34	1.41
35	0.0029	0.40	0.21	1.86
100	0.016	0.46	1.45	2.17

[a]Pulse system, 3-μL pulses into 60 cm^3 min^{-1}; carrier gas: mixture of He and H_2.
[b]0.4 g of Rh, T = 453 K. From Z. Paál, unpublished results.
[c]0.31 g of Pd, T = 648 K. After Ref. 60.
[d]0.4 g of Pt, T = 633 K. After Ref. 50.

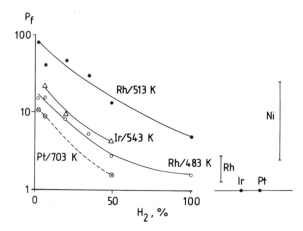

FIGURE 10 Depth of hydrogenolysis of n-hexane, expressed in terms of P_f [31], as a function of the hydrogen content of the carrier gas. Pulse system, 0.06 g of Ir black, 0.11 g of Rh black, and 1 g of Pt black (From Z. Paál, unpublished data. On the right-hand side, P_f values from Ref. 31 are given.)

it also shows a marked dependence on hydrogen pressure [113]. This is illustrated in Fig. 11. Although each reactant has its own "fingerprint-like" pattern, there is a general tendency—that splitting in the middle of the molecule is favored by increasing p_{H_2}. Whenever there is a possibility of splitting a C_5 unit from the molecule, this usually predominates at low p_{H_2} and its ω value drops nearly to random hydrogenolysis (curves 1, 2, and 3 in Fig. 11 for n-hexane, n-heptane, and n-octane, respectively) at hydrogen pressures where C_5 cyclization (the other possible reaction of a C_5-cyclic surface species) has its maximum yield (arrows in Fig. 11). The C_5 fragment may undergo secondary cyclization, but the maximum of this reaction is at lower p_{H_2} than that of C_5 cyclization of the parent molecule [46]. The degradative reaction of an adsorbed "C_5 unit" is shown in Scheme 9.

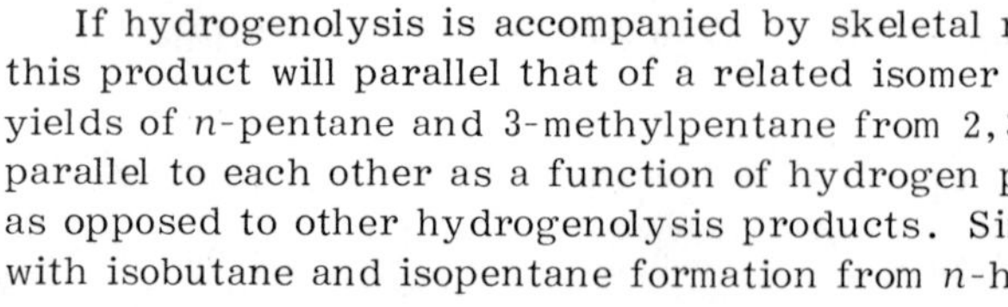

Scheme 9

If hydrogenolysis is accompanied by skeletal rearrangement, the yield of this product will parallel that of a related isomer [114]. For example, the yields of n-pentane and 3-methylpentane from 2,3-dimethylbutane change parallel to each other as a function of hydrogen pressure and temperature, as opposed to other hydrogenolysis products. Similar trends were observed with isobutane and isopentane formation from n-hexane [47] (Scheme 6).

E. Homologation

Several metals under low hydrogen pressure may produce hydrocarbons with more C atoms than in the parent molecule [7,115-120]. This "homologation ability" parallels the activity of the metal shown in ethane hydrogenolysis

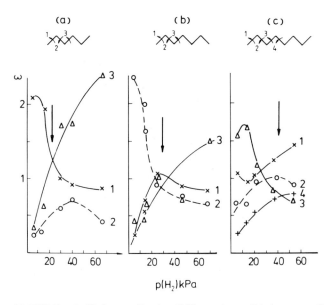

FIGURE 11 Values of ω in different positions as a function of hydrogen pressure with (a) n-hexane, (b) n-heptane, and (c) n-octane reactants. For conditions, see Fig. 2. (After Ref. 113; used by permission of Akadémiai Kiadó.)

[117]. The products are predominantly aromatics. The yield of benzene was maximum for a H_2/n-pentane ratio between 0.5:1 and 2:1 at 570 K [115]. This reaction was interpreted [116,117] in terms of carbene addition to adsorbed olefins: Both are rather dehydrogenated species. Scheme 10 also shows the possibility of simultaneous degradation of buildup products.

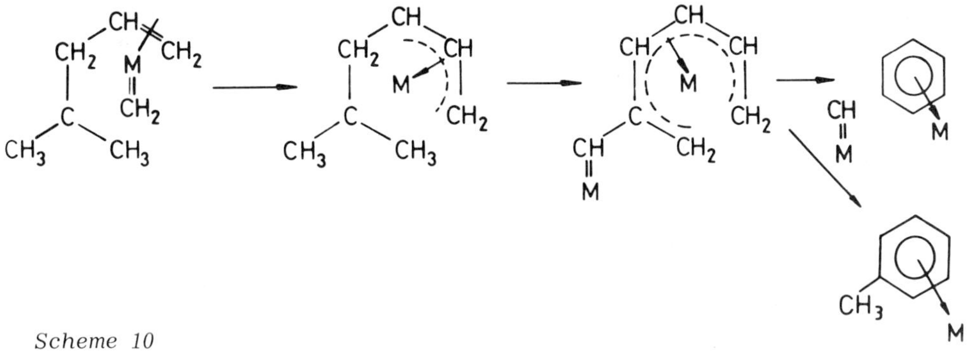

Scheme 10

The optimum hydrogen pressure is even lower than that of hydrogenolysis (Fig. 12). Under these conditions, fragments react with each other on the surface rather than desorbing. Increasing hydrogen pressure favors degradation. Chain lengthening occurs mostly in the end position; however, on Pt and Pd, addition to secondary C atoms was also observed [118-122]. Radiotracer experiments verified that C_1 species active in homologation are formed even on Pt by *deep* fragmentation of the parent molecule [119], which occurs at low hydrogen pressures and higher temperatures (Section II.D).

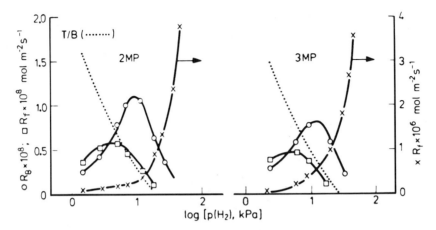

FIGURE 12 Rate of formation of benzene (○), toluene (□), and fragments (×) as a function of the hydrogen pressure. Benzene/toluene ratio is given by a dotted line. Catalyst: 0.04 g of 8.7% $NiSiO_2$; dispersion: 16% prepared prepared by impregnation; T = 603 K, p_{HC} = 4 kPa. (After Ref. 102; used by permission of Academic Press, Inc.)

F. Cyclic Hydrocarbons

Reactions in the Ring

Alicyclic hydrocarbons may undergo essentially the same reactions as alkanes. However, both the cyclic character and the number of C atoms influence considerably both reactivities and reaction directions. The great variety of transformations of various small (C_3-C_5), medium (C_6-C_8), and large (>C_8) rings has been discussed extensively [14, 18].

Cyclopentanes will be used to illustrate the reactions of small rings. Their conversion usually increases very markedly with increasing hydrogen pressure, and *ring opening* becomes predominant [100, 121]. Three arguments will be given that this represents a reaction of C-C bond rupture essentially different from hydrogenolysis over Pt, Pd, Ir, and Rh; certain features also appear on Ru and Ni. First, there is a temperature range for the first four metals where alkanes without degradation are produced in ample amounts from methylcyclopentane but fragmentation (and also that of alkanes) is still a minor reaction (Table 12). No such differences can be seen for other group VIII metals and rhenium [28]. Second, the energy of activation of ring opening is considerably lower for Pd, Pt, Rh, and Ni than that of hydrogenolysis [18, 51, 121]. The third and most relevant feature is the different hydrogen sensitivity of hydrogenolysis and C_5 ring opening. Figure 13 illustrates this for Ir, but data for Rh [26] and Pt [122] have also been published.

Ring opening may occur in various positions: It can be "selective" (hindered in position *a* next to the alkyl substituent) or "nonselective" (when positions *a*, *b*, and *c* are all equivalent). The latter reaction was attributed to disperse catalysts, whereas the former occurred on large crystallites [123-125]. With four Pt catalysts (Pt black, Pt/C, Pt/SiO_2, and Pt/Al_2O_3), the actual *b/a* selectivity values depended on the dispersity; more disperse metals gave less selective ring opening. Hydrogen effects were superimposed on this particle-size effect. More hydrogen caused more selective opening of methylcyclopentane (expressed in terms of the *b/a* ratio, i.e., the ratio 2-methylpentane to *n*-hexane [84, 126]).

TABLE 12 Ratio of C_6 Products to Fragments on Different Metal Blacks

Metal[a]	$T(K)$[b]	$C_6/<C_6$	2MP/nH[c]
Co	573	0.042	1.9
Ni	543	0.316	2.8
Ru	543	0.076	1.7
Rh	453	12.0	6.5
Pd	648	5.82	2.9
Os	513	0.043	0.5
Ir	468	18.1	32.4
Pt	603	30.0	9.8

[a] Pulse system, 3-μL pulses into 60 ml min^{-1} H_2 carrier gas. After Ref. 28.
[b] The temperatures were selected in a way that the number of C_6 molecules underwent *fragmentation* should be similar (5 to 6 × 10^{17} molecules per square meter of metal). With Pd and Pt this value was lower (∿3 × 10^{17}), with Os and Ir somewhat higher (∿10^{18}).
[c] 2-Methylpentane/n-hexane ratio (b/a) from methylcyclopentane feed.

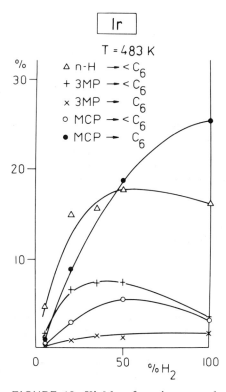

FIGURE 13 Yields of various products as a function of the hydrogen content of the carrier gas. Catalyst: 0.06 g of Ir black; pulse system, 3-μL hydrocarbon pulses (From Z. Paál, unpublished data.)

Some b/a ratios for group VIII metals in H_2 are shown in Table 12. An ethyl substituent decreased selectivity less than did a methyl substituent [18]. Lower temperatures (= higher hydrogen coverages) gave higher selectivities [121,123]. All these facts can be interpreted in terms of a flat-lying, less dehydrogenated intermediate for selective ring opening, similar to that proposed for ring closure. Another, more dissociated intermediate, perhaps with a fluctional olefinic bond [22], may be responsible for nonselective opening.

The opening of smaller rings also showed hydrogen effects. The selectivity of isobutane to n-butane from methylcyclopropane is discussed in Section III. The cyclobutane ring behaved very similarly to the cyclopentane ring: b/a selectivities increased with increasing hydrogen pressure [127,128]; temperature effects on the selectivity [123] as well as the difference between methyl and ethyl substituents were also observed.

Alkylcyclopentanes may form aromatics by *ring enlargement*. This reaction has its hydrogen optimum at rather low p_{H_2} [86]. It probably represents a peculiar sort of "C_3-cyclic" isomerization involving a rather dehydrogenated intermediate, like that published by Kane and Clarke [129] for the reaction 1,1-dimethylcyclopentane → toluene. The presence of a quaternary C atom facilitates this process [130]. Even in mechanisms describing this process as ring opening and reclosure [131], no intermediate desorption is assumed. With higher hydrocarbons, several parallel directions of ring enlargement are possible, as reflected by the aromatic composition. For example, increasing the hydrogen pressure from 1 bar to 27 bar, the fraction of $(m + p)$-xylenes, even more that of o-xylene, increased on (acidic) Pt/Al_2O_3, whereas ethylbenzene prevailed on alkaline catalyst [132]. The metal-catalyzed ring enlargement has to be distinguished from $C_5 \rightarrow C_6$ ring transformation, assumed to be the main route of aromatization on bifunctional catalysts (see Chapter 25). This reaction takes place on acidic centers but not on metallic centers and may involve more dehydrogenated species (e.g., methylcyclopentadiene) [133].

The $C_4 \rightarrow C_5$ ring enlargement showed similar hydrogen dependence to that observed for $C_5 \rightarrow C_6$ ring expansion [14]. This indicates that the method of primary adsorption may be more important here than the larger driving force toward the formation of a more stable aromatic ring. Other reactions of small cycles, such as dehydrogenation and fragmentation, do not differ from those of alkanes, and therefore are not discussed separately.

Several hydrogen effects on the various reaction routes of larger cyclic hydrocarbons were summarized in our earlier review [14]. Selectivities to aromatics, to form cyclo-olefins or bicyclic products, to ring opening, and so on, all depend largely on the amount of hydrogen available. For example, cyclo-octane gave at 633 K 37% bicyclic, 40% octane, 1.5% cyclo-octene, and 6.5% aromatics in hydrogen; the corresponding selectivities in helium were: 8, 2.5, 40, and 15% [14].

Reactions in the Side Alkyl Groups

The *loss of side groups* from cycloalkanes is similar to common hydrogenolysis reactions and usually occurs with much lower rates than ring opening. Typical C_6-alkane/cyclopentane ratios are between 10 and 100, increasing markedly with p_{H_2}.

The side groups may undergo cis-trans isomerization at relatively low temperatures where only this reaction occurs [19]. This *configurational isomerization* of dialkylcycloalkanes did not start in the absence of hydrogen on Pt/C [134], Pd/C, Ir/C, Rh/C, Ru/C, and Os/C [135], in a flow apparatus, with a gas diluent added in a large excess. If this gas contained between 0 and 3% H_2, a rapid increase in the rate was reported; above 20 to 30% H_2 in the gas mixture, the rate reached saturation. The reaction rate had a broad maximum at a hydrogen excess of about 250 to 500 times (Pt/C, T=380 to 420 K).

The possible mechanisms have been reviewed (19,136). It is not excluded that a changeover of mechanisms takes place as hydrogen coverage grows and that a less dehydrogenated intermediate will prevail under hydrogen-rich conditions. The reaction is somewhat related to epimerization of cyclic alcohols (see Chapter 19).

An interesting hydrogen effects was reported in *xylene isomerization* over Pt/Al_2O_3. Dermietzel et al. [137] reacted p-xylene with 10% of ^{14}C-labeled m-xylene and 1% o-xylene added. The specific radioactivities of the three isomers were monitored as functions of the space velocity in hydrogen and nitrogen flow. In hydrogen the following reaction was found:

$$o\text{-xylene} \rightleftharpoons m\text{-xylene} \rightleftharpoons p\text{-xylene}$$

In nitrogen, C_9 aromatics and toluene appeared and the course of specific activities pointed to the validity of a triangular reaction scheme. Here m-xylene was not the only intermediate and the following scheme was suggested:

$$o\text{-xylene} \rightleftharpoons m\text{-xylene} \rightleftharpoons p\text{-xylene}$$
$$\diagdown \quad 1,2,4\text{-trimethylbenzene} \quad \diagup$$

G. Catalyst Deactivation

One of the products of catalytic hydrocarbon reactions is a deeply dehydrogenated species that remains on the catalyst surface as "coke precursor." Their accumulation is the cause of "coke deactivation" of the catalysts. Since this is just a peculiar sort of dehydrogenation reaction, it is proper that we close this section with this reaction. As the phenomenon is of enormous importance, it is investigated extensively; here only a few examples can be mentioned to illustrate hydrogen effects.

It is likely that several types of coke precursors are formed: C_1 species (CH_x surface radicals) [138,139], open-chain surface polyenes [14,62], or unsaturated C_5 cyclics [140]. Their formation is slowed down enormously in the presence of hydrogen (Table 13). The degree of deactivation (as expressed in terms of remaining activity) also depends on the reactant: *cis*-hexatriene forms much more benzene than does any other n-C_6 hydrocarbon and deactivates the catalyst to the lowest degree. This served as the basis of developing of the polyene model of coking [14,62]. It was assumed that dienes or trienes that are in elongated configuration on the surface cannot desorb or isomerize to *cis* isomers (to give benzene) when the hydrogen supply is insufficient. Then they undergo polymerization/condensation to give carbonaceous precursors. The resulting polymers produce an infrared band at 1580 cm^{-1} characteristic of C-C vibrations in hydrogen-deficient polyaromatic ring structures such as graphites of low crystallinity [139]. Such surface structures were characteristic for adsorption at 623 K, where the amount of retained hydrogen must have been small. The rate of carbon accumulation on the surface was much higher with any open-chain C_6 reactant than with benzene.

Earlier we suggested [14] the direct formation of six-ring systems during coking. However, C_5-ring closure of dienes and trienes (Scheme 5) is not excluded, either; then the further development of surface coke will be closely related to the route suggested by Parera et al. [140].

TABLE 13 Deactivation of Pt Black by Hydrocarbons of Various Unsaturation[a]

Reactant	Remaining activity of the catalyst (%)[b] after passing 30 μL of hydrocarbon if the carrier gas is:		
	He	5% H_2 + 95% He	20% H_2 + 80% He
n-Hexane	9.4	26.5	56.5
1-Hexene	1.8	—	—
1,5-Hexadiene	—	36.4	83.0
1,4-Hexadiene	3.6	40.7	—
1,3-Hexadiene	1.4	45.0	69.0
2,4-Hexadiene			
trans-trans	2.2	33.0	55.5
cis-trans	2.1	31.0	62.0
1,3,5-Hexatriene			
trans	5.6	23.5	90.5
cis	25.7	99.2	97.5

[a]For conditions, see Table 5.
[b]Expressed as (benzene % from final pulse/benzene % from first pulse) × 100.
Source: After Ref. 62.

The rate of carbon accumulation in ambients containing various amounts of hydrogen has been studied using ^{14}C-labeled 3-methylpentane [54]. A material balance was calculated considering the radioactivity of introduced hydrocarbon pulse, on the one hand, and the sum of the effluent plus the radioactivity swept off by a single subsequent air pulse from the catalyst. (No radioactivity was observed in the outflow of the second air pulse.) The extent of possible surface blocking of a Pt-black sample is shown in Table 14. With polycondensed aromatic species lying over the (111) surface, the 1:1 C:Pt stoichiometry would be realistic. These values would correspond to 0.13 C_6 unit per Pt atom after one pulse into helium and 0.035 C_6/Pt in 5% H_2. These data compare well with those of Luck et al. [141], who reported an abrupt increase in surface carbonization up to 0.06 C_6 unit per Pt atom during the first moment of reaction, even in hydrogen excess (0.65-kPa 2-methylpentane in 99.35-kPa H_2). This may be similar to the process described by Lietz et al. [138] involving the incorporation of C_1 species into the surface, which attenuates hydrogenolysis activity and produces "steady-state carbonized" catalyst [142]. We believe that further carbonization may occur via polymerization/condensation of larger units. After a conversion of 300 molecules per surface Pt, the value of carbonization increased up to 0.25 C_6/Pt at 633 K and 0.15 C_6/Pt at 623 K, while at 573 K the initial value showed hardly any changes. This comparison also shows a slower deactivation in higher hydrogen pressure and demonstrates that temperature can critically alter the hydrogen coverage in this range to the extent that the ability of hydrogen to prevent polymerization-type coking would be lost between 573 and 623 K.

The percentage of activity loss shown in Table 14 is less than the number of uncovered Pt atoms. This means that reactions take place on partially carbonized catalysts which retain almost all their dehydrogenation activity. Also, a three-dimensional carbon island may start to build up [142]. If reactions were performed on this partially carbonized catalysts at a higher

TABLE 14 Material Balance of Radioactive 3-Methylpentane Pulses Introduced onto Pt Black Under Different Atmospheres and the Remaining Catalytic Activity[a]

	Carrier gas	
	He	5% H_2
Surface atoms × 10^{15}	3,630	3,630
Input (10^{15} molecules)	13,800	13,800
Product (10^{15} molecules)	518 (3.8%)	1,524 (11.0%)
Retained (10^{15} molecules)	468 (3.4%)	124 (0.9%)
Percent of surface atoms blocked (one C atom per surface Pt atom)	77	20
Remaining free surface (%)	23	80
Remaining activity (%)[b]		
Hydrogenolysis	30	81
Isomers plus MCP[c]	25	68
Benzene plus olefins	68	97

[a]Catalyst: 0.4 g of Pt Black, 1.1 $m^2 g^{-1}$
[b]Activity in second pulse divided by activity in first pulse, in the same carrier gas as the deactivation.
[c]Methylcyclopentane.
Source: Adapted from Ref. 54.

hydrogen pressure than that used for deactivation, the activity drop is disproportionately high, probably because hydrogen and hydrocarbon compete for the remaining few free active sites. This would be equivalent to a more steep descending branch of the bell-shaped curve on carbonized catalysts.

If the coke precursors are not too deeply dehydrogenated, they may be removed by hydrogenation of the catalyst, usually at temperatures higher than that of their deposition. However, above a certain degree of dehydrogenation (if they are formed above 673 K from n-C_6 hydrocarbons), they cannot be hydrogenated off the catalyst surface [139]. About half of the deposit formed from 2-methylpentane at 573 K could be hydrogenated off at 623 K; this was not possible with residues formed from methylcyclopentane [141]. If various hydrocarbons were injected onto Pt black in the absence of hydrogen, deactivation caused by cyclopentane and methylcyclopentane was appreciably higher than that induced by hexane isomers. Of the latter, neohexane deactivated the least. Deactivation shifted the selectivity toward olefin formation. With hydrogenation, the activity level could be raised somewhat (e.g., from 3% to 6% of the original level) and also, olefin selectivity decreased from, for example, 50% to 25%, compared to 0.5% on the regenerated catalyst. The corresponding isomerization selectivities were 6 and 22% (36% after regeneration). In these experiments carried out at 633 K [93], a temperature where appreciable removal of coke precursors by hydrogen cannot be expected [141], the deactivating agent was cyclopentane (the most efficient agent of all C_5 and C_6 alkanes and cyclopentanes).

On industrial supported catalysts, increasing the hydrogen coverage of the catalyst usually reduces the steady-state level of unsaturated molecules on the surface, thus also reducing deactivation by coking [143]. Increasing the hydrogen pressure from atmospheric pressure to 1.5 MPa caused a conspicuous slowing down of deactivation of model alumina-supported Pt and PtRe catalysts at 770 K, for the reasons noted above [143,144].

Both isomerization and hydrogenolysis of n-hexane showed a negative hydrogen exponent of about -1.5 on a 0.35% Pt/Al_2O_3 in the 8-30-bar p_{H_2} range ($p_{n-hexane}$ = 0.35 bar, T = 723 K). Upon the n-hexane accumulating a carbonaceous overlayer (by heating the catalyst at 800 K in n-hexane of 0.5 bar and H_2 of 10 bar for 1 h), both exponents decreased to about -1 [109]. This may be in accordance with the smaller ensembles available for reactions on a partially carbonized catalyst.

Subsequent hydrogenation at reaction temperature reactivated alumina-supported Pt and PtRe catalysts fairly well after reacting methylcyclohexane (of a few torr pressure) in excess H_2 over them. At the initial stage of deactivation (ca. 65% activity loss) the regeneration was complete at 650 K; after 85% activity loss, only 40% of the original activity could be regained. These are much higher values than those attained with unsupported metals at lower temperatures. Reactivation was never complete when the reaction took place at 700 K [145]. More rapid deactivation of Pt/Al_2O_3 was observed at lower hydrogen pressures [146]. The initial rapid deactivation period (E_a = 33 kJ mol^{-1}) and a subsequent slow deactivation period (E_a = 164 kJ mol^{-1}) could be clearly distinguished at all hydrogen pressures. This latter deactivation period was defined as causing irreversible deactivation, whereas initial deactivation was not only reversible but its extent could be correlated to hydrogen partial pressure by a simple equilibrium equation.

III. PERMANENT EFFECTS OF VARIOUS HYDROGEN TREATMENTS

It has been discussed in Chapters 10 to 13 that heating metal catalysts in hydrogen could bring about various changes. They undergo sintering with loss of surface area and reconstruction of exposed faces, hydrogen may be transferred between metal and support, and various metal-support interactions can be induced. These changes may and do alter catalytic properties (see Chapter 13). In what follows we deal with such effects, emphasizing their catalytic aspects.

A. High-Temperature Hydrogen Treatment of Supported Metal Catalysts

Den Otter and Dautzenberg reported an unexpected drop in hydrogenolysis activity of Pt/Al_2O_3 when it was treated in hydrogen between 773 and 973 K [147]. The problem was studied in detail by Menon and Froment [148,149]. They reduced Al_2O_3-, SiO_2-, and TiO_2-supported Pt catalysts between 673 and 873 K. High-temperature reduction gave catalysts with almost totally suppressed hydrogenolysis activity. The minimum was at 823 K with Pt/Al_2O_3 [148] and at 873 K with Pt/SiO_2 [149]. One typical plot of n-pentane and n-hexane transformations over Pt/SiO_2 treated at various temperatures in hydrogen is shown in Fig. 14. The phenomena were interpreted in terms of self-inhibition by one of the reactants rather than in terms of SMSI. Temperature-programmed desorption experiments indicate the presence of a high-temperature hydrogen peak on Pt with suppressed hydrogenolysis activity; this may be responsible for the activity drop, discussed also in Chapter 4.

Parera et al. [150] studied the phenomena mentioned with alumina-supported Pt, Re, and PtRe catalysts under high-pressure conditions simulating catalytic reforming. Increasing the prereduction temperature

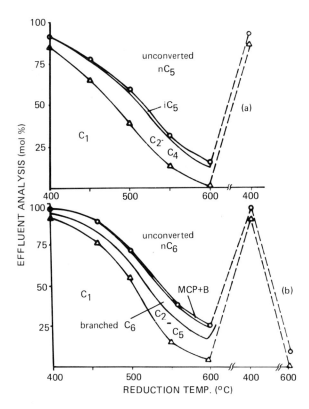

FIGURE 14 Effect of reduction temperature on reactions of (a) n-pentane and (b) n-hexane on 0.49% Pt/SiO_2 (prepared by ion exchange, dispersion: 63.5%). The catalyst is identical to that described in Ref. 153. Pulse reactor, T = 673 K. (After Ref. 149; used by permission of Elsevier Science Publ. Inc.)

from 753 K to 823 K continuously decreased the production of methane. This resulted in a 5% increase of liquid product yield, as opposed to a 10% increase with presulfided catalyst. Aromatic production had a minimum at a pretreatment temperature of 773 K with PtRe and 823 K with Pt. The yield of C_9 aromatics increased with increasing prehydrogenation temperatures and with sulfidation because the loss of alkyl groups was suppressed to a greater extent than dehydrocyclization. Thus the effect of high-temperature hydrogenation was compared with that of controlled deactivation by C, S, or by adding inactive alloying elements such as Cu, Au, and Sn. However, in a flow system and at elevated pressures the effects were not as marked as in a pulse system at atmospheric pressure, as reported by Menon and Froment [148,149].

The experiments of Margitfalvi et al. [47] show that high-temperature hydrogen treatment decreased the overall activity of a Pt/Al_2O_3 catalyst only slightly. However, the selectivity changed considerably: Aromatization increased at the cost of hydrogenolysis and isomerization (Table 15).

B. Structure Sensitivity on Supported Pt, Pd, and Rh Pretreated in Hydrogen

An excellent systematic study carried out in the group led by Burwell and Butt was intended to elucidate how the structure, hydrogen chemisorption,

TABLE 15 Role of High-Temperature Hydrogen Treatment on the Activity and Selectivity of a 0.5% Pt/Al_2O_3 Catalyst

Pretreatment temperature (K)	Rate[b]	Selectivity (%)[a]			
		Hydrogenolysis	Isomers	MCP[c]	Benzene
773[d]	2.12	11.8	47.2	10.3	9.0
903[d]	2.0	7.0	41.9	11.1	12.5
773[e]	—	13.5	37.2	8.1	28.5
903[e]	—	8.5	28.5	8.6	37.2

[a] Does not give 100%. The rest consists of various unsaturated products.
[b] n-Hexane conversion rate, in mol g_{cat}^{-1} s^{-1} × 10^5. T = 753 K.
[c] Methylcyclopentane.
[d] Measured at zero conversion.
[e] Measured at 20% conversion.
Source: After Ref. 48.

hydrogen-oxygen titration stoichiometry metal particle sizes, and catalytic activity of supported Pt, Pd, and Rh catalysts change under the effect of controlled pretreatment. The ultimate aim was to obtain information on the possible surface structures and their catalytic properties. Excellent reviews by each of the leading scientists of the group have appeared [151,152]; therefore, only a short summary of their work is given. We include as much as we do because the systematic and purpose-oriented character of this work is almost unparalleled.

Large-pore SiO_2 (14 nm) and Al_2O_3 (12 nm) supports were chosen for catalyst preparation. Ion exchange with ammin complexes of Pt, Pd, and Rh was the preferred method of catalyst preparation [153-155]. Also, impregnation of SiO_2 with H_2PtCl_6 [153] and with Rh carbonyls [155] was used. Al_2O_3 was impregnated with $Pt(NH_3)_4(NO_3)_2$ to avoid the incorporation of chlorine [156]. Care was taken that the supports were free of alkaline and sulfate impurities. The catalysts were then stored in air for years and became oxidized. Their standard pretreatment consisted of oxygen flow at 573 K (300°C) for 0.5 h, abbreviated by the authors as O_2, 300, 0.5. (Although temperatures are expressed in kelvin in this chapter, here we adopt the notation used by the authors.) Then the catalyst was subjected to various hydrogen treatments.

The pretreatment O_2, 300, 0.5; H_2, 300, 1; Ar 450, 1; cool in Ar was called the "standard" pretreatment. In other pretreatments the temperature of hydrogen flow (usually for 1 h), T_p, varied. The percentage of platinum atoms exposed, D_t, was measured by hydrogen-oxygen titration. H_2, 25 or H_2, 100 gave higher D_t values than those given by the standard pretreatment, probably because the surface became rough or faceted. This was observed with a catalyst of D_t = 6.3% and not found with D_t = 40 or 81. Obviously, the smaller particles there could not undergo surface roughening.

The pretreatment H_2, 450; cool in H_2 resulted in the incorporation of "excess" hydrogen into SiO_2-supported catalysts with D_t = 6.3, 40, and 81. This "excess" hydrogen could be liberated by heating in Ar at 723 to 773 K [157].

Essentially similar results were obtained with Pt/Al_2O_3 [156], Pd/SiO_2 [154], and Rh/SiO_2 [155], except for variations on the extent of oxidation during storage, the conditions of oxygen removal, and small variations of D_h after various treatments.

The extensive x-ray diffraction (XRD) studies of these catalysts [158-161], also using small-angle diffraction and EXAFS, revealed that they contained mostly perfect equiaxed crystals (except for the lowest dispersion, as noted in Chapter 11). Some hydride formation was observed with Pd/SiO_2, and this itself was also structure sensitive. Larger particles form hydride [154,160], whereas smaller ones remain metallic [162]. For further aspects of hydride formation and activity, we refer to Chapter 14.

The catalytic properties of samples subjected to such a careful analysis are of considerable interest. Hydrogenation of propene [163], ring opening of cyclopropane and methylcyclopropane [154,155,163-165], and H-D exchange in cyclopentane [154,165,166] and 2,2-dimethylbutane [167] were used as model reactions. These experiments may give insight into possible surface structures brought about by hydrogen treatments as well as the possible structure sensitivity of the model reactions.

As to isotope exchange, that of neohexane [167] presents more marked effects as a function of hydrogen pretreatment. The turnover number of exchange was almost independent of catalyst dispersion (except for the lowest dispersion) after standard pretreatment; H_2, 100, decreased, H_2, 450 always increased turnover numbers (Fig. 15). The ratio of exchanged t-butyl groups to exchanged ethyl groups was around 2 with Pt/SiO_2 (increasing slightly with increasing pretreatment temperature) and around 0.4 with Pd/SiO_2. The presence of a quaternary carbon atom hinders 1,3-adsorption, and therefore exchange in the t-butyl group typically produces d_1 species, whereas species up to d_5 were found in the ethyl group. The distribution of d_2 to d_5 in ethyl was nearly uniform over Pt/SiO_2 and showed a monotonic increase from d_2 to d_5 over Pd. The rate of $-CH_2$-exchange in cyclopentane was about 40 to 80 times larger on Pt/SiO_2 and 160 times larger on Pd/SiO_2 than that into the single methylene group in neohexane. The latter requires either its single adsorption on C-3, -CHM-, or 3,4-adsorption of the molecule $-CHM-CH_2M$. Such adsorption modes are hindered by the isobutyl group of the molecule, and this hindrance may also be valid on edges and corners.

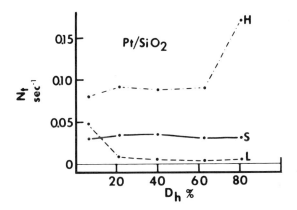

FIGURE 15 Turnover number, N_t (s^{-1}) as a function of dispersion, D_h, determined by hydrogen chemisorption of Pt/SiO_2. T = 373 K. S, standard pretreatment: O_2, 300, 0.5; He, 300, 0.25; H_2, 300, 1; He or Ar, 450, 1, cool in He or Ar. H_2, 300, 1 was replaced by H_2, 100, 1, in the case of line L; and by H_2, 450, 1, cool in H_2, in the case of line H. (After Ref. 167; used by permission of Academic Press, Inc.)

The authors put forward as a tentative explanation the possibility that "flat" low-index planes are, in fact, not perfectly flat. Knowing the several types of reconstructions induced by hydrogen, we may be less cautious and regard these data as other, direct catalytic evidence of probable surface reconstruction under the effect of high-temperature treatment, H_2, 450. On the other hand, H_2, 100 may smooth the surfaces, thus making the turnover frequencies lower (Fig. 15). This reasoning is in accordance with the linear correlation of turnover numbers with the mean vibrational amplitude of Pt atoms as calculated from XRD data in three reactions: propene hydrogenation, and ring hydrogenolysis of cyclopropane and methylcyclopropane [163]. The correlation against the edge + corner atoms per total exposed atoms calculated from ideal crystal geometry is less perfect (not linear).

Another reaction widely investigated on these well-characterized catalysts was ring opening (hydrogenolysis) of methylcyclopropane [155,164,165]. Although this reaction proceeds with high rates at low temperatures (like propene hydrogenation), both its turnover number and its selectivity (isobutane/n-butane ratio) may change in a wide range as a function of catalyst dispersion and pretreatment. With Pt/SiO_2, at a typical reaction temperature of 273 K, both isobutane and n-butane turnover frequencies show broad minima as a function of reduction temperature at T_p 420 to 470 K [164]. The curves characteristic of catalysts of low and high dispersion cross each other: Low-temperature pretreatment gives high turnover numbers with catalysts of low dispersity; high-temperature pretreatment enhances

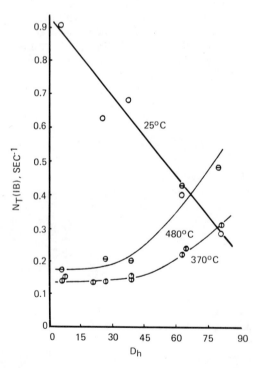

FIGURE 16 Turnover number, N_T, of isobutane formation at T = 273 K as a function of the dispersion, D_h, of Pt/SiO_2 catalysts pretreated in hydrogen at different temperatures (see the text and Fig. 15 for details of pretreatment). The numbers on the curves mean the temperature of final reduction in Celsius. (After Ref. 164; used by permission of Academic Press, Inc.)

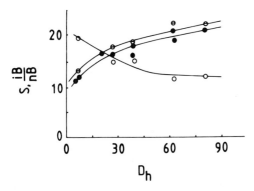

FIGURE 17 Structure sensitivity of selectivity, isobutane to n-butane ratio, at T = 273 K as a function of dispersion, D_h, for Pt/SiO_2 catalysts pretreated in hydrogen at different temperatures: ○, 298 K; ●, at 643 K; ⊖, 753 K. (After Ref. 164; used by permission of Academic Press, Inc.)

the activity of disperse samples (Fig. 16). The turnover frequencies are much lower for n-butane; therefore, ring opening is selective: The selectivity isobutane/n-butane changes between 10 and 23. Again, their change with dispersion parallels the variations in turnover numbers with Pt/SiO_2 pretreated at various temperatures (Fig. 17). The qualitative picture for Pt/Al_2O_3 samples is the same [165]. No essential differences are seen between high- and low-temperature pretreated Pd/SiO_2 [154]. The turnover numbers for n-butane and isobutane have broad maxima at a dispersity D_h between 40 and 60, with a much lower selectivity isobutane/n-butane ratio, between 3 and 4. The turnover frequencies for the formation of both butanes increase monotonically with dispersion over Rh/SiO_2, Rh being about three times more structure sensitive than Pt [155]. The selectivity value showed a marked decrease as a function of dispersity. Comparative values for the three metals are shown in Fig. 18. A third type of pattern was observed with Rh/SiO_2 as far as the catalyst pretreatment temperature is concerned. After a minimum on Pt and a slight increase on Pd, the turnover numbers for butanes did not change with increasing pretreatment temperatures for catalysts up to 50% dispersity; a slight increase appeared at D_h = 58 and a marked increase from a value of 0.05 (H_2, 25) to 0.80 (H_2, 300 and H_2, 450) with the most disperse catalyst of D_h = 108. (This higher-than-100% value comes from the fact that the H/Rh stoichiometry may not be equal to unity [155].)

The authors discuss their results in terms of catalyst geometry, electronic factors (similar to those mentioned in Chapter 13 for SMSI), and surface stoichiometry. At least on Pt and Rh catalysts, low-temperature hydrogen treatment may not be sufficient to remove all oxygen [153,155,156]; that may be the reason for the low activity of these catalysts at higher dispersion. For low-dispersion samples, the treatment may be sufficient to remove O_{ads}. The authors argue [164] that doing so may create many vacancies which anneal above 470 K, and hence the activity will be minimal. At higher temperature, hydrogen penetration of "excess" into subsurface layers (as shown in Chapters 1 and 11 and suggested by Burwell [152]) may loosen the surface atoms, thus creating imperfect flat surfaces, for example (111), with small islands of atoms of (111) symmetry on them [152]. They may be responsible for enhanced activity in neohexane H-D exchange (Fig. 15) and also for the increase of activity in methylcyclopropane ring opening. This phenomenon

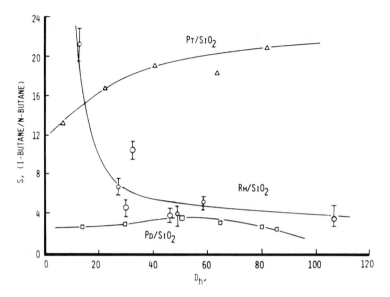

FIGURE 18 Selectivity, isobutane to n-butane, as a function of dispersion of Pt/SiO_2, Pd/SiO_2, and Rh/SiO_2, standard pretreatment (see Fig. 15). (After Ref. 155; used by permission of Academic Press, Inc.)

is more marked with more disperse Pt catalysts, whereas random vacancies may be more efficiently formed on larger particles of lower dispersity. That may be the reason for the opposite change of activity (Fig. 16). With Pd, hydride formation—eventually only in the upper few layers [168]—would cause a more profound surface reconstruction; thus any other local surface structure created by hydrogen would be "washed off."

As far as selectivity is concerned, the authors discuss the reaction in terms of hydrogenolysis mechanism (i.e., they assume the formation of 1,1,3-triadsorbed species). Therefore, they reject primary adsorption via the tertiary C atom, since this would lead to *its* diadsorption and to n-butane. If we apply the dual-site mechanism with flat-lying intermediate to this reaction [14,50] (see also Sections II.B and II.F), this contradiction can be eliminated. Then the first adsorption on the tertiary C atom would keep the molecule flat on the surface and ring opening would occur, probably involving not too deeply dissociated intermediates, predominantly to isobutane. This is the situation with Pt. Here larger metal-on-metal islands at higher pretreatment temperature may also increase the selectivity, whereas higher surface vacancy concentration at high-dispersity catalysts pretreated at 298 K (H_2, 25) decreases the number of continuous doublets suitable for flat-lying hydrogenolysis (Fig. 17). This suggestion cannot explain lower selectivity on Pd, but the comparison of Pt and Pd blacks in methylcyclopentane ring opening (Table 13) shows the same trend there, too, in spite of the enormous reactivity differences. On Rh, the selectivity drop at higher dispersity may be attributed to geometric reasons—fewer continuous sites for "flat" adsorption.

In Chapter 19 it is shown that both "flat" and "two-point" adsorption of three- and four-membered heterocycles, oxiranes and oxetanes, is possible, in their ring opening on metals, with hydrogen coverage being one of the factors regulating the ratio of different adsorption types.

C. Catalytic Properties of Unsupported Metals Pretreated at Different Temperatures

In Chapter 11 we described two Pt-black samples sintered at 473 K and 633 K, respectively. The latter showed much larger crystallites (not longer transparent in a lattice resolution electron microscope), its specific surface was lower, and the amount of surface impurities was higher than with the former sample, where a lattice resolution image could be obtained (Fig. 5 in Chapter 11). One may wonder how these changes in morphology and composition are reflected in the catalytic activity of these samples.

The overall activity of Pt-473 (or Pt-200) per unit surface was higher than that of Pt-633 (or Pt-360), as shown for n-heptane in Fig. 19. The overall conversion, as well as the yields of the individual products, show similar bell-shaped hydrogen dependence in both cases (like Fig. 2) and are higher per unit surface with Pt-473.

The same characteristics as used in Sections II.B and II.C for comparing selectivities have been calculated for these samples and compared with data obtained with Pt/SiO$_2$ (Euro Pt-1 [169]) under the same conditions. Toluene/C$_5$-cyclic ratios in Table 16 are comparable to the benzene/methylcyclopentane ratios listed in Tables 3 and 4. Since C$_5$ cyclics may react further to give skeletal isomers, the ratio of toluene to C$_7$ saturated products sometimes seems more appropriate for characterizing the ratios of the corresponding deeply and slightly dehydrogenated surface intermediates. Both ratios decrease initially with increasing p_{H_2}; and this trend is more marked with the toluene/C$_7$ saturated ratio. Of the two unsupported samples, Pt-473 gives relatively more aromatics at higher hydrogen pressures. In this

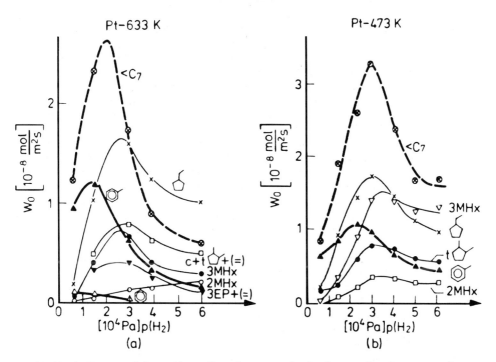

FIGURE 19 Rates of formation of various products from n-heptane as a function of temperature, over Pt-black samples pretreated at different temperatures. For conditions, see Fig. 2. [(a) Adapted from Ref. 45; (b) from H. Zimmer, unpublished data.]

TABLE 16 Ratio of Aromatics to Saturated Products on Different Pt Catalysts[a]

Catalyst	Ratio[c]	p_{H_2} (torr)[b]				
		50	120	250	320	500
Pt black, 473 K	T/C$_5$ cyclics	1.35	0.63	0.35	0.30	0.28
	T/C$_7$ sat.	0.78	0.42	0.19	0.15	0.13
Pt black, 633 K	T/C$_5$ cyclics	2.5	0.51	0.37	0.21	0.09
	T/C$_7$ sat.	0.89	0.53	0.17	0.12	0.06
Pt/SiO$_2$[d]	T/C$_5$ cyclics	0.51	0.47	0.57	0.90	1.32
	T/C$_7$ sat.	0.41	0.36	0.37	0.45	0.51
Pt/Al$_2$O$_3$	T/C$_5$ cyclics	0.75	0.90	1.21	1.60	1.42
	T/C$_7$ sat.	0.59	0.65	0.79	0.89	0.74

[a]Static-circulation system; $p_{heptane}$ = 10 torr; T = 603 K; conditions and catalysts were described in Ref. 59.
[b]±10%.
[c]T, toluene; C$_5$ cyclics, ethylcyclopentane + cis- and trans-dimethylcyclopentanes + ethylcyclopentene; C$_7$ sat., C$_5$ cyclics plus isoheptanes.
[d]EuroPt-1 as described in Ref. 169.

respect it is closer to Pt/SiO$_2$. The relative extent of aromatization *increases* on the latter at higher hydrogen pressures. This may be due to the better "hydrogen economy" of the latter sample. Here higher pressures may enhance spillover, which decreases the actual hydrogen coverage on the metallic sites, thus making them more suitable for aromatization. Hydrogen on the support in the vicinity of metal particles may serve as a reservoir preventing extensive coking. Also, the migration of coke precursor by a spillover-like mechanism is not excluded [140]. Toluene formation is enhanced at higher p_{H_2} on Pt/Al$_2$O$_3$, perhaps due to the start of a bifunctional mechanism.

Another ratio, that of isomers to C$_5$ cyclics, may be characteristic of this hydrogen economy. After an initial drop, it remains constant on Pt-633 and on carbonized Pt/SiO$_2$, while it increases continuously on Pt-473 and supported Pt samples (Table 17). As indicated also in Section II.G, if hydrocarbons and hydrogen compete for free surface sites, the situation will be unfavorable for hydrocarbons on carbonized catalysts. Perhaps adsorbed hydrogen may be attached to carbonaceous residues (as proposed in Chapters 16 and 23). This process is not sufficient to remove the residues but is enough to stop, for example, the opening of surface cyclopentanes; they appear as cyclic products. With cleaner catalysts, isomer formation increases monotonically with hydrogen pressure.

The enhanced toluene formation on Pt-473 is also manifested with C$_5$-cyclic reactants. Ethylcyclopentane produces mainly olefins on Pt-633. They can be regarded as the first products of ring enlargement according to the mechanism by Kane and Clarke [129] which we accepted for ring enlargement in Section II.F. Toluene appears at most in traces over Pt-633 but in appreciable amounts over Pt-473; its hydrogen dependence is analogous to that reported previously for the methylcyclopentane→benzene reaction [14,100]. Enhanced aromatization is observed with 1,1-dimethylcyclopentane feed (as predicted by the literature [130]) and the hydrogen dependence is different from that found with ethylcyclopentane. The superiority of Pt-473 is seen here, too (Fig. 20).

TABLE 17 Isomer/C_5 Cyclic Ratios from n-Heptane as a Function of Hydrogen Pressure[a]

Catalyst	Isomers/cyclopentanes if the hydrogen pressure (torr) is:			
	50	120	220	480
Pt-black-633	1	0.54	0.58	0.53
Pt-black-473	0.74	0.50	0.82	1.18[b]
Pt/Al_2O_3	0.28	0.38	0.54	0.96
Pt/SiO_2	0.29	0.32	0.57	1.56
Pt/SiO_2 (carb.)[c]	—	0.38	0.30	0.48

[a] Static-circulation system, T = 600 K, p/n-heptane/ = 9 torr. After Ref. 54 and H. Zimmer, unpublished data.
[b] 510 torr.
[c] Partly carbonized catalyst (treated with 1-torr n-heptane at 600 K for 1 min, without hydrogen).

A selective ring opening of methylcyclopentane was seen over most metals in Section II.F; selectivity increases with hydrogen pressure [126]. Pt-633 with larger particles shows a more conspicuous increase than Pt-473. The hydrogen dependence of the b/a selectivity on the latter sample is more or less parallel to that observed with Pt/SiO_2 (Fig. 21).

The examples enumerated give evidence on the marked effect of the temperature of pretreatment of Pt-black catalysts. It cannot be decided at present whether these differences are caused mainly by particle-size effects or by the different amounts of impurities. Since the crystallite sizes (14 and 30

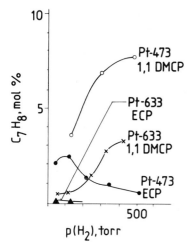

FIGURE 20 Yield of toluene (C_7H_8), mol %, from 1,1-dimethylcyclopentane and ethylcyclopentane over two Pt-black samples, as a function of hydrogen pressure. T = 603 K. For conditions, see Fig. 2. (From H. Zimmer, unpublished data.)

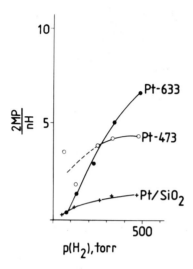

FIGURE 21 Ratio of 2-methylpentane to n-hexane (2MP/nH) formed from methylcyclopentane over various Pt catalysts, as a function of hydrogen pressure. For conditions, see Fig. 2. (From H. Zimmer, unpublished data.)

TABLE 18 Activity and Selectivity of Group VIII Metals in Neopentane Reactions

Catalyst	Treatment[a]	$T(K)$[b]	Reaction rate	Selectivity (%)	
				Hydro-genolysis	Isomeri-zation
Pt-film	O_2, 245, 0.25; H_2, 245, 0.5	516	1.3×10^{-2}[d]	12.3	87.7
	O_2, 245, 0.25; H_2, 460, 13	516	4×10^{-5}	9.6	90.4
Pd-film	O_2, 290, 0.25; H_2, 290, 0.6	562	8.1×10^{-4}	71.4	28.6
	O_2, 290, 0.25; H_2, 500, 13.5	562	4.4×10^{-4}	32.5	67.5
Pd-powder	standard[c]	559	9.2×10^{-9}[e]	92.0	8.0
	O_2, 300, 0.25; H_2, 700, 12	647	4.3×10^{-9}	78.9	21.0
Ir-powder	standard[c]	462	2×10^{-8}	97	3
	O_2, 320, 0.25; H_2, 850, 18	481	3.4×10^{-9}	92	8
Rh-powder	standard[c]	467	1.9×10^{-9}	100	0
	O_2, 300, 0.25; H_2, 900, 12	540	1.6×10^{-8}	98	2

[a]First letter: gas atmosphere, followed by the temperature of treatment in °C; final number: duration of treatment in hours.
[b]Reaction temperature; static-circulation system, typical HC/H_2 ratio: 1:10. Films were deposited on mica.
[c]"Standard" pretreatment with powders only: O_2, 300, 0.25; H_2, 300, 1; H_2, 400, 1.
[d]Rate on metal films is expressed as turnover frequencies, molecule reacted per surface atom per second.
[e]Initial rate on metal powders is expressed as mol reacted per g catalyst per second.
Source: After Ref. 173.

nm, respectively [170]; see also Table 1 in Chapter 11) are well above the ranges where particle-size effects are expected [15] and because large differences were found in the surface compositions of these two samples [171,172], the latter reason seems to be more likely. The parallelism demonstrated between carbonized Pt/SiO$_2$ and Pt-633 supports this opinion.

That high-temperature hydrogen treatment can, indeed, alter the selectivity and activity of Group VIII metals independently of support effects, as postulated in [148,149], was recently confirmed by Karpinski et al. [173]. They reacted neopentane on Pt, Pd, Rh, and Ir powders and films. High-temperature hydrogen treatment resulted in the considerable decrease of total activity and in a somewhat increased isomerization selectivity. This latter result was most marked with Pd and least marked with Pt, which exhibited the highest isomerization selectivity (Table 18). It can be seen that the "high-temperature" treatment was carried out at different temperatures with various metals. Pd-powder, after standard pretreatment, showed 81% isomerization selectivity; it proved to be inactive up to 773 K after H$_2$, 900, 12.

IV. CONCLUSIONS

Hydrogen effects are numerous in skeletal reactions of hydrocarbons and may have different causes. Theoretically, as expected on the basis of the equations in Chapter 15, surface hydrogen concentration regulates the degree of dehydrogenation of hydrocarbons adsorbed on the surface. This simple picture, although we believe it to be qualitatively valid quite generally, fails to be quantitative except for the simplest hydrocarbons. There are various reasons for this. Even hydrogen itself is delocalized on the surface (Chapters 1 and 2); also, each hydrocarbon occupies a patch of different size on the surface. Thus the hydrogen coverage optimum for one reactant may be far from that for another. Further, the coverage—pressure correlation usually cannot be expressed analytically in the case of coadsorption with hydrocarbons. The presence of carbonaceous residues may complicate the picture further: They can be combined with hydrogen [142] and may also be sources of hydrogen in TPD experiments [142,173].

The possibility of spillover represents another sink or source for hydrogen on metals. Even on clean surfaces, various types of hydrogen may exist and coexist. Infrared spectroscopy gives evidence (see Chapter 7) that atom-top hydrogen may be present before multiple sites are saturated. TPD is a good tool to identify these types but cannot answer the simple question of how much of them is present on the surface contacting with gas-phase hydrogen (and eventually, hydrocarbon reactants and products of its conversion). Although proposals were put forward on the possible role of multiple-bonded and atom-top hydrogen in C$_5$-cyclic reactions, and we believe that the explanations offered are realistic, they are far from being proven absolutely.

Still, some general trends seem to take shape if we compare the results obtained in very different experimental setups such as single crystals and disperse catalysts at subatmospheric or atmospheric pressures as well as industrial samples studied at elevated pressures (Chapters 16 and 25). Accordingly, every reaction has an optimum hydrogen pressure at which its yield (rate) is maximal. It seems to us that this picture is valid for catalysts carbonized to an extent corresponding to steady-state activity. The bell-shaped (or volcano) activity response to hydrogen pressure may be a general feature of such systems.

ACKNOWLEDGMENTS

The author is grateful to Professor Pál Tétényi for several stimulating and valuable discussions during many years of fruitful common work. He thanks Professors R. L. Burwell, Jr. and John B. Butt for sending unpublished data and advice relating to Section III.B. Many unpublished data are thankfully acknowledged to Dr. Helga Zimmer. Thanks are due Mrs. Rózsa Lörincz for typing and Mrs. Edit Rockenbauer-Fülöp for technical assistance in preparing the chapter.

REFERENCES

1. N. D. Zelinsky, B. A. Kazansky, and A. F. Plate, *Ber. Dtsch. Chem. Ges.*, *B66*:1415 (1933).
2. N. D. Zelinsky, B. A. Kazansky, and A. F. Plate, *Ber. Dtsch. Chem. Ges.*, *B68*:1869 (1935).
3. B. A. Kazansky and A. F. Plate, *Ber. Dtsch. Chem. Ges.*, *B69*:1862 (1936).
4. Yu. K. Yuryev and P. Ya. Pavlov, *Zh. Obshch. Khim.*, 7:97 (1937).
5. B. A. Kazansky, A. L. Liberman, T. F. Bulanova, V. T. Aleksanyan, and Kh. E. Sterin, *Dokl. Akad. Nauk SSSR*, 95:77 (1954); B. A. Kazansky, A. L. Liberman, V. T. Aleksanyan, and Kh. E. Sterin, *Dokl. Akad. Nauk SSSR*, 95:281 (1954).
6. N. I. Shuikin, N. G. Berdnikova, and S. S. Novikov, *Dokl. Akad. Nauk SSSR*, 89:1029 (1953); N. I. Shuikin, *Adv. Catal.*, 9:783 (1957).
7. C. O'Donohoe, J. K. A. Clarke, and J. J. Rooney, *J. Chem. Soc. Chem. Commun.*, 648 (1979); *J. Chem. Soc. Faraday Trans. 1*, 76:345 (1980).
8. G. A. Mills, H. Heinemann, T. H. Milliken, and A. G. Oblad, *Ind. Eng. Chem.*, 45:134 (1953).
9. Y. Barron, D. Cornet, G. Maire, and F. G. Gault, *J. Catal.*, 2:162 (1963); J. R. Anderson and N. R. Avery, *J. Catal.*, 2:542 (1963).
10. J. R. Anderson, *Adv. Catal.*, 23:1 (1973).
11. J. H. Sinfelt, *Adv. Catal.*, 23:91 (1973).
12. J. K. A. Clarke and J. J. Rooney, *Adv. Catal.*, 25:125 (1976).
13. G. A. Somorjai, *Adv. Catal.*, 26:268 (1977).
14. Z. Paál, *Adv. Catal.*, 29:273 (1980).
15. F. G. Gault, *Adv. Catal.*, 30:1 (1981).
16. B. A. Kazansky, in *Mechanisms of Hydrocarbon Reactions* (F. Márta and D. Kalló, eds.), Akadémiai Kiadó, Budapest (1975), p. 15.
17. H. Pines, *The Chemistry of Catalytic Hydrocarbon Conversions*, Academic Press, New York (1981).
18. Z. Paál and P. Tétényi, *Catalysis*, Vol. 5 (G. C. Bond and G. Webb, eds.), Royal Society of Chemistry, London (1982), p. 80.
19. O. V. Bragin and A. L. Liberman, *Transformations of Hydrocarbons on Metal Containing Catalysts* (in Russian), Khimiya, Moscow (1981).
20. G. V. Isagulyants, M. I. Rozengart, and Yu. G. Dubinsky, *Catalytic Aromatization of Aliphatic Hydrocarbons* (in Russian), Nauka, Moscow (1983).
21. V. Ponec, in *The Chemical Physics of Solid Surfaces and Heterogeneous Catalysis*, Vol. 4 (D. A. King and D. P. Woodruff, eds.), Elsevier, Amsterdam (1982), p. 365; G. A. Somorjai, ibid., p. 217.
22. E. H. Van Broekhoven and V. Ponec, *Prog. Surf. Sci.*, 19:351 (1985).
23. J. C. Rohrer, H. Hurwitz, and J. H. Sinfelt, *J. Phys. Chem.*, 65:1458 (1961).
24. Z. Paál, *Magy. Kém. Foly.*, 75:478 (1969).

25. Z. Paál and P. Tétényi, *Dokl. Akad. Nauk SSSR*, *201*:1119 (1971).
26. Z. Paál and P. G. Menon, *Catal. Rev.-Sci. Eng.*, *25*:229 (1983).
27. J. F. Taylor and J. K. A. Clarke, *Z. Phys. Chem. (Neue Folge)*, *103*: 216 (1976).
28. Z. Paál and P. Tétényi, *Nature*, *267*:234 (1977).
29. A. A. Balandin, *Adv. Catal.*, *19*:1 (1969).
30. D. A. Dowden, *J. Chem. Soc.*, 242 (1950); for a review, see, e.g., V. Ponec, in *Electronic Structure and Reactivity of Metal Surfaces* (E. G. Derouane and A. A. Lucas, eds.), Plenum Press, New York (1976), p. 590.
31. R. Montarnal and G. Martino, *Rev. Inst. Fr. Pet.*, *32*:367 (1977).
32. Z. Paál, *Faraday Discuss. Chem. Soc.*, *72*:82 (1981).
33. M. W. Vogelzang, M. J. P. Botman, and V. Ponec, *Faraday Discuss. Chem. Soc.*, *72*:33 (1981).
34. D. O. Hayward in *Chemisorption and Reactions on Metallic Films*, Vol. 1 (J. R. Anderson, ed.), Academic Press, London (1971), pp. 225-326; see in particular, Fig. 13 on p. 248.
35. S. L. Kiperman, *Bulg. Acad. Sci. Commun. Dept. Chem.*, *16*:22 (1983).
36. A. Frennet, G. Liénard, A. Crucq, and L. Degols, *J. Catal.*, *53*:150 (1978).
37. L. Guczi, B. S. Gudkov, K. Matusek, A. Sárkány, K. M. Sharan, and P. Tétényi, *Kém. Közl.*, *43*:187 (1975).
38. Z. Paál, G. Székely, and P. Tétényi, *J. Catal.*, *58*:108 (1979).
39. J. Margitfalvi, M. Hegedüs, P. Szedlacsek, and F. Nagy, *Acta Chim. Hung.*, *119*:213 (1985).
40. P. G. Menon, J. C. de Deken, and G. F. Froment, *J. Catal.*, *95*:313 (1985).
41. P. Biloen, J. N. Helle, H. Verbeek, F. M. Dautzenberg, and W. M. H. Sachtler, *J. Catal.*, *63*:112 (1980).
42. G. Leclercq, L. Leclercq, and R. Maurel, *Bull. Soc. Chim. Belg.*, *88*:599 (1979).
43. F. Garin and F. G. Gault, *J. Am. Chem. Soc.*, *97*:4466 (1975).
44. P. Parayre, V. Amir-Ebrahimi, F. G. Gault, and A. Frennet, *J. Chem. Soc. Faraday Trans. 1*, *76*:1704 (1980).
45. H. Zimmer, Z. Paál, and P. Tétényi, *Acta Chim. Acad. Sci. Hung.*, *111*:513 (1982).
46. H. Zimmer, M. Dobrovolszky, P. Tétényi, and Z. Paál, *J. Phys. Chem.*, *90*:4758 (1986).
47. E. Ruckenstein and D. B. Dadyburjor, in *Growth and Properties of Metal Clusters* (J. Bourdon, ed.), Elsevier, Amsterdam, 1980, p. 435.
48. J. Margitfalvi, P. Szedlacsek, M. Hegedüs, and F. Nagy, *Appl. Catal.*, *15*:69 (1985).
49. B. S. Gudkov, L. Guczi, and P. Tétényi, *J. Catal.*, *74*:207 (1982); L. Guczi, A. Frennet, and V. Ponec, *Acta Chim. Hung.*, *112*:127 (1983).
50. Z. Paál, M. Dobrovolszky, and P. Tétényi, *J. Catal.*, *45*:189 (1976).
51. P. Tétényi, L. Guczi, Z. Paál, and A. Sárkány, *Kém. Közl.*, *47*:391 (1977).
52. P. Biloen, F. M. Dautzenberg, and W. M. H. Sachtler, *J. Catal.*, *50*:77 (1977).
53. H. Verbeek and W. M. H. Sachtler, *J. Catal.*, *42*:257 (1976).
54. Z. Paál, M. Dobrovolszky, and P. Tétényi, *J. Catal.*, *46*:65 (1977).
55. S. M. Davis, F. Zaera, and G. A. Somorjai, *J. Catal.*, *85*:206 (1984).
56. Z. Paál and P. Tétényi, *Kém. Közl.*, *37*:129 (1973).
57. Yu. V. Fomichov, I. G. Gostunskaya, and B. A. Kazansky, *Izv. Akad. Nauk SSSR Ser. Khim.*, 1112 (1968).
58. Z. Paál, *J. Catal.*, *91*:181 (1985).

59. Z. Paál, H. Zimmer, and P. Tétényi, *J. Mol. Catal.*, 25:99 (1984).
60. Z. Paál and P. Tétényi, *Appl. Catal.*, 1:9 (1981).
61. V. S. Fadeev, I. V. Gostunskaya, and B. A. Kazansky, *Dokl. Akad. Nauk SSSR*, 189:788 (1969).
62. Z. Paál and P. Tétényi, *J. Catal.*, 30:350 (1973).
63. P. C. Stair and G. A. Somorjai, *J. Phys. Chem.*, 66:2036 (1977).
64. H. Zimmer, V. V. Rozanov, A. V. Sklyarov, and Z. Paál, *Appl. Catal.*, 2:51 (1982).
65. V. V. Rozanov and A. V. Sklyarov, *Kinet. Katal.*, 19:1533 (1978).
66. G. Twigg, *Proc. R. Soc. London*, A178:106 (1941).
67. Z. Paál and P. Tétényi, *Acta Chim. Acad. Sci. Hung.*, 55:273 (1968).
68. Z. Paál and P. Tétényi, *Acta Chim. Acad. Sci. Hung.*, 54:175 (1967).
69. F. M. Dautzenberg and J. C. Platteeuw, *J. Catal.*, 19:41 (1970).
70. B. A. Kazansky, A. L. Liberman, G. V. Loza, and T. V. Vasina, *Dokl. Akad. Nauk SSSR*, 128:1188 (1959).
71. B. H. Davis, *J. Catal.*, 29:398 (1973); 46:348 (1977).
72. V. Amir-Ebrahimi, A. Choplin, P. Parayre, and F. G. Gault, *Nouv. J. Chim.*, 4:431 (1980).
73. V. P. Lebedeva and V. M. Gryaznov, *Izv. Akad. Nauk SSSR Ser. Khim.*, 611 (1981).
74. Z. Paál, M. Dobrovolszky, J. Völter, and G. Lietz, *Appl. Catal.*, 15:33 (1985).
75. J. R. H. Van Schaik, R. P. Dessing, and V. Ponec, *J. Catal.*, 38:273 (1975).
76. J. K. A. Clarke, A. F. Kane, and T. Baird, *J. Catal.*, 64:200 (1980).
77. T. Koscielski, Z. Karpinski, and Z. Paál, *J. Catal.*, 77:539 (1982).
78. A. L. Liberman, *Kinet. Katal.*, 5:128 (1964).
79. O. Finlayson, J. K. A. Clarke, and J. J. Rooney, *J. Chem. Soc. Faraday Trans. 1*, 80:191 (1984).
80. F. G. Gault, V. Amir-Ebrahimi, F. Garin, P. Parayre, and F. Weisang, *Bull. Soc. Chim. Belg.*, 88:475 (1979).
81. S. A. Krasavin, M. S. Kharson, M. M. Kostyukovsky, O. V. Bragin, and S. L. Kiperman, *Izv. Akad. Nauk SSSR Ser. Khim.*, 1231 (1982).
82. S. A. Grin', N. A. Gayday, T. Yu. Sergeeva, B. S. Gudkov, N. I. Koltsov, and S. L. Kiperman, *Kinet. Katal.*, 26:907 (1985).
83. H. Zimmer and Z. Paál, *Proceedings of the 8th International Congress on Catalysis, West Berlin*, Vol. 3 (1984), p. 417.
84. T. G. Olferyeva, S. A. Krasavin, and O. V. Bragin, *Izv. Akad. Nauk. SSSR Ser. Khim.*, 605 (1981).
85. S. M. Csicsery, *Adv. Catal.*, 28: 293 (1979).
86. L. A. Erivanskaya, G. A. Shevtsova, A. Khalima-Mansur, and A. F. Plate, *Kinet. Katal.*, 15:810 (1974).
87. L. A. Erivanskaya, V. P. Lozhkina, L. M. Korovina, T. V. Antipina, and A. F. Plate, *Neftekhimiya*, 15:341 (1975).
88. F. E. Shephard and J. J. Rooney, *J. Catal.*, 3:129 (1964).
89. Y. Barron, G. Maire, J. M. Muller, and F. G. Gault, *J. Catal.*, 5:428 (1966).
90. Z. Paál and P. Tétényi, *Acta Chim. Acad. Sci. Hung.*, 72:277 (1972).
91. B. A. Kazansky, V. S. Fadeev, and I. V. Gostunskaya, *Izv. Akad. Nauk. SSSR Ser. Khim.*, 677 (1971).
92. V. S. Fadeev, I. V. Gostunskaya, and B. A. Kazansky, *Dokl. Akad. Nauk SSSR*, 199:622 (1971).
93. Z. Paál, to be published.
94. H. C. de Jongste and V. Ponec, *Bull. Soc. Chim. Belg.*, 88:453 (1979).
95. M. A. McKervey, J. J. Rooney, and N. E. Sammann, *J. Catal.*, 30:330 (1973).

96. V. Amir-Ebrahimi and F. G. Gault, *J. Chem. Soc. Faraday Trans. 1,* 76:1735 (1980).
97. K. Foger and J. R. Anderson, *J. Catal.,* 54:318 (1978).
98. Guo Xiexian, Yang Yashu, Den Maicum, Li Huimin, and Lin Zhiyin, *J. Catal.,* 99:218 (1986).
99. P. Parayre, V. Amir-Ebrahimi, and F. G. Gault, *J. Chem. Soc. Faraday Trans. 1,* 76:1723 (1980).
100. Z. Paál and P. Tétényi, *J. Catal.,* 29:175 (1973).
101. Z. Paál, M. Dobrovolszky, and P. Tétényi, *Proceedings of the 3rd International Conference on Heterogeneous Catalysis, Varna, 1975,* Bulgarian Academy of Sciences, Sofia (1978), p. 350.
102. A. Sárkány, *J. Catal.,* 89:14 (1984).
103. A. Sárkány, *J. Catal.,* 97:407 (1986).
104. L. Guczi, A. Sárkány, and P. Tétényi, *J. Chem. Soc. Faraday Trans. 1,* 70:1971 (1974).
105. A. Sárkány, L. Guczi, and P. Tétényi, *Acta Chim. Acad. Sci. Hung.,* 96:27 (1978).
106. Z. Karpinski, *Nouv. J. Chim.,* 4:561 (1980).
107. A. Sárkány, K. Matusek, and P. Tétényi, *J. Chem. Soc. Faraday Trans. 1,* 73:1699 (1977).
108. E. Kikuchi and Y. Morita, *J. Catal.,* 15:217 (1969).
109. E. G. Christoffel and Z. Paál, *J. Catal.,* 73:30 (1982).
110. G. Leclercq, L. Leclercq, and R. Maurel, *J. Catal.,* 50:87 (1977).
111. J. L. Carter, J. A. Cusumano, and J. H. Sinfelt, *J. Catal.,* 20:223 (1971).
112. Z. Paál and P. Tétényi, *React. Kinet. Catal. Lett.,* 12:131 (1979).
113. H. Zimmer, P. Tétényi, and Z. Paál, *Acta Chim. Hung.,* 124:13 (1987).
114. P. Tétényi and Z. Paál, *Acta Chim. Hung.,* 119:83 (1985).
115. A. Sárkány and P. Tétényi, *J. Chem. Soc. Chem. Commun.,* 527 (1980).
116. A. Sárkány, S. Pálfi, and P. Tétényi, *React. Kinet. Catal. Lett.,* 15:1 (1980).
117. A. Sárkány, S. Pálfi, and P. Tétényi, *Acta Chim. Acad. Sci. Hung.,* 111:633 (1982).
118. A. Sárkány, *J. Chem. Soc. Faraday Trans. 1,* 82:103 (1986).
119. Z. Paál, M. Dobrovolszky, and P. Tétényi, *J. Chem. Soc. Faraday Trans. 1,* 80:3037 (1984).
120. M. A. Dobrovolszky, Z. Paál, and P. Tétényi, *Acta Chim. Hung.,* 119:95 (1985).
121. A. Sárkány, J. Gaál, and L. Tóth, *Proceedings of the 7th International Congress on Catalysis, Tokyo,* Part A (1980), p. 291.
122. Z. Paál, K. Matusek, and P. Tétényi, *Acta Chim. Acad. Sci. Hung.,* 94:119 (1977).
123. G. Maire, G. Plouidy, J. C. Prudhomme, and F. G. Gault, *J. Catal.,* 4:556 (1965).
124. J. R. Anderson, R. J. MacDonald, and Y. Shimoyama, *J. Catal.,* 20:147 (1971).
125. C. Corolleur, S. Corolleur, and F. G. Gault, *J. Catal.,* 24:385 (1972).
126. O. V. Bragin, Z. Karpinski, K. Matusek, Z. Paál, and P. Tétényi, *J. Catal.,* 56:219 (1979).
127. Z. Paál, M. Dobrovolszky, and P. Tétényi, *React. Kinet. Catal. Lett.,* 2:97 (1975).
128. Z. Paál, *Proceedings of the 6th International Congress on Catalysis, London,* Vol. 1 (1976), p. 127.
129. A. F. Kane and J. K. A. Clarke, *J. Chem. Soc. Faraday Trans. 1,* 76:1640 (1980).

130. B. A. Kazansky, A. L. Liberman, G. V. Loza, and T. V. Vasina, *Dokl. Akad. Nauk SSSR*, *128*:1188 (1959).
131. V. Amir-Ebrahimi and F. G. Gault, *J. Chem. Soc. Faraday Trans. 1*, *77*:1813 (1981).
132. B. H. Davis, *Proceedings of the 8th International Congress on Catalysis, West Berlin*, Vol. 2 (1984), p. 469.
133. J. M. Parera, J. N. Beltramini, C. A. Querini, E. E. Martinelli, E. J. Churin, P. E. Aloe, and N. S. Figoli, *J. Catal.*, *99*:39 (1986).
134. O. V. Bragin, Tao Lung-Hsiang, and A. L. Liberman, *Kinet. Katal.*, *8*:98 (1967).
135. O. V. Bragin, Tao Lung-Hsiang, and A. L. Liberman, *Kinet. Katal.*, *8*:931 (1967).
136. M. Bartók and coauthors: *Stereochemistry of Heterogeneous Metal Catalysis*, Wiley, Chichester, West Sussex, England (1985).
137. J. Dermietzel, F. Bauer, M. Rösseler, W. Jockisch, H. Franke, J. Klempin, and H. J. Barz, *Isotopenpraxis*, *12*:57 (1976).
138. G. Lietz, J. Völter, M. Dobrovolszky, and Z. Paál, *Appl. Catal.*, *13*:77 (1984).
139. A. Sárkány, H. Lieske, T. Szilágyi, and L. Tóth, *Proceedings of the 8th International Congress on Catalysis, West Berlin*, Vol. 2 (1984), p. 613.
140. J. M. Parera, N. S. Figoli, J. N. Beltramini, E. J. Churin and R. A. Cabrol, *Proceedings of the 8th International Congress on Catalysis, West Berlin*, Vol. 2 (1984), p. 593.
141. F. Luck, S. Aeiyach, and G. Maire, *Proceedings of the 8th International Congress on Catalysis, West Berlin*, Vol. 2 (1984), p. 695.
142. S. M. Davis, F. Zaera, and G. A. Somorjai, *J. Catal.*, *77*:439 (1982); for reviews, see G. A. Somorjai and F. Zaera, *J. Phys. Chem.*, *86*:3070 (1982); R. J. Koestner, M. A. Van Hove, and G. A. Somorjai, *J. Phys. Chem.*, *87*:203 (1983).
143. V. K. Shum, J. B. Butt, and W. M. H. Sachtler, *J. Catal.*, *99*:126 (1986).
144. V. K. Shum, J. B. Butt, and W. M. H. Sachtler, *J. Catal.*, *96*:371 (1985).
145. L. W. Jossens and E. E. Petersen, *J. Catal.*, *76*:265 (1982).
146. L. W. Jossens and E. E. Petersen, *J. Catal.*, *73*:377 (1982).
147. G. J. Den Otter and F. M. Dautzenberg, *J. Catal.*, *53*:116 (1978).
148. P. G. Menon and G. F. Froment, *J. Catal.*, *59*:138 (1979).
149. P. G. Menon and G. F. Froment, *Appl. Catal.*, *1*:31 (1981).
150. J. M. Parera, E. L. Jablonski, R. A. Cabrol, N. S. Figoli, J. C. Musso, and R. J. Verderone, *Appl. Catal.*, *12*:125 (1984).
151. R. B. Butt, *Appl. Catal.*, *15*:161 (1985).
152. R. L. Burwell, Jr., *Langmuir*, *2*:2 (1986).
153. T. Uchijima, J. M. Herrmann, Y. Inoue, R. L. Burwell, Jr., J. B. Butt, and J. B. Cohen, *J. Catal.*, *50*:464 (1977).
154. R. Pitchai, S. S. Wong, N. Takahashi, J. B. Butt, R. L. Burwell, Jr., and J. B. Cohen, *J. Catal.*, *94*:478 (1985).
155. Z. Karpinski, T.-K. Chuang, H. Katsuzawa, J. B. Butt, R. L. Burwell, Jr., and J. B. Cohen, *J. Catal.*, *99*:184 (1986).
156. M. Kobayashi, Y. Inoue, N. Takahashi, R. L. Burwell, Jr., J. B. Butt, and J. B. Cohen, *J. Catal.*, *64*:74 (1980).
157. R. Pitchai, unpublished results; cited after R. L. Burwell, Jr., personal communication.
158. S. R. Sashital, J. B. Cohen, R. L. Burwell, Jr., and J. B. Butt, *J. Catal.*, *50*:479 (1977).

159. R. K. Nandi, R. Pitchai, S. S. Wong, J. B. Cohen, R. L. Burwell, Jr., and J. B. Butt, *J. Catal.*, *70*: 298 (1981).
160. R. K. Nandi, P. Georgopoulos, J. B. Cohen, J. B. Butt, R. L. Burwell, Jr., and D. H. Bilderback, *J. Catal.*, *77*: 421 (1982).
161. R. K. Nandi, F. Molinaro, C. Tang, J. B. Cohen, J. B. Butt, and R. L. Burwell, Jr., *J. Catal.*, *78*: 289 (1982).
162. M. Boudart and H. S. Hwang, *J. Catal.*, *39*: 44 (1975).
163. P. H. Otero-Schipper, W. A. Wachter, J. B. Butt, R. L. Burwell, Jr., and J. B. Cohen, *J. Catal.*, *50*: 494 (1977).
164. P. H. Otero-Schipper, W. A. Wachter, J. B. Butt, R. L. Burwell, Jr., and J. B. Cohen, *J. Catal.*, *53*: 414 (1978).
165. S. S. Wong, P. H. Otero-Schipper, W. A. Wachter, Y. Inoue, M. Kobayashi, J. B. Butt, R. L. Burwell, Jr., and J. B. Cohen, *J. Catal.*, *64*: 84 (1980).
166. Y. Inoue, J. M. Herrmann, H. Schmidt, R. L. Burwell, Jr., J. B. Butt, and J. B. Cohen, *J. Catal.*, *53*: 401 (1978).
167. V. Eskinazi and R. L. Burwell, Jr., *J. Catal.*, *79*: 118 (1983).
168. D. D. Eley and E. I. Pearson, *J. Chem. Soc. Faraday Trans. 1*, *74*: 223 (1978).
169. G. C. Bond and P. B. Wells, *Appl. Catal.*, *18*: 221, 225 (1985); J. W. Geus and P. B. Wells, *Appl. Catal.*, *18*: 231 (1985); A. Frennet and P. B. Wells, *Appl. Catal.*, *18*: 243 (1985).
170. Z. Paál, P. Tétényi, L. Kertész, A. Szász, and J. Kojnok, *Appl. Surf. Sci.*, *14*: 101 (1982-83).
171. Z. Paál, P. Tétényi, D. Prigge, X. Zh. Wang, and G. Ertl, *Appl. Surf. Sci.*, *14*: 307 (1982-83).
172. Z. Paál and D. Marton, *Appl. Surf. Sci.*, *26*: 161 (1986).
173. Z. Karpinski, W. Juszczyk, and J. Pielaszek, *J. Chem. Soc. Faraday Trans. 1*, *83*: 1293 (1987).
174. T. S. Wittrig, P. D. Szuromi, and W. H. Weinberg, *J. Chem. Phys.*, *76*: 3305 (1982).
175. P. Tétényi, J. Király, and L. Babernics, *Acta Chim. Acad. Sci. Hung.*, *29*: 35 (1961).
176. P. Tétényi and L. Babernics, *Acta Chim. Acad. Sci. Hung.*, *35*: 419 (1963).
177. J. Pérez Pariente, Ph.D. Thesis, Madrid, 1984.

18 | Hydrogen Effects in Organic Hydrogenations

ÁRPÁD MOLNÁR

József Attila University, Szeged, Hungary

GERARD V. SMITH

Southern Illinois University at Carbondale, Carbondale, Illinois

I.	INTRODUCTION	499
II.	HYDROGEN AVAILABILITY	500
	A. Pressure Effects	500
	B. Effect of Temperature	504
	C. Effect of Catalyst Quantity and Metal Loading	505
	D. Other Factors Influencing Hydrogen Availability	506
III.	HYDROGEN SPILLOVER	509
	A. Use of Hydrogen Spillover in Hydrogenations	509
	B. Reverse Spillover	509
	C. Phenomena Interpreted by Hydrogen Spillover	510
IV.	EFFECT OF VARIOUS TYPES OF HYDROGEN IN HYDROGENATIONS	510
	A. Surface Hydrogens	510
	B. Hydrogen on Raney Nickel	512
	C. Absorbed Hydrogen	513
V.	MORPHOLOGICAL CHANGES AND HYDROGENATION	513
	A. Dissolution of Metals	513
	B. SMSI	514
	C. Miscellaneous Phenomena	514
VI.	CONCLUDING REMARKS	515
	REFERENCES	515

I. INTRODUCTION

Experimental chemists working with metal hydrogenation catalysts often encounter phenomena which cannot be interpreted easily. These phenomena may be called "hydrogen effects." The term indicates that hydrogen, besides taking part stoichiometrically in addition, affects the outcome of the transformations in some known or unknown way. The most readily identifiable hydrogen effects are the effects of the direct experimental variables (pressure, temperature, catalysts) on the selectivity or stereochemistry of the transformations of different organic functional groups by altering hydrogen availability. This is the topic of Section II.

Section III deals with spillover and reverse spillover, with emphasis on the use of spillover hydrogen in the hydrogenation of alkenes.

Developments in experimental techniques have made it possible to identify different types of hydrogen on or in metal catalysts. Section IV presents experimental observations in which the presence of these different types of hydrogen have already been correlated with transformations taking place.

It is known that certain conditions of catalyst pretreatment lead to catalysts with special features. Moreover, the catalytic process itself can alter the morphology of the catalysts. As a result, the actual working catalyst may have characteristics substantially different from those before or after reaction. Examples of these phenomena are presented in Section V.

II. HYDROGEN AVAILABILITY

In the interpretation of the mechanism of hydrogenation of organic compounds with multiple bonds, the Horiuti-Polanyi mechanism [1] is widely used (Scheme 1). Since the relative rates of the consecutive steps are a function of the

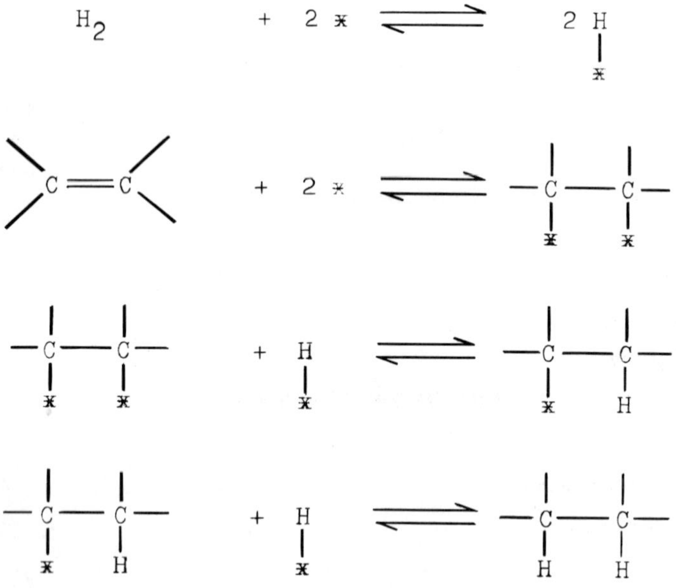

$$H_2 + 2* \rightleftharpoons 2 H{-}* \quad (1)$$

$$\!\!>\!\!C{=}C\!\!<\! + 2* \rightleftharpoons {-}\underset{*}{C}{-}\underset{*}{C}{-} \quad (2)$$

$$-\underset{*}{C}{-}\underset{*}{C}{-} + H{-}* \rightleftharpoons -\underset{*}{C}{-}\underset{H}{C}{-} \quad (3)$$

$$-\underset{*}{C}{-}\underset{H}{C}{-} + H{-}* \rightleftharpoons -\underset{H}{C}{-}\underset{H}{C}{-} \quad (4)$$

Scheme 1

concentration of adsorbed hydrogen, all the factors influencing the hydrogen availability (hydrogen partial pressure, catalyst quantity, metal structure, metal loading, support, temperature, surface modifiers, additives, diffusion phenomena associated with pores, solvent, and agitation in the case of liquid-phase hydrogenations) have an effect on the transformation by changing not only the rate-determining step, but also the geometry-determining step.

A. Pressure Effects

Stereochemistry of Hydrogenation of Olefins

In the hydrogenation of the isomeric dimethylcyclohexenes ([1] and [2]) and 2-methylmethylenecyclohexane [3] on Pt (Scheme 2), the stereochemistry of

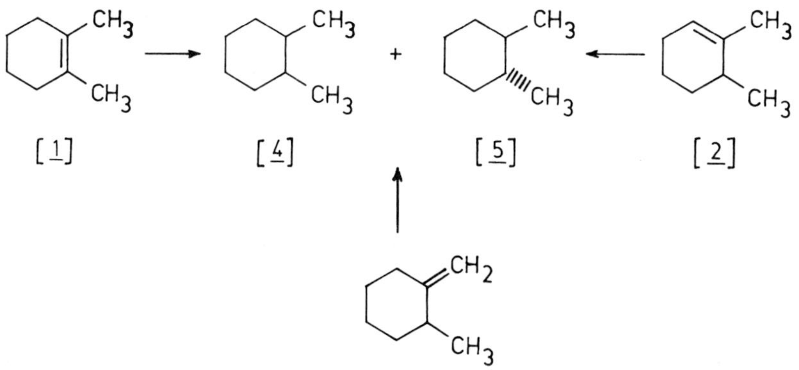

Scheme 2

the hydrogen addition, that is, the ratio of cis- and trans-1,2-dimethylcyclohexanes ([4] and [5]), depends on the pressure of hydrogen. [1] is converted mainly to the thermodynamically less stable cis compound [2,3], but the trans compound is also formed, as a result of isomerization through the half-hydrogenated state (step 3 in Scheme 1). If the pressure increases, the amount of cis isomer also increases (Table 1), because the higher pressure increases the rate of addition relative to isomerization.

In contrast, for compounds [2] and [3], a hydrogen pressure increase results in a decrease of the cis isomer (Tables 1 and 2). At low hydrogen hydrogen pressures the product controlling reaction in the Horiuti-Polanyi mechanism is the formation of the half-hydrogenated intermediate (step 3 in Scheme 1), while at high hydrogen pressures it is the adsorption of the alkene (step 2). As a result of this, the ratio of the isomeric products approaches unity, because the steric effects in the adsorption of the alkenes are less pronounced than in the transition state leading to the half-hydrogenated state. Therefore, for the hydrogenation of alkyl-substituted methylenecyclohexanes, this general trend toward a decrease in the quantity of the less stable isomer can be interpreted as a result of the shift of the product-determining step from (3) to (2) [5-7].

Similar phenomena were observed in the hydrogenation of substituted cyclopentenes [8], substituted t-butylcyclohexenes [5-7], 4-t-butylcyclohexenyl methyl ether [9], and α- and β-pinenes [6,10] (see Table 7).

TABLE 1 Percent cis-1,2-Dimethylcyclohexane in Hydrogenation of [1] and [2][a]

| Hydrogen pressure | % cis | |
(× 10^5 Pa)	[1]	[2]
1	81.8	76.6
4	83.3	70.6
40	88.9	68.8
150	93.4	71.1

[a]PtO_2, acetic acid, 298 K, conversion: 100%.
Source: Ref. 2.

TABLE 2 Percent cis Isomer in Hydrogenation of Methyl-methylenecyclohexanes[a]

	Hydrogen pressure ($\times 10^5$ Pa)		
	1	3	50
2-Methyl	70	69	69
3-Methyl	28	35	43
4-Methyl	73	70	66

[a]PtO_2, acetic acid, 298 K, conversion: 100%.
Source: Ref. 4.

Stereochemistry of Hydrogenation of Aromatics

The hydrogenation of disubstituted benzenes gives widely changing isomer ratios, depending on the substrate and the catalyst. Both the increase and decrease in the rates of hydrogenation and cis/trans ratios were observed as a function of hydrogen pressure on Pt and Rh [11-14]. Siegel et al. found [12-14] that the corresponding cycloalkenes are intermediates in the hydrogenation of aromatics. For example, the concentrations of the intermediate olefins gave maximum curves during the hydrogenation of 1,4-di-t-butylbenzene.

Because of the greater number of elementary steps and the presence of desorbed intermediates, the application of the Horiuti-Polanyi mechanism to aromatics is much more complicated than its application to olefins. Moreover, the relative rates of formation and disappearance of the intermediate cycloalkenes are different for the different substrates on the different metals, depending on the hydrogen coverage. Also, there is a delicate balance between the rate of hydrogenation of the starting aromatic compound and the intermediate cyclohexene which determines the ratio of the cis- and trans-disubstituted cyclohexanes. Consequently, it is impossible to give a clear-cut, general interpretation of the effect of the pressure of hydrogen on the stereochemistry of hydrogenation of aromatic compounds.

A similar complex reaction scheme is operative in the hydrogenation of cresols and diphenols, during which the corresponding substituted cyclohexanones and cyclohexanediones can be intermediates. Increasing hydrogen pressures result in increasing quantities of the cis isomer (Table 3). m-Cresol exhibits the most pronounced change, while p-cresol exhibits no change.

Stereochemistry of Hydrogenation of Carbonyl Compounds

In the hydrogenation of 2-methylcyclopentanone on Pd, there is a change from the more stable trans isomer in excess to the cis isomer in excess as the hydrogen pressure increases [17]. The excess of the cis isomer is also characteristic of hydrogenations on Raney Ni, where the product-controlling step is the adsorption of the starting compound. This was interpreted by Mitsui et al. [18] to indicate that the hydrogen transfer from the catalyst surface to the half-hydrogenated state became easier and the product-determining step shifted toward the adsorption of the substrate on the catalyst. This interpretation is supported by reaction kinetic measurements with cycloalkanones on Pt/SiO_2 and Ru/SiO_2 [19]. At low hydrogen pressures, when the rate-determining step is the reaction of the adsorbed species, the reaction order in ketone is approaching zero, while at high pressures (rate-

TABLE 3 Isomer Distribution (cis/trans) in Hydrogenation of Cresols and Dihydroxybenzenes[a]

	Hydrogen pressure ($\times 10^5$ Pa)	
	1	100
o-Cresol	74/26	82/18[b]
1,2-Dihydroxybenzene	75/25	81/19
m-Cresol	40-45/55-60	73/27
1,3-Dihydroxybenzene	49/51	68-72/28-32
p-Cresol	67/33	67/33[b]
1,4-Dihydroxybenzene	52/48	68/32

[a] Rh-Pt, acetic acid, room temperature.
[b] 142×10^5 Pa.
Source: Refs. 15 and 16.

determining step is the adsorption of the substrate) the reaction order is approaching unity.

The same phenomenon, that is, an increase in quantity of the less stable isomer (isomenthol) by increasing pressure was found in the hydrogenation of menthone on Rh/C and Pt/C (Table 4). On the other hand, in the hydrogenations of the alkyl-substituted cyclohexanones, an increase in hydrogen pressure always results in an increase in the quantity of the more stable equatorial alcohol (trans isomer from the 2- and 4-substituted compounds and cis isomer from the 3-substituted cyclohexanones) (Table 5). No explanation of this has been offered. In the case of the isomeric cyclohexanediones at higher pressure, there is always a higher amount of the *cis*-diol formed (Table 6).

Isomerizations During Addition

Because of the reversibility of the formation of the half-hydrogenated state (step 3 in Scheme 1) in the Horiuti-Polanyi mechanism, the ratio of isomerization versus addition can be changed by changing the hydrogen partial

TABLE 4 Isomer Distribution in Hydrogenation of Menthone[a]

Catalyst	Temperature (K)	Hydrogen pressure ($\times 10^5$ Pa)	Menthol/isomenthol	Conversion (%)
Rh/C	378	5	37/63	87
	378	10	33/67	49
	423	7	58/42	97
	423	35	43/57	99
Pt/C	423	7	49/51	95
	423	35	41/59	98

[a] Solvent: 0.7 M NaOH.
Source: Ref. 20.

TABLE 5 Hydrogenation of Alkylcyclohexanones[a]

	Hydrogen pressure ($\times 10^5$ Pa)	% Axial alcohol	Conversion (%)
2-Methyl	1	70	88
	70	55	61
3-Methyl	1	88	100
	85	57	100
3-t-Butyl	1	94	100
	110	56	100
4-Methyl	1	84	100
	80	60	100
4-t-Butyl	1	94	100
	80	58	90

[a]Rh/C, ethanol, room temperature.
Source: Ref. 18.

pressure. Although the isomerization is not a hydrogen-consuming reaction, step 3 cannot take place without hydrogen. A good example of this is found in the hydrogenations of cis- and trans-cyclododecene [21]. In vacuo, on a prereduced Pt catalyst isomerization takes place mainly with an initial ratio of isomerization to hydrogenation of 5.3, while at 3.04×10^5 Pa the main process is addition, and the ratio of isomerization to hydrogenation is 0.47.

On the basis of isotopic distribution during tritiation of octadecenoates on Ni/SiO$_2$, it was concluded that cis-trans isomerization and double-bond migration take place on different sites influenced by the occupation of the catalyst surface by hydrogen atoms [22]. Sites with low hydrogen coverage (NiH sites) catalyze hydrogen abstraction and, therefore, double-bond migration, while sites with high hydrogen coverage (NiH$_2$) catalyze cis-trans isomerization.

B. Effect of Temperature

The hydrogen availability can be changed by altering the reaction temperature as well, but it is difficult to predict and to interpret the effect of temperature on stereoselectivity because the change in temperature can also affect variables

TABLE 6 Cis/trans Ratios in Hydrogenation of Cyclohexanediones[a]

	Hydrogen pressure ($\times 10^5$ Pa)	
	1	100
1,2-	76/24	82/18
1,3-	56-58/42-44	72/28
1,4-	59/41	61/39

[a]Rh-Pt, acetic acid, room temperature.
Source: Ref. 16.

TABLE 7 Pressure and Temperature Effect in Hydrogenation of α- and β-Pinene

α-Pinene		β-Pinene	
Pt/C, ethanol, 293 K		PtO_2, acetic acid, 300 K	
p (× 10^5 Pa)	% cis	p (× 10^5 Pa)	% cis
1	90	1	85
30	97.5	80 - 100	90
100	98.5		
Pd/C, propionic acid, 10^5 Pa		Borohydride-reduced Pt/C, ethanol, 0.1 N acetic acid, 10^5 Pa	
T(K)	% cis	T(K)	% cis
273	80	273	94
325	64.5	298	86
400	47		

Source: Refs. 6, 10, and 25.

other than hydrogen availability. For example, the advantageous effect on selectivity of lowering the reaction temperature in hydrogenations of acetylenic compounds [23,24] may be connected to a change in the adsorption coefficient of acetylenes relative to olefins.

Detailed investigation was made of the temperature dependence of hydrogenation of α-pinene. In propionic acid on Pd/C it was found that there is a decrease of the cis isomer as the temperature increases [10]. The same was observed in the transformation of β-pinene on Pt/C catalyst [25] (Table 7). In Mitsui's interpretation [18] the decreasing selectivity is a result of the change of hydrogen availability. That is, as the temperature increases, the hydrogen availability decreases and step 4 in the Horiuti-Polanyi mechanism (Scheme 1) becomes the product-controlling process. As a result of this, the possible half-hydrogenated states may come into equilibrium.

On the basis of this, the increasing hydrogen pressure and the higher temperature should have the opposite effect on product distribution. In fact, as the pressure is raised there is an increase in the cis/trans ratio for both α- and β-pinene as a result of higher hydrogen availability [6,10] (Table 7).

A similar increase in the amount of the more stable isomer with increasing temperature was observed in the hydrogenation of different cyclic ketones (menthone [20] (Table 4), 2-carbethoxycycloalkanones [26], 2-methylcyclohexanone [27]), and aromatics [28]. Although the correlation between the temperature change and the hydrogen availability in these transformations is not clear, nevertheless, the change of selectivity similarly shows a trend opposite to that produced by an increase in pressure observed by Mitsui et al. [17] during the hydrogenation of 2-methylcyclopentanone.

C. Effect of Catalyst Quantity and Metal Loading

Observations indicate that on palladium the stereoselectivity of hydrogenation of the olefinic double bond is controlled by the amount of catalyst and metal loading. By investigating the hydrogenation of 2-methylmethylenecyclohexane

Mitsui et al. arrived at the following conclusions [29,30]. At low reactant/ catalyst ratios (large amount of catalyst) the hydrogen available is spread over a large surface area, decreasing its probability of interaction with the substrate. Since the product-controlling reaction of hydrogenation is hydrogen transfer to the half-hydrogenated state, the preceding reactants and intermediates become equilibrated with one another and result in decreased selectivity. Under such experimental conditions, which favor reversible adsorption, isomerization occurs between the two epimeric half-hydrogenated intermediates, and the rapid migration of the exo-cyclic double bond into the ring can be observed during the hydrogenation of 2-methylmethylenecyclohexane. As a result of this equilibrium it is the thermodynamically more stable trans isomer which is formed preferentially.

When a small amount of catalyst is used (large reactant/catalyst ratio, resulting in high hydrogen availability), more cis isomer is formed. Under such conditions the hydrogen transfer to the adsorbed species is relatively easy, so it is the initial adsorption of the olefin that becomes the product-controlling step in the Horiuti-Polanyi mechanism.

The foregoing reasoning originally was put forward by House et al. [31] to interpret the stereochemistry of hydrogenation of an unsaturated polycyclic lactone and was also used by Augustine to explain the change of stereoselectivity in hydrogenation of $\Delta^{1(9)}$-octalone-2 upon changing the catalyst quantity and metal loading of palladium catalysts [32,33]. Similarly, this interpretation explains the increasing amount of the 5β isomer as a result of increasing catalyst quantity or metal loading in the hydrogenation of cholestenone on Pd [34].

The change of regioselectivity upon changing the catalyst quantity was also observed in the hydrogenation of *endo*-2-acetoxy-7-isopropylidenebicyclo[2.2.1]-5-heptene [35]. At low catalyst quantities the reduction of the endocyclic double bond takes place with high selectivity, while larger amounts of catalyst lead to more random reductions.

D. Other Factors Influencing Hydrogen Availability

Effect of Catalyst Prereduction

In the hydrogenation of different unsaturated compounds on palladium catalysts the results depend on whether the catalysts are prereduced.

During the transformation of 1,4-butynediol on 0.2% Pd on Al_2O_3 both the selectivity and the rate of hydrogen uptake exhibited substantial differences if the catalyst was prereduced or preequilibrated with the substrate [34]. This was explained by assuming that the preadsorption of substrate affects the hydrogen adsorption and causes changes in both the rate- and selectivity-determining steps.

Depending on the experimental conditions in the hydrogenation of 7-acetoxynorbornadiene on Pd/C, large differences in both the rate of hydrogenation and the isomeric ratio were also observed. On the other hand, the pretreatment of Pt/C resulted in only slight changes (Table 8).

Mitsui et al. observed similar changes in the hydrogenation of methylcyclohexanones on PdO [18]. Moreover, on the unreduced catalyst, the ratio of the isomeric alcohols gradually changes as the hydrogenation proceeds. This phenomenon is presumed due to very strong adsorption of the ketones, which prevents hydrogen adsorption on the catalyst. During the progress of the hydrogenation, as a result of a decrease in ketone concentration, the hydrogen supply increases and the ratio of isomeric alcohols changes. The exceptionally strong adsorption of ketones may totally hinder reduction [38]. On the other hand, gradual surface reconstruction, as discussed in Section V.C (p. 514) may

TABLE 8 Transformations of 7-Acetoxynorbornadiene[a]

	k_{anti}/k_{syn}	Saturated (%)	Conversion to olefin (%)	Syn/anti
Pd/C	3.42	31	49	84:16
Pd/C, prereduced	0.58	23	47	56:44
Pt/C	1.1	29	39	92.3:7.7
Pt/C, prereduced	1.8	21	48	91.7:8.3

[a]Ethanol, room temperature, 1×10^5 Pa.
Source: Refs. 36 and 37.

account for gradually changing product ratios during hydrogenations over incompletely reduced palladium catalysts.

Selectivity of Hydrogenation of Hydrocarbons with Multiple Unsaturations

The selective formation of olefins in the partial hydrogenation of hydrocarbons with multiple unsaturation (dienes, acetylenes) is an interesting and important process. By investigating the deuteration of 1,2-propadiene and 1,2-butadiene on palladium, the most selective metal in these transformations, Olivier and Wells observed a decrease in selectivity for olefin formation with an increase in deuterium pressure [39]. Similarly, in the hydrogenation of acetylene there is a rapid rate increase of hydrogenation when the reaction diverts from selective ethene formation to ethane formation [40]. At sufficiently high initial hydrogen pressures a completely nonselective reaction takes place, which led the authors to conclude that high surface hydrogen concentrations provide conditions for the formation of [6] or [9] from the allenes or acetylene, respectively. These species, they propose, are responsible for the formation of the saturated hydrocarbons, which decrease selectivity. In their interpretation [39], chemisorbed hydrogen at a critical coverage can modify the electronic structure of palladium such that during the hydrogenation of acetylene the vinylic intermediate [7] is able to form an additional metal bond [8] (Scheme 3). This new bond activates the methylene group for hydrogen atom attack, producing [9].

The hydrogenation of acetylene as an impurity in ethene streams obtained from steam cracking is an important industrial process. Making detailed studies in this field, Margitfalvi et al. [41] and Moses et al. [42] arrived at the conclusion that the decreasing selectivity is due to high hydrogen coverage. They found that under their experimental conditions it is a carbonaceous deposit which results in the increase in surface hydrogen concentration. The surface species leading to the formation of ethane is formulated as [10] on the basis of the observation of Kesmodel [43]. Pd black and supported palladium behave similarly in their selectivity of hydrogenation as a result of the effect of carbonaceous deposits.

All factors removing excess chemisorbed hydrogen result in increased selectivity. An oxidation-reduction treatment of the used catalyst leads to better selectivity. Higher oxidation temperatures further improve the selectivity because carbonaceous deposits can be removed more completely at higher temperatures. Higher temperatures of evacuation after oxidation-reduction treatments have the same effect because the concentration of hydrogen, either adsorbed or chemisorbed, is further decreased.

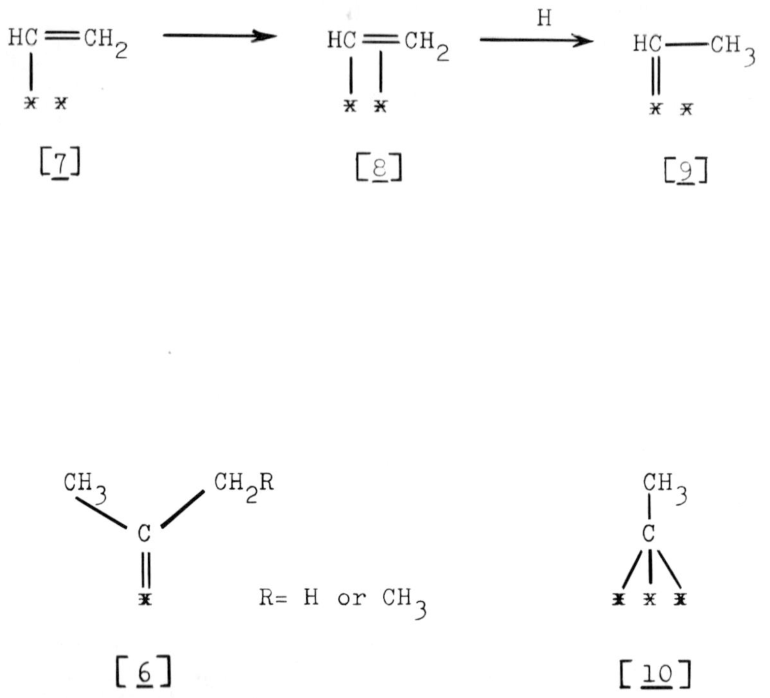

Scheme 3

Effect of Additives

During hydrogenation of alkylcyclohexanones Mitsui observed that in the presence of an alkaline additive (NaOH), the amount of the more stable isomer (trans alcohol from 2- and 4-, and cis alcohol from the 3-substituted cyclohexanones) increases on Pt, Pd, and Rh [18]. They interpreted this to mean that the alkaline additive decreases hydrogen availability, and steps 3 and 4 (Scheme 1), particularly the latter, are retarded. Step 4 becomes the product-controlling step, and of the two epimeric half-hydrogenated states, the one leading to equatorial alcohol is favored. On Raney Ni the additive has the opposite effect. This indicates that the product-determining step of transformation is not the hydrogen transfer to the substrate but the adsorption of the substrate on the catalyst.

In the hydrogenation of acetylenes additives play an important role, modifying catalysts for high selectivity. The best known modified palladium catalyst is Lindlar's catalyst, palladium-lead on calcium carbonate [44]. Investigating the effect of Pb as a modifier in the hydrogenation of acetylene, Palczewska et al. found [45-47] that Pb, among others, strongly suppresses the dissociative adsorption of hydrogen. The amount of both adsorbed and chemisorbed hydrogen decreases with increasing surface concentration of Pb. The selectivity of palladium can be further improved by using CO in the stream. Surface studies [47,48] and other types of investigations [49-51] indicate that CO blocks the surface for hydrogen (and ethene) adsorption and reduces the surface hydrogen coverage. The effect of both CO and Pb is to lower the hydrogen accessibility for hydrogenation. Additionally, it was found that alloying of Pd with Cu leads to dissociated hydrogen depletion on the catalyst surface [51].

III. HYDROGEN SPILLOVER

A. Use of Hydrogen Spillover in Hydrogenations

The early hypothesis that hydrogenation occurs by spilt-over hydrogen was challenged (see detail in Ref. 52) as an explanation for the support-enhanced catalytic activity in the hydrogenation of ethylene and benzene in flow systems. However, recent results by Antonucci et al. [53] seem to prove the existence of spillover on diluted catalysts. They observed that the activity of Pt/Al_2O_3 in the hydrogenation of benzene increased with the degree of dilution, with γ-alumina giving a saturation-type curve at a ratio of 50:1 alumina to catalyst. This indicates that atomic hydrogen cannot migrate beyond a certain distance.

Later, Gardes et al. [54] developed a static hydrogenation technique, which has recently been improved [55], for demonstrating the use of spilt-over hydrogen to hydrogenate unsaturated compounds. In their system a supported metal catalyst and the support are kept in hydrogen at 573 K. After removal of the catalyst the support is able to hydrogenate unsaturated compounds. Detailed investigations with hydrogen-ethene mixtures on alumina activated with Ni/Al_2O_3 revealed [56] that spilt-over hydrogen and molecular hydrogen are needed simultaneously to hydrogenate ethene. An interesting observation is an induction period in the transformation of the first dose of ethene-hydrogen mixture on different oxides activated with Pt. On the basis of this induction period, it was concluded that the spilt-over hydrogen is an inhibitor in the hydrogenation of ethene [57-61].

In addition to ethene [54-62], Teichner's group studied the transformations of butenes [57,58]; benzene, cyclohexene, and cyclohexadienes [60,61,63,64]; and acetylene [63]. A detailed analysis of these results is given in Chapter 12.

By using a fluidized-bed technique in which a Pt/γ-Al_2O_3 pellet, sitting on a fritted disk, activates with hydrogen different supports which are in a fluidized state above it, Lau and Sermon could hydrogenate ethene at 473 K [65]. γ-Alumina, silica + alumina, and MoO_3 were active in the transformation; silica was not.

The enhanced reaction rate of ethene hydrogenation in the Pt/γ-alumina system during spillover was interpreted as altering the initial reaction conditions and not as utilizing the extra hydrogen formed by spillover [66].

Unlike the foregoing activation procedures, in which spillover is induced by atomic hydrogen produced by metals, hydrogen plasma alone is able to bring about spillover activation of silica. This activated silica can also be used to hydrogenate ethene [67].

B. Reverse Spillover

Bond and Sermon [68,69] used 1-pentene as a titrant in an alkene titration measurement to study the state of adsorbed hydrogen. On a Pt/SiO_2 catalyst which has first been saturated by hydrogen, the only product from the addition of 1-pentene is pentane. Later, pentane formation decreases markedly and double-bond isomerization takes place. It was concluded that the hydrogen chemisorbed on the metal reacts with the titrant. Later, when hydrogen coverage on the metal decreases below a certain value, isomerization becomes the main reaction (Scheme 1). The low quantity of pentane observed under the latter circumstances is formed as a result of reverse spillover.

Augustine et al. developed a pulsed alkene titration method for investigation of reverse spillover [70,71]. Presaturation of either Pt or Pd with hydrogen results in the storage of a large amount of hydrogen in the support by spillover. This hydrogen can be used to saturate 1-butene and 1-pentene by reverse spillover. During this reverse process hydrogen migrates to the active sites of the catalyst and reacts with alkenes. As the hydrogen is removed by the reaction with olefin, more hydrogen migrates to the active sites from the spillover reservoir. This process takes time and depends on the catalysts used. It occurs at a lower rate with Pd than with Pt.

C. Phenomena Interpreted by Hydrogen Spillover

Vannice and Chou [72] observed a 5- to 10-fold increase in catalytic activity of Pd/TiO_2 and $Pd/SiO_2 + Al_2O_3$ in the hydrogenation of benzene when the catalysts were reduced at low temperatures (448 K instead of 673 K). Because of the slow spillover at this low temperature they postulated that hydrogen migrates to special sites present at the metal-support interface. The enhanced rate of hydrogenation of benzene on a $Ru-Pd/Al_2O_3$ catalyst is considered to be due to the migration of spilt-over hydrogen from Pd to Ru [73]. Depending on the reduction temperature, some $Ni-Cu/Al_2O_3$ catalysts exhibited a similar activity change as well as different selectivities in the hydrogenation of crotonaldehyde. Here, again, spillover was evoked to interpret the results [74].

In the hydrogenation of benzene, Pt and Pd zeolites show higher relative activities compared to metal blacks and supported metals [75]. The authors conclude that additional hydrogenation surface is created by spillover. The hydrogenation of toluene on Pt/HY zeolite was also interpreted as hydrogen spillover [76]; for the hydrogenation of benzene on Rh [77] and Pt zeolites [78], spillover effects were ruled out.

Decreasing selectivities during acetylene hydrogenations on Pd catalysts are attributed to hydrogen spillover [50,51,79]. Spillover is believed to cause higher hydrogen concentrations in the vicinity of the Pd-support interface. This hydrogen can react with the ethene adsorbed on support sites and result in decreased selectivity. Surface polymers and carbonaceous deposits formed under the experimental conditions might facilitate hydrogen migration from Pd to the support. Using the same reasoning, some type of reactive surface polymer may facilitate spillover from the support to the Pd and furnish hydrogen atoms for ethene hydrogenation. Both processes bring about the formation of larger amounts of ethane. In contrast, added copper increases the selectivity by providing desorption sites for spillover hydrogen which would otherwise lead to ethene hydrogenation [51,79].

Double-spillover effects are proposed to take place in the solid-state hydrogenation of alkyl-substituted phenols [80]. To participate in hydrogenation, activated hydrogen must first migrate from the metal to the support, then migrate between the support and the organic reagent.

Sachtler et al. used the hydrogen spillover effect to interpret the enantioselective hydrogenation of carbonyl compounds [81]. A dual-site hydrogenation catalyst was prepared by depositing either tartaric acid or Ni tartrate complexes on Ni/SO_2 [81,82]. Hydrogenation requires hydrogen migration from the adsorption sites (Ni metal) to the addition sites (tartrate complex). Distinction between the sites was made by hydrogenating methyl acetoacetate. Support for the authors' explanation was drawn from the fact that the Ni tartrate complex itself was inactive in hydrogenation, whereas an enantiomeric excess was achieved by the modified Ni/SiO_2.

IV. EFFECT OF VARIOUS TYPES OF HYDROGEN IN HYDROGENATIONS

A. Surface Hydrogens

Using a pulsed thermokinetic technique, two different types of chemisorbed hydrogen were identified in the transformations of ethene on Ni/Al_2O_3 [83].

They result from a fast and a slow dissociation process. The one formed in a fast dissociation is active only in hydrogenolysis. The rate of formation of the other depends on the coverage by adsorbed hydrogen, and that is the species that takes part in the hydrogenation of ethene. This hydrogen is held in the interstices of the Ni(100) planes, where it migrates from other more accessible sites.

A possible way to demonstrate the participation of chemisorbed hydrogen in hydrogenations is to use Pd or Pd alloy membranes. Atomic hydrogen may be supplied to a reaction on one side of the membrane by diffusing it through from the other side. A direct correlation was found between the rate of hydrogenation of olefins and the rate of diffusion of chemisorbed hydrogen [84-86]. The high selectivity of hydrogenation of cyclopentadiene on some palladium membrane catalysts was partially attributed to a low, uniform hydrogen concentration on entire catalyst [86,87] (see further discussion in Section V).

Infrared (IR) investigations led to recognition of the activity of surface hydrogens in the hydrogenation of benzene. It was observed that the IR band of preadsorbed hydrogen on platinum (2125 cm^{-1}) was extinguished when more ethene was added than preadsorbed hydrogen. In contrast, when the added ethene was less than the preadsorbed hydrogen, the band disappeared and then reformed [88]. When this hydrogen was removed by evacuation at room temperature, the irreversibly adsorbed hydrogen, which remained on the surface, did not react with benzene [89]. It was concluded that only reversibly and weakly adsorbed hydrogen is active in the hydrogenation of benzene on platinum. The same conclusion was reached by Davis and Somorjai in the transformations of cyclohexene on a stepped Pt (223) surface at 423 K [90]. They, as well as others [91-93], pointed out that at high reactant pressures (above 13.3 Pa) hydrogenation becomes the main reaction instead of benzene formation because of the high surface concentration of reactive, weakly adsorbed hydrogen.

The presence of different surface hydrogens was also demonstrated by Kochloefl et al. [94]. They measured the quantity of adsorbed hydrogen on supported nickel catalysts by both oxygen and olefin titrations. The former method always gave higher values. On the basis of this difference the authors concluded that there are different surface hydrogens, some of which cannot react with olefins because of their low energies or because of steric hindrance. Both the support and the method of preparation strongly affect the hydrogen adsorption capacity of the catalysts.

By hydrogenating ethene over Pt/SiO$_2$ catalysts the turnover number was observed to go through a maximum for particles between 0.6 and 1.1 nm. At lower particle sizes no hydrogenation occurs, which demonstrates a minimum particle-size requirement for olefin hydrogenation [95]. It was reasoned that the chemisorption of hydrogen is governed by this size, because corner positions are thought to be necessary to chemisorb hydrogen dissociatively [96].

On the basis of the suggestion of Siegel et al. [97], hydrogens bound to coordinatively highly unsaturated metal atoms (^{3}M sites) are supposed to take part in the hydrogenation of olefinic double bonds. By the single-turnover olefin hydrogenation process, Augustine and Warner were able to distinguish two ^{3}M sites on platinum catalysts [98]. Hydrogen adsorbs reversibly on one of the sites and can be removed after a short sweep-off time, while in the other ^{3}M site, irreversibly adsorbed hydrogen is swept off at a much slower rate. Both types of hydrogen have the same reactivity in 1-butene hydrogenation. At the same time, hydrogens attached to ^{2}M sites are active primarily in the isomerization reaction and can catalyze addition only under special circumstances. The transformations of (+)-apopinene, a special probe molecule, make it possible to determine certain surface features of palladium catalysts, and in this way to differentiate hydrogens attached to different types of surface metal atoms [99,100].

The activity of different types of hydrogen detected in temperature-programmed desorption (TPD) measurements was investigated in the hydrogenations of different hydrocarbons. A maximum rate of hydrogenation of benzene on an Fe catalyst was observed at 453 K and a TPD peak was also found at this temperature [101]. The authors supposed that type H(IV) (loosely adsorbed) species [102] might be involved in the surface reaction. A linear correlation was observed between the height of the low-temperature TPD peak and the benzene hydrogenation rate at 323 K on Pt/Al_2O_3 [103]. This hydrogen, which desorbs at about 293 to 303 K, is weakly bound hydrogen. Tsuchiya and co-workers made detailed studies in this field. They investigated the hydrogenation of different olefins (ethene [104], propene [105], 1-butene [106], and 2-butenes [107]) on platinum black and found that two of the types of chemisorbed hydrogens react with the olefins. These two were presumed to be present on the surface in the form of hydrogen atoms chemisorbed on top of platinum atoms (γ peak) and in the bridged form of molecular hydrogen (β peak). In the hydrogenation of benzene on different supported Re catalysts a correlation was found between the maximum of the low-temperature desorption peak and the catalytic activity [108]. On the basis of surface studies [109] hydrogen chemisorbed in the molecular form is supposed to be active in benzene hydrogenation on Re.

Other experimental techniques made it possible to detect different forms of hydrogen on metal surfaces. By means of adsorption and TPD measurements, as well as the static-capacitor method, three forms of hydrogen were identified on Pd, Pt, and Ni [110-114]: the atomic, electronegatively polarized β^- form; the atomic, electropositively polarized β^+ species (hydrogen adatoms); and the reversibly adsorbed, positively polarized, molecular α form. By monitoring the surface potential changes during the adsorption of hydrogen and ethene, Duś was able to study the reactivities of the different hydrogen species with respect to ethene at 195 and 298 K. On Ni and Pd films, the hydrogen adsorbed as the β^- form is active in hydrogenation [115,116]. In contrast, on Pt only the β^+ form of hydrogen is able to hydrogenate ethene admolecules [114]. At the same time, ethene molecules adsorbed in the second surface layer are also reactive toward gas-phase hydrogen.

Mechanistic studies of hydrogenations in the liquid phase also led to the recognition of different active forms of hydrogen [117]. The results of such studies reveal the most favorable experimental conditions (metal, solvent, pressure) in the formation of hydrogen species which are the most active in the hydrogenation of certain functional groups.

B. Hydrogen on Raney Nickel

A detailed review article published recently by Fouilloux [118] deals with the problems of different hydrogen species on Raney Ni. Here we discuss only the information concerning the activity of different types of hydrogen on Raney Ni in hydrogenations.

During its preparation Raney Ni is saturated with hydrogen, and it was observed long ago that there is a relationship between the activity of the catalyst and its hydrogen content. Kokes and Emmett studied the hydrogenation of ethene at 195 K [119] and found that the activity of Raney Ni decreased in proportion to the quantity of hydrogen evolved as a result of heat treatment. Electric or thermal removal of hydrogen originally held by a Raney Ni plate led to a decrease in activity for hydrogenation of acetone [120]. After the hydrogen removal, the catalyst was able to readsorb hydrogen, and the activities before and after readsorption were practically the same. The correlation between increasing temperatures of heat treatment versus decreasing activities was found to be valid in the hydrogenation of nitrobenzene [121,122] and eugenol [122]. Heat treatment at temperatures higher than 473 K resulted in catalysts with no activity [122].

After investigating the hydrogenation of cyclohexene and adiponitrile, Orchard et al. [123] concluded that the activity of a Raney Ni-Co catalyst is related to the hydrogen chemisorbed on the catalyst. The decreasing activity with increasing quantity of Co suggests that Co reduces the quantity of chemisorbed hydrogen which is active in hydrogenation.

TPD measurements on Raney Ni have identified the surface hydrogen associated with a low-temperature desorption peak to be involved primarily in hydrogenation of eugenol and nitrobenzene at room temperature and atmospheric pressure [124].

On the basis of neutron inelastic spectroscopy measurements in connection with the hydrogenation of benzene [125], it was concluded that strongly adsorbed hydrogen does not take part in the reaction, and it is only the loosely bound hydrogen that is active in hydrogenation [118]. A similar conclusion was reached by Brendel et al. [126] in the hydrogenation of acetone. By using thiourea for sulfur poisoning, the activity of Raney Ni ceases at a relatively low sulfur coverage (25%) in spite of the fact that a large quantity of hydrogen is present on the nickel. This reactive, weakly and reversibly bound hydrogen is considered to be a linear (or terminal) species bound to one superficial nickel atom [127].

C. Absorbed Hydrogen

A detailed discussion of the participation of the hydride phase of metals in hydrocarbon conversions is given in Chapter 14. Hydrogen occlusion and its role in the selectivity of the hydrogenation of 1,3-butadiene is discussed by Grant et al. [128] and Wells [129], assuming a cavity model. On a highly ordered zone of the metal surface, both hydrogen and diene can chemisorb competitively. The high surface concentration of the strongly adsorbed diene relative to hydrogen ensures the selective formation of olefin. The other, so-called cavitated zone contains a disordered arrangement of metal atoms and occludes hydrogen. At the junction of the two zones the hydrogen supply is sufficient to convert the diene directly to the saturated compound. The measured diminishing extent of hydrogen occlusion in the sequence Ir > Os > Ru > Rh > Pt > Co coincides with the increasing selectivity of metals in the hydrogenation of 1,3-butadiene [129].

V. MORPHOLOGICAL CHANGES AND HYDROGENATION

A. Dissolution of Metals

The solubility of platinum metals during hydrogenations has long been a problem (see, e.g., Ref. 130). A detailed study of the effect of reaction conditions on palladium dissolution during the hydrogenation of 2,4-dinitrotoluene on Pd/C was made by Bird and Thompson [131]. They found that palladium could be rendered insoluble if the metal could be maintained as the β-hydride phase during the catalytic reaction. The experimental conditions for this are sufficiently high partial pressure of hydrogen and kinetic control of the reaction; that is, the rate of hydrogen arrival at the catalyst surface is faster than the rate of hydrogen consumption by the surface reaction.

During the hydrogenation of vinylacetylene [132], Boitiaux et al. refer to the observation that an important part of the catalyst is lost, but they do not give any exact reference. Dissolution and fracturing of bulk palladium catalysts were observed in the hydrogenation of terminal acetylenes [133]. During hydrogenations of phenylacetylene and 1-octyne over amorphous Pd-Si and Pd-Ge alloys and over Pd foil the consumption of the first mole of hydrogen or deuterium resulted in the solution turning gray and turbid. This turbidity disappeared exactly at the complete consumption of the starting acetylene compounds, after which a black deposit appeared on the walls of the reaction flask. Using a rapidly cooled pure Pd catalyst, fracturing of the sample was observed. The authors suggest that the metal dissolved by complexation with terminal acetylenes. The remarkable effect of this is the increased activity of the catalyst for olefin hydrogenation. Moreover, the rapidly cooled catalysts show a larger rate increase for isomerization than for addition. The exposure of new, ordered surface is believed to occur as a result of dissolution and fracturing.

B. SMSI

The SMSI effect, that is, the sharp drop of CO and hydrogen chemisorption capacity of supported metals following high-temperature reduction, results in a decrease in activity for hydrogenation of organic compounds.

Most work deals with titania [134-137] and silica [138-142] as supports, but there are data for ceria [135], molybdena [143], and magnesia [144]. The metals used are platinum [134-136,139,144], rhodium [134,136], iridium [136], palladium [138], nickel [140-143], and copper [137]. Mainly, the hydrogenation of benzene was used as a test reaction for catalytic activity [134-136,138-144] but hydrogenation of ethene and styrene was also studied [136]. The largest decrease in activity for the hydrogenation of benzene was observed on titania and silica. Only a relatively small effect was exhibited by Pt/CeO_2 [135] and Pt/MgO [144].

There is only one observation of SMSI in a non-group VIII metal. This was on copper [137]. After reduction at 773 K, Cu/TiO_2 showed decreased activity for hydrogenation of 2-methylbutanal, while the activity of Cu/SiO_2 was unaffected by the same reduction conditions.

Unlike platinum metals and Ni on silica, Ni on molybdena, prepared from Ni molybdate, does not exhibit a correlation between the decreasing activity with increasing reduction temperature and the resulting increasing hydrogen adsorption capacity [143]. Molybdenum formed during high-temperature reduction covers the Ni surface, and although it chemisorbs hydrogen, it is not active for the hydrogenation of benzene.

C. Miscellaneous Phenomena

Palczewska et al. studied the effect of the palladium β-hydride phase on the activity of acetylene hydrogenation [145]. The β-hydride phase was produced and decomposed in Pd films in seven consecutive cycles. As a result of this treatment the films were totally reconstructed to emphasize the Pd(111) plane. The rearranged catalysts displayed an enhanced catalytic activity; a tenfold increase in acetylene conversion was observed. A similar treatment of reduced PdO_2 resulted in increased isomerization activity relative to addition activity in the hydrogenation of (+)-apopinene [100].

A platinum-on-silica catalyst reduced above 573 K lost its ethene hydrogenation activity but retained hydrogen chemisorption capacity [146]. The dispersion of the metal into clusters below the critical size for ethene chemisorption explains the phenomenon.

An increased catalytic activity of 0.5% Pd/Al$_2$O$_3$ in the hydrogenation of a C$_2$-C$_4$ olefin-diene mixture at 353 K was observed [147]. This was a result of redispersion which occurred during high-temperature hydrogen treatment and reduction and during the reaction.

Differences in turnover frequencies in hydrogenations of propene on Pt/SiO$_2$ catalysts with different pretreatments were attributed either to inhibition by strongly bound hydrogen or to formation of less active surface morphology (surface reconstruction) [148].

VI. CONCLUDING REMARKS

Usually, dihydrogen is essential for hydrogenation, that is, the conversion of π bonds to σ bonds by the addition of two hydrogens. Exceptions to this do exist: for example, metal-catalyzed disproportionation (hydrogen transfer) reactions, such as the conversion of cyclohexene to benzene and cyclohexane. But generally the source of hydrogen is dihydrogen, and it is dissociated by surface sites on the catalysts. On metals it is believed to be dissociated into hydrogen atoms and on metal oxides, and possibly on metal sulfides, into ions. We have seen that the effects of these forms of hydrogen on catalytic surfaces can be either chemical or physical. Sometimes one effect is misinterpreted as the other and leads to strange conclusions. Sometimes it is not clear which is occurring, so it is best to refer to it simply as a "hydrogen effect."

Probably the most insidious hydrogen effects are those involving the movement of hydrogen to, from, and over the catalytic surface. These may be physical effects in the case of rate-limiting pore diffusion during liquid-phase hydrogenations, or chemical effects in the case of migration of hydrogen atoms across a catalytic surface or from one surface to another (e.g., spillover).

The pore diffusion, or intraparticle migration, problem has been examined by several workers [149-152], and its characteristics are generally understood, although not always easily identifiable. The surface migration problem is neither well understood nor easily identifiable. For example, are hydrogen atoms able to migrate across all sites with approximately equal ability? Probably not, as evidenced by hydrogen spillover requirements. It does seem probable, though, that hydrogen atoms may migrate across metal surfaces. Yet, evidence exists that such surface migration is not easy [104-107, 153,154] and that hydrogen adsorbed on one type of site (e.g., terraces) does not readily migrate to other sites (e.g., vertices) [98]. Therefore, until we fully understand surface migration of hydrogen during hydrogenation we will not be able to describe the complete mechanism.

REFERENCES

1. J. Horiuti and M. Polanyi, *Trans. Faraday Soc.*, 30:663, 1164 (1934).
2. S. Siegel and G. V. Smith, *J. Am. Chem. Soc.*, 82:6082 (1960).
3. S. Siegel, P. A. Thomas, and J. T. Holt, *J. Catal.*, 4:73 (1965).
4. S. Siegel, M. Dunkel, G. V. Smith, W. Halpern, and J. Cozort, *J. Org. Chem.*, 31:2802 (1966).
5. S. Siegel and B. Dmuchovsky, *J. Am. Chem. Soc.*, 84:3132 (1962).
6. S. Siegel, M. Foreman, and D. Johnson, *J. Org. Chem.*, 40:3589 (1975).
7. R. L. Augustine, J. F. Van Peppen, and F. Yaghmai, *Prep. Can. Symp. Catal.*, 6:256 (1979).
8. S. Siegel and B. Dmuchovsky, *J. Am. Chem. Soc.*, 86:2192 (1964).

9. Y. Takagi and S. Teratani, *J. Catal.*, *34*:490 (1974).
10. W. Cocker, P. V. R. Shannon, and P. A. Staniland, *J. Chem. Soc.*, C 41 (1966).
11. S. Siegel, G. V. Smith, B. Dmuchovsky, D. Dubbell, and W. Halpern, *J. Am. Chem. Soc.*, *84*:3136 (1962).
12. S. Siegel, *Adv. Catal.*, *16*:124 (1966).
13. S. Siegel and N. Garti, in *Catalysis in Organic Syntheses* (G. V. Smith, ed.), Academic Press, New York (1977), p. 9
14. S. Siegel, J. Outlaw, Jr., and N. Garti, *J. Catal.*, *58*:370 (1979).
15. F. Zymalkowski and G. Strippel, *Arch. Pharm.*, *297*:727 (1964).
16. F. Zymalkowski and G. Strippel, *Arch. Pharm.*, *298*:604 (1965).
17. S. Mitsui, H. Saito, S. Sekiguchi, Y. Kumagai, and Y. Senda, *Tetrahedron*, *28*:4751 (1972).
18. S. Mitsui, H. Saito, Y. Yamashita, M. Kaminga, and Y. Senda, *Tetrahedron*, *29*:1531 (1973).
19. P. Geneste, M. Bonnet, and M. Rodriguez, *J. Catal.*, *57*:147 (1979).
20. J. Solodar, *J. Org. Chem.*, *41*:3461 (1976).
21. G. V. Smith, *J. Catal.*, *5*:152 (1966).
22. I. Heertje, G. K. Koch, and W. J. Wösten, *J. Catal.*, *32*:337 (1974).
23. K. Sporka, J. Hanika, V. Ruzicka, and B. Vostry, *Collect. Czech. Chem. Commun.*, *37*:52 (1972).
24. C. A. Henrick, *Tetrahedron*, *33*:1845 (1977).
25. C. A. Brown, *J. Am. Chem. Soc.*, *91*:5901 (1969).
26. G. Bernáth, Gy. Göndös, and L. Gera, *Acta Phys. Chem. (Szeged)*, *20*:139 (1974).
27. R. J. Wicker, *J. Chem. Soc.*, 2165 (1956).
28. P. N. Rylander and D. R. Steele, *Engelhard Ind. Tech. Bull.*, *3*:91 (1962).
29. S. Mitsui, K. Gohke, H. Saito, A. Nanbu, and Y. Senda, *Tetrahedron*, *29*:1523 (1973).
30. S. Mitsui, M. Shionoya, K. Gohke, F. Watanabe, S. Imaizumi, and Y. Senda, *J. Catal.*, *40*:372 (1975).
31. H. O. House, R. G. Carlson, H. Muller, A. W. Noltes, and C. D. Slater, *J. Am. Chem. Soc.*, *84*:2614 (1962).
32. R. L. Augustine, *J. Org. Chem.*, *28*:152 (1963).
33. R. L. Augustine, *Adv. Catal.*, *25*:56 (1976).
34. I. Jardine, R. W. Howsam, and F. J. McQuillin, *J. Chem. Soc.*, C 260 (1969).
35. C. H. DePuy and P. R. Story, *J. Am. Chem. Soc.*, *82*:627 (1960).
36. B. Franzus, W. C. Baird, E. I. Snyder, and J. H. Surridge, *J. Org. Chem.*, *32*:2845 (1967).
37. W. C. Baird, B. Franzus, and J. H. Surridge, *J. Org. Chem.*, *34*:2944 (1969).
38. E. Breitner, E. Roginski, and P. N. Rylander, *J. Org. Chem.*, *24*:1855 (1959).
39. R. G. Olivier and P. B. Wells, *J. Catal.*, *47*:364 (1977).
40. G. C. Bond and P. B. Wells, *J. Catal.*, *5*:65 (1966).
41. J. Margitfalvi, L. Guczi, and A. H. Weiss, *J. Catal.*, *72*:185 (1981).
42. J. M. Moses, A. H. Weiss, K. Matusek, and L. Guczi, *J. Catal.*, *86*:417 (1984).
43. L. L. Kesmodel, L. H. Dubois, and G. A. Somorjai, *J. Chem. Phys.*, *70*:2180 (1979).
44. H. Lindlar, *Helv. Chim. Acta*, *35*:446 (1952).
45. W. Pałczewska, I. Szymerska, and M. Krawczyk, *Proceedings of the 5th International Symposium, Varna*, Part I (1983), p. 357.

46. W. Pałczewska, A. Jablonski, Z. Kaszkur, G. Zuba, and J. Wernisch, J. Mol. Catal., 25:307 (1984).
47. W. Pałczewska, I. Ratajczykowa, I. Szymerska, and M. Krawczyk, Proceedings of the 8th International Congress on Catalysis, West Berlin, Vol. 4 (1984), p. 173.
48. I. Ratajczykowa and I. Szymerska, Proceedings of the 5th International Symposium, Varna, Part II (1983), p. 413.
49. W. T. McGown, C. Kemball, and D. A. Whan, J. Catal., 51:173 (1978).
50. A. Sárkány, L. Guczi, and A. H. Weiss, Appl. Catal., 10:369 (1984).
51. S. LeViness, V. Nair, A. H. Weiss, Z. Schay, and L. Guczi, J. Mol. Catal., 25:131 (1984).
52. P. A. Sermon and G. C. Bond, Catal. Rev.-Sci. Eng., 8:211 (1973).
53. P. Antonucci, N. van Truong, N. Giordano, and R. Maggiore, J. Catal., 75:140 (1982).
54. G. E. E. Gardes, G. M. Pajonk, and S. J. Teichner, J. Catal., 33:145 (1974).
55. D. Maret, G. M. Pajonk, and S. J. Teichner, Stud. Surf. Sci. Catal., 17:215 (1983).
56. D. Bianchi, G. E. E. Gardes, G. M. Pajonk, and S. J. Teichner, J. Catal., 38:135 (1975).
57. S. J. Teichner, A. R. Mazabrard, G. Pajonk, G. E. E. Gardes, and C. Hoang-Van, J. Colloid. Interface Sci., 58:88 (1977).
58. M. Lacroix, G. M. Pajonk, and S. J. Teichner, Bull. Soc. Chim. Fr., I-101 (1981).
59. M. Lacroix, G. M. Pajonk, and S. J. Teichner, Bull. Soc. Chim. Fr., I-87 (1981).
60. D. Bianchi, M. Lacroix, G. Pajonk, and S. J. Teichner, J. Catal., 59:467 (1979).
61. M. Lacroix, G. Pajonk, and S. J. Teichner, Proceedings of the 7th International Congress on Catalysis, Tokyo (1980), p. 279.
62. M. Lacroix, G. Pajonk, and S. J. Teichner, React. Kinet. Catal. Lett., 12:369 (1979).
63. M. Lacroix, G. M. Pajonk, and S. J. Teichner, Bull. Soc. Chim. Fr., I-265 (1981).
64. M. Lacroix, G. M. Pajonk, and S. J. Teichner, Bull. Soc. Chim. Fr., I-258 (1981).
65. M. S. W. Lau and P. A. Sermon, J. Chem. Soc. Chem. Commun., 891 (1978).
66. H. Saltsburg and M. E. Mullins, Stud. Surf. Sci. Catal., 17:295 (1983)
67. J. P. Nogier, J. L. Bonardet, and J. P. Fraissard, Stud. Surf. Sci. Catal., 17:233 (1983).
68. G. C. Bond and P. A. Sermon, React. Kinet. Catal. Lett., 1:3 (1974).
69. P. A. Sermon and G. C. Bond. J. Chem. Soc. Faraday Trans., 72:745 (1976).
70. R. L. Augustine and R. W. Warner, J. Org. Chem., 46:2614 (1981).
71. R. L. Augustine, K. P. Kelly, and R. W. Warner, J. Chem. Soc. Faraday Trans. 1, 79:2639 (1983).
72. M. A. Vannice and P. Chou, Proceedings of the 8th International Congress on Catalysis, West Berlin, Vol. 5 (1984), p. 99.
73. K. Nakano and K. Kusunoki, Proceedings of the 3rd Pacific Chemical Engineering Congress, Vol. 2 (1983), p. 262.
74. H. Noller and W. M. Lin, J. Catal., 85:25 (1984).
75. V. N. Romannikov, K. G. Ione, and L. A. Pedersen, J. Catal., 66:121 (1980).
76. J. Bandiera, C. Naccache, and B. Imelik, J. Chim. Phys., 75:406 (1978).
77. G. del Angel, B. Coq, R. Dutarte, F. Fajula, F. Figueras, and C. Leclerq, Stud. Surf. Sci. Catal., 17:301 (1983).

78. M. Primet, E. Garbowski, M. V. Mathieu, and B. Imelik, *J. Chem. Soc. Faraday Trans. 1*, 76:1953 (1980).
79. A. H. Weiss, S. LeViness, V. Nair, L. Guczi, A. Sárkány, and Z. Schay, *Proceedings of the 8th International Congress on Catalysis, West Berlin*, Vol. 5 (1984), p. 591.
80. R. Lamartine and R. Perrin, *Stud. Surf. Sci. Catal.*, 17:251 (1983).
81. W. M. H. Sachtler and L. J. Bostelaar, *Stud. Surf. Sci. Catal.*, 17:207 (1983).
82. A. Hoek and W. M. H. Sachtler, *J. Catal.*, 58:276 (1979).
83. J. T. Richardson and H. Friedrich, *J. Catal.*, 37:8 (1975).
84. R. S. Yolles, B. J. Wood, and H. Wise, *J. Catal.*, 21:66 (1971).
85. B. J. Wood and H. Wise, *J. Catal.*, 5:135 (1966).
86. V. M. Gryaznov, V. S. Smirnov, and M. G. Slin'ko, *Proceedings of the 7th International Congress on Catalysis, Tokyo* (1980), p. 224.
87. V. M. Gryaznov, V. S. Smirnov, and M. G. Slin'ko, *Proceedings of the 6th International Congress on Catalysis, London* (1976), p. 894.
88. Y. Soma, *J. Catal.*, 59:239 (1979).
89. J. M. Basset, G. Dalmai-Imelik, M. Primet, and R. Mutin, *J. Catal.*, 37:22 (1975).
90. S. M. Davis and G. A. Somorjai, *J. Catal.*, 65:78 (1980).
91. M. Primet, J. M. Basset, M. V. Mathieu, and M. Prettre, *J. Catal.*, 28:368 (1973).
92. J. D. Clewley, J. F. Lynch, T. B. Flanagan, *J. Catal.*, 36:291 (1975).
93. L. T. Dixon, R. Barth, R. J. Kokes, and J. W. Gryder, *J. Catal.*, 37:368 (1975).
94. K. Kochloefl, A. Nuemeier, and O. Bock, *Proceedings of the 7th International Congress on Catalysis, Tokyo* (1980), p. 1476.
95. A. Masson, B. Bellamy, G. Colomer, M. M'Bedi, P. Rabette, and M. Che, *Proceedings of the 8th International Congress on Catalysis, West Berlin*, Vol. 4 (1984), p. 333.
96. K. Christman and G. Ertl, *Surf. Sci.*, 60:365 (1976).
97. S. Siegel, J. Outlaw, Jr., and N. Garti, *J. Catal.*, 52:102 (1978).
98. R. L. Augustine and R. W. Warner, *J. Catal.*, 80:358 (1983).
99. G. V. Smith, O. Zahraa, A. Molnár, M. M. Khan, B. Rihter, and W. E. Brower, *J. Catal.*, 83:238 (1983).
100. G. V. Smith, A. Molnár, M. M. Khan, D. Ostgard, and N. Yoshida, *J. Catal.*, 98:502 (1986).
101. R. Badilla-Ohlbaum, H. J. Neuberg, W. F. Graydon, and M. J. Phillips, *J. Catal.*, 47:273 (1977).
102. Y. Amenomiya and G. Pleizier, *J. Catal.*, 28:442 (1973).
103. P. C. Aben, H. van der Eijk, and J. M. Oelderik, *Proceedings of the 5th International Congress on Catalysis, Miami* (1972), p. 717.
104. S. Tsuchiya and M. Nakamura, *J. Catal.*, 50:1 (1977).
105. S. Tsuchiya, M. Nakamura, and N. Yoshioka, *Bull. Chem. Soc. Jpn.*, 51:981 (1978).
106. S. Tsuchiya and N. Yoshioka, *J. Catal.*, 87:144 (1984).
107. S. Tokitaka, H. Imamura, and S. Tsuchiya, *Bull. Chem. Soc. Jpn.*, 58:82 (1985).
108. Kh. M. Minachev, V. I. Avaev, R. V. Dmitriev, and M. A. Ryashentseva, *Izv. Akad. Nauk SSSR Ser. Khim.*, 1456 (1984).
109. R. Ducros, J. J. Ehrhardt, M. Alnot, and A. Cassuto, *Surf. Sci.*, 55:509 (1976).
110. D. D. Eley and E. J. Pearson, *J. Chem. Soc. Faraday Trans. 1*, 74:223 (1978).
111. R. Duś, *Surf. Sci.*, 42:324 (1974).

112. R. Duś and F. C. Tompkins, *J. Chem. Soc. Faraday Trans. 1*, 71:930 (1975).
113. R. Duś, *J. Catal.*, 42:334 (1976).
114. R. Duś and W. Lisowski, *Surf. Sci.*, 85:183 (1979).
115. R. Duś, *Surf. Sci.*, 50:241 (1975).
116. R. Duś, W. Lisowski, and A. Leszczynski, *Rocz. Chem.*, 50:553 (1976).
117. D. V. Sokolskii, *Kinet. Katal.*, 18:1223 (1977).
118. P. Fouilloux, *Appl. Catal.*, 8:1 (1983).
119. R. J. Kokes and P. H. Emmett, *J. Am. Chem. Soc.*, 82:4497 (1960).
120. I. Nakabayashi, T. Hisano, and T. Terazawa, *J. Catal.*, 58:74 (1979).
121. N. I. Popov, D. V. Sokolskii, I. S. Svets, L. A. Kolomytsev, and S. I. Kan, *Zh. Fiz. Khim.*, 47:1725 (1973).
122. A. Tungler, J. Petró, T. Máthé, J. Heiszman, F. Buella, and Z. Csűrös, *Acta Chim. Acad. Sci. Hung.*, 89:151 (1976).
123. J. P. Orchard, A. D. Tomsett, M. S. Wainwright, and D. J. Young, *J. Catal.*, 84:189 (1983).
124. J. Heiszman, J. Petró, A. Tungler, T. Máthé, and Z. Csűrös, *Acta Chim. Acad. Sci. Hung.*, 86:117 (1975).
125. A. J. Renouprez, G. Clubnet, and H. Jobic, *J. Catal.*, 74:296 (1982).
126. A. Brendel, P. Fouilloux, G. A. Martin, and P. Bussiere, *J. Chim. Phys.*, 72:665 (1975).
127. A. J. Renouprez, P. Fouilloux, G. Coudurier, D. Tochetti, and R. Stockmeyer, *J. Chem. Soc. Faraday Trans. 1*, 73:1 (1977).
128. J. Grant, R. B. Moyes, and P. B. Wells, *J. Catal.*, 51:355 (1978).
129. P. B. Wells, *J. Catal.*, 52:498 (1978).
130. J. Llopis, *Catal. Rev.-Sci. Eng.*, 2:161 (1968).
131. A. J. Bird and D. T. Thompson, in *Catalysis in Organic Syntheses* (W. H. Jones, ed.), Academic Press, New York (1980), p. 61.
132. J. P. Boitiaux, J. Cosyns, and G. Martino, *Stud. Surf. Sci. Catal.*, 11:355 (1982).
133. Á. Molnár, G. V. Smith, and M. Bartók, *J. Catal.*, 101:67 (1986).
134. P. Meriaudeau, H. Ellestad, and C. Naccache, *Proceedings of the 7th International Congress on Catalysis, Tokyo* (1980), p. 1464.
135. P. Meriaudeau, J. F. Dutel, M. Dufaux, and C. Naccache, *Stud. Surf. Sci. Catal.*, 11:95 (1982).
136. P. Meriaudeau, O. H. Ellestad, M. Dufaux, and C. Naccache, *J. Catal.*, 75:243 (1982).
137. F. S. Delk II and A. Vavere, *J. Catal.*, 85:380 (1984).
138. R. L. Moss, D. Pope, B. J. Davis, and D. H. Edwards, *J. Catal.*, 58:206 (1979).
139. G. A. Martin, R. Dutartre, and J. A. Dalmon, *React. Kinet. Catal. Lett.*, 16:329 (1981).
140. G. A. Martin and J. A. Dalmon, *React. Kinet. Catal. Lett.*, 16:325 (1981).
141. H. Praliaud and G. A. Martin, *J. Catal.*, 72:934 (1981).
142. G. A. Martin and J. A. Dalmon, *J. Catal.*, 75:233 (1982).
143. M. P. Astier, M. L. Lacroix, and S. J. Teichner, *J. Catal.*, 91:356 (1985).
144. J. Goldwasser, C. Bolivar, C. R. Ruiz, B. Arenas, S. Wanke, H. Royo, R. Barrios, and J. Giron, *Proceedings of the 8th International Congress on Catalysis, West Berlin*, Vol. 5 (1984), p. 195.
145. A. Janko, W. Palczewska, and I. Szymerska, *J. Catal.*, 61:264 (1980).
146. L. Gonzales-Tejuca, K. Alka, S. Namba, and J. Turkevich, *J. Phys. Chem.*, 81:1399 (1977).
147. O. Leonte, M. Birjega, N. Popescu-Pogrion, C. Sárbu, D. Macovei, P. Pausescu, and M. Georgescu, *Proceedings of the 8th International Congress on Catalysis, West Berlin*, Vol. 2 (1984), p. 683.

148. E. Rorris, J. B. Butt, R. L. Burwell, Jr., and J. B. Cohen, *Proceedings of the 8th International Congress on Catalysis, West Berlin*, Vol. 4 (1984), p. 321.
149. R. Ciola and R. L. Burwell, Jr., *J. Phys. Chem.*, *65*:1158 (1961).
150. J. Newham and R. L. Burwell, Jr., *J. Phys. Chem.*, *66*:1431 (1962).
151. H. H. Kung, R. J. Pellet, and R. L. Burwell, Jr., *J. Am. Chem. Soc.*, *98*:5603 (1976).
152. G. W. Roberts, in *Catalysis in Organic Syntheses* (P. N. Rylander and H. Greenfield, eds.), Academic Press, New York (1976), p. 1.
153. S. Tsuchiya, Y. Amenomiya, and K. J. Cvetanovic, *J. Catal.*, *19*:245 (1970).
154. H. Verbeek and W. M. H. Sachtler, *J. Catal.*, *42*:257 (1976).

19 | Hydrogen Effects in Catalytic Transformations of Oxygenated Carbon Compounds on Metals

MIHÁLY BARTÓK

József Attila University, Szeged, Hungary

I.	INTRODUCTION	521
II.	ALCOHOLS	522
	A. Isotope Exchange	522
	B. Epimerization	522
	C. Dehydrogenation	523
	D. Dehydration	525
	E. Hydrogenolysis	525
	F. Amination	526
III.	DIOLS	527
IV.	UNSATURATED ALCOHOLS	527
V.	AMINO ALCOHOLS	529
VI.	OXO COMPOUNDS	529
VII.	CARBOXYLIC ACIDS AND THEIR DERIVATIVES	530
VIII.	OXYGEN-CONTAINING HETEROCYCLES	530
	A. Oxiranes and Oxetanes	530
	B. Oxolanes and Oxanes	537
	C. 1,3-Dioxacycloalkanes	538
IX.	CONCLUDING REMARKS	539
	REFERENCES	540

I. INTRODUCTION

Experimental observations indicative of the special role of hydrogen in heterogeneous metal catalysis (apart from its participation in hydrogenation reactions) were made as early as some 60 years ago [1], even for compounds containing C-O bonds [2], and systematic work was begun to investigate a whole series of metal-catalyzed organic chemical reactions [3-9]. It should be noted that study of the role of hydrogen first extended to the isomerization of hydrocarbons of various types [4,8]. In this chapter we survey the literature relating to the effects of hydrogen in certain transformations on metals of organic compounds containing C-O bonds.

II. ALCOHOLS

A. Isotope Exchange

The fundamental studies by Patterson and Burwell [10,11] are well known; we have already reviewed these briefly [12]. The exchange reaction of methanol was investigated with deuterium on Raney Ni at 308 K [13]. This exchange reaction proceeded with low activation energy (14.63 kJ mol^{-1}). The rate of the reaction decreased linearly with decrease in the initial deuterium pressure. Under the conditions applied, only the hydrogen of the OH group underwent exchange.

B. Epimerization

Stoichiometrically, there is no need for hydrogen for the epimerization of alcohols, but on metal catalysts its participation can justifiably be presumed. The epimerization borneol $\rightleftharpoons$ isoborneol on Ni and Cu catalysts was described as early as 1941. The reaction route depicted in Scheme 1 was assumed in order to interpret the process [14,15].

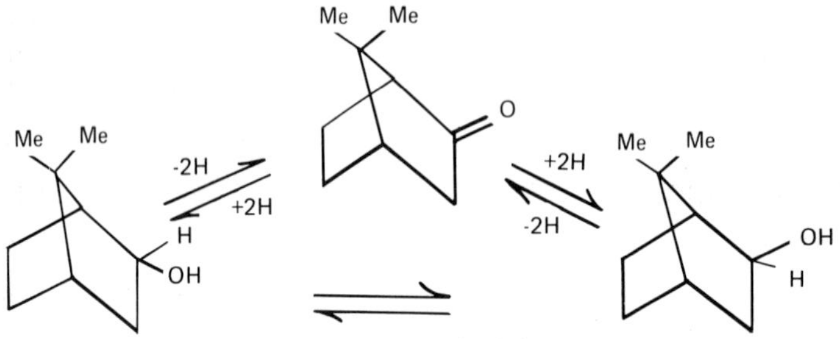

Scheme 1

In our view the lower two reactions in this scheme occur by an S_N2 mechanism, with the participation of chemisorbed hydrogen (associative mechanism). Under hydrogen pressure on a reduced Ni catalyst, the epimerization takes place in both directions [16]. The substituted cycloalkanols have otherwise proved to be good models for study of the configuration of the OH group.

As an example, the epimerization of 4-*t*-butylcyclohexanol has been studied on a Raney Ni catalyst in the presence of hydrogen [17,18]. Neither the mechanism of the reaction nor the role of hydrogen were dealt with. We consider that the epimerization takes place via a *t*-butylcyclohexanone intermediate (i.e., the first dehydrogenation step is followed by hydrogenation). The hydrogen present therefore exerts a significant effect on both reactions, despite the fact that hydrogen may be regarded as an astoichiometric component as far as the overall reaction is concerned. In 1956, Wicker [19] put forward the general rule that "substituted cyclohexanols in general can be isomerized by heating them with hydrogenation catalysts such as nickel or platinum, in an atmosphere of hydrogen." He therefore drew attention to the role of hydrogen. The isomerization was interpreted in terms of the following reaction steps:

$$\text{trans-Alcohol} \quad \xrightleftharpoons[]{-2H} \quad \rightleftharpoons \quad \rightleftharpoons \quad \xrightleftharpoons[]{+2H} \quad \text{cis-Alcohol} \tag{1}$$

Hydrogen uptake occurs via a "half-hydrogenated state."

The hydrogenation and the addition of deuterium to 2-methylcyclopentanone and 2-methylcyclohexanone were studied on Pt, Ni and Rh catalysts, as was the isomerization of the corresponding cycloalkanols [20], which was considered not to involve a cycloalkanone intermediate. [The possibility of direct isomerization was first suggested by Burwell [21] in connection with the racemization of (R)-2-butanol on Cu.]

The isomerization of 2-methylcycloalkanols took place only in hydrogen (deuterium), demonstrating its hydroisomerization nature [20] (hydroisomerization: isomerization that does not occur in the absence of hydrogen or its isotopes). The presumed surface species [1] in the case of 2-methylcyclopentanol is shown as follows:

$$\tag{2}$$

The epimerization of 2-, 3-, and 4-alkylcyclohexanols in the presence of hydrogen also takes place in the kinetic regime on Ru/Al_2O_3 under selected experimental conditions [22]. The reaction rate increases in proportion not only to the amount of catalyst, but also to hydrogen pressure up to 10^6 Pa. The reaction rate is no longer dependent on the hydrogen pressure at higher values than this. Configurational isomerization cannot be observed in an inert atmosphere. Epimerization also occurs in the case of ethers [23], which proves that the presence of the OH group is not vital; thus, besides proceeding via oxo compounds, epimerization may be a consequence of a direct configurational change on the C atom.

The 1,4-dimethoxycyclohexanes undergo epimerization on Pt/C only in the presence of hydrogen [23]. As far as the participation of hydrogen in the reaction is concerned, this differs from that observed for the dialkylcyclohexanes. We hold the view that for the 1,4-dimethoxycyclohexanes the reaction mechanism is a dissociative one.

The addition of hydrogen to the enol ethers has been studied on platinum metals [24, 25]. The hydrogenation was accompanied by hydrogenolysis involving cleavage of the C-O bond. The role of ionically adsorbed hydrogen in the process is stressed. As the temperature is raised, the quantity of ionically adsorbed hydrogen on the surface of the catalyst decreases, and the extent of hydrogenolysis decreases proportionally [24]. We consider that the hydrogen plays a role not only in addition via half-hydrogenated states, but also in the accompanying epimerization.

C. Dehydrogenation

The first experimental observation relating to the effect of hydrogen in the dehydrogenation of alcohols probably dates from 1911, when a study was made of the dehydrogenation of octanol-1 in a stream of hydrogen [2]. In the dehydrogenation of butanol-1 on a Cu catalyst, in the absence of hydrogen the catalyst was deactivated [26]. At not too high temperatures, the hydrogen took part as a reactant in the process, for it promoted the transformation of the aldehyde formed in the dehydrogenation, into the corresponding ester. The results supported the earlier finding that in certain reactions hydrogen exhibits a favorable effect, either by inhibiting side reactions which otherwise cause poisoning of the catalyst, or by removing reaction products from the surface that are otherwise desorbed only with difficulty [1].

In the absence of adsorbed hydrogen, on Pt/C cyclohexanol is converted to cyclohexanone and phenol in two parallel processes [27]. In the presence of adsorbed hydrogen, the products are cyclohexane, benzene, and cyclohexanediol.

For study of the hydrogen effect, the cyclohexanol ⇌ cyclohexanone system on various metals in the presence of hydrogen and nitrogen proved to be a good model [8, 28-31] (Scheme 2), as transformations requiring hydrogen and

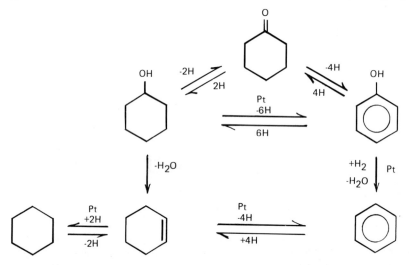

Scheme 2

transformations producing hydrogen prevail on the different metals. The hydrogen effect is quite clearly demonstrated by the experimental observation that in a nitrogen atmosphere (i.e., in the absence of hydrogen), not only do the activities of the catalysts decrease, but there is also a change in the selectivity of the transformation [8, 30].

The reaction route outlined in Scheme 2 was based on the results of the investigations in hydrogen and nitrogen atmospheres and radiotracer studies [30].

The product composition depends on the amount of hydrogen present and on the ability of the catalyst to utilize the available hydrogen for one or another reaction type [28]. The activities of the various group VIII metals were compared in the dehydrogenation of cyclohexanol. These metals may be classified into three classes. Class 1 comprises Os, Co, Ru, Re, and Fe. Here the main process in the absence of hydrogen is dehydrogenation to cyclohexanone. The extent of participation of hydrogen in the dehydrogenation reaction depends on the metal. Its main effect is to enhance the fragmentation reactions. In class 2, consisting of Ni, Pd, and Pt, the main reaction is aromatization (formation of both phenol and benzene). Hydrogen depresses the conversion, mainly through decrease of the yield of aromatics. Class 3 metals (Rh and Ir) display the most striking hydrogen effects. Hydrogen greatly increases the activity, leading to more fragmentation and more aromatization. Both phenol and benzene may be observed as aromatic products; their ratio depends on the ability of the catalyst to bring about hydrogenolytic cleavage of the phenolic OH group.

D. Dehydration

When the alcohols are dehydrogenated on Pd film [32], on Ni [33-43], on Pt [44,45], on Cu [44], and on Rh and Ir [45] in the presence of hydrogen, ethers are formed, too. The rate of the ether formation reaction is much lower in the absence of hydrogen [35,37,44]. As an example, 2-propanol reacted on Rh with a conversion of 38% in hydrogen and of 13% in helium [45]. Hydrogen favored ether formation, with a selectivity of 81%, compared with 62% in helium; at the same time, a propane selectivity of 29% in hydrogen corresponded to one of 38% in helium. Hydrogen retained by the catalyst was more active in the hydrogenation of propene than was the abundant gaseous hydrogen. Higher temperatures (less surface hydrogen present) similarly favored propane formation.

The catalyst support does not catalyze ether formation [44]. Hydrogen chemisorbed on the metals plays a dual role: It participates in the development of the active sites, and it allows the mechanism of ether formation, as in

$$CH_3-CH_2-CH_2-CHO + C_4H_9OH \longrightarrow C_3H_7-\overset{OH}{\underset{H}{C}}-O-C_4H_9 \xrightarrow{2H} C_4H_9OC_4 + H_2O \quad (3)$$

Aldehyde condensation products are formed during the transformation; in the absence of hydrogen, these deactivate the catalyst [37].

E. Hydrogenolysis

The hydrogenolysis of enantiomeric benzyl alcohols and their methyl ethers has been studied on Raney Ni and on Pd containing various amounts of hydrogen [Pd(H) and Pd($H_{0.4}$)] in hydrogen and helium atmospheres [46]. On all three catalysts, the same stereoselectivity was observed in hydrogen and in helium. On Ni the hydrogenolysis proceeded with retention of the configuration, whereas on the Pd catalysts, inversion occurred:

$$\underset{R^2}{\overset{R^1}{\underset{|}{Ph-C-H}}} \xleftarrow[-HOX]{2H, Ni} \underset{R^2}{\overset{R^1}{\underset{|}{Ph-C-OX}}} \xrightarrow[-HOX]{2H, Pd} \underset{R^2}{\overset{R^1}{\underset{|}{H-C-Ph}}} \quad (4)$$

X = H

These measurements indicated [46] that the Khan mechanism [47] does not hold, as hydrogenolysis is also observed in helium on the two Pd catalysts. Because of the rapid hydrogen diffusion, the hydrogen on the Pd attacks from the surface of the catalyst. In the mechanisms given by Mitsui et al. [46], the substrate is bound to the surface via its oxygen atom on Ni, and via its carbon atom on Pd. The latter is subject to question, and further experimental evidence is required.

F. Amination

The amination of alcohols on Cu catalysts is an important industrial process [48]. Cu is relatively quickly deactivated as a consequence of surface nitride formation. In the presence of hydrogen, however, less deactivation occurs (e.g., during 1000 h the activity decreases by only 2%) [49]. Provision of the optimum hydrogen partial pressure proved to be a factor enhancing the selectivity in the preparation of ethylenediamine from monoethanolamine on a Ni catalyst [50].

In the amination of ethylene glycol with dimethylamine on Cu [51]:

$$\begin{array}{c} CH_2-CH_2 \\ | \quad\quad | \\ OH \quad OH \end{array} + HNMe_2 \xrightarrow[-H_2O]{H_2, Cu/Al_2O_3} \begin{array}{c} CH_2-CH_2 \\ | \quad\quad | \\ OH \quad NMe_2 \end{array} + \begin{array}{c} CH_2-CH_2 \\ | \quad\quad | \\ NMe_2 \quad NMe_2 \end{array} \quad (5)$$

$$[2] \quad\quad\quad\quad [3]$$

the selectivity of formation of the main product [2] attains 70%. The conversion proved to be apparently independent of the partial pressure of hydrogen. However, the selectivity of formation of [3] could be raised to 55% by performing the reaction in nitrogen instead of in hydrogen.

Cyclohexanol, too, undergoes amination in the presence of ammonia and hydrogen, yielding aniline [27]. The mechanism of the amination depends on the partial pressure of hydrogen. The sequence of reaction steps at low hydrogen concentration is as follows:

$$\text{cyclohexanol} \rightarrow \text{cyclohexanone} \rightarrow \text{cyclohexylamine} \rightarrow \text{aniline} \quad (6)$$

At higher hydrogen concentration, the process takes place as follows:

$$\text{cyclohexanol} \rightarrow \text{cyclohexylamine} \rightarrow \text{aniline} \quad (7)$$

The latter is more favorable from the aspect of aniline production. In this case the reaction mechanism of the process cyclohexanol → aniline is as follows [27]:

(8)

The recognition and investigation of the hydrogen effect led to the development of an industrially applicable procedure for aniline production.

III. DIOLS

Studies of the catalytic transformations of diols on metals (dehydrogenation, dehydration, and hydrogenolysis) revealed the occurrence of epimerization [52,53]; this probably takes place through hydroxy-oxo compound intermediates (dehydrogenation followed by hydrogen addition), and it therefore seems apparent that the transformation is influenced by the presence of hydrogen, although its role has not been examined.

The effect of adsorbed hydrogen in the process 1,3-diol → oxo compound + H_2O has similarly not been studied, but an astoichiometric role may be attributed to hydrogen in the transformation, which proceeds with high selectivity [54-56]. Emphasis is laid on the role of hydrogen transfer in the multistep process [54,55,57].

In the transformation of 1,2-diols on Cu, which is known to involve a 1,2-bond shift mechanism (see also the discussion of hydrocarbon isomerization in Chapter 17), the experimental results can similarly be interpreted in terms of hydrogen abstraction and subsequent hydrogen addition [58].

IV. UNSATURATED ALCOHOLS

Besides hydrogenation, metal catalysts are known to catalyze the isomerization of unsaturated alcohols to oxo compounds [59,60]. Detailed studies have been made of the isomerization of α,β-unsaturated alcohols to aldehydes, mainly on Pt and Pd catalysts [59-61].

The role of the supports, particularly hydrogen, has been investigated through measurement of the kinetics of the reaction. Isomerizations involving monomolecular and bimolecular mechanisms take place; the latter has been demonstrated on metals when hydrogen is required. When plotted as a function of the ratio of hydrogen and unsaturated alcohol, the reaction rate exhibited a maximum. The possibilities of two assumptions were considered in efforts to explain this phenomenon: (a) hydrogen plays a role in the step determining the reaction rate, and (b) the concentration of catalytically active surface atoms increases in the presence of hydrogen. Kinetic analysis supported the latter view [61]. A check with deuterium confirmed this, for deuterium was not found in the aldehyde.

For study of the effect of hydrogen, experiments were also carried out on Pt/T (T = thermolyte, a type of firebrick), Pt/C, Ni/Al, and Cu/Al catalysts in the presence of helium as carrier gas [62]. In helium the Pt/T effectively lost its activity (Fig. 1). The transformation of crotyl alcohol is slow, and butyraldehyde is not even formed in traces; the presence of hydrogen is therefore determining as concerns the occurrence of isomerization.

In the case of Cu/Al, the rate of transformation was not influenced by the carrier gas, but the composition of the product did change. It appears that on this catalyst, too, hydrogen plays a role in the isomerization to butyraldehyde, but this is of a different nature from that for Pt/T.

Unsaturated alcohols undergo isomerization to oxo compounds on Pt, probably with the participation of chemisorbed hydrogen, similarly to the isomerization of olefins [63] — that is, by an intermolecular mechanism, probably through π-allyl-type complexes (Scheme 3). Hydrogen transfer plays a considerable role in the isomerization of unsaturated alcohols on Cu and Zn catalysts of Raney type [64].

Ni behaves differently from supported Pt catalysts in the isomerization of unsaturated alcohols, despite the fact that similar to the platinum metals, it chemisorbs hydrogen and is very active in the hydrogenation of the carbon-carbon double bond.

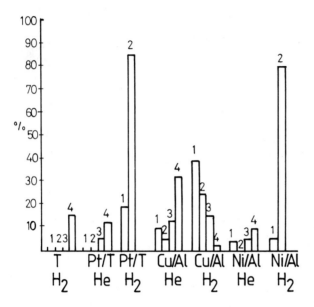

FIGURE 1 Maximum yields of the main products from crotyl alcohol at 373 to 573 K, on various catalysts, in the presence of hydrogen and helium carrier gases. 1, butyraldehyde; 2, 1-butanol; 3, crotonaldehyde; 4, allylcarbinol + methylvinylcarbinol.

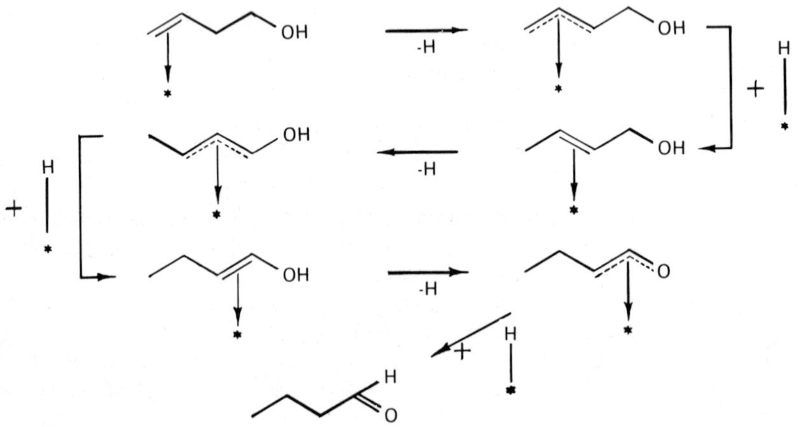

Scheme 3

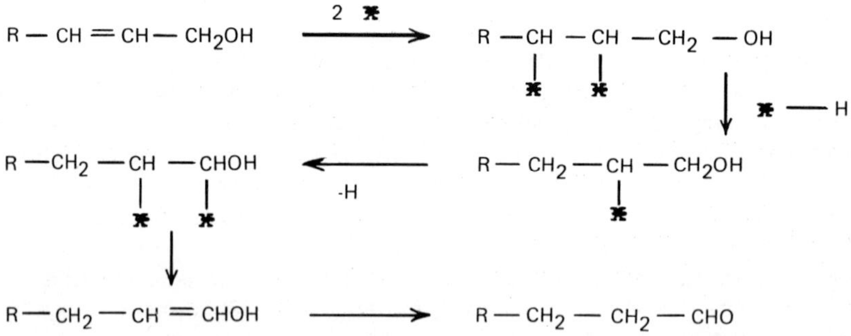

Scheme 4

In the absence of hydrogen, the isomerization of α, β-unsaturated alcohols on Raney Ni is slow [65,66]. The isomerization proceeds via the half-hydrogenated state [67] (Scheme 4), as confirmed by a stereochemical study of the process [66].

V. AMINO ALCOHOLS

In the preparation of piperazine from ethanolamine on Co/γ-Al$_2$O$_3$, it has been observed that the role of hydrogen is decisive [68]. Optimization of the partial pressure of hydrogen similarly proved to be a factor improving the selectivity in the preparation of ethylenediamine from monoethanolamine on Ni [50].

Hydrogen transfer has an important role in the transformation of 1,3-amino alcohols to ketones [69,70]. In the light of this, Mannich condensation of ketones has been utilized for α methylation of ketones [71].

VI. OXO COMPOUNDS

Hydrogen or deuterium is a reactant in the H-D exchange reaction of oxo compounds; the general rules relating to exchange reactions hold for these processes [12]. Certain individual steps in the H-D exchange and the hydrogenation are identical with those in the mechanism of double-bond migration (isomerization) in olefins [72]. The addition-abstraction mechanism is characterized by preformation of a metal-hydrogen bond, followed by its interaction with the olefin to form σ-bonded alkyl intermediates (see Scheme 5).

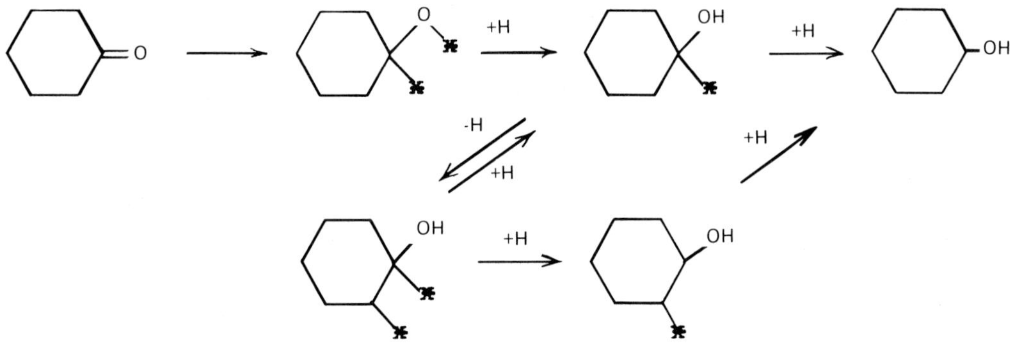

Scheme 5

Probably, the reverse reactions occur in cyclohexanol dehydrogenation (Scheme 2). Other authors stress the significant role of transfer hydrogenation (hydrogen transfer) in the hydrogenation of oxo compounds [73].

Investigations dealing with the hydrogenation of glucose, fructose, and their mixtures (invert sugar) to sorbitol and mannitol are of practical importance [74]. Experiments have been carried out on Raney Ni, Ru, and Rh catalysts as a function of hydrogen pressure in the interval 3.5×10^5 to 7×10^9 Pa [74]. Ru proved to be the most active in the hydrogenation; glucose was not hydrogenated on Rh, and fructose was not hydrogenated on Raney Ni. The rate of the hydrogenation reaction exhibited a maximum curve as a function of the hydrogen pressure in certain cases, while in other cases (at higher temperature) it increased sharply above 4.2×10^6 Pa. In the case of fructose, when both sorbitol and mannitol are formed, the selectivity of the hydrogenation reaction proved independent of the hydrogen pressure.

Conclusions as to the mechanism of the reaction were drawn from studies of the kinetics of the hydrogenation: the hydrogenation of glucose on Raney Ni and of fructose on Ru can be explained by a mechanism involving a surface reaction between unadsorbed sugar and chemisorbed hydrogen.

VII. CARBOXYLIC ACIDS AND THEIR DERIVATIVES

As far as the transformation of this type of compounds on metals is concerned, very few experimental observations have been reported in connection with the role of hydrogen.

The effects of adsorbed oxygen and hydrogen on the dehydrogenation and dehydration of formic acid on Pt have been studied [75]. Pretreatment with oxygen or hydrogen has a considerable influence on the selectivities of the two reactions, by changing the surface state of the Pt (oxygen pretreatment favors dehydrogenation). It is probable that chemisorbed hydrogen has a large effect on the transformations of cyclic carboxylic acid esters on Cu, where hydrogen abstraction and hydrogen addition are assumed [76].

VIII. OXYGEN-CONTAINING HETEROCYCLES

The isomerization of cyclic compounds containing one ring oxygen atom (cyclic ethers = oxacycloalkanes) or two ring oxygen atoms (1,3-dioxacycloalkanes) on metals has been studied widely (for reviews see Refs. 12, 77, and 78). In the case of the oxacycloalkanes, the general nature of the isomerization to carbonyl compounds has been successfully demonstrated from the oxiranes to the oxanes [79-82]:

$$(CH_2)_n \underset{CH_2}{\overset{O}{\diagup}} C \underset{H}{\overset{Me}{\diagup}} \xrightarrow{M, H_2} (CH_2)_{n-1} \underset{CH_2}{\overset{Me}{\diagup}} C \underset{Me}{\overset{O}{\diagdown}} \qquad (9)$$

$$n = 0, 1, 2, 3$$

Despite the fact that the studies have been performed in the presence of hydrogen, there has been hardly any mention of the hydroisomerizational nature of the isomerization (i.e., the determining role of hydrogen). The reason for this is that the aim of the investigations of the oxacycloalkanes in the presence of hydrogen was not the study of the isomerization, but the study of the hydrogenolysis to alcohols.

However, attention was drawn to the role of hydrogen in the isomerization of epoxides as long ago as 1958 [79], and the determining role of hydrogen was later proved for the oxetanes [83-87], the oxolanes and oxanes [82,88,89], and the 1,3-dioxacycloalkanes [90] too.

A. Oxiranes and Oxetanes

From kinetic studies of the isomerization and hydrogenolysis of ethylene oxide on Ni, Pd, and Pt, Tenma and Kwan [79] concluded that "a mechanism was suggested that both isomerization and hydrogenolysis of ethylene oxide are initiated by the ring-opening addition of adsorbed hydrogen atom. . . ."

Cornet et al. considered that the isomerization and hydrogenolysis of epoxides are parallel reactions, the relative rates of which depend on the

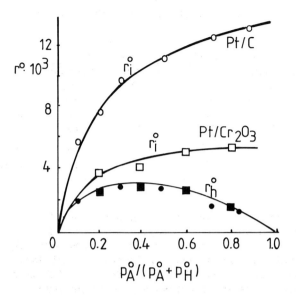

FIGURE 2 Dependence of the initial reaction rates for isomerization r_i^o and for hydrogenolysis r_h^o (in mol h^{-1} g_{cat}^{-1}) on the composition of the reaction mixture at total atmospheric pressure. (From Ref. 96; used by permission of Academic Press.)

nature of the metal, and that the presence of hydrogen or deuterium is necessary for the isomerization [80, 91-94].

For 1,2-epoxybutane, the ratio of isomerization and hydrogenolysis on Pt catalysts depends on the partial pressure of hydrogen. The rate of isomerization increases with the partial pressure of hydrogen [95-97] (Fig. 2). The role of hydrogen is reflected by the kinetic equation

$$r_i^o = \frac{k_i K_A^i p_A^o (1 + C_H p_H^o)}{1 + K_A^i p_A^o} \quad (10)$$

where

r_i^o = initial rate of isomerization reaction
k_i = rate constant of isomerization reaction
K_A^i = adsorption coefficient of 1,2-epoxybutane
p_A^o = initial partial pressure of 1,2-epoxybutane
C_H = empirical coefficient
p_H^o = initial partial pressure of hydrogen

The experimental data referred to above indicate that the rates of hydrogenolysis and isomerization are influenced by variation in the hydrogen coverage of the Pt surface, and that the secondary alcohol is formed primarily by an associative mechanism [91, 82], while the ketone is formed by a dissociative mechanism [98, 99]. Accordingly, it may be assumed that any change in the hydrogen coverage of the surface will act in different ways on these two reactions. For this reason we carried out detailed investigations in this respect in a catalytic microreactor [100].

TABLE 1 Initial Distribution of Products on Pt/C in a State Microreactor System

Compound	Temperature (K)	Initial distribution of products (mol %)				b/a
		Aldehyde (a)	1-Alcohol (a)	Ketone (b)	2-Alcohol (b)	
Methyloxirane	373	0	13	3	84	6.7
	473	0	11	3	86	8.0
2-Methyloxetane	373	0	13	1	86	6.7
	473	0	75	5	20	0.33

The regioselectivities of the transformations of methyloxirane and 2-methyloxetane display different behavior. In 2-methyloxetane, the less sterically hindered C-O bond is cleaved at 373 K, whereas the more sterically hindered one is cleaved at elevated temperatures. In the case of methyloxirane, the regioselectivity remains unchanged (Table 1). The rate of formation of 1-propanol from methyloxirane increases monotonously as the pressure of hydrogen is raised (Fig. 3), whereas at 473 K it passes through a maximum (Fig. 4). This suggests that the associative mechanism predominates at 373 K, while the dissociative mechanism prevails at 473 K. In the case of the associative mechanism the regioselectivity is governed by the stereochemical factors, and cleavage of the less sterically hindered bond is favored. In contrast, at higher temperatures a dissociative mechanism occurs. In the latter case, the transformation begins with the splitting of a C-H bond, and the higher reactivity of the secondary C-H bond will be the factor determining the selectivity. However, the dissociative mechanism is competitive only at low hydrogen pressure; above a hydrogen pressure of 20 kPa, the dissociative mechanism is suppressed and the associative mechanism comes into the foreground. This is why 1-propanol is hardly formed in the investigations in the microreactor (Scheme 6). These experimental results also support the

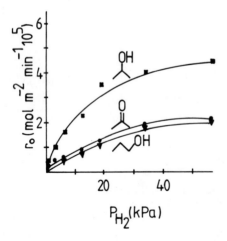

FIGURE 3 Initial rates of appearance of products from methyloxirane at 373 K over Pt on carbon catalyst at different hydrogen pressures. $P_{oxirane}$, 2.0 kPa; catalyst, 5×10^{-3} g. (From Ref. 100; used by permission of Academic Press.)

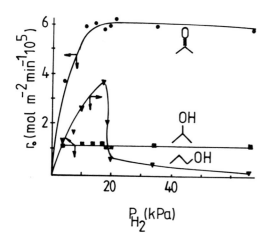

FIGURE 4 Initial rates of appearance of products from methyloxirane at 473 K over Pt on carbon catalyst at different hydrogen pressures. $P_{oxirane}$, 2.0 kPa; catalyst, 5×10^{-3} g. (From Ref. 100; used by permission of Academic Press.)

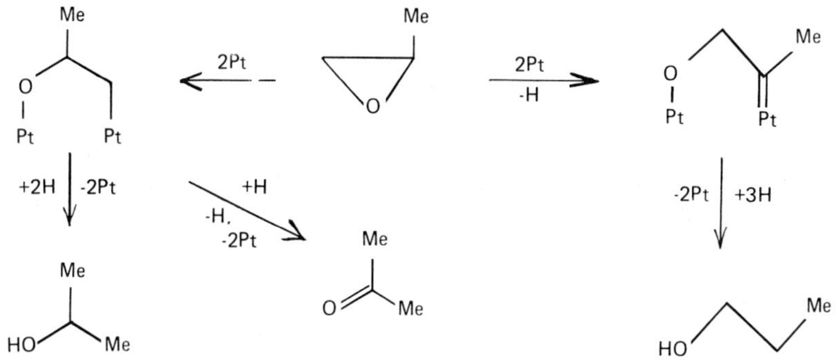

Scheme 6

earlier observation (see Chapter 17) that a higher hydrogen pressure is required for the same degree of hydrogen coverage at higher temperature.

For 2-methyloxetane at 373 K, the less sterically hindered bond is cleaved. At 473 K, the dissociative mechanism becomes predominant (Fig. 5). In this case the dissociative mechanism is suppressed to a much lower extent by increase of the hydrogen pressure than in the case of methyloxirane, and thus the associative mechanism will no longer be competitive at higher temperature. It is probable that in the four-membered ring the C-H and C-O bonds are not cleaved simultaneously, and it is therefore less sensitive to change in the hydrogen coverage. Since the rates of formation of the two alcohols display similar curves as functions of the hydrogen pressure, it is likely that both are formed by a dissociative mechanism and that the difference in reactivity of the C-H bonds is the factor controlling the regioselectivity.

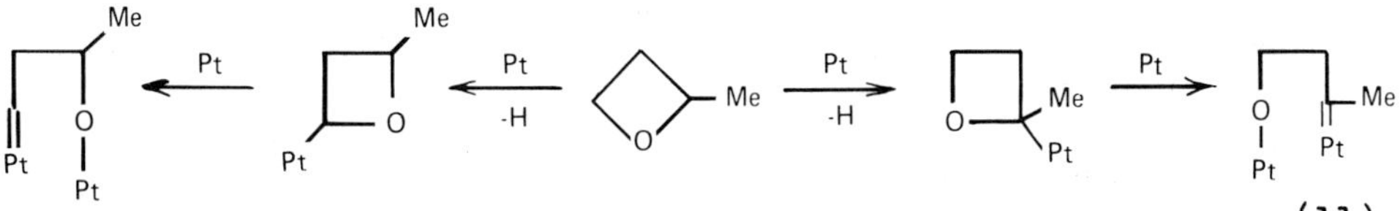

(11)

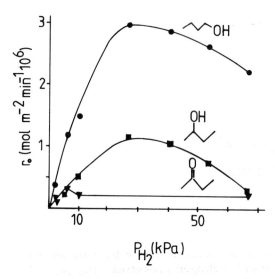

FIGURE 5 Initial rates of appearance of products from 2-methyloxetane at 473 K over Pt on carbon catalyst. $p_{oxetane}$, 2.0 kPa; catalyst, 5×10^{-3} g. (From Ref. 100; used by permission of Academic Press.)

The rate of formation of ethyl methyl ketone from 2,3-dimethyloxirane isomers exhibited a maximum as a function of the hydrogen pressure; this maximum was particularly sharp for the cis isomer (Fig. 6). This shows that a relatively narrow range of surface hydrogen coverage is optimum for ketone formation. It also means that the mechanisms of transformation of methyloxirane and the 2,3-dimethyloxiranes are not completely identical; in the case of methyloxirane, the alcohol and the ketone are probably formed from a common surface species, and the selectivity depends on the temperature-controlled hydrogen coverage (Scheme 6). For 2,3-dimethyloxirane, however, the two products are derived from different surface species (Scheme 7).

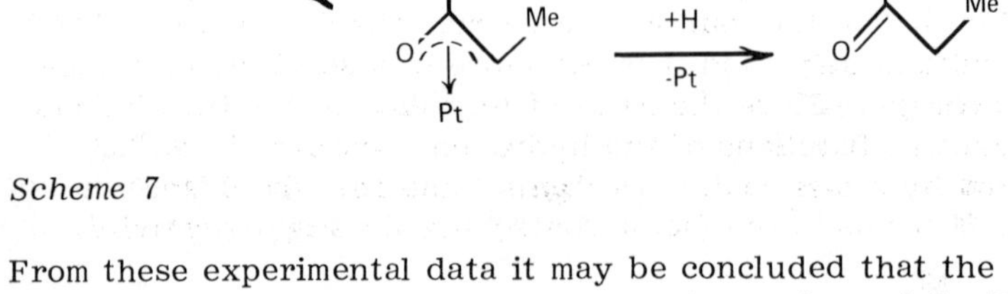

Scheme 7

From these experimental data it may be concluded that the 2,3-dimethyloxiranes are adsorbed in a flat-lying manner on the surface of the Pt catalyst, while methyloxirane undergoes two-point adsorption [98].

Studies of the kinetics of transformation of oxiranes on Pt at 373 K in the hydrogen pressure interval 6 to 26 kPa [79] revealed that the rate of hydrogenolysis is described by the formula

$$r = kp_{H_2}^n p_{oxirane}^o \quad (n = 0.5) \tag{12}$$

The mechanism to be seen in Scheme 8 was proposed for the reaction.

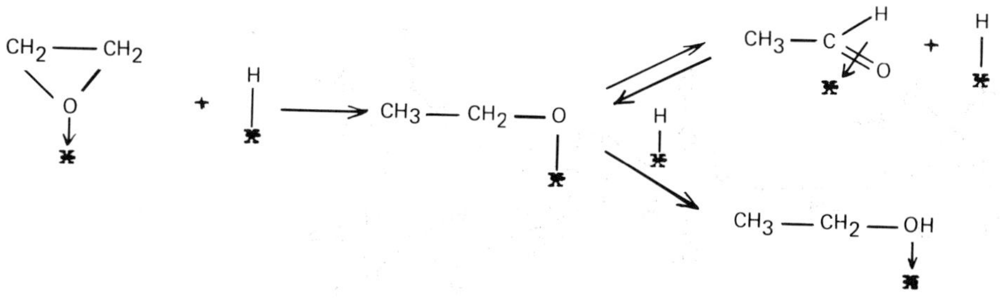

Scheme 8

Davidová and Kraus investigated the kinetics of transformation of ethyloxirane on Pt/C [96]. The interpretation of the results is hampered by the fact that the catalyst they used behaved as a bifunctional catalyst [97,99]. Other authors [80,91,92] studied the transformations of the dimethyloxiranes on various metals and suggested the mechanism outlined in Scheme 7. The effect of change of the hydrogen pressure was examined in the case of trans-2,3-dimethyloxirane at 293 K on Pd film [91]. The rate of isomerization did not vary regularly with the hydrogen pressure, but for the rate of hydrogenolysis it was found that n = -0.4.

The question arises as to which of the two mechanisms outlined in Scheme 7 and 8 can be regarded as correct. Does the mechanism depend on the structure of the oxirane and on the reaction conditions? Is the observed negative hydrogen order characteristic of the trans-2,3-dimethyloxirane reactant or of the Pd catalyst? The experimental data [91,92,100,101] suggest that the mechanism proceeds as in Scheme 8 for methyloxirane, and as in Scheme 7 for the dimethyloxiranes. We concluded that the Scheme 8 mechanism involves two-point adsorption, and that in Scheme 7 flat-lying adsorption.

The most recent findings indicate that the "two-point adsorption" of the oxiranes consists of the adsorption of the oxygen to the surface through two

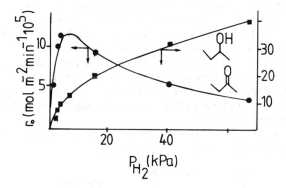

FIGURE 6 Initial rates of appearance of products from cis-2,3-dimethyloxirane at 373 K over Pt on carbon catalyst. $p_{oxirane}$, 1.4 kPa; catalyst, 5×10^{-3} g. (From Ref. 100; used by permission of Academic Press.)

electron pairs, whereas in the "flat-lying adsorption" the oxygen atom is bound to the surface via only one nonbonding electron pair.

The transformation of the oxiranes as a function of the hydrogen pressure on Cu has also been examined [101]. For methyloxirane, mainly the bond adjacent to the methyl group is cleaved (Scheme 9).

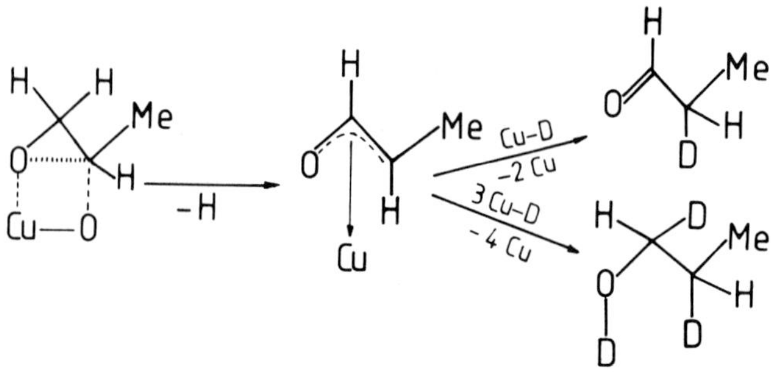

Scheme 9

The product composition and the reaction rate are strongly dependent on the hydrogen pressure. The experimental results permit the following conclusions [101]:

1. At low hydrogen pressure chiefly propanal, and at high hydrogen pressure chiefly 1-propanol is formed.
2. Above a hydrogen pressure of 10 kPa, the rates of formation of propanal and 1-propanol vary in opposite directions, but their sum is practically constant.
3. Below a hydrogen pressure of 3 kPa, hardly any 1-propanol is formed, and the rate of formation of propanal increases with increase of the pressure of hydrogen.
4. Propanal is also formed in the absence of hydrogen, but at a lower rate.
5. Acetone is formed only in the presence of hydrogen.
6. Two types of isomerization reaction probably take place side by side: one type occurs on copper oxide (electrophilic catalysis), the product being exclusively propanal, and there being no need for hydrogen; the rate of the other type increases with the hydrogen pressure, and it becomes the main reaction at higher hydrogen pressure.

It has already been mentioned that the main product is propanal at low hydrogen pressure and 1-propanol at high hydrogen pressure. It was proposed by Sabatier and Durand [102] that on Ni the oxirane is first isomerized to aldehyde, which is then hydrogenated to primary alcohol. On Cu, the desorbed aldehyde is not readsorbed on the oxirane-covered surface, and thus 1-propanol is not formed through the hydrogenation of propanal. It is probable, however, that both compounds are formed from a common surface species, which is desorbed from the surface in the form of aldehyde or alcohol, depending on the surface hydrogen coverage.

The isomerization to propanal on Cu can be interpreted on the basis of Scheme 9, while acetone is formed as in the following equation:

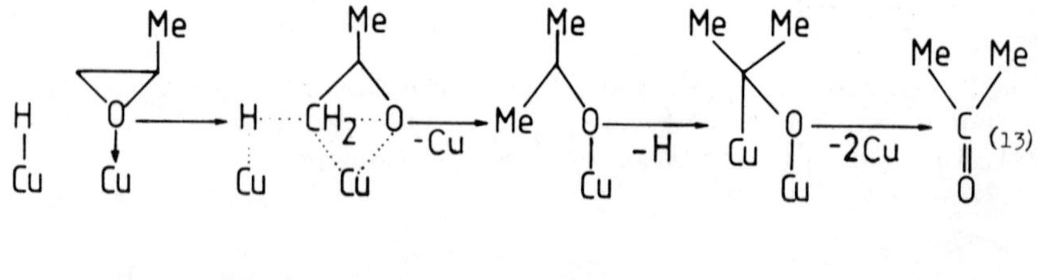

(13)

Our experimental results demonstrate that the surface of the catalyst is oxidized through the deoxygenation of the oxiranes in the course of the transformation, and Lewis acid site/basic site pairs are produced [103,104]. Propanal is produced predominantly on these centers. The number of active centers of this type, and hence the rate of formation of propanal, are maximum on the partially oxidized surface. Acetone is formed on the reduced metal surface. Both reactions may be regarded as hydroisomerizations.

The metal-catalyzed transformations of epoxides are also of practical significance. Epoxides produced from terminal olefins can be converted to secondary or primary alcohols, depending on the catalyst and the hydrogen pressure. Mainly the primary alcohols are of great importance. Hydroisomerization is also of practical significance, particularly for those industrially valuable ketones that are difficult to prepare by other means (e.g., cyclododecanone) [105].

Finally, it should be mentioned that intermolecular hydrogen transfer reactions may take place on metal catalysts in the absence of hydrogen (i.e., when hydrogenation and hydrogenolysis cannot occur). A number of transformations of this nature are known not only for hydrocarbons, but also for compounds containing C-O bonds (e.g., 1,2-epoxycyclohexan-3-ol on Pd) [106].

$$\text{epoxycyclohexanol} \xrightarrow{Pd} \text{cyclohexanediol} + \text{catechol} \qquad (14)$$

It is to be noted that study of the hydroisomerization of oxetanes (i.e., the role of hydrogen) led to the first experimental evidence of the 1,3-bond shift isomerization reaction—as opposed to the common 1,2-bond shift—in the case of 2,2,4,4-tetramethyloxetane on metals [87].

B. Oxolanes and Oxanes

Investigation of the role of hydrogen effects in these types of compounds has been carried out by means of H-D exchange, epimerization, isomerization to ketones, and the hydrogenation of furans.

Work on the H-D exchange of oxolanes [107,108] revealed the steric arrangement of the five-membered ring on the surfaces of metals, and provided further experimental evidence of the rollover mechanism [109], shown by the following in the case of cyclopentane:

$$\text{(rollover mechanism scheme)} \qquad (15)$$

Detailed studies of the H-D exchange of five-membered cyclic ethers have been made on Rh and Pd [108]. The first step in the process is the formation of an O-metal bond; this bond remains unchanged in the event of a single exchange, but is broken in multiple exchanges. An α,γ-diadsorbed species is produced in the single exchange.

The simultaneous adsorption of the oxygen and the vicinal carbon atom (formation of an α,β-diadsorbed species) leads to hydrogenolysis. Multiple exchange on Rh proceeds via α,β-diadsorbed species, whereas on Pd the rollover mechanism is favored.

For the 2,5-dimethyloxolane isomers, the H-D exchange was accompanied by epimerization (more on Pd than on Rh), which was similarly explained by the rollover mechanism. Hydrogen or deuterium plays a determining role in this process, too, despite the fact that attention was not drawn to this specially. The hydroisomerizational character of the isomerization of five- and six-membered cyclic ethers to ketones (for a survey, see Ref. 78) has been confirmed experimentally [82,110] (Table 2).

Studies have been made on Pd of the decarbonylation and hydrogenation of 5-methylfurfural [111], and of the hydrogenation and hydrogenolysis of 2-methylfuran [112]. Kinetic examinations of the deuterium isotope effect showed that hydrogen participated in the rate-determining step of hydrogenation. The latter process is again of practical importance: Methyl propyl ketone is produced from furfural.

C. 1,3-Dioxacycloalkanes

It has been found that the transformations of 1,3-dioxolanes and 1,3-dioxanes on Pt involve isomerization to esters [90,113,114].

$$[4] \xrightarrow{Pt, H_2, R=Me} [5] \quad (16)$$

$$[6] \xrightarrow{Pt, H_2, R=Me} [7] \quad (17)$$

Studies relating to the role of the hydrogen effect have produced the following findings [90,113,115-118] (see Table 3):

1. Both the conversion and the ester selectivity are considerably lower in the absence of hydrogen.

TABLE 2 Conversion and Selectivity of Transformations of Oxacycloalkanes in the Presence and Absence of H_2[a]

	Me-oxirane		Me-oxetane		Me-oxolane		Me-oxane	
	a	b	a	b	a	b	a	b
Pt/CS + H_2	100	34	85	22	68	57	53	49
Pt/CS - H_2	42	2	33	8	16	6	10	10

a, conversion (%); b, ketone selectivity (mol %); CS, Cab-O-Sil.

TABLE 3 Variation in Extent of Ester Formation with Hydrogen Content[a]

	Feed			
	2-Methyl-1,3-dioxolane [4]		2-Methyl-1,3-dioxane [6]	
% H_2 in N_2	Conversion (%)	Selectivity of ethylacetate [5] (mol %)	Conversion (%)	Selectivity of propylacetate [7] (mol %)
0	0	0	0	0
7	2	72	3	52
20	4	91	5	80
50	6	61	10	62
100	8	31	18	27

[a]See Eqs. (16) and (17).

2. Isomerization to esters can be optimized via the hydrogen coverage of the catalyst.
3. At an appropriate hydrogen/reactant ratio, the selectivity of ester formation can be raised to more than 90%

Under hydrogen-deficient conditions at higher temperature, the reaction takes place by a dissociative mechanism, that is, in the course of the σ,π-bond shift, hydrogen abstraction accompanies ring opening, with subsequent hydrogen addition.

(18)

IX. CONCLUDING REMARKS

In this chapter we have surveyed the experimental data demonstrating the role of hydrogen in the metal-catalyzed transformations of organic compounds containing C-O bonds. The experimental results discussed reveal that in contrast with hydrocarbons (see Refs. 119 to 126 and Chapters 16 to 18), for organic compounds containing oxygen atoms, research involving systematic studies of the hydrogen effects is still in the comparatively early stages.

After new types of metal catalysts have been developed and new methods of investigation have been introduced, study of the metal-catalyzed transformations of various organic compounds, including the hydrogen effects, will assume greater importance. Such research will be an integral part of the efforts to discover the laws relating to metal-catalyzed reactions. It may also make a major contribution to the new organic chemical technology needed to produce low-volume, high-value-added, high-technology products, collectively called "specialities," which are much sought after by the present-day chemical industry.

REFERENCES

1. E. F. Armstrong and T. P. Hilditch, *Proc. R. Soc.*, 97:262 (1920).
2. R. H. Pickard and J. Kenyon, *J. Chem. Soc.*, 99:56 (1911).
3. Ya. T. Eidus, *Astoichiometric Components of Catalytic Reactions* (in Russian), Nauka, Moscow (1975).
4. M. Bartók, *Acta Phys. Chem. (Szeged)*, 21:79 (1975).
5. M. Bartók, *React. Kinet. Catal. Lett.*, 3:115 (1975).
6. O. V. Bragin and A. L. Liberman, *Transformations of Hydrocarbons on Metal Containing Catalysts* (in Russian), Khimiya, Moscow (1981).
7. S. L. Kiperman, *Bulg. Acad. Sci. Commun. Dept. Chem.*, 16:22 (1983).
8. Z. Paál and P. G. Menon, *Catal. Rev.-Sci. Eng.*, 25:229 (1983).
9. G. C. Bond and P. B. Wells, *Adv. Catal.*, 15:92 (1964).
10. W. R. Patterson and R. L. Burwell, Jr., *J. Am. Chem. Soc.*, 93:833 (1971).
11. W. R. Patterson and R. L. Burwell, Jr., *J. Am. Chem. Soc.*, 93:839 (1971).
12. M. Bartók and coauthors, *Stereochemistry of Heterogeneous Metal Catalysis*, Wiley-Interscience, Chichester, West Sussex, England (1985).
13. H. A. Smith and B. B. Stewart, *J. Am. Chem. Soc.*, 82:3898 (1960).
14. S. Yamada, *Bull. Chem. Soc. Jpn.*, 16:187 (1941).
15. S. Yamada, *Bull. Chem. Soc. Jpn.*, 16:239 (1941).
16. E. G. Peppiatt and R. J. Wicker, *J. Chem. Soc.*, 3122 (1955).
17. E. L. Eliel and S. H. Schroeter, *J. Am. Chem. Soc.*, 87:5031 (1965).
18. E. L. Eliel, S. H. Schroeter, T. J. Brett, F. J. Biros, and J.-C. Richer, *J. Am. Chem. Soc.*, 88:3327 (1966).
19. R. J. Wicker, *J. Chem. Soc.*, 2165 (1956).
20. D. Cornet and F. G. Gault, *J. Catal.*, 7:140 (1967).
21. R. L. Burwell, *J. Am. Chem. Soc.*, 70:2865 (1948).
22. L. Kh. Freidlin, E. F. Litvin, V. V. Yakubenok, and S. J. Shcherbakova, *Izv. Akad. Nauk SSSR Ser. Khim.*, 1791 (1979).
23. V. I. Tatarinova, O. V. Bragin, and A. L. Liberman, *Izv. Akad. Nauk SSSR Otd. Khim. Nauk*, 1756 (1970).
24. Y. Takagi and S. Teratani, *J. Catal.*, 34:490 (1974).
25. S. Teratani, Y. Takagi, K. Tanaka, and Y. Muramatsu, *J. Catal.*, 63:102 (1980).
26. B. N. Dolgov, V. V. Mazurek, and V. A. Krol', *Zh. Obshch. Khim.*, 28:2395 (1958).
27. J. T. Richardson and W.-C. Lu, *J. Catal.*, 42:275 (1976).
28. M. Dobrovolszky, P. Tétényi, and Z. Paál, *J. Catal.*, 74:31 (1982).
29. J. Manninger, Z. Paál, and P. Tétényi, *J. Catal.*, 48:442 (1977).
30. J. Manninger, Z. Paál, and P. Tétényi, *Acta Chim. Acad. Sci. Hung.*, 97:439 (1978).
31. M. Dobrovolszky, P. Tétényi, and Z. Paál, *Acta Chim. Acad. Sci. Hung.*, 107:343 (1981).
32. J. F. Hemidy and F. G. Gault, *Bull. Soc. Chim. Fr.*, 1710 (1965).
33. H. Pines and P. Steingaszner, *J. Catal.*, 10:60 (1968).
34. H. Pines and T. P. Kobylinski, *J. Catal.*, 17:375 (1970).
35. T. P. Kobylinski and H. Pines, *J. Catal.*, 17:384 (1970).
36. H. Pines and T. P. Kobylinski, *J. Catal.*, 17:394 (1970).
37. J. Hensel and H. Pines, *J. Catal.*, 24:197 (1972).
38. H. Pines, J. Hensel, and J. Šimonik, *J. Catal.*, 24:206 (1972).
39. J. Šimonik and H. Pines, *J. Catal.*, 24:211 (1972).
40. H. Pines and J. Šimonik, *J. Catal.*, 24:220 (1972).
41. E. Licht, Y. Schächter, and H. Pines, *J. Catal.*, 31:110 (1973).
42. E. Licht, Y. Schächter, and H. Pines, *J. Catal.*, 34:338 (1974).

43. E. Licht, Y. Schächter, and H. Pines, J. Catal., 55:191 (1978).
44. M. Bartók, F. Notheisz, and I. Török, Acta Phys. Chem. (Szeged), 17: 101 (1971).
45. E. Licht, Y. Schächter, and H. Pines, J. Catal., 61:109 (1980).
46. S. Mitsui, S. Imaizumi, and Y. Esashi, Bull. Chem. Soc. Jpn., 43:2143 (1970).
47. A. M. Khan, F. J. McQuillin, and I. Jardine, J. Chem. Soc., C136 (1967).
48. A. Baiker and J. Kijenski, Catal. Rev.-Sci. Eng., 27:653 (1985).
49. A. Baiker and W. Richarz, Stud. Surf. Sci. Catal., 7:1428 (1981).
50. C. M. Barnes, Ind. Eng. Chem. Prod. Res. Dev., 20:399 (1981).
51. J. Runeberg, A. Baiker, and J. Kijenski, Appl. Catal., 17:309 (1985).
52. Á. Molnár and M. Bartók, React. Kinet. Catal. Lett., 4:315 (1976).
53. Á. Molnár and M. Bartók, React. Kinet. Catal. Lett., 4:425 (1976).
54. Á. Molnár and M. Bartók, Acta Chim. Acad. Sci. Hung., 89:393 (1976).
55. M. Bartók and Á. Molnár, Kém. Közl., 45:335 (1976).
56. Á. Molnár and M. Bartók, J. Catal., 72:322 (1981).
57. Á. Molnár and M. Bartók, Acta Phys. Chem. (Szeged), 25:161 (1979).
58. M. Bartók and Á. Molnár, J. Chem. Soc. Chem. Commun., 1178 (1980).
59. J. Šimonik and L. Beránek, J. Catal., 24:348 (1972).
60. M. Kraus, Collect. Czech. Chem. Commun., 37:460 (1972).
61. M. Beránek and L. Kraus, Collect. Czech. Chem. Commun., 37:3778 (1972).
62. M. Bartók and I. Török, Acta Chim. Acad. Sci. Hung., 88:35 (1976).
63. G. H. Twigg, Trans. Faraday Soc., 35:934 (1939).
64. G. Eadon and M. Y. Shiekh, J. Am. Chem. Soc., 96:2288 (1974).
65. J. V. N. Vara Prasad, A. G. Samuelson, and C. N. Pillai, J. Catal., 75:1 (1982).
66. J. V. N. Vara Prasad and C. N. Pillai, J. Catal., 88:418 (1984).
67. J. Horiuti and M. Polányi, Nature (London), 132:819 (1933).
68. A. A. Balandin, E. I. Karpeiskaya, V. A. Ferapontov, A. A. Tolstopyatova, and L. S. Gorshkova, Dokl. Akad. Nauk. SSSR, 165:99 (1965).
69. M. Bartók, G. Sirokmán, and Á. Molnár, J. Mol. Catal., 14:379 (1982).
70. Á. Molnár, G. Sirokmán, and M. Bartók, J. Mol. Catal., 19:25 (1983).
71. G. Sirokmán, Á. Molnár, and M. Bartók, Appl. Catal., 7:133 (1983).
72. Y. Takagi, S. Teratani, and K. Tanaka, J. Catal., 27:79 (1972).
73. Y. Saito and Y. Ogino, J. Catal., 55:198 (1978).
74. J. Wisniak and R. Simon, Ind. Eng. Chem. Prod. Res. Dev., 18:50 (1979).
75. L. Riekert, Z. Elektrochem., 63:198 (1959); 66:207 (1962).
76. M. Bartók and Á. Molnár, J. Catal., 79:485 (1983).
77. M. Bartók and K. L. Láng, in The Chemistry of Functional Groups (S. Patai, ed.), Wiley, Chichester, West Sussex, England (1980), Suppl. E, Chap. 14.
78. M. Bartók, in The Chemistry of Functional Groups (S. Patai, ed.), Wiley, Chichester, West Sussex, England (1980), Suppl. E, Chap. 15.
79. S. Tenma and T. Kwan, Shokubai, 15:11 (1958).
80. D. Cornet, Y. Gault, and F. G. Gault, Proceedings of the 3rd International Congress on Catalysis, Amsterdam, 1964, North-Holland, Amsterdam (1965), p. 1184.
81. M. Bartók, I. Török, and I. Szabó, Acta Chim. Acad. Sci. Hung., 76:417 (1973).
82. M. Bartók, Acta Chim. Acad. Sci. Hung., 88:395 (1976).
83. M. Bartók and K. Kovács, Acta Chim. Acad. Sci. Hung., 55:49 (1968).
84. M. Bartók and B. Kozma, Acta Chim. Acad. Sci. Hung., 55:61 (1968).
85. M. Bartók and K. Kovács, Acta Chim. Acad. Sci. Hung., 56:369 (1968).

86. M. Bartók, K. Kovács, and N. I. Shuikin, *Acta Chim. Acad. Sci. Hung.*, 56:393 (1968).
87. M. Bartók, *J. Chem. Soc. Chem. Commun.*, 139 (1979).
88. M. Bartók, D.Sc. dissertation, Szeged, Hungary (1972).
89. M. Bartók and F. Notheisz, *Mechanism of the Platinum Heterogeneous Catalytic Isomerization of Oxacycloalkanes.* III. Intern. Katalyse-Konf. der DDR. Reinhardsbrunn (1974), Kurzref. 52. p.
90. M. Bartók, J. Apjok, R. A. Karakhanov, and K. Kovács, *Acta Chim. Acad. Sci. Hung.*, 61:315 (1969).
91. G. Sénéchal and D. Cornet, *Bull. Soc. Chim. Fr.*, 773 (1971).
92. G. Sénéchal, J. C. Duchet, and D. Cornet, *Bull. Soc. Chim. Fr.*, 783 (1971).
93. J. C. Duchet and D. Cornet, *Bull. Soc. Chim. Fr.*, 1141 (1975).
94. J. C. Duchet and D. Cornet, *Bull. Soc. Chim. Fr.*, 1135 (1975).
95. G. Davidova and M. Kraus, *Proceedings of the All-Union Conference on Mechanism of Catalytic Transformations*, Moscow, Vol. 1 (1979), p. 90.
96. G. Davidova and M. Kraus, *J. Catal.*, 61:1 (1980).
97. M. Kraus and G. Davidova, *J. Catal.*, 68:252 (1981).
98. M. Bartók, F. Notheisz, and Á. G. Zsigmond, *J. Catal.*, 63:364 (1980).
99. M. Bartók, F. Notheisz, and J. T. Kiss, *J. Catal.*, 68:249 (1981).
100. M. Bartók, F. Notheisz, Á. G. Zsigmond, and G. V. Smith, *J. Catal.*, 100:39 (1986).
101. F. Notheisz, Á. Molnár, Á. G. Zsigmond, and M. Bartók, *J. Catal.*, 98:131 (1986).
102. P. Sabatier and J. F. Durand, *C.R.*, 182:826 (1926).
103. I. E. Wachs and R. J. Madix, *J. Catal.*, 53:208 (1978).
104. I. E. Wachs and R. J. Madix, *Appl. Surface Sci.*, 1:303 (1978).
105. F. A. Tsernyskova and D. V. Mushenko, *Neftekhim.*, 16:250 (1976).
106. J. E. Lyons, in *Catalysis in Organic Synthesis* (P. N. Rylander and H. Greenfield, eds.), Academic Press, New York (1976).
107. J. M. Forrest, R. L. Burwell, Jr., and B. K. C. Shim, *J. Phys. Chem.*, 63:1017 (1959).
108. J. C. Duchet and D. Cornet, *J. Catal.*, 44:57 (1976).
109. R. L. Burwell, Jr. and K. Schrage, *Discuss. Faraday Soc.*, 41:215 (1966).
110. M. Bartók and F. Notheisz, *J. Catal.*, 68:209 (1981).
111. Zh. G. Yuskovets, N. V. Nekrasov, M. V. Shimanskaya, and S. L. Kiperman, *Kinet. Katal.*, 22:1214 (1981).
112. Zh. G. Yuskovets, N. V. Nekrasov, M. M. Kostyukovskii, M. V. Shimanskaya, and S. L. Kiperman, *Kinet. Katal.*, 25:1361 (1984).
113. M. Bartók and J. Apjok, *Acta Phys. Chem. (Szeged)*, 21:49 (1975).
114. M. Bartók and J. Apjok, *Acta Phys. Chem. (Szeged)*, 21:69 (1975).
115. J. Apjok, M. Bartók, and K. Kovács, *Acta Chim. Acad. Sci. Hung.*, 69:445 (1971).
116. M. Bartók and J. Apjok, *Kém. Közl.*, 45:359 (1976).
117. M. Bartók and J. Czombos, *J. Chem. Soc. Chem. Commun.*, 106 (1981).
118. M. Bartók and J. Czombos, *J. Chem. Soc. Chem. Commun.*, 978 (1981).
119. F. G. Ciapetta and J. B. Hunter, *Ind. Eng. Chem.*, 45:147 (1953).
120. H. Pines and A. W. Shaw, *Adv. Catal.*, 9:569 (1957).
121. P. B. Wells and G. R. Wilson, *J. Catal.*, 90:70 (1967).
122. S. Siegel, *Adv. Catal.*, 16:124 (1964).
123. M. Boudart, *Adv. Catal.*, 20:153 (1969).
124. J. R. Anderson, *Adv. Catal.*, 23:1 (1973).
125. J. H. Sinfelt, *Adv. Catal.*, 23:91 (1973).
126. J. K. A. Clarke and J. J. Rooney, *Adv. Catal.*, 25:125 (1976).

20 | Role of Hydrogen in CO Hydrogenation

CALVIN H. BARTHOLOMEW

Brigham Young University, Provo, Utah

I. INTRODUCTION	543
II. ROLE OF HYDROGEN AS A REACTANT IN CO HYDROGENATION ON GROUP VIII METALS	544
A. Role of Hydrogen in CO Hydrogenation Kinetics	544
B. Effects of Hydrogen Adsorption Kinetics and Energetics on the Activity and Selectivity of Group VIII Metals in CO Hydrogenation	548
III. EFFECTS OF PREPARATION AND PRETREATMENT ON CO HYDROGENATION ACTIVITY/SELECTIVITY OF GROUP VIII METALS	549
IV. EFFECTS OF SUPPORTS AND PROMOTERS ON CO HYDROGENATION ACTIVITY AND SELECTIVITY OF GROUP VIII METALS	555
A. Effects of Supports in CO Hydrogenation	555
B. Effects of Promoters in CO Hydrogenation	560
V. TECHNOLOGICAL IMPLICATIONS	561
VI. SUMMARY	561
REFERENCES	562

I. INTRODUCTION

CO hydrogenation on group VIII metal catalysts comprises an important class of catalytic reactions, including methanation, Fischer-Tropsch synthesis, and alcohol synthesis. Hydrogen plays two important roles in these catalytic processes. (a) It is typically a reducing agent in the activation of synthesis catalysts, and this treatment determines the initial chemical states of the catalyst in these reactions. (b) Hydrogen is a reactant in CO hydrogenation; thus its adsorption kinetics and energetics as well as its interactions with surface CO/carbon species strongly affect the activity/selectivity behavior of a given catalyst.

    The objective of this chapter is to discuss the role of hydrogen in CO hydrogenation processes, especially in methanation and Fischer-Tropsch (FT) synthesis. Alcohol synthesis will not be treated here. In view of their commercial significance in FT synthesis and methanation, cobalt, nickel, and iron catalysts will be emphasized. Since the chemical states of the catalysts used in these processes are determined in large part by the preparation and activation procedures, the effects of these procedures on the activity/ selectivity behavior of syngas conversion catalysts will also be considered. Since in this short chapter it will not be possible to review the voluminous literature dealing with the catalysis, kinetics, and mechanisms of CO hydrogenation on metal catalysts, only the more recent articles are cited.

In Chapter 5 it was demonstrated that support, promoter, and pretreatment can greatly change the kinetics and energetics of hydrogen adsorption on cobalt, iron, and nickel catalysts. Since hydrogen adsorption is one of the important steps in CO hydrogenation, these changes could greatly affect the availability on the surface of reactant hydrogen, hence affecting the rate of conversion and product selectivity. Accordingly, in this chapter we will consider the relationship of the changes in hydrogen adsorption properties to changes in catalytic properties for CO hydrogenation on the same cobalt, iron, and nickel catalysts.

The understanding of how hydrogen participates in CO hydrogenation reactions and of the relationship between the chemical/physical and catalytic properties has important technological implications. For example, it enables the tuning of product distributions through careful design of the chemical/ physical properties of the catalyst. These technological implications and their potential applications are also discussed in this chapter.

II. ROLE OF HYDROGEN AS A REACTANT IN CO HYDROGENATION ON GROUP VIII METALS

A. Role of Hydrogen in CO Hydrogenation Kinetics

The chemistry of methanation and Fischer-Tropsch (FT) synthesis processes [1-5] can be described by the following set of reactions:

$$CO + 3H_2 = CH_4 + H_2O \tag{1}$$

$$CO + 2H_2 = -CH_2- + H_2O \tag{2}$$

$$CO + H_2O = CO_2 + H_2 \tag{3}$$

$$2CO = C + CO_2 \tag{4}$$

Reaction (1) is the formation of methane, Reaction (2) the synthesis of hydrocarbons heavier than methane, Reaction (3) the water-gas-shift reaction, and Reaction (4) the Boudouard reaction resulting in deposition of carbon. Reactions (1) and (2), the principal reactions in methanation and FT synthesis, involve hydrogen as a reactant, while Reaction (3) produces hydrogen. Reaction (4), although generally slow compared to the other three reactions, occurs more rapidly at low H_2/CO feed ratios and is thus also influenced by hydrogen.

The most widely accepted mechanisms for Reactions (1) and (2) [6-10] involve the following sequence of elementary steps (where M = a metal surface site):

$$H_2 + 2M \leftrightarrow 2M-H \tag{5}$$

$$CO + M \leftrightarrow CO-M \tag{6}$$

$$CO-M + M \rightarrow C-M + O-M \tag{7}$$

$$C\text{-}M + H\text{-}M \rightarrow CH\text{-}M + M \quad (8)$$

$$CH\text{-}M + H\text{-}M \rightarrow CH_2\text{-}M + M \quad (9)$$

$$CH_2\text{-}M + H\text{-}M \rightarrow CH_3\text{-}M + M \quad (10)$$

$$CH_3\text{-}M + H\text{-}M \rightarrow CH_4 + 2M \quad (11)$$

$$CH_3\text{-}M + CH_2\text{-}M \rightarrow CH_3CH_2\text{-}M + M \quad (12)$$

$$CH_3CH_2\text{-}M + M \rightarrow CH_2=CH_2 + M\text{-}H \quad (13)$$

$$CH_3CH_2\text{-}M + H\text{-}M \rightarrow CH_3\text{-}CH_3 + 2M \quad (14)$$

$$O\text{-}M + 2H\text{-}M \rightarrow H_2O + 3M \quad (15)$$

Steps (5) and (6) involve reversible dissociative and associative adsorptions of hydrogen and CO, respectively. In step (7) adsorbed CO is dissociated to atomic, carbidic carbon, which is hydrogenated by adsorbed atomic hydrogen in step 8; further hydrogenation of CH_x species occurs in steps (9) to (11). Step (12) represents the polymerization process leading to C_{2+} hydrocarbons in which methylene group monomers are added to alkyl groups serving as active centers, while steps (11), (13), and (14) represent termination by either hydrogen addition to form normal alkanes [steps (11) and (14)] or β elimination of hydrogen to form α-olefins [step (13)]. Not shown are other secondary reactions involving hydrogenation of olefins, hydrogenolysis, and isomerization.

Generally, the kinetics of reaction [6,7,9,10] are consistent with a fast, quasi-equilibrium for the adsorption of reactants and with either step (7), (8), or (9) limiting the rate of reaction. However, it has been shown that the rate determining step can shift with changes in temperature and reactant concentration [10].

Reaction kinetics of CO hydrogenation on supported nickel [10-14], cobalt, [11,15,16], and iron [11,17-19] have been reported in a number of papers. Representative kinetic parameters for CO hydrogenation on these catalysts are summarized in Table 1. It is evident from these data that the hydrogen reaction order for CO hydrogenation on base metal catalysts is typically about 0.5 to 1.0, while that for CO is either negative or near zero. Hence the rate of reaction is generally proportional to the concentration of hydrogen in the gas phase. If the rate-determining step involves hydrogen adsorption [Eq. (5)] or hydrogenation of atomic carbon or a CH_x intermediate [Eqs. (8), (10), and (12)], it follows that the reaction rate will be proportional to the fractional coverage of hydrogen on the surface. This observation has important implications considering that the physical and chemical properties of catalysts can significantly affect hydrogen coverage during reaction (see Sections II.B and III).

In Fischer-Tropsch synthesis hydrocarbon product selectivities are determined by the ability of a catalyst to catalyze chain propagation versus chain termination steps (see Fig. 1). The distribution of hydrocarbon products in Fischer-Tropsch synthesis is generally described by a chain polymerization kinetics model ascribed to Anderson, Schulz, and Flory [5,7,20], henceforth referred to as the Anderson-Schulz-Flory (ASF) model. The ASF product distribution is mathematically represented by the following two equations:

$$\frac{W_n}{n} = (1 - \alpha)^2 \alpha^{n-1} \quad (16)$$

$$\alpha = \frac{r_p}{r_p + r_t} \quad (17)$$

TABLE 1 Kinetic Parameters for CO Hydrogenation on CO, Fe, and Ni Catalysts

Catalysts	x (CO)[a]	y (H$_2$)[a]	E_{act}[b] kJ mol^{-1}	Ref.
5% Ni/Al$_2$O$_3$	-0.31	0.77	105	11
2, 10% Ni/SiO$_2$	Neg. to pos.	0.6 to 0.8	84 to 105	12
5% Ni/Al$_2$O$_3$	-1.0 to 0.0	0.6 to 1.0	70 to 130	10
5% Ni/SiO$_2$	-0.7 to 0.4	0.2 to 0.9	80 to 115	13
18% Ni/Al$_2$O$_3$	-1.0 to 0.5	-0.5 to 1.0	103	14
2% Co/Al$_2$O$_3$	-0.5	1.22	113	11
3-15% Co/Al$_2$O$_3$	—	—	96 to 146	15
6.2% Co/Al$_2$O$_3$	-1.5 to 0.5	-0.5 to 1.0	--	16
Promoted Fe	0	1	70	17
15% Fe/Al$_2$O$_3$	-0.05	1.14	89	11
Fused, promoted Fe	0 to 1	1	85	18
	-0.26 to 0.25[c]	0.67 to 1.15[c]	100[c]	19

[a] $r_{CH_4} = k[p_{CO}]^x [p_{H_2}]^y$
[b] Activation energy.
[c] Reaction order and activation energies were determined for each carbon product from C_1 to C_8 at reaction times of 14 and 96 h.

where n is the number of carbon atoms in the product, W_n the weight fraction of product containing n carbon atoms, α the chain growth probability, r_p the chain propagation rate, and r_t the chain termination rate. Generally, the value of α is obtained by a least-squares linear regression of the logarithmic form of Eq. (16), the slope and intercept yielding α:

$$\ln \frac{W_n}{n} = \ln (1 - \alpha)^2 + (n - 1)\ln \alpha \qquad (18)$$

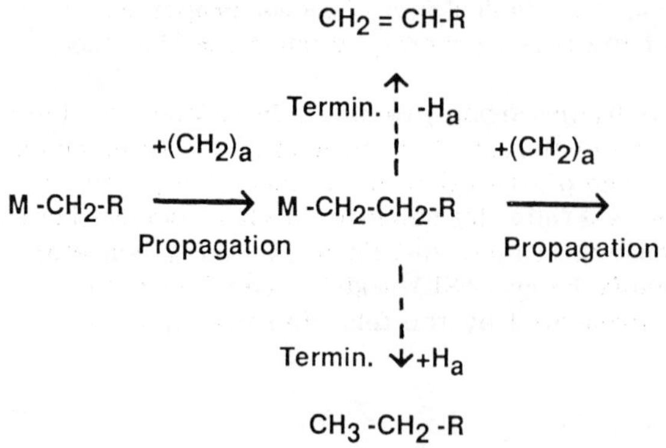

```
                        CH₂ = CH-R
                            ↑
                   Termin.  ¦ -Hₐ
           +(CH₂)ₐ          ¦        +(CH₂)ₐ
  M-CH₂-R ─────────→ M-CH₂-CH₂-R ─────────→
           Propagation      ¦        Propagation
                            ¦
                   Termin.  ↓ +Hₐ
                        CH₃-CH₂-R
```

FIGURE 1 Chain growth and termination in Fischer-Tropsch synthesis. Solid arrows indicate propagation; dashed arrows indicate termination.

TABLE 2 Power-Law Rate Expressions[a] for the Synthesis of C_1 Through C_4 Hydrocarbons[b] on Ru/Alumina

C_n	A [atm$^{(m-n)}$ s^{-1}]	m	n	E_a (kcal mol^{-1})	% Dev.[c]
C_1	1.3×10^9	1.35	-0.99	28	5.6
$C_2^=$	2.5×10^8	0.74	-0.68	28	5.7
C_2	1.6×10^6	1.34	-0.81	25	11.3
$C_3^=$	2.3×10^7	0.82	-0.58	25	4.2
C_3	1.4×10^3	1.39	-0.55	18	5.8
$C_4^=$	3.8×10^6	0.70	-0.44	24	9.6
C_4	8.7×10^3	1.14	-0.47	19	4.8

[a] $N_{Cn} = A \exp(-E_a/RT)\, (p_{H_2})^m (p_{CO})^n$.
[b] Reaction conditions: T = 448 to 548 K; P = 1 to 10 atm; H_2/CO = 1 to 3.
[c] Average deviation between predicted and observed rates.
Source: Ref. 9.

There are relatively few data available on the kinetics of propagation and termination. In fact, Wojchiechowski and Taylor [21] state that these rates cannot be determined experimentally from steady-state data. Nevertheless, Fu et al. [20] have demonstrated how rates of propagation and termination can be estimated for CO hydrogenation on catalysts in which the degree of branching is negligibly small. Their results show that rates of propagation and termination have different concentration and temperature dependencies, thus providing a basis for explaining why propagation probabilities (α values) change with H_2/CO and temperature. For example, the higher activation energy for termination relative to that for propagation explains the decrease in the average molecular weight of the hydrocarbon product for 10% Co/alumina with increasing temperature. Since the rate of termination increases more significantly relative to that for propagation with increasing H_2/CO ratio, it follows that the termination rate is more highly dependent on hydrogen concentration relative to that of propagation.

In a detailed study of the kinetics of CO hydrogenation on a 1% Ru/alumina catalyst, Kellner and Bell [9] determined the rate expressions and kinetic parameters for formation of paraffins and olefins in the range C_1 to C_{10}. Their kinetic data for C_1 to C_4 hydrocarbons, summarized in Table 2, indicate that hydrogen reaction orders are significantly higher (in the range 1.1 to 1.3) for formation of paraffins relative to olefins (0.7 to 0.8) but reasonably independent of carbon number. Moreover, the activation energies for formation of olefins are larger than those for formation of paraffins, indicating that the olefin/paraffin ratio should increase with increasing temperature. However, it was found experimentally that olefin/paraffin ratios increase through a maximum, as the olefins formed at higher temperature are readily hydrogenated. Thus it is clear that the selectivity for olefins is dependent on both temperature and hydrogen concentration.

Based on the reaction mechanism presented in Eqs. (5) to (15), Kellner and Bell [9] derived the following expression for the propagation probability α in terms of the rates of propagation and termination:

$$\alpha = \frac{k_p \theta_{CH_2}}{k_p \theta_{CH_2} + k_{to} \theta_v + k_{tp} \theta_H} \tag{19}$$

where k_p is the rate constant for propagation; k_{to} and k_{tp} are rate constants for termination to olefins and paraffins, respectively; θ_{CH_2} is the coverage of methylene species; θ_H is the coverage of hydrogen atoms; and θ_v is the coverage of vacant sites. From this expression it is evident that the rate of propagation is proportional to the coverage of methylene species, while the rates of termination to olefins and paraffins are proportional to the fraction of vacant sites and the hydrogen atom coverage. Thus, the reaction model predicts that both the rate of termination and the molecular weight of the hydrocarbon product are highly influenced by the coverage of hydrogen on the surface during reaction.

One of the yet unsettled mechanistic controversies in CO hydrogenation regards the role of hydrogen in the CO dissociation reaction [Eq. (7)] on methanation and FT catalysts. While some authors propose that CO dissociation is assisted by hydrogen [12,22], others [6,7,10] argue that hydrogen simply facilitates the process by hydrogenating the products of the dissociation [Eqs. (8) and (15)]. Evidence in favor of a hydrogen-assisted CO dissociation on iron was recently presented by Bianchi and Bennett [22]. They argued that the hydrogen-assisted dissociation explains concentration transients of carbon, carbon dioxide and water. Nevertheless, there are clearly alternative explanations for these transients. Moreover, the fact that (a) CO dissociation occurs readily on cobalt, nickel, iron and many other group VIII metals at relatively low temperature (e.g., 450 to 550 K), that (b) the carbon formed in dissociation is hydrogenated at rates comparable with CO dissociation [6,7,10], argue against the need for postulating a hydrogen-assisted dissociation.

B. Effects of Hydrogen Adsorption Kinetics and Energetics on the Activity and Selectivity of Group VIII Metals in CO Hydrogenation

Although it is true that hydrogen adsorbs rapidly during CO hydrogenation on most methanation and FT catalysts there are some important exceptions to this trend. These exceptions may occur in CO hydrogenation on catalysts involving either highly activated hydrogen adsorption or strongly bound hydrogen states.

In Chapter 5 data were presented providing evidence that hydrogen adsorption kinetics and energetics on group VIII metals can be dramatically affected by the choice of calcination pretreatment, reduction temperature, support, and promoter. For example, it was shown that precalcination of Fe/silica at 473 K (see Fig. 10 in Chapter 5), reduction of Ni/titania at or above 673 K (Figs. 5 to 7 in Chapter 5), interaction between nickel and alumina in Ni/alumina catalysts of low loading (Figs. 3 and 4 in Chapter 5), and promotion of Fe/silica with potassium (Fig. 10 in Chapter 5) all cause the appearance of new, highly activated adsorption states of high binding energy. The appearance of these new adsorption states was attributed in the first three cases to decoration of metal crystallites by support moieties and in the last case to decoration of metal crystallites by potassium oxide species. In most of these catalyst systems, the new high-temperature adsorption states are filled and/or desorbed only at temperatures in or above the range typical of CO hydrogenation on these catalysts (500 to 675 K). Accordingly, the kinetics of hydrogen adsorption may limit the rate of CO hydrogenation and lead to low hydrogen atom surface coverages during reaction in some of these catalyst systems. Moreover, the support moieties and/or

20. Role of Hydrogen in CO Hydrogenation

promoters present on the metal surface block sites for hydrogen adsorption and thus physically lower the hydrogen surface coverage during reaction.

Based on the discussion in Section II.A, it is expected that a decrease in hydrogen surface coverage during CO hydrogenation on group VIII metals due to kinetic, chemical, or physical limitations would lead to (a) lower rates of reaction, since the rate is proportional to hydrogen coverage, and (b) significantly higher selectivities to olefins, since the olefin/paraffin ratio is greater at lower hydrogen coverages. Indeed, the rates of CO hydrogenation on Fe/silica are lowered by precalcination and promotion with potassium [23] and are lower in Ni/alumina and Co/alumina catalysts of low loading relative to those of high loading [15,24]. Nevertheless, specific rates of CO hydrogenation on Ni/titania, reduced at progressively higher temperatures, increase [24,25] rather than decrease. In this case, hydrogen coverage is *increased* because the TiO_x decorant (a) increases the binding energy of hydrogen [26] (see Chapter 5), (b) decreases the binding energy of CO and its ability to compete with hydrogen for surface sites or to inhibit the reaction [26], and (c) increases the rate of CO dissociation [27].

In all of these catalyst systems, the above-mentioned effects lead to higher olefin selectivities relative to untreated, unpromoted, or conventional catalysts. Indeed, olefin/paraffin ratios are significantly higher in CO hydrogenation on precalcined Fe/silica relative to uncalcined Fe/silica [23] and on potassium-promoted Fe/silica relative to unpromoted Fe/silica [23]. Moreover, Ni/titania catalysts reduced at 673 K or higher as well as Ni/alumina and Ni/silica catalysts of low metal loadings produce significantly higher fractions of C_{2+} hydrocarbons than do Ni/silica and Ni/alumina catalysts of moderate or high loadings [24,25]. Thus it is clear that changes in hydrogen adsorption kinetics, energetics, and coverage due to crystallite decorants may account for many, maybe even most of the interesting modifications in CO hydrogenation activity and selectivity of group VIII metals as a result of changes in preparation, pretreatment, support, and promoter level. Examples of these phenomena are presented in greater detail in Sections III and IV.

III. EFFECTS OF PREPARATION AND PRETREATMENT ON CO HYDROGENATION ACTIVITY/SELECTIVITY OF GROUP VIII METALS

The effects of preparation on the CO hydrogenation activity and selectivity of nickel on different supports were investigated by Bartholomew et al. [24]. Their data, summarized in Fig. 2, show that specific activities and methane selectivities for Ni/silica are very similar to those for unsupported nickel. However, Ni/alumina and Ni/titania catalysts are significantly more active; moreover, those prepared by controlled pH precipitation [24,28,29] are significantly more active and selective for C_{2+} hydrocarbons relative to those prepared by impregnation. Hadjigeorghiou and Richardson [30] found similarly that coprecipitated Ni/alumina catalysts, activated by a short calcination at 773 K followed by reduction at 773 K, were remarkably selective for C_{2+} hydrocarbons at 498 K (75% yield of C_{2+} hydrocarbons at a conversion of 35%, 20 to 30% yield at 100% conversion). The method of calcination was found to significantly affect activity and selectivity; moreover, activity and C_{2+} hydrocarbon selectivity increased with increasing reduction temperature up to 900 K. The same investigators [30] observed similar behavior for nickel on other supports when prepared and pretreated in the same manner. However, precipitated, precalcined Ni/ThO_2 catalysts (reduced at 773 K) were found to have the highest selectivities for C_{2+} hydrocarbons (70 wt % C_2 to C_8 hydrocarbons at 100% conversion). The authors speculated that the nickel-support

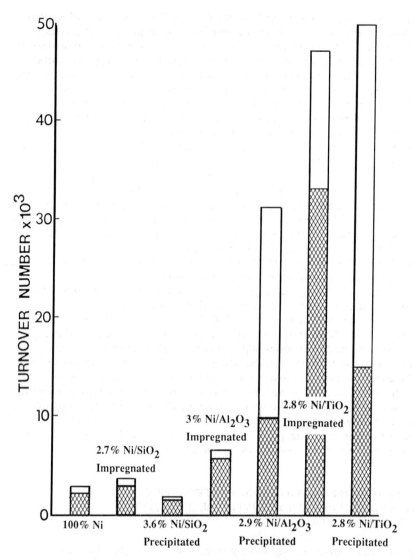

FIGURE 2 Effects of support and preparation on methane turnover frequency of nickel at 525 K: Shaded bar is proportional to the CH_4 turnover frequency; unshaded bar denotes the C_{2+} hydrocarbon turnover frequency; total bar length is CO turnover frequency. (From Ref. 24.)

interface is maximized by the precipitation preparation and calcination pretreatment and that the interface promotes CO "adlineation" (simultaneous bonding of the carbon atom to the metal and the oxygen atom to the support at the interface to facilitate CO dissociation) leading to the observed high selectivities for postmethane hydrocarbons.

Reuel and Bartholomew [15] found that the effects of catalyst preparation on activity/selectivity behavior of cobalt varied with support. For example, 3% Co/silica prepared by controlled pH-precipitation was observed to be less active and less selective for C_{2+} hydrocarbons than 3% Co/silica prepared by impregnation; on the other hand, 3% Co/alumina prepared by precipitation was a factor of 6 more active and produced a factor of 2 less methane than 3% Co/alumina prepared by impregnation. They also found the by increasing reduction temperature for Co/alumina and Co/titania catalyst from 650 or 675 K to 800 K, the average carbon number of the hydrocarbon product increased by 10 to 15%.

Castner and Santilli [31] found the order of activity of supported cobalt catalysts precalcined at 783 to 813 K, namely Co/silica > Co/titania > Co/alumina, to be different from that observed by Reuel and Bartholomew [15]. The precalcined Co/alumina contained three different cobalt species, Co_3O_4, Co(II), and Co aluminate, which could be reduced at 673 to 700 K, 700 to 973 K, and >1123 K, respectively; precalcined Co/silica, on the other hand, contained only the easily reducible Co_3O_4. These workers also observed that the amount of cobalt exposed to the surface in Co/titania and Co/alumina decreased when these catalysts were reduced at high temperatures (753 K and 973 K, respectively).

These interesting changes in activity/selectivity behavior observed in these four studies of nickel and cobalt [15,24,30,31] might be explained in a number of ways; however, the results are most consistent with variations in the degree of metal crystallite decoration by support moieties with variations in reduction temperature, precalcination treatment, and preparation method. Indeed, the ESCA data of Castner and Santilli provide direct evidence of either decoration of cobalt by support-derived species or diffusion of cobalt into the support during reduction at high temperatures. Further evidence for decoration effects is documented for these same catalysts in Chapter 5.

Lohrengel et al. [32] studied the effects of increasing reduction temperature from 573 K to 773 K on the adsorption and catalytic properties of a precipitated Fe/Mn/Zn/Cu/K CO catalyst. While BET surface area and pore-size distributions were little affected by the increasing reduction temperature, the heat of adsorption of hydrogen was increased 30% while that of ethylene was decreased by a factor of 2. Specific activity decreased four to five times and the selectivity toward olefins and short-chain hydrocarbons was significantly enhanced. Since ESCA studies revealed no significant difference in surface composition, the authors attributed the differences in behavior to formation of different catalytic surface compounds at the two different reduction temperatures. Nevertheless, the increase in the heats of adsorption for hydrogen and the decrease in the heat of adsorption for ethylene are consistent with an increasing degree of contamination of the iron surface with Mn, Cu, Zn, and K oxides, assuming that this occurs without a change in overall surface composition, due to a more uniform spreading of these contaminants over the surface in the form of small aggregates during reduction at the higher temperature (see the proposed model in Fig. 3).

Dictor and Bell [19] studied the influence of reduction temperature on the physical and activity/selectivity properties of a fused iron catalyst. They observed a marked increase in BET surface area with increasing reduction temperature which they attributed to the formation of large pores upon more severe reduction. Nevertheless, the surface composition/structure of the catalyst was not totally independent of the extent of reduction, since methane selectivity decreased and hydrogen reaction order increased with increasing reduction temperature and/or reaction time.

Similar but more dramatic changes in activity and selectivity due to precalcination of Fe/K/silica catalysts (reduced 36 h at 723 K) were observed by Rankin and Bartholomew [23]. Their data, summarized in Table 3 and illustrated in Figs. 4 and 5, indicate that the activity and activation energy of Fe/silica decrease while selectivity for light olefins increases with increasing temperature of precalcination and increasing potassium content. While the product distribution of 15% Fe/3% K/silica precalcined at 373 K is typical of FT synthesis on iron and consistent with the Anderson-Schulz-Flory model

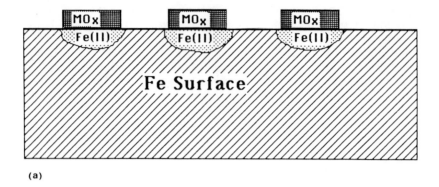

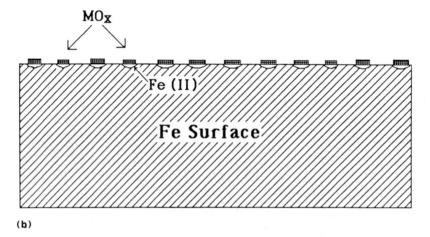

FIGURE 3 Proposed model for effects of reduction temperature on promoted iron catalyst. (a) Reduction at 300°C; iron crystallites are covered with large aggregates of Mn, Zn, Cu oxides. (b) Reduction at 500°C; promoter oxide aggregates breakup into small moieties which interrupt the surface ensembles for reaction. Some alloying of Fe with Cu may occur, reducing activity further. However, the overall surface composition is the same as in part (a). (Model based on Ref. 32.)

(Fig. 4), the product of the same catalyst precalcined at 473 K consists only of methane, ethylene, and propylene. The significantly lower activation energy suggests that a different mechanism is operative on the catalyst calcined at the higher temperature. The authors attributed this dramatic change in catalytic behavior to increases in the adsorption activation energy for hydrogen and hence a decrease in hydrogen surface coverage with increasing precalcination temperature (see Chapter 5). These changes in the hydrogen adsorption properties may be the result of greater silica or potassium silicate decoration as a result of the precalcination at higher temperature [15] (Chapter 5).

TABLE 3 Effects of Potassium Level and Calcination Temperature on Activity and Selectivity of K-Promoted Fe/SiO$_2$ Catalysts[a]

Catalyst	Calcin. temp. (K)	Reactor temp. (K)	$N_{CO} \times 10^3$ (s^{-1})[b]	CO_2 select.[c]	Hydrocarbon selectivity[d] (wt %)				Olefin Paraffin[e]	Activation energy (kJ mol^{-1})[f]
					C_1	C_2-C_4	C_{5+}	C_{OH}		
15% Fe/SiO$_2$	373	498	3.2	47.0	30.2	50.9	5.7	13.2	0.33	93
	473	473	1.6	5.8	26.5	46.0	23.1	4.4	1.1	110
15% Fe/0.2% K/SiO$_2$	473	473	0.27	30.4	19.0	81.0	0	0	8.2	104
15% Fe/1% K/SiO$_2$	473	473	0.12	25.6	18.5	81.6	0	0	321	72
15% Fe/3% K/SiO$_2$	473	473	0.063	2.4	20.8	79.2	0	0	>400	32
15% Fe/3% K/SiO$_2$	373	473	5.9	14.4	15.2	32.1	50.6	2.1	1.5	124

[a]H_2/CO = 2, 1 atm.
[b]CO turnover frequency: molecules CO reacted per hydrogen adsorption site per second.
[c]Mole percent of converted CO appearing as CO_2.
[d]Weight percent hydrocarbon groups in the hydrocarbon product.
[e]C_2 to C_4 fraction.
[f]Activation energy for CO conversion determined from CO turnover frequencies in the temperature range 473 to 523 K.
Source: Adapted from Ref. 23.

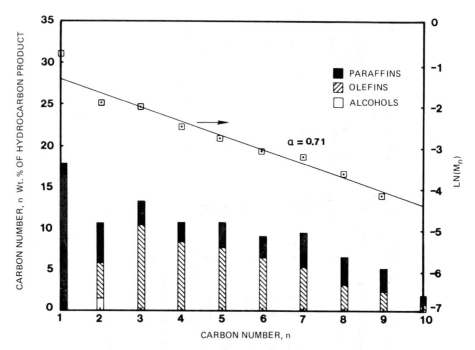

FIGURE 4 Hydrocarbon product distribution and Anderson-Schulz-Flory plot. 15% Fe/3% K/silica calcined at 373 K; reaction at 498 K, $H_2/CO = 2$, 1 atm. (From Ref. 23.)

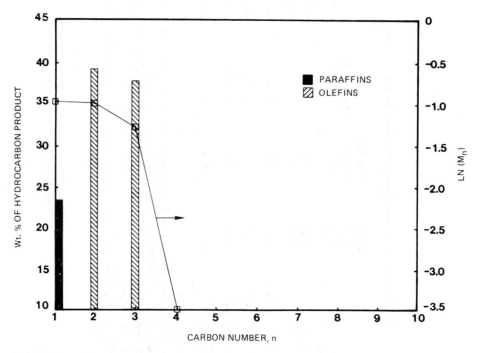

FIGURE 5 Hydrocarbon product distribution and Anderson-Schulz-Flory plot. 15% Fe/3% K/silica calcined at 473 K; reaction at 498 K, $H_2/CO = 2$, 1 atm. (From Ref. 23.)

IV. EFFECTS OF SUPPORTS AND PROMOTERS ON CO HYDROGENATION ACTIVITY AND SELECTIVITY OF GROUP VIII METALS

A. Effects of Supports in CO Hydrogenation

The influence of support and dispersion on CO hydrogenation activity/selectivity properties of nickel were first detailed in papers by Vannice and Garten [25] and Bartholomew et al. [24]. Their results, summarized in Tables 4 and 5 and in Fig. 2, show increasing activity in the order Ni/silica > Ni/alumina > Ni/titania and decreasing activity with increasing dispersion. Ni/titania was found to produce a higher fraction of C_{2+} hydrocarbons relative to the other catalysts. Moreover, Bartholomew et al. [24] found that C_{2+} yields of Ni/silica and Ni/alumina catalysts increased with increasing dispersion (see Table 4 and Fig. 6). These results, initially interpreted in terms of electronic interactions between the support and metal crystallites, are nevertheless consistent with an increasing degree of crystallite decoration by support moieties with increasing dispersion and in the order Ni/silica > Ni/alumina > Ni/titania (see Chapter 5 for supporting evidence).

Recent studies confirm that supports can greatly influence the activity and selectivity of nickel in CO hydrogenation [33,34]. For example, Turlier et al. [34] investigated the influence of support on nickel reducibility and the activity/selectivity behavior of nickel in CO hydrogenation. Their reported sequence of activity for CO conversion was titania > zirconia > thoria > unsupported > alumina, chromia, magnesia, silica > ceria. Since activity correlated to some extent with the ease of reducibility of nickel, the authors ascribed these effects to the presence of unreduced nickel patches of variable concentration on the nickel crystallites. Nevertheless, C_{2+}

TABLE 4 Specific Activities of Ni/TiO$_2$ Relative to Other Ni Catalysts for the CO-H$_2$ Reaction[a]

Catalyst	N_{CO} (s^{-1} × 10^3)		N_{CH4} (s^{-1} × 10^3)	
	H$_2$ (fresh)	H$_2$ (used)	H$_2$ (fresh)	H$_2$ (used)
1.5% Ni/TiO$_2$	500	1142	231	528
1.5% Ni/TiO$_2$ (100)[b]	16		7.4	
10% Ni/TiO$_2$	1607	2500	196	305
10% Ni/TiO$_2$ (100)[b]	90		11	
5% Ni/η-Al$_2$O$_3$	19.6	44	16.4	37
8.8% Ni/η-Al$_2$O$_3$	10.7	128	7	85
42% Ni/α-Al$_2$O$_3$	58	109	23.8	43
30% Ni/α-Al$_2$O$_3$	32.3	35	16.6	18
16.7% Ni/SiO$_2$	14.8	47	10.7	34
20% Ni/graphite	43.1	79	27.8	51
Ni powder	30	18	26.6	16

[a]Reaction conditions: 548 K (275°C), 103 kPa (1 atm), H$_2$/CO = 3.
[b]Calculated assuming 100% nickel metal exposed.
Source: Ref. 25.

TABLE 5 Effects of Support and Preparation on CO Hydrogenation Activity/Selectivity Properties

Catalyst	Percentage dispersion[a]	Turnover number[b] × 10^3 (s^{-1}) at 525 K		CH_4 yield[c]
		N_{CO}	N_{CH_4}	
100% Ni (INCO)	0.05	2.8	2.2	0.79
2.7% Ni/SiO_2 (impregnated)	16	3.9	2.6	0.66
3.6% Ni/SiO_2 (precipitated)	26	1.7	1.4	0.82
3% Ni/Al_2O_3 (impregnated)	22	6.4	5.7	0.89
2.9% Ni/Al_2O_3 (precipitated)	22	31	9.8	0.32
2.8% Ni/TiO_2 (impregnated)	11(18)[d]	47(29)[d]	33(20)[d]	0.70
2.8% Ni/TiO_2 (precipitated)	12	50	15	0.30

[a]Percentage Ni exposed based on H_2 adsorption.
[b]Molecules CO converted or CH_4 produced per nickel site per second based on H_2 adsorption.
[c]Mole fraction of converted CO appearing as methane.
[d]Percentage dispersion or turnover number based on transmission electron microscopy.
Source: Ref. 24.

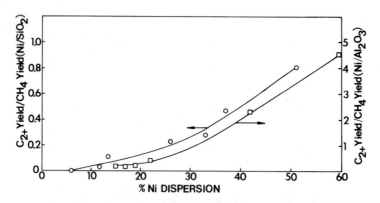

FIGURE 6 Ratio of C_{2+} hydrocarbon and methane yields versus percent nickel dispersion. ○, Ni/silica at 525 K; □, Ni/alumina at 500 K. (From Ref. 24.)

TABLE 6 Turnover Frequencies and Activation Energies for Conversion of CO Hydrogenation on Cobalt Catalysts

Catalyst	$N_{CO} \times 10^3$ (s^{-1})[a] at 225°C	E_{CO}[b] (kJ mol^{-1})
100% CO	5.8	95
Co/SiO$_2$		
3%	5.5	67
10%	7.5[c]	69
Co/Al$_2$O$_3$		
1%	d	d
3%	2.8[c]	96
10%	12	98
15%	63[c]	146
Co/TiO$_2$		
3%	25	96
10%	38[c]	142
Co/MgO		
3%	d	d
10%	0.13[c]	164
Co/C (Spheron)		
3%	0.64[c]	153
10%	5.9	121

[a]Turnover frequency for CO conversion (to hydrocarbons and CO_2), i.e., the number of CO molecules converted per catalytic site (based on total H_2 uptake) per second at 1 atm, H_2/CO = 2, 225°C. These data were measured within a few minutes of initial reaction and hence correspond to initial activities.
[b]Activation energy for CO conversion based on the temperature dependence of N_{CO} at three or four different temperatures.
[c]Extrapolated values; in most cases the extrapolation was over a small (25 to 50°C) range of temperature.
[d]Inactive up to 400°C.
Source: Ref. 15.

selectivity varying from 0.6 (Ni/Cr$_2$O$_3$) to 0.1 (Ni/silica) did not correlate with reducibility. Unfortunately, their model fails to explain why some catalysts are more and some less active than completely reduced, unsupported nickel. It appears that the crystallite decoration model is more consistent with these results, since some support decorants might promote activity while other support contaminants would decrease activity.

The effects of support, loading, and dispersion on the CO hydrogenation activity/selectivity properties of cobalt were studied by Reuel and Bartholomew [15] and Lu and Bartholomew [35]. Reuel and Bartholomew [15] observed variations in the initial specific activity spanning three orders of magnitude for cobalt supported on alumina, silica, titania, magnesia, and carbon (see Table 6); the order of decreasing activity from their data is Co/titania > Co/silica > Co/alumina > Co/carbon > Co/magnesia. For a given cobalt/support system (e.g., cobalt/alumina) both initial and steady-state activities were found to increase with increasing loading; in fact, a linear increase in steady-

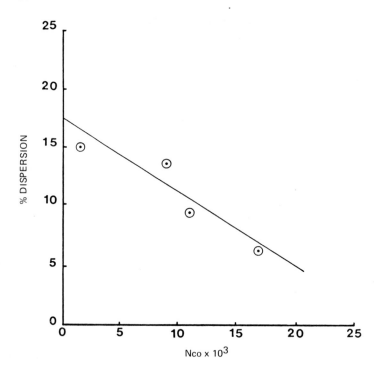

FIGURE 7 Percentage dispersion versus CO turnover frequency at 473 K, 1 atm for 3, 10, 15, 25% Co/alumina catalysts after 20 h of reaction. (Adapted from Ref. 35.)

state activity with decreasing dispersion was observed (see Fig. 7). Reuel and Bartholomew also observed a linear increase in the average carbon number of the hydrocarbon product with ln[dispersion] for the same group of catalysts. Similarly, Fu and Bartholomew found that the average carbon number of the hydrocarbon product of Co/alumina increased linearly with decreasing dispersion (see Fig. 8).

While these results might be explained by a primary structure sensitivity (i.e., variations in activity due to changes in site geometry with changes in crystallite size) [9,35], the application of a geometric model to explain these data is not consistent with studies of CO adsorption on single-crystal cobalt. Indeed, the work of Bridge et al. [36] and Prior et al. [37] indicates that CO does not adsorb dissociatively on the smooth basal plane of cobalt [i.e., Co(0001)] but does dissociate on the stepped planes [i.e., Co(1012)]. Since CO dissociation is one of the necessary elementary steps in CO hydrogenation (see Section II.A), these results predict higher CO hydrogenation activity for stepped surfaces. However, it was recently observed in this laboratory [38] that CO does *not* dissociate on 1% Co/alumina, a catalyst having no measurable activity for CO hydrogenation (see Table 6) but does adsorb dissociatively on the highly active 10 and 15% Co/alumina catalysts (Table 6). Thus the trend of decreasing activity with increasing dispersion for the alumina-supported cobalt is opposite to that predicted from the single-crystal results, since the more highly dispersed 1% Co/alumina should contain metal clusters with a higher incidence of step planes, corners, and edges relative to the catalysts of higher loading. This observation leads to the conclusion that secondary structure sensitivity due to preferential contamination of small crystallites by support moieties or carbon is probably responsible for the

20. *Role of Hydrogen in CO Hydrogenation*

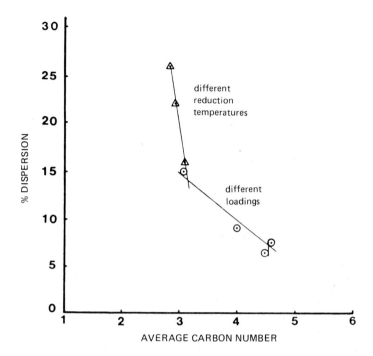

FIGURE 8 Correlations of dispersion with average carbon number for Co/alumina catalysts of different loadings and reduction temperatures. △, Different reduction temperatures; ○, different loadings.

lower activity of 1% Co/alumina. Preferential contamination of the low loading catalysts with carbon during reaction is less likely, since the same trend is observed for both initial and steady state rates on Co/alumina catalysts.

The Fe/carbon system is one of the more interesting and unique metal/support systems in CO hydrogenation catalysis. Studies by Vannice et al. [39], Jung et al. [40], and Jones et al. [41] indicate that Fe/C catalysts have typically higher activities and markedly higher olefin selectivities than do unpromoted Fe/alumina and Fe/silica catalysts (see Table 7). As in the

TABLE 7 Comparison of Steady-State Activities and Selectivities of Iron Catalysts[a]

Catalyst	Turnover frequency[b] $\times 10^3$ (s^{-1})	Olefin/paraffin[c] (wt ratio)
Fe	4.0	0.61
10% Fe/carbon	2.8	4.1
15% Fe/silica	0.36	1.2
15% Fe/alumina	0.21	0.72

[a]1 atm, 498 K, $H_2/CO = 2$, 24 h.
[b]Molecules of CO converted per catalytic site per second.
[c]C_2 to C_4 fraction.
Source: Ref. 41.

case of Co/alumina, the activity of Fe/C decreases with increasing dispersion. Vannice et al. [39] and Jung et al. [40] have attributed the unusual activity/ selectivity behavior to promotion of iron by the carbon support, analogous to promotion by potassium. However, the question arises why this high activity and high olefin selectivity are observed in Fe/C catalysts of low dispersions, in which the crystallites are sufficiently large (4 to 40 nm) to preclude intimate contact of the carbon support with a significant fraction of the iron surface. Again, decoration of the metal by the support could account for these effects. In fact, Snel [42] designed a catalyst of high activity and high olefin selectivity comparable to those of Fe/C by introducing carbon to the iron surface through partial degradation of iron citrate complexes involving the precursor $Fe_2O_3C_x$. This result strongly suggests that carbon contamination of the iron surface by the support is responsible for the unique properties of Fe/C catalysts.

B. Effects of Promoters in CO Hydrogenation

A number of recent studies indicate that potassium promotion affects the adsorption and activity/selectivity properties of iron in many significant ways [23, 43-46]. Benziger and Madix [43] found that potassium enhances CO and H_2 binding strengths and the amount of CO dissociated relative to the clean surface. Rankin and Bartholomew [23] observed similarly that CO was more easily dissociated on potassium-promoted Fe/silica and that the binding energy for hydrogen and the activation energy for hydrogen adsorption were increased on Fe/K/silica relative to those for unpromoted Fe/silica; moreover, they observed a dramatic shift of hydrogen adsorption states to higher temperature and a further large increase in the activation energy for hydrogen adsorption when Fe/K/silica was calcined at 473 K prior to reduction.

Arakawa and Bell [44] reported that potassium promotion of Fe/alumina causes a decrease in iron dispersion, an increase in CO adsorption strength, a decrease in the CO conversion turnover frequency, an increase in average molecular weight and olefin-to-paraffin ratio of the product, and an increase in water-gas-shift activity. Rankin and Bartholomew [23] observed similarly that potassium decreased the dispersion of Fe/silica by blocking the surface for hydrogen and CO adsorption. They also observed an increase in the average molecular weight and olefin/paraffin ratio of the product for potassium-promoted catalysts precalcined at only 373 K (see Table 3); however, the selectivity of catalysts precalcined at 478 K was dramatically shifted by potassium addition toward production of light olefins; indeed, a 15% Fe/3% K/ silica catalyst precalcined at 473 K produced only methane, ethylene, and propylene (see Table 3). Moreover, the specific activity of Fe/silica calcined at 373 K *increased* with addition of potassium in contradiction to the results of Arakawa and Bell, while the activity of the same catalyst calcined at 473 K was dramatically lowered by addition of potassium (see Table 3).

Bonzel and Krebs [45] found that K_2CO_3 promotion of an iron foil caused a decrease in the rate of methane formation, an increase in the rate of carbon deposition, and a shift in selectivity toward longer hydrocarbon chains. Dwyer [46] found that the nature and amount of carbon deposited on iron powder was greatly influenced by potassium; in fact, while the surface and bulk of the unpromoted iron catalyst was carbided, there was no evidence of a carbonaceous deposit on the iron surface. On the other hand, a heavy, multilayer hydrocarbonaceous deposit was observed on the potassium-promoted catalyst, following reaction.

Considered as a whole, the data from these previous studies suggest that potassium affects the selectivity for CO hydrogenation by changing the kinetics and energetics of the adsorption of the reactants, H_2 and CO, thereby affecting their relative surface coverages during reaction. Indeed, the significant increase in the activation energy for hydrogen adsorption, together with the increasing rate of CO dissociation, leads to a carbon-rich, hydrogen-poor surface. It follows that a lower H/C ratio favors the formation of hydrogen-poor, olefinic products.

Rankin and Bartholomew [23] speculated that in addition to potassium oxides, silica or potassium silicate species partially cover the surface of Fe/K/silica catalysts precalcined at higher temperatures (e.g., 473 K), leading to even lower H/C surface coverage ratios and hence lower activities and higher olefin selectivities. Thus it appears that either promoter or support moieties can be used to control the relative surface coverages of hydrogen and carbon during reaction and therewith the product distribution.

V. TECHNOLOGICAL IMPLICATIONS

The data presented in this chapter provide evidence that the activity and selectivity properties of group VIII metals in CO hydrogenation can be greatly influenced by the presence on the metal surface of reduced support moieties and/or promoter species. The relative concentration of these adspecies is largely a function of the preparation and pretreatment of the catalyst. Upon introduction to the surface these adspecies affect the relative rates and strengths of hydrogen and CO adsorptions and thereby the relative surface coverages of hydrogen and CO or carbon during reaction; changes in the H/C surface coverage ratio can lead to very significant changes in product distribution. For example, it was previously shown that TiO_x adspecies increase hydrogen adsorption strength relative to that of carbon monoxide on nickel, leading to high rates of CO conversion as a result of less CO inhibition, while in the case of potassium promotion, surface coverage of hydrogen is markedly decreased relative to that of carbon, leading to a higher fraction of olefinic products.

Thus, by controlling the degree of support and/or promoter decoration, it is possible to control, within the constraints of the Anderson-Schulz-Flory model, the hydrocarbon product distribution. There are a number of options for maximizing the yield of a specific product group (e.g., gasoline or diesel fuel): (a) use of pretreatments, support decorants, or promoters that maximize olefin production, followed by oligomerization to gasoline or diesel fuel [47]; (b) choice of pretreatments, support decorants, and/or promoters that maximize heavy molecular weight products which can then be hydrocracked to diesel fuel and gasoline [47]; and (c) use of shape-selective supports to limit chain growth of heavy waxes [48]. Clearly, additional research and development activities are needed to develop better understanding and control of the effects of support decorants and promoters on product distribution; nevertheless, the general principles and approach, as outlined in this chapter, should be clear.

VI. SUMMARY

In this short chapter we have focused on the role of hydrogen as a reactant in CO hydrogenation. The following conclusions follow from our discussion regarding this role.

1. The hydrogen reaction order in CO hydrogenation on cobalt, iron, and nickel catalysts is typically 0.5 to 1.0. Hence the reaction rate is

generally proportional to the concentration of hydrogen in the gas phase and to the fractional coverage of hydrogen on the surface.

2. Mechanistically, it follows that if the rate-determining step involves either hydrogen adsorption or hydrogenation of atomic carbon or CH_x intermediates, the reaction rate will be proportional to the fractional coverage of hydrogen on the surface. Since CO hydrogenation rates are generally proportional to hydrogen surface coverage, it appears that one of these two steps is usually rate determining or co-rate determining.

3. In Fischer-Tropsch synthesis, hydrocarbon product selectivities are determined by the ability of a catalyst to catalyze chain propagation versus chain termination steps. Rates of chain termination are generally more dependent on hydrogen concentration than are chain propagation rates. Hydrogen reaction orders are also higher for the production of paraffins (1.1 to 1.3) than for olefins (0.7 to 0.8).

4. Although hydrogen adsorption during reaction is rapid on most synthesis catalysts, there are a number of important exceptions to this trend as a result of highly activated hydrogen adsorption or the existence of high-temperature hydrogen adsorption states. In either of these cases the surface coverage of hydrogen during reaction can be significantly lower relative to the case in which the adsorption process is facile.

5. Activity and selectivity in CO hydrogenation on group VIII metals is determined in large part by the coverage of hydrogen relative to carbon monoxide. This relative coverage of reactants on group VIII metals during CO hydrogenation is dependent in turn on the energetics and kinetics of hydrogen and CO adsorptions.

6. The kinetics and energetics of hydrogen and CO adsorptions (and hence their relative surface coverages) are greatly influenced by the concentrations of support and promoter adspecies, which depend in turn on at least three important catalyst variables: (a) preparation, (b) pretreatment, and (c) the relative concentrations of active phase, support, and promoter. Generally, the concentration of adspecies is enhanced by aqueous impregnations or precipitations, high-temperature reduction, easily reducible supports such as titania, and highly mobile additives such as potassium oxide.

7. It follows that the activity and selectivity of CO hydrogenation catalysts can be tuned, within the limitations of Anderson-Schulz-Flory kinetics, toward a specific product group (e.g., light olefins) through careful manipulation of these three catalyst variables. This chapter provides examples of how catalyst preparation, pretreatment, and additive variables influence the activity/selectivity properties.

REFERENCES

1. H. H. Storch, N. Golumbic, and R. B. Anderson, *The Fischer-Tropsch and Related Synthesis*, Wiley, New York (1951).
2. H. A. Pichler and A. Hecktor, *Kirk-Othmer Encyclopedia of Chemistry and Technology*, 2nd ed., Vol. 4 (1964), pp. 446-489.
3. M. A. Vannice, *Catal. Rev.-Sci. Eng.*, *14*:153 (1976).
4. M. E. Dry, in *Catalysis Science and Technology* (J. R. Anderson and M. Boudart, eds.), Springer-Verlag, New York (1981), pp. 159-256.
5. R. B. Anderson, *The Fischer-Tropsch Synthesis*, Academic Press, New York (1984).
6. P. Biloen and W. M. H. Sachtler, *Adv. Catal.*, *30*:165 (1981).
7. A. T. Bell, *Catal. Rev.-Sci. Eng.*, *23*:203 (1981).
8. C. K. Rofer-DePoorter, *Chem. Rev.*, *81*:447 (1981).
9. C. S. Kellner and A. T. Bell, *J. Catal.*, *70*:418 (1981).

10. E. L. Sughrue and C. H. Bartholomew, *Appl. Catal.*, *2*:239 (1982).
11. M. A. Vannice, *J. Catal.*, *37*:462 (1975).
12. S. V. Ho and P. Harriott, *J. Catal.*, *64*:262 (1980).
13. R. Z. C. van Meerten, J. G. Vollenbrock, M. H. J. M. De Croon, P. F. M. T. Van Nisselroy,, and J. W. E. Coenen, *Appl. Catal.*, *3*:29 (1982).
14. J. Klose and M. Baerns, *J. Catal.*, *85*:105 (1984).
15. R. C. Reuel and C. H. Bartholomew, *J. Catal.*, *85*:78 (1984).
16. A. O. I. Rautavouma and H. S. van der Baan, *Appl. Catal.*, *1*:247 (1981).
17. M. E. Dry, *Ind. Eng. Chem. Prod. Res. Dev.*, *15*:282 (1976).
18. H. E. Atwood and C. W. Bennett, *Ind. Eng. Chem. Process Des. Dev.*, *18*:163 (1979).
19. R. A. Dictor and A. T. Bell, *Appl. Catal.*, *20*:145 (1986).
20. L. Fu, J. L. Rankin, and C. H. Bartholomew, *C1 Mol. Chem.*, *1*:369 (1986).
21. B. W. Wojchiechowski and P. D. Taylor, *Can. J. Chem. Eng.*, *61*:98 (1983).
22. D. Bianchi and C. O. Bennett, *J. Catal.*, *86*:433 (1984).
23. J. L. Rankin and C. H. Bartholomew, *J. Catal.*, *100*:533 (1986).
24. C. H. Bartholomew, R. B. Pannell, and J. L. Butler, *J. Catal.*, *65*:335 (1980).
25. M. A. Vannice and R. L. Garten, *J. Catal.*, *56*:236 (1979).
26. G. B. Raupp and J. A. Dumesic, *J. Catal.*, *96*:597 (1985).
27. C. H. Bartholomew and C. K. Vance, *J. Catal.*, *91*:78 (1985).
28. J. A. van Dillen, J. W. Geus, L. A. M. Hermsna, and J. van der Meivben, *Proceedings of the 6th International Congress on Catalysis, London* (1976).
29. J. T. Richardson and R. J. Dubus, *J. Catal.*, *54*:207 (1978).
30. G. A. Hadjigeorghiou and J. T. Richardson, *Appl. Catal.*, *21*:11, 27, 47 (1986).
31. D. G. Castner and D. S. Santilli, in *Catalytic Materials: Relationship Between Structure and Reactivity* (T. E. Whyte et al., eds.), *ACS Symposium Series 248*, American Chemical Society, Washington, D.C. (1984), pp. 39-56.
32. G. Lohrengel, M. R. Dass, and M. Baerns, *Preparation of Catalysts II*, Elsevier, Amsterdam (1979).
33. R. Burch and A. R. Flambard, *J. Catal.*, *85*:16 (1984).
34. P. Turlier, H. Praliaud, P. Moral, G. A. Martin, and J. A. Dalmon, *Appl. Catal.*, *19*:287 (1985).
35. L. Fu and C. H. Bartholomew, *J. Catal.*, *92*:376 (1985).
36. M. E. Bridge, C. M. Comrie, and R. M. Lambert, *Surf. Sci.*, *67*:393 (1977).
37. K. A. Prior, K. Schwaba, and R. M. Lambert, *Surf. Sci.*, *77*:193 (1970).
38. W. H. Lee and C. H. Bartholomew, presented at the *Spring Meeting of the California Catalysis Society*, Richmond, Calif. (1986).
39. M. A. Vannice, P. L. Walker, H-J. Jung, C. Moreno-Castilla, and O. P. Mahajan, *Proceedings of the 7th Congress on Catalysis, Tokyo*, Paper A31 (1980).
40. H. J. Jung, P. L. Walker, Jr., and M. A. Vannice, *J. Catal.*, *75*:416 (1982).
41. V. K. Jones, L. R. Neubauer, and C. H. Bartholomew, *J. Phys. Chem.*, *90*:4832 (1986).
42. R. Snel, presented at the *1985 Annual Meeting of the American Chemical Society, Division of Petroleum Chemistry* (1985).
43. J. B. Benziger and R. J. Madix, *Surf. Sci.*, *94*:119 (1980).

44. H. Arakawa and A. T. Bell, *Ind. Eng. Chem. Process Des. Dev.*, *22*:97 (1983).
45. H. P. Bonzel and H. J. Krebs, *Surf. Sci.*, *109*:L527 (1981).
46. D. J. Dwyer, *Preprints ACS Div. Petr. Chem.* (Aug. 26-31, 1985), p. 715.
47. *Chem. Eng. News* (Jan. 7, 1985), p. 45.
48. D. L. King, J. A. Cusumano, and R. L. Garten, *Catal. Rev.-Sci. Eng.*, *23*:233 (1981).

21 | Effect of Hydrogen on the Activity of Oxide Catalysts

JÓZSEF ENGELHARDT and JÓZSEF VALYON

Central Research Institute for Chemistry of the
Hungarian Academy of Sciences, Budapest, Hungary

I.	INTRODUCTION	565
II.	SURFACE CHEMISTRY OF OXIDES	566
	A. Reduced Molybdena-Alumina	566
III.	HYDROGEN ADSORPTION ON OXIDES	567
	A. Molecular Adsorption	567
	B. Dissociative Adsorption	568
IV.	EFFECT OF HYDROGEN ON CATALYTIC ACTIVITY	572
	A. Zinc Oxide	572
	B. Cobalt Oxide	572
	C. Molybdena-β-Titania	575
	D. Reduced Molybdena-Alumina	576
V.	CONCLUSIONS	580
	REFERENCES	580

I. INTRODUCTION

The large number of publications concerning adsorptive and catalytic properties of metal oxides and the effort to keep this survey brief and informative forced the authors to focus on a few oxides only, those for which the effect of adsorbed hydrogen on catalytic activity has been clearly established—namely on zinc oxide, cobalt oxide, and molybdena-alumina.

A relevant review of the catalytic function of hydrogen bound to surfaces of oxides was given by Hall [1]. According to the picture that emerged for adsorption and activation of hydrogen, molecules of H_2 interact with the strong electric field between the electropositive and electronegative surface centers developed upon dehydration or reduction of the catalyst. The strong polar Lewis acid/Lewis base pair sites tend to dissociate H_2, or at lower temperature the same or related sites adsorb H_2 as molecules. In some cases hydrogen bound to the surface desorbs at elevated temperature and/or at low pressure in the form of H_2. The cocatalytic and inhibiting effects of this "reversible type" of adsorbed hydrogen are discussed in this chapter.

Today, ZnO can be considered as the best characterized oxide catalyst [2-5]. Chromia and supported chromia have also been studied extensively; however, with the exception of the results reported by Selwood [6] as to the promotion and poisoning effects of hydrogen adsorbed by ruby (a dilute solution of chromia in α-alumina) on H_2-D_2 exchange and on magnetic parahydrogen conversion, respectively, no well-documented experiment has been carried out with chromia which would demonstrate the effect of adsorbed hydrogen on catalytic properties.

Redox properties of oxides are not closely related to our subject. However, surface chemistry of reduced molybdena-alumina catalysts will be discussed briefly. We do that because both the extent of reduction and the number of oxide ion vacancies formed in reaction with hydrogen affect significantly the form and the amount of hydrogen adsorption. We lay more stress on molybdena-alumina because most of the results demonstrating the promoting or inhibiting effects of adsorbed hydrogen were obtained with these catalysts.

II. SURFACE CHEMISTRY OF OXIDES

The surface layer of an oxide prepared in aqueous solution is composed solely of hydroxide ions. The removal of oxide ions of the surface by dehydration,

$$2O_sH^- \rightarrow O_s^{2-} + H_2O(g) \tag{1}$$

creates coordinatively unsaturated sites (CUS) (subscript s designates an O atom in the surface layer), resulting in a pronounced change in the adsorptive and catalytic properties of the oxide.

The general idea that surface atoms are in some way unsaturated is old [7]. Using the concept of coordinative unsaturation to describe catalytic sites, significant progress was achieved recently in understanding the adsorption and reaction mechanism on oxides [8].

Increasing the oxidation number of surface ions of altervalent metals (metals with a variety of possible oxidation states) can permit the surface of an oxide to become covered with oxygen without the formation of CUS ions, yet preserving electrical neutrality. Reducing this surface layer with hydrogen (or with some other reducing agent), CUS and concomitantly, metal ions in lower-valence states are formed:

$$M^{n+}O_s^{2-} + H_2(g) \rightarrow M^{(n-2)+} + H_2O(g) \tag{2}$$

Surface properties of supported oxide catalysts, including reducibility, depend strongly on the nature of the support and on other factors, such as the method of preparation, loading, and so on.

A. Reduced Molybdena-Alumina

Molybdena-alumina catalysts are usually prepared by reacting alumina with a solution of $(NH_4)_6Mo_7O_{24}$ [9]. Upon calcination, ammonium paramolybdate decomposes, ammonia is evolved, and polymolybdic acid reacts with basic $(O_sH)^-$ groups of alumina. Polyanions maintain their integrity on the wet and dried catalysts; that is, the surface of the support becomes partly covered by molybdena polyanion patches [10,11].

Millman et al. [12], Hall and Massoth [13], and Hall and LoJacono [14] studied the reduction process and depicted changes that occurred in terms of anion vacancies and new $(O_SH)^-$ groups formed during reduction. The system was described by the following experimental variables: H_C, hydrogen consumed in reduction; □/Mo, anion vacancies produced by removal of water; H_R, reversibly and H_I, irreversibly chemisorbed hydrogen; and e/Mo, the average extent of reduction. The value of e/Mo can be calculated from the amount of oxygen consumed when the catalyst is reoxidized, or can be taken as the amount of hydrogen consumed in the reduction, corrected for reversibly chemisorbed hydrogen, H_R.

Only a part of hydrogen introduced during reduction can be removed as H_2 by evacuation at the reduction temperature (H_R). Another part of hydrogen remains on the catalyst under such conditions (H_I). The distinction between the two types of adsorbed hydrogen seems to be more or less arbitrary, as the amount of hydrogen released strongly depends on desorption conditions. Infrared (IR) spectroscopic measurements showed that H_I could be identified with alumina hydroxyl groups [12]. Strong IR and nuclear magnetic resonance (NMR) evidence was provided that H_R is also linked with alumina hydroxyls [12,15].

III. HYDROGEN ADSORPTION ON OXIDES

Above -148°C, both H_2-D_2 equilibration and ortho-para conversion of hydrogen proceed on dehydrated alumina, but ortho-para conversion only was found fast enough to be measured below -148°C [16]. This behavior, observed many times with other catalysts, too, can be interpreted by the different activation mechanisms of hydrogen at low and high temperatures. Dissociative mechanisms may generally be favored at higher temperatures, while at low temperatures associative mechanisms might prevail [17]. H_2-D_2 exchange was assumed to require a dissociative mechanism, while allotropic conversion may proceed when molecular H_2 interacts with paramagnetic surface sites.

Dehydration and/or reduction usually produces several types of surface sites; for example, sites of different degrees of coordinative unsaturation, or in the case of altervalent metal oxides, surface ions of different oxidation numbers. As one might expect, more than one form of adsorbed hydrogen is often observed on the same catalyst. (Five different types of hydrogen appear on alumina; some of them involve the same surface sites [18]. More than one form of adsorbed hydrogen was reported for zinc oxide [2-5], chromium oxide [6,8,19,20], cobalt oxide [21], and molybdenum oxide/aluminum oxide [22,23]). The way hydrogen is adsorbed and activated on these catalysts depends on the experimental conditions: the temperature, the pressure of hydrogen, and the nature of sorption sites.

A. Molecular Adsorption

Molecular adsorption of hydrogen was described on dehydrated alumina [1], ZnO [3-5], Co_3O_4 [24], and chromia and supported chromia [8,19,20,25] at -196°C, as distinct from the dissociative adsorption prevailing on most metals (see Chapter 2). At low coverage weak molecular chemisorption of hydrogen on ZnO can be characterized with a heat of adsorption of 9.2 to 10.5 kJ mol^{-1}. Direct spectroscopic evidence for the molecular adsorption of hydrogen is available only for ZnO. Bands in the IR spectra were attributed to the stretching vibrations of the perturbed H_2, D_2, or HD molecules. A more substantial shift of these bands toward lower frequencies was reported than for bands of species adsorbed physically [26,27]. Adsorbed hydrogen could be displaced selectively by molecular nitrogen at low temperature (-196°C) [28].

Hydrogen or deuterium was described to be adsorbed irreversibly on dehydrated chromia at -196°C. Hydrogen or deuterium adsorbed could be liberated by deuterium or hydrogen, respectively. The complete absence of HD during displacement suggests that hydrogen was adsorbed molecularly [19]. Similar to that observed with ZnO, displacement of hydrogen adsorbed on chromia by nitrogen took place as well. It was suggested that at low temperature, hydrogen molecules are adsorbed by some form of polarization interaction [1, 4, 26] probably between Cr^{3+} (CUS) and O^{2-} (CUS) [19, 20].

Molecular adsorption of hydrogen at -196°C on molybdena-alumina catalysts was characterized by the value of the separation factor (S) for H_2 and D_2 [22]. S is the ratio of adsorption equilibrium constants. It was determined by the method of Kokes et al. [29], which can be applied quite correctly if adsorption isotherms for H_2 and D_2 follow Henry's law or can be described by the Langmuir expression. Data for adsorption of hydrogen on the nonuniform surface of molybdena-alumina fitted accurately the Freundlich equation. Consequently, one type of average S value can be obtained by the Kokes treatment. S values reflect differences in the average strength of the sites for the different samples up to the point where dissociation of H_2 occurs. Values of S larger than unity are predicted for molecular adsorption, but not for hydrogen adsorbed dissociatively.

Almost identical S values were obtained for calcined molybdena-alumina and for the alumina support. H_2 chemisorbed at -196°C could be eluted by D_2 pulses, indicating that molecular adsorption took place. In this respect calcined molybdena-alumina and dehydrated chromia show similar behavior. The number of Cr^{3+} ions was altered by changing the degree of dehydration of chromia [19]; the surface chemistry of molybdena-alumina, however, was varied by controlling the extent of reduction [13, 14]. S increased from 1.5 to about 2.3 with the extent of reduction of molybdena-alumina, up to about e/Mo = 1.0 (i.e., in this range adsorption of hydrogen should be molecular).

When the extent of reduction was e/Mo $\approx$ 1.2 or higher, the ratio of the chemisorption of D_2 to H_2 at -196°C decreased to about 1. This value is above 2 for the oxidized catalyst [23]. Upon H_2 chemisorption new alumina $(O_sH)^-$ groups were detected on the molybdena-alumina by NMR [30]. Unlike chemisorption on calcined or slightly reduced catalysts, H_2 adsorption was inhibited by prechemisorbed O_2 or NO [23, 30]. These results and the observation that H_2-D_2 exchange also took place suggested that adsorption of hydrogen on the extensively reduced molybdena-alumina is dissociative even at a temperature as low as -196°C.

A molecular mechanism is proposed for low-temperature H_2-D_2 equilibration on reduced supported chromia [25].

B. Dissociative Adsorption

Dissociative adsorption of hydrogen was concluded to be heterolytic on ZnO [31, 32], chromia [19], Co_3O_4 [33], and reduced molybdena-alumina [23]. This might occur as follows [8]:

$$H_2(g) + M^{n+}O_s^{2-} \rightarrow \begin{bmatrix} H^- & H^+ \\ M^{n+} & O_s^{2-} \end{bmatrix} \rightarrow M^{n+}H^-(O_sH)^- \qquad (3)$$

Heterolytic dissociative adsorption is isomeric with the reductive adsorption of hydrogen [8]:

$$H_2(g) + 2(M^{n+}O_s^{2-}) \to 2\left[M^{(n-1)+} \begin{array}{c} H^+ \\ O_s^{2-} \end{array}\right] \to 2\left(M^{(n-1)+}(O_sH)\right) \quad (4)$$

Here, H_2 is adsorbed as two protons and the electrons reduce one or two M^{n+} ions. Reduction of altervalent metal oxides occurs probably in subsequent steps expressed by Eqs. (4) and (1).

Unambiguous spectroscopic evidence for heterolytic dissociative adsorption of H_2 has been reported for ZnO only. Two types of adsorbed hydrogen were observed by Kokes and Dent [2,3]. One is formed rapidly (type I), the other more slowly (type II). Type I hydrogen could be removed by evacuation at room temperature, while type II remained on the catalyst at room temperature even after evacuation for several hours. Type I adsorption produces two IR active species having stretching vibrations at 845 to 850 cm^{-1} and 817 cm^{-1} [34], assigned to Zn-H and O-H, respectively. Type II adsorbed hydrogen results in IR inactive species. A heterolytic dissociative mechanism for adsorption of hydrogen on $Zn^{2+}O^{2-}$ pair sites was deduced for both types of adsorption. This model for adsorption has been refined further by Murphy et al. [35]. Another form of adsorbed hydrogen was observed upon adsorption at "high temperature" (220 to 320°C); this form, however, appears to varying extents at room temperature as well [36].

The greater part of the hydrogen adsorbed on Co_3O_4 at room temperature corresponds to the type II adsorption described on ZnO [21]. This hydrogen does not take part in H_2-D_2 exchange and hydrogenation. Rapid exchange was observed, however, even at -78°C when a mixture of H_2 and D_2 was passed over the catalyst. A small fraction of coordinatively unsaturated surface sites must adsorb H_2 or D_2 rapidly and reversibly, making exchange possible.

Adsorption of hydrogen on dehydrated microcrystalline α-chromia and amorphous chromia at -196°C was found to be molecular. However, passing pulses of H_2 or D_2 at -161°C over chromia containing preadsorbed D_2 or H_2, respectively, HD was released. No isotopic exchange has been observed with surface hydroxyl groups on slightly dehydroxylated chromia even at 0°C [19]. These $(O_sH)^-$ groups do not differ from those generated according to Eq. (3). Nevertheless, activation of hydrogen at -161°C has been interpreted by heterolytic dissociative chemisorption of hydrogen. It was assumed that isotopic exchange with $(O_sH)^-$ groups is unobservable due to the very small fraction of the surface where dissociative adsorption and activation of D_2 can occur.

There has been considerable discussion on the nature of hydrogen bound to reduced molybdena-alumina. In addition to that adsorbed molecularly, at least three forms of surface-bound hydrogen have to be considered: (a) hydrogen adsorbed dissociatively at -196°C on catalysts reduced extensively; (b) H_R, hydrogen adsorbed reversibly at high temperature; (c) H_I, hydrogen adsorbed irreversibly upon reduction.

a. The increasing strength of electric field generated by reduction of centers available for molecular chemisorption is considered at least a significant factor in the dissociation, if it is heterolytic [1,22]. Hall and Millman [23] assumed that the rare species involved in dissociative chemisorption of hydrogen are $Mo^{3+}O^{2-}$ sites, and that dissociation is heterolytic. The same Mo^{3+} sites were supposed to chemisorb O_2 and NO [37,38].

b. Reversibly held hydrogen (H_R) was also considered to be heterolytically cleaved upon adsorption. Infrared and NMR data supported this notion [12,15]. For each H_R molecule removed by evacuation, one hydroxyl group was lost. A search was made to find an IR band attributable to Mo-H, but without success [12].

Homolytic cleavage and reductive chemisorption of hydrogen was not considered as a possible method of adsorption either at -196°C or at high temperature. It was argued that strong magnetic interaction with paramagnetic low-valence molybdenum species, which would form necessarily near the adsorbed hydrogen if adsorption were reductive, would probably prevent the detection of ^{1}H by NMR [15,30].

Another mechanism for the activation of hydrogen is its oxidative addition to M^{n+} ions. This can be represented, for example, as

$$H_2(g) + 2M^{n+} \rightarrow 2\left(M^{(n+1)+}H^-\right) \qquad (5)$$

Oxidative chemisorption is favored by oxides of altervalent metals in which the metal is in a low oxidation state [19]. Molybdena-alumina was considered to be difficult to reduce below an *average* valence state of Mo^{4+}, although the presence of low-valent ions in small quantities could not be excluded. Cr^{2+} has long been known as a major species on a reduced chromia catalyst [39-41]. The existence of Mo^{2+} ions on reduced molybdena-alumina was substantiated recently [42]. Until now no formal evidence for oxidative chemisorption on oxide catalysts has been reported; however, oxidative chemisorption as a possible way for the activation of hydrogen cannot be rejected entirely.

H_R removed by evacuation can be readsorbed. The amount of hydrogen adsorbed at 65 kPa as a function of time and temperature can be seen in Fig. 1. Each curve for hydrogen uptake was determined on identically pretreated samples. The catalyst was reduced to e/Mo = 1.3 and the amount of H_R removed by evacuation at 500°C was 2.5×10^{20} H atoms per gram of catalyst.

Similar to that observed on ZnO at room temperature, two types of hydrogen adsorption can be observed above 300°C: one adsorbing rapidly in the initial short period, and a second wich adsorbs slowly for a long time. No fast adsorption was observed below 300°C and on the alumina support alone.

Four different forms of hydrogen adsorbed reversibly were found by Engelhardt [44] depending on the temperature of adsorption. The catalyst reduced with hydrogen for 2 h at 550°C (e/Mo $\approx$ 2.0) was eluted with nitrogen for 2 h at the same temperature to remove H_R, then cooled to the temperature of second hydrogen treatment (0, 120, 300, 450, and 550°C). After a short evacuation, the catalyst was treated with hydrogen for 15 min and cooled to 0°C. The temperature-programmed desorption method was applied by using nitrogen as carrier gas. Desorptograms (Fig. 2) show four desorption peaks at 75, 115, 340, and above 560°C. Desorption of hydrogen above 450°C was observed in all experiments, even if the catalyst was not treated with hydrogen for a second time. The exact nature of the four different forms of hydrogen is not yet known.

c. H_I is presumably held as OH since it may be desorbed as H_2O (not as H_2) on raising the temperature [13].

Homolytic cleavage and adsorption of hydrogen can take place in special cases without changing the valence state of the metal ion. From magnesium oxide, not water, but hydrogen is evolved above 300°C [45]. O_S^- ions are formed and no CUS is generated. As a reverse process, the homolytic dissociative adsorption of hydrogen occurs above 25°C on O_S^- ions.

$$H_2(g) + 2O_S^- \rightarrow 2(O_SH)^- \qquad (6)$$

Hydrogen was supposed to be activated for H_2-D_2 exchange on magnesium oxide by a similar dissociative mechanism at -196°C without annihilating active sites [46].

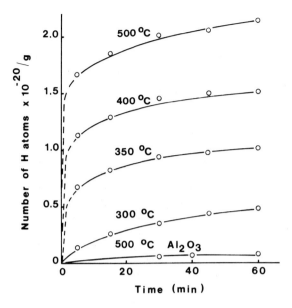

FIGURE 1 Adsorption of hydrogen at 65 kPa as a function of time on a reduced molybdena-alumina. (From Ref. 43.)

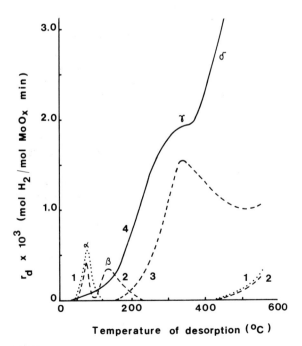

FIGURE 2 Temperature-programmed desorption of hydrogen. Reduced molybdena-alumina was treated with hydrogen for 15 min at (1) 0°C, (2) 120°C, (3) 300°C, and (4) any temperature between 450 and 550°C. (From Ref. 44.)

IV. EFFECT OF HYDROGEN ON CATALYTIC ACTIVITY

Oxide systems are usually active in hydrocarbon reactions which do not require skeletal rearrangement, such as hydrogenation or isomerization of olefins (e.g., reactions listed in Chapters 18 and 24. In what follows, we briefly summarize the hydrogen effects of some typical olefin reactions, as well as in H_2-D_2 exchange.

A. Zinc Oxide

Hydrogenation of ethylene at room temperature over zinc oxide was studied by Kokes and Dent [2,3]. It was found that hydrogenation occurred at a rate comparable to the initial rates of both type I and type II chemisorption of hydrogen. A freshly degassed catalyst was exposed to deuterium overnight, then type I deuterium was removed by brief evacuation and the reaction was carried out with $C_2H_4 + H_2$. Although the turnover on the surface was approximately four times as much as the amount of deuterium adsorbed on type II, less than 2% of the ethane stemmed from this deuterium. From this it follows that type II chemisorption was inert in the hydrogenation.

Type I hydrogen species were primarily responsible for room-temperature hydrogenation. Type II hydrogen, however, modified the catalyst and promoted hydrogenation. This promotional effect did not persist for a second hydrogenation experiment with the same catalyst, even though only about one-fourth of type II hydrogen was removed in the first run. No sufficient explanation for the promotional effect of chemisorbed hydrogen has been given.

B. Cobalt Oxide

Isomerization of Butene

Isomerization of butene at room temperature over Co_3O_4 catalyst was found to be promoted by both hydrogen chemisorbed and hydrogen in the gas phase [47,48]. The cobalt oxide sample was first heated at 400°C for 2 h in circulating air, evacuated at the same temperature for 5 h, and then cooled in vacuum to room temperature. The oxide sample was kept then in hydrogen (8 kPa) for 1 or 2 days and evacuated at room temperature for 15 h.

Isomerization of *cis*-2-butene was carried out in a closed recirculation system. Second and further runs were performed under identical conditions on the same catalyst evacuated each time at room temperature for 30 min before a following run. A very rapid cis-trans isomerization took place at the beginning of the first run (Fig. 3). After a short time the fast isomerization was followed by a slow one, and the fast process did not reappear in successive runs. The high initial activity of the catalyst suggested that hydrogen chemisorbed during pretreatment was acting as active species, which was displaced with butene, resulting in the rapid decrease of isomerization rate.

When isomerization was carried out on a catalyst pretreated with D_2 instead of H_2, a one-to-one mixture of d_1- and d_2-butane was found in the products in addition to a small amount of HD and D_2 [48]. The amount of butane molecules was of the same magnitude as the number of active sites as estimated by CO poisoning of ethylene hydrogenation.

Experiments were carried out for isomerization of *cis*-2-butene (4 kPa) in which in the course of the second run, hydrogen (0.4 kPa) was introduced into the reactor together with *cis*-2-butene [47]. Isomerization was enhanced by hydrogen (Fig. 4).

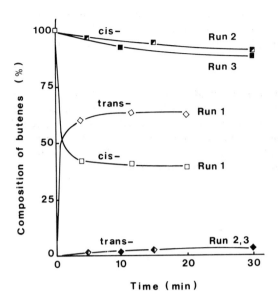

FIGURE 3 Isomerization of *cis*-2-butene at room temperature in successive runs over cobalt oxide with prechemisorbed hydrogen. (From Ref. 47.)

A close correlation was found between the amount of hydrogen prechemisorbed and the initial (high) isomerization activity of the catalyst [48]. The activity could be controlled by the amount of hydrogen chemisorbed (i.e., by the pretreatment conditions). For example, 1- and 15-h evacuations at room temperature after hydrogen treatment resulted in 10% and 50% reductions of the initial isomerization rate, respectively, compared to the rate obtained after a 0.5-h evacuation. It is to be noted that the initial rate of isomerization in the second and successive runs was smaller by more than

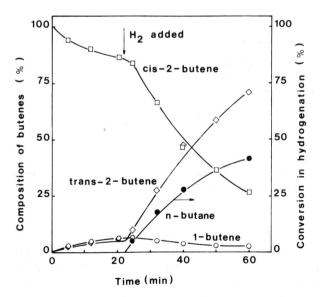

FIGURE 4 Effect of hydrogen addition on the isomerization of *cis*-2-butene. (From Ref. 47).

95% if butene contacted the catalyst for about 1.5 min only in the first run. The period of rapid isomerization, as well as the rate of the subsequent slow one were found to be independent of the previous chemisorption of hydrogen.

C_2H_4-C_2D_4 *Exchange*

In the presence of hydrogen, a linear correlation was observed [47] between hydrogen partial pressure and the rate of C_2H_4-C_2D_4 exchange reaction over catalyst without hydrogen chemisorbed previously. To interpret this promoting effect of hydrogen, exposed cobalt atoms to which a hydride anion and an olefin molecule can coordinate simultaneously were suggested as catalytically active sites. This can be represented as shown in Scheme 1.

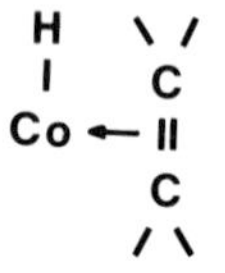

```
    H   \ /
    |    C
   Co ← ||
         C
        / \

   I.
```

Scheme 1

The enhancement of activity by hydrogen was ascribed to the increase in the stationary amount of hydrido-cobalt species, which is determined by the ratio of the rates of hydrogen chemisorption and consumption of hydrogen by the olefins. Both exchange reaction and isomerization of butene were concluded to proceed via an alkyl intermediate [48]. The results obtained for transformation of butene on catalyst holding chemisorbed deuterium were consistent with Scheme 2. Splitting of butene from alkyl intermediate

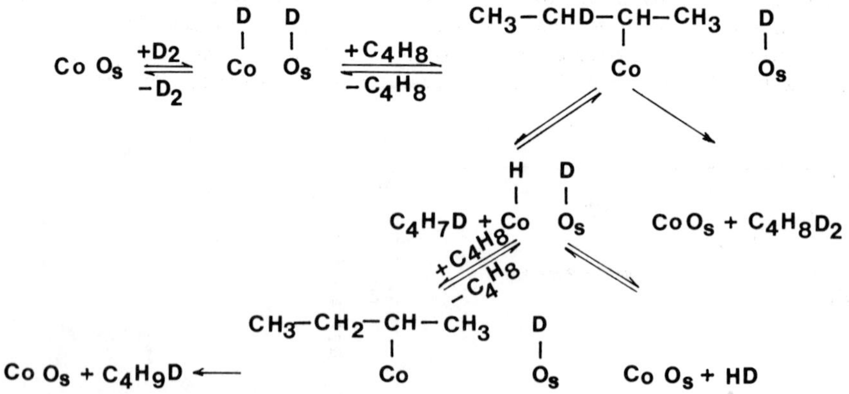

```
                      D  D              CH3–CHD–CH–CH3          D
                      |  |                       |              |
   Co Os  ⇌(+D2/−D2)  Co Os  ⇌(+C4H8/−C4H8)      Co             Os
                                                ⇅  ↘
                                    H  D
                                    |  |
                         C4H7D + Co Os              CoOs + C4H8D2
                                 ↗+C4H8
                                 −C4H8
                                 ⇅
                   CH3–CH2–CH–CH3    D
                              |      |
   CoOs + C4H9D ←              Co    Os    Co Os + HD
```

Scheme 2

may result in both isomer restoration (with surface hydrogen-butene hydrogen exchange; this is the route for C_2H_4-C_2D_4 exchange) and the formation of new isomers.

H_2-D_2 *Exchange*

Irreversible adsorption of hydrogen on calcined cobalt oxide takes place even at -196°C, and increases with increasing temperature [24]. H_2-D_2 equilibration was carried out at -196°C on cobalt oxide dehydrated only and on samples pretreated with hydrogen at a pressure of 28 kPa for 10 min at -196, -78, and 0°C. The catalysts were cooled to -196°C and evacuated for 3 min at the same temperature [48]. The rate of H_2-D_2 equilibration decreased in the sequence noted above, which corresponds to increasing hydrogen content. As opposed to butene isomerization and C_2H_4-C_2D_4 exchange, H_2-D_2 equilibration is inhibited by hydrogen chemisorbed.

C. Molybdena-β-Titania

The variation of catalytic activity with the degree of reduction of molybdena/β-titania in reactions of olefins was described by Tanaka et al. [49]. The catalyst was reduced at 500°C. The extent of reduction was estimated from the amount of hydrogen consumed for reducing the oxidized catalyst.

Metathesis of propene and isomerization of cis-2-butene were carried out in a closed recirculation system at room temperature and 60 kPa (Fig. 5). Catalysts with an average valence state of e/Mo = 0.2 to 1.4 (state I) gave, preferentially, olefin metathesis, with little hydrogen scrambling of the olefins. The catalyst of e/Mo = 1.4 to 2.0 (state II) was active for isomerization in the presence of hydrogen only. Catalysts with an e/Mo value higher than 2.0 (state III) were active in isomerization in the absence of hydrogen. Hydrogenation of olefins was also observed over catalysts of states II and III. The more reduced catalyst shows higher activity for hydrogenation.

This example demonstrates how difficult it is to distinguish whether hydrogen influencing catalytic activity acts as a cocatalyst or as a reducing agent. In the case of MoO_x/β-TiO_2, there is only a narrow range of valence state (e/Mo = 1.4 to 2.0) where isomerization of olefins requires hydrogen in the gas phase (i.e., apparently hydrogen is a cocatalyst). However, at a higher extent of reduction, isomerization proceeds in the absence of hydrogen. Is it not possible that in the course of isomerization, with hydrogen present in the gas phase, the catalyst is simply reduced from state II to state III? For the same catalyst a homolytic dissociative chemisorption of hydrogen was substantiated by Tanaka et al. [50]. Why should not this chemisorption be reductive? Moreover, it cannot be precluded that on

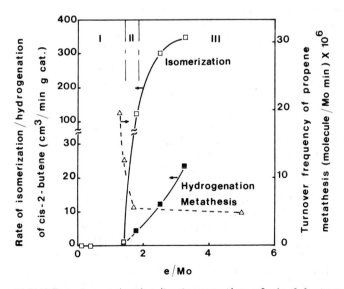

FIGURE 5 Isomerization/hydrogenation of cis-2-butene + H_2 (1:1) and metathesis of propene at room temperature on MoO_x/β-TiO_2 catalyst. (From Ref. 49.)

catalysts in state III, isomerization is promoted by chemisorbed hydrogen, that is, the consumption of hydrogen used for estimating the extent of reduction may include hydrogen chemisorbed on the catalyst.

D. Reduced Molybdena-Alumina

Isomerization of n-Butenes

Different mechanisms of isomerization of n-butenes on unreduced molybdena-alumina and on molybdena-alumina reduced with hydrogen reflect different types of surface sites, intermediates, and chemistry.

Over a freshly oxidized catalyst, isomerization (double bond and cis-trans transformation) was found to occur predominantly via nonmetathetic reaction route on proton acid sites [51]. With increasing duration of contact with the reactant, the metathetic cis-trans interconversion of 2-butenes becomes the prevailing reaction, while the rate of acidic-type isomerization decreases.

Over catalysts reduced with hydrogen, nonmetathetic isomerization was found to proceed by a hydride insertion mechanism via Mo-2-butyl intermediate formed on coordinatively unsaturated sites of the catalyst [52]. This was accompanied by cis-trans transformation via metathesis. The extent of reduction as well as the chemisorption of hydrogen on the catalyst in any state of reduction influences the participation of the two routes in the overall isomerization process [53,54].

Butene isomerization experiments were carried out at 40°C in a closed recirculation system over molybdena-alumina (9.1 wt % MoO_3 on γ-Al_2O_3) reduced with hydrogen for 2 h at 550°C (e/Mo $\simeq$ 2.0) [55]. The catalyst was evacuated at 550°C for 1 h to remove hydrogen adsorbed reversibly. This standard catalyst was treated again with hydrogen. For studying the effect of adsorbed hydrogen on the catalytic activity two catalysts were prepared. One was contacted with hydrogen at 40°C, the other at 550°C for 10 min each. Before the admission of butene, hydrogen was pumped off from the reactor at 40°C for 2 min.

Initial rates of isomerization reaction over the standard catalyst (column 1) and over that treated with hydrogen at 40°C (column 2) are shown in Table 1. Rates increased significantly as a result of adsorbed hydrogen. A further increase (except for the rate of 1-butene conversion) was observed when 1% of hydrogen was admixed to the gas phase (column 3). In subsequent isomerization runs the high initial reaction rates of the first runs were not achieved, although the rates were higher over catalysts pretreated with hydrogen and with hydrogen in the gas phase than over the standard catalyst.

In the reaction of cis-2-C_4H_8 + cis-2-C_4D_8, carried out at 60°C in a pulse system, it was found that both nonmetathetic transformation and cis-trans interconversion via metathesis were enhanced by hydrogen preadsorbed at 60°C. Over catalyst holding hydrogen preadsorbed at 550°C, however, metathesis played no significant role in cis-trans transformation, while the initial overall rate of isomerization was several times higher than over standard catalyst (see Fig. 6).

In the course of isomerization over the standard catalyst, the amount of butane formed was identical to that produced over a catalyst, holding hydrogen bound at 550°C. When isomerization was carried out on a catalyst with pre-chemisorbed D_2 instead of H_2, butene-d_1 was found in the products besides a small amount of butane. These data show that hydrogen chemisorbed at 550°C takes part in the isomerization process at room temperature, but can be hardly removed by butene in the form of butane. A possible interpretation of the foregoing results is that hydrogen chemisorbed affects the concentration of surface intermediates.

TABLE 1 Initial Rates ($r \times 10^{12}$ mol cm_{cat}^{-2} s^{-1}) of n-Butene Isomerization on Reduced Molybdena-Alumina[a]

	(1)[b]	(2)[c]	(3)[d]
cis → trans	68.4	102.1	158.5
cis → 1	2.7	4.8	7.3
trans → cis	42.4	55.3	68.3
trans → 1	2.3	4.7	5.7
1 → cis	3.4	10.2	6.5
1 → trans	3.7	15.4	9.8

[a]n-Butene isomerization was carried out in closed recirculation system at 40°C. The catalysts contained 9.1 w/w % MoO_3 on γ-Al_2O_3. Initial pressure of reactant was 106 kPa.
[b]Standard catalyst. The catalyst was reduced with hydrogen for 2 h and evacuated for 1 h at 550°C.
[c]Standard catalyst, treated with hydrogen at 40°C for 10 min.
[d]The same as footnote c, but isomerization of n-butene was carried out in the presence of 1% H_2 in the gas phase.
Source: Ref. 55.

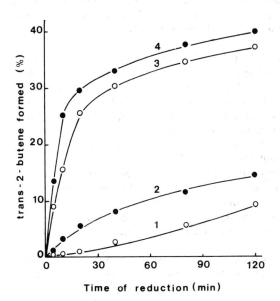

FIGURE 6 Concentration of *trans*-butene formed in the first 30 s of the reaction of 100% *cis*-2-butene (curves 1, 3) and 99% *cis*-2-butene, 1% H_2 (curves 2, 4) over molybdena-alumina reduced with hydrogen at 550°C for different times and evacuated at 550°C (curves 1 and 2) or at 40°C (curves 3 and 4) for 1 h, reaction temperature 40°C. (From Ref. 54.)

Hydrogen chemisorbed at low temperature seems to promote the formation of Mo-alkyl species which are common intermediates for isomerization and metathesis; hydrogen chemisorbed at high temperature, however, inhibits the transformation of Mo-alkyl to Mo-alkylidene. Metathesis reaction requires Mo-alkylidene intermediates. It was suggested that surface species such as II in Scheme 3 may exist on a molybdena-alumina reduced and evacuated at high temperature. Hydrogen at low temperature can be bound in position

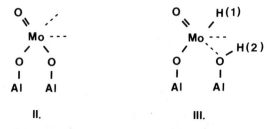

Scheme 3

(1), and at high temperature in both positions (1) and (2), forming species such as III. Hydrogen in position (1) enhances the formation of Mo-alkyl species. Hydrogen in Mo-alkyl species close to the molybdenum ion (2) may be attracted by the oxygen in the sublayer. If hydrogen abstraction is completed, Mo-alkylidene may be formed. When there are already OH groups in the sublayer close to the molybdenum, as in species III, hydrogen abstraction from Mo-alkyl, consequently, transformation to Mo-alkylidene species, is hindered [56].

Metathesis of Propene

Similarly to that of *n*-butenes, metathesis of propene over reduced molybdena-alumina was found to be accelerated by hydrogen adsorbed at low temperature and by hydrogen in the gas phase, but hindered by hydrogen bound at high temperature [44,56].

Metathesis of propene was studied at 40°C in the flow reactor at atmospheric pressure [44]. Molybdena-alumina (9.1 wt % MoO_3 on γ-Al_2O_3) was reduced with hydrogen, evacuated at the same temperature for 1 h (e/Mo $\simeq$ 2, standard catalyst), then treated with hydrogen (100 kPa) at various temperatures for 15 min. The reactor was cooled rapidly to 40°C in hydrogen and the feed of propene was started without any evacuation.

After an initial increase, the catalytic activity decreased continuously in each case. Maximum rates of metathesis reaction are plotted in Fig. 7 (curve b) as a function of the temperature of hydrogen treatment. Up to 180°C this treatment resulted in an increase, above this temperature in a decrease of the rate of metathesis compared to the rate obtained on standard catalyst (line c in Fig. 7). The amount of hydrogen consumed by the catalyst in hydrogen treatment was determined in separate experiments and is plotted in Fig. 7 (curve a).

As shown in Fig. 2, hydrogen adsorbed at low temperature can be removed more easily than that adsorbed at higher temperature. It was assumed that hydrogen at low temperature is bound in position (1) of species III and enhances the formation of Mo-alkyl species, one of the intermediates for metathesis [56]. At higher temperatures, presumably, a greater fraction of hydrogen is bound in position (2) of species III, hindering the transformation of Mo-alkyl to Mo-alkylidene species, which are the direct intermediates for metathesis.

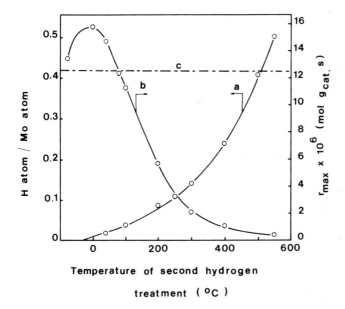

FIGURE 7 (a) Amount of hydrogen consumed by reduced molybdena-alumina during second hydrogen treatment for 15 min; (b) rate of propene metathesis at 40°C over reduced molybdena-alumina treated with hydrogen at different temperatures; and (c) over standard catalyst. (From Ref. 44.)

Hydrogenation of Olefins

Two catalysts retaining different amounts of hydrogen (H_R) were prepared by reducing molybdena-alumina (8 w/w % Mo on γ-Al_2O_3) with hydrogen at 500°C for 2 h and evacuating it for 1 h at 500°C or at room temperature. These catalysts were tested in hydrogenation of ethylene and 2-butenes in a closed recirculation system [43].

Hydrogenation of ethylene was carried out at -78°C with an equimolar mixture of ethylene and hydrogen in the system at a total starting pressure of 30 kPa. The rate of hydrogenation was approximately one order of magnitude higher on the catalyst with more H_R. The same results were obtained in the activity tests in the second runs, indicating that hydrogen adsorbed at a high temperature was not removed in the first run by ethylene. The activity of catalyst originally holding less hydrogen did not increase in the course of reaction, although the catalyst certainly adsorbed hydrogen at the low reaction temperature.

Reactions of 2-butenes were studied at 0°C at an initial pressure of 106 kPa; the ratio of 2-butenes to hydrogen was 2:3. The effect of hydrogen bound to the catalyst at high temperature on the catalytic activity in hydrogenation (and isomerization) of *cis*-2-butene is demonstrated by Fig. 8. The rates of both hydrogenation and isomerization of 2-butenes were significantly higher over the catalyst holding more H_R than were the rates on the other catalyst tested.

Similar experiments were carried out with catalyst reduced with hydrogen for different amounts of time [54]. Hydrogenation activity of the catalyst increases with increasing extent of reduction. However, the catalyst reduced for 10 min and holding hydrogen retained at 500°C is more active than that reduced for 2 h but evacuated at 500°C for 1 h to remove the hydrogen chemisorbed. These results show that new sites for olefin hydrogenation were formed by hydrogen chemisorbed at high temperature.

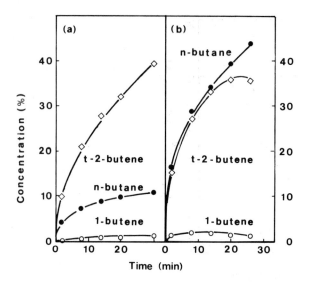

FIGURE 8 Concentration of products formed at 0°C in reaction of cis-2-butene and hydrogen (molar ratio = 2:3) over molybdena-alumina reduced with hydrogen at 500°C and evacuated for 1 h at 500°C (a) or at 0°C (b). (From Ref. 43.)

V. CONCLUSIONS

The results given above demonstrate that hydrogen chemisorbed on oxides might play an important role in influencing their catalytic properties. Reduced molybdena-alumina presents the most striking example. Without hydrogen chemisorbed at high temperature, it is highly active in olefin metathesis, and this activity is even increased by hydrogen chemisorbed at low temperatures. The same catalyst with the same extent of reduction, but holding hydrogen chemisorbed at high temperature, is completely inactive in olefin metathesis but very active in isomerization and hydrogenation of olefins. Studying the adsorption and activation of hydrogen on oxide catalysts may help chemists to learn more about catalytically active sites and reaction mechanisms.

REFERENCES

1. W. K. Hall, *Acc. Chem. Res.*, 8:257 (1975).
2. A. L. Dent and R. J. Kokes, *J. Phys. Chem.*, 73:3772 (1969).
3. R. J. Kokes and A. L. Dent, in *Advances in Catalysis*, Vol. 22 (D. D. Eley, H. Pines, and P. B. Weisz, eds.), Academic Press, New York (1972), p. 1.
4. R. J. Kokes, *Acc. Chem. Res.*, 6:226 (1973).
5. C. S. John, in *Catalysis*, Vol. 3 (C. Kemball and D. A. Dowden, eds.), Chemical Society, London (1980), p. 169.
6. P. W. Selwood, *J. Am. Chem. Soc.*, 88:2676 (1966).
7. H. S. Taylor, *Chem. Rev.*, 9:1 (1931).
8. R. L. Burwell, G. L. Haller, K. C. Taylor, and J. F. Read, in *Advances in Catalysis*, Vol. 20 (D. D. Eley, H. Pines, and P. B. Weisz, eds.), Academic Press, New York (1969), p. 1.
9. F. E. Massoth, in *Advances in Catalysis*, Vol. 27 (D. D. Eley, H. Pines, and P. B. Weisz, eds.), Academic Press, New York (1978), p. 265.

10. W. K. Hall, in *Proceedings of the 4th Climax International Conference on the Chemistry and Uses of Molybdenum* (H. F. Barry, P. C. H. Mitchell, eds.), Climax Molybdenum Company, Golden, Colo. (1982), p. 224.
11. L. Wang and W. K. Hall, *J. Catal.*, 77:232 (1982).
12. W. S. Millman, M. Crespin, A. C. Cirillo, Jr., S. Abdo, and W. K. Hall, *J. Catal.*, 60:404 (1979).
13. W. K. Hall and F. E. Massoth, *J. Catal.*, 34:41 (1974).
14. W. K. Hall and M. LoJacono, in *Proceedings of the 6th International Congress on Catalysis*, Vol. 1 (G. C. Bond, P. B. Wells, and F. G. Tomkins, eds.), Chemical Society, London (1977), p. 246.
15. A. C. Cirillo, Jr., F. R. Dollish, and W. K. Hall, *J. Catal.*, 62:379 (1980).
16. F. H. VanCauwelaert and W. K. Hall, *Trans. Faraday Soc.*, 66:454 (1970).
17. M. Boudart, in *Topics in Applied Physics*, Vol. 4, *Interactions on Metal Surfaces* (R. Gomer, ed.), Springer-Verlag, New York (1975), p. 275.
18. Y. Amenomiya, *J. Catal.*, 22:109 (1973).
19. R. L. Burwell, Jr. and K. S. Stec, *J. Colloid Interface Sci.*, 58:54 (1977).
20. S. R. Ely and R. L. Burwell, Jr., *J. Colloid Interface Sci.*, 65:244 (1978).
21. K. Tanaka, H. Nihira, and A. Ozaki, *J. Phys. Chem.*, 74:4510 (1970).
22. W. S. Millman, F. H. VanCauwelaert, and W. K. Hall, *J. Phys. Chem.*, 83:2764 (1979).
23. W. K. Hall and W. S. Millman, in *Proceedings of the 7th International Congress on Catalysis, Tokyo, 1980*, Part B (T. Seiyama and K. Tanabe, eds.), Elsevier, New York (1981), p. 1304.
24. Y. Shigehara and A. Ozaki, *J. Catal.*, 21:78 (1971).
25. V. Indovina, A. Cimino, and M. Inversi, *J. Phys. Chem.*, 82:285 (1978).
26. C. C. Chang, L. T. Dixon, and R. J. Kokes, *J. Phys. Chem.*, 77:2634 (1973).
27. W. C. Conner, Jr. and R. J. Kokes, *J. Catal.*, 36:199 (1975).
28. C. C. Chang and R. J. Kokes, *J. Phys. Chem.*, 77:2640 (1973).
29. R. J. Kokes, A. L. Dent, C. C. Chang, and L. T. Dixon, *J. Am. Chem. Soc.*, 94:4429 (1972).
30. A. C. Cirillo, Jr., J. M. Dereppe, and W. K. Hall, *J. Catal.*, 61:170 (1980).
31. T. Okuhara, T. Kondo, and Ken-ichi Tanaka, *J. Phys. Chem.*, 81:808 (1977).
32. A. L. Dent and R. J. Kokes, *J. Phys. Chem.*, 73:3781 (1969).
33. S. Siegel, *J. Catal.*, 30:139 (1973).
34. R. P. Eichens, W. A. Pliskin, and M. J. D. Low, *J. Catal.*, 1:180 (1962).
35. W. R. Murphy, T. F. Veerkamp, and T. W. Leland, *J. Catal.*, 43:304 (1976).
36. A. Baranski and J. Galuszka, *J. Catal.*, 44:259 (1976).
37. W. S. Millman and W. K. Hall, *J. Catal.*, 59:311 (1979).
38. W. S. Millman and W. K. Hall, *J. Phys. Chem.*, 83:427 (1979).
39. F. S. Stone, *Chimia*, 23:490 (1969).
40. J. B. Peri, *J. Phys. Chem.*, 78:588 (1974).
41. R. Merryfield, M. McDaniel, and G. Parks, *J. Catal.*, 77:348 (1982).
42. J. Valyon and W. K. Hall, *J. Catal.*, 84:216 (1983).
43. J. Valyon, J. Engelhardt, and D. Kallo, *Acta Chim. Hung.*, 124:83 (1987).
44. J. Engelhardt, *Recl. Trav. Chim. Pays-Bas*, 96:M101 (1977).
45. W. Gieseke, H. Nagerl, and F. Freund, *J. Phys. Chem.*, 78:758 (1974).
46. M. Boudart, A. Delbouille, E. G. Derouane, V. Indovina, and A. B. Valters, *J. Am. Chem. Soc.*, 94:6622 (1972).
47. T. Fukushima and A. Ozaki, *J. Catal.*, 32:376 (1974).

48. T. Fukushima and A. Ozaki, *J. Catal.*, *41*:82 (1976).
49. K. Tanaka, K. Miyahara, and Ken-ichi Tanaka, *Bull. Chem. Soc. Jpn.*, *54*:3106 (1981).
50. K. Tanaka, Ken-ichi Tanaka, and K. Miyahara, *Chem. Lett.*, 943 (1978).
51. J. Goldwasser, J. Engelhardt, and W. K. Hall, *J. Catal.*, *70*:275 (1981).
52. J. Engelhardt, J. Goldwasser, and W. K. Hall, *J. Catal.*, *76*:48 (1982).
53. J. Engelhardt, J. Goldwasser, and W. K. Hall, *J. Mol. Catal.*, *15*:173 (1982).
54. J. Engelhardt, J. Kallo, and I. Zsinka, *J. Catal.*, *88*:317 (1984).
55. J. Engelhardt and D. Kallo, *J. Catal.*, *71*:209 (1981).
56. J. Engelhardt and D. Kallo, *Acta Chim. Hung.*, *119*:249 (1985).

22 | Reactivity of Hydrogen in Sulfide Catalysts

RICHARD B. MOYES

University of Hull, Hull, North Humberside, England

I.	INTRODUCTION	583
II.	THEORIES OF SULFIDE ACTIVITY	585
	A. Importance of Edge Sites	585
	B. Monolayer Model	585
	C. Intercalation Model	586
	D. Contact Synergy Model	586
	E. Remote Control Theory	587
	F. Defect Control Theory	587
	G. Co-Mo-S Model	588
III.	HYDROGEN SORPTION BY SULFIDE CATALYSTS	588
	A. Molybdenum Sulfide System	588
	B. Supported Molybdenum Sulfide	593
	C. Hydrogen Sorption on Tungsten Sulfide	595
	D. Hydrogen Sorption on Co_9S_8	596
	E. Supported and Promoted Molybdenum Catalysts	596
	F. Temperature-Programmed Desorption Studies	597
IV.	KINETICS OF DESULFURIZATION AND RELATED REACTIONS	599
V.	PREDICTIVE VALUE OF HYDROGEN SORPTION MEASUREMENTS	602
	REFERENCES	604

I. INTRODUCTION

The reaction of sulfide catalysts with hydrogen is a complex process, and hydrogen transfer by these catalysts is of great importance in a number of industrial processes. These catalysts will become increasingly important as the fossil fuel supply turns to heavier petroleum fractions or to coal as the basic feedstock. For a review of sulfide catalysts in general, and these aspects in particular, the book of Weisser and Landa covers the literature in a comprehensive way up to 1973 [1]. More recent reviews by Mitchell [2,3] cover much of the literature since this book was published.

Sulfide catalysts find extensive use as hydrotreating catalysts. Hydrotreating is a general term that includes hydrocracking, hydrogenation, and hydrogenolysis. The four most important reactions in the latter category are hydrodesulfurization (HDS), hydrodenitrogenation (HDN), hydrodeoxygenation (HDO), and hydrodemetallation (HDM). All of these reactions involve hydrogen, and it is for this reason that the purpose of this chapter is to examine the role of hydrogen in sulfide catalysts. Most work has been done on the cobalt-molybdenum-alumina sulfide system (Co-Mo/alumina) as a hydrogenation and hydrodesulfurization catalyst, and most examples will derive from this system. This is not to say that other systems are not equally important, merely that the majority of published work is based on this system. Similar work on the corresponding tungsten sulfide analogs reveals similarities, but other sulfides can be quite unlike these systems.

It is natural that in order to pursue basic research on the nature and activity of sulfide catalysts, attempts should be made to simplify the system and reduce it to its component parts. It is known that under reaction conditions [4] the Co-Mo/alumina system contains some MoS_2 and some Co_9S_8 which are the thermodynamically stable materials expected in these reaction conditions. As a result, studies are to be found on cobalt sulfide catalysts and on molybdenum sulfide catalysts both as powders or supported on alumina or other supports. Although these studies suggest lines of thought in developing models of catalyst activity, it must be stressed that the whole working catalyst is much more than the component parts, partly because of the synergic effects found when cobalt is used to promote the action of molybdenum and partly because of the support effects given to the system by the high area support and its interaction with the active phases. A related but less clear feature is the role of sulfur stoichiometry and the extent to which this is affected by use as an HDS catalyst, with subsequent formation of equilibrium sulfur compositions. Furthermore, industrial catalysts contain specific promoters for specific purposes [3], and these may also have effects on hydrogen reactivity. It has long been known that preparative conditions strongly affected the activity of the resulting catalyst, and understanding this feature, in terms of forming particles of the correct size, active materials, and avoiding unwanted formation of compounds with the support, has taken a great deal of research effort.

In this chapter, the second in the book that does not deal with metallic catalysts, we look briefly at the structure of the sulfides and the explanations of their reactivity in terms of such structures. It will be convenient to divide catalysts into the following classes: (a) powdered, pure sulfides; (b) the effects of supporting the pure materials; and (c) the effects of promoters. Industrial catalysts are always produced as the oxide form [4], which is then reduced and sulfided. The extent to which the materials become sulfided and the effect that this has on the activity remains a matter of debate, as Massoth has pointed out [5].

The hydrogen reactivity can also be subdivided into (a) the reactions by which oxide forms are converted into active sulfides (i.e., the sulfidation step), and (b) hydrogen reduction that may be present at the surface level. There is extensive literature on the first process, but this appears to have little direct relevance to the hydrogen sorption properties of the final sulfide catalysts.

The chapter bears little relevance to previous ones because of the totally different catalyst structures involved. One may note, however, that inelastic neutron scattering studies (see Chapter 6) and infrared spectroscopy (see Chapter 7) may supply useful information here, too.

II. THEORIES OF SULFIDE ACTIVITY

Many studies have centered on activity for thiophene hydrodesulfurization using the cobalt-molybdenum-alumina catalyst system. This provides a useful beginning, but there is some study of the equally important nickel-tungsten alumina catalyst, or related materials using iron as the group VIII metal [3]. Theories of sulfide activity must successfully explain a number of experimental facts, such as those summarized by Pirotte et al. [6]. These are that there appear, from kinetic studies, to be two types of active site, those responsible for hydrogenation and those responsible for hydrogenolysis. These are sometimes differentiated also in terms of relative acidity; the sites can be separated in terms of their susceptibility to poisoning by bases [7], where the hydrogenolysis site is more easily poisoned than the hydrogenation site. Molybdenum sulfide and tungsten sulfide share the same layered structure, so that structural effects are thought to be similar. Theories usually use the fact that MoS_2 is a layer lattice, although such theories must also account for the activity of other sulfides, which are of different structure. Most theories are based on adsorption of the organic compound through the sulfur atom at a sulfur vacancy. Finally, there is a maximum in the activity of the synergic mixtures of group VIII and group VI sulfides which requires explanation.

A. Importance of Edge Sites

A number of reaction mechanisms show the reactant adsorbed at a sulfur-deficient molybdenum site. These sites are seen as being exposed at the edge of the layer, with the basal plane presenting a continuous layer of sulfur. A very useful account of the nature of active sites for hydrogenation in MoS_2 has been given by Tanaka and Okuhara [8] in explaining the activity of MoS_2 single crystals for hydrogenation and exchange. This picture gives a careful description of the edge sites where molybdenum is exposed, or, in other terms, where coordinatively unsaturated sites are available at varying degrees of coordination (Figs. 1 and 2).

This picture is drawn with reference to experiments in which the number of edge sites was varied by cutting wafers of single-crystal MoS_2. In bulk MoS_2 each molybdenum is surrounded by six sulfur atoms, but at the edges, one, two, or three of these atoms may be absent, giving rise to ^1Mo, ^2Mo, or ^3Mo sites, respectively. These sites are seen as the basis of the activity of sulfide catalysts by providing active sites for coadsorption of the hydrocarbon and the necessary hydrogen for hydrogenation. The explanation implies that the basal plane is ineffective as a hydrogenation catalyst (although attack on the basal plane by hydrogen at elevated temperatures could "dig holes" in the basal plane and thus provide more sites). The reported experiments did not include cobalt promotion or hydrodesulfurization. The importance of edge sites had long been recognized, and this led to attempts to characterize them by specific gas chemisorption; of O_2 at a sufficiently low temperature [9], of strongly retained hydrogen [10], and of NO [11]. These results give similar values which themselves can be related to activity.

B. Monolayer Model

The monolayer theory suggests that the activity results from the formation of a monolayer of oxysulfide on the alumina surface [12,13]. The oxide form of the catalyst has an "epitaxial" layer of molybdenum on the surface, bound to two oxygen atoms from the alumina, and having two further atoms on the exposed surface. These atoms can be sulfided in the sulfidation step. The

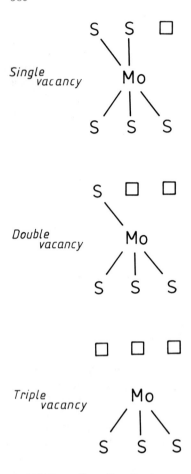

FIGURE 1 Coordinatively unsaturated sites on MoS$_2$ edges.

position of the cobalt is not clear, but is considered to be in the alumina, possibly as a spinel. Massoth [14] has modified the theory slightly to account for sulfur content and steric effects. Most would now accept as unlikely that a monolayer of sulfide covers the surface. It does describe the precursor oxide structure, and has colored views of the sulfided oxide state.

C. Intercalation Model

The intercalation model proposes intercalation of nickel between the WS$_2$ layers in the WS$_2$ crystal, in some cases within the van der Waals gap, in others, at edges [15,16]. The similarity of the molybdenum and tungsten sulfide structures led to the assumption that a similar theory would hold for molybdenum sulfide. There is some difficulty in understanding how such large amounts of nickel as those present in the most active materials can be accommodated in the structure, unless the particles are very small and so most of the molybdenum is at the edges of the crystallites.

D. Contact Synergy Model

The contact synergy model concentrates attention on the contact between the two phases thought to be active in the cobalt-molybdenum-alumina catalyst,

monolayer model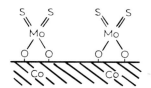

contact synergy model

intercalation model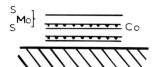

CoMoS model

FIGURE 2 Diagrammatic representation of some theories explaining synergy between cobalt and molybdenum supported on alumina.

notably Co_9S_8 and MoS_2 [17,18]. Hydrogen is thought to dissociate at the contact point and be fed by a type of spillover process onto the MoS_2 (for spillover phenomena, see Chapter 12).

E. Remote Control Theory

The remote control theory expands the contact synergy model in that it suggests that cobalt sulfide controls the activity of MoS_2 by providing spillover hydrogen, which then creates active MoS_2 sites. This theory predicts that the synergy curve for hydrogenation should be different from the curve for hydrodesulfurization, as is found in practice [6].

F. Defect Control Theory

Aoshima and Wise [19] and Wentrcek and Wise [20] have experimented with a series of single-crystal MoS_2 catalysts which strongly support the suggestion that control of sulfur defects controls the activity of the catalyst. This occurs partly through the provision of defects that act as hydrodesulfurization sites, and partly as control of the amount and nature of the adsorbed hydrogen. It is clear that surface stoichiometry during reaction reaches an equilibrium point which is different from that of the pure sulfides alone, or of the catalyst in its original state.

G. Co-Mo-S Model

The Co-Mo-S model is based on Mössbauer studies of cobalt-molybdenum-alumina catalysts [21-23] which clearly demonstrate that activity is related to preparative method, intimate contact between the two sulfides being extremely important. This activity is also related to the Mössbauer emission spectrum (MES) of the cobalt in the catalysts, which is not that for Co_9S_8. Cobalt is recognized as being present as "Co-Mo-S" which appears to provide the activity and is seen as a surface phase. Cobalt is also present in a spinel form in the alumina surface. At higher Co/(Co + Mo) ratios, signals arising from Co_9S_8 appear, but as these increase in intensity, the activity for HDS falls off.

This theory has been buttressed by a series of investigations by other laboratories which lend substantial support to it by finding the same results and by extending the ideas to related systems [24,25]. The MES studies do not support the earlier models; for example, Co_9S_8 is not necessary for promotion, as required by the remote control theory.

In a recent review [26], the evidence for the Co-Mo-S model is carefully examined and updated. Results for the Co-Mo/alumina system, unsupported powdered systems, and carbon-supported systems are drawn together to give a picture of the genesis of active catalysts and the role of preparative conditions. The molybdenum oxide layer must be prepared as a monolayer, in order to maximize the edge-site concentration after sulfiding. The cobalt must be added so as to make as much of the Co-Mo-S state as possible. The Co-Mo-S is in MoS_2-like crystallites of very small size, present as a single sheet on the support. The cobalt is thought to occupy edge sites on the MoS_2 structure rather than intercalating, as no van der Waals gap between layers is said to exist. Direct confirmation of the cobalt site has been obtained by use of infrared and analytical electron microscopy [27,28]. Co-Mo-S is not a stoichiometric phase and can be looked upon as cobalt supported on an active MoS_2 support, where interaction involving the nearby molybdenum has been shown by EPR measurements [29]. As far as the effect of hydrogen is concerned, the valence state of the Co in Co-Mo-S is altered by the H_2/H_2S ratio in the gas in contact with the catalyst. Thus the active site is seen as the *cobalt* in Co-Mo-S, which is associated with nearby molybdenum.

The Co-Mo-S model brings together a number of previously unexplained features of sulfide catalysts and explains them. It is still being investigated, but it provides an extensive and eminently reasonable view of the promotion of sulfide catalysts. Curiously, there is little comment on the ability of the phase to interact with hydrogen in the active form.

III. HYDROGEN SORPTION BY SULFIDE CATALYSTS

A. Molybdenum Sulfide System

The molybdenum sulfide system has attracted the greatest effort, although some work on tungsten, cobalt, and other sulfides also exists. As is noted above, molybdenum sulfide has a layer lattice, which can crystallize in one of two forms, hexagonal and rhombohedral. An extensive review by Wilson and Yoffe [30] discusses the physical properties of the transition metal dichalcogenides, including structural and electrical properties. MoS_2 is available as a naturally occurring mineral, molybdenite, and is a commonly used source. It can be made by reduction of MoS_3 with hydrogen or H_2/H_2S mixtures, or by heating in an inert gas, by reduction of ammonium tetrathiomolybdate [31], or by a nonaqueous metathesis reaction [32]. As a result, some confusion can exist in published work about the form of MoS_2 used in some studies. It is now clear that those studies which use MoS_2 produced by hydrogen reduction will be at least partly covered with S-H groups, but these surfaces are not particularly stable [33] and will react on air exposure to form oxides at the exposed edge planes and into the bulk crystal [34]. In discussing hydrogen uptake, it is convenient to express the amount adsorbed as a hydride of the form MoS_2H_x, as this appears more meaningful.

An early study by Badger et al. [35] is a convenient starting point. Molybdenum disulfide made by the reduction of the trisulfide was cleaned in hydrogen at 673 K and evacuated for 48 h at the same temperature before the adsorption measurements were carried out. As Fig. 3 shows (open circles), the material began to adsorb hydrogen under 1 atm pressure at 413 K, the amount adsorbed then rose abruptly and leveled off. On cooling in a hydrogen atmosphere the adsorbed hydrogen was retained (dashed line). The amount of hydrogen sorbed extrapolated to the axis was 26 ml g^{-1} or about 0.37 mol of hydrogen per mole of MoS_2. The surface area was given as 16.6 m^2 g^{-1}, which means that as long as taking the BET measurement did not involve a change in the surface, more hydrogen was adsorbed than was sufficient to cover the surface of the powder. It is tempting to suggest that this hydrogen has been intercalated in the van der Waals gap between the repeating MoS_2 layers of the MoS_2 structure. The authors state that adsorption was slow; several hours were necessary to obtain equilibrium.

As part of a study of the exchange of H_2S with deuterium, Wilson et al. [36] studied the exchange of hydrogen with deuterium catalyzed by MoS_2 powder. This powder was made by reduction of ammonium thiomolybdate in a 1:1 $H_2:H_2S$ mixture. The analysis of the material gave Mo = 59.1% and

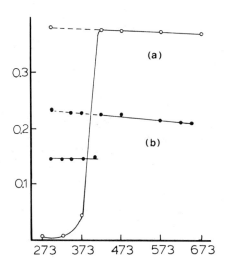

FIGURE 3 (a) Isobars of hydrogen sorption on MoS_2. (b) Ordinate, moles of hydrogen per mole of MoS_2; abscissa, temperature (K). (Open points redrawn from Ref. 35; solid points redrawn from Ref. 45.)

S = 36.7% (MoS$_2$, Mo = 59.94%, S = 40.06%). The authors noted that H$_2$/D$_2$ reaction mixtures gave some H$_2$S in the gas phase, suggesting sulfur excess, although they do not comment on this possibility. The surface area was very low, 2.16 m^2 g^{-1}. Exchange began at 423 K, which the authors note was the same temperature as that required for adsorption by Badger et al. [35]. For a particular temperature, exchange with H$_2$S occurred at a similar rate. The mechanism suggested is given mostly in terms of H$_2$S reacting at the surface, that is,

H$_2$S(g) → HS(a) + H(a)

D(a) + HS(a) → HS(a) + HDS(g)

Ratnasamy and Fripiat [37] reported S-H groups on the surface of MoS$_2$ on the basis of IR measurements. The samples chosen were MoS$_2$(i) from BDH Ltd. with a surface area of 29.6 m^2 g^{-1} and MoS$_2$(ii) obtained by reduction of MoS$_3$. The spectra shown were for hydrogen adsorbed on MoS$_2$(ii) at 298 K, with desorption at 373 and 473 K. A broad strong band around 2500 cm^{-1} was observed at 298 K. This shifted to 2525 cm^{-1} at 373 K and 2516 cm^{-1} at 473 K. Evacuation at 573 K gave rise to three bands centered on 2454, 2532, and 2649 cm^{-1} which persisted, although with a diminished intensity, even after evacuation at 673 K. The bands from MoS$_2$(i) were said to be essentially the same. Two bands were observed at room temperature, at 2500 and 2640 cm^{-1}, which gradually disappeared with increasing temperature, probably due to removal of S-H groups as H$_2$S. The bands were assigned to S-H stretching by comparison with those from mercaptans (2550 to 2650 cm^{-1}), thiophenol (2550 cm^{-1}), gaseous H$_2$S (2668 cm^{-1}), and from other compounds containing S-H groups. The observation of S-H groups present at room temperature appears to be the first evidence of the surface sulfhydryl groups present on reduced sulfide surfaces.

Stevens and Edmonds [38] studied the adsorption of hydrogen on MoS$_2$ which had been ground in air or in heptane. Air-ground material was shown to be of low surface area (10 m^2 g^{-1}) with 95% edge planes exposed. Heptane-ground material had higher surface area (52 m^2 g^{-1}) and had 72% basal plane exposure. ESCA examination showed substantial similarities between the two samples, except that the edge-plane sample contained more oxygen than did the basal-plane sample. Hydrogen adsorption on the two samples was quite different. As with Badger et al. [35], hydrogen adsorption on edge-plane material began around 430 K and increased to a maximum at 500 K. On the basal plane sample 480 K was needed to begin adsorption and the amount adsorbed was much less. This was explained in terms of hydrogen adsorption as an activated process, occurring more readily at the edge than at the basal plane. Edge-site adsorption was completely reversible and no products were detected during the adsorption. The adsorption of thiophene was also studied; it behaved similarly, but was more weakly adsorbed. The catalysts were also tested for CS$_2$ and for thiophene HDS activity. Edge-plane material was more active in hydrogenation, basal plane in HDS.

Wise and co-workers have examined the relationship of catalytic activity and defect structure in MoS$_2$. In an early study [19, 20] a single-crystal wafer of molybdenite was used. Sulfur could be removed from the wafer by hydrogen treatment above 900 K, and could be replaced by H$_2$S treatment at 700 K. The process could be monitored by resistance measurements. High-conductivity material was the result of hydrogen treatment and was an HDS catalyst of less activity than one of low conductivity, produced by H$_2$S treatment. Furthermore, the hydrogenation activity of the latter material was different in that butane, rather than butene, was produced from butanethiol. This n-type catalyst is the preferred HDS catalyst, and the authors state that the differences between the ends of the conductivity range are due to the availability of surface-sorbed hydrogen, a parameter controlled by the electronic properties of the catalyst. In a later paper [39] the properties were shown to be modified by the addition of cobalt. Hou and Wise [40] later showed that sulfur excess in molybdenum sulfide powders gave rise to a second form of adsorbed hydrogen, distinguishable by temperature-programmed desorption. Wise [41] found that a sample made from ammonium thiomolybdate and pretreated at 800 K in H_2/H_2S mixtures showed a change in fractional hydrogen coverage with the H_2/H_2S ratio in the gas mixture. The maximum sorption found corresponded to a stoichiometry of $MoS_2H_{0.039}$.

Blake et al. [42,43] and Eyre [44] prepared the two crystalline forms of MoS_2 (stoichiometry $MoS_{2.0 \pm 0.05}$) by heating MoS_3 in nitrogen to 1273 K for different periods. No hydrogen adsorption was thereafter measurable up to 600 K, but hydrogen-deuterium exchange was observed at 423 K, with an energy of activation of 63 kJ mol^{-1}. This implies rapid dissociation of hydrogen, probably at edge sites. Use of the catalyst for hydrodesulfurization altered the exchange energy of activation to 180 kJ mol^{-1}, no doubt because the sites available were filled by sulfur removed from the thiophene. There were no large differences in the exchange behavior of the two forms of MoS_2.

Wright et al. reported a study of hydrogen sorption on MoS_2 in 1980 [10]. Samples of MoS_2 were made by hydrogen reduction of MoS_3 at 673 K and examined thereafter without exposure to air. Degassing overnight at 573 K was found to give a material which still contained some hydrogen; the stoichiometry was $MoS_2H_{0.011}$ as measured by H_2/D_2 exchange. This material was then capable of sorbing hydrogen reversibly at 573 K to a stoichiometry of $MoS_2H_{0.067}$ at 1 atm pressure. The process was slow. Adsorption of H_2 on MoS_2 made from the elements at high temperature did not occur. Single-point BET-surface-area measurements for the hydridic material gave a value of 8.4 m^2 g^{-1} when either N_2 or CO_2 was the adsorbate. Thus the "surface area" as measured by hydrogen sorption was more than three times the BET value if the assumption is made that one hydrogen atom occupies the same area as one nitrogen molecule. The values differ, of course, according to whether edge planes or basal planes are the most exposed, and allowance is made for this in the calculation. These results are compared with those of Badger et al. [35], where the ratio of hydrogen to BET areas is about 7. The authors then went on to examine the inelastic neutron scattering spectra (INS) of the three materials MoS_2, $MoS_2H_{0.011}$, and $MoS_2H_{0.067}$. In the latter two samples, a peak at 640 ± 40 cm^{-1} is observed. Analysis of the relative intensities of the fundamental and overtones suggested, from the data available, that the hydrogen atom is most likely to be surrounded by sulfur atoms rather than being attached to the metal. The argument was then developed comparing MoS_2 with TaS_2, which also absorbs quantities of hydrogen and in which the hydrogen is thought to be attached to the metal.

Some preliminary neutron diffraction data are given. The hydrogen-containing material had a different c-axis measurement (MoS_2 = 6.15 Å, $MoS_2H_{0.011}$ = 6.389 Å, and after deuterium sorption = 6.372 Å). This expansion was later confirmed by Sampson et al. [45] and Vasudevan et al. [46] by x-ray diffraction (XRD) measurements and is taken to mean that the hydrogen occupies sites in the van der Waals gap of the MoS_2 crystal. The mechanism of adsorption was not elucidated, but some agreement was expressed with the idea that hydrogen creates active centers for further hydrogen adsorption. These centers could well be edge sites, and readier adsorption occurs at catalysts with a predominance of edge sites.

Condensible material (e.g., H_2S) was not detected. Thus it is reasonable to conclude that hydrogen was sorbed and not reacting with the surface to produce H_2S in the gas phase. Such H_2S could have been adsorbed [36] but no INS evidence for SH_2 groups was found, so it was reasonable to assume that no adsorbed H_2S was present.

Moyes and co-workers continued [47] to examine MoS_2H_x, made by reduction of ammonium tetrathiomolybdate directly, in detail. This material also gives a stoichiometry of $MoS_2H_{0.067}$ under 1 atm of hydrogen. The surface area measured by nitrogen adsorption with or without exposure to air was found to be around 30 m^2 g^{-1} for a number of samples. This value gives good agreement between the BET and the hydrogen surface area. This led to the view that sorption of hydrogen at pressures up to 1 atm served only to cover the existing crystallites with S-H groups without intercalation. These results fitted a model in which the sulfur was attached to the hydrogen, with the hydrogen above the sulfur plane, suggesting a picture of the sulfide surface covered with S-H groups, analogous to the O-H groups on the surface of hydrated silica or of alumina. The basal plane material is removed by pumping at 400 K; the edge-plane hydrogen was not removed rapidly by pumping. The latter form was measured by deuterium exchange, giving a value of 0.011 mol of hydrogen per mole of MoS_2, about 0.16 of the surface area. Sorption was slow, particularly when compared with the rate of exchange, again pointing to the importance of the abundance of edge sites.

Sampson et al. [45] and Vasudevan et al. [46] examined hydrogen sorption on MoS_2 in a differential scanning calorimeter and in a high-pressure INS cell. Samples were made by nonaqueous precipitation and resulfiding and compared with results from thermal decomposition. Stoichiometries were $MoS_{1.8}$ and $MoS_{2.06}$, respectively. Isobars similar to those of Badger et al. [35] were measured, but with adsorption of 0.22 mol of hydrogen per mole of MoS_2, while 0.142 mol was adsorbed at lower temperatures (Fig. 3, solid circles). Here, partial formation of a second form of MoS_2 at high pressures with considerable expansion of the c-axis lattice spacing, from 12.6 to 15.6 Å, was calculated from XRD measurements. A theoretical calculation gave an expansion of 2.2 Å if the S-H bonding is perpendicular to the MoS_2 sulfur layers. High-pressure adsorption was associated with a new INS peak at 400 cm^{-1} which appears only at higher temperatures.

Work by Blackburn and Sermon [48] and by others has shown that the intercalation of hydrogen is possible under less usual conditions. Blackburn and Sermon used spillover from platinum to assist the adsorption and intercalation processes. MoS_2 alone adsorbed hydrogen under 1 atm pressure, to the extent of forming $MoS_2H_{0.22}$, a value close to that found by Sampson et al. [45]. Addition of platinum raised the amount adsorbed to the stoichiometry $MoS_2H_{0.84}$. The temperature used for these studies was 325 K. Forming an electrode from MoS_2 and connecting it in an electrochemical cell with dilute H_2SO_4 as an electrolyte also gives rise to hydridic materials. In this way Schollhorn et al. [49] has made a hydride of stoichiometry $MoS_2H_{0.5}$. Vogdt et al. [50] have used γ-γ correlation spectroscopy to distinguish between ordinary and protonated MoS_2. This work shows that there are clear differences between MoS_2 single crystal and MoS_2H_x identified by the technique. Claims are made for the detection of anion vacancies and hydrogen uptake.

To sum up: Three processes of hydrogen sorption can be identified on unsupported molybdenum sulfide. The first is dissociation at edge sites to produce adsorbed H. This is a rapid process, as the exchange of H_2 with D_2 occurs very rapidly and at lower temperatures than adsorption. The concentration of sites depends on the preparation method, but in hydrogen-reduced material of surface area 30 m^2 g^{-1}, edge sites constitute some 16%

of the surface. The high temperature, stoichiometric, well-crystallized material seems to be free of surface sulfhydryl groups, no doubt due to the reduced number of edge sites. In contrast, the stoichiometric material made by reduction with hydrogen probably has a "rag" structure [51] with a high concentration of edge sites, which act as centers for the dissociation of the hydrogen, which then "spills over" onto the basal-plane sites. The dissociated hydrogen then spreads across the surface of the crystal, forming S-H bonds [52], until a complete layer of sulfhydryl groups is present. This process is complete at a stoichiometry of $MoS_2H_{0.07}$ for the 30-m^2 g^{-1} example. At higher pressures and temperatures, or when catalyzed by platinum, dissociated hydrogen diffuses into and along the van der Waals gap in the bulk crystal to form sulfhydryl groups within the structure, apparently by a slow process which finally approaches the stoichiometry MoS_2H, although $MoS_2H_{0.2}$ stoichiometry is obtained more rapidly. Alternatively, exfoliation of the structure may occur, with the layers opening out at the edges under pressure. The first two of these processes are complete at atmospheric pressure, and the most strongly held material is adsorbed on the edge sites of the MoS_2 crystallites, while the weaker material, removed by pumping, is adsorbed on the basal planes of sulfur. Increase in temperature and pressure initiates the process of slow absorption into the MoS_2 structure, which is complete only at pressures of 50 to 70 atm.

These considerations apply to the stoichiometric material. Nonstoichiometric sulfur can increase hydrogen uptake. It was only at relatively high temperatures that Wise found reduction of the sulfide to occur. The molybdenum sulfide with excess sulfur must contain sites of a different type, perhaps as hydrogen ions [20].

B. Supported Molybdenum Sulfide

Supported and promoted catalysts have been the subject of extensive research in attempting to relate hydrotreating activity to preparation. Only those features that affect the hydrogen sorption will be considered here.

Screen [53] examined the hydrogen adsorption properties of alumina-supported MoS_2. The catalysts were made by thermal decomposition of ammonium tetrathiomolybdate (ATM) supported on Alon C in an inert gas stream. Preparation of MoS_2 powders by the same method gave MoS_2 of stoichiometry very close to $MoS_{2.00}$, and avoided questions of the degree of sulfidation raised by sulfiding oxide precursors. Adsorption of hydrogen at 450 K was slow, but measurable and faster at 573 K. There was some adsorption on the support, but none on unsupported MoS_2 made by the same method. The loading of MoS_2 on the support was varied and the graph of the amount of hydrogen adsorbed versus loading went through a maximum between 6 and 9% w/w (Fig. 4).

The monolayer of MoS_2 on the alumina used was calculated to be about 15% w/w, the exact value depending on the basis of the calculation. The composition of the hydride corresponding to the 6% loading at a pressure of 300 torr in the gas phase (1 torr = 133 N m^{-2}) was $MoS_2H_{0.294}$, a value close to that found later by Blackburn and Sermon on bulk MoS_2 at 1 atm pressure [48]. Treatment with H_2/H_2S mixtures only increased the speed at which equilibrium was attained; similar total amounts were adsorbed. Adsorption rates showed an energy of activation of 63 ± 8 kJ mol^{-1}. Hydrogen-deuterium exchange occurred, but was complicated by the presence of the large pool of hydrogen present as surface hydroxyls. At 433 K, exchange of surface hydrogen with gaseous deuterium was slow, but was rapid at 480 K. Exchange did not occur on the alumina alone at 430 K, and occurred very slowly at 480 K. Thus MoS_2 can catalyze spillover of hydrogen from the

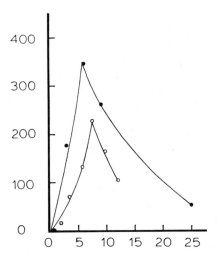

FIGURE 4 Variation of amount of sorbed hydrogen by alumina-supported MoS_2 at different loadings. [Solid points redrawn from Ref. 53, micromole (g $MoS_2)^{-1}$; open points redrawn from Ref. 54, micromole (g catalyst)$^{-1}$.]

alumina support at temperatures greater than 450 K. This observation restricts experimentation with deuterium on alumina-supported Co-Mo catalysts, unless it can be shown that the alumina surface hydroxyls are not involved. In measurements of the rate of adsorption combined with the rate of exchange, it was found that exchange was about 10 times faster than adsorption at 480 K. Adsorption was seen, as before, as a dual-site process, with one site catalyzing fast exchange with dissociation, followed by slower diffusion to a second site.

Screen [53] rationalized the curve of quantity of hydrogen adsorbed against loading (Fig. 4) by suggesting that in the lowest loadings, 1% and 3%, the molybdenum sulfide underwent strong interaction with the support at relatively few strong adsorption sites. This suggestion is supported by the anomalous behavior of the 1% catalyst in HDS and hydrogenation reactions. This strong interaction reduced the edge-site concentration available for the dissociation of hydrogen. As the loading was increased, the importance of the strong interaction became less and the MoS_2 was able to form small crystallites on the surface. These grew with loading until they reached the optimum surface-to-volume ratio, after which the amount of hydrogen adsorbed per gram of MoS_2 fell.

Blake and Moyes [47], following Screen, showed that MoS_2 supported on Vulcan carbon behaved in a similar way to MoS_2 supported on alumina; that is, the amount of hydrogen adsorbed approached a stoichiometry of $MoS_2H_{0.29}$.

In addition to the work on MoS_2 powders outlined above, Wright et al. [10] had examined alumina-supported MoS_2 similar to that examined by Screen, except that the support used was a Filtrol alumina of higher surface area. X-ray diffraction from the supported materials showed broad peaks from which crystallite sizes could be estimated. These sizes were estimated to be 64 × 75 Å before hydriding and 106 × 55 Å after hydriding, showing an increase in the c direction on hydriding. Again, evidence is found for the formation of S-H surface groups on the MoS_2 crystallites. Inelastic neutron scattering studies of this supported material were complicated by the dilution of effects by alumina, particularly by the surface hydroxyls. The spectra gave peaks at 686 and 960 cm^{-1} (compare $MoS_2H_{0.067}$, 662 cm^{-1}). The 686-cm^{-1}

peak was thought to be that from surface S-H on the MoS_2, while that at 960 cm^{-1} was from surface alumina hydroxyls.

Reddy et al. [54] examined the adsorption of hydrogen on a series of sulfided, γ-alumina-supported molybdenum catalysts. The loadings were 2%, 4%, 6%, 8%, 10%, and 12%. Very large amounts of hydrogen were adsorbed at 573 K (the equilibrium pressure is not given) and the curve of amount adsorbed versus loading was similar to that found by Screen et al. [53] (Fig. 4). A maximum of 245 μmol per gram of catalyst was found at 8% loading, corresponding to a stoichiometry of $MoS_2H_{0.98}$. As the adsorption of hydrogen was compared with that of oxygen, the edge plane-to-basal plane area ratio can be calculated and is about 1:10, and the authors conclude that only the surface is hydrogen covered; there is no intercalation.

Earlier, Massoth [55] had obtained data from an alumina-supported molybdenum oxide which had been sulfided and reduced at 673 K. This showed the ratio of hydrogen to molybdenum to be 1.26:1, and was interpreted as a surface covered with Mo-SH groups.

These catalysts were made by sulfiding molybdenum oxide, unlike Screen, so the agreement of the results suggests that the preparative route through the oxide gives results similar to that by decomposition of ammonium tetrathiomolybdate. Reddy and co-workers rationalized their results for the variation of hydrogen uptake with loading in the following terms. Hydrogen adsorption occurs through dissociation at the edge-plane sites and diffuses onto the basal-plane sites. The area of basal plane on the microcrystallites grows with loading up to about 8%. Around this value of loading, the patches of MoS_2 begin to form multilayer crystallites in the c direction rather than continuing to spread across the alumina surface. The authors attribute the fall in the amount adsorbed per gram of catalyst to the relative fall in the basal surface area. But their results are plotted per gram of catalyst. It would appear that the surface area would grow gently, as more edge plane area is added, so the amount of hydrogen adsorbed should be constant or grow slightly. In fact, the amount falls per gram of catalyst with loading, so there must be some process that reduces the MoS_2 surface area to hydrogen. This implies that the crystals must grow in all directions allowed, so that the surface-to-volume ratio decreases. This would reinforce their contention that the external surface of the crystallite is available to hydrogen, whereas the van der Waals gap is not.

In summary, supporting molybdenum disulfide increases the access of hydrogen to the microcrystallites, up to a point. The total adsorption then falls. The maximum in the curve corresponds to MoS_2H approximately.

C. Hydrogen Sorption on Tungsten Sulfide

In general terms, hydrogen sorption on tungsten sulfide (WS_2) resembles molybdenum disulfide (MoS_2). WS_2 has a similar crystalline form and can be made from ammonium thiotungstate by a similar reaction sequence to that used for MoS_2 [56]. Like MoS_2, WS_2 takes up hydrogen, although there are some minor disagreements about the extent of this uptake, based on questions about the time needed for equilibrium to be reached. Early results by Donath [57] were improved by Friz [58], who demonstrated that the amount adsorbed at constant pressure rose steadily with temperature up to 523 K, with a stoichiometry of $WS_2H_{0.0084}$. Decrue and Susz [56] found higher uptakes, corresponding to $WS_2H_{0.04}$ at 383 K, while Gonikberg and Levitskii [59] found that at high pressure and 673 K the stoichiometries ranged from $WS_2H_{0.09}$ at 10 bar to $WS_2H_{0.16}$ at 97 bar. These figures do not suggest that intercalation is occurring [60]. Wright and co-workers examined the WS_2/hydrogen system [60] using a WS_2 preparation which involved reduction of the sulfide precursor in hydrogen. The product was $WS_{2.09}$ by gravimetric analysis. Hydrogen sorption at 673 K was measured, and a limiting coverage gave a stoichiometry of $WS_2H_{0.048}$. Unlike Friz, the authors found more adsorption on cooling to 423 K, at which temperature the stoichiometry corresponded to $WS_2H_{0.056}$, which was about 70% of the one-point BET surface area measurement of 20 m^2 g^{-1}.

The inelastic neutron scattering spectrum was similar to that from MoS_2, but in this case peaks at 2470 cm^{-1} and 2679 cm^{-1} were ascribed to an S-H stretching vibration. The hydrogen vibration parallel to the surface was at 694 cm^{-1}, and was accompanied, as for MoS_2, by harmonics at 1380 and 2074 cm^{-1}. The spectrum was more intense and the features were better resolved than in the corresponding MoS_2 spectra. The analysis of these peaks suggested, as before, an S-H group perpendicular to the sulfur plane, but these results made possible the testing of different models for the position of hydrogen in or on the crystal. Arguments were proposed which show that the hydrogen was unlikely to be attached to tungsten. Related measurements on adsorbed hydrogen sulfide shows that H_2S adsorbs at the surface vacancies, where it dissociates.

D. Hydrogen Sorption on Co_9S_8

Unlike molybdenum sulfide and tungsten sulfide, cobalt sulfide (Co_9S_8) has a cubic lattice. Hoodless et al. [61] have made a study of deuterium adsorption and exchange on carefully characterized crystalline material made by H_2/H_2S treatment of Co_3O_4. Surface areas were 4 to 7 m^2 g^{-1}, and adsorption and exchange were examined in the temperature range 473 to 673 K. At 673 K adsorption was slow, and the first adsorption led to the appearance of H_2S in the gas phase. Repetition of the measurements showed less adsorption on the reduced surface, but reversible behavior. The amount of sulfur lost as H_2S was small, so that the material remained within the composition range for Co_9S_8. There was some hydrogen retained from the preparation step equivalent to $Co_9S_8H_{0.005}$. Coverage with hydrogen in the reproducible conditions was about twice this amount.

Deuterium exchange was more rapid than adsorption, and rates were measured in the range 473 to 583 K. This difference was explained in a manner similar to the same difference on molybdenum sulfide; exchange occurs rapidly at a primary adsorption site (metal atom or ion), while slow uptake is associated with transport of the hydrogen to sulfide ions.

The salient points which emerge are that cobalt adsorbs hydrogen to a much smaller extent than molybdenum or tungsten, and that this adsorption process is slow and occurs only at much higher temperatures. Some reduction of the cobalt sulfide also occurs. The relevance of these results to catalyst activity is shown in Section III.F, where temperature-programmed experiments are considered. The static adsorption experiments give good agreement with the dynamic experiments reported there.

E. Supported and Promoted Molybdenum Catalysts

Samuel and Yeddanapali [62] studied the chemisorption of hydrogen on a catalyst described as MoS_2-Al_2O_3, but, in fact, a commercial cobalt-molybdenum-alumina catalyst. They report multiple kinetic stages in the adsorption process on analysis by use of the Elovich equation. The temperature range 450 to 550 K was used and the sorption divided into three kinetic groups: instantaneous adsorption, rapid adsorption, and slow adsorption.

Stevens and Edmonds [38] found that hydrogen adsorption on sulfided CoMo catalysts occurred only after sulfidation of the oxide precursor. The behavior was similar to that of basal-plane MoS_2 powders, but the extent of adsorption was much greater. There was some formation of H_2S.

Screen [53] noted that promotion with cobalt did not increase the amount of hydrogen adsorbed, but merely increased the speed of adsorption. Reddy and co-workers found that cobalt promotion did not alter the extent of adsorption [54]. They found reasonable agreement between commercial catalysts and their laboratory analogs and between promoted and unpromoted catalysts.

Further evidence for surface sulfhydryl groups on cobalt molybdenum alumina catalysts is given by Maternova [63], who measured surface sulfhydryl groups by reaction with silver nitrate in pyridine solution. The reaction was thought to be

$$*\text{-S-H} + Ag^+ \rightarrow *\text{-S-Ag} + H^+$$

However, Ag^+ adsorbed exceeded H^+ liberated in the most active catalysts. The surface S-H groups were found to correlate with activity for thiophene HDS, and the author claims this as evidence for the importance of the reduction of sulfhydryl groups at active catalyst surfaces as a step in the HDS process:

$$Mo^{(n)+} + (1/2)H_2 \rightarrow Mo^{(n-1)+} + H_2S$$

The results show that large numbers of S-H groups are present, 0.35 mmol per gram of catalyst. If such groups were entirely on the molybdenum part of the catalyst and an 8% loading is present, the stoichiometry would be $MoS_2H_{1.2}$.

F. Temperature-Programmed Desorption Studies

Extensive studies have been made of the molybdenum oxide-alumina system by temperature-programmed reduction (TPR), in order to characterize the oxide before sulfidation. As this work refers to the structure of the oxide, it will not be discussed here. Arnoldy et al. [64] have also examined the sulfidation step in the preparation of the supported catalyst by temperature-programmed sulfidation. Little work, however, appears to have been done on the temperature-programmed desorption (TPD) of hydrogen from molybdenum sulfide and promoted materials. Yarochkin et al. [65] studied the TPD of hydrogen from Co-Mo/alumina catalysts and found that the catalytic activity for thiophene HDS was in good agreement with the amount of weakly bound hydrogen (amount desorbed up to 773 K). It is not clear whether the authors performed the experiments on the oxide precursors or on the sulfided catalysts. As a relationship between activity and hydrogen uptake was shown by Dicks and co-workers [66], and activity was also related to pore volume from pores of radius 37.5 to 55 Å, the value of Yarochkin's observations may be questioned. The work of Dicks and co-workers concerned measurements on the oxide precursors and may explain the results of Yarochkin et al. if their measurements were made on the oxide forms.

Hou and Wise [40] examined the desorption of hydrogen from sites on MoS_{2+x}, made from MoS_3 by reduction, and also titrated the acidity of the sites with ammonia. This work supported the view that hydrogen sorption was related to the sulfur stoichiometry. In a similar study, Fulstow [67] examined the adsorption of hydrogen and of ammonia on MoS_2 made from ATM. This work showed that two distinct desorption peaks were obtained from each adsorbate. The hydrogen desorption peaks were centered on 440 and 590 K, and were of similar area in the fully reduced catalyst. Treatment with H_2S at 723 K and subsequent TPD of hydrogen gave an increased area to the higher-temperature peak, in agreement with Wise [41]. The presence of preadsorbed ammonia removed the higher-temperature hydrogen desorption peak from MoS_2.

Yerofeyev and Kaletchits [68] examined a range of Co-Mo/alumina catalysts and the separate sulfided oxides of cobalt and molybdenum supported on alumina. Hydrogen adsorbed at 298, 373, 473, or 573 K was desorbed in a temperature-programmed experiment to a maximum of 873 K. The adsorption at 298 and 373 K gave rise to two peaks, centered on 443 and 873 K, with calculated energies for hydrogen desorption of 34.7 and 83.3 kJ mol^{-1}, respectively. Hydrogen adsorbed at 473 and 573 K gave rise to three peaks, centered on 443 and 773 K, and with further material desorbed while the sample was being held at its highest temperature, 873 K. The peak that arose in this isothermal regime was accompanied by the emission of some H_2S, and the authors suggested that this hydrogen is in the form of S-H groups. The 443-K peak was attributed to molecular hydrogen, because the desorption was found to be first order. The higher-temperature peaks were attributed to dissociated hydrogen, because the desorption behavior was second order. Thiophene adsorption was investigated similarly. Co-adsorbed hydrogen affected thiophene HDS, and the authors suggested that the hydrogen desorbed around 773 K is that which hydrogenates butene, while the latter peak reflects the H_2S formation step. The authors noted that full conversion of the oxides to the corresponding sulfides had not occurred.

Hoodless et al. [69] examined a set of cobalt-promoted MoS_2 powders. These were made by homogeneous coprecipitation and the activity for thiophene HDS was obtained. The materials were air exposed, and so had to be reduced, resulfided and rereduced, and finally cooled in hydrogen. The desorptogram from MoS_2 showed an envelope containing three peaks, with maxima at 350, 500, and 725 K. The first peak appeared only when the sample temperature in flowing hydrogen fell below room temperature. The peak shape suggests that the first maximum is sharp and overlays the second, broader one. This second (500 K) peak is relatively large in area compared with the other two. The third, high-temperature peak is much smaller, and it was noted that the area of this peak depended on the reduction temperature. The authors report a desorption of hydrogen which corresponds to a stoichiometry of $MoS_2H_{0.054}$, close to their value of $MoS_2H_{0.067}$ obtained in earlier studies.

An explanation of the series of peaks might well be that the first peak, of weakly held material, corresponds to molecular adsorption. The broader second peak contains the basal-plane material, which has to migrate to the edge sites to recombine as molecules. The relatively small amount of edge material is represented by the highest-temperature peak, which requires the greatest thermal energy for desorption, as it is the most strongly bound hydrogen.

The picture is substantially changed by the incorporation of cobalt. Again at the Co/(Co + Mo) ratio of 0.167 two peaks are observed, at 500 and 725 K. The latter now predominates and appears at the temperature corresponding to that from pure Co_9S_8. This change of relative area continues as the Co content increases through the conventional range of catalysts with high synergic effects [i.e., Co/(Co + Mo) ratios of 0.286 to 0.444]. The relative area of this peak then declines as more Co is added. Thus the desorptograms in the series reflect the activity for HDS of thiophene up to a Co/(Co + Mo) ratio of 0.5. The peak at 500 K is apparent throughout the series but is approximately constant. Thus cobalt appears to occupy the edge sites on the crystals of MoS$_2$, a view suggested by Topsøe et al. [27], and this cobalt alters the hydrogen sorption behavior profoundly.

It has been said that addition of cobalt assists the dissociation of hydrogen [70], a suggestion that appears to be borne out in this case. It can be seen that the hydrogen retention by MoS$_2$ is useful for hydrogenation reactions, which occur at low temperatures, but hydrogen has been desorbed at those temperatures at which MoS$_2$ is active for HDS (600 to 700 K). The presence of cobalt appears to correspond with a maximum in the TPD profile at temperatures appropriate for the HDS reaction, so that cobalt makes hydrogen available in sufficient quantities at the high temperatures needed for reaction.

IV. KINETICS OF DESULFURIZATION AND RELATED REACTIONS

As before, much of the information available relates to the cobalt-molybdenum sulfide system, and as this is the industrial catalyst of choice in many cases, it will serve to exemplify many of the problems. Thiophene is often used as a test reactant, because of the relative difficulty in desulfurizing the aromatic molecule. Delmon [70] has pointed out some of the difficulties in comparing studies of sulfide catalysts. The maximum synergic effect occurs at different Co/(Co + Mo) ratios for the hydrodesulfurization, hydrogenation, and isomerization reactions. The synergic effect is amplified by increased pressure. The reactivity in HDS is improved by pretreatment in hydrogen, while the hydrogenation activity is depressed; the opposite is true when pretreatment in H$_2$/H$_2$S mixtures is used. Transient effects are observed when operating conditions are changed. All these factors make comparison of the work of different authors very difficult. One feature of the HDS catalyst is generally agreed--that the HDS and hydrogenation sites appear to be different.

Satterfield and Roberts [71] studied the intrinsic kinetics of thiophene hydrogenolysis in a differential reactor. They were able to correlate the rate of thiophene disappearance with a Langmuir-Hinshelwood rate expression, which included terms for the retardation of the rate by H$_2$S and by the reactant thiophene.

$$\text{rate of HDS} = k \frac{K_T p_T K_H p_H}{(1 + K_T p_T + K_S p_S)^2}$$

Here k is the rate constant, K the adsorption constant, and the subscripts T, H, and S refer to thiophene, H$_2$, and H$_2$S, respectively.

The reaction was thought to involve two types of site. One of these was available to, and was competed for, by thiophene and by H$_2$S. The other site was available only to hydrogen. The rate-limiting step was seen as the combination of adsorbed thiophene with adsorbed hydrogen. The authors admitted that the range of hydrogen pressure variation was small, but Frye and Mosby had found that there was a first-order dependence on hydrogen pressure over a wide range of pressure [72].

Massoth [73] also examined thiophene HDS on sulfided molybdena-alumina, using a similar Langmuir-Hinshelwood model but with particular reference to the order in hydrogen. A "fair" correlation was found with a rate expression that was first order in thiophene and in hydrogen and inhibited by thiophene and by H$_2$S. A "superior" correlation was found by introducing a hydrogen inhibition term into an expression of the form

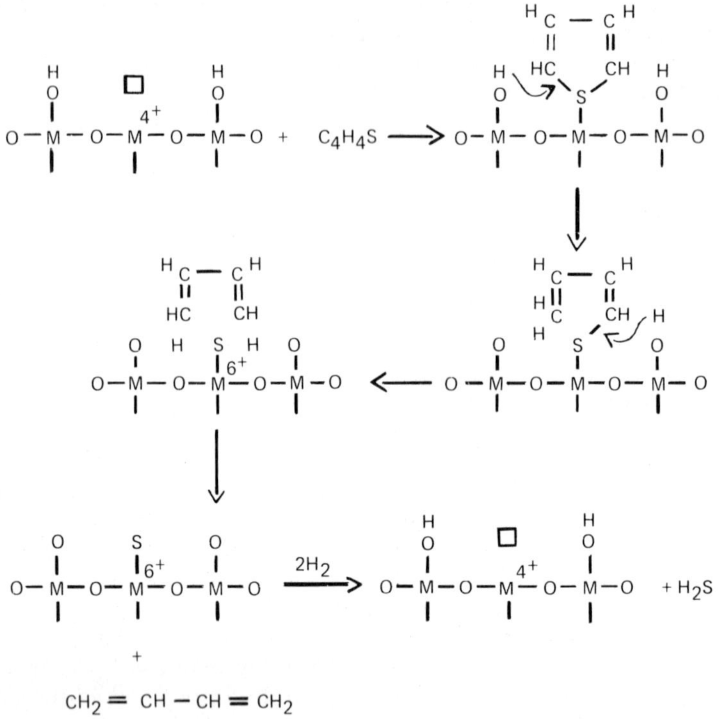

FIGURE 5 Vacancy-site mechanism for thiophene hydrodesulfurization. (From Ref. 12.)

$$\text{rate of HDS} = \frac{k_T p_T p_H}{(1 + K_T p_T + K_S p_S)(1 + K_H p_H)}$$

As before, surface reaction between adsorbed species were found to be the rate-limiting steps, using the dual-site mechanism suggested by Lipsch and Schuit [12].

Thiophene was adsorbed at sulfur vacancy; nearby there was an S-H group (Fig. 5). Hydrogen performed two functions in this mechanism, that of assisting cleavage of the C-S bond and that of removing the sulfur as H_2S, reforming the sulfur vacancy. The wide range of hydrogen partial pressures necessary to demonstrate the difference between single and dual sites unequivocally was not undertaken. The author also considered molecularly adsorbed hydrogen. Lee and Butt [74] used selective poisoning with pyridine to show that the hydrogenolysis of thiophene and the hydrogenation of butene occurred at different sites on Co-Mo/alumina catalysts. They were unable to find kinetic evidence for dissociative adsorption of hydrogen, but suggested that at lower temperatures the major HDS reaction involved weakly adsorbed hydrogen, with increasing contributions from more strongly adsorbed hydrogen at higher temperatures. Ozimek and Radomyski [75] showed, by use of product analyses, that thiophene was adsorbed at two sites on the catalyst and reacted with gas-phase hydrogen and not with adsorbed hydrogen. Their catalyst was a Co-Mo/alumina preparation. Massoth and MuraliDhar [76] reviewed the kinetic evidence as far as hydrogen was concerned and concluded that separate sites were involved and that on one of these hydrogen was molecularly adsorbed. They were aware, however, that other studies had shown that hydrogen could readily be dissociated at sulfide surfaces, but pointed out that this may not be the hydrogen involved in HDS.

The possibility that molecular hydrogen is involved is reflected in some hydrogenation studies. Okuhara et al. [77] observed that hydrogen molecularity was conserved in the hydrogenation of olefins and of butadiene on MoS_2, and Blake et al. [42] found that deuterium addition to thiophene and to butadiene occurred pairwise. In an investigation of the HDS of thiophene with deuterium catalyzed by MoS_2 powders, they also found that two pools of "hydrogen" existed on the surface with different H/D ratios, a finding repeated in an investigation of the HDS of tetrahydrothiophene [43]. One explanation of these two pools was that the first existed as hydrogen adsorbed on edge sites, while the second consisted of basal-plane hydrogen.

Recently, Vrinat [78] has reviewed the kinetics of the HDS process. He lists eight different forms of the rate equation for thiophene HDS and similar numbers for dibenzothiophene. Like Delmon, he cautioned the reader against comparisons of data obtained by widely differing methods. In spite of the range of results, he considered that the hydrogen dependence still required investigation, particularly at elevated pressures. His comment on the influence of H_2 partial pressure was that despite some experimental findings, the current view was that the dual-site mechanism held sway, with hydrogen on one site and thiophene on the other. He quotes the paper of Devanneaux and Maurin [79] as good evidence for the involvement of dissociated hydrogen in the thiophene and benzothiophene HDS reactions. For dibenzothiophene, however, molecularly adsorbed hydrogen was preferred.

In summary, it appears that even for the relatively simple thiophene system, there is no clear picture of the kinetic role of hydrogen and its relation to mechanism. This may have arisen because the multiplicity of sites postulated by some investigators had not been compared with the modes of hydrogen sorption on MoS_2 suggested by investigators and summarized in Fig. 5. The positions in which hydrogen can be attached are often very like those occupied by cobalt in the picture drawn by Topsoe and Clausen [80].

The hydrogen may be an S-H group (Fig. 6, a), or it may occupy a sulfur vacancy as in (b), where most authors consider initial dissociation of hydrogen

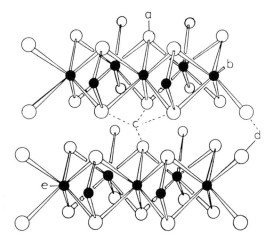

FIGURE 6 Representation of sites for hydrogen adsorption on MoS_2: (●) Mo, (○) S. (From Ref. 80.)

molecules is likely to occur. It may be contained in the van der Waals gap between the layers of the crystal (c), or be an S-H group at an edge (d). Finally, it may be attached to molybdenum with its full complement of sulfur, creating an Mo(V) species (e). The evidence for some of these sites is given in Section III, most of it garnered from experiments on solid MoS_2 exposed to gas-phase hydrogen. The behavior of the MoS_2/hydrogen system in solvents has yet to be investigated. It will be recalled that the higher molecular mass sulfur compounds are usually desulfurized in a catalyst/liquid/gas system. As shown by temperature-programmed desorption, the introduction of cobalt massively changes the hydrogen behavior. Further extensive studies are clearly needed to elucidate the active forms of hydrogen adsorption.

V. PREDICTIVE VALUE OF HYDROGEN SORPTION MEASUREMENTS

In discussing the measurements of hydrogen sorption, reference has been made to relationships with activity for HDS or for hydrogenation reactions, by Stevens and Edmonds [38], Wise and co-workers [19, 20, 39-41], and Moyes and co-workers [42, 43], among others.

Screen [53] found a parallel relationship between the amount of hydrogen sorbed (Fig. 4, open circles) and thiophene HDS (Fig. 7) and butadiene hydrogenation (Fig. 8). The paper by Reddy et al. [54] found no such relationship, but instead an activity for HDS which, when plotted against uptake, initially increased and then leveled off. Oxygen chemisorption gave a better correlation with HDS activity. For hydrogenation, however, there was a reasonable correlation with uptake, implying a different role for hydrogen in hydrogenation reactions. Fulstow [67] has examined the interrelationship of the adsorption of ammonia, oxygen, hydrogen, and H_2S and finds

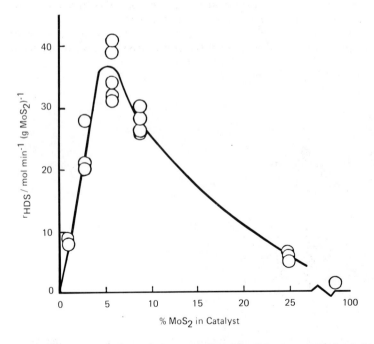

FIGURE 7 Variation of the activity for thiophene HDS at 513 K with MoS_2 loading on alumina. (Compare Fig. 4.) (From Ref. 53.)

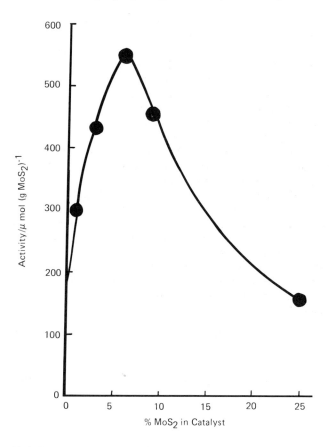

FIGURE 8 Variation of the activity for buta-1,3-diene hydrogenation at 450 K with MoS_2 loading on alumina. (Compare Fig. 4.) (From Ref. 53.)

O, NH_3, and H each occupy a similar number of sites, and this is half the number occupied by H_2S. All this evidence seems to point to the need to recognize the possible sites for adsorption outlined in Fig. 6, and to find ways to differentiate between them. Edge-site concentrations appear to be measured by H_2-D_2 exchange at low pressures and temperatures, or by TPD, although the well-established oxygen adsorption measurement is unlikely to be displaced. On the other hand, the availability of hydrogen at the surface level at higher temperatures and hydrogen partial pressures may be important. The studies discussed in Section III show how difficult it is to ensure that the results are reproducible and comparable between investigators. Some studies suggest that hydrogen sorption measures basal-plane area, and basal-plane area is probably more easily measured by the standard BET method.

A number of papers have been devoted to deuterium exchange with thiophene and other sulfur compounds. Early work by Kieran and Kemball [81,82] did much to establish views about desulfurization mechanisms, particularly the form in which thiophene adsorbs. As this chapter is concerned with hydrogen in sulfide catalysis, there is no space to discuss more than peripheral aspects of HDS mechanisms, but Zdrazil [83] has reviewed the situation recently. Later, John and co-workers suggested that exchange reactions of sulfur-containing compounds were essentially the same whether catalyzed by MoS_2 or by partially sulfided Co-Mo/alumina catalysts [84].

In the same vein, Smith and co-workers [85] and Behbahany et al. [86] have desulfurized thiophene using deuterium and have related the exchange patterns in the products to activity. A paper by Mikovsky and Silvestri [87] suggested that intramolecular HDS occurs by formation of H_2S from the β-hydrogen atoms in thiophene. The evidence to support this lay in the deuterium content of the product H_2S, which was below that expected. The catalyst used was a Co-Mo/alumina one, and as in Section III.B, it is necessary to ensure that the surface hydroxyls were not involved, although the authors report that the catalyst was thoroughly purged with deuterium. Gudkov et al. [88] made similar experiments on Ni-Mo/alumina, Co-Mo/alumina, and Ni-W/alumina catalysts. In this case not only were experiments with C_4H_4S/D_2 mixtures done, but the reverse, C_4D_4S/H_2, mixtures were also investigated. The results show that external hydrogen must be involved rather than the intramolecular mechanism suggested by Mikovsky et al. The necessity of extermal hydrogen for HDS of thiophene is also suggested by Morávek and Kraus [89], who found that a $Co-Mo/Al_2O_3$ catalyst lost its activity after heating at 773 K for 8 h, which probably caused the desorption of hydrogen.

In summary, there seems to be a general agreement that hydrogen sorption relates to hydrogenation activity, but that HDS is more complex and there is no simple relationship with activity. Hydrogen sorption and exchange studies, however, shed light on possible mechanisms.

ACKNOWLEDGMENTS

The author is grateful to the British Gas Corporation, the SERC, and AERE, Harwell for generous support of some of the studies reviewed here.

REFERENCES

1. O. Weisser and S. Landa, *Sulphide Catalysts, Their Properties and Applications*, Pergamon Press, Oxford (1973).
2. P. C. H. Mitchell, in *Catalysis*, Vol. 1 (C. Kemball, ed.), Chemical Society, London (1977), p. 204.
3. P. C. H. Mitchell, in *Catalysis*, Vol. 4 (C. Kemball and D. A. Dowden, eds.), Chemical Society, London (1981), p. 175.
4. P. C. H. Mitchell, in *The Chemistry of Some Hydrodesulphurisation Catalysts Containing Molybdenum*, Climax Molybdenum Co. Ltd., London (1967).
5. F. E. Massoth, *Adv. Catal.*, 27:265 (1978).
6. D. Pirotte, J. M. Zabala, P. Grange, and B. Delmon, *Bull. Soc. Chim. Belge.*, 90:1239 (1981).
7. P. Desikan and C. H. Amberg, *Can. J. Chem.*, 42:843 (1963).
8. K. Tanaka and T. Okuhara, in *Proceedings of the Climax 3rd International Conference on the Chemistry and Uses of Molybdenum*, Ann Arbor, Mich. (H. F. Barry and P. C. H. Mitchell, eds.) (1980), p. 170.
9. S. J. Tauster, T. A. Pecoraro, and R. R. Chianelli, *J. Catal.*, 63:515 (1980).
10. C. J. Wright, C. Sampson, D. Fraser, R. B. Moyes, P. B. Wells, and C. Riekel, *J. Chem. Soc. Faraday Trans. 1*, 76:1585 (1980).
11. N-Y. Topsoe and H. Topsoe, *Bull. Soc. Chim. Belge.*, 90:1311 (1981).
12. J. M. J. G. Lipsch and G. C. A. Schuit, *J. Catal.*, 15:179 (1969).
13. G. C. A. Schuit and B. C. Gates, *AIChE J.*, 19:41 (1973).
14. F. E. Massoth, *J. Catal.*, 36:164 (1975).
15. R. J. H. Voorhoeve and J. C. M. Stuiver, *J. Catal.*, 23:243 (1971).

16. A. L. Farragher and P. Cossee, *Proceedings of the 5th International Congress on Catalysis* (J. W. Hightower, ed.), North-Holland, Amsterdam (1972), p. 1301.
17. G. Hagenbach, P. Courty, and B. Delmon, *J. Catal.*, *31*:264 (1973).
18. P. Grange and B. Delmon, *J. Less-Common Met.*, *36*:353 (1974).
19. A. Aoshima and H. Wise, *J. Catal.*, *34*:145 (1974).
20. P. R. Wentrcek and H. Wise, *J. Catal.*, *45*:349 (1976).
21. B. S. Clausen, S. Morup, H. Topsoe, and R. Candia, *J. Phys. (Paris) Colloq.*, *37*:C6-249 (1976).
22. C. Wivel, R. Candia, B. S. Clausen, S. Mørup, and H. Topsøe, *J. Catal.*, *68*:453 (1981).
23. H. Topsøe, B. S. Clausen, R. Candia, R. Wivel, and S. Mørup, *J. Catal.*, *68*:433 (1981).
24. M. Breysse, B. A. Bennett, D. Chadwick, and M. Vrinat, *Bull. Soc. Chim. Belge.*, *90*:1271 (1981).
25. S. Gobolos, Q. Wu, J. Ladriere, F. Delannay, and B. Delmon, *Bull. Soc. Chim. Belge.*, *93*:687 (1984).
26. H. Topsøe, R. Candia, N-Y. Topsøe, and B. S. Clausen, *Bull. Soc. Chim. Belge.*, *93*:783 (1984).
27. N-Y. Topsøe, H. Topsøe, O. Sorensen, B. S. Clausen, and R. Candia, *Bull. Soc. Chim. Belge.*, *93*:727 (1984).
28. H. Topsøe, N-Y. Topsøe, O. Sorensen, R. Candia, B. S. Clausen, S. Kallesøe, and E. Pederson, in *Proceedings of the ACS National Meeting, American Chemical Society, Washington, D.C.* (1983).
29. E. G. Derouane, E. Pederson, B. S. Clausen, Z. Gabelica, R. Candia, and H. Topsøe, *J. Catal.*, *99*:253 (1986).
30. J. A. Wilson and A. D. Yoffe, *Adv. Phys.*, *18*:211 (1969).
31. E. Y. Rode and B. A. Lebedev, *Russ. J. Inorg. Chem.*, *6*:608 (1961).
32. R. R. Chianelli and M. B. Dines, *Inorg. Chem.*, *17*:2758 (1978).
33. F. T. Eggertson and R. M. Roberts, *J. Phys. Chem.*, *63*:1981 (1959).
34. B. S. Parekh and S. W. Weller, *J. Catal.*, *47*:100 (1977).
35. E. H. M. Badger, R. H. Griffiths, and W. B. S. Newling, *Proc. R. Soc. London*, *A197*:184 (1949).
36. R. L. Wilson, C. Kemball, and A. K. Galwey, *Trans. Faraday Soc.*, *58*:583 (1962).
37. P. Ratnasamy and J. J. Fripiat, *Trans. Faraday Soc.*, *66*:2897 (1970).
38. G. C. Stevens and T. Edmonds, *J. Less-Common Met.*, *54*:321 (1977).
39. P. R. Wentrcek and H. Wise, *J. Catal.*, *50*:80 (1976).
40. P. Hou and H. Wise, *J. Catal.*, *78*:469 (1982).
41. H. Wise, in *Proceedings of the Amax 5th International Conference on the Chemistry and Uses of Molybdenum* (A. G. Sykes and P. C. H. Mitchell, eds.), Pergamon Press, Oxford (1986), p. 145.
42. M. R. Blake, M. Eyre, R. Moyes, and P. B. Wells, in *Proceedings of the 7th International Congress on Catalysis, Tokyo* (T. Seiyama and K. Tanabe, eds.) (1980), p. 591.
43. M. R. Blake, M. Eyre, R. B. Moyes, and P. B. Wells, *Bull. Soc. Chim. Belge.*, *90*:1293 (1981).
44. M. Eyre, Ph.D. thesis, University of Hull (1979).
45. C. Sampson, J. M. Thomas, S. Vasudevan, and C. J. Wright, *Bull. Soc. Chim. Belge.*, *90*:1215 (1980).
46. S. Vasudevan, J. M. Thomas, C. J. Wright, and C. Sampson, *J. Chem. Soc. Chem. Comm.*, 418 (1982).
47. M. R. Blake and R. B. Moyes, unpublished work.
48. A. Blackburn and P. Sermon, *J. Chem. Technol. Biotechnol.*, *A33*:120 (1983).

49. R. Schollhorn, M. Kumpers, and D. Plorin, *J. Less-Common Met.*, 58:55 (1978).
50. C. Vogdt, T. Butz, A. Lerf, and H. Knozinger, in *Proceedings of the Amax 5th International Conference on the Chemistry and Uses of Molybdenum* (A. G. Sykes and P. C. H. Mitchell, eds.), Pergamon Press, Oxford (1986), p. 95.
51. R. R. Chianelli, E. B. Prestridge, T. A. Pecoraro, and J. P. de Neufville, *Science*, 203:1105 (1979).
52. D. Chadwick and M. Breysse, *J. Catal.*, 71:226 (1981).
53. C. T. Screen, Ph.D. thesis, University of Hull (1981).
54. M. B. Reddy, K. V. R. Chary, V. S. Subrahmanyam, and N. K. Nag, *J. Chem. Soc. Faraday Trans. 1*, 81:1655 (1985).
55. F. E. Massoth, *J. Catal.*, 36:164 (1975).
56. J. Decrue and B. Susz, *Helv. Chim. Acta*, 39:619 (1956).
57. E. E. Donath, *Adv. Catal.*, 8:245 (1956).
58. H. Friz, *Z. Elektrochem.*, 54:538 (1950).
59. M. G. Gonikberg and I. I. Levitskii, *Izv. Akad. Nauk SSSR Otd. Khim. Nauk*, 1170 (1960).
60. C. J. Wright, D. Fraser, R. B. Moyes, and P. B. Wells, *Appl. Catal.*, 1:49 (1981).
61. R. C. Hoodless, R. B. Moyes, and P. B. Wells, *J. Catal.*, submitted for publication
62. P. Samuel and L. M. Yeddanapali, *J. Appl. Chem. Biotechnol.*, 24:777 (1974).
63. J. Maternova, *Appl. Catal.*, 6:61 (1983).
64. P. Arnoldy, J. A. M. van Heijkant, G. D. de Bok, and J. A. Moulijn, *J. Catal.*, 92:35 (1985).
65. V. N. Yarochkin, V. I. Sokolova, G. A. Berg, and V. I. Kuz'min, *Kinet. Katal.*, 16(2):476 (1975).
66. A. L. Dicks, R. L. Ensell, T. R. Phillips, A. K. Szczepura, M. Thorley, A. Williams, and R. D. Wragg, *J. Catal.*, 72:266 (1981).
67. A. N. Fulstow, Ph.D. thesis, University of Hull (1986).
68. V. I. Yerofeyev and I. V. Kaletchits, *J. Catal.*, 86:55 (1984).
69. R. C. Hoodless, R. B. Moyes, and P. B. Wells, *Bull. Soc. Chim. Belge.*, 93:673 (1984).
70. B. Delmon, in *Proceedings of the Climax 3rd International Conference on the Chemistry and Uses of Molybdenum, Ann Arbor* (H. F. Barry and P. C. H. Mitchell, eds.) (1980), p. 73.
71. C. N. Satterfield and G. W. Roberts, *AIChE J.*, 14:159 (1968).
72. C. G. Frye and J. F. Mosby, *Chem. Eng. Prog.*, 63(9):66 (1967).
73. F. E. Massoth, *J. Catal.*, 47:316 (1977).
74. H. C. Lee and J. B. Butt, *J. Catal.*, 49:320 (1977).
75. B. Ozimek and B. Radomyski, *Chem. Stosow.*, 19:97 (1975); 20:20 (1976).
76. F. E. Massoth and G. MuraliDhar, in *Proceedings of the Climax 4th International Conference on the Chemistry and Uses of Molybdenum, Ann Arbor, Mich.* (H. F. Barry and P. C. H. Mitchell, eds.) (1982), p. 343.
77. T. Okuhara, T. Kondo, and K. Tanaka, *Chem. Lett.*, 717 (1976).
78. M. L. Vrinat, *Appl. Catal.*, 6:137 (1983).
79. J. Devanneaux and J. Maurin, *J. Catal.*, 69:202 (1981).
80. H. Topsøe and B. S. Clausen, *Cat. Rev.-Sci. Eng.*, 26:395 (1984).
81. P. Kieran and C. Kemball, *J. Catal.*, 4:380 (1965).
82. P. Kieran and C. Kemball, *J. Catal.*, 4:394 (1965).
83. M. Zdrazil, *Appl. Catal.*, 4:107 (1982).
84. C. S. John, J. G. Williamson, L. V. F. Kennedy, and J. Kelvin Tyler, *J. Chem. Soc. Faraday Trans. 1*, 76:1356 (1980).

85. G. V. Smith, C. C. Hinkley, and F. Behbahany, *J. Catal.*, *30*:218 (1973).
86. F. Behbahany, Z. Sheikhrezai, M. Djalali, and S. Salajegheh, *J. Catal.*, *63*:285 (1980).
87. R. J. Mikovsky, A. J. Silvestri, *J. Catal.*, *34*:324 (1974).
88. B. S. Gudkov, N. A. Gayday, L. Beranek, and S. L. Kiperman, *Collect. Czech. Chem. Commun.*, *49*:2400 (1984).
89. V. Moravek and M. Kraus, *Collect. Czech. Chem. Commun.*, *50*:2159 (1985).

V
TECHNOLOGICAL IMPLICATIONS

23 | Hydrogen Effects in Industrial Catalysis

P. G. MENON

*Dow Chemical (Nederland) B.V., Terneuzen, The Netherlands,
and Chalmers University of Technology, Gothenburg, Sweden*

I.	INTRODUCTION	611
II.	FIVE MAJOR AREAS	613
III.	SOME MORE SUBTLE HYDROGEN EFFECTS IN CATALYSIS	615
	REFERENCES	619

I. INTRODUCTION

During the last four decades large quantities of hydrogen have become available, thanks to the development of both large-scale and highly economical means of hydrogen manufacture from coal, oil, and natural gas, and the by-product hydrogen from modern petroleum refining processes. This has triggered rapid expansion of the uses of hydrogen in several industrial applications, including its extensive captive use in petroleum and petrochemical processing. World production of hydrogen was barely 6000 metric tons in 1938, but it soared to nearly 45 million metric tons in 1980, and it is still increasing at the rate of 6 to 8% per year.

The importance of hydrogen as a source of energy was recognized by Jules Verne as early as 1870. In his "The Mysterious Island" he wrote: "I believe that water will one day be employed as fuel, that hydrogen and oxygen which constitute it, used singly or together, will furnish an inexhaustible source of heat and light, of an intensity of which coal is not capable. . . . I believe, then, that when the deposits of coal are exhausted we shall heat and warm ourselves with water. Water will be the coal of the future." Although that scenario is not yet fully realized, it is no longer just science fiction. Liquid hydrogen and liquid oxygen are indeed used to fuel the manned space flights. The outstanding merits of hydrogen as a fuel are well known, both for direct burning and for internal-combustion engines. Nontoxic, odorless, and tasteless by itself, hydrogen is almost completely benign to the environment, because it produces no smoke, no fumes, no CO or CO_2, and much less NO_x than that produced by gasoline combustion. The absence of soot and acidic combustion products (which cause "acid rain," with its disastrous consequences to forests, vegetation, and ancient monuments) is welcome not only to the environment but also to the engine, because it will prolong engine life and reduce engine-maintenance costs. Several comprehensive volumes have been published: *Hydrogen Technology for Energy* [1], *Hydrogen for Energy Distribution* [2], *Hydrides for Energy Storage* [3], and *Catalysis on the Energy Scene* [4]. Hence these important aspects of hydrogen in the overall energy/fuels situation will not be discussed in this book.

In the world of industrial catalysis, hydrogen plays a very important role. Conversely, catalysts are required for most uses or applications of hydrogen in industry, because molecular hydrogen has first to be dissociated into atoms or radicals before it can react with most other molecules. In the case of chemical or biological reactions in an aqueous phase, the hydrogen-ion concentration or pH of the solution is often a vital process variable. Such aspects of hydrogen in industrial catalysis are too well known and too varied to be covered in a book like the present one.

The history of industrial catalysis has been reviewed briefly by Heinemann [5]. Catalysis as an art was practiced in brewing alcoholic beverages and making soap, perhaps even in prehistoric times. The birth of catalyzed chemical industries is usually identified with the processes for sulfuric acid and nitric acid in the nineteenth century and the hydrogenation of fats and oils early in this century. The development of ammonia synthesis and methanol synthesis, hydrogenation of coal (1927), and so on, paved the way for the emergence of catalyst technology as we know it today. The advent of motor cars and airplanes and the energy-intensive life-style of the last half-century have led to (and were made possible by) the phenomenal growth of the chemical, petroleum, and petrochemical industries. For developed economies such as those of the United States, Western Europe, and Japan, current estimates are that about 20% of the gross national product is the result of the use of catalysts, directly or indirectly. Industrial catalysis plays a major role in satisfying the diverse needs of modern societies, particularly their food production, energy production/conversion/conservation, defense technologies, environmental protection, and health and medical care. The recent Pimentel Committee Report in the United States, which underlined the pivotal role of chemistry in the development of present-day economy and quality of life, has cited catalysis as one of the most crucial areas for top research priority in the coming decades: "Ultimately, new catalyst systems will result that will lay the foundations for the development of new chemical technologies" [6].

Catalysis involving hydrogen directly or indirectly constitutes a very important part of modern developments in industrial catalysis. For example, much of petroleum refining may be regarded as processes for increasing the H/C atomic ratio from 1.5-1.9 in feedstocks to the 1.8-2.1 needed for most transportation fuels (Fig. 1). Making hydrogen from coal, oil, or natural

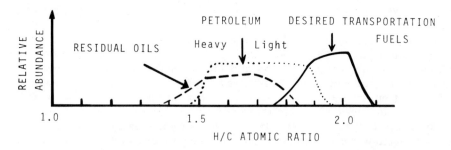

FIGURE 1 Much of petroleum processing may be looked upon as efforts to raise the H/C ratio of the available feedstocks.

TABLE 1 Some Major Hydrogenation and Dehydrogenation Processes

A. Hydrogenation of:
Benzene to cyclohexane
Phenol to cyclohexanone (for caprolactam)
Nitrobenzene to aniline
Edible oils to shortenings
Inedible oils to saturated fatty acids (rubber additives)
Fatty acids to fatty alcohols
Diolefins and acetylenes to mono-olefins
Dinitrotoluenes to toluene diamines (for toluene di-isocyanate)

B. Dehydrogenation of:
Butanes to butene/butadiene
Butene to butadiene
n-Paraffins to α-olefins (for biodegradable detergents)
Ethylbenzene to styrene

---

gas is in itself a major process industry, since it is the starting point for all ammonia- and urea-based fertilizers. A comprehensive volume, *Hydrogen Manufacture by Electrolysis, Thermal Decomposition and Unusual Techniques*, was published in 1978 [7]. Quite apart from the hydrogenation of nitrogen to ammonia and of carbon monoxide to methanol, higher alcohols, hydrocarbons, and oxygenates (cf. Fischer-Tropsch synthesis), hydrogenations and dehydrogenations form the basis of several other major process industries. A representative list of these is given in Table 1. Several processes in pharmaceuticals and fine chemicals also are based on highly selective (partial) hydrogenation and dehydrogenation. However, such aspects of hydrogen in industrial catalysis are well known and readily accessible in several monographs and reviews on those topics. Hence they will not be discussed further here. The purpose of Part V of this book is to draw attention to a few typical cases of more subtle and less known aspects and effects of hydrogen in industrial catalysis.

II. FIVE MAJOR AREAS

As mentioned above and as shown in Table 1, selective catalytic hydrogenation is a very important chemical engineering unit process today. In petrochemical complexes, such processes are used to hydrogenate diolefins and acetylenes to mono-olefins in the ethylene, propylene, butene/butadiene, and pyrolysis gasoline streams. The role of hydrogen in the selectivity for competitive hydrogenation in these and many other reactions is highlighted in Chapter 24. The strong chemical reaction engineering flavor of this chapter is intended to be complementary to the chemical catalysis aspects of hydrogen effects in organic hydrogenations discussed in Chapter 18, and in selective/partial hydrogenation of acetylenes, where the formation, stability, and hydrogen-transfer efficiency of metal hydrides could be so decisive, as pointed out in Chapter 14. This entire area has become of special importance in recent years in the context of several large chemical/petrochemical firms trying to transfer a good part of their business from bulk chemicals and intermediates to low-volume, high-value-added, high-technology "specialties."

Catalytic cracking is often considered to be the heart of a modern refinery [8]. The use of zeolitic cracking catalysts, introduced by Mobil in the 1960s, totally revolutionized catalytic cracking. But the importance of hydrogen transfer in the overall cracking process has remained practically buried in the patent literature; it has seldom been discussed, understood, or appreciated by the catalysis community. To bridge this gap in our understanding of catalysis on zeolites, Chen and Haag of Mobil have made a valuable contribution in Chapter 27, where they have also discussed several industrial applications of hydrogen transfer on zeolites. Interesting examples presented there include the use of erionite and ZSM-5 for enhancing the octane number (RON) of gasoline in catalytic reforming (Mobil's Selectoforming and M-Forming processes) and of ZSM-5 for the same purpose in fluid catalytic cracking (FCC). Still another innovative application of hydrogen transfer is in Davison's tailor-made catalysts for octane boosting in FCC. This consists of the incorporation of appreciable quantities of ultrastable Y (USY) zeolites in these catalysts. The mechanism of octane improvement by USY catalysts proposed by Magee et al. [9,10] is as follows. In USY, the acid sites are more widely dispersed and are of stronger acidity than those in conventional rare-earth exchanged Y (REY) zeolites. Consequently, cracking over USY yields a product significantly richer in olefins at the expense of paraffins, and slightly richer in aromatics. Since both these factors will minimize the hydrogen-transfer reaction

olefins + naphthenes = paraffins + aromatics

fewer paraffins are produced ultimately. Since the hydrogen transfer is impeded, there is also less likelihood of aromatics condensing to form polynuclear aromatic coke precursors and finally, coke. Thus less coke and an improved yield of aromatics are secondary benefits. Furthermore, the strong acid sites enhance cracking over hydrogen transfer. Hence large alkyl aromatics would more likely dealkylate, producing aromatics and olefins in the gasoline or LPG range, rather than ending up as high-molecular-weight components in the light-cycle oil range or polynuclear aromatic coke precursors.

Hydrocarbon reactions at high temperatures are often accompanied by deposition of carbonaceous residues ("coke") on catalysts. An effective way to minimize this and give the catalyst an economically acceptable life is to use hydrogen under pressure in the process (e.g., in catalytic reforming, hydrodesulfurization/hydrotreating, hydrocracking, even many dehydrogenations). Catalytic reforming is a fascinating network of hydrogenation/dehydrogenation reactions on the metal sites and carbenium-ion reactions on the acid sites of the catalyst in a classical and most instructive example of bifunctional catalysis. In fact, it was the study of catalytic reforming mechanisms which led to the recognition of polyfunctional heterogeneous catalysis as such; this recognition, in turn, has led to many major improvements in both the catalysts and process for reforming and to several entirely new processes as well. In these developments of the last three decades, there are several subtle roles of hydrogen under pressure. These are brought out elegantly in Chapter 25.

The interaction of hydrogen with metals is of vital importance to the subject matter of this book. The theoretical or physicochemical aspects of this interaction have been dealt with in some detail in Chapter 1 and later in Chapter 3. This interaction is also the basis for using hydrogen as a tool for characterization of catalysts, discussed in Chapter 4. Hydrogen-metal interactions also figure prominently in several other chapters. Going one step further, one can consider hydrogen *in* metals. The importance of this topic can be appreciated from the latest International Symposium on Hydrogen in Metals, held at Belfast, North Ireland, March 26-29, 1985. The diffusion of hydrogen through metal membranes can have an impact on catalytic processes and catalyst technology. Attention is drawn to this in Chapter 26.

23. Hydrogen Effects in Industrial Catalysis

The last major example chosen is still more remote from the general theme of heterogeneous catalysis: Ziegler-Natta polymerization has generally remained a domain of polymer scientists rather than of catalysis scientists, although Natta himself was a most distinguished and successful heterogeneous catalysis scientist before his epoch-making discovery of propylene polymerization in 1955. Hydrogen is used for a very unusual purpose in olefin polymerization: to control the molecular-weight distribution in the polymer. Perhaps for the very first time in the literature, this role of hydrogen in olefin polymerization has been dissected and critically examined in Chapter 28 of this book.

III. SOME MORE SUBTLE HYDROGEN EFFECTS IN CATALYSIS

Chapters 24 to 28 thus focus attention on five important and rather unusual technological aspects of hydrogen in industrial catalysis. There are a host of other phenomena involving hydrogen, the technological implications of which for the near future are not yet so obvious. Some of these are mentioned briefly below.

The spillover of hydrogen from a metal to its support has been covered in Chapter 12. The mobility/diffusion of active hydrogen as atoms or radicals on the surface of heterogeneous catalysts is of great importance for the activity and selectivity of these catalysts and for the kinetics and mechanism of reactions on them. Two specific examples may be cited here to illustrate that hydrogen spillover is beginning to make its impact on industrial catalysis.

1. Using 0.6% Pt/Al_2O_3 reforming catalysts of varying Cl content, Parera and co-workers [11] have found that the hydrogen spillover under catalytic reforming conditions was maximum for the catalyst containing 0.9 wt % Cl, and it was precisely this catalyst which had the most stable activity and minimum coke deposition. In industrial reforming operations also, often 0.9 wt % is considered as the optimum value for the Cl content of this catalyst.
2. In catalytic reforming, the bimetallic $Pt-Re/Al_2O_3$ catalyst is well known to form less coke and has a better activity stability, enabling it to operate at higher temperatures and lower pressures than the monometallic Pt/Al_2O_3 catalyst. Barbier et al. [12] have found that the spillover of hydrogen in the bimetallic catalyst is maximum for a weight ratio Pt/Re = 1, which is also the metal ratio in commercial catalysts. These are two examples of empirical trials and practical experience leading to an optimum catalyst, where the rationalization came only later: In both cases, maximum hydrogen spillover apparently keeps the support surface relatively free of polyolefinic coke precursors.

The influence of reduction temperature on the properties of the catalyst has received much attention since the SMSI controversy (see Chapter 13) started in 1978. Except in Fischer-Tropsch synthesis and the new C_1 chemistry (see Chapter 20), the industrial potential of SMSI was, however, rated very low, as indicated by the ready publication from industries in 1978-1984 before taking patents. Some change seems to be occurring here. Two recent patents issued to Engelhard [13] and Chevron [14] suggest that even in Pt/Al_2O_3 and $Pt-Re/Al_2O_3$ reforming catalysts, improvements in C_{5+} yields and selectivity can be obtained by a sequential reduction of the catalyst: first at a low temperature of 120 to 260°C and then at a higher temperature of 370 to 600°C.

Ammonia synthesis is perhaps the most important single use of hydrogen in industry. The coadsorption of nitrogen and hydrogen on iron and their mutual interactions have been discussed in Chapter 1. Although well-defined crystal planes are dealt with there, several of the conclusions drawn from such surface-science studies could have some relevance to industrial catalysis of ammonia in modern 1500-metric ton/day single loops. An extreme case of the hydrogen effect in catalysis will be of hydrogen as a catalyst poison in ammonia synthesis, discussed recently by A. Nielsen [15]. Experimental results of H. Nielsen [16] indicate that a blocking of the catalyst surface by chemisorbed hydrogen may occur at low temperatures. Below 380°C hydrogen is so strongly chemisorbed on the synthesis catalyst that it competes with nitrogen for active sites on the surface. This becomes important in the context of recent efforts (in the wake of the energy crisis) to lower the temperature and pressure in the ammonia synthesis process. While the most economical H_2/N_2 ratio for ammonia synthesis is in the range 2.5 to 3.0 for temperatures above 300°C, for synthesis at lower temperatures and pressures a much lower H_2/N_2 ratio may represent the optimum condition in the synthesis loop. Furthermore, the occurrence of a maximum in the steady-state rate of formation of ammonia with increasing mole fraction of hydrogen in the H_2-N_2 synthesis gas has been reported by Jain et al. [17] at 400°C, 2.38 MPa total pressure, using a gradientless recycle reactor. A maximum of this type with increasing hydrogen pressure is well known in hydrocarbon reactions studied in the laboratory (cf. Chapter 17) and under typical industrial process conditions (cf. Chapter 25). But its occurrence in such an industrially important inorganic reaction as ammonia synthesis lends support to the generality of this phenomenon involving hydrogen for catalysis as a whole.

The importance of hydrogen as an integral component of catalytic systems has been emphasized in the Preface of this book. Under the conditions of a catalytic reaction, the surface of the catalyst will be covered to a greater or less extent with reactive species or intermediates, the relative amounts of which in hydrocarbon reactions will be regulated by hydrogen (see Chapter 17). Thus carbonaceous surface species with varying amounts of hydrogen and with or without oxygen have been reported on a variety of catalysts for several technologically important reactions, such as methanation, methanol synthesis, steam reforming of hydrocarbons, Fischer-Tropsch synthesis, water-gas-shift reaction, and the Mobil methanol-to-gasoline process. Menon et al. [18] have recently pointed out that these surface species need not be mutually exclusive; instead, they may be in a state of equilibrium on the catalyst surface, as shown in Fig. 2. Is the CH_yO intermediate on the surface in an enolic form or a ketonic form? Is the value of y in these species 1, 2, or 3? Does the attachment of hydrogen to the carbon occur before or after the dissociation of CO into C and O on the surface in the case of Fischer-Tropsch and methanol synthesis, methanation, and other $CO + nH_2$ reactions? These all remain as very relevant questions. But the answers to these questions need not be unequivocal since they may depend very much on the nature of the catalyst used, the concentration of the reactants, and the severity of the prevailing experimental conditions. The fact that a particular surface species in the equilibrium of Fig. 2 is not detected in a specific investigation can have at least two causes [19]: (a) its concentration is below the level of detection of the analytical technique used; and (b) it is readily converted into another actually detected form during the experimental manipulation (e.g., raising the temperature in temperature-programmed desorption or reaction, or shifting the equilibrium during reactive scavenging of a particular species from the catalyst surface).

The recognition of a unifying general equilibrium as of Fig. 2, existing for so many industrially important catalytic processes, may perhaps help to

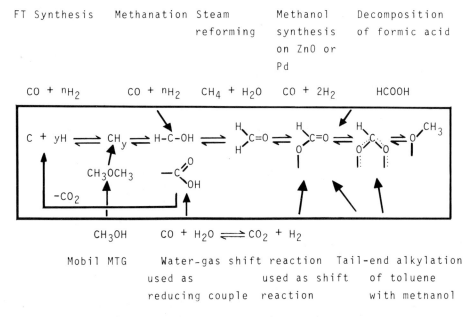

FIGURE 2 Hydrogen-containing surface species in equilibrium, possibly common for several major catalytic processes. (From Ref. 18.)

resolve some of the conflicting evidences and controversies regarding their mechanisms as reported in the literature. In any case, the availability of an overall hydrogen pool on the catalyst surface and the hydrogen balance in the various surface species should have a vital role in determining the activity, selectivity, and operational stability (life) of the catalysts in these major catalytic processes.

Recent work in catalysis and surface science provides more direct experimental support for the existence of carbonaceous overlayers on catalyst surfaces and the vital importance of the H/C ratio of these overlayers in determining the performance of the catalysts (cf. Somorjai [20]). Whether or not the carbonaceous overlayer is an active reaction intermediate or a deactivating coke is often determined by its exact location on the catalyst surface, its hydrogen content, and its nature/structure under actual reaction conditions. For instance, during catalytic reforming on Pt/alumina or Pt-Re/alumina bifunctional catalysts, coke on the metal surface is formed mainly from hydrogenolysis, whereas coke on the alumina surface results predominantly from hydrocracking and such carbenium-ion reactions. The reforming catalyst can tolerate even 6 to 8 wt % coke on the large alumina surface (180-200 m^2/g), but a small quantity of it deposited on the metal surface (only 1 to 3 m^2/g catalyst) by hydrogenolysis can rapidly deactivate the catalyst [21]. Special pretreatments of the catalyst such as chloriding and sulfiding and other special start-up procedures are often directed to one goal: minimizing the hydrogenolytic coke on the metal surface. Similarly, in steam reforming of methane to produce synthesis gas using Ni/α-alumina catalysts, large quantities of reactive carbon or coke will be present on the catalyst surface under reaction conditions, while a deactivated catalyst will contain both carbidic and graphitic carbon not only on the surface of the catalyst but also in the subsurface layers or bulk of Ni crystallites. The presence of subsurface or bulk carbon in such a deactivated catalyst was indicated both by the very slow combustion of carbon in the catalyst at 893 to

TABLE 2 AES and SIMS Data on the Characteristics of Carbonaceous Deposits Formed from Ethylene on Rh, Ir, and Pt Foils

Metal	T (K)	$\theta_c{}^a$	$\%(a+g)^b$	h^c	g^d
Rh	325	0.4	10	0.70	0.011
	525	0.7	30	0.32	0.020
	775	1.2	70	0.10	0.015
Ir	325	0.5	30	0.68	0.014
	525	1.2	50	0.49	0.024
	775	1.4	80	0.32	0.030
Pt	325	0.5	35	0.77	0.012
	525	1.2	60	0.47	0.031
	775	1.8	90	0.20	0.047

[a] Approximate carbon coverage in monolayers, determined from Auger spectra.
[b] Fraction (amorphous + graphitic) carbon, determined from Auger spectra.
[c] Hydrogen content parameter, determined from SIMS; h = 0.8 for adsorbed methane, and about zero for dehydrogenated graphitic carbon.
[d] Graphitization parameter, determined from SIMS; g varies between about zero for a low coverage of adsorbed hydrocarbons and 0.05 for graphitic carbon on Pt.
Source: Ref. 23.

993 K over several hours and an XPS study of the depth-composition profile of the catalyst (de Deken et al. [22]; see also the discussion on temperature-programmed surface reaction in Section IV.C of Chapter 4).

Quantitative determination of hydrogen in carbonaceous overlayers formed on metal catalyst surfaces has recently been achieved by Niemantsverdriet and van Langeveld [23]. By an elegant combination of Auger electron spectroscopy (AES) and secondary ion mass spectrometry (SIMS), these authors distinguished four types of carbonaceous layers on metal surfaces. AES indicates whether the carbon can be categorized as molecular/carbidic or amorphous/graphitic. Combination with data on hydrogen content and degree of polymerization from SIMS enables further conclusions to be made on the state of the carbon deposits. Typical results for deposits formed from ethylene on foils of Rh, Ir, and Pt are given in Table 2. The carbonaceous deposit formed on Rh at 525 K is a mixture of mainly carbidic carbon and some hydrocarbon fragments, while that formed at 775 K has a low hydrogen content and low degree of graphitization and hence contains mainly hydrogen-depleted amorphous and carbidic carbon. Carbon deposit on Pt at 775 K is largely graphitic in nature; on Ir at the same temperature, it is characterized by a lower degree of graphitization and a higher hydrogen content and hence has a more amorphous character than that on Pt. The deposits on Pt and Ir at 525 K are very similar and their hydrogen content is practically the same; still, the higher value of g suggests that there is more graphitic carbon on Pt. Such experimental evidence for the simultaneous existence of different carbonaceous species with varying hydrogen content on metal catalyst surfaces broadly supports the type of generalization proposed in Fig. 2. Furthermore, the data of Table 2 also show that, in general, the deposits formed on Rh contain significantly less hydrogen than those on Pt and Ir. This is in agreement with the relatively high activity of Rh for breaking C-H bonds, reported by Merta and Ponec [24]. For Ir and Pt substrates, the degree of polymerization of carbon increases with increasing reaction temperature, while for Rh it remains at a low level. Pt is more sensitive than Ir for graphite formation, which is in agreement with work on alumina-supported Pt and Ir catalysts by Carter et al. [25]. The AES/SIMS data of Table 2 show that the tendency of the substrate metals to promote graphitization decreases in the order Pt, Ir, Rh, which is indeed the order of increasing hydrogenolytic activity for these metals [26]. Paál and Marton [27] have recently employed the SIMS technique to determine the hydrogen content and depth-composition profile of various Pt black samples with and without carbonaceous overlayers, when subjected to different pretreatments; the presence of hydrogen not only on the surface but also in the subsurface layers of Pt could be shown by this technique. Such quantitative comparisons of the hydrogen content and nature of the carbonaceous overlayer or deposit on catalyst surfaces may thus provide a direct rationalization for the intrinsic catalytic activity and the actual performance of catalysts, both in laboratory and industry.

REFERENCES

1. D. A. Mathis, *Hydrogen Technology for Energy*, Noyes Data Corp., Park Ridge, N.J. (1976).
2. *Hydrogen for Energy Distribution*, Institute of Gas Technology, Chicago (1978).
3. A. F. Andresen and A. J. Maeland (eds.), *Hydrides for Energy Storage*, Pergamon Press, Elmsford, N.Y. (1978).
4. S. Kaliaguine and A. Mahay (eds.), *Catalysis on the Energy Scene*, Elsevier, Amsterdam (1985).
5. H. Heinemann, in *Catalysis—Science and Technology*, Vol. 1 (J. R. Anderson and M. Boudart, eds.), Springer, Berlin (1981), p. 1.
6. *Chem. Eng. News* (Jan. 2, 1984), p. 8; (Oct. 14, 1985), p. 9.
7. M. S. Casper (ed.), *Hydrogen Manufacture by Electrolysis, Thermal Decomposition and Unusual Techniques*, Noyes Data Corp., Park Ridge, N.J. (1978).
8. P. B. Venuto and E. Habib, *Catal. Rev.-Sci. Eng.*, *18*:1 (1978).
9. J. S. Magee and R. E. Ritter, *Symposium on Octane in the 1980's*, American Chemical Society, Division of Petroleum Chemistry (1978), p. 1057.
10. J. S. Magee, W. E. Cornier, and G. M. Wolterman, *6th Katalistiks Symposium, Munich* (1985).
11. J. M. Parera, N. S. Figoli, E. L. Jablonski, M. R. Sad, and J. L. Beltramini, in *Catalyst Deactivation* (B. Delmon and G. F. Froment, eds.), Elsevier, Amsterdam (1980), p. 571.
12. J. Barbier, H. Charcossett, G. de Periera, and J. Reviére, *Appl. Catal.*, *1*:71 (1981).
13. S. O. Oyekan (to Engelhard), U.S. Patent 4,539,307 (1985).
14. L. A. Field (to Chevron), U.S. Patent 4,539,304 (1985).
15. A. Nielsen, *Catal. Rev.-Sci. Eng.*, *23*:43 (1981).
16. H. Nielsen, unpublished results from Haldor Topsoe, quoted by A. Nielsen in *Catal. Rev.-Sci. Eng.*, *23*:43 (1981).
17. A. K. Jain, C. Li, P. L. Silveston, and R. R. Hudgins, *Chem. Eng. Sci.*, *40*:1029 (1985).
18. P. G. Menon, J. C. de Deken, and G. F. Froment, *J. Catal.*, *95*:313 (1985).

19. P. G. Menon, *J. Mol. Catal.*, *39*:383 (1987).
20. G. A. Somorjai, *Chemistry in Two Dimensions: Surfaces*, Cornell University Press, Ithaca (1981).
21. P. G. Menon, unpublished results.
22. J. de Deken, P. G. Menon, G. F. Froment, and G. Haemers, *J. Catal.*, *70*:225 (1981).
23. J. W. Niemantsverdriet and A. D. van Langeveld, *Fuel*, *65*:1396 (1986); A. D. van Langeveld and J. W. Niemantsverdriet, *Surf. Interface Anal.*, *9*:215 (1986).
24. R. Merta and V. Ponec, *Proc. IVth Internat. Congr. Catal.*, 1968, Moscow, *2*:53 (1971).
25. J. L. Carter, G. B. McVicker, W. Weissman, W. S. Kmak, and J. H. Sinfelt, *Appl. Catal.*, *3*:327 (1982).
26. B. C. Gates, J. R. Katzer, and G. C. A. Schuit, *The Chemistry of Catalytic Processes*, McGraw-Hill, New York, p. 209 (1979).
27. Z. Paál and D. Marton, *Applied Surf. Sci.*, *26*:161 (1986).

24 | Selectivity in Competitive Hydrogenation Reactions

NILS-HERMAN SCHÖÖN

Chalmers University of Technology, Gothenburg, Sweden

I.	CONCEPT OF SELECTIVITY	622
II.	HYDROGEN EFFECT IN SOME CHEMICAL PROCESSES	623
	A. Different Hydrogenation Processes	623
	B. Hydroformylation	629
	C. Reforming	630
	D. Fischer-Tropsch Synthesis and Other Carbon Monoxide Hydrogenation Processes	630
	E. Hydrodesulfurization	631
	F. Hydrodenitrogenation	631
	G. Coal Hydrogenation	632
III.	POSSIBLE EXPLANATIONS OF THE HYDROGEN EFFECT ON THE SELECTIVITY	632
	A. Different Reaction Orders in Hydrogen Explained by Various Rate-Determining Steps of the Surface Reactions	633
	B. Hydrogen Effect on the Selectivity Explained by the Influence of Slow Adsorption and Desorption (No Rate-Determining Step)	637
	C. Hydrogen Effect on the Selectivity Caused by a Change of the Surface of the Catalyst	639
	D. Hydrogen Effect on the Selectivity Explained by the Influence of Slow Pore Transport	639
	E. Hydrogen Effect on the Selectivity Explained by the Influence of Slow Film Transport	641
	F. Hydrogen Effect on the Selectivity in Homogeneous Hydrogenations	642
IV.	DIFFERENT WAYS TO UTILIZE THE HYDROGEN EFFECT TO MAXIMIZE THE SELECTIVITY IN CONSECUTIVE REACTIONS	642
	A. Traditional Control	642
	B. Periodic Operation	643
	C. Determination of Hydrogen Concentration to Optimally Control the Hydrogenation	644

	D. Influence of the Catalyst Geometry on the Selectivity in Liquid-Phase Hydrogenation	645
V.	CONCLUDING REMARKS	646
	NOMENCLATURE	647
	REFERENCES	648

I. CONCEPT OF SELECTIVITY

The ability to selectively favor one among various competitive chemical reactions is by far the most fascinating characteristic of a catalyst. This ability is usually more difficult to explain than other properties of the catalyst. In heterogeneous catalysis, the selectivity action is associated with the chemical composition and structure of the solid surface and is also influenced by the type of supporting material. Permanently adsorbed compounds on the surface may also have a great influence on the selectivity. The variables mentioned may be called the catalyst variables, and they define the *intrinsic selectivity* of the catalyst.

The two main factors influencing intrinsic selectivity are the *mechanistic* factor and the *thermodynamic* factor [1]. The mechanistic factor is related to the kinetic properties of the different steps involved in the surface processes, while the thermodynamic factor reflects the relative adsorption ability of the catalyst with respect to different compounds when the surface reactions are the rate-determining step.

The porosity of the solid catalyst is a catalytic variable which does not influence the intrinsic selectivity but restricts the accessibility of the active surface to the reactants. A high porosity often gives rise to a slow intraparticle transport, which may also result in an effect on the selectivity. This effect can be modified by the process variables, such as the composition of the reaction mixture and the temperature. When using a catalyst with molecule sieving properties, the restricted diffusion in the very narrow pores is influenced very little by these process variables. The pore transport limitation is a desired effect in this case and is the origin of *shape selectivity* [2].

In homogeneous catalysis, with some reactant in the gas phase and the catalyst in the liquid phase, the selectivity will be very much influenced by the resistance against the transport of the gas reactant across the liquid film close to the phase boundary [3-9]. This effect arises from the fact that the chemical reactions start in the liquid film, which means that these reactions are closely coupled with the diffusion of the reactants and also very much dependent on the concentration gradients created as a result of this coupling. This effect on the selectivity can only be controlled by the process variables: for example, the temperature, the concentration of catalyst, the partial pressure of the reactant, and the rate of the mechanical stirrer. A similar effect on the selectivity can also appear in a slurry gas-liquid process with a finely divided solid catalyst. The condition for this effect is that the small catalyst particles can penetrate the liquid film surrounding the gas bubbles and that the chemical reactions start in this film [10,11].

The term "selectivity" also has more chemically related aspects, and three classes of selectivity can be recognized here. Complex organic compounds normally have more than one reactive functional group. The ability of the catalyst to discriminate among the reactive groups is often referred to as *chemoselectivity*. At a given reactive group there may be several orientations, and this defines the problem of *regioselectivity*. Besides difference in orientation, the groups in the product can differ in spatial arrangement. The control of the catalyst action in this respect is referred to as *diastereoselectivity* and *enantioselectivity*. Enantioselectivity refers to the ability of the catalyst to discriminate between mirror-image isomers or enantiomers.

Besides these selectivity definitions related to organic chemistry, it should be noted that there is a need to define special selectivity terms for different industrial chemical processes. For example, the fat hydrogenation industry uses four different selectivity terms to describe the selectivity properties of this process [12].

The term "selectivity," according to the IUPAC's Commission for Definitions, Terminology and Symbols in Colloid and Surface Chemistry [13], is used to describe the relative rates of two or more competing reactions. In the literature the meaning of the term is very diverse, and is often related to distribution of the products or to the yield at the end of the process. With this meaning, reaction engineering factors will also influence the selectivity. This means that the selectivity of a batch process is not always the same as in a continuous process. In the continuous process this type of selectivity is influenced by the residence-time distribution, the mixing mechanism, the manner of adding reactants, the recirculation flow from the outlet to the inlet of the reactor, and many other factors [14].

"Selectivity" in the more strict definition and as related to product distribution are two quite different concepts but have certain points in common. A further discussion of the difference between selectivity and yield is not fruitful here, since there exist many types of yields in the literature [15]. In general, high selectivity corresponds to high yield, and conversely. This means that if hydrogen has an effect on the selectivity, it will also have an effect on most types of selectivities and yields. In the present review of the literature on the hydrogen effect, all types of selectivities will therefore be included, and only if there is a risk of misunderstanding will the meaning of the selectivity of interest be given.

II. HYDROGEN EFFECT IN SOME CHEMICAL PROCESSES

The very large consumption of hydrogen in the chemical industry is a good measure of the importance played by hydrogenation reactions. In many of these processes the catalyst is selective enough that side reactions are of minor interest; or if this is not the case, the side reactions and the main reaction are influenced by hydrogen to the same extent, so that no increased selectivity can be attained by a proper choice of the hydrogen pressure. There are, however, many chemical processes of industrial interest where hydrogen has a clear effect on the competition between these reactions. Some examples of this effect are given below.

A. Different Hydrogenation Processes

Hydrogenation of Hydrocarbons and Related Compounds

The selectivity of ethylene formation from acetylene catalyzed by alumina-supported noble group VIII metals is a well-known example of the influence of hydrogen on the selectivity [16-19]. The selectivity decreases with increasing hydrogen pressure. The same effect of hydrogen pressure is found for the initial selectivity in the hydrogenation of 1,3-butadiene with the same type of catalysts [20]. A very high hydrogen effect on selectivity is also found in the hydrogenation of unsaturated aldehydes, as, for example, the hydrogenation of 2-ethyl-2-hexenal in the presence of nickel as catalyst [21].

The hydrogen effect on the selectivity in acetylene hydrogenation in the presence of precious metal catalysts, as referred to above, is found only when hydrogen is in excess. If, instead, the initial ratio of hydrogen and acetylene is as low as unity, the products formed are quite different. If we disregard the greater tendency of coke formation, the most striking difference is that, besides the C_2 hydrocarbons, a sizable fraction of 1-butene, trans-but-2-ene, cis-but-2-ene, butane, and butadiene is also formed [16, 17, 19].

Sometimes the selectivity in the hydrogenation of an alkyne or an alkadiene with palladium catalysts is affected very drastically by the hydrogen. On increasing the hydrogen pressure beyond a critical point, the selectivity is diminished abruptly [22]. The same effect is reported for the hydrogenation of methylacetylene over unsupported alloys of nickel and copper [23]. The mechanism of the selectivity break is not understood, but various explanations have been put forward [22]

Often a high selectivity in alkadiene and alkyne hydrogenation may be attributed to an unusually low value of surface coverage of hydrogen atoms [24], but the ability to catalyze these hydrogenations selectively is also more or less related to the extent of hydrogen occlusion in the group VIII metals [22, 25]. For example, the order of hydrogen occlusion in Ir, Os, Ru, Rh, and Pt was found to be the reverse of the order of selectivity.

Stereochemistry of the Hydrogenation of Cyclic Olefins

The hydrogenation of disubstituted cyclohexenes and of substituted methylene-cyclohexenes is generally observed to yield a mixture of cis and trans isomers of the corresponding cyclohexanes. The isomer composition was found to be a function of the hydrogen pressure when the hydrogenations were performed with platinum dioxide [26, 27]. The hydrogen effect was most striking in the hydrogenation of 1,2-dimethylcyclohexene, and the cis isomer increased from about 80% at atmospheric pressure to over 95% at very high hydrogen pressure.

A different stereochemical pattern is found in the hydrogenation of 2-, 3-, or 4-alkyl-substituted methylene-cyclohexanes over a platinum catalyst [26, 28, 29]. With these compounds an increased pressure of hydrogen gives an increased proportion of the more stable saturated isomer in the product (i.e., the formation of trans- 1, 2-, cis-1, 3-, and trans-1, 4-dialkylcyclohexane is favored).

Other examples are the hydrogenations of 4-tert-butylmethylenecyclohexane and 4-tert-butyl-1-methylcyclohexene in the presence of platinum dioxide. The cis-trans proportion of saturated stereoisomers was found to be affected by the hydrogen pressure in quite different ways. In hydrogenation of 4-tert-butylmethyl compounds the cis/trans ratio of the saturated isomers decreased strongly at increasing hydrogen pressure, whereas in hydrogenation of the 4-tert-butyl-1-methyl compound this ratio increased at increasing hydrogen pressure [29].

Hydrogenation of Aromatic Compounds

The selectivity problems are intricate and subtle in the hydrogenation of aromatic compounds, and the hydrogen effect on the selectivity is often very strong.

One typical example is the liquid-phase hydrogenation of benzene on a ruthenium catalyst [30]. On increasing the hydrogen pressure, the yield of cyclohexene increases and reaches a sharp maximum, and then decreases upon further increase of the hydrogen pressure. This maximum was explained by the assumption that the reaction order with respect to hydrogen could be different for the two consecutive reactions and for the shunt hydrogenation reaction of benzene directly to cyclohexane [30].

The same behavior was found in the hydrogenation of 1,4-di-*tert*-butylbenzene over rhodium on alumina, where 1,4-di-*tert*-butylcyclohexene is formed as an intermediate. The yield of this compound was found to increase with increasing hydrogen pressure from 0.34 atm to 1.35 atm, but decreased upon increasing the hydrogen pressure from 7.7 atm to 150 atm [31]. The differing hydrogen effect on the selectivity at high and low hydrogen pressure was explained by Siegel et al. [32] on the assumption of a two-site catalyst model.

The hydrogen pressure influences not only the selectivity of the consecutive hydrogenation reactions of aromatic compounds, but also the parallel formation of the cis and trans isomers. In the hydrogenation of o-xylene in the presence of palladium dioxide, for example, the cis/trans ratio of the saturated stereoisomers is very much affected by the hydrogen pressure. The cis/trans ratio passes a minimum value on increasing the hydrogen pressure [33].

The ratio of the cis/trans isomers of 1,4-di-*tert*-butylcyclohexane was also influenced by the hydrogen pressure when hydrogenation of 1,4-di-*tert*-butylbenzene occurred in a slurry reactor with rhodium on alumina as a catalyst [31]. However, when performing the hydrogenation in a cross-flow catalytic reactor, this cis/trans ratio was much higher and, moreover, the selectivity of the cis/trans formation was very much influenced by the hydrogen pressure [34]. This surprising effect was explained by the competition of the different modes of adsorption of 1,4-di-*tert*-butylbenzene, together with the particular concentration profiles in the porous plates of the cross-flow catalyst. This explanation is commented on further in Section IV.D.

Enantio-Differentiating Hydrogenation with a Modified Solid Catalyst

One of the most sophisticated selectivity properties of a catalyst is the ability to influence the distribution of enantiomers in a chemical reaction. In heterogeneous hydrogenation reactions this type of selectivity is obtained with a Raney nickel catalyst which has been modified with an adsorbed optically active compound, as, for example, an amino acid [35]. The conventional theory of the hydrogenation presupposes that an adsorption precedes the reaction. If this is the case, the proportions of the two optical forms (si and re faces) of the adsorbed reactant should be independent of the hydrogen pressure and the optical yield should not be influenced by the hydrogen pressure [35]. Quite contrary to this theory, the hydrogen pressure was shown to have a strong effect on the optical yield in the asymmetric hydrogenation of methyl acetoacetate in the presence of a catalyst of silica-supported nickel modified with tartaric acid [36]. Similar hydrogen pressure effects on the optical yield have been reported by other authors [37-39].

Hydrogenation of Fatty Acid Glycerides in Vegetable Oils

The greatest importance of the hydrogen effect on selectivity is found in the hardening process for edible oils and fats [40-43]. The initial raw oils consist of a mixture of trienoic, dienoic, monoenoic, and saturated fatty acid glycerides. These oils are to be modified by hydrogenation to give a liquid-solid mixture with very specific rheologic properties at 8 to 25°C and with high stability against oxidation. This stability can be brought about if the content of trienoic acids in the glycerides is strongly reduced by hydrogenation. This hydrogenation should be performed in such a way that the formation of saturated fatty acids in the glycerides is avoided. The desired rise in melting point is, instead, attained by favoring the translocation of the alkene bonds of the dienoic and monoenoic acids in the glycerides. The desired plasticity of the hydrogenated fat is thus regulated by the cis/trans ratio of the fatty acids and by the partition of the carbon-carbon double bonds along the fatty acid molecule. This partition of the double bonds does not exist in the original raw oil but is a result of migration of the double bonds during the hydrogenation.

The hydrogenation of vegetable oils consists mainly of consecutive reactions, where trienoic acids give dienoic acids, and so on. Parallel to these reactions, different shunt reactions also proceed (i.e., trienoic acids are hydrogenated directly to monoenoic acids and dienoic acids are hydrogenated directly to stearic acid). In addition to these reactions, parallel isomerization reactions proceed as mentioned above. All these reactions are very dependent on the hydrogen pressure. As a general rule the selectivity in the consecutive reactions decreases with increasing hydrogen pressure. Also, the isomerization reactions are decreased upon increasing the hydrogen pressure [40-43]. The same behavior is valid for methyl esters of fatty acids in liquid-phase hydrogenation [40-43] and in vapor-phase hydrogenation [44].

A good example of the influence of the hydrogen pressure on the selectivity in fat hydrogenation is derived from Pihl et al. [45] in Fig. 1, which shows the content of monoenoic acids in glycerides in hydrogenation of cottonseed oil in the presence of an unsupported nickel catalyst at different hydrogen pressures from 0.5 to 7.3 bar. The hydrogenations were performed at 160°C and every series at the same pressure comprises 10 experimental values, which for the sake of clarity are not given in Fig. 1.

Hydrogenation with Unsupported and Supported Homogeneous Catalysts and with Cluster Compounds

The applicability of homogeneous hydrogenation is limited to those processes in which heterogeneous catalysts are unsuitable, when for example side reactions such as disproportionation, hydrogenolysis, and isomerization have to be suppressed. The homogeneous catalysis is especially preferred for asymmetric hydrogenation and for reactions that present particular problems of selectivity. The rate of homogeneously catalyzed hydrogenation reactions is found to depend on hydrogen pressure in a manner similar to the corresponding heterogeneous processes [46]. The hydrogen effect of the selectivity seems to have been of minor interest to research, since the homogeneous catalysts often give high selectivity and high activity, and there has thus been no need to improve these properties further by changing the hydrogen pressure. Moreover, hydrogenation and isomerization reactions with homogeneous and heterogeneous metal catalysts are related mechanistically [47], and it is therefore reasonable to suppose that the intrinsic selectivity depends on the hydrogen pressure in the same way. It is expected that this similarity is also valid in hydrogenations with supported transition-metal complexes.

A notable hydrogen effect on the selectivity was, however, found in the hydrogenation of some steroids. It is known that 3-oxo-1,4-diene steroids are hydrogenated to the corresponding 3-oxo-4-enes using $RhCl(PPh_3)_3$ as a catalyst [48,49]. Although the hydrogenation with this catalyst proceeds selectively, it was not possible to hinder the reaction when proceeding further to give the corresponding saturated ketones. A more selective action was found with the corresponding ruthenium-complex catalyst in the hydrogenation of 1,4-androstadiene-3,17-dione to give the desired product 4-androstene-3,17-dione (I) instead of the androstane-3,17-dione (II) formed in parallel

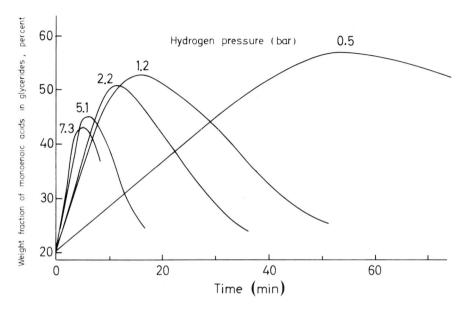

FIGURE 1 Content of monoenoic acids in glycerides formed in hydrogenation of cottonseed oil at different hydrogen pressures. Temperature, 160°C; catalyst load, 0.02% of unsupported nickel. (From Ref. 45.)

[50]. Most interesting is the finding that the fraction of compound I increases relative to compound II when increasing the hydrogen pressure. The high hydrogen pressure also resulted in an increased reaction rate.

Hydrogenation of olefins in the presence of polymer-bound metal clusters has proved to be dependent on the hydrogen pressure to a varying extent. At the present stage of progress in this particular research field, the hydrogen effect on the selectivity seems to be of minor interest. However, a hydrogen effect of that type was reported in the hydrogenation of terminal alkynes to olefins by the unsupported iron cluster (η^5-C_5H_5)$_4$Fe$_4$(μ_3-CO)$_4$ [51].

Electrocatalytic Hydrogenation

Electrocatalysis is that branch of electrochemistry where adsorbed reaction intermediates on the electrode surface are controlled by chemisorptive forces. This situation is restricted to transition metal electrodes. Hydrogenations performed with mercury or lead or similar metals are, instead, electrochemical hydrogenations. The reaction mechanism in electrochemical hydrogenations includes a short-lived radical cation or similar species which undergoes subsequent reaction in solution rather than on the electrode surface.

Acetylenic compounds are electrocatalytically hydrogenated on Pt, Pd, Ni, and Co cathodes. The reactions are not very selective, and a mixture of ethylenic and saturated compounds are formed [52]. Increased selectivity is not brought about primarily by varying the hydrogen pressure, but rather, by changing the potential of the catalyst. The mechanism of the electrocatalytic hydrogenation of acetylenic compounds has been shown to be principally the same as the catalytic process [52]. This result implies that hydrogen has the same effect on the selectivity in the electrocatalytic hydrogenation as in the catalytic hydrogenation of these compounds. This conclusion may be valid for all types of electrocatalytic hydrogenations. It is therefore of little advantage to use electrocatalytic methods instead of common catalytic methods.

A quite different situation is, of course, valid for the much broader field of electrochemical hydrogenations and reductions.

Transfer Hydrogenation (or Hydrogen Transfer)

All of the hydrogenation reactions discussed until now have been of the type where the hydrogen is supplied as molecular hydrogen. An alternative type of hydrogenation is represented by the reaction where the hydrogen is supplied by a donor molecule DH_2 which itself undergoes dehydrogenation during the course of the process, according to

$$A + DH_2 \rightarrow AH_2 + D$$

Very often the donor molecules are present as solvent, since it is an advantage if they are in excess over the unsaturated reactant to be hydrogenated. This type of hydrogenation, called transfer hydrogenation, is heterogeneously or homogeneously catalyzed. An essential feature of the transfer hydrogenation is the fact that there is no change in the overall unsaturation in the reaction system. Transfer hydrogenation also occurs as a slow side reaction during a normal hydrogenation. This was demonstrated by Coussemant and Jungers [53] in the hydrogenation of phenol, where phenol was hydrogenated by transfer of hydrogen from cyclohexanol.

It should be mentioned in this connection that a general mechanism for ordinary hydrogenations on metals has been put forward, where the reaction is interpreted as a hydrogen transfer between an adsorbed hydrocarbon and the adsorbed olefin, instead of being regarded as an addition of hydrogen directly to the latter [54].

The hydrogen effect on the selectivity in a transfer hydrogenation corresponds to the effect of the donor molecule on the selectivity. This hydrogen effect can be found by changing the concentration of the donor molecule or by comparing the effect of donor molecules with differing abilities to donate hydrogen molecules.

The influence on selectivity of different donor molecules was demonstrated in hydrogenation of methyl linoleate, with indoline and isopropyl alcohol as hydrogen donating agents, in the presence of $RuH_2(PPh_3)_4$ as a catalyst [55].

A hydrogen effect was also found in hydrogenation of 1,3-butadiene in the presence of cerium dioxide as a catalyst, and with different cyclohexadiene isomers as hydrogen donor molecules [56]. The hydrogen effect on the cis-trans isomerization of the butene formed in this hydrogenation was, however, found to be rather small. Hydrogen transfer on zeolite catalysts is so important that Chapter 27 is completely devoted to this topic.

Biohydrogenation

It has long been known that enzyme systems activate molecular hydrogen and hydrogenate a number of substances in the same way as in heterogeneous catalysis [57]. Selective hydrogenation of linolenic and linoleic acid has been performed in the presence of, among others, the enzyme hydrogenase from rumen organism [58-60]. Little attention was, however, given to the selectivity aspects and factors corresponding to what has here been called the hydrogen effect.

Miscellaneous Industrial Hydrogenation Processes

There are several industrial processes where functional groups other than unsaturated carbon-carbon bonds are to be hydrogenated, and where the hydrogenation consists of many competitive side reactions. One important type of reaction is the hydrogenation of nitro groups and nitrile groups. Well-known examples of industrial interest are the hydrogenation of nitroarenes [61], adiponitrile, benzonitrile [62], and butyronitrile [63]. To this type belongs also the hydrogenation of chloronitroarenes. These hydrogenation processes consist of a number of consecutive reactions. Little, however, is reported about the effect of hydrogen on the selectivity in these processes. It is important to note that a partial hydrogenation, to obtain some of the intermediate compounds in these processes, cannot be achieved industrially by regulation of the hydrogen concentration, but has to be done by moderating the catalyst or by using some chemical roundabout route.

The hydrogen effect on selectivity is of more interest in the hydrogenation of alkylanthraquinones to alkylanthrahydroquinones, which is an important step in the production of hydrogen peroxide according to the anthraquinone process. A certain overhydrogenation to alkyltetrahydroanthrahydroquinones is desirable and is partly controlled by the hydrogen pressure [64].

In hydrogenation of mixtures of glucose and fructose, the hydrogen concentration in the liquid phase was shown to have a clear effect on the ratio of the sorbitol and mannitol formed [65].

In partial hydrogenation of phenols, the overhydrogenation is surmounted primarily by a proper choice of the catalyst. This has been the leading problem, and the role played by hydrogen on this selectivity has been noticed very little. However, in hydrogenation of thymol over Raney nickel, the selectivity of the intermediately formed ketone was very strongly influenced by the hydrogen concentration [53,66].

Unsaturated aldehydes such as acrolein, crotonaldehyde, cinnamaldehyde, and 2-ethylhexenal can be partially hydrogenated to unsaturated alcohols or to saturated aldehydes. The selectivity is controlled primarily by the choice of catalyst, but hydrogen has a great effect on the selectivity in at least the route giving saturated aldehydes [21].

Hydrogenolysis of unsaturated fatty acid glycerides in vegetable oils to give unsaturated fatty alcohols is also a process that is very much influenced by the hydrogen concentration. The unsaturation of the alcohol is largely preserved despite the hydrogenolysis reaction if nickel is avoided as a catalyst.

B. Hydroformylation

Many of the processes discussed above are related to fine chemical manufacture. Hydroformylation is an example of a large-scale process in which hydrogen is an important reactant without being critical for the selectivity at normal reaction conditions. In the production of butyraldehyde by hydroformylation of propene, using rhodium or cobalt homogeneous catalysts, one important selectivity aspect is related to the possibility of increasing the formation of n-butyraldehyde and decreasing the formation of i-butyraldehyde (i.e., making the normal to branched selectivity as high as possible). This selectivity is controlled primarily by the type of catalyst, but much interest has also been directed toward the possibility of increasing this selectivity by proper choice of the H_2/CO ratio. This ratio has, however, an upper limit, due to the fact that the formation of propane is favored at high hydrogen pressure.

There are different observations reported about the hydrogen effect on the normal to branched selectivity. The apparently contradictory results may be explained by the fact that quite different catalysts have been used.

In an ordinary gas-sparged reactor and with a PPh_3-modified Rh catalyst, the selectivity was shown to be unaffected by an increased H_2/CO ratio [67]. On the other hand, the selectivity was found to be favored by increasing this ratio in the presence of a $P(OPh)_3$-modified catalyst [68]. When using a SLP catalyst with $HRh(CO)(PPh_3)_3$ in PPh_3 as a liquid phase, the selectivity was found to be independent of the hydrogen pressure but increased with decreasing carbon monoxide pressure [69]. With the same SLP catalyst, but with a solvent instead of PPh_3 as the liquid phase, the selectivity was found to increase with increasing H_2/CO ratio [70]. Many authors have reported an increased propane formation when increasing the H_2/CO ratio [68,70-72].

The selectivity in hydroformylation of 1- and 2-pentene in the presence of the cluster catalysts $Co_3(CO)_9(\mu_3-CC_6H_5)$ and $Co_4(CO)_8(\mu_2-CO)_2(\mu_4-PC_6H_5)_2$ was shown to be positively affected by the increase in hydrogen pressure at a constant H_2/CO ratio [51].

C. Reforming

Catalytic reforming is one of the basic refining processes in the petroleum industry. Despite the fact that the dehydrogenation of naphthenes and the dehydrocyclization of paraffins result in a net production of hydrogen, the pressure of hydrogen has an important positive role here: to prevent the formation of coke on the surface of the catalyst and thereby make possible the desired isomerization and ring-closure reactions. The coke formation involves polymerization and cyclization of polyolefins to give polynuclear compounds which undergo dehydrogenation, aromatization, and further polymerization. These processes occur at low hydrogen pressure under otherwise typical reforming reaction conditions. These side reactions can be markedly reduced by increasing the hydrogen pressure. This suppression of the coke formation has been shown to increase the rate of dehydrocyclization of n-alkanes [73] and also the rate of isomerization of these hydrocarbons to give branched hydrocarbons [73]. Moreover, the rates of isomerization and dehydroisomerization of cycloalkanes to give aromatic compounds are also increased under these conditions [74]. All these studies were performed in the presence of Pt on Al_2O_3.

If, however, the hydrogen pressure is increased beyond about 10 atm, removal of the carbonaceous species from the metal fraction of the catalyst is no longer a rate-determining step, and therefore the dehydrogenation and dehydrocyclization reactions do not increase upon further increasing the pressure. An increasing hydrogen pressure will, instead, promote the hydrogenation reactions at the expense of the dehydrogenation reactions. This means that dehydrogenation and dehydrocyclization will decrease upon further increasing the hydrogen pressure, and the rate of these two processes will thus reach a maximum. These and other aspects of hydrogen in catalytic reforming are discussed in Chapter 25.

D. Fischer-Tropsch Synthesis and Other Carbon Monoxide Hydrogenation Processes

Fischer-Tropsch Synthesis

The number of possible products that can be formed in the Fischer-Tropsch synthesis is very large, and the primary objective has been to find a specific, robust, and cheap catalyst in order to get some narrow fractions of main products of economic interest. Besides the concern with the catalyst and the process design, the H_2/CO ratio has been one of many important variables studied. In the liquid-phase variation of the Fischer-Tropsch synthesis, it was found [75] that it is more difficult to synthesize preferentially low-boiling hydrocarbons from a hydrogen-rich gas, because of a lower operating temperature and the higher alkalization of iron catalysts which is required in the liquid process. This higher hydrogen partial pressure gives a low olefin content. When hydrogen-rich gases, corresponding to a H_2/CO ratio of 1.25 to 2, are used, the main products will thus be slack and hard paraffins.

Methanol Synthesis

The hydrogenation of carbon monoxide in gas phase gives a broad spectrum of products with alkanes, alkenes, alkynes, ethers, alcohols, aldehydes, ketones, and acids [76], and the selectivity of the different reactions is very much dependent on the catalyst. However, the conditions that favor alcohol formation at the expense of hydrocarbons are, among others, a low H_2/CO ratio. The catalyst used in industrial production of methanol gives a very high selectivity with minor formation of methane and dimethyl ether. The formation of methane increases rapidly at temperatures above 400°C. The possibility of further increasing the selectivity of methanol formation by changing the H_2/CO ratio has not been reported in the literature.

E. Hydrodesulfurization

Hydrogen is consumed in many parallel reactions in the hydrodesulfurization process [77]. The main process involves the hydrogenolysis reactions that result in cleavage of C-S bonds. Under industrial conditions the hydrogenolysis also results in cleavage of C-C bonds. Parallel to these reactions, hydrogen takes part in hydrogenation of unsaturated hydrocarbons. Other important reactions are demetallization, cracking reactions, and coke formation. Many of these reactions increase linearly with increasing hydrogen pressure. This is not always the case for the hydrodesulfurization reactions. For example, in desulfurization of a Kuwait atmospheric residuum, Beuther and Schmid [78] found that the rate of hydrodesulfurization does not increase with increasing hydrogen pressure beyond a pressure of 70 bar. This observation means that hydrogen may have a strong negative effect on the hydrodesulfurization selectivity under reaction conditions prevailing in a common industrial process (i.e., an increased hydrogen pressure increases only the rate of the other processes, not the hydrodesulfurization). It should, however, be noted that Cecil et al. [77] found a quite different hydrogen effect on the selectivity in question when they studied the hydrosulfurization of an unidentified Middle Eastern residuum. In this case the reaction rate of the hydrodesulfurization was proportional to the hydrogen pressure.

F. Hydrodenitrogenation

The hydrogen effect on the hydrodenitrogenation reactions may be illustrated by the kinetic studies performed with different model substances, such as pyridine and quinoline. For example, Rosenheimer and Kiovsky [79] reported a second-order reaction in hydrogen in the hydrodenitrogenation of pyridine, whereas other authors found first-order kinetics with respect to hydrogen, or 1.5 order [80]. The reaction of the opening of the heterocyclic ring was, on the other hand, found to be zero order in hydrogen [80,81]. If we assume that other reactions proceeding simultaneously with the hydrodenitrogenation reactions are on the average first order in hydrogen, we consequently find that the hydrogen effect on the hydrodenitrogenation selectivity may be positive if the observation made by Rosenheimer and Kiovsky is correct. On the other hand, there is no hydrogen effect on the ring-opening reactions.

G. Coal Hydrogenation

The future supply of petroleum is difficult to prophesy, but sooner or later we will reach the stage of evolution where hydroliquefaction of coal for production of gasoline and oil will compete economically with petroleum. The hydrogen effect on the selectivity will then be one of many important properties to study. Even now, Shah et al. [82] have shown that the product distribution of hydrocarbons is very dependent on the hydrogen pressure in a catalytic liquefaction study of Big Horn Coal in a segmented bed reactor with a Gulf-patented catalyst. Since other variables besides hydrogen pressure were changed at the same time, it is difficult to decide how much of the change of product distribution really is a function of the hydrogen pressure.

III. POSSIBLE EXPLANATIONS OF THE HYDROGEN EFFECT ON THE SELECTIVITY

The engineering aspects of the effect of hydrogen on the selectivity in hydrogenation reactions are related to the possibility of controlling the hydrogen concentration in some way so as to increase the selectivity. To achieve this, the influence of the hydrogen concentration must be expressed with a mathematical model. The usual method is to express the rates of the competitive reactions with power equations. The reaction orders with respect to hydrogen in the rate equations will then attain different values if the selectivity is dependent on the hydrogen concentration. The reaction having the highest order will thus be favored at high hydrogen concentration, and conversely.

The reaction order with respect to hydrogen can be expected to be either 1/2 or 1, according to the two most common reaction mechanisms for hydrogenation surface reactions. In reality the reaction order found experimentally is, however, often quite different from these values. In hydrogenation of the carbon-carbon double bond of 2-ethyl-2-hexenal in the presence of nickel, the reaction order in hydrogen was found to be -0.8 [21]. Other, not so extreme values of this reaction order have been observed for this hydrogenation process [88]. It should also be mentioned that hydrogenations of nitro compounds are often reported to be zero order in hydrogen [83,84]. Many hydrogenation reactions have been reported to depend very much on the hydrogen concentration, corresponding to a reaction order of 1.5 or more [20]. Despite these examples of extreme values of the reaction order, it is important to note that the reaction order in hydrogen is between 1/2 and 1 for most hydrogenation reactions.

The selectivity in consecutive hydrogenation reactions has been shown to have the greatest hydrogen influence, and efforts have been made to maximize the selectivity by an optimal control of the hydrogen concentration in these processes. For most of the consecutive reactions, the reaction order for the second reaction is greater than that for the first one, which means that a decreasing hydrogen concentration favors the net formation of the intermediate product at increasing conversion.

In hydrogenation of acetylene in the presence of nickel, Komiyama and Inoue reported quite the reverse property [85]. The hydrogenation of acetylene was shown to be more dependent on the hydrogen pressure than was the consecutive hydrogenation of the formed ethylene. This implies that the selectivity of ethylene formation is favored by a high hydrogen pressure. In the presence of a palladium catalyst the ethylene selectivity is, however, favored by a low hydrogen pressure [20], like most other consecutive hydrogenation reactions.

24. *Selectivity in Hydrogenation Reactions*

The difference in reaction orders in hydrogen between the consecutive hydrogenation reactions is very dependent on the type of unsaturated compound to be hydrogenated and the type of catalyst used. In hydrogenation of 2-ethyl-2-hexenal in the presence of nickel, this difference in reaction order was shown to be as large as 1.6 and corresponds to a very high hydrogen effect on the selectivity. Commonly, the difference is of the order of 0.5, which is found in hydrogenations of fatty acids in glycerides [45], among others.

The influence of hydrogen on the selectivity in consecutive hydrogenation reactions is not only of interest from a process-control viewpoint. It is also important to understand the background of this effect in order to be able to increase the selectivity by a proper change of catalyst variables. In a first attempt to understand this effect on the selectivity, we shall formulate rate equations for a simple heterogeneous hydrogenation reaction, using the common Langmuir-Hinshelwood formalism for different possible reaction pathways.

A. Different Reaction Orders in Hydrogen Explained by Various Rate-Determining Steps of the Surface Reactions

The monounsaturated reactant A is assumed to be hydrogenated according to the reaction

$$A + H_2 \rightarrow B$$

We assume that the reaction is preceded by an adsorption of the reactants, where this adsorption includes only one active site for the adsorption of A, according to

$$A + * \underset{k_{-1}}{\overset{k_1}{\rightleftharpoons}} A* \tag{1}$$

and two active sites for the dissociative adsorption of hydrogen, according to

$$H_2 + 2* \underset{k_{-2}}{\overset{k_2}{\rightleftharpoons}} 2H* \tag{2}$$

The hydrogenation is then assumed to proceed either via the formation of a half-hydrogenated compound or without this formation. The formation of the half-hydrogenated compound may be written as

$$A* + H* \underset{k_{-3}}{\overset{k_3}{\rightleftharpoons}} AH* + * \tag{3a}$$

or

$$* + A* + H_2 \underset{k'_{-3}}{\overset{k'_3}{\rightleftharpoons}} AH* + H* \tag{3b}$$

This half-hydrogenated compound may thereafter be assumed to react according to some of the following alternatives, which for the sake of simplicity are assumed to be irreversible:

$$AH* + H* \xrightarrow{k_4} B* + * \tag{4a}$$

$$AH* + 2H* \xrightarrow{k'_4} B* + H* + * \tag{4b}$$

$$* + AH* + H_2 \xrightarrow{k''_4} B* + H* \tag{4c}$$

The adsorbed product B then desorbs according to

$$B* \underset{k_{-5}}{\overset{k_5}{\rightleftharpoons}} B + * \tag{5}$$

The desorption of B is sometimes very rapid, and the formation of B* in reactions (4a) to (4c) is then replaced by a direct formation of B in these reactions.

If the half-hydrogenated intermediate compound is not formed, we have instead the following two alternatives:

$$A* + 2H* \xrightarrow{k_6} B* + 2* \tag{6}$$

or

$$A* + H_2 \xrightarrow{k_7} B* \tag{7}$$

Rate equations have been derived for different possible rate-determining steps (3a), (3b), (4a), (4b), (4c), (6), and (7). The rate equations are in principle of the same form, with a numerator consisting of the kinetic term and a denominator with the adsorption terms. The rate equations are given in Table 1 and the parameters and other symbols included are explained in the section "Nomenclature." The rate equations are derived with the assumption that all the reaction species compete mutually for the same active sites. If this competition is restricted in some way, the corresponding rate equations will differ mainly in the denominator, in comparison to the rate equations without this restriction. If, moreover, the adsorption of A includes two active sites instead of one site as in Eq. (1), the rate equation will be different but the reaction order in hydrogen will not be changed.

As seen from Table 1, it is possible to find reaction pathways that result in rate equations where the reaction order in hydrogen is between 1/2 and 3/2. In most rate equations the influence of the hydrogen concentration is very complicated, since this concentration is also found in the denominator of the rate equation.

The high reaction order 3/2 is found only when the hydrogenation of the half-hydrogenated intermediate is assumed to be the rate-determining step, and when H_2 or 2H* is included in this reaction step. The more probable hydrogenation agent H* in this step does not give such a high reaction order. An objection may be raised against the assumption that H_2 is included in the reaction in the presence of a metal catalyst. Despite this objection, the same explanation of the reaction order 3/2 in hydrogen was given by Bond et al. [1] for the hydrogenation of acetylene in the presence of platinum, and by Hatziantoniou et al. [86] for the hydrogenation of nitrobenzene in the presence of palladium.

TABLE 1 Structure of the Rate Equation When the Surface Reaction Is the Rate-Determining Step (Mutual Competition of Active Sites)[a]

Rate-determining step	Step preceding the rate-determining step	Numerator of the rate equation	Denominator of the rate equation
(3a)	—	$k_3 K_A K_H^{1/2} C_A C_{H_2}^{1/2}$	$(1 + K_A C_A + K_B C_B + K_H^{1/2} C_{H_2}^{1/2})^2$
(3b)	—	$k_3' K_A C_A C_{H_2}$	$(1 + K_A C_A + K_B C_B + K_H^{1/2} C_{H_2}^{1/2})^2$
(4a)	(3a)	$k_4 K_A K_{AH} K_H C_A C_{H_2}$	$[1 + K_A C_A + K_B C_B + K_H^{1/2} C_{H_2}^{1/2} (1 + K_A K_{AH} C_A)]^2$
	(3b)	$k_4 K_A K_{AH}' C_A C_{H_2}$	$[1 + K_A C_A + K_B C_B + K_H^{1/2} C_{H_2}^{1/2} (1 + K_A K_{AH}' K_H^{-1} C_A)]^2$
(4b)	(3a)	$k_4' K_A K_{AH} K_H^{3/2} C_A C_{H_2}^{3/2}$	$[1 + K_A C_A + K_B C_B + K_H^{1/2} C_{H_2}^{1/2} (1 + K_A K_{AH} C_A)]^3$
	(3b)	$k_4' K_A K_{AH}' K_H^{1/2} C_A C_{H_2}^{3/2}$	$[1 + K_A C_A + K_B C_B + K_H^{1/2} C_{H_2}^{1/2} (1 + K_A K_{AH}' K_H^{-1} C_A)]^3$
(4c)	(3a)	$k_4'' K_A K_{AH} K_H^{1/2} C_A C_{H_2}^{3/2}$	$[1 + K_A C_A + K_B C_B + K_H^{1/2} C_{H_2}^{1/2} (1 + K_A K_{AH} C_A)]^2$
	(3b)	$k_4'' K_A K_{AH}' K_H^{-1/2} C_A C_{H_2}^{3/2}$	$[1 + K_A C_A + K_B C_B + K_H^{1/2} C_{H_2}^{1/2} (1 + K_A K_{AH}' K_H^{-1} C_A)]^2$
(6)	—	$k_6 K_A K_H C_A C_{H_2}$	$(1 + K_A C_A + K_B C_B + K_H^{1/2} C_{H_2}^{1/2})^3$
(7)	—	$k_7 K_A C_A C_{H_2}$	$1 + K_A C_A + K_B C_B + K_H^{1/2} C_{H_2}^{1/2}$

[a] The adsorption of A is assumed to include only one active site.

It should be noted that molecular hydrogen has also been assumed to take part in hydrogenation of methyl linoleate with copper [87], in hydrogenation of 2-ethyl-2-hexenal with nickel sulfide [88], in hydrogenation of glucose in the presence of Raney nickel [89], and in hydrogenation of ethylene on alumina after hydrogen spillover [90]. Moreover, it was proved experimentally that molecular hydrogen was the active agent in the hydrogenation of olefins and butadiene in the presence of molybdenum sulfide [91].

In hydrogenation of methyl oleate in vapor phase in the presence of nickel on alumina, the reaction order was also 3/2 in hydrogen. After comparing the fit of different rate equations to the data using statistical methods, it was obvious that the reaction pathway involving 2H∗ was more probable than that with H_2 [92].

It follows from the rate equations in Table 1 that a zero-order reaction in hydrogen, as found experimentally in hydrogenation of nitro compounds [83, 84], is found only if hydrogen is adsorbed very strongly. Since this is not very probable, the only explanation of this result based on the reaction model given here is that the adsorption of A is the rate-determining step.

Some Notes on Studying the Rate Equations

The rate equations given in Table 1 are very complicated and it is a difficult task to show which of them describe the experimental result best. In order to be able to discriminate between the different rate equations, the experimental runs have to be designed carefully by means of the design theories by Box and Hill [93] and by Box and Hunter [94]. The result must also be analyzed using proper statistical methods before any conclusions can be drawn. Discrimination between rival reaction models is often performed with the initial rate method, but this method suffers from several weak points and cannot be recommended [95]. Moreover, the parameters of the rate equation should not be determined from a linear plot of some function of the inverse reaction rate versus some function of the independent variable, unless special precautions are taken against the quite different statistical situation introduced by this transformation of the original rate equation [96-100].

Since the rate equations in Table 1 contain many parameters related to the adsorption properties of the reaction species, it is often recommended that these parameters be determined in separate experiments, in order to have only very few parameters remaining to be determined from the kinetic experiments. A determination of the adsorption parameters in separate experiments under conditions that do not exactly imitate the reaction conditions often gives data which are not valid under real reaction conditions. Kabel and Johanson's study [101] of the dehydration of ethanol over an organic ion exchanger in acid form is often referred to as a good example in this respect, where the adsorption data were determined separately and then used successfully in interpreting the kinetic result [102]. Due to the much more complicated reaction conditions in hydrogenation reactions, with a tendency toward coke formation, among others, it is not easy to make this type of study. Lidefelt et al. [87] found, however, good agreement between adsorption parameter values determined separately and from kinetic experiments in vapor-phase hydrogenations of methyl linoleate in the presence of copper on alumina as a catalyst.

Other Aspects of the Interpretation of the Hydrogen Effect on the Selectivity

The most important aim in studying the hydrogen effect of the selectivity in consecutive hydrogenation reactions is not necessarily to find the reaction

24. Selectivity in Hydrogenation Reactions

pathways and the rate equations that explain this property. If the consecutive reactions really proceed according to two different mechanisms, it is of more interest to know why these mechanisms are so different. It is also of interest to understand why the first of the two consecutive reactions is generally less hydrogen dependent than the second reaction.

In hydrogenation of a molecule with two carbon-carbon double bonds, it is easy to understand that a conjugated double bond may react with a mechanism other than a single double bond. If the double bonds are not conjugated, the first hydrogenation reaction may start with an initial conjugation preceding the hydrogenation, which may explain the difference in the reaction mechanism between the hydrogenations of the two double bonds.

It may perhaps be easier to explain the hydrogen effect on the selectivity by starting from a quite opposite argument. The assumption can instead be that the two hydrogenation reactions proceed with the same reaction mechanism, which means that the reactions are kinetically equal with respect to the reaction orders in the unsaturated compound and in hydrogen. The hydrogen effect on the selectivity must then be found by assuming that some step of the catalytic process other than the surface reactions can influence the global process in such a way that this initiates a hydrogen effect on the selectivity.

B. Hydrogen Effect on the Selectivity Explained by the Influence of Slow Adsorption and Desorption (No Rate-Determining Step)

Niklasson et al. [103] showed by mathematical simulation of the consecutive reactions

$$A + H_2 \to B \tag{8}$$

$$B + H_2 \to C \tag{9}$$

that a slow adsorption can result in a hydrogen effect on the selectivity of B even when the rates of the two surface reactions are first order in hydrogen.

The net adsorption rates of A, B, and C were assumed to follow (here given for the reactant A only)

$$r_A = k_A p_A \theta - \frac{k_A \theta_A}{K_A} \tag{10}$$

where θ is the fraction of vacant active sites.

For hydrogen the corresponding equation is

$$r_{H_2} = k_H p_{H_2} \theta^2 - \frac{k_H \theta_H^2}{K_H} \tag{11}$$

For the surface reactions we have, moreover,

$$r_{AB} = k_{AB} \theta_A \theta_H^2 \tag{12}$$

and

$$r_{BC} = k_{BC} \theta_B \theta_H^2 \tag{13}$$

If stationary conditions prevail and no step of the process is considered to be rate determining, it is possible to calculate the fractional coverages of A, B, C, and H and also the fraction of vacant active sites for a given set of parameter values from the following system of equations:

$$k_A p_A \theta - \frac{k_A \theta_A}{K_A} = k_{AB} \theta_A \theta_H^2 \tag{14}$$

$$k_B p_B \theta - \frac{k_B \theta_B}{K_B} = k_{BC} \theta_B \theta_H^2 - k_{AB} \theta_A \theta_H^2 \tag{15}$$

$$k_C p_C \theta - \frac{k_C \theta_C}{K_C} = -k_{BC} \theta_B \theta_H^2 \tag{16}$$

$$k_H p_{H_2} \theta^2 - \frac{k_H \theta_H^2}{K_H} = k_{AB} \theta_A \theta_H^2 + k_{BC} \theta_B \theta_H^2 \tag{17}$$

It is then possible to calculate the rate of reaction from Eqs. (12) and (13). The reaction rates can also be formulated as

$$r_{AB} = k'_{AB} p_A p_{H_2}^\alpha \tag{18}$$

and

$$r_{BC} = k'_{BC} p_B p_{H_2}^\beta \tag{19}$$

From the calculated values of r_{AB} and r_{BC} according to Eqs. (12) and (13) it is possible to determine the reaction orders α and β in Eqs. (18) and (19). A simulation was performed with the values given in Table 2. This simulation gave the reaction order values $\alpha = 0.78$ and $\beta = 0.96$ with a standard deviation of 0.06, which shows that hydrogen has an effect on the selectivity. The reason for this effect is that the slow adsorption of A does not compensate for the decrease of θ_A due to the rapid surface reaction giving a decreased ratio of θ_A/θ_B. If the adsorption rate constants are increased by a factor of 100 without changing the other parameter values, α and β will both attain unit value. Due to the lack of adsorption kinetic data valid under reaction condi-

TABLE 2 Parameter Values Used in the Simulation

Rate constants	Equilibrium constants	Range of partial pressures
$k_A = 10^3$	$K_A = 10^3$	$p_A = (0.05\text{-}0.5) \times 10^{-3}$
$k_B = 10^3$	$K_B = 10^3$	$p_B = (0.05\text{-}0.5) \times 10^{-3}$
$k_C = 2 \times 10^3$	$K_C = 1$	$p_C = 10^{-3}$
$k_H = 10^2$	$K_H = 2 \times 10^2$	$p_{H_2} = (1\text{-}10) \times 10^{-3}$
$k_{AB} = 10$		
$k_{BC} = 10$		

C. Hydrogen Effect on the Selectivity Caused by a Change of the Surface of the Catalyst

In most hydrogenations in gas phase, the surface of the catalyst is readily covered with carbonaceous deposit. At low hydrogen pressure, this deposit may be assumed to cover the most active sites, whereas the less active and perhaps the more selective sites are uncovered. When increasing the hydrogen pressure, new and less selective sites will then be exposed, which may thus be a conceivable explanation of the hydrogen effect on the selectivity in consecutive hydrogenation reactions.

Opposite to this theory is the observation that carbonaceous deposit results in increased hydrogen coverage [104,105]. Since a high hydrogen coverage gives a low selectivity if the hydrogen effect originates from different reaction orders in hydrogen of the surface reactions, a high degree of carbonaceous coverage may, consequently, result in a low selectivity.

Siegel et al. [32] have stated the hypothesis that two different types of sites may exist on the surface of the catalyst. The activity of these sites depends on the hydrogen pressure, and the proportion of the sites is related to the history of the catalyst, in particular to the conditions of its exposure.

D. Hydrogen Effect on the Selectivity Explained by the Influence of Slow Pore Transport

Wheeler [106] showed, in his now classical analysis, that the selectivity of the formation of B in the consecutive reactions A → B → C decreases upon increasing the chemical reaction rate relative to the maximum possible physical transport rate of the reactants in a porous catalyst. A corresponding effect can also occur for the consecutive hydrogenation reactions (8) and (9). An increased chemical reaction rate can be obtained here by increasing only the hydrogen concentration. If the two consecutive reactions are both first order in the respective unsaturated reactants and hydrogen, there will be no hydrogen effect on the selectivity at low hydrogen concentration. This concentration is so low that the physical transport in the pores does not influence the rate of the global process. When increasing the hydrogen concentration, the reaction rate will increase and the transport in the pores will gradually influence the global rate. Parallel to this increase in the hydrogen concentration, the selectivity of the formation of B will decrease. If the global rates of the two consecutive reactions are written as power rate equations, this decreasing selectivity can be evaluated as different apparent reaction orders in hydrogen for the two reactions; and since the selectivity decreases with increasing hydrogen concentration, the reaction order will be greater for the second reaction. This difference in reaction orders was found by Hell et al. [107] in efforts to understand the hydrogen effect on the selectivity in hydrogenation of fatty acids in glycerides by mathematical simulation of the chemical processes in the porous catalyst. The hydrogen effect on the selectivity in the similar liquid-phase hydrogenation of methyl linoleate was explained by Tsuto et al. [108] as a combined effect of nonequilibrium adsorption and a slow pore transport. That the transport in the catalyst pores really has an effect on the selectivity in fat hydrogenation was experimentally proved by Coenen et al. [109] in 1964 and later by Beasly et al. [110]. The pore transport problems in fat hydrogenation were also studied by van der Plank et al. [111] and Bern et al. [112].

The creation of the apparent reaction orders α and β in hydrogen due to the influence of the pore transport effect on the two consecutive reactions, which are first order in hydrogen, can be illustrated by mathematical simulation of the process with the classical single-pore model. The material balances of the three species A, B, and H_2 in the pore may be written as

$$\frac{d^2 C_A}{dx^2} = \frac{4}{dD_A} k''_{AB} C_A C_{H_2} \tag{20}$$

$$\frac{d^2 C_B}{dx^2} = -\frac{4}{dD_B} (k''_{AB} C_A C_{H_2} - k''_{BC} C_B C_{H_2}) \tag{21}$$

$$\frac{d^2 C_{H_2}}{dx^2} = \frac{4}{dD_H} (k''_{AB} C_A C_{H_2} + k''_{BC} C_B C_{H_2}) \tag{22}$$

with the common boundary values

$$\frac{dC_A}{dx} = \frac{dC_B}{dx} = \frac{dC_{H_2}}{dx} = 0 \tag{23}$$

at the pore end $x = L$, and

$$C_A = C_A^o$$
$$C_B = C_B^o$$
$$C_{H_2} = C_{H_2}^o \tag{24}$$

at the pore mouth $x = 0$. The system of three differential equations was solved using a numerical code called COLSYS [113], and the molar fluxes of A and B at the pore mouth ($x = 0$) were calculated and correlated with the hydrogen concentration at this boundary in order to determine the reaction orders α and β.

The calculated reaction orders α and β are given in Fig. 2 at various hydrogen concentrations and under reaction conditions corresponding to a liquid-phase hydrogenation similar to fat hydrogenation, apart from the fact that the hydrogen concentration spans over a much larger concentration interval. As seen from Fig. 2, α and β are equal to unity at low hydrogen concentration and the difference β-α increases to a maximum of about 0.6 at the hydrogen concentration 11 mol m^{-3} at the pore mouth. The location of this maximum is, of course, dependent on the reaction and transport conditions, and appears at much lower hydrogen concentration when the maximum pore transport rate is lower than in the simulation given here. A maximum in the difference β-α also appears at conditions corresponding to a gas-phase hydrogenation.

On further increasing the hydrogen concentration, the reaction orders α and β approach the limiting value 0.5. The hydrogen effect on the selectivity originating from the pore transport limitation does not exist at very low and very high hydrogen concentrations, but somewhere between these values the hydrogen effect has a maximum corresponding to α = 0.8 and β = 1.4

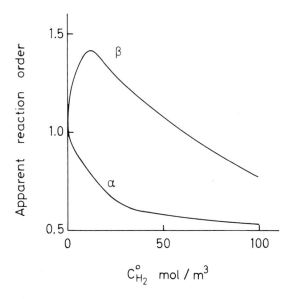

FIGURE 2 Apparent reaction orders α and β as a function of the hydrogen concentration at the pore mouth. $D_H/D_A = D_H/D_B = 71.43$; $k''_{BC}/k''_{AB} = 7.143 \times 10^{-2}$; $L^2(k''_{AB}/D_A d) = 0.1082$ m^3 mol^{-1}; $C^o_A = C^o_B = 1000$ mol m^{-3}.

E. Hydrogen Effect on the Selectivity Explained by the Influence of Slow Film Transport

In liquid-phase hydrogenations the influence of a slow physical step in the porous catalyst may easily be eliminated if the particle size is reduced. However, there is a risk that one jumps out of the frying pan into the fire by this measure. It is a well-known fact that finely divided solid catalyst particles can penetrate the diffusion liquid film surrounding the gas bubbles [10,11]. If, moreover, these particles are active, the chemical reaction starts in this liquid film, and if the activity of the particles is very high, the chemical reaction will proceed exclusively in this film. This coupling between the chemical reactions and the physical transport step in the liquid film may be expected to result in a falsification of the reaction orders in hydrogen for the two consecutive reactions (8) and (9), of the same type as that found for the corresponding coupling of these processes in the porous catalyst.

The magnitude of the apparent reaction orders in hydrogen can be illustrated with a simulation of the same type as that given in Section III.D. The two simulations differ, among other things due to the fact that that the boundary conditions are different. Another difference is the fact that the reactants are diffusing countercurrently in the liquid film and not cocurrently as in the porous catalyst. The result of one typical simulation is given in Fig. 3, where the catalyst activity per volume of liquid is assumed to be the same in the liquid film and in the bulk liquid. As can be seen from Fig. 3, the penetration of catalyst particles into the liquid film surrounding the bubbles creates a hydrogen effect on the selectivity reflected by a difference between α and β. The hydrogen effect at very high hydrogen concentration is the reverse of that at low concentration. The reverse hydrogen effect ($\beta < \alpha$) is explained by the fact that B reaches a maximum in the liquid film and is also depleted in this film. In the present simulation the

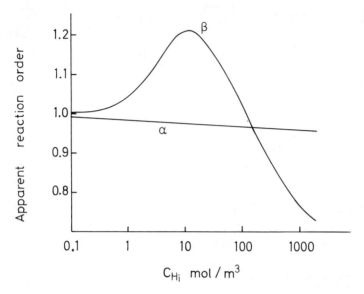

FIGURE 3 Apparent reaction orders α and β as a function of the hydrogen concentration in the liquid at the gas-liquid interface. $C_{Ab} = 1000$ mol m^{-3}; $C_{Bb} = 100$ mol m^{-3}; $k_L a = 0.3$ s^{-1}; $a = 300$ m^{-1}; $D_H/k_L = 1.6 \times 10^{-5}$ m; $D_H = 1.6 \times 10^{-8}$ m^2 s^{-1}; $D_A = D_B = 5 \times 10^{-10}$ m^2 s^{-1}; $k''_{BC}/k''_{AB} = 0.4$; $k''_{AB} = 0.1$ m^3 mol^{-1} s^{-1}. Hatta number $\gamma_{AB} = 1.29$.

hydrogen concentration was chosen too high compared to real conditions, to demonstrate that the hydrogen effect on the selectivity can turn over. This phenomenon may occur at much lower hydrogen concentration with another set of parameter values chosen.

F. Hydrogen Effect on the Selectivity in Homogeneous Hydrogenations

The hydrogen effect on the selectivity discussed in Section III.E, caused by the penetration of the catalyst into the liquid film surrounding the gas bubbles, is of the same type as may occur in gas-liquid processes with a homogeneous catalyst in the liquid phase. This phenomenon seems to have been noticed very little in interpreting the results from kinetic studies of homogeneous hydrogenations. The influence of a slow physical transport on the selectivity in consecutive reactions in gas-liquid processes in general has, however, been studied by many authors [5-9].

IV. DIFFERENT WAYS TO UTILIZE THE HYDROGEN EFFECT TO MAXIMIZE THE SELECTIVITY IN CONSECUTIVE REACTIONS

A. Traditional Control

The traditional control implies that the hydrogen concentration in the reactor is varied so that the selectivity is always maximal at every moment during the hydrogenation. For most consecutive hydrogenation reactions, where the reaction order in hydrogen is larger for the second reaction ($\beta > \alpha$), the selectivity is favored by a strongly decreasing hydrogen concentration with increasing conversion if the reaction time is short [114]. For slower hydrogenation processes the optimal hydrogen concentration will, instead, be constant and low. In liquid-phase hydrogenation processes with very reactive components, it is difficult to attain a high hydrogen concentration in the liquid phase in the beginning of the hydrogenation, since the mass transport resistances allow only a slow compensation of the rapid consumption of hydrogen. In a practical situation, the hydrogen concentration in many liquid hydrogenations increases instead with increasing conversion, which is quite contrary to the optimal concentration course. An optimal hydrogen concentration is in this case more or less restricted by the difficulty of cooling the reaction mixture, since a high hydrogen concentration at the start of the process also means a rapid temperature rise caused by rapid evolution of the heat of reaction. The loss in selectivity is, however, rather small in the beginning of the hydrogenation, since the concentration of the intermediate product B is still relatively low compared to the starting reactant A. It is therefore more important to see that the hydrogen concentration reaches the proper optimal level after this initial reactive period of the process.

B. Periodic Operation

Periodic operation means a periodic varying of the feed composition or feed temperature, or both simultaneously. If the reaction system fulfills certain proper dynamic properties, which are rather difficult to specify, the periodic variation (if optimally performed) can provide a product selectivity not obtainable from conventional optimal steady-state operation [115].

Numerous theoretical works have been published about periodic operation, but only a few experimental applications have been presented. Some of these experimental works concern the effort to increase yield and selectivity in hydrogenation processes.

Based on the fact that the ethylene selectivity in hydrogenation of acetylene in the presence of a nickel catalyst is a convex function of the hydrogen partial pressure, Bilimoria and Bailey [116] experimentally studied the possibility of increasing this selectivity by varying the hydrogen flow rate in a bang-bang fashion. The result was, however, not very encouraging and no improvement of the ethylene selectivity was obtained. The stationary kinetic behavior of the process was not the right basis on which to judge the conditions for successful utilization of periodic operation. It is obvious that the dynamics of the gas phase/catalyst surface interaction cannot be neglected when determining a periodic control strategy.

A more successful study of increasing the selectivity by periodic operation was given by Al-Taie and Kershenbaum [117] in hydrogenation of 1,3-butadiene in the presence of nickel on kieselguhr in a tube reactor. It was shown that as much as 20% improvement in the yield of butenes can be attained with periodic operation, depending on the choice of amplitude, cycle time, and residence time. The improvement in yield and selectivity were compared with the optimum steady-state values. The increased yield and selectivity were explained by postulating that the kinetics of adsorption and desorption processes play an important role in the system dynamics. The main cause of the improvement was thought to be the resonance induced by the adsorption-desorption kinetics during periodic operation.

The influence of forced cycling of the composition of the feed on the Fischer-Tropsch synthesis was studied by Feimer et al. [118]. This type of processing did not influence the total hydrocarbon product distribution.

C. Determination of Hydrogen Concentration to Optimally Control the Hydrogenation

An optimal control of the hydrogen concentration according to the traditional control strategy presupposes that it is possible to measure the hydrogen concentration at the catalyst surface during the process. In liquid-phase hydrogenations this determination is more difficult than in gas-phase hydrogenations, due to the many different transport resistances for the hydrogen transport from the gas phase to the catalyst surface. Andersson and Berglin [119] constructed a fast-response hydrogen analog to the well-known membrane-covered oxygen electrode, which makes possible rapid measurement of the hydrogen concentration in the bulk liquid. Figure 4 shows the hydrogen concentration in the bulk liquid determined with this hydrogen electrode at different positions in an industrial hydrogenation tank reactor during hydrogenation of anthraquinones with different types of hydrogen supply. This determination of the hydrogen concentration also makes it possible to calculate directly the volumetric mass transfer coefficient $k_L a$ for the transport of hydrogen across the liquid film close to the gas bubbles.

Much more important in this respect is the concentration or activity of hydrogen at the surface of the catalyst. This activity is easily measured if the liquid contains free protons. This measurement is based on the fact that an electrochemical exchange occurs between adsorbed species on the catalyst surface and the solution, giving a potential difference relative to

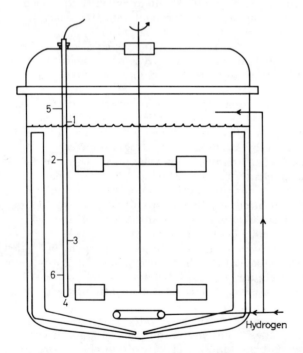

FIGURE 4 Determination of the hydrogen concentration in the bulk liquid during industrial hydrogenation of alkylanthraquinones in a tank reactor ($10\ m^3$) with continuous liquid and gas flows. The hydrogen concentrations at levels 1, 2, 3, 4, 5, and 6 were 7.48, 6.28, 6.43, 6.58, 7.48, and 6.51, mol m^{-3}, respectively. When hydrogen was supplied only from the top of the reactor, the hydrogen concentration at level 6 was 3.35 mol m^{-3}. Equilibrium hydrogen concentration is 7.48 mol m^{-3}. Power input from the stirrer per volume of gas and liquid is 2900 W m^{-3}. (From Ref. 119.)

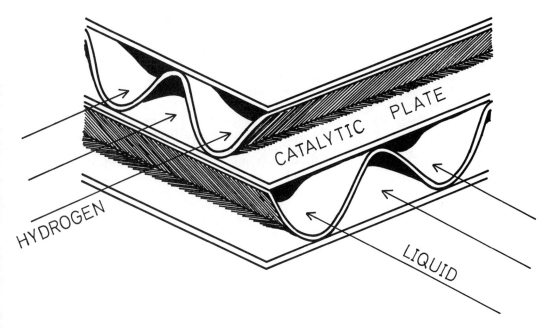

FIGURE 5 Section of the cross-flow catalyst. The thickness of the porous catalytic plates is 0.2 mm. (From Ref. 125.)

the solution. This potential difference can be measured by a metal wire, which does not take part in any electrochemical reaction during the hydrogenation, and to which colliding catalyst particles impart their potential. This method of following a liquid-phase hydrogenation was first described by Müller and Schwabe [120] in 1928. The method has been used extensively by Sokolskii [121] in hydrogenation of a wide variety of substances. This method was also used by Andersson [122,123] to study the mass transfer resistance for the transport of hydrogen across the liquid film close to the catalyst particles in a slurry reactor. The method can also be used to determine the rate-determining step, including the dissociation of H_2 in hydrogenation processes [64,123].

D. Influence of the Catalyst Geometry on the Selectivity in Liquid-Phase Hydrogenation

Many liquid-phase hydrogenations are performed at high pressure in trickle-bed reactors [124]. The availability of the catalytic material is relatively low in this reactor. Moreover, the resistance against the transport of hydrogen to the catalyst surface can be high due to the fact that the liquid film surrounding the pellets is thick and the flow rate in this film is low. The availability of the catalytic material is much greater in reactors where the catalyst consists of thin porous plates. If, in addition, hydrogen and liquid are supplied from each side of the plates, hydrogen does not need to pass across a thick liquid film before it comes into contact with the active catalyst. Two examples of this type of reactor are the cross-flow catalyst reactor [125] and the cell reactor [126]. In the cross-flow catalyst reactor, the porous plates are separated by corrugated planes. These planes are placed at a 90° angle to the corrugated planes immediately above and below. If the gas and the liquid are also separated by 90°, this will make it possible to supply gas and liquid to different sides of every catalytic plate (Fig. 5). In the cell reactor the thin active plates are mounted in a special rack like a filter press.

The fact that hydrogen and liquid reactants are not transported cocurrently in the porous catalytic structure of these plates may give rise to a pronounced hydrogen effect on the selectivity. De Vos et al. [34] found that the formation of cis-1,4-di-*tert*-butylcyclohexane was strongly supported, especially at elevated hydrogen pressure in liquid-phase hydrogenation of 1,4-di-*tert*-butylbenzene in cyclohexane in a cross-flow catalyst reactor with rhodium as the active metal. This increased selectivity was difficult to explain, but one theory was the possibility for different competition behavior of vertically and horizontally adsorbed 1,4-di-*tert*-butylbenzenes, together with the particular concentration profiles in the porous plates of the cross-flow catalyst.

Both the cross-flow catalyst reactor and the cell reactor are characterized by their very low pressure drop compared to the common trickle-bed reactor. The same favorable property is also valid for the monolithic catalyst reactor (i.e., a catalyst with the same geometry as used in catalysts for cleaning exhaust gas from cars). In this reactor, gas and liquid compete for the same channels [86]. Due to the very specific mass transfer conditions for hydrogen in this reactor, the hydrogen effect on the selectivity can easily be controlled by the gas flow rate in the channels.

V. CONCLUDING REMARKS

Hydrogen has been shown to have a very important influence on the selectivity in many industrial chemical processes. This hydrogen effect is most often found in consecutive hydrogenation reactions. The selectivity of the net formation of the intermediate product in these reactions is, moreover, favored by a low hydrogen concentration. This property has been given various interpretations:

1. The consecutive surface reactions proceed according to two reaction mechanisms, giving rise to different reaction orders in hydrogen.
2. Different rate-determining steps control the two consecutive reactions.
3. An increased hydrogen concentration decreases the coke layer blocking the active sites and exposes new active metal sites which are assumed to be less selective.
4. The adsorption and desorption steps are slow in comparison to the surface reactions.
5. The transport in the porous catalyst is a slow process, but not too slow, and the chemical reactions are dependent on both the inward supply of the reactants and the outward transport of the intermediate product.
6. The catalyst particles are small enough to penetrate the liquid film surrounding the gas bubbles in liquid-phase hydrogenations. The chemical reactions can start in this film and are intimately coupled with the transport processes in the film.
7. In liquid-phase hydrogenations in the presence of a homogeneous catalyst, the reaction can start in the liquid film close to the gas bubbles, an effect of the same type as that above in the slurry process.

If the hydrogen concentration is to be controlled in a traditional way, by slow changes of the hydrogen concentration relative to the time constants of the reaction and transport steps, the optimal hydrogen concentration to maximize the selectivity may be independent of the origin of the hydrogen effect observed in kinetic studies at stationary conditions. If the hydrogen concentration control instead includes more rapid changes, the optimal control function will be dependent on the kinetic details of the reaction and transport processes.

ACKNOWLEDGMENT

The sustained support, over a long period, of the National Swedish Board for Technical Development for research on different aspects of hydrogenation reactions is gratefully acknowledged.

NOMENCLATURE

a	interfacial area of the gas-liquid mixture
A, B, C	starting, intermediate, and final compounds, respectively, in consecutive reactions
c	concentration
d	pore diameter
D	molecule taking part in hydrogen transfer
D_A, D_B, D_H	diffusivities of A, B, and H_2, respectively
k	rate constant
k_3, k_3'	rate constants in reactions (3a) and (3b), respectively
k_4, k_4', k_4''	rate constants in reactions (4a), (4b), and (4c), respectively
k_5, k_6, k_7	rate constants of reactions (5), (6), and (7), respectively
k_A, k_B, k_C, k_H	adsorption rate constants for the adsorption of A, B, C, and H_2, respectively
k_{AB}, k_{BC}	rate constants of reactions (12) and (13), respectively
k_{AB}', k_{BC}'	rate constants of reactions (18) and (19), respectively
k_{AB}'', k_{BC}''	rate constants referred to unit surface area in Eqs. (20) to (22) and in Figs. 1 and 2
$k_L a$	volumetric mass transfer coefficient in liquid for the transport of hydrogen
K	adsorption equilibrium constant
K_A	equilibrium constant equal to k_1/k_{-1}
K_B	equilibrium constant equal to k_{-5}/k_5
K_C	adsorption equilibrium constant for adsorption of C
K_H	equilibrium constant equal to k_2/k_{-2}
K_{AH}	equilibrium constant equal to k_3/k_{-3}
K_{AH}'	equilibrium constant equal to k_3'/k_{-3}'
L	pore length
p_A, p_B, p_C, p_{H_2}	partial pressures of A, B, C, and H_2, respectively
r_A, r_H	net rates of adsorption of A and H_2, respectively
r_{AB}, r_{BC}	reaction rates according to Eqs. (12) and (13)
x	length coordinate
*	active site

Greek letters

α, β	apparent reaction orders in hydrogen
γ_{AB}	Hatta number = $(k_{AB}'' C_{Bb} D_H)^{1/2}/k_L$
θ	fraction of free active sites
θ_A, θ_B, θ_H	fraction of active sites occupied by A, B, and H, respectively

Superscript

o   conditions at the pore mouth

Additional Subscripts

b   conditions in the bulk liquid
i   conditions at the interface between gas and liquid

REFERENCES

1. G. C. Bond, D. A. Dowden, and N. Mackenzie, *Trans. Faraday Soc.*, 54: 1537 (1958).
2. P. B. Weisz, *Ann. Rev. Phys. Chem.*, 21:193 (1970).
3. J. C. van de Vusse, *Chem. Eng. Sci.*, 21:631 (1966).
4. J. C. van de Vusse, *Chem. Eng. Sci.*, 21:645 (1966).
5. M. Teramoto, T. Nagayasu, T. Matsui, K. Hashimoto, and S. Nagata, *J. Chem. Eng. Jpn.*, 2:186 (1969).
6. K. Onda, E. Sada, T. Kobayashi, and M. Fujine, *Chem. Eng. Sci.*, 25: 753 (1970).
7. K. Onda, E. Sada, T. Kobayashi, and M. Fujine, *Chem. Eng. Sci.*, 25: 761 (1970).
8. K. Onda, E. Sada, T. Kobayashi, and M. Fujine, *Chem. Eng. Sci.*, 25: 1023 (1970).
9. K. Onda, E. Sada, T. Kobayashi, and M. Fujine, *Chem. Eng. Sci.*, 27: 247 (1972).
10. E. Alper, B. Wichtendahl, and W.-D. Deckwer, *Chem. Eng. Sci.*, 35: 217 (1980).
11. O. Stenberg and N.-H. Schöön, *Chem. Eng. Sci.*, 40:2311 (1985).
12. J. W. E. Coenen, *J. Am. Oil Chem. Soc.*, 53:382 (1976).
13. IUPAC, *Pure Appl. Chem.*, 4:71 (1976); *Adv. Catal.*, 26:351 (1977).
14. K. R. Westerterp, W. P. M. van Swaaij, and A. A. C. M. Beenackers, *Chemical Reactor Design and Operation*, Wiley, New York (1984).
15. K. G. Denbigh, *Chem. Eng. Sci.*, 14:24 (1961).
16. G. C. Bond and P. B. Wells, *J. Catal.*, 4:211 (1965).
17. G. C. Bond and P. B. Wells, *J. Catal.*, 5:65, 419 (1966).
18. G. C. Bond and P. B. Wells, *J. Catal.*, 6:397 (1966).
19. G. C. Bond, G. Webb, and P. B. Wells, *J. Catal.*, 12:157 (1968).
20. G. C. Bond and P. B. Wells, *Adv. Catal.*, 15:91 (1964).
21. D. J. Collins, D. E. Grimes, and B. H. Davis, *Can. J. Chem. Eng.*, 61: 36 (1983).
22. P. B. Wells, in *Surface and Defect Properties of Solids*, Vol. 1, Chemical Society, London (1972), p. 236.
23. R. S. Mann and K. C. Khulbe, *Can. J. Chem.*, 46:623 (1968).
24. G. C. Bond, J. J. Phillipson, P. B. Wells, and J. M. Winterbottom, *Trans. Faraday Soc.*, 62:433 (1966).
25. P. B. Wells, *J. Catal.*, 52:498 (1978).
26. S. Siegel and G. V. Smith, *J. Am. Chem. Soc.*, 82:6082 (1960).
27. S. Siegel and B. Dmuchovsky, *J. Am. Chem. Soc.*, 86:2192 (1964).
28. S. Siegel, *Adv. Catal.*, 16:123 (1966).
29. S. Siegel and B. Dmuchovsky, *J. Am. Chem. Soc.*, 84:3132 (1962).
30. C. U. I. Odenbrand and S. T. Lundin, *J. Chem. Technol. Biotechnol.*, A31:660 (1981).
31. S. Siegel, J. Outlaw, Jr., and N. Garti, *J. Catal.*, 58:370 (1979).
32. S. Siegel, J. Outlaw, Jr., and N. Garti, *J. Catal.*, 52:102 (1978).

33. S. Siegel and N. Garti, in *Catalysis in Organic Syntheses 1977* (G. V. Smith, ed.), Academic Press, New York (1977), p. 9.
34. R. de Vos, G. Smedler, and N.-H. Schöön, *Ind. Eng. Chem. Process Des. Dev.*, *25*:197 (1986).
35. Y. Izumi, *Adv. Catal.*, *32*:215 (1983).
36. Y. Nitta, F. Sekine, T. Imanaka, and S. Teranishi, *Chem. Lett.*, 541 (1981).
37. E. I. Klabunovskii, A. A. Vedenyapin, Yu. M. Talanov, and N. P. Sokolova, *Kinet. Catal.*, *16*:595 (1975).
38. V. I. Neypokoev, Yu. I. Petrov, and E. I. Klabunovskii, *Izv. Akad. Nauk SSSR Ser. Khim.*, 113 (1976).
39. E. N. Lipgart, Yu. I. Petrov, and E. I. Klabunovskii, *Kinet. Catal.*, *12*: 1326 (1971).
40. E. N. Frankel and H. J. Dutton, in *Topics in Lipid Chemistry*, Vol. 1 (F. D. Gunstone, ed.), Logos Press, London (1970), p. 161.
41. H. J. Dutton, in *Progress in Chemistry of Fats and Other Lipids*, Vol. 9, Part 3 (R. T. Hohman, ed.), Pergamon Press, Oxford (1968), p. 351.
42. J. W. E. Coenen, *Chem. Ind.*, 709 (1978).
43. R. J. Dutton, *Chem. Ind.*, 9 (1982).
44. J.-O. Lidefelt, *J. Am. Oil Chem. Soc.*, *60*:588 (1983).
45. M. Pihl and N.-H. Schöön, *Acta Polytech. Scand. Chem. Ind. Metall. Ser.*, *100*:4 (1971).
46. S. Carra and E. Santacesaria, *Catal. Rev.-Sci. Eng.*, *22*:75 (1980).
47. G. Webb, in *Catalysis*, Vol. 2, Chemical Society, London (1978), p. 143.
48. A. J. Birch and K. A. M. Walker, *J. Chem. Soc.*, *C*:1894 (1966).
49. C. Djerassi and J. Gutzwiller, *J. Am. Chem. Soc.*, *88*:4537 (1966).
50. S. Nishimura and K. Tsuneda, *Bull. Chem. Soc. Jpn.*, *42*:852 (1969).
51. C. U. Pittman, Jr. and R. C. Ryan, *CHEMTECH*, *8*(3):170 (1978).
52. M. D. Birkett, A. T. Kuhn, and G. C. Bond, in *Catalysis*, Vol. 6, Chemical Society, London (1983), p. 61.
53. F. Coussemant and J. C. Jungers, *Bull. Soc. Chim. Belge.*, *59*:295 (1950).
54. S. J. Thomson and G. Webb, *J. Chem. Soc. Chem. Commun.*, 526 (1976).
55. T. Nishiguchi, T. Tagawa, H. Imai, and K. Fukuzumi, *J. Am. Oil Chem. Soc.*, *54*:144 (1977).
56. H. Shima and T. Yamaguchi, *J. Catal.*, *90*:160 (1984).
57. T. I. Taylor, in *Catalysis*, Vol. 5 (P. H. Emmett, ed.), Reinhold, New York (1957), p. 386.
58. C. E. Polan, J. J. McNeill, and S. B. Tove, *J. Bacteriol.*, *88*:1056 (1964).
59. C. R. Kepler, K. P. Hirons, J. J. McNeill, and S. B. Tove, *J. Biol. Chem.*, *241*:1350 (1966).
60. C. R. Kepler and S. B. Tove, *J. Biol. Chem.*, *242*:5686 (1967).
61. R. L. Augustine, *Catal. Rev.-Sci. Eng.*, *13*:285 (1976).
62. H. Greenfield, *Ind. Eng. Chem. Prod. Res. Dev.*, *6*:142 (1967).
63. H. Greenfield, *Ind. Eng. Chem. Prod. Res. Dev.*, *15*:156 (1976).
64. T. Berglin and N.-H. Schöön, *Ind. Eng. Chem. Process Des. Dev.*, *20*: 615 (1981).
65. J. Wisniak and R. Simon, *Ind. Eng. Chem. Prod. Res. Dev.*, *18*:50 (1979).
66. J. C. Jungers and F. Coussemant, *J. Chim. Phys. (Fr.)*, *47*:139 (1950).
67. A. Hershman, K. K. Robinson, J. H. Craddock, and J. F. Roth, *Ind. Eng. Chem. Prod. Res. Dev.*, *8*:372 (1969).
68. Y. Matsui, H. Taniguchi, K. Terada, T. Anezaki, and M. Iriuchijima, *Bull. Jpn. Pet. Inst.*, *19*:62 (1977).
69. L. A. Gerritsen, Ph.D. thesis, Delft University of Technology (1979).
70. K. L. Olivier and F. B. Booth, *Hydrocarbon Process.*, *49*(4):112 (1970).

71. D. M. Fenton and K. L. Olivier, *CHEMTECH*, 2:220 (1972).
72. B. Cornils, R. Payer, and K. C. Traenckner, *Hydrocarbon Process.*, 54(6):83 (1975).
73. J. C. Rohrer, H. Hurwitz, and J. H. Sinfelt, *J. Phys. Chem.*, 65:1458 (1961).
74. J. H. Sinfelt and J. C. Rohrer, *J. Phys. Chem.*, 65:978 (1961).
75. H. Kölbel and M. Ralek, *Catal. Rev.-Sci. Eng.*, 21:225 (1980).
76. P. J. Denny and D. A. Whan, in *Catalysis*, Vol. 2, Chemical Society, London (1978), p. 46.
77. R. R. Cecil, F. X. Mayer, and E. N. Cart, *American Institute of Chemical Engineers Meeting, Los Angeles* (1968), cited in G. C. A. Schuit and B. C. Gates, *AIChE J.*, 19:417 (1973).
78. H. Beuther and B. K. Schmid, *Proceedings of the 6th World Petroleum Congress, Hamburg*, Sec. 3 (1964), p. 297.
79. M. O. Rosenheimer and J. R. Kiovsky, *Am. Chem. Soc. Div. Pet. Chem. Prepr.*, 12:B147 (1967).
80. M. J. Ledoux, in *Catalysis*, Vol. 7, Chemical Society, London (1985), p. 125.
81. D. G. Gavin, in *Catalysis*, Vol. 5, Chemical Society, London (1982), p. 220.
82. Y. T. Shah, D. C. Cronauer, H. G. McIlvried, and J. A. Paraskos, *Ind. Eng. Chem. Process Des. Dev.*, 17:288 (1978).
83. G. J. K. Acres and B. J. Cooper, *J. Appl. Chem. Biotechnol.*, 22:769 (1972).
84. P. B. Kalantri and S. B. Chandalla, *Ind. Eng. Chem. Process Des. Dev.*, 21:186 (1982).
85. H. Komiyama and H. Inoue, *J. Chem. Eng. Jpn.*, 1:142 (1968).
86. V. Hatziantoniou, B. Andersson, and N.-H. Schöön, *Ind. Eng. Chem. Prod. Res. Dev.*, 25:964 (1986).
87. J.-O. Lidefelt, J. Magnusson, and N.-H. Schöön, *J. Am. Oil Chem. Soc.*, 60:600 (1983).
88. G. Smedler, private communication.
89. P. H. Brahme and L. K. Doraiswamy, *Ind. Eng. Chem. Process Des. Dev.*, 15:130 (1976).
90. B. Bianchi, G. E. E. Gardes, G. M. Pajonk, and S. J. Teichner, *J. Catal.*, 38:135 (1975).
91. T. Okuhara, K. Tanaka, and K. Miyahara, *J. Chem. Soc. Chem. Commun.*, 42 (1976).
92. J.-O. Lidefelt, J. Magnusson, and N.-H Schöön, *J. Am. Oil Chem. Soc.*, 60:603 (1983).
93. G. E. P. Box and W. J. Hill, *Technometrics* 9:57 (1967).
94. G. E. P. Box and W. G. Hunter, *Proceedings of the IBM Scientific Computing Symposium on Statistics* (1965), p. 113.
95. M. Boudart, *Chem. Eng. Sci.*, 22:1387 (1967).
96. G. E. P. Box and W. J. Hill, *Technometrics*, 16:385 (1974).
97. D. J. Pritchard and D. W. Bacon, *Technometrics*, 19:109 (1977).
98. D. J. Pritchard, J. Downie, and D. W. Bacon, *Technometrics*, 19:227 (1977).
99. D. J. Egy, *Ind. Eng. Chem. Fundam.*, 21:337 (1982).
100. G. E. P. Box, D. M. Steinberg, and W. J. Hill, *Ind. Eng. Chem. Fundam.*, 23:267 (1984).
101. R. L. Kabel and L. N. Johanson, *AIChE J.*, 8:621 (1982).
102. S. W. Weller in *Chemical Reaction Engineering Reviews* (H. M. Hulburt, ed.), *Advances in Chemistry Series 148*, American Chemical Society, Washington, D.C. (1975), p. 26.

103. C. Niklasson, B. Andersson, and N.-H. Schöön, *Ind. Eng. Chem. Res.*, 26, No. 7 (1987).
104. J. M. Moses, A. H. Weiss, K. Matusek, and L. Guczi, *J. Catal.*, 86:417 (1984).
105. M. Boudart, A. W. Aldag, and M. A. Vannice, *J. Catal.*, 18:46 (1970).
106. A. Wheeler, *Adv. Catal.*, 3:249 (1951).
107. M. Hell, L.-E. Lundqvist, and N.-H. Schöön, *Acta Polytech. Scand. Chem. Ind. Metall. Ser.*, 100:59 (1971).
108. K. Tsuto, P. Harriott, and K. B. Bischoff, *Ind. Eng. Chem. Fundam.*, 17:199 (1978).
109. J. W. E. Coenen, H. Boerma, B. G. Linsen, and B. De Vries, *Proceedings of the 3rd International Congress on Catalysis*, Vol. 2, North Holland, Amsterdam (1964), p. 1387.
110. P. F. Beasley, A. J. Bird, and M. C. Sweeney, in *Characterisation of Catalysts* (J. M. Thomas and R. M. Lambert, eds.), Wiley, Chichester, West Sussex, England (1980), pp. 62-65.
111. P. van der Plank, B. G. Linsen, and H. J. van den Berg, *Proceedings of the 5th European, 2nd International Symposium on Chemical Reaction Engineering*, Elsevier, Amsterdam (1972), p. B6.
112. L. Bern, M. Hell, and N.-H. Schöön, *J. Am. Oil Chem. Soc.*, 52:182 (1975).
113. U. Ascher, J. Christiansen, and R. D. Rusell, *ACM Trans. Math. Software*, 17:209 (1981).
114. L.-E. Lundqvist and N.-H. Schöön, *Acta Polytech. Scand. Chem. Ind. Metall. Ser.*, 100:37 (1971).
115. J. E. Bailey and F. J. M. Horn, *Ber. Bunsenges. Phys. Chem.*, 73:274 (1969).
116. M. R. Bilimoria and J. E. Bailey, in *Chemical Reaction Engineering—Houston* (V. W. Weekman, Jr., and D. Luss, eds.), *Advances in Chemistry Series 65*, American Chemical Society, Washington, D.C. (1978), p. 526.
117. A. S. Al-Taie and L. S. Kershenbaum, in *Chemical Reaction Engineering—Houston* (V. W. Weekman, Jr., and D. Luss, eds.), *Advances in Chemistry Series 65*, American Chemical Society, Washington, D.C. (1978), p. 512.
118. J. L. Feimer, P. L. Silveston, and R. R. Hudgins, *Can. J. Chem. Eng.*, 63:86 (1985).
119. B. Andersson and T. Berglin, *Chem. Eng. J.*, 24:201 (1982).
120. E. Müller and K. Schwabe, *Z. Electrochem.*, 34:170 (1928).
121. D. W. Sokolskii, *Hydrogenation in Solution*, Akademii Nauka Kazakhskoi, Alma Ata (1962).
122. B. Andersson, *Chem. Eng. Sci.*, 37:93 (1982).
123. B. Andersson, *AIChE J.*, 28:333 (1982).
124. Y. T. Shah, *Gas-Liquid-Solid Reactor Design*, McGraw-Hill, New York (1979).
125. R. de Vos, V. Hatziantoniou, and N.-H. Schöön, *Chem. Eng. Sci.*, 37:1719 (1982).
126. V. Hatziantoniou, B. Andersson, T. Larsson, N.-H. Schöön, L. Carlsson, S. Schwarz, and K.-B. Widéen, *Ind. Chem. Eng. Process Des. Dev.*, 25:143 (1986).

25 | Hydrogen and Catalytic Reforming

JEAN-PAUL BOURNONVILLE and JEAN-PIERRE FRANCK

Institut Français du Pétrole, Rueil-Malmaison, France

I.	INTRODUCTION	653
	A. Hydrogen and Catalytic Reforming	654
II.	HYDROGEN AND AGING	656
	A. Hydrogen Pressure and Sintering	656
	B. Hydrogen Pressure and Coke Fouling	658
III.	HYDROGEN AND DEHYDROCYCLIZATION	660
IV.	HYDROGEN AND SELECTIVITY	669
V.	CONCLUSION	674
	REFERENCES	674

I. INTRODUCTION

In the course of catalytic reforming a mixture of paraffins, naphthenes, and aromatics, containing on average 6 to 10 carbon atoms and called naphtha, is transformed into a base for high-octane fuels [1-6]. The reaction is carried out under hydrogen partial pressure in the presence of a bifunctional catalyst having a metallic and an acid function, thus promoting hydro-dehydrogenation reactions as well as molecular rearrangements [1-3]. The metallic function is provided by platinum (between 0.20 and 0.60 wt % of the catalyst), which is today commonly promoted by a few hundred to a few thousand ppm of various additives, such as rhenium, iridium, tin, and germanium. The acid function is provided by a high-specific-area alumina of which the intrinsic acidity is enhanced by 0.8 to 1.3 wt % chlorine [1-6].

The improvement in antiknock properties, measured by the research octane number (RON) of the reformate, is obtained through the production of aromatic hydrocarbons with a RON higher than 100, as illustrated in Fig. 1 in the case of hydrocarbons with seven carbon atoms. Therefore, the most sought-after reactions in catalytic reforming are those which lead to the formation of aromatic hydrocarbons. These are: (a) the dehydrogenation of naphthenes, and (b) the dehydrocyclization of paraffins.

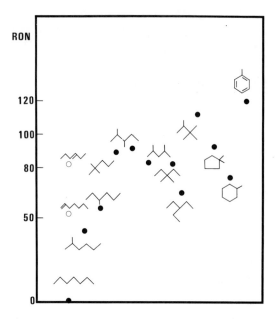

FIGURE 1 Research octane number (RON) for seven-carbon-atom hydrocarbons.

In the absence of cracking side reactions there is, therefore, an increase in the average C/H ratio of the hydrocarbon mixture, and hydrogen is produced in an amount that can be up to 2 to 4 wt % of the feedstock being treated. Most of this hydrogen production is sent to various hydrotreating units in the refinery. But in the meantime, 3 to 6 mol of hydrogen per mole of naphtha being treated is recycled to the catalytic reforming unit to maintain a constant hydrogen partial pressure.

The role of this hydrogen partial pressure is what we deal with in this chapter. We tried to collect industrial or industrial-like data for an understanding from the process point of view. As our purpose was not to make an exhaustive survey of the literature dealing with catalysis by metals, we have considered actual industrial equilibrium catalysts and only made reference to some of the fundamental works. Reactions under laboratory and industrial conditions may not differ much, provided that laboratory experiments are performed with representative catalysts, that is, with the right composition as far as alumina, metals, and chlorine are concerned, but also with highly dispersed metallic phase and at least 2 wt % carbon, under industrial-like operating conditions, particularly with regard to the hydrogen partial pressure.

A. Hydrogen and Catalytic Reforming

In view of the dehydrogenation nature of the reactions to be carried out, it is clear from the thermodynamic point of view that they must not involve an increase in the hydrogen partial pressure [1,2]. On the other hand, a reaction scheme for the dehydrocyclization of paraffins that is the most probable mechanism in industrial conditions which better matches kinetic studies and industrial observations is reported in Fig. 2. Although it is complex and probably involves at least three reaction paths [7-9], nevertheless the final step of the transformation is a dehydrogenation reaction: hence

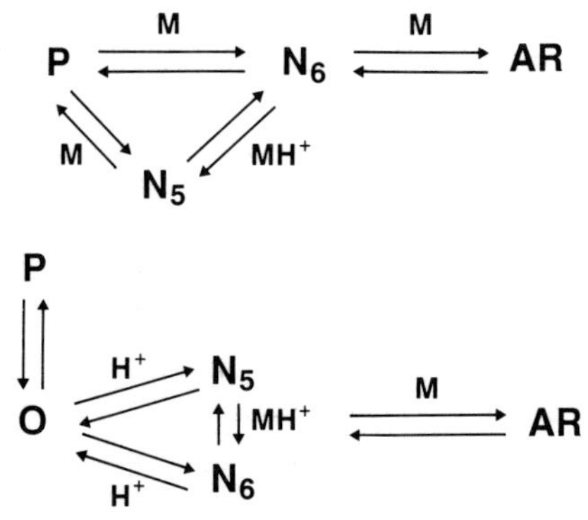

25. Hydrogen and Catalytic Reforming

FIGURE 2 Dehydrocyclization mechanism. (From Refs. 7, 8, and 9.)

the well-known negative reaction order reported for the influence of the pressure [7,10,11]. It does not exclude other possible mechanisms demonstrated in specific cases (see Chapter 17 and references therein). From the industrial point of view, it is common to find relationships such as those shown in Fig. 3 which illustrate that for a given RON, the yield of reformate C_5^+ is higher when the pressure (expressed, oddly enough, as overall pressure, not in terms of hydrogen pressure) is lower [12,13]. Why, therefore, maintain hydrogen pressure between 5 and 15 bar, and often more, during catalytic reforming operations?

In fact, it has already been well established that aging of the catalyst, measured by the temperature required to maintain the RON of the reformate constant, proceeds even more quickly under low-pressure conditions (Fig. 4). The relationships shown in Figs. 3 and 4, drawn up according to the overall pressure, generally for a constant hydrogen-to-naphtha ratio (H2/HC), help to illustrate the influence of the ratio between hydrogen and hydrocarbon pressures. Moreover, it has been shown that in the absence of hydrogen, the rate of dehydrocyclization could be nil after a given working time [15].

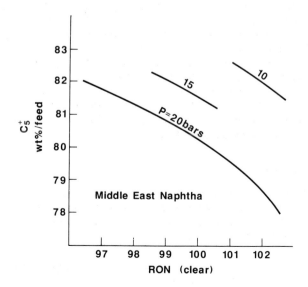

FIGURE 3 Effect of pressure on reformate yield.

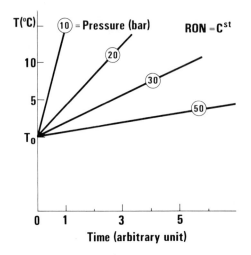

FIGURE 4 Typical influence of pressure on stability. (From Ref. 14.)

First we examine the effect of the pressure and, more precisely, the effect of the hydrogen pressure on the aging of the catalytic reforming catalysts. Then we deal with its influence on the dehydrocyclization of paraffins and the selectivity of the transformation.

II. HYDROGEN AND AGING

While in use, all reforming catalysts, however efficient they may be, age—in other words, lose their activity. This aging is reflected by a drop in the RON of the reformate and by a decrease in the production of hydrogen. To maintain a constant reformate octane number, the average operating temperature is gradually increased.

Unfortunately, this remedy is not without its drawbacks. On the one hand, there is a limit to the temperature the plant can withstand, and on the other hand, this increase in the operating temperature generally entails an increase in the rate of the cracking side reactions (hydrocracking and hydrogenolysis), and thus an increase in the production of gases from methane to butane, which are generally less valuable than reformate and hydrogen.

The acknowledged causes for this aging are carbon deposits on the catalyst as well as sintering of the metallic phase. To these one might also add accidental poisoning by various impurities carried by the feed. As the aging of reforming catalysts has, on the whole, been largely dealt with elsewhere [1-6,13,16-18], we will limit ourselves to illustrating the influence of hydrogen pressure on aging through sintering and coke fouling.

A. Hydrogen Pressure and Sintering

Deactivation of reforming catalysts by sintering or recrystallization of the metallic phase is a long recognized phenomenon [19, 20]. In 1968, Somorjai [21] examined what was known about platinum. A survey has been done more recently [22], which can be outlined as follows:

1. The recrystallization rate depends on the temperature, the carrier, the metal concentration, and the atmosphere of treatment.

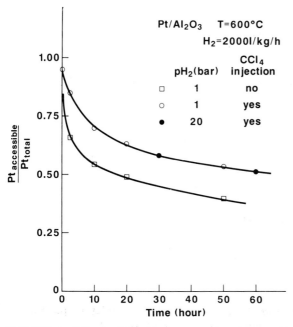

FIGURE 5 Effect of hydrogen pressure and chlorine addition on platinum sintering. (From Ref. 22.)

2. Two sintering mechanisms can be distinguished involving either particle migration or the removal of atoms or clusters from particles which after diffusion in the gaseous phase or on the solid settle again on another crystallite [23,24] (see also Chapter 10). Our own work in this field has allowed us to determine, among other results, the action of chlorine as well as the influence of hydrogen pressure [22,25].

In a reducing atmosphere, the results shown in Fig. 5 demonstrate that the presence of chlorine inhibits platinum sintering and that the hydrogen pressure neither inhibits nor increases the rate of sintering. These results can be interpreted by an increase in crystallite adhesion to the surface of the alumina or by greater difficulty in the removal of an atom or an aggregate from a crystallite and moving it elsewhere.

One can say that interaction between the platinum moiety and the support is enhanced by increasing the acidity and so reducing its mobility on the surface, or that the withdrawing of electrons decreases the volatility of the surface atoms and prevents their escaping from the crystallite. This increase in interaction between the metal and the support in the presence of chlorine can be highlighted by means of infrared spectroscopy study of the carbon monoxide chemisorbed onto the metal [26]. The greater resistance of platinum impregnated on acid carriers to hydrogen sulfide poisoning is also an indication of the change in the interaction between the metal and the support [27-29]. Similarly, the results in Fig. 6 show that the rate of toluene hydrogenation with platinum impregnated on chlorinated alumina is lower in the presence of hydrogen sulfide when the acidity of the catalyst is neutralized [30].

Therefore, it is necessary to maintain the chlorine content and then the acidity of a reforming catalyst when on stream in order to slow down the sintering of the metallic phase but also to minimize poisoning by hydrogen

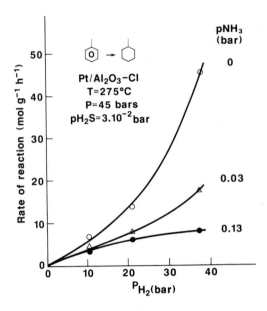

FIGURE 6 Effect of acidity poisoning on hydrogenation activity of platinum on chlorinated alumina in the presence of H_2S. (From Ref. 20.)

sulfide and ammonia (Fig. 6), especially as the hydrogen pressure is lower. This increase in interaction between metal and support may also be enhanced through the modification of platinum by various additives, such as iridium, rhenium, germanium, tin, lead, indium, ruthenium, and rhodium [22]. This is shown on Fig. 7 for platinum modified by indium, rhenium, and tin.

From the industrial point of view, this means that for a given catalytic reforming catalyst, the rate of sintering depends only on the temperature and the chlorine content of the catalyst, not on the operating pressure. So we can assume that the influence of pressure as reported on Fig. 4 deals mainly with the inhibition by hydrogen of aging by coke fouling rather than by sintering.

B. Hydrogen Pressure and Coke Fouling

The positive influence of hydrogen pressure on the stability of catalytic reforming catalysts has been also recognized for a long time [31]. Some authors claim that hydrogen intervenes with a negative apparent order superior or equal to 2 on the rate of aging through coke fouling [32].

To illustrate the influence of hydrogen pressure, most of the data generally appear in relation to the recycle ratio (i.e., the hydrogen-to-hydrocarbon molar ratio at the inlet of catalytic reforming reactors). In this respect, Figure 8 shows the deactivation rate expressed in the number of Celsius degrees by which the operating temperature has to be increased daily in order to maintain constant the RON of the reformate, according to the recycle ratio.

The results presented in Fig. 9 demonstrate that the rate of coke deposit is roughly divided by 7 when the hydrogen-to-hydrocarbon ratio is increased from 4 to 8. In both sets of results reported in Figs. 8 and 9, the effect is extremely important below $H_2/HC = 10$; the fact that total pressure is quite low in both cases helps explain this phenomenon. Due to the almost continuous improvement in catalytic reforming catalysts, today most reforming units are

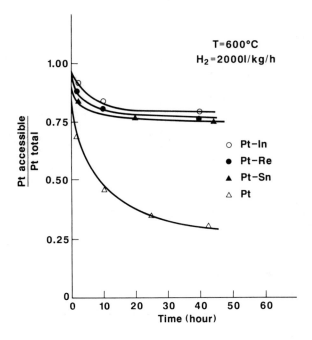

FIGURE 7 Effect of various additives on platinum sintering. (From Ref. 22.)

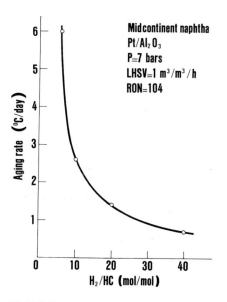

FIGURE 8 Catalyst aging versus hydrogen-to-hydrocarbon ratio. (From Ref. 31.)

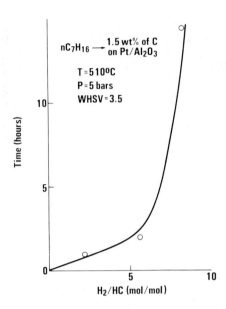

FIGURE 9 Rate of coke deposit versus hydrogen-to-hydrocarbon ratio. (From Ref. 33.)

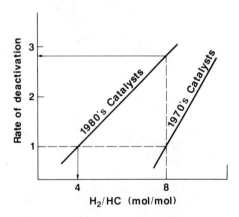

FIGURE 10 Relative rate of deactivation versus hydrogen-to-hydrocarbon ratio. (From Ref. 34.)

operated with hydrogen-to-hydrocarbon ratios of around 4, whereas they were operated at around 8 about 10 years ago (Fig. 10). Regarding the structure of the coke formed on reforming catalysts, it has been established by means of laser Raman spectroscopy that an increase in the H_2/HC ratio has an effect similar to the addition of iridium or rhenium to platinum [35] (i.e., an increase in the hydrogenation function of the catalyst) [13,22,36,37].

III. HYDROGEN AND DEHYDROCYCLIZATION

In view of the reaction scheme for the dehydrocyclization of paraffins in Fig. 2, one can only expect that the effect of hydrogen pressure on the rate

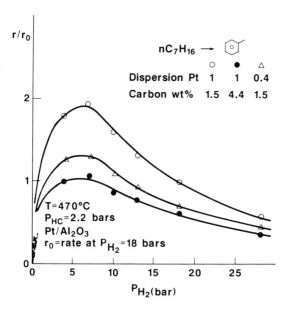

FIGURE 11 Rate of dehydrocyclization versus hydrogen pressure. (From Ref. 10.)

of reaction, should exist, will be complex. The results presented in Fig. 11 illustrate that the relationship between the n-heptane dehydrocyclization rate and the hydrogen pressure is actually complex whatever the stage of aging of the catalyst might be. When the hydrogen pressure decreases at constant hydrocarbon pressure, the rate of dehydrocyclization increases up to a maximum and then tends toward zero for low hydrogen pressures.

The existence of a maximum for the rate of dehydrocyclization versus hydrogen pressure is interpreted according to a double effect of hydrogen [7]:

1. At low pressures, hydrogen acts by limiting the concentration of highly dehydrogenated species on the catalyst surface, whence the formation of coke.
2. At high pressures, through adsorption, hydrogen reduces the number of vacant sites that are needed for cyclization as well as the concentration of dehydrogenated intermediaries necessary for cyclization.

The results reported in Fig. 12 show that when the n-heptane partial pressure varies, the shape of the curve giving the rate of dehydrocyclization versus hydrogen pressure does not change. One should note, however, that the increase in n-heptane pressure leads to a shift in the maximum rate toward higher hydrogen pressures as well as to a reduction in the value of the highest rate, thus illustrating the competition between hydrogen and hydrocarbons on the active sites of the catalyst. The same is observed under laboratory conditions (see Chapter 15). Tests carried out at atmospheric pressure with a hydrogen pressure of between 0.10 and 0.95 bar enabled us to demonstrate that even under these conditions, when the catalyst was very unstable, the hydrogen had a positive effect on the activity [10].

It seems, therefore, that the dehydrocyclization rate tends toward zero together with the hydrogen pressure. From the results shown in Fig. 12, it is possible to draw the curves giving the dehydrocyclization rate versus

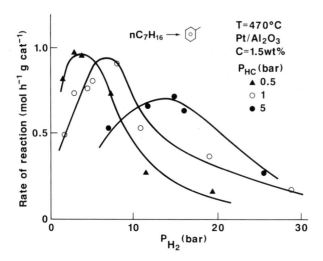

FIGURE 12 Rate of dehydrocyclization versus hydrogen pressure for various hydrocarbon pressures. (From Ref. 10.)

hydrocarbon pressure for different hydrogen pressures (Fig. 13). The curves obtained are once again with a maximum; they have been extrapolated to zero, as it is obvious that there can be no reaction when the hydrocarbon pressure is nil. As for hydrogen pressure, the apparent order of reaction for hydrocarbon pressure can be positive, negative, or nil, depending on the relative values of hydrogen and hydrocarbon pressures.

From the point of view of formal kinetics it is possible to account for these results by writing the overall rate of reaction according to the generally acknowledged reaction scheme for dehydrocyclization [7-10]. According to this scheme, the overall rate of reaction is equal to the sum of the rates of the three independent reaction paths. These three paths are mainly different from the point of view of the cyclization step and they are generally termed the metallic monofunctional path (MFM), the metallic bifunctional path (BFM), and the acid bifunctional path (BFA).

The metallic monofunctional path takes place on the metallic phase only and we believe that it involves a cyclohexanic intermediate which is very rapidly dehydrogenated into aromatic. Some claim, however, that the paraffin should dehydrogenate to triolefin before cyclization (see Chapter 17). The metallic bifunctional path involves a cyclopentanic intermediate formed on the metal; this intermediate undergoes ring expansion on the acid function before dehydrogenating. The acid bifunctional path is supposed to take place from olefins or diolefins produced by the dehydrogenation of paraffins.

For all these paths, cyclization is the probable rate determining step. The equilibrium between adsorption and desorption is very fast and the dehydrogenation reactions on the surface of the catalyst are also very fast. Moreover, bifunctional isomerization, which leads to ring expansion in the metallic bifunctional path, is supposed to be faster than cyclization on metal; this is actually correct for low metal loading. The hypothesis put forth to account for the effect of hydrogen pressure [7,10] can also be expected to account for the shape of the curve giving the rate of reaction versus the hydrogen pressure (Fig. 11).

Figure 14 shows the possible reaction scheme for dehydrocyclization with cyclization on metal (MFM + BFM). The corresponding rate equation can be derived from dehydrogenation of paraffin P, which requires two vacant sites

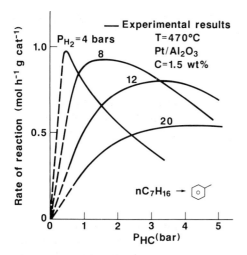

FIGURE 13 Rate of dehydrocyclization versus hydrocarbon pressure for various hydrogen pressures. (From Ref. 10.)

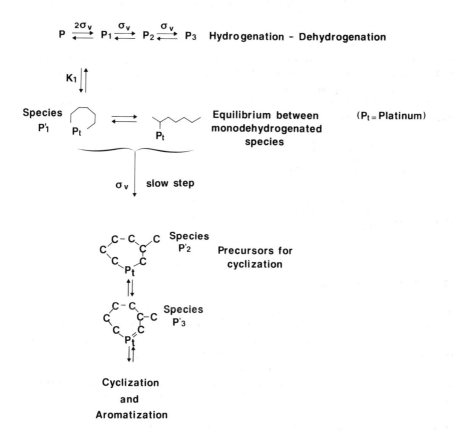

FIGURE 14 Mechanism for paraffin dehydrocyclization on metal. (From Ref. 11.)

σ_V to adsorb the first dehydrogenated intermediate P_1 and the hydrogen atom produced [11]. This first dehydrogenation step can be represented as

$$P + 2\sigma_V \underset{}{\overset{K_1}{\rightleftarrows}} \sigma_{P_1} + \sigma_H \quad (1)$$

The intermediate P_1 is then dehydrogenated into P_2 adsorbed onto the same metal atom as P_1. This step requires a new vacant site to adsorb the hydrogen atom produced and dehydrogenation continues in the same way according to

$$\sigma_{P_i} + \sigma_V \underset{}{\overset{k_{i+1}}{\rightleftarrows}} \sigma_{P_{i+1}} + \sigma_H$$

In the mechanism reported in Fig. 14 it is considered that among the P_1 dehydrogenated species in equilibrium which are also in equilibrium with the other P_2, P_3, P_i species, only a P_1' intermediate is converted into a P_2' cyclization precursor. This step is more difficult than the transformation of P_2' into P_3' which is then cyclized. Taking into account the dehydrogenated intermediates up to P_3, the following rate equation is obtained:

$$V = kK_1'K_1 \frac{P_P(b_H P_{H_2})^{5/2}}{\{(b_H P_{H_2})^{3/2} + (b_H P_{H_2})^2 + P_P K_1[(b_H P_{H_2}) + K_2(b_H P_H)^{1/2} + K_2 K_3]\}^2} \quad (2)$$

The calculation of the abscissa P_{HM} and the ordinate V_M corresponding to the maximum rate as well as their study toward the limits showed that such an expression correctly matches the experimental results. There is, therefore, a set of hydrogen partial pressure (P_{HM}) and hydrocarbon partial pressure (P_{PM}) for which a maximum rate (V_M) is obtained.

$$P_{PM} = \frac{(b_H P_{HM})^{3/2} + 3(b_H P_{HM})^2}{K_1[(b_H P_{HM}) + 3K_2(b_H P_{HM})^{1/2} + 5K_2 K_3]} \quad (3)$$

$$V_M = kK_1' \frac{[(b_H P_{HM}) + 3(b_H P_{HM})^{3/2}][(b_H P_{HM}) + 3K_2(b_H P_{HM})^{1/2} - 5K_2 K_3]}{[(2 + 3K_2)(b_H P_{HM}) + (4K_2 + 8K_2 K_3)(b_H P_{HM})^{1/2} + 6K_2 K_3 + 4(b_H P_{HM})^{3/2}]^2} \quad (4)$$

In Eqs. (1) to (4) we have

P_{H_2}, P_P = hydrogen and paraffin partial pressures
P_1, P_2, P_3, . . . , P_i = dehydrogenated species chemisorbed on the metal and having lost 1, 2, 3, . . . , i hydrogen atoms
σ_{P_1}, σ_{P_2}, σ_{P_i} = fraction of the metallic sites occupied by the corresponding intermediates
σ_H = fraction of the metallic sites occupied by hydrogen
σ_V = vacant fraction of the metallic sites

So it appears that the overall rate of reaction and the rate for reaction paths involving cyclization on metal respond in a similar way to the hydrogen partial pressure, which means that for cyclization on acid (BFA) a reaction

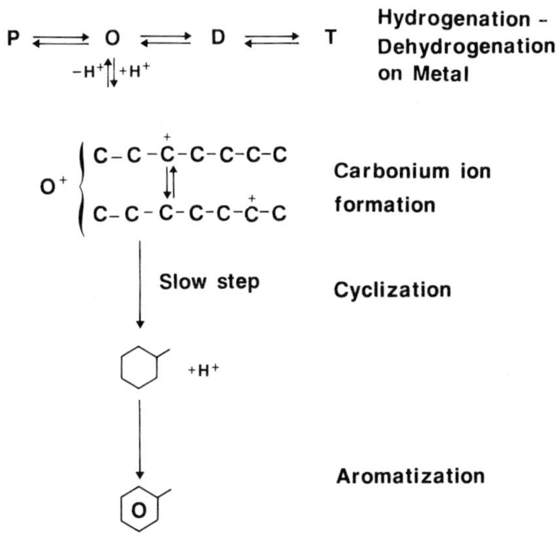

```
P ⇌ O ⇌ D ⇌ T      Hydrogenation –
   -H⁺↑↓+H⁺              Dehydrogenation
                         on Metal

       ⎧ C-C-Ċ⁺-C-C-C-C
  O⁺  ⎨        ↕           Carbonium ion
       ⎩ C-C-C-C-C-Ċ⁺-C     formation

       │ Slow step         Cyclization
       ↓
       ⌬  +H⁺

       ↓                   Aromatization
       ⌬
```

FIGURE 15 Mechanism for paraffin dehydrocyclization on acid. (From Ref. 11.)

scheme symmetrical to that for cyclization on metal can be admitted [11]. Such a scheme is shown in Fig. 15. According to this scheme, paraffin is dehydrogenated on the metal into mono-olefins (O), diolefins (D), and triolefins (T), together in equilibrium with the paraffin in the vapor phase.

The unsaturated hydrocarbons react with the protons H^+ of the support and give adsorbed ions (O^+), (D^+), and (T^+) on the acid sites which remain in equilibrium with the corresponding compounds in the vapor phase. Among the carbonium ions (O^+) only a few ($O^{+'}$) are likely to be cyclized. This cyclization, which is the final step of the transformation, is achieved by means of an electrophilic reaction on a saturated carbon followed by the removal of a proton. The dehydrogenation of the naphthene is then carried out on platinum. One can compare the rate equation with that observed for the cyclization on the metallic phase only as long as an acid site takes part in the mechanism (i.e., nucleophilic assistance during the ring closure).

As already mentioned, the aging of the catalyst does not change the overall effect of the hydrogen partial pressure and so has no influence on the competition between hydrogen and hydrocarbons. Moreover, an experiment carried out using a series of normal paraffins [38] enabled us to establish that the competition between hydrogen and paraffins was independent of the molecular weight of the molecule under consideration (Fig. 16). One might therefore conclude that the three reaction paths mentioned in the literature about the dehydrocyclization of paraffins, among which two involve cyclization on the metallic phase, coexist on a coked bifunctional catalytic reforming catalyst. Moreover, these three reaction paths respond in a similar way to the variations in hydrogen partial pressure regardless of the molecular weight of the reagent. Only a change in the metallic phase carried out by means other than coking or sintering can alter the competition between hydrogen and hydrocarbons.

Figure 17 shows the results obtained at 470°C, using various bimetallic catalysts with 0.6% platinum by weight, a second metal-to-platinum atomic ratio of 1/3, and containing 3% coke by weight, when the hydrogen pressure varies from 2 to 28 bar at constant n-heptane partial pressure (2.2 bar). One may note the following:

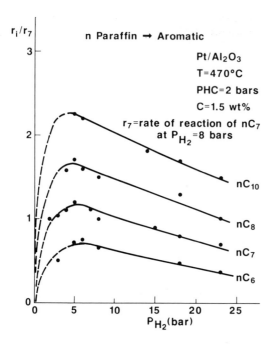

FIGURE 16 Rate of dehydrocyclization versus hydrogen pressure and paraffin molecular weight. (From Ref. 38.)

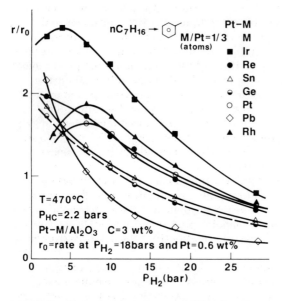

FIGURE 17 Rate of dehydrocyclization for monometallic and bimetallic catalysts. (From Ref. 11.)

1. Compared to monometallic catalysts, bimetallic catalysts are more or less active depending on the operating conditions.
2. The shape of the curves giving the dehydrocyclization rate versus the hydrogen partial pressure is virtually the same for all the catalysts but with a shift of the maximum rate toward low pressures. (The maximum rate is obtained at p_{H_2} = 5 bar for Pt-Ir, about 2 bar for Pt-Re, and lower than 1 bar for Pt-Sn, Pt-Ge, and Pt-Pb instead of 7 bar for platinum alone and Pt-Rh.)

In view of the presence of the second metal, these results might be explained in the following ways: (a) through a change in the amount of coke deposited on the metallic phase, as has been demonstrated by Barbier et al. in the case of model catalysts [39], or (b) through a change in the adsorption properties of the platinum, which is a hypothesis all the more likely since the mere addition of a second metal leads to a shift in the maximum dehydrocyclization rate.

Modification of platinum, rhodium, and nickel by tin with respect to hydrogen adsorption has been clearly demonstrated [40-44]. The temperature-programmed desorption (TPD) of hydrogen demonstrates that two types of adsorbed hydrogen exist in the presence of tin, of which one is in strong interaction with the metallic phase. One might say that the presence of tin creates a storage of activated hydrogen in strong interaction with the metallic phase.

After dissociation on the metal, the activated hydrogen can also, through spillover, be stabilized into the surface of oxides [45,46]. Then these activated spillt-over species can compete with hydrocarbons on catalytic sites as well as react with coke precursors more efficiently than can molecular hydrogen [47-49]. This phenomenon, already demonstrated with platinum, has also been observed with supported iridium [50]. After dissociation on the metal, hydrogen spillover is favored by the presence of chlorine [51,52], water [53], or oxides such as tin oxide [47]. Therefore, a catalytic reforming catalyst contains all the components required to promote hydrogen spillover and a cleaning effect. All of these results dealing with the effect of hydrogen partial pressure on the paraffin dehydrocyclization rate highlight the major role of hydrogen activation by means of the metallic phase and how wrong one can be to think only in terms of total pressure.

After examining Figs. 12 and 13 it becomes clear that the same rate of reaction can be achieved under very different operating conditions, particularly as far as the stability of the catalyst is concerned. What is more, the structure sensitivity generally reported for paraffin dehydrocyclization on clean surfaces [54] disappears on coked and stabilized catalysts, on which the only phenomenon is competitive chemisorption and hydrogen spillover. Thus the same rate of n-heptane dehydrocyclization is achieved on two Pt/Al_2O_3 catalysts of similar metallic area as measured by H_2/O_2 titration [55] but with very different dispersions (40% and 100%) [1]. When using as a reagent, methylcyclopentane which is supposed to be an intermediate during dehydrocyclization (Fig. 2), taking the place of n-paraffin, shows a maximum dehydrocyclization rate on the basis of hydrogen partial pressure.

One can see from Fig. 18 that contrary to what was noted when using n-heptane (Fig. 17), the hydrogen pressure corresponding to the maximum rate does not depend on the catalyst formula; furthermore, the deactivation is very fast as soon as the hydrogen partial pressure is lower than 5 bar [56]. These results can, nevertheless, be interpreted without contradicting those presented previously.

In the transformation of methylcyclopentane the rate-limiting step no longer occurs on the metallic phase but rather, on the acid function of the catalyst

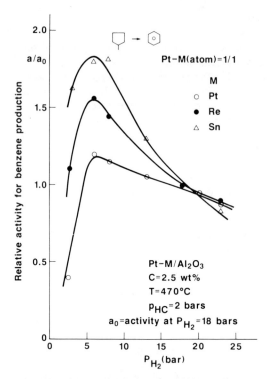

FIGURE 18 Effect of hydrogen pressure on methylcyclopentane aromatization. (From Ref. 56.)

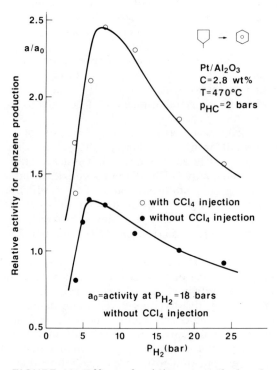

FIGURE 19 Effect of acidity on methylcyclopentane aromatization.

(i.e., ring expansion). Thus the rate of reaction no longer depends on hydrogen-hydrocarbon competition, and only the overall activity of the catalyst depends on the metallic phase. By enhancing the acidity of catalysts through injecting carbon tetrachloride [57], the results obtained confirm that the isomerization of methylcyclopentane into cyclohexane is probably the rate-determining step of the reaction (Fig. 19). Therefore, the chlorine content of a catalytic reforming catalyst when on-stream affects not only the stability but also the activity and selectivity. One must finally note that in terms of the absolute rate of reaction, the appearance of an aromatic is faster when using methylcyclopentane than when using n-heptane under similar operating conditions [10, 56].

As far as deactivation is concerned, when the hydrogen partial pressure is lower than 5 bar, one might assume that it is due to the presence of very efficient cyclopentadienic-type coke precursors [31, 58]. In the same way, the relative activities of the various metallic phases studied can be interpreted on the basis of a different toxicity and the amount of covering of the metallic phases by the same cyclopentadienic coke precursors [36].

IV. HYDROGEN AND SELECTIVITY

Together with the reactions promoted to improve antiknock properties as well as to ensure maximum production of reformate, side-cracking reactions cannot be avoided. These cracking reactions are generally called hydrocracking and hydrogenolysis.

The hydrocracking reactions involve the two functions of the catalyst, metal and acid, and lead primarily to the formation of liquefied petroleum gas. As for the hydrogenolysis reactions, they involve only the metallic function of the catalyst and lead essentially to the formation of fuel gas. These two reactions are hydrogen consuming, but they respond differently to variations in hydrogen partial pressure.

Whereas as far as hydrogen pressure is concerned, the apparent reaction order is always positive in the case of hydrocracking (2), it can be either negative or positive in the case of hydrogenolysis, depending on the range of hydrogen pressure under consideration [4, 59, 60].

The results shown in Fig. 20 for n-hexane hydrogenolysis on iridium-impregnated silica demonstrate that just as in the case of dehydrocyclization, the relationship between the rate of reaction and the reagent's partial pressure is far from simple. When the hydrogen partial pressure increases from zero at a given hydrocarbon partial pressure, the rate of hydrogenolysis increases, reaches a maximum, and then decreases toward zero at high hydrogen pressures.

When the hydrocarbon partial pressure increases, the value corresponding to the maximum specific activity decreases and the maximum occurs at increasingly higher hydrogen pressures, demonstrating the competition between the hydrogen and the hydrocarbon in the same way as for dehycrocyclization (Fig. 12).

From the results shown in Fig. 20 it is possible to obtain, as in Fig. 21, evolution of the specific activity versus the n-hexane pressure at constant hydrogen pressure. Once again volcano curves are obtained whose maxima depend on the hydrogen pressure, exactly as in the case of dehydrocyclization. This complex relationship between hydrogen and hydrocarbon is in agreement with the reaction scheme generally acknowledged for the hydrogenolysis. The paraffin is adsorbed onto the catalyst while dehydrogenation proceeds simultaneously and then continues to lose hydrogen atoms, leading to a dehydrogenated chemisorbed entity C_6H_x that then undergoes carbon-carbon breaking, which is the slowest step of the reaction [59, 60].

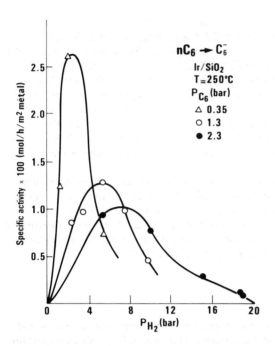

FIGURE 20 Hydrogenolysis specific activity versus hydrogen pressure. (From Ref. 60.)

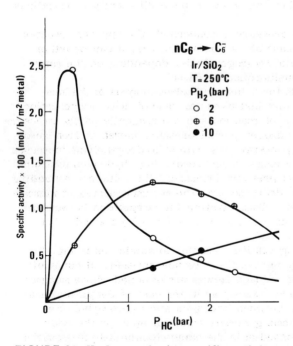

FIGURE 21 Hydrogenolysis specific activity versus hydrocarbon pressure. (From Ref. 60.)

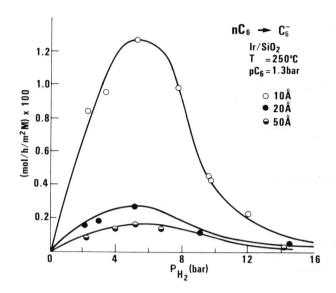

FIGURE 22 Influence of crystallite size on hydrogenolysis. (From Ref. 64.)

At first the hydrogen enhances the rate of reaction, and one might think that it takes part in the carbon-carbon bond breakage; but when it starts to compete on the surface of the catalyst with the dehydrogenated intermediates, of which the stationary concentration diminishes as the hydrogen pressure increases, its effect is inverted.

The hydrocarbon effect is similar: When the pressure of the n-hexane increases, the stationary concentration of the reactive species C_6H_x increases. At the same time the rate of hydrogenolysis increases, then decreases when intermediates other than C_6H_x, which are more or less dehydrogenated, become exceedingly numerous on the surface of the catalyst.

As far as the dispersion of the metallic phase is concerned, the paraffin hydrogenolysis has often been described as a demanding reaction, in other words, sensitive to the structure of the catalyst, but this is generally a result of experiments carried out at atmospheric pressure and therefore far from industrial operating conditions [61-63]. The results in Fig. 22 show that, in fact, the demanding character of the reaction is noticeably reduced when the hydrogen partial pressure increases. The reaction is most sensitive to the structure around the maximum rate, especially in the case of small crystallites (1 to 2 nm). It therefore seems that the demanding nature of the hydrogenolysis as it is generally defined depends on the operating conditions under which the reaction is studied and apparently does not occur in catalytic reforming conditions.

It was demonstrated in the same study [65] that the selectivity of the hydrogenolysis does not change with hydrogen partial pressure. For example, with 1-nm crystallites, the molecule is cut mainly in the middle, whereas it is generally cut at the ends with 5-nm crystallites, whatever the hydrogen pressure might be. Although the overall reaction scheme is probably the same for both sizes of crystallite, a hypothesis that is borne out by the similarity of the kinetic curves, the absence of variation of the selectivity for a given size of crystallite when the hydrogen pressure varies can only be explained by a change in the chemisorption process or in the nature of the reactive intermediate in relation to the size of the crystallites [64].

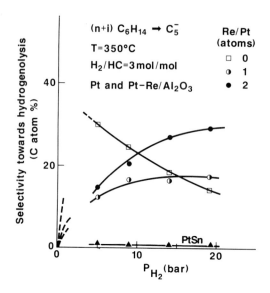

FIGURE 23 C_6 paraffins hydrogenolysis with various bimetallic catalysts.

The results shown in Fig. 23 illustrate once again the possibility of modifying the properties of platinum by means of various additives and the consequences of hydrogen activation [66]. Whereas optimum selectivity in hydrogenolysis is obtained at a hydrogen pressure of 0 to 5 bar with a monometallic Pt/Al_2O_3 catalyst, no maximum is actually observed when using a bimetallic $PtRe/Al_2O_3$ catalyst at between 0 and 20 bar.

Moreover, a $PtRe/Al_2O_3$ catalyst is more active for hydrogenolysis than a Pt/Al_2O_3 catalyst when hydrogen pressure is around 20 bar, whereas the inverse occurs at hydrogen pressures near 10 bar. The conclusions dealing with the selectivities of these two types of formula will depend, therefore, on the operating conditions under investigation.

One can also see from Fig. 23 that the hydrogenolysis activity of $PtRe/Al_2O_3$ catalysts increases noticeably when the atomic Re/Pt ratio changes from 1, as it is for most PtRe industrial reforming catalysts [4], to 2. This result is in agreement with the generally reported optimum reformate yield, while the atomic Re/Pt ratio is equal to 1 [67,68]. One must note the low hydrogenolysis activity of PtSn catalysts, hence the best selectivity, as reported in technical and scientific literature, for these types of bimetallic catalysts [69-71].

To illustrate better the relationship between the metallic phase and hydrogen pressure, one may, as in Fig. 24, compare the rates of aromatization and ring opening (a reaction very similar to hydrogenolysis) versus the hydrogen partial pressure for various catalyst formulas when studying the transformation of methylcyclopentane [57]. As before, we obtain volcano curves. At high hydrogen pressure, the Pt, PtRe, and PtSn catalysts have roughly the same selectivity. When the hydrogen pressure decreases, the Pt and PtRe catalysts behave in the same way, whereas the PtSn formula is far more selective, both intrinsically and when compared to the other two types of catalysts.

Whatever the effects put forward, be they geometric or electronic [72-74], to explain the modifications of the hydrogenolyzing properties of platinum through the addition of rhenium, iridium, tin, and so on, one must take into account the hydrogen chemisorption onto these metallic phases. From the

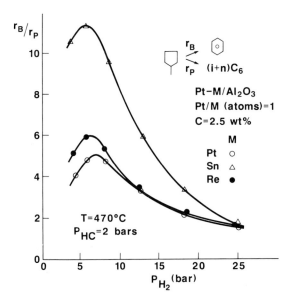

FIGURE 24 Influence of hydrogen pressure on selectivity during methylcyclopentane aromatization.

industrial point of view, these results enable us to understand why Pt and PtRe are said to give the same selectivity when the hydrogen partial pressure is in the range of 10 to 20 bar [5,75], whereas PtSn catalysts are reported to be far more selective at very low hydrogen pressures and generally with continuous regeneration of the catalyst [5,76].

Instead of examining selectivity under fixed operating conditions, another approach consists of examining the evolution of selectivity during operation while the temperature is gradually increased to keep constant the RON of the reformate, of which the yield generally decreases. As, on the one hand, the different reactions involved do not have the same apparent activation energy [2] and, on the other hand, it may depend on the hydrogen pressure [10] and the catalyst formula [1], a study has been undertaken to tackle this evolution of selectivity with temperature and hydrogen pressure [77].

For this purpose, the apparent activation energy for dehydrocyclization of n-heptane as well as that for the overall cracking reactions have been determined for different metallic formulas according to the hydrogen partial pressure. Figure 25 shows the ratio of the apparent activation energy for dehydrocyclization (E_{a1}) to the apparent activation energy for overall cracking reactions (E_{a2}) versus hydrogen partial pressure for different types of catalysts. One may note, first, that this ratio is always lower than 1 in the case of the monometallic catalyst, whereas it is always higher than 1 for the PtSn catalyst. In other words, whatever the hydrogen pressure may be, it is impossible to obtain a constant yield of reformate with a Pt/Al_2O_3 catalyst when the catalyst deactivates, but it is possible with a $PtSn/Al_2O_3$ catalyst.

As far as platinum-iridium and platinum-rhenium are concerned, the reformate yield can be constant with aging of the catalyst if the hydrogen pressure is higher than 10 bar in the case of platinum-iridium and 5 bar in the case of platinum-rhenium. In the course of industrial runs an almost constant production of reformate and hydrogen is actually observed with bimetallic platinum-rhenium catalysts [5,75].

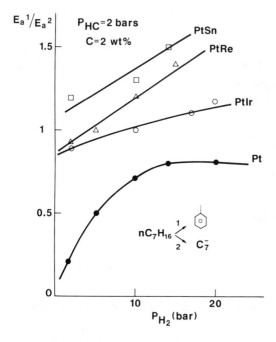

FIGURE 25 Influence of hydrogen pressure on the selectivity maintenance for various bimetallic catalysts. (From Ref. 77.)

V. CONCLUSION

It is clear that in catalytic reforming, hydrogen does not act only as a cleaning agent of the metallic phase in order to prevent deactivation through coking. By means of the competition between the hydrogen and hydrocarbons it is possible to find the best operating conditions for each metallic phase. From the scientific point of view, the activation of hydrogen, or the hydrogen spillover, or even the competition between hydrogen and hydrocarbons, are evidently powerful tools to characterize the metallic phases. An understanding of the mechanism by which metallic phases are modified, especially the platinum-based ones, will certainly be an efficient guide to further progress in the field of catalysis by metals.

REFERENCES

1. J. P. Franck, in *Catalyse par les métaux* (B. Imelik, G. A. Martin, and A. J. Renouprez, eds.), CNRS, Paris (1984), p. 401.
2. G. Martino, in *Catalyse de Contact* (J. F. Le Page, ed.), Technip, Paris (1978), p. 575.
3. B. C. Gates, J. R. Katzer, and G. C. A. Schuit, *Chemistry of Catalytic Processes*, McGraw-Hill, New York (1979), p. 184.
4. J. H. Sinfelt, in *Catalysis Science and Technology* (J. R. Anderson and M. Boudart, eds.), Springer-Verlag, New York (1981), p. 257.
5. D. M. Little, *Catalytic Reforming*, Penn Well Books, Tulsa, Okla. (1985).
6. M. D. Edgar, in *Applied Industrial Catalysis*, Vol. 1 (B. E. Leach, ed.), Academic Press, New York (1983), p. 123.

7. J. C. Rohrer, H. Hurwitz, and J. H. Sinfelt, *J. Phys. Chem.*, 65:1458 (1961).
8. B. A. Kazansky and A. L. Liberman, *Proceedings of the 5th World Petroleum Congress, New York*, Sec. 4, Paper 3 (1959), p. 29.
9. W. L. Callender, S. G. Brandenberger, and W. K. Meerbott, *Proceedings of the 5th International Congress on Catalysis, Miami, 1972*, Vol. 2, North-Holland, Amsterdam (1973), p. 1265.
10. C. Alvarez-Herrera, thesis, Poitiers (1977).
11. G. Abolhamd, thesis, Paris (1980).
12. J. P. Franck and M. Berthelin, unpublished results.
13. W. L. Nelson, *Oil Gas J.* (Aug. 2, 1971), p. 76.
14. J. P. Franck and G. Martino, in *Deactivation and Poisoning of Catalyst* (J. Oudar and H. Wise, eds.), Marcel Dekker, New York (1985), p. 205.
15. R. K. Herz, E. E. Petersen, W. Gillespie, and G. A. Somorjai, *72nd AICHE Annual Meeting, San Francisco*, Paper A1 (1979).
16. F. G. Ciapetta, R. M. Dobres, and R. W. Baker, in *Catalysis*, Vol. 6 (P. H. Emmett, ed.), Reinhold, New York (1958), p. 495.
17. F. G. Ciapetta and D. N. Wallace, in *Catalysis Reviews*, Vol. 5 (H. Heineman, ed.), Marcel Dekker, New York (1972), p. 67.
18. W. N. N. Knight and M. L. Peniston-Bird, in *Modern Petroleum Technology* (G. D. Hobson, ed.), Applied Science Publishers, Barking, Essex, England (1973), p. 327.
19. G. A. Mills, S. Weller, and E. B. Cornelius, *Proceedings of the 2nd International Congress on Catalysis, Paris 1960*, Vol. 2, Technip, 2, Paris (1961), p. 2221.
20. R. A. Herman, S. F. Adler, M. S. Goldstein, and R. M. Debaun, *J. Phys. Chem.*, 65:2189 (1961).
21. G. A. Somorjai, in *X-Ray and Electron Methods of Analysis* (H. Van Ophen and W. Porrich, eds.), Plenum Press, New York (1968), Chap. 6.
22. J. P. Bournonville, thesis, Paris (1979).
23. E. Ruckenstein and B. Pulvermacher, *J. Catal.*, 29:224 (1973); 35:115 (1974); 37:416 (1975).
24. P. C. Flynn and S. E. Wanke, in *Catalysis Reviews* (H. Heinemann, ed.) Marcel Dekker, New York (1975), p. 93 and references therein.
25. J. P. Bournonville and G. Martino, in *Catalyst Deactivation* (B. Delmon and G. F. Froment, eds.), Elsevier, Amsterdam (1980), p. 159.
26. M. Primet, J. M. Basset, M. V. Matthieu, and M. Pretre, *J. Catal.*, 29:213 (1973).
27. R. A. Della Betta and M. Boudart, *Proceedings of the 5th International Congress on Catalysis, Miami, 1972*, Vol. 2, North-Holland, Amsterdam (1973), p. 1239.
28. F. Figueras, R. Gomez, and M. Primet, *Adv. Chem. Ser.*, 121:480 (1973).
29. P. Gallezot, J. Datka, J. Massardier, M. Primet, and B. Imelik, *Proceedings of the 6th International Congress on Catalysis, London, 1976*, Vol. 2, Chemical Society, London (1977), 696.
30. J. B. Marin Gil, thesis, Poitiers (1978).
31. C. G. Myers, W. H. Lang, and P. B. Weisz, *Ind. Eng. Chem.*, 53:299 (1961).
32. *Oil Gas J.* (April 9, 1973), p. 88.
33. J. P. Bournonville and J. P. Franck, unpublished results.
34. J. P. Franck, Institut Français du Pétrole, industrial results.
35. D. Espinat, H. Dexpert, E. Freund, and G. Martino, *Appl. Catal.*, 16:343 (1985).
36. D. Espinat, thesis, Rueil-Malmaison (1982).
37. J. P. Bournonville, J. P. Franck, and G. Martino, *7th Indian National Symposium on Catalysis*, Wiley Eastern, New Delhi (1985).

38. F. Ammour, thesis, Rueil-Malmaison (1986).
39. J. Barbier, G. Corro, and Y. Zhang, *Appl. Catal.*, *13*:245 (1985).
40. L. Lin, J. Tsang, J. Wu, and P. Chiang, *Proceedings of the 7th International Congress on Catalysis, Tokyo, 1980*, Vol. 1, Elsevier, Amsterdam (1981), p. 1466.
41. J. Völter, H. Lieske, and C. Lietz, *React. Kinet. Catal. Lett.*, *16*:87 (1981).
42. B. Kuznetsov and Y. Yermakov, *Kinet. Katal.*, *23*:519 (1982).
43. J. P. Candy, O. A. Ferretti, G. Mabilon, and J. P. Bournonville, *J. Chem. Soc. Chem. Commun.*, 1197 (1985).
44. O. A. Feretti, thesis, Rueil-Malmaison (1986).
45. P. A. Sermon and G. C. Bond, *Catal. Rev.-Sci. Eng.*, *8(2)*:211 (1973).
46. G. C. Bond, in *Spillover of Adsorbed Species* (G. M. Pajonk, S. J. Teichner, and J. E. Germain, eds.), Elsevier, Amsterdam (1983), p. 1.
47. D. L. Trimm, *Appl. Catal.*, *5*:293 (1983).
48. C. A. Bernardo and D. L. Trimm, *Carbon*, *17*:115 (1979).
49. S. J. Tauster and R. M. Koros, *J. Catal.*, *27*:307 (1972).
50. S. J. Tauster and S. C. Fung, *J. Catal.*, *55*:29 (1978).
51. E. J. Nowak, *J. Phys. Chem.*, *73*:3790 (1969).
52. J. M. Parera, N. S. Figoli, E. L. Jablonski, M. R. Sad, and J. N. Beltramini, in *Catalyst Deactivation* (B. Delmon and G. F. Froment, eds.), Elsevier, Amsterdam (1980), p. 571.
53. R. B. Levy and M. Boudart, *J. Catal.*, *32*:304 (1974).
54. M. Boudart, in *Advances in Catalysis*, Vol. 20 (D. D. Eley, H. Pines, and P. B. Weisz, eds.), Academic Press, New York (1969), p. 153.
55. J. E. Benson and M. Boudart, *J. Catal.*, *4*:705 (1965).
56. A. Abdeladim, thesis, Rueil-Malmaison (1982).
57. P. Gauthier, J. P. Bournonville, and J. P. Franck, unpublished results.
58. J. Barbier, L. Elassal, N. S. Gnep, M. Guisnet, W. Molina, Y. R. Zhang, J. P. Bournonville, and J. P. Franck, *Bull. Soc. Chim. Fr.*, *9/10*:250 (1984).
59. A. Cimino, M. Boudart, and H. S. Taylor, *J. Phys. Chem.*, *58*:796 (1954).
60. J. P. Boitiaux, thesis, Paris (1976).
61. Y. Barron, G. Maire, D. Cornet, and F. G. Gault, *J. Catal.*, *2*:152 (1963).
62. M. Boudart, A. Aldag, J. E. Benson, N. A. Dougharty, and C. G. Harkins, *J. Catal.*, *6*:92 (1966).
63. R. Maurel, G. Leclercq, and L. Leclercq, *Bull. Soc. Chim. Fr.*, 491 (1972).
64. J. P. Boitiaux, G. Martino, and R. Montarnal, *3rd Symposium France-USSR on Catalysis, Villeurbanne* (1976).
65. J. P. Boitiaux, G. Martino, and R. Montarnal, *C. R. Acad. Sci. (Paris)*, *280*:1451 (1975).
66. J. P. Bournonville and J. P. Franck, unpublished results.
67. U.S. Patent 3,415,737 (1968).
68. E. L. Pollitzer, V. Haensel, and J. C. Hayes, *Proceedings of the 8th World Petroleum Congress, Moscow*, Vol. 4 (1971), p. 255.
69. Franch Patent 2,031,984 (1969).
70. H. Verbeek and W. M. H. Sachtler, *J. Catal.*, *42*:257 (1976).
71. P. Biloen, J. N. Helle, H. Verbeek, F. M. Dautzenberg, and W. M. H. Sachtler, *J. Catal.*, *63*:112 (1980).
72. L. R. Moss, *Catalysis*, Vol. 1 (C. Kemball, senior reporter), Chemical Society, London (1977), p. 37.
73. D. A. Dowden, *Catalysis*, Vol. 2 (C. Kemball and D. A. Dowden, Senior Reporters), Chemical Society, London (1978), p. 1.

74. H. Charcosset, *Rev. Inst. Fr. Pét.*, 34(2):239 (1979).
75. T. R. Hugues, R. L. Jacobson, K. R. Gibson, L. G. Schornack, and J. R. McCabe, *Oil Gas J.* (May 17, 1976), p. 121.
76. B. J. Cha, A. Vidal, R. Huin, and H. Van Landeghem, *API Division of Refining, Proceedings of the 38th Midycar Meeting, Washington*, Vol. 53 (1973), p. 138.
77. S. Marty, thesis, Rueil Malmaison (1986).

26 | Metal Membranes for Hydrogen Diffusion and Catalysis

JOHN PHILPOTT and DUNCAN R. COUPLAND

Catalytic Systems Division—Engineered Products, Johnson Matthey, Royston, Hertfordshire, England

I.	INTRODUCTION	679
II.	DIFFUSION OF HYDROGEN IN METALS	681
III.	PALLADIUM MEMBRANES	682
	A. Palladium/Rare Earth	683
	B. Palladium/Boron	685
	C. Industrial Applications	686
IV.	POISONING AND SURFACE TREATMENTS	690
	A. Chemical Reactions	690
	B. Chemisorption	690
V.	PERMEABLE MEMBRANES AS CATALYSTS	691
	REFERENCES	693

I. INTRODUCTION

The different aspects of the interaction of hydrogen with metal catalyst surfaces have been examined from various angles in several chapters in this book. In the present chapter we consider the diffusion of hydrogen into metals through permeable metallic membranes and in some interesting applications for catalysis. The discussion here will be very brief since part of the relevant material has already been covered in greater detail in Chapters 14, 18, and 24. The emphasis here is on the development of rugged hydrogen-permeable membranes and their uses for preparing ultrapure hydrogen for both laboratory and field use and, quite recently, in novel molecularly engineered catalytic applications.

The major requirement for ultrapure hydrogen in industrial usage is in the manufacture of silicon chips, the basic components in most of today's electronic equipment. Other applications for dry hydrogen at moderate pressure without the need for expensive compressors from on-site generators are on remote meteorological sites for filling balloons and on large power stations for cooling alternators.

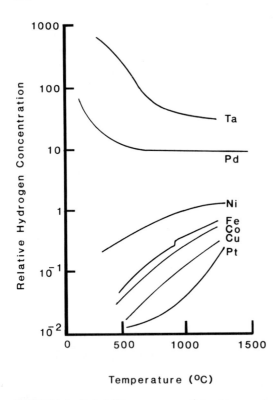

FIGURE 1 Solubility curves of hydrogen in some endothermic and exothermic occluding metals at atmospheric pressure. (From Ref. 18.)

Hydrogen reactions with solid metals are of two types, exothermic and endothermic. The distinction between the two is defined by the relative positions of the metals in the periodic table. The metals in groups III, IV, and V react exothermally, while those in groups VI, VII, and VIII react endothermally. In general, the exothermic reactions produce stable hydrides, and the endothermic reactions produce low-stability hydrides. Of the metals that react endothermally, palladium is notable in its formation of the stable alloy PdH_2. Commonly, endothermic reacting metals absorb only small quantities of hydrogen, the solubility increasing with temperature (Fig. 1). Conversely, exothermic-reacting metals absorb large volumes of hydrogen gas, but solubility falls with increasing temperature. Palladium stands out from this general pattern since it is an endothermally reacting metal, being in group VIII of the periodic table, but behaves toward hydrogen exothermally. This is discussed in detail in Chapter 14.

The reversible reactions of metals with gaseous hydrogen take place in the form of a series of discrete steps. The reaction is initiated by the absorption of atomic or molecular hydrogen onto the metal surface. If the gas is molecular, dissociation occurs, followed by diffusion of ions into the metal lattice to form solid solutions. As a result of the small size of the hydrogen atom, the solid solutions are interstitial rather than substitutional, with the hydrogen occupying octahedral and/or tetrahedral interstices, depending on the lattice structure of the matrix metal. The degree of filling of these sites is dependent on the partial pressure of the hydrogen and the ambient temperature.

26. Metal Membranes for Hydrogen Diffusion

For the majority of metals the dissolution of hydrogen results in distortion of the metal lattice and leads to brittleness. However, exceptions do exist in cerium and palladium, both of which suffer lattice expansions without significant distortions. The topic of diffusion of hydrogen in metals has been reviewed extensively by Hempelmann [1].

II. DIFFUSION OF HYDROGEN IN METALS

Measurements of the diffusivity of gases in metals have been widely reported. They all clearly show that hydrogen has a diffusivity 15 to 20 orders of magnitude higher than that for oxygen and for nitrogen, in the same metal. One consequence of this high mobility is the existence of ordering transitions below ambient temperatures. These ordering reactions provide an explanation for the hysteresis of solubility of hydrogen in palladium during heating and cooling cycles (see Fig. 2 in Chapter 14).

The process of diffusion through a metal can be mathematically defined. A simple equation expressing permeability through a membrane is: $J = -D \, \Delta c(A)$ (Fick's law), where J is the hydrogen diffusion coefficient, A the surface area in contact with the gas, D the thickness of the metal membrane, and Δc the concentration difference between the high- and low-pressure sides. Clearly, the permeability J is a function of D and Δc, both of which will vary with temperature. For normal endothermally reacting metals, increasing temperature will benefit both hydrogen mobility and solubility, whereas for palladium and other exothermally reacting metals, increasing temperature will cause reductions in solubility. Figure 1 shows the latter metals to have such high solubilities that at equal temperatures they have higher solubilities for hydrogen than do the endothermic metals.

As noted previously, nearly all exothermic metals form strong compounds with hydrogen, often several in number. These compounds are brittle and promote weakness in any form of structural component. Such compounds are

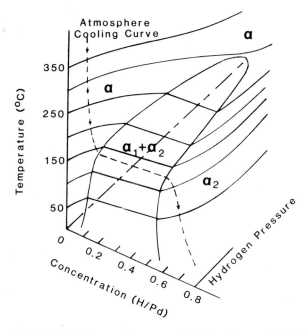

FIGURE 2 Phase diagram of the hydrogen/palladium system. (From Ref. 18.)

also susceptible to [3] oxidation and [2] poisoning, which restrict permeabilities; fortunately, palladium, the endothermic metal with the highest shown solubility of its type, does not form distinct compounds except over a restricted range of composition, temperature, and gas pressure. If this "no go" zone, as shown in Fig. 2, is avoided, palladium remains ductile.

The diffusivity [5] of hydrogen in silver/palladium alloys has been shown to change little with silver content up to approximately 25 at % at ambient temperature. However, above 25% silver content diffusivity falls rapidly. Set against this is the effect of solubility, which is approximately opposite to that of diffusivity. The net effect is to cause permeability to vary little with silver content at ambient temperature.

Additionally, the permeability of cold-worked and annealed silver/palladium changes little at ambient temperature. Cold working has three effects: (a) [7] to decrease the hydrogen diffusity (possibly due to hydrogen trapping at lattice imperfections); (b) to increase, with increasing alloy element addition, the activation energy of diffusion [6]; and (c) to increase the solubility of hydrogen. The overall result is that at ambient temperature very little difference exists in the permeabilities of hydrogen through cold-worked and annealed silver/palladium alloys.

III. PALLADIUM MEMBRANES

Because of its special properties, palladium is the most important element for hydrogen purification membranes. These properties include the resistance to embrittlement, high hydrogen solubility, and mobility of the proton. In combination, these properties provide a commercially useful hydrogen permeability which can be exploited in hydrogen diffusion membranes, as in Fig. 3.

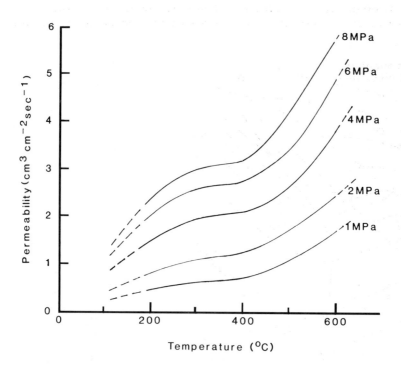

FIGURE 3 Permeabilities of hydrogen in 25% Ag-Pd foil of 0.0025 cm thickness. (From Ref. 18.)

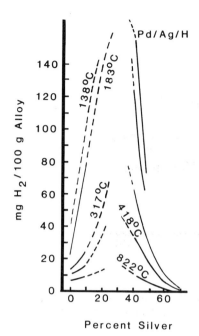

FIGURE 4 Comparison of hydrogen solubilities in silver/palladium alloys at a pressure of 0.1 MPa. (After Ref. 26.)

However, the miscibility gap must be avoided; otherwise, the repeated precipitation and re-solutioning of the hydrogen-rich phase, which occurs below the annealing temperature, will prove seriously deleterious to the metal structure. Severe strains are set up by these phase changes and result in almost catastrophic distortion of the membranes, which if constrained, eventually puncture. In practice, it was very difficult to avoid the gap; consequently, commercial membranes had to be maintained at their operating temperature of 300°C at all times when in the presence of hydrogen.

These difficulties were finally overcome following the efforts of Hunter in 1955 [4] utilizing alloying as a means of lowering the temperature at which immiscibility of the PdH_x occurs, to below room temperature. Silver additions, in the range 20 to 25 at %, were found (Fig. 4) to produce considerable increases in the solubility of hydrogen in palladium, while removing the damaging phase precipitation. As might be expected from theories of diffusion, this increased solubility also produces increased permeability (Figs. 5 and 6).

With the object of enhancing the hydrogen permeability of palladium alloys, numerous alloying additions other than silver have been considered [8,9]. The majority of the binary systems can be dismissed, since reduction in permeability results, as can be seen from Table 1. However, further consideration must be given to rare earth/palladium alloys and also, because of implications for the mechanisms of hydrogen diffusion, palladium/boron alloys.

A. Palladium/Rare Earth

Alloying rare earth metals with palladium was first considered because of the large range of palladium-rich solid solutions [14]. The size difference between atomic palladium and rare earth metals ensures significant solid solution strengthening and relatively high work-hardening rates. The increased inter-

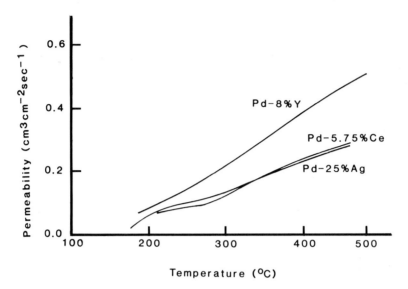

FIGURE 5 Comparison of permeabilities of alloys at a pressure of 0.34 MPa (all values normalized to a membrane thickness of 0.01 cm). (From Ref. 17.)

stitial spacing resulting from this size mismatch was also expected to promote rapid hydrogen diffusion. At the same time, it was found [15] that the α_1/α_2 miscibility gap (see Fig. 2) was suppressed to below room temperature for rare earth levels between 6 and 8%. Measurements of the permeability of hydrogen in Pd-Y and Pd-Ce [16] indicated significant improvement over silver/palladium (Fig. 5), although serious problems were encountered with poisoning of the Pd-Ce alloy. From the permeability and solubility determina-

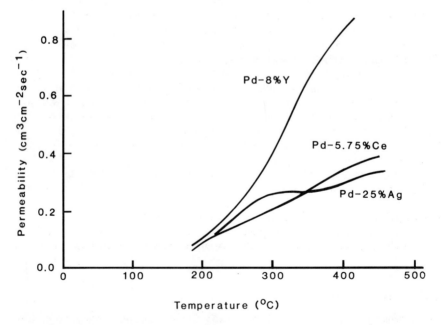

FIGURE 6 Comparison of the permeabilities of palladium alloys at a differential pressure of 0.68 MPa (all values normalized to a membrane thickness of 0.0025 cm). (From Ref. 17.)

TABLE 1 Comparison of Hydrogen Permeability Values for Palladium Binary Alloys

Alloy	Max. Permeability ($cm^3\ cm^{-2}\ s^{-1}$)	Alloy	Max. Permeability ($cm^3\ cm^{-2}\ s^{-1}$)
90 Pd-10 Y	5.38	99.5 Pd-0.5 B	1.35
77 Pd-23 Ag	2.48	95 Pd-5 Ru	0.47
92.3 Pd-7.7 Ce	2.24	90 Pd-10 Ni	0.27
95 Pd-5 Au	1.52	Pd/Fe	Low
60 Pd-40 Cu	1.52	Pd/Pt	Low
Pd	1.43	Pd/Rh	Low

Source: After Ref. 8.

tions the diffusivity values were calculated, which indicated (Table 2) that the rare earth additions did expand the lattice interstices and hence speed diffusion rates.

These alloys have improved mechanical properties and enhanced permeabilities even relative to the silver/palladium alloys. At the present time problems in mechanically fabricating these alloys, and their sensitivity to surface poisoning, prevent their commercial exploitation. If these problems can be overcome, a promising future could exist for these alloys as hydrogen membranes.

B. Palladium/Boron

Boron is believed to enter palladium on interstitial sites, and it was originally predicted that this propensity would lead to an expansion of octahedral interstices in the lattice with consequent increase in the hydrogen diffusion rate and hence in permeability [10,11]. In practice, at low temperatures diffusion coefficients were found to decline steadily with boron content from that of pure palladium [12]. This reduction in diffusion indicates that hydrogen mobility in palladium is via the tetrahedral holes. Using standard shielding models [13] it can be deduced that the blocking of diffusion stems from the distortion of the tetrahedral holes by the presence of boron atoms in the interstitial position. The activation energy of the hydrogen diffusion appears to be raised sufficiently by this distortion to block diffusion up to at least 70°C.

TABLE 2 A Comparison of Diffusivity Values for Selected Palladium Binary Alloys

Alloy	E 1 [kcal (g atom)$^{-1}$]	D ($cm^2\ s^{-1} \times 10^3$)	Ref.
Pd	5.17	5.7	19
Pd-20% Ag	6.8	10.9	19
Pd-25% Ag	7.12 ± 0.2	10.4 ± 0.7	17
Pd-5.75% Ce	7.35 ± 0.2	10.9 ± 0.7	17
Pd-8% Y	7.31 ± 0.25	12.2 ± 0.8	17

FIGURE 7 Hydrogen purifier unit housed in a tough plastic shelter, with door temporarily removed, on the roof of the STC factory at Harlow, Essex, where it has been operating successfully for over six years purifying hydrogen for silicon chip manufacture.

C. Industrial Applications

Although palladium foil has been used occasionally to diffuse hydrogen for experimental purposes where exceptionally pure hydrogen was needed, it was not until the properties of the silver/palladium alloys were established that commercial applications for hydrogen purification by membrane diffusion became practicable. Since that time many thousands of pieces of laboratory hydrogen purification equipment have been built to purify supplies of commercial hydrogen contained in gas cylinders, to a much higher purity level than that of the original cylinder gas. Increasingly, modern manufacturing techniques rely on control of the purity of the process materials used to ensure a high standard of product quality and low rejection rates. Hydrogen purified by diffusion through a palladium/silver alloy membrane can be guaranteed to purity levels better than 0.5 ppm total impurities. To meet the needs of these modern techniques, a range of small to medium-size standard hydrogen diffusion units are available which can supply purified gas at 1 liter min^{-1} up to 100 m^3 h^{-1} (Figs. 7 to 11). Figure 7 shows hydrogen diffusion equipment housed in a tough plastic tent on the roof of a factory building, from where it has successfully supplied very high purity hydrogen to the factory beneath for over 5 years with very little call for maintenance. In the interests of safety, diffusion equipment is often installed outside a factory building, where with some protection against the weather it will operate successfully for long periods. The purified hydrogen is used in the manufacture of silicon chips, where it acts as a carrier to transport small quantities of vaporized chemical compounds needed to "dope" the chip. The combination of specific elements in the doping compounds with the silicon areas exposed on the

FIGURE 8 Laboratory hydrogen purifier unit EP20 for purification of commercial-quality hydrogen. The unit is self-contained and need only be connected to a source of impure commercial hydrogen and an electrical supply to provide hydrogen output at 99.99995% purity.

masked chip creates the electronic circuits of the chip. The necessary electrical controls are mounted inside the factory close to the point of use for the hydrogen. Diffusion equipment is used not only for gas purification but also for hydrogen recovery from hydrogen-rich gas streams.

Because the silver/palladium membrane is permeable only to hydrogen, the diffused gas is very pure, although at startup this purity may be degraded by the need to sweep out moisture from the pure gas pipework and also to reduce easily reducible oxides on the internal pipework surfaces. In general, stainless steel pipework is used for pure hydrogen services. Welded junctions are preferred over compression fittings to keep to a minimum the risk of leaks and, equally important, the effects of back diffusion of impurities into the pure gas. To facilitate maintenance, diffusion equipment is usually offered in modular units that can be interconnected to provide the required pure gas output; and the controls are usually automatic, based on the use of a microprocessor.

As an extension of the commercial application, a number of pieces of hydrogen generator equipment have been designed. These provide on-site production of hydrogen in places remote from hydrogen supplies and in places where self-sufficiency is of particular importance. The generating process

FIGURE 9 G4M field-version hydrogen generator for on-site generation of hydrogen for meteorological balloon filling. The balloon has just been released from the filling cradle and is about to lift the radar reflector and sonde equipment. The hydrogen is supplied simultaneously from the hydrogen generator in the center and the extra storage trailer on the right.

FIGURE 10 Small, containerized hydrogen generator for remote meteorological stations. The model shown is now in operation at the Halley Bay base of the British Antarctic Survey.

FIGURE 11 Full hydrogen generator system on the Central Electricity Power Station site at the Isle of Grain. The hydrogen is used for alternator cooling and the picture shows the fuel storage tanks where the equivalent of a 6-month hydrogen supply is stored. The hydrogen generator is sheltered by a Dutch barn type of structure, and beyond are the low-pressure hydrogen storage vessels.

is based on the decomposition of a simple organic liquid, usually a methanol-water mixture, and extracting hydrogen from the resulting gas stream. The gas extracted can either be used immediately or fed to a low-pressure storage vessel. Generators of this type are in use on meteorological stations, where the hydrogen they make is used for balloon filling (Figs. 9 and 10); on power stations, where the hydrogen is used to cool the alternators (Fig. 11); and in chemical processes.

A major advantage of these generators is that small quantities of fuel can be used to store compactly large volumes of hydrogen obtainable from it. For example, two 200-liter drums of methanol-water fuel will produce one month's supply of hydrogen (assuming that six 500-g balloons are released every day). Hydrogen is therefore available constantly, with no risk of interruption in the supply line to the remote location. Generating hydrogen on-site from methanol ensures security of supply under most conditions, since methanol is available worldwide and its ease of storage makes it an ideal source for hydrogen generation.

The outputs of these generators range from 1 to 500 m^3 h^{-1}. They are compact, freestanding, and need only minimum protection from weather conditions. Many are in use in extreme climatic conditions, such as in Australia, Antarctica, and the deserts of Saudi Arabia. A measure of their inherent robustness is that they are in service for the supply of hydrogen for meteorological balloons in adverse climates usually considered unsuitable for the reliable operation of a small mobile chemical plant. Furthermore, these

furthermore, designed to operate unattended, while meeting stringent safety requirements, and over temperature ranges of -40°C to +55°C.

The use of hydrogen generation and purification provides a ready source of hydrogen gas at a potentially catalytic surface. This offers the opportunity to carry out novel hydrogenation processes. Although work in this area is still in its infancy, the use of special palladium alloys to promote selective hydrogenation, and to obtain very high product yields from hydrogenation reactions, has already been described. A review of the work to date forms Section V (see also Chapter 14).

IV. POISONING AND SURFACE TREATMENTS

Undoubtedly the palladium/silver membrane is presently the most widely used for production of ultrapure hydrogen. The feed gas for all palladium alloy membranes can be relatively low in hydrogen, but caution has to be exercised in respect to the composition of the feed gas, to avoid membrane poisoning. Poisoning causes blocking of catalytically active sites on the alloy surface and hence retards the dissociation of hydrogen, thus reducing permeability. Two types of poisoning have been recognized. These result from either chemical reaction or chemisorption of unsaturated hydrocarbons.

A. Chemical Reactions

The most common poisons are gas-borne reactive species such as sulfur, arsenic, chlorine, iodine, and mercury. As an example of the effect of sulfur, 1600 ppm of hydrogen sulfide caused a 50% reduction in diffusion rate in only 16 h at 500°C. It is unlikely that heating in air could regenerate such a badly damaged membrane.

The detrimental effect of chlorine and iodine is believed to result from interaction with ferrous materials, creating volatile chlorides and iodides. The poisoning effect of ferric chloride has been demonstrated by washing silver/palladium membranes in a dilute solution of the salt, resulting in significant reduction of permeability [18].

B. Chemisorption

It is well known that active unsaturated hydrocarbons such as acetylene, ethylene, propylene, and butylene at high concentrations in the hydrogen gas feed stream cause surface poisoning. In extreme cases this brings about total blockage to hydrogen transfer [4]. The effect of minor levels of such species is unclear, however.

Temperature-programmed reduction (TPR) techniques have recently been used to assess the effect of the exposure of thin palladium film to ethylene [20]. They indicated that at temperatures between 100 and 200°C a minor structural change occurred indicative of some reaction, and the process was found to be reversible. The reaction was found to be the production of a phase that was identified as PdC_x, with x lying in the range 0.10 to 0.15. Total blockage to hydrogen transfer [4] was removed by the simple expedient of exposure of surfaces to air at 450°C.

Hydrocarbons such as methane, benzene, and cyclohexane mixed with hydrogen have not been shown to be detrimental to the surface properties of silver/palladium. Similarly, the chemisorption of carbon dioxide and carbon monoxide were reported as not believed directly to cause poisoning of these membranes.

The long-term effect of any hydrocarbon, whether saturated or unsaturated, on the behavior of silver/palladium is uncertain. However, it is well known that the solubility of carbon in palladium is quite significant, expecially at elevated temperatures [21, 22], and more recent work [23] points to the possibility of formation of a PdC_x phase with as much as 11.5 at % of carbon in the lattice after reaction treatment at 373 K. On heating at about 460 K in hydrogen, methane was produced. It may be suggested that carbon, occupying interstitial sites in the same way as boron, might reduce hydrogen permeability if sufficiently high concentrations are achieved.

V. PERMEABLE MEMBRANES AS CATALYSTS

Work done over the past 15 to 20 years has shown that palladium alloys, in addition to having a high permeability to hydrogen while restricting the flow of all other gases, have important catalytic activity for hydrogenation and dehydrogenation reactions. This catalytic activity is usually achieved by

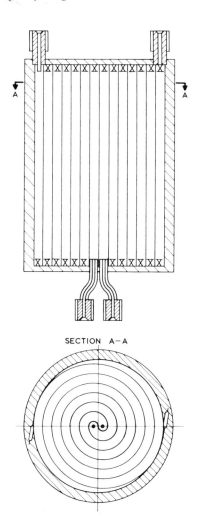

FIGURE 12 This reactor incorporates the membrane catalyst in the form of a plate coiled as a double spiral the edges of which are fixed into the wall of the reactor. (From Ref. 25.)

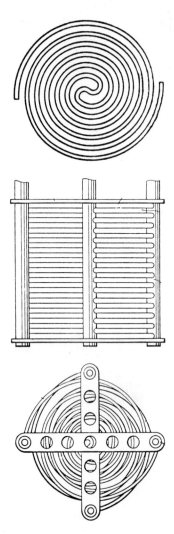

FIGURE 13 An alternative arrangement consists of a flat double spiral of thin-walled palladium alloy tube. The inside of the reactor is filled with these spirals, arranged so as to maximize their number. The inlets and outlets for the diffusion tubes consist of four header pipes arranged perpendicularly to the spirals. (From Ref. 25.)

reaction over supported bulk catalysts. The beneficial effect of utilizing membranes as catalytic surfaces, where hydrogen is removed or provided through the membrane, has recently been highlighted in the work of Gryaznov and co-workers [24, 25] in the Soviet Union.

It is well known that the hydrogenation rate and selectivity of the foregoing reactions depend on the concentration ratio of the hydrogen and hydrogenated substance on the catalyst surface (see Chapters 18 and 24). Conventionally, this ratio is related only to the initial mixture content and will not be constant along the catalyst bed. However, for a membrane catalyst the situation is very different, since the hydrogen reaction takes place on the catalyst where adsorbed reactant can combine with atomic hydrogen flowing to it from the opposite membrane surface. The rate of hydrogen diffusion can be controlled and hence the rate of reaction and the degree of completeness to which it is taken, allowing incomplete hydrogenation to be performed satisfactorily. Apart from these advantages the complexity of chemical plant can be reduced, since separation of reactants and catalyst is maintained when using membranes, whereas the very nature of supported or Raney catalysts ensures some reactant contamination from this source. Also, the corrosion resistance of the membranes in combination with significant strength allows combination of previously separate reaction stages, such as selective hydrogenation plus esterification, as in the one-stage production of vitamin K_4, described by Gryaznov [25]. It is believed that by using membrane catalysts which are permeable to hydrogen, rather than conventional catalysts, very pure chemicals and pharmaceuticals can be produced. Many hydrogenation reactions have been attempted [24, 25], with results which indicate that improved results are obtained using membranes.

Dehydrogenation reactions have also been performed successfully. The membrane catalyst acts to deplete the reactants of hydrogen (see also Chapter 17). Yields are reported to be increased using this technique, especially if the released hydrogen is oxidized. Membrane catalysts are also discussed in Chapters 14 and 17.

If the two processes of hydrogenation and dehydrogenation are performed on opposite sides of the same membrane catalyst, energy savings can also arise, since the processes of hydrogen adsorption and desorption are exothermic and endothermic, respectively. Two alternative arrangements of membrane catalysts, using plate and tube membranes, are illustrated schematically in Figs. 12 and 13. In both, the membrane catalyst forms part of the reactor in that it subdivides the reactor into two compartments. Hydrogen is introduced into one of these compartments for it to diffuse through the membrane to react with the hydrocarbon molecules absorbed on the surface of the membrane in the other compartment. These and other considerations open up several fascinating possibilities for membrane catalysts in industrial processes.

REFERENCES

1. R. Hempelmann, *J. Less-Common Met.*, 101:69 (1984).
2. R. Sherman and H. K. Birnbaum, *J. Less-Common Met.*, 105:339 (1985).
3. Atlantic Refining Co., U.S. Patent 2,773,561 (1956).
4. J. B. Hunter, *Symposium on the Production of Hydrogen*, American Chemical Society Meeting, New York (1963).
5. H. Zuchner, *Z. Naturforsch. Teil*, A25:1490 (1970).
6. H. Buchold, G. Sieking and E. Wicke, *Ber. Bunsenges. Phys. Chem.*, 80:446 (1976).
7. Y. Sakamoto, S. Hirata and H. Nashikawa, *J. Less-Common Met.*, 88:387 (1982).
8. A. G. Knapton, *Platinum Met. Rev.*, 21:44 (1977).
9. G. J. Grashoff, C. E. Pilkington, and C. W. Corti, *Platinum Met. Rev.*, 27(4):157 (1983).
10. A. S. Darling, British Patent 956,176 (1964).
11. D. L. McKinley, U.S. Patent 3,439,474 (1969).
12. K. D. Allard, E. B. Flanagan, and E. Wicke, *J. Phys. Chem.*, 74(2):298 (1979).
13. J. Zimon, *Principles of the Theory of Solids*, Cambridge University Press, Cambridge (1964).
14. M. L. H. Wise, J. P. G. Farr, and I. R. Harris, *J. Less-Common Met.*, 41:115 (1975).

15. J. R. Hirst, M. L. H. Wise, D. Fort, J. P. G. Farr, and I. R. Harris, J. Less-Common Met., 49:193 (1976).
16. D. Fort and I. R. Harris, J. Less-Common Met., 45:247 (1976).
17. D. T. Hughes and I. R. Harris, J. Less-Common Met., 61:9 (1978).
18. A. S. Darling, Symposium on the Less Common Means of Separation, Institute of Chemical Engineers (1963), p. 103.
19. G. Bohmholdt and E. Wicke, Z. Phys. Chem. (Neue Folge), 56:133 (1965).
20. S. B. Ziemecki and G. A. Jones, J. Catal., 95:691 (1985).
21. G. L. Selman, P. J. Ellison, and A. S. Darling, Platinum Met. Rev., 14:14 (1970).
22. J. Stachurski, J. Chem. Soc. Faraday Trans. 1, 81:2813 (1985).
23. J. Stachurski and A. Frackiewicz, J. Less-Common Met., 108:249 (1985).
24. N. D. Fomin, V. M. Gryaznov, A. P. Mischenko, A. P. Maganjuk, V. N. Kulakov, V. P. Polyakova, N. R. Roschan, and E. M. Savickij, British Patent 2,056,043 (1981).
25. V. M. Gryaznov, Platinum Met. Rev., 30:68 (1986), and references therein.
26. A. Sieverts, E. Jutisch, and A. Metz, Z. Allgem. Chem., 92(4):329 (1915).

27 | Hydrogen Transfer in Catalysis on Zeolites

N. Y. CHEN and W. O. HAAG

Mobil Research and Development Corporation, Princeton, New Jersey

I.	SCOPE OF STUDY	696
II.	NATURE OF CARBENIUM IONS	698
III.	ROLE OF HYDROGEN TRANSFER IN STERICALLY UNRESTRICTED ACID-CATALYZED REACTIONS	699
	A. Paraffin Isomerization and Disproportionation	699
	B. Bimolecular Paraffin Cracking, Including "Hydrocracking" with Catalyst Containing a Weak Metal Function	700
	C. Paraffin Alkylation	703
	D. Conjunct Polymerization	704
	E. Aromatization	705
	F. Coke Formation	705
IV.	CLASSIFICATION OF ZEOLITES	705
	A. Eight-Membered Oxygen Ring Systems	706
	B. Ten-Membered Oxygen Ring Systems	706
	C. Twelve-Membered Oxygen Ring Systems	707
V.	MAJOR EFFECTS OF ZEOLITES ON HYDROGEN TRANSFER REACTIONS	707
	A. Steric Inhibition or Spatial Constraint	707
	B. Superactivity	709
	C. Other Subtle Effects	710
VI.	COMMERCIAL IMPLICATIONS	711
	A. Catalytic Cracking of Gas Oils	711
	B. Catalytic Dewaxing	713
	C. Octane Boosting	713
	D. Aromatization	715
	E. Methanol to Gasoline	717
	F. Olefin Polymerization	718
VII.	CONCLUSIONS	720
	REFERENCES	720

I. SCOPE OF STUDY

The transfer of hydrogen from one molecule to another molecule occurs in many organic reactions:

$$AH + B \rightarrow A + HB$$

In general, the attacking molecule B can be either a cation, an anion, or a free radical (Table 1). In this chapter, only the cationic hydrogen transfer reactions involving carbocations or carbenium ions are discussed. Since the hydrogen is transferred as a negative species, the reaction is also referred to as hydride transfer.

Carbenium ions play a key role as intermediates in acid-catalyzed reactions. Hydride transfer by the carbenium ions is of great importance in many reactions, such as catalytic cracking, alkylation, isomerization (Fig. 1a), and hydrocracking of paraffins and the aromatization of paraffins and olefins.

However, not all carbenium ion reactions involve intermolecular hydrogen transfer. For example, the reaction of olefins in isomerization (Fig. 1b), oligomerization, polymerization, and cracking, and the alkylation of aromatics with olefins do not involve intermolecular hydrogen transfer, nor do the classical bifunctional reactions, using highly active metal hydrogenation catalysts (e.g., Ni, Pt), in hydroisomerization and hydrocracking of paraffins. As shown in Fig. 1c for hydroisomerization, the bifunctional reaction mechanism involves two types of catalytic sites: the dehydrogenation/hydrogenation site and the acid site. The olefinic intermediate formed at the dehydrogenation sites (Pt) migrates to the protonic acid sites forming the carbenium ion, where it undergoes monomolecular skeletal isomerization. These intermediate product molecules then migrate back to the hydrogenation site to form the saturated species. With pentane and hexane as feed, the olefinic intermediates can be intercepted by the hydrogenation catalyst component and their cracking reactions minimized. High selectivity approaching 100% can be obtained for the products that have an isomer distribution which is close to equilibrium. With larger paraffins as feed, the isomerization selectivity is less than 80% because of the greater cracking rate of the intermediate olefins. These reactions are outside the scope of this study.

In contrast to these catalytic systems, some bifunctional catalysts rely on their hydrogen transfer activity to function properly. For example, hydrogen transfer plays an important role in hydrocracking with catalysts containing a weak dehydrogenation/hydrogenation function, such as NiS_x, WS_x, and so on. This conclusion is based on the observation that when reaction

TABLE 1 Classification of Hydrogen Transfer Reactions

1. Cationic = hydride transfer

    $AH + B^+ \rightarrow A^+ + HB$

2. Anionic = proton transfer

    $AH + B:^- \rightarrow A:^- + HB$

3. Free radical = hydrogen atom transfer (hydrogen abstraction)

    $AH + B\cdot \rightarrow A\cdot + HB$

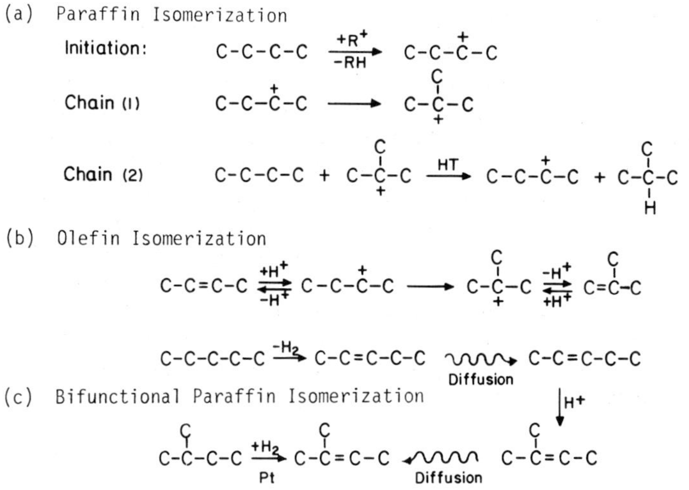

(a) Paraffin Isomerization

Initiation: C-C-C-C $\xrightarrow[-RH]{+R^+}$ C-C-$\overset{+}{C}$-C

Chain (1):   C-C-$\overset{+}{C}$-C $\longrightarrow$ C-$\overset{\overset{C}{|}}{\underset{+}{C}}$-C

Chain (2):   C-C-C-C + C-$\overset{\overset{C}{|}}{\underset{+}{C}}$-C $\xrightarrow{HT}$ C-C-$\overset{+}{C}$-C + C-$\overset{\overset{C}{|}}{\underset{\underset{H}{|}}{C}}$-C

(b) Olefin Isomerization

C-C=C-C $\underset{-H^+}{\overset{+H^+}{\rightleftarrows}}$ C-C-$\overset{+}{C}$-C $\longrightarrow$ C-$\overset{\overset{C}{|}}{\underset{+}{C}}$-C $\underset{+H^+}{\overset{-H^+}{\rightleftarrows}}$ $\overset{\overset{C}{|}}{C}$=C-C

(c) Bifunctional Paraffin Isomerization

C-C-C-C-C $\xrightarrow{-H_2}$ C-C=C-C-C $\overset{\text{Diffusion}}{\rightsquigarrow}$ C-C=C-C-C $\downarrow H^+$

C-$\overset{\overset{C}{|}}{C}$-C-C $\underset{Pt}{\xleftarrow{+H_2}}$ C-$\overset{\overset{C}{|}}{C}$=C-C $\overset{\text{Diffusion}}{\leftsquigarrow}$ C-$\overset{\overset{C}{|}}{C}$=C-C

FIGURE 1 Mechanisms of isomerization: (a) acid-catalyzed paraffin isomerization via hydride transfer (HT); (b) acid-catalyzed olefin isomerization; (c) paraffin isomerization with acid (H^+) and noble metal (Pt) catalysts via olefin intermediate.

products of various types of hydrocracking catalysts are compared, there is a marked difference in their product distribution. The products from hydrocracking with a catalyst containing a weak dehydrogenation/hydrogenat function bear a striking resemblance to that of acid cracking in terms of carbon number distribution and isomer distribution.

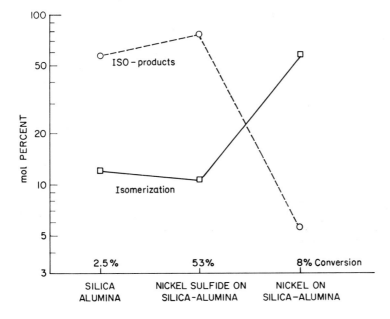

FIGURE 2 Hydrocracking of n-decane at 288°C, 8 LHSV, 82 atm, $H_2/HC = 10$. Isomerization: % selectivity to isomeric decanes; iso-products: mol % branched products in total cracked product. (Data from Ref. 1.)

Figure 2 compares the cracking and hydrocracking of n-decane over silica-alumina, sulfided nickel on silica alumina, and nickel on silica-alumina. With nickel on silica-alumina, classical dual-functional hydrocracking occurs, typified by a high isomerization selectivity and a predominance of linear cracked products. By contrast, despite the large difference in conversion, the products obtained with the nickel sulfide on silica-alumina catalyst are quite similar to those from silica-alumina: The isomerization selectivity is low and the cracked products consist predominantly of C_4 to C_7 isoparaffins [1]. This is evidence that the weak metal function (Ni_3S_2) does not actively participate in the main reaction pathway. The bifunctionality of this reaction may be considered trivial. The major reactions take place on the monofunctional acid sites via carbenium ion intermediates and intermolecular hydrogen transfer, and the role of metal is primarily to saturate the final products and to prevent catalyst aging by keeping the olefinic and diolefinic intermediates at a low concentration. These reactions are discussed in more detail later.

II. NATURE OF CARBENIUM IONS

Carbenium ions are the principal reaction intermediates in acid-catalyzed reactions carried out either in solution or on heterogeneous solid surfaces. Although much less is known of the nature of these ions on solid surfaces than in solution, the application of carbenium ion mechanisms to interpret reaction kinetics and product distribution is quite satisfactory and indicates a strong similarity for both systems.

Depending on the nature of the substituents on the central carbon atom, carbenium ions may be classified into primary, secondary, and tertiary carbenium ions (Fig. 3). They can be formed by hydride abstraction from paraffins or by proton addition to olefins. It can be assumed that in the presence of Brønsted acid sites, an equilibrium is rapidly established between the olefins and their corresponding carbenium ions, relative to the reactions of carbenium ions, such as skeletal isomerization or cracking.

It is well known that in sterically unrestricted systems, hydrogen transfer to and from tertiary carbenium ions occurs much more readily than from secondary carbenium ions, and primary carbenium ions are usually not involved. This is commonly explained by the difference in the heat of formation of these different types of carbenium ions (Table 2) [2]. Inferred by the heat of formation is the relative stability of these ions [3]. Thus the tertiary ions are more stable than the primary ions by 32 to 34 kcal mol^{-1}, and the secondary ions are more stable than the primary ions by 16 to 18 kcal mol^{-1}. It should be kept in mind that these values are valid only for free gas-phase carbenium ions. The magnitude of these differences in their stability in catalytic systems containing negative counterions has not been quantified and remains as one of the unsolved problems.

Nevertheless, tertiary ions are more stable than the other ions and hence are present in larger concentrations. But the relation between concentration and reactivity remains obscure. For our discussion it is assumed that in

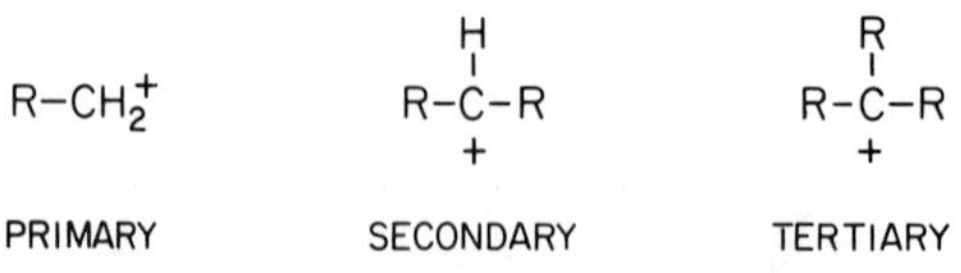

```
                        H                  R
                        |                  |
     R-CH₂⁺           R-C-R              R-C-R
                        +                  +

     PRIMARY         SECONDARY          TERTIARY
```

FIGURE 3 Types of carbenium ions.

TABLE 2 Heat of Formation of Carbenium Ions

Carbenium ions	ΔH_{298} (kcal mol^{-1})
Methyl	261
Ethyl	219
Propyl	208
Secondary propyl	192
Primary butyl	201
Primary isobutyl	199
Secondary butyl	183
Tertiary butyl	167

Source: Data from Ref. 2.

unrestricted systems, primary ions are so unstable that their equilibrium concentration is very low; if formed at all, they are rapidly converted to secondary and tertiary ions by intramolecular hydride shift. On the other hand, the involvement of the less stable secondary and perhaps even primary carbenium ions must be invoked in acid-catalyzed reactions occurring in small-pore and medium-pore zeolites, which sterically cannot accommodate highly substituted molecules.

III. ROLE OF HYDROGEN TRANSFER IN STERICALLY UNRESTRICTED ACID-CATALYZED REACTIONS

A. Paraffin Isomerization and Disproportionation

It is now well accepted that the acid-catalyzed paraffin isomerization proceeds via carbenium ion intermediates. Initiation usually requires the presence of impurities such as olefins, alcohols, or alkyl halides, which are readily converted by Brønsted acids to carbenium ions. After the initiation, isomerization takes place in a chain reaction via intramolecular skeletal rearrangement of the carbenium ions, followed by intermolecular hydrogen transfer between the reactant and the carbenium ion (Fig. 1a).

When the skeletal isomerization of carbenium ions involves a change in the number of side chains, it probably occurs via protonated cyclopropane intermediates, as shown in Fig. 4a. When no change in the degree of branching is involved, a simple 1,2-Wagner-Meerwein shift of an intact alkyl group is the most likely reaction path (Fig. 4b).

The isomerization reaction of butane and higher paraffins is accompanied by disproportionation reactions, which produce both higher- and lower-molecular-weight paraffins. It is believed [3] that disproportionation occurs when a proton is lost by a tertiary carbenium ion; the olefin produced condenses with a neighboring carbenium ion to produce a dimer ion. Isomerization and β scission of the dimer ion lead to disproportionation products (Fig. 5). Again, hydrogen transfer is an essential part of the reaction.

Isomerization and disproportionation reactions of normal and branched paraffins are readily catalyzed by aluminum halides. Sulfuric acid and silica-aluminas can isomerize only branched paraffins and only molecules containing a tertiary hydrogen are produced, because these catalysts only catalyze tertiary-tertiary hydrogen transfer reactions [4].

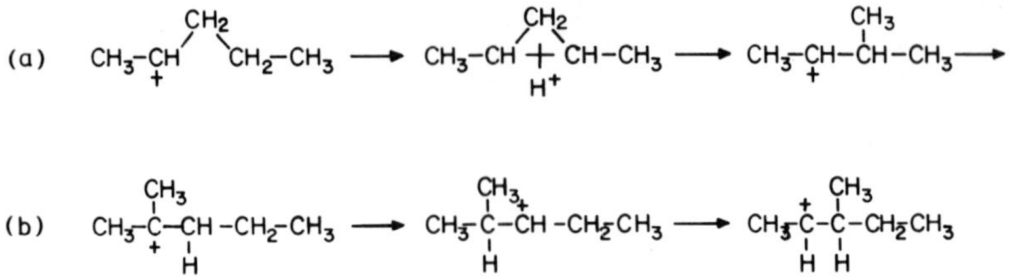

(a) [Structure: CH₃-CH(+)-CH₂-CH₂-CH₃ with CH₂ bridge → CH₃-CH···CH-CH₃ with CH₂ bridge and H⁺ → CH₃-CH(+)-CH(CH₃)-CH₃ →]

(b) [Structure: CH₃-C(CH₃)(+)(H)-CH-CH₂-CH₃ → CH₃-C(CH₃)-CH(+)-CH₂CH₃ with H → CH₃-C(CH₃)(H)-C(H)(+)-CH₂CH₃]

FIGURE 4 Skeletal rearrangement of carbenium ions: (a) isomerization of n-pentyl to t-pentyl carbenium ion via a protonated cyclopropane intermediate; (b) isomerization of 2-methyl- to 3-methylpentyl carbenium ion via 1,2-methyl shift.

B. Bimolecular Paraffin Cracking, Including "Hydrocracking" with Catalyst Containing a Weak Metal Function

Carbenium ions are also the intermediates in catalytic cracking reactions. The cracking of olefins with six or more carbon atoms is a monomolecular reaction; smaller olefins usually crack via dimerization/cracking steps. In the hydrocracking of paraffins with high-activity-metal-containing catalysts, olefins are produced as intermediates that are likewise cracked by the acid component. Intermolecular hydrogen transfer is not involved in these reactions (Fig. 6a and b). On the other hand, the acid-catalyzed cracking of paraffins, except under some conditions (discussed later), generally involves a bimolecular hydrogen transfer reaction between the reactant molecule and a product carbenium ion (Fig. 6c). In other words, as in the case of paraffin isomerization, hydrogen transfer reactions are the essential part of the cracking reaction.

The carbenium ion, once formed, can undergo intramolecular hydride shifts and skeletal isomerization to form the more stable tertiary carbenium ion and/or crack by β scission to an olefin and a smaller carbenium ion. The observed product distribution, which varies with the reaction condition and the nature of the catalyst, is a reflection of relative rate of these reactions.

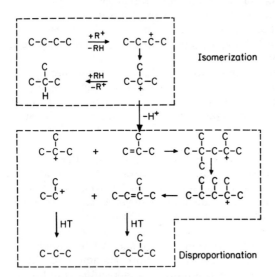

FIGURE 5 Mechanism of isomerization of n-butane of isobutane, accompanied by disproportionation to propane and pentanes.

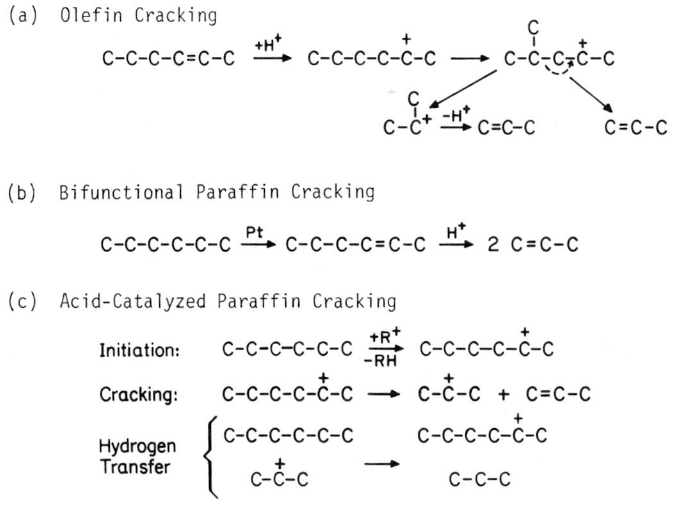

27. Hydrogen Transfer in Catalysis on Zeolites

(a) Olefin Cracking

C-C-C-C=C-C $\xrightarrow{+H^+}$ C-C-C-C-$\overset{+}{C}$-C ⟶ C-$\overset{C}{\underset{|}{C}}$-$\overset{+}{C}$-$\overset{\cdot}{C}$-C

C-$\overset{C}{\underset{|}{C}^+}$ $\xleftarrow{}$ ↓

C-$\overset{C}{\underset{|}{C^+}}$ $\xrightarrow{-H^+}$ C=C-C   C=C-C

(b) Bifunctional Paraffin Cracking

C-C-C-C-C-C $\xrightarrow{Pt}$ C-C-C-C=C-C $\xrightarrow{H^+}$ 2 C=C-C

(c) Acid-Catalyzed Paraffin Cracking

Initiation: C-C-C-C-C-C $\xrightarrow[-RH]{+R^+}$ C-C-C-C-$\overset{+}{C}$-C

Cracking: C-C-C-C=$\overset{+}{C}$-C ⟶ C-$\overset{+}{C}$-C + C=C-C

Hydrogen Transfer:
{ C-C-C-C-C-C   C-C-C-C-$\overset{+}{C}$-C
  C-$\overset{+}{C}$-C   ⟶   C-C-C }

FIGURE 6 Mechanisms of cracking: (a) acid-catalyzed cracking of olefins; (b) bifunctional cracking of paraffins via olefin intermediate; (c) acid-catalyzed cracking of paraffins via bimolecular hydrogen abstraction.

The following is a list of observations that can be attributed to the participation of hydrogen transfer in bimolecular cracking reactions:

1. *Hydrogen transfer determines the reactivity or crackability of a paraffin molecule.* In paraffin cracking the rate-determining step involves a hydrogen transfer reaction between a product carbenium ion and a reactant paraffin molecule. It follows that the reactivity or crackability of a paraffinic molecule is determined by its rate of hydrogen transfer, which is inversely proportional to the relative strength of the different types of C-H bond in a paraffinic molecule: Primary C-H bonds are the strongest, tertiary C-H bonds are the weakest. Thus branched paraffins are intrinsically more reactive than a straight-chain paraffin. The reaction mechanism involved is shown in Fig. 7, using 2-methylpentane and n-hexane as illustrative examples. In the case of 2-methylpentane, the rate-determining step involves breaking of a weak tertiary C-H bond, while in the case of n-hexane, the hydrogen associated with the stronger secondary C-H bond is transferred to the carbenium ion. Therefore, it is not surprising to find that 2-methylpentane is intrinsically more reactive and cracks about 2.5 times faster than n-hexane. By statistically correcting for the number of secondary and tertiary hydrogen atoms in these two molecules, one calculates that the tertiary carbon-hydrogen bond is broken about 16 times faster than the secondary carbon-hydrogen bond.

(a) C-C-C-$\overset{C}{\underset{C}{|}}$-H + C$^+$ $\xrightarrow{fast}$ C-$\overset{C}{\underset{+}{C}}$-C-C-C + HC

(b) C-C-C-C-$\overset{C}{\underset{|}{C}}$-H + C$^+$ $\xrightarrow{slow}$ C-C-C-C-$\overset{+}{C}$-C + HC

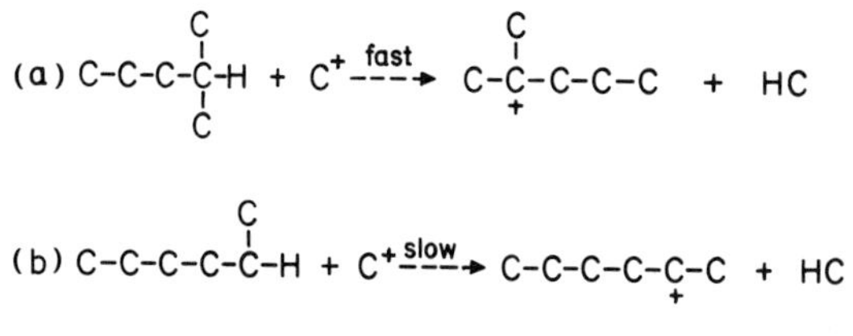

FIGURE 7 Relative rates of hydride transfer from tertiary (a) and secondary (b) carbon atoms.

2. *Saturated cracked products are the result of hydrogen transfer.*
The stoichiometry of paraffin cracking requires the formation of 1 mol of paraffin per mole of paraffin cracked. The number of moles of olefins produced depends on the number of secondary reactions, such as cracking or polymerization. The cracked paraffins are formed from olefins and their corresponding carbenium ions via hydrogen transfer. Thus the relative concentration and reactivity of these carbenium ions determine the nature of the paraffinic product.

In practice, the number of moles of paraffin formed per mole of paraffin cracked often exceeds 1. This is the result of hydrogen transfer reactions between the olefinic products, which produce hydrogen-rich paraffins and hydrogen-poor polyolefins, aromatics, and coke. These reactions are discussed in more detail later.

3. *Hydrogen transfer leads to high ratios of i-C_4/n-C_4 and i-C_5/n-C_5.*
It is generally found that in cracking or hydrocracking (with weak hydrogenation function) of paraffins at moderate temperatures, where the products are determined by kinetics rather than thermodynamics, the ratio of iso- to n-butane and that of iso- to n-pentane greatly exceed their thermodynamic equilibrium values.

As shown in Fig. 8, the rate of formation of a cracked paraffin depends on the concentration of the precursor carbenium ion and its rate constant for hydrogen transfer. The high iso-to-normal ratio is primarily the result of the higher concentration of the tertiary carbenium ions. The value of the ratio of rate constants, k_i/k_n, is not known; it may be assumed that the more stable tertiary carbenium ion is less reactive than the secondary carbenium ion, and k_i/k_n should be < 1. Thus the selective formation of isoparaffinic cracked products is probably a carbenium ion concentration effect, because the equilibrium between secondary and tertiary ions strongly favors the tertiary ions.

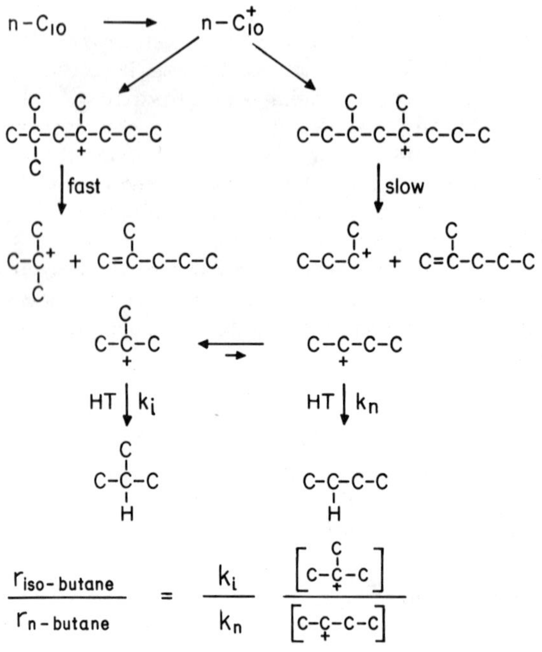

FIGURE 8 Formation of isobutane and n-butane from the cracking of decane.

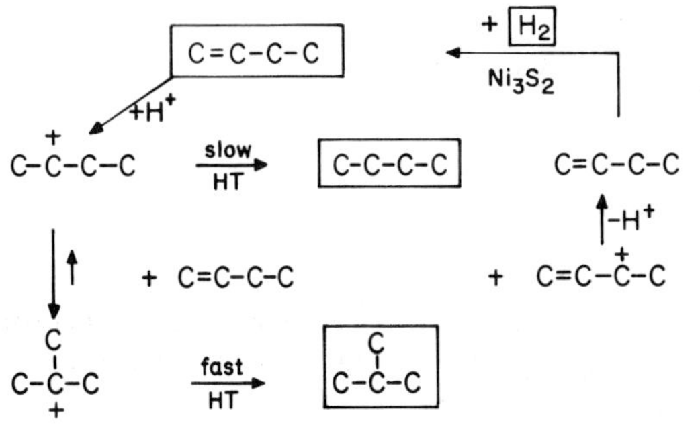

FIGURE 9 Hydroisomerization of 1-butene to isobutane.

A similar explanation can be given for the selective hydroisomerization of n-butenes to isobutane over an acidic catalyst containing NiS_x. Here the weak metal function selectively hydrogenates the butadiene produced by the hydrogen transfer reaction to butenes and prevents the catalyst from rapid deactivation. This is shown schematically in Fig. 9.

C. Paraffin Alkylation

Hydrogen transfer plays a number of key roles in the alkylation of an isoparaffin, usually isobutane, with an olefin, such as ethene, propene, and butene. For example, the tertiary butyl carbenium ion intermediate is formed from isobutane by hydrogen transfer reaction with a carbenium ion. The hydrogen transfer reaction not only activates the reactant isoparaffin and propagates the alkylation reaction, but also produces the saturated alkylation product as a result.

The relative rate of hydrogen transfer reaction also determines the isomer distribution of the alkylate. Since tertiary carbenium ions are present in larger concentrations than are secondary or primary carbenium ions, the alkylation of isobutane with ethene yields exclusively 2,3-dimethylbutane as the primary product, rather than the expected 2,2-dimethylbutane, which does not contain a tertiary C-H bond.

The reaction product can be clearly understood by comparing the relative rates of isomerization and hydrogen transfer of the carbenium ion intermediates (Fig. 10).

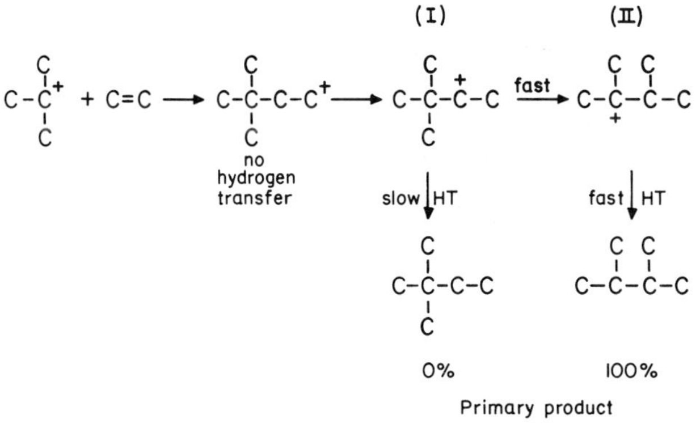

FIGURE 10 Reaction pathway for the alkylation of isobutane with ethene.

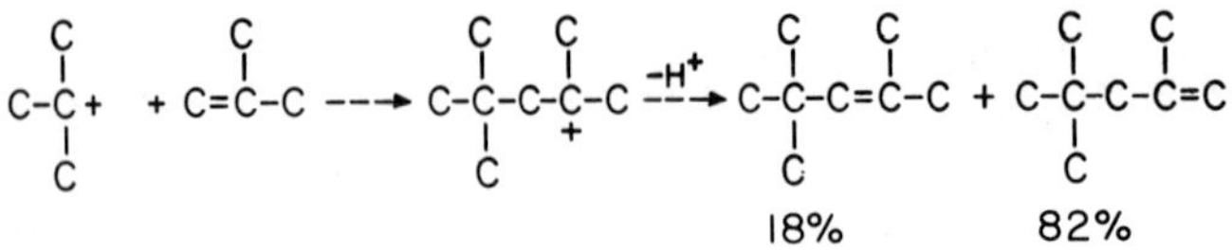

$$\underset{\underset{C}{\overset{C}{|}}}{C-C+} + C=\overset{C}{\overset{|}{C}}-C \longrightarrow \underset{\underset{C}{\overset{C}{|}}}{C-\overset{C}{\overset{|}{C}}-C-\overset{C}{\overset{|}{C}}-C} \xrightarrow{-H^+} \underset{\underset{C}{\overset{C}{|}}}{C-\overset{C}{\overset{|}{C}}-C=\overset{C}{\overset{|}{C}}-C} + \underset{\underset{C}{\overset{C}{|}}}{C-\overset{C}{\overset{|}{C}}-C-\overset{C}{\overset{|}{C}}=C}$$

　　　　　　　　　　　　　　　　　　　　　　　18%　　　　　82%

FIGURE 11 Dimerization of isobutene.

Isobutane is unique because it has no secondary C-H bonds. Its alkylation products can be predicted from known mechanisms. Higher isoparaffins, on the other hand, form much more complex alkylation products. For example, isopentane forms a t-amyl ion which contains a secondary C-H bond. It loses a proton more readily to form an olefin, which either alkylates with another isopentane or self-polymerizes.

Alkylation with higher olefins, such as pentene, produces complex alkylates not only because of the large number of isomers present, but also because hydrogen transfer occurs more readily than with lower olefins.

D. Conjunct Polymerization

Olefin oligomerization and polymerization reactions occur very readily over acid catalysts. The polymer product distributions from C_3 to C_7 olefins have been studied in detail by Schmerling and Ipatieff [5] and were shown to obey some basic rules governing the carbenium ion mechanism. Specifically, double-bond and skeletal isomerization of carbenium ions occur rapidly to form preferentially the most stable tertiary carbenium ion. Expulsion of a proton from a primary carbon atom of a carbenium ion is easier than expulsion of a proton from a secondary carbon: The released proton attacks another olefin and the reaction propagates in a chain fashion. For example, consistent with these rules, the dimerization of isobutene yields 18% 2,4,4-trimethylpentene-2 and 82% 2,4,4-trimethylpentene-1 (Fig. 11).

At higher temperatures or longer contact times, however, other side reactions, including hydrogen transfer reactions, can predominate over the acid-catalyzed oligomerization reaction of olefins [6]. Under these conditions, termed conjunct polymerization, the product comprises not only olefinic polymers but also saturated and highly unsaturated cyclic hydrocarbons, resulting from hydrogen transfer reactions. For example, in addition to olefinic products, paraffins, cycloparaffins, cycloolefins, and aromatics were formed when propene was polymerized at 330 to 370°C using 90% phosphoric acid (Table 3).

TABLE 3 Products from Polymerization of Propene over Phosphoric Acid

	Composition (wt %)
Paraffins	15
Olefins	63
Cycloparaffins	10
Cycloolefins	6
Aromatics	6

Source: Ref. 6.

The importance of hydrogen transfer in acid catalysis of olefin polymerization was also recognized by Schmerling [5] and others [7].

Ipatieff et al. [8,9] polymerized ethylene over phosphoric acid catalysts at 250 to 330°C and pressures of about 50 atm. Paraffinic, olefinic, naphthenic, and aromatic hydrocarbons were obtained. No aromatics were found in thermally produced polymers and the catalyst must, therefore, have hydrogen transfer properties.

E. Aromatization

As discussed earlier, in catalytic cracking, hydrogen transfer reactions forming excess saturated cracked products coproduce unsaturated products, including aromatics. Similarly, in conjunct polymerization, hydrogen-rich paraffins and hydrogen-poor aromatics are coproduced. The reaction occurs not only between olefins but also between olefins and paraffins, and cycloolefins and naphthenes. For example, Pines and his co-workers [10] studied the reaction of monocyclic olefins and diolefins over an alumina/hydrogen chloride catalyst and found, for example, that ethylcyclohexene was converted to aromatic hydrocarbons and cycloparaffins.

The reaction stoichiometry is such that 4 mol of olefins and/or naphthenes produce 1 mol of aromatics. In practice, this reaction contributes to the formation of aromatics in catalytic cracking, and as will be shown later, it assumes major importance in zeolite catalysis.

F. Coke Formation

The formation of "coke" over acidic catalysts involves a complex set of reactions. From the foregoing discussion, there is little doubt that reactions involving hydrogen transfer among hydrogen-poor molecules play a major role in "coke" formation. It is reasonable to expect that successive hydrogen transfer reactions will lead to the formation of condensed polyunsaturated and cyclic molecules—principal components of "coke."

Coking is the major cause of deactivation of cracking catalysts. For this reason, current catalytic cracking processes operate in a cyclic mode, alternating between reaction and coke burning.

To avoid frequent catalyst regeneration while maintaining the characteristics of acid cracking reactions, hydrocracking catalysts have been developed by incorporating a weak hydrogenation function. Operating in a hydrogen atmosphere, the weak hydrogenation function hydrogenates the coke precursors without actively participating in the main reaction.

IV. CLASSIFICATION OF ZEOLITES

Zeolites are characterized by their unique, uniform pore/channel systems. Available zeolites cover a pore size range of about 4 to 13 Å, similar to the critical dimensions of many hydrocarbon molecules present in petroleum. Thus, as pointed out by Weisz [11], the transport of reactants and products and their intermediates in zeolites falls in a regime beyond the Knudson diffusion regime, known as the configurational diffusion regime (Fig. 12). Configurational diffusion occurs in situations where the structural dimensions of the catalyst approach those of the diffusing molecules. In this case, even subtle changes in the dimensions of molecules can result in large changes in diffusivity. Zeolites of interest to catalysis may be divided into three major groups according to their pore/channel systems.

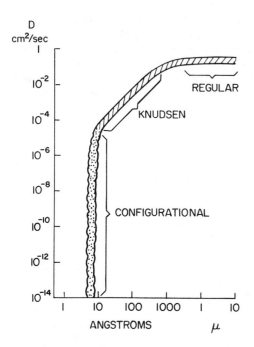

FIGURE 12 Diffusivity and size of aperture (pore): the classical regions of regular and Knudsen and the new regime of configurational diffusion. (Data from Ref. 11.)

A. Eight-Membered Oxygen Ring Systems

These include most of the earliest known shape-selective zeolites, such as Linde A, erionite, and chabazite. Other members of this group include ZK-5, high-silica analogs of Linde A, zeolite alpha, ZK-4, ZK-21, and ZK-22, and several other less common natural zeolites.

The pore-channel systems of these zeolites also contain "supercages," which are much larger than the size of the pore openings. These zeolites sorb only straight-chain molecules, such as n-paraffins and olefins. Therefore, they cannot accommodate molecules with methyl branches, except perhaps in the supercages.

B. Ten-Membered Oxygen Ring Systems

These include ZSM-5; ZSM-11; ZSM-22; Theta-1, which is isostructural with ZSM-22; ZSM-23; ZSM-48; and laumontite. Except laumontite, which has puckered 10-membered oxygen rings in its structure, almost all medium-pore zeolites of interest to catalysis are synthetic in origin. Among the zeolites in this group, only ZSM-5 and ZSM-11 have bidirectional intersecting channels; the others have nonintersecting unidirectional channels.

These zeolites sorb straight-chain and slightly branched-chain molecules and exclude highly branched molecules. Molecules diffuse typically in the configurational diffusion regime. For example, o-xylene diffuses more than three orders of magnitude slower than its para isomer in ZSM-5. Shown in Fig. 13 are data for several aromatic hydrocarbons, obtained by the gravimetric adsorption method [12]. Also included are diffusivity data for three hexane isomers obtained from the relationship between crystal size and observed activity [13]. Again, the various isomers differ by a factor of $>10^3$ in diffusivity.

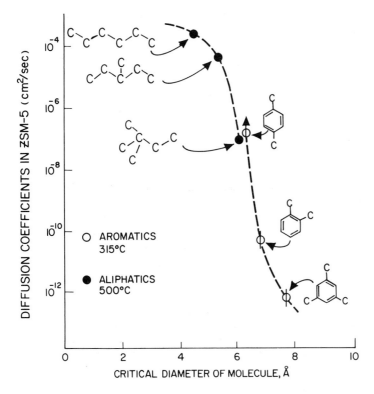

FIGURE 13 Diffusion coefficients of representative paraffins and aromatics in the 10-membered-ring zeolite, HZSM-5. (Data from Refs. 12 and 13.)

Detailed analysis of the composition of cracked products from the hydrocracking and hydroisomerization of n-paraffins over ZSM-5 [14] suggests that reactions involving tertiary carbenium ion intermediates do not take place in ZSM-5. The relatively high yield of C_3 hydrocarbons, typical of medium- and small-pore zeolites, is a consequence of β scission and hydrogen transfer with secondary and perhaps primary propyl carbenium ions.

C. Twelve-Membered Oxygen Ring Systems

These include the well-known faujasites and mordenite. Other members of this group include Linde L; mazzite, which is isostructural with ZSM-4 and Zeolite Omega; ZSM-12, and ZSM-20.

The pore/channel systems of some of these zeolites, faujasites, and ZSM-20 also contain "supercages," which are much larger than the size of the pore openings. Others contain straight channels. They sorb most of the hydrocarbon molecules present in petroleum gas oils.

V. MAJOR EFFECTS OF ZEOLITES ON HYDROGEN TRANSFER REACTIONS

A. Steric Inhibition or Spatial Constraint

Hydrogen transfer between two molecules involves the alignment of a reaction complex which is considerably larger than the individual molecules. In a sterically restricted environment, such as the pores of a zeolite, the formation

TABLE 4 Molecular Cross Sections of Reactants and Hydrogen Transfer Transition States (Å)

	Reactant	Transition state
n-Hexane	3.9 × 4.3	4.9 × 6.0
3-Methylpentane	4.4 × 5.8	6.0 × 7.0
Nominal pore size of ZSM-5		5.4 × 5.6

of such a reaction complex could be sterically constrained. The extent of steric effects may range from complete inhibition, as in some small-pore zeolites, to subtle changes in product distribution due to the change in relative rates of various competing reactions in a sterically restricted system. Following are some examples of steric inhibition effects.

Relative Rate of Cracking n-Hexane and 3-Methylpentane: The Constraint Index

The Constraint Index test [15] is a useful diagnostic test devised to distinguish medium-pore zeolites from large- and small-pore zeolites. The numerical value determined by this test is approximately the ratio of the cracking rate constants for n-hexane and 3-methylpentane. As discussed earlier, in sterically unrestricted systems, 3-methylpentane cracks 2.5 times faster than n-hexane (i.e., these systems have a constraint index of 0.4). By contrast, n-hexane cracks at a rate of up to about 10 times faster than 3-methylpentane over medium-pore zeolites such as ZSM-5. As shown previously [13], this is a result of steric inhibition of 3-methylpentane cracking, since its hydrogen transfer transition state has a larger cross section than that of n-hexane (Table 4).

Because of spatial constraints on the bimolecular cracking reaction, a second type of cracking mechanism, a monomolecular path, has been observed. While it occurs with all amorphous and crystalline acid catalysts examined at high temperatures, it can dominate the cracking reaction in medium-pore zeolites. This monomolecular cracking mechanism involves direct protonation of the paraffin molecule to form a penta-coordinated carbonium ion (Fig. 14). Unlike cracking by the classical bimolecular reaction, the paraffinic products are not formed by hydrogen transfer reactions, and their distribution depends strongly on the structure of the reactant (e.g., 3-methylpentane cracks exclusively to methane and ethane, 2,4-dimethylpentane to methane

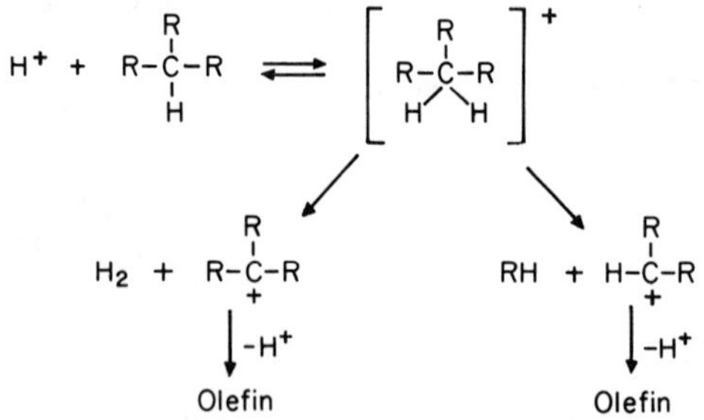

FIGURE 14 Monomolecular paraffin cracking mechanism via protonated penta-coordinated carbonium ion.

and isobutane, and 2,3-dimethylbutane to methane and propane) [16].
Significantly, hydrogen is also produced as a primary cracking product.
The mechanism advanced [16] provides a rationale for this acid-catalyzed
formation of hydrogen at high temperatures, and of the reverse reaction,
the "hydrogenation" activity of acid catalysts in the absence of metals. The
formation of hydrogen can be considered to be a result of hydride transfer
from a paraffin to a proton.

Since the monomolecular reaction is not spatially constrained by the available intracrystalline space as are the bimolecular reaction intermediates, the size difference between n-hexane and 3-methylpentane should be partially compensated by the difference in their chemical reactivity. This rationale is consistent with the experimental observation that the relative rate of cracking n-hexane to 3-methylpentane in ZSM-5 approaches unity with increasing temperature [15,17].

Nonaging Properties of Medium-Pore Zeolites

As discussed earlier, catalyst aging due to "coke" buildup involves a series of hydrogen transfer reactions among olefins and aromatics to form condensed polyunsaturated and cyclic carbonaceous molecules. Earlier, Venuto and Hamilton [18] described the aging of faujasite, a large-pore zeolite, as a "reverse" molecular shape selectivity, because the bulky molecules trapped inside the supercages are too large to escape through the pore openings.

The channel structures of medium-pore zeolites that do not contain supercages are believed to be a major reason for their low hydride transfer activity and hence their nonaging properties [19]. As an example, Fig. 15 compares the activity for methanol conversion as a function of reaction time for a number of zeolites; ZSM-5 is by far the most stable catalyst.

B. Superactivity

Compared to amorphous silica-alumina catalyst, zeolites have about the same amount of BET surface area and aluminum concentration, yet some zeolites

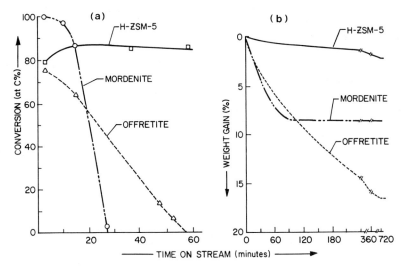

FIGURE 15 Conversion of methanol (a) and catalyst weight gain due to coking (b) as a function of reaction time for various zeolite catalysts. (Data from Ref. 20.)

have been shown to have several orders of magnitude higher intrinsic acid activity [21, 22].

A number of factors probably contribute to the observed superactivity: (a) catalytic acidity in zeolites is derived from protons associated with the tetrahedral aluminum in the crystalline framework, and zeolites have higher concentrations of accessible tetrahedral aluminum than of amorphous silica-aluminas; (b) the Bronsted acid sites in zeolites have higher specific intrinsic activities than those in amorphous silica-aluminas; and (c) the hydrocarbon sorption equilibrium in zeolites is more favorable in zeolite than in amorphous solids, thus the reactant concentration is higher.

Among these three factors, we believe that higher reactant concentration plays a significant role in the superactivity found with zeolites. High reactant concentrations not only increase the reaction rate but also favor bimolecular reactions, such as hydrogen transfer and paraffin cracking, over monomolecular reactions, such as olefin isomerization and cracking.

For monomolecular reactions such as the cracking of n-octene, HY and REY can be 30 times more active than an amorphous silica-alumina. The same zeolite cracks a paraffinic molecule, such as n-decane, over 1000 times faster than an amorphous silica-alumina, because of the effect of higher reactant concentration on bimolecular reactions.

C. Other Subtle Effects

Several zeolites, such as ZSM-4 and mordenite, show remarkable activity and selectivity for the isomerization and disproportionation of light paraffins.

For example, mordenite is known to have a higher selectivity for paraffin isomerization than that of amorphous silica-alumina or other large-pore zeolites. Guisnet et al. [23] and Hilaireau et al. [24] found that hydrogen mordenite also catalyzes the disproportionation of n-butane to produce equimolar yields of propane and pentanes. Recently, Chen reported the remarkable activity of ZSM-4 for these reactions, and showed that it is at least 20 times more active than mordenite [25].

The only feature of these two zeolites that is different from those of other large-pore zeolites is the shape of their channels. They are straight and do not contain either intersections or supercages. Whether this feature provides a specific site geometry that promotes interaction of carbenium ions with reactant molecules remains to be determined.

However, higher local reactant concentration could favor the relative rate of isomerization over cracking, even though both require bimolecular hydrogen transfer reactions. Figure 16 shows the reaction pathways of these two reactions. The isomerized tertiary carbenium ion can either crack by β scission (a), or abstract hydrogen from another paraffin molecule to yield the isomer (b). Thus, the relative rate of isomerization and cracking (i.e., the isomerization selectivity) depends on the hydrocarbon reactant concentration.

$$r_{Cr} = k_{Cr}[C^+]$$

$$r_{Iso} = k_{HT}[C^+][HC]$$

$$\frac{r_{Iso}}{r_{Cr}} = k[HC]$$

The effective hydrocarbon concentration is that prevailing near the active site where the carbenium ion is generated, that is, in the zeolite pore, $[HC]_S$. Using a Langmuir adsorption equation, its concentration is

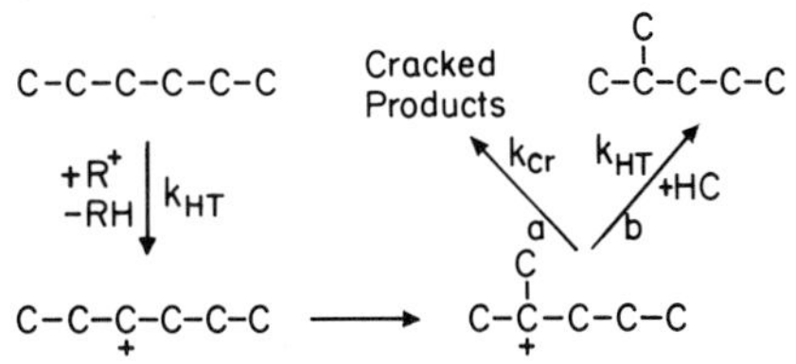

27. Hydrogen Transfer in Catalysis on Zeolites

```
C-C-C-C-C-C        Cracked        C
                   Products       |
                              C-C-C-C-C
  +R⁺ |                  ↖k_cr  k_HT↗
  -RH | k_HT                  ↘    ↗+HC
      ↓                      a\  /b
                                C
                                |
C-C-C-C-C-C   ⟶   C-C-C-C-C
      +                +
```

FIGURE 16 Acid-catalyzed paraffin isomerization versus cracking.

$$[HC]_S = \frac{K[HC]}{1 + K[HC]} \simeq K[HC]$$

where $[HC]_S$ and $[HC]$ are the sorbed and the fluid-phase concentrations, respectively. The isomerization selectivity thus is given by

$$S_{Iso} = \frac{r_{Iso}}{r_{Cr}} = \frac{k_{HT}}{k_{Cr}} K \frac{[HC]}{1 + K[HC]}$$

For a low sorption equilibrium value, the selectivity is a linear function of $[HC]$ and depends on the sorption constant K.

Steric inhibition, on the other hand, can override the concentration effect of some zeolites. For example, the medium-pore zeolite ZSM-5 has a poorer selectivity for isomerization than does mordenite, because its hydride transfer activity (k_{HT}) is low as a result of steric inhibition.

VI. COMMERCIAL IMPLICATIONS

A. Catalytic Cracking of Gas Oils

The very large impact of zeolites on gas-oil cracking has been well recognized over the years [11]. As discussed earlier, hydrogen transfer is a principal reaction in catalytic cracking and is responsible for the formation of isoparaffins and aromatics. The outstanding selectivity of zeolites in gas-oil cracking for the production of gasoline was recognized long before their reaction pathways were known [26,27].

Figure 17 shows a comparison of the composition of gasoline produced over the zeolite catalyst and the amorphous catalyst. The gasoline produced by zeolite cracking contains more aromatics and paraffins and fewer olefins and naphthenes than that produced by the amorphous catalyst. This difference is consistent with the expected higher rate of hydrogen transfer in the zeolite catalyst [28,29], which is clearly reflected in a shift of hydrogen according to

olefins + naphthenes → paraffins + aromatics

The hydrogen transfer reaction "stabilizes" the gasoline product toward further cracking by converting the reactive olefinic and naphthenic molecules to the more stable paraffinic and aromatic molecules.

The excellent hydrogen transfer activity of large-pore zeolites also contributes to increasing the gasoline yield by higher single-pass gas-oil conversion. With the amorphous silica-alumina catalyst, the gasoline yield begins to reach an asymptotic value at conversions above about 60% conversion. The zeolite cracking catalyst, with its higher hydrogen transfer activity, reduces the concentration of olefin, which leads to overcracking, and enables

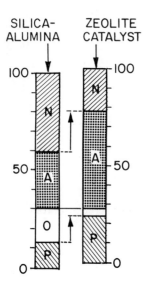

FIGURE 17 Shift of the compound classes (P, paraffins; O, olefins; N, naphthenes; A, aromatics) in the gasoline product from cracking the same hydrotreated petroleum fraction over amorphous silica-alumina and over zeolite-cracking catalyst. (Data from Ref. 11.)

much deeper conversion without losing gasoline selectivity. This can be understood by the following reaction scheme:

$$C_1 \xrightarrow{k_0} a_1 C_2 + a_2 C_3$$

$$C_2 \xrightarrow{k_2} C_3$$

where

C_1 = gas oil charged
C_2 = gasoline
C_3 = C_4's, dry gas and coke
a_1, a_2 = mass ratios of C_2/C_1 and C_3/C_1, respectively
k_0, k_2 = rate constants for gas-oil cracking and gasoline recracking, respectively

Weekman and Nace [30] showed that for an isothermal vapor-phase plug flow reactor with negligible interparticle diffusion, the foregoing reactions can be modeled to account for the effect of process variables on the conversion and gasoline yields in terms of modified reaction rate constants, K_0, K_1, and K_2, with $K_1/K_0 = a_1$, the initial gasoline selectivity; and the relative rate of gasoline overcracking to gas-oil cracking given by the ratio K_2/K_0. Thus, for a given initial gasoline selectivity, an increase in the overcracking ratio reduces the gasoline selectivity at high gas-oil conversions, as shown in Fig. 18.

Gas-oil cracking over zeolites also produces less "coke" and dry gases. Hydrogen transfer reactions are major contributors to coke and dry gas production. However, the reduction in overcracking preserves the gasoline fraction such that both coke and dry gas are significantly reduced. An

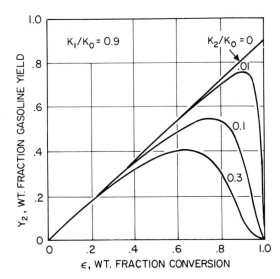

FIGURE 18 Effect on selectivity of varying gasoline-to-gas oil cracking ratios at constant initial selectivity. (Data from Ref. 30.)

example of this was shown by Venuto and Habib [31], based on published data [32], and is illustrated in Table 5.

B. Catalytic Dewaxing

The application of shape-selective zeolites in catalytic dewaxing of gas oils and lube oils has gained worldwide acceptance in recent years. Catalytic dewaxing of gas oils increases the supply of low-pour-point jet fuels and other distillate fuels by selectively converting normal and slightly branched paraffins from the oil to improve its fluidity. Catalytic dewaxing of lube oils replaces the traditional solvent dewaxing processes.

These selective cracking processes make use of the size exclusion and spatioselective properties of medium-pore zeolites to achieve their objectives through the subtleties of hydrogen transfer reactions discussed earlier. Thus the rate of paraffin cracking is found to change with respect to the chain length and the degree of branching. For example, Fig. 19 shows the relative rate of cracking of each component when a mixture of C_5 to C_7 hydrocarbons was passed over HZSM-5 at 1.4 LHSV, 35 atm, and 340°C.

C. Octane Boosting

The selective paraffin cracking reaction also finds application in improving the octane rating of gasoline products. The Selectoforming process [33] operates in conjunction with a naphtha reforming process. It upgrades the octane rating of the reformate and produces LPG (primarily propane) as the major by-product. The catalyst uses a small-pore zeolite, erionite. It cracks only the low-octane straight-chain paraffins in the reformate. To assure catalyst stability, a weak metal function was incorporated into the catalyst to saturate olefins and prevent coke formation but not to hydrogenate the aromatics present in the feed.

The M-Forming process [34] is another post-reforming process. It uses ZSM-5 as the catalyst. For octane boosting, its cracking selectivity is a most

TABLE 5 Comparison of Yield Structure for Fluid Catalytic Cracking of Waxy Gas Oil over Commercial Equilibrium Zeolite and Amorphous Catalysts

Yields at 80 vol % conversion	Amorphous, high alumina	Zeolite XZ-25	Change from amorphous
Hydrogen (wt %)	0.08	0.04	-0.04
$C_1 + C_2$'s (wt %)	3.8	2.1	-1.7
Propene (vol %)	16.1	11.8	-4.3
Propane (vol %)	1.5	1.3	-0.2
Total C_3's	17.6	13.1	-4.5
Butenes (vol %)	12.2	7.8	-4.4
i-Butane (vol %)	7.9	7.2	-0.7
n-Butane (vol %)	0.7	0.4	-0.3
Total C_4's	20.8	15.4	-5.4
C_5-390 at 90% ASTM gasoline (vol %)	55.5	62.0	+6.5
Light fuel oil (vol %)	4.2	6.1	+1.9
Heavy fuel oil (vol %)	15.8	13.9	-1.9
Coke (wt %)	5.6	4.1	-1.5
Gasoline octane no., R + O	94	89.8	-4.2

Source: Ref. 31.

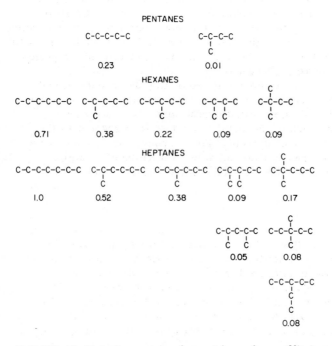

FIGURE 19 Relative rates of cracking of paraffins over ZSM-5 at 1.4 LHSV, 35 atm, 340 C. (Data from Ref. 17.)

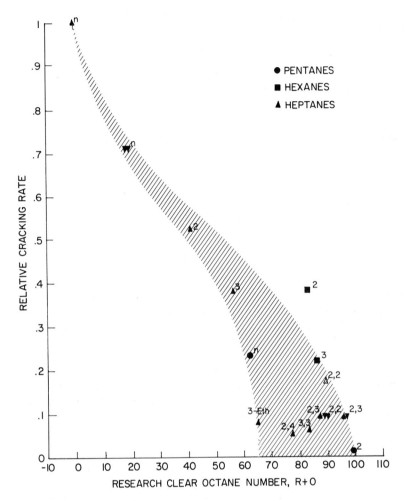

FIGURE 20 Relationship between octane rating and cracking rate of paraffins. (Data from Ref. 35.)

desirable feature, because the octane rating of paraffinic molecules increases as the cracking rate decreases. Figure 20 shows the relationship between the relative cracking rate of C_5 to C_7 paraffins and their octane ratings.

Use of ZSM-5 as an adjunct to a conventional gas-oil cracking catalyst is another application of its selective cracking properties. It boosts the octane rating of the cracked gasoline one to three numbers by selectively cracking the low-octane component to C_3 and C_4 olefins and paraffins [36-39]. The light olefins can be fed to an isobutane alkylation unit to produce additional high-octane gasolines, raising the octane pool of the refinery.

D. Aromatization

With medium-pore zeolites, light olefins undergo rapid isomerization, oligomerization, and interconversion reactions at low temperatures with remarkable selectivity and catalyst stability not found with large-pore zeolites [40,41].

As shown in Table 6, with medium-pore zeolites such as HZSM-5, the olefinic product from different feeds is substantially independent of the feed

TABLE 6 Comparison of Pentene Product Isomers, Obtained with HZSM-5 Catalyst, with Equilibrium at 275°C

Pentene isomer (%)	Feed					Equilibrium
	Ethene	Propene	Pentene Mix	1-Hexene	1-Decene	
1-Pentene	2	2	2	2	2	2
2-Methyl-1-butene	18	16	18	18	17	24
3-Methyl-1-butene	1	2	2	2	1	2
trans-2-Pentene	11	10	11	12	13	9
cis-2-Pentene	5	5	5	5	6	7
2-Methyl-2-butene	63	65	62	61	61	56

Source: Data from Ref. 39.

molecules. This suggests a rapid equilibration of olefins to a composition determined by the temperature and pressure of the system [42].

At longer contact times or increased temperatures, a shift in product distribution toward BTX aromatics and light paraffins occurs (Table 7), as hydrogen transfer and cracking reactions become appreciable.

Aromatization reactions of olefins become significant at about 370°C at atmospheric pressures over HZSM-5. Paraffin aromatization, on the other hand, requires a higher temperature. The reaction products are again found to be independent of feed composition and reflect the equilibrium concentra-

TABLE 7 Conversion of Propene over HZSM-5 at 390°C, 1 atm, 0.5 WHSV

Carbon no.	Product (wt %)				
	Olefin	Isoparaffin	n-Paraffin	Aromatics	Total
1	—	—	0.2	—	0.2
2	—	—	0.8	—	0.8
3	0.4	—	23.3	—	23.7
4	0.7	17.1	9.1	—	26.9
5	0.1	7.2	1.7	—	9.0
6	0.1	1.0	0.2	2.0	3.3
7	0.1	0.9	0.2	10.7	11.9
8	—	—	—	12.8	12.8
9	—	—	—	6.3	6.3
10	—	—	—	2.1	2.1
11+	—	—	—	3.0	3.0
Total	1.4	26.2	35.5	36.9	100.0

Source: Data from Ref. 39.

TABLE 8 Hydrocarbon Product Distribution in Long-Term Aging Test

	Cycle 1		Cycle 6		Cycle 9	
Time on stream in cycle (h)	73	314	94	333	77	316
Cumulative charge, wt of MeOH/wt of conv. cat.	117	501	4162	4545	7123	7504
Hydrocarbon composition (wt %)						
Methane	2.1	1.8	0.9	0.7	0.9	0.8
Ethane	1.3	0.4	0.6	0.2	0.4	0.3
Ethene	0.0	0.0	0.0	0.0	0.0	0.0
Propane	11.6	4.6	6.6	3.0	5.3	3.9
Propene	0.1	0.2	0.2	0.2	0.2	0.3
n-Butane	5.9	2.6	3.7	2.0	3.1	2.3
Isobutane	8.5	8.5	9.3	6.4	8.9	8.0
Butenes	0.5	1.1	1.0	1.1	1.2	1.4
Total C_4^-	30.0	19.2	22.3	13.6	20.0	17.0
n-Pentane	2.7	1.3	1.7	1.2	1.6	1.2
Isopentane	10.1	11.3	11.4	10.1	11.2	10.6
Pentenes	0.7	2.4	1.9	2.8	2.3	2.9
Cyclopentane	0.3	0.2	0.3	0.3	0.3	0.3
C_6^+ nonaromatic	20.0	38.1	31.5	43.6	35.3	41.1
Benzene	0.4	0.2	0.2	0.2	0.2	0.2
Toluene	4.4	1.7	2.0	1.8	2.1	1.8
Ethylbenzene	0.9	0.5	0.6	0.6	0.6	0.6
p- and m-Xylenes	9.5	6.7	7.4	7.1	7.2	6.2
o-Xylene	2.7	1.7	2.0	1.8	2.0	1.7
Trimethylbenzenes	8.4	7.3	8.2	7.3	7.3	6.5
Methylethylbenzenes	2.3	2.5	2.8	2.8	2.8	2.6
Propylbenzenes	0.1	0.1	0.1	0.2	0.2	0.3
1,2,4,5-Tetramethylbenzene	3.6	4.2	4.2	3.6	3.2	3.2
1,2,3,5-Tetramethylbenzene	1.1	0.2	0.4	0.2	0.4	0.1
1,2,3,4-Tetramethylbenzene	0.4	0.1	0.2	0.1	0.1	0.1
Other C_{10} benzenes	1.5	1.6	1.9	1.9	1.9	1.8
C_{11} alkylbenzenes	0.8	0.6	0.8	0.7	0.7	0.7
Naphthalenes	0.1	0.1	0.1	0.1	0.2	0.1
Unknowns	0.0	0.0	0.0	0.0	0.4	0.7
Total C_5^+	70.0	80.8	77.7	86.4	80.0	83.0
Total aromatics	36.1	27.5	31.0	28.5	29.4	26.7

Source: Ref. 45.

tions of the intermediate olefins [42]. They are formed in a consecutive reaction pathway via cracking and hydrogen transfer reactions. The yield of aromatics is subject to the stoichiometric constraint of the carbon/hydrogen balance. For example, for olefinic feeds 3 mol of paraffins is produced per mol of aromatics [43].

M2-forming, a generic name, has been coined to describe the reactions producing BTX aromatics from nonaromatic hydrocarbons over the medium-pore zeolites [44].

E. Methanol to Gasoline

Mobil's methanol-to-gasoline process (MTG) has attracted worldwide attention. Its discovery has been hailed as the first new route in more than 40 years for the production of gasoline from coal [45]. The chemistry of methanol/ dimethyl ether to hydrocarbons reactions over medium-pore zeolites has been comprehensively reviewed by Chang [46].

Table 8 shows some typical hydrocarbon products made from methanol in a fixed-bed reactor during a long-term aging test. In methanol conversion, where methylation of aromatic rings is an important reaction, there is sharp cutoff of the polyalkylbenzenes at a size no larger than 1,2,4,5-tetramethylbenzene.

The primary hydrocarbon products from methanol are light olefins. Other than for the effect of water and oxygenates, the role of hydrogen transfer in the conversion of olefins to other hydrocarbons is essentially analogous to that described previously.

F. Olefin Polymerization

Olefin polymerization over acid catalysts is well known. The production of polymer gasoline from olefins using phosphoric acid catalysts is practiced commercially. As discussed above, this and similar polymerization processes are always accompanied by hydrogen transfer reactions which yield polyunsaturated and aromatic products (conjunct polymerization) leading to short catalyst life. These problems are avoided by using medium-pore zeolites such as ZSM-5, which sterically inhibit these hydrogen transfer reactions. The use of ZSM-5 to produce gasoline and distillate from light olefins led to the development of Mobil's Olefin to Gasoline and Distillate Process [47,48].

In this process, light olefins are converted to higher-molecular-weight iso-olefins. The iso-olefins in the gasoline boiling range have a high octane number; those in the diesel fuel range have a high cetane number and a very low pour point (Table 9). The process is highly flexible and can be designed to produce from 100% gasoline to 80% distillate-20% gasoline. Feeds are normally propylene or butenes, but could range from ethylene to C_{12} and higher olefinic naphtha.

The process is based on the chemistry of the olefin interconversion reaction mentioned earlier, in which a pseudo-equilibrium of olefins is established by a series of polymerization and cracking steps [41], as shown in Fig. 21. At moderate temperatures ($\leq 175°C$), olefins constitute over 98% of the total

TABLE 9 Typical MOGD Distillate Product

	Raw	Hydrotreated
Specific gravity	0.79	0.78
Bromine no.	79	4.0
Hydrogen (wt %)	14.33	15.14
Pour point (°C)	<-51	<-51
Cetane no.	33	52
Sulfur (wt %)	<0.002	<0.002
Distillation (D2887) (°F)		
20%	396	390
50%	440	457
90%	633	648

Source: Ref. 47.

At low temperature: True Oligomerization

$$C_3 \longrightarrow C_6 \longrightarrow C_9 \longrightarrow C_{12} \cdots$$

At higher temperature (> 150 °C): Random Oligomerization

$$C_3 + C_6 \rightleftharpoons C_9 \rightleftharpoons C_4 + C_5$$

$$C_3 + C_4 \rightleftharpoons C_7$$

$$C_4 + C_4 \rightleftharpoons C_8 \rightleftharpoons C_3 + C_5$$

FIGURE 21 Reaction scheme for olefin oligomerization and random oligomerization (olefin interconversion).

product obtained with ZSM-5. This is in sharp distinction to the polymerization of olefins with non-shape-selective catalysts, such as phosphoric acid, whereby the polymerization is invariably accompanied by hydrogen transfer which yields paraffins and aromatics (Table 3). Shape-selective rate retardation of the bimolecular hydrogen transfer is again operating with the medium-pore zeolite ZSM-5.

At higher temperature, some hydrogen transfer does occur, forming first cyclo-olefins and paraffins, and then aromatics and paraffins [49], as shown in Fig. 22. Very little coke is formed, leading to long catalyst life. Large-pore zeolites and amorphous catalysts undergo severe hydrogen transfer and produce heavy aromatics and coke in a very short time under these conditions.

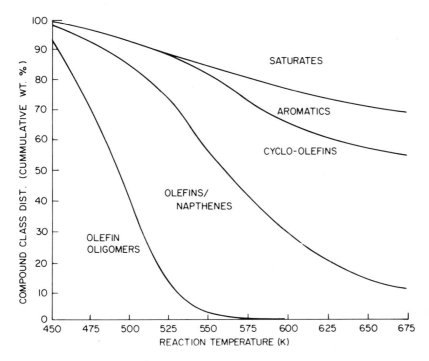

FIGURE 22 Propylene conversion over HZSM-5 at 54 atm, 1 WHSV. Effect of temperature on compound class distribution. (Data from Ref. 49.)

VII. CONCLUSIONS

Hydrocarbon conversions over solid acid catalysts occur through tri-coordinated carbenium ion intermediates. Typical carbenium reactions are: isomerization, addition to another olefin (oligomerization) or aromatic (alkylation), β scission (cracking), and finally, hydride abstraction from another molecule (hydrogen transfer), the subject of this chapter. As a result, isoparaffins crack more readily than normal paraffins, and isobutane and isopentane predominate over their n-isomers in the cracked products.

Relative to amorphous catalysts such as silica-alumina, large-pore zeolite catalysts such as zeolite Y exhibit enhanced hydrogen transfer activity. This is the result of a higher reactant concentration in the zeolite pores, which favors the bimolecular hydrogen transfer over monomolecular isomerization or cracking reactions. This important fact is responsible for the superactivity of zeolites and for the phenomenal success of zeolite catalysts in gas-oil cracking, where hydrogen transfer leads to greatly enhanced gasoline yields.

With medium-pore zeolites such as ZSM-5, however, the hydrogen transfer activity is greatly reduced, below that of amorphous catalysts. This results from a steric inhibition of the bulky bimolecular transition state for hydride transfer in the narrow pores of these zeolites. Thus olefin conversions such as polymerization/cracking can be carried out with a previously unattainable lack of hydrogen transfer side reactions such as conjunct polymerization and coking. Applications of these principles to a wide variety of commercial processes have been discussed.

One of the unanswered problems of catalytic carbenium ion chemistry is the relative stability of primary, secondary, and tertiary carbenium ions. While the relative stability of free carbenium ions in the gas phase is well known, in acid catalysis these ions are always associated with a negatively charged counter ion that forms an ion pair with varying degrees of covalent bonding between the carbenium ion and the anion. To establish the relative stability and hence the relative concentration of interconverting carbenium ions at various temperatures remains a challenging task that would lead to a quantitative kinetic understanding of activity and selectivity phenomena in acid catalysis.

REFERENCES

1. G. E. Langlois and R. F. Sullivan, *Refining Petroleum for Chemicals*, Advances in Chemistry Series 97, American Chemical Society, Washington, D.C. (1970), pp. 38-67.
2. F. P. Lossing and G. P. Semeluk, *Can. J. Chem.*, 48:955 (1970).
3. H. Pines, *The Chemistry of Catalytic Hydrocarbon Conversions*, Academic Press, New York (1981), p. 5.
4. H. Pines, *The Chemistry of Catalytic Hydrocarbon Conversions*, Academic Press, New York (1981), p. 23.
5. L. Schmerling and V. N. Ipatieff, *Adv. Catal.*, 2:21 (1950).
6. V. N. Ipatieff and H. Pines, *Ind. Eng. Chem.*, 28:684 (1936).
7. F. G. Ciapetta, *Ind. Eng. Chem.*, 37:1210 (1945).
8. V. N. Ipatieff and H. Pines, *Ind. Eng. Chem.*, 27:1364 (1935).
9. V. N. Ipatieff, R. E. Schaad, and W. B. Shaney, *Science of Petroleum*, Vol. 4, Part 2, Oxford University Press, Oxford (1953), p. 14.
10. H. Pines, R. C. Olberg, and V. N. Ipatieff, *J. Am. Chem. Soc.*, 74:4872 (1952).
11. P. B. Weisz, *Chem. Technol.*, 3:498 (1973).

12. D. H. Olson, G. T. Kokotailo, S. L. Lawton, and W. M. Meier, *J. Phys. Chem.*, 85:2238 (1981).
13. W. O. Haag, R. M. Lago, and P. B. Weisz, *Faraday Discuss. Chem. Soc.*, 72:317 (1982).
14. J. Weitkamp, P. A. Jacobs, and J. A. Martens, *Appl. Catal.*, 8:123 (1983).
15. V. J. Frilette, W. O. Haag, and R. M. Lago, *J. Catal.*, 67:218 (1981).
16. W. O. Haag and R. M. Dessau, *Proceedings of 8th International Congress on Catalysis*, Vol. 2 (1984), p. 305.
17. N. Y. Chen and W. E. Garwood, *J. Catal.*, 52:453 (1978).
18. P. B. Venuto and L. A. Hamilton, *Ind. Eng. Chem. Prod. Res. Dev.*, 6:190 (1967).
19. L. D. Rollmann and D. E. Walsh, *J. Catal.*, 56:139 (1979).
20. P. Dejaifve, A. Auroux, P. C. Gravelle, J. C. Vedrine, Z. Gabelica, and E. G. Derouane, *J. Catal.*, 70:123 (1981).
21. J. N. Miale, N. Y. Chen, and P. B. Weisz, *J. Catal.*, 6:278 (1966).
22. P. B. Weisz and J. N. Miale, *J. Catal.*, 4:527 (1965).
23. M. Guisnet, N. S. Gnep, C. Bearez, and F. Chevalier, *Catalysis by Zeolites* (B. Imelik et al., 3ds.), Elsevier, Amsterdam (1980), p. 77.
24. P. Hilaireau, C. Bearez, F. Chevalier, G. Perot, and M. Guisnet, *Zeolites*, 2:69 (1982).
25. N. Y. Chen, *Proceedings of the 7th International Zeolite Conference* (Y. Murakami, A. Iijima, and J. W. Ward, eds.), Elsevier, Amsterdam (1986), p. 653.
26. S. C. Eastwood, R. D. Drew, and F. D. Hartzell, *Oil Gas J.*, 60(44):152 (1962).
27. C. J. Plank, E. J. Rosinski, and W. P. Hawthorne, *Ind. Eng. Chem. Prod. Res. Dev.*, 3:165 (1964).
28. P. B. Weisz, *Ann. Rev. Phys. Chem.*, 21:175 (1970).
29. S. C. Eastwood, C. J. Plank, and P. B. Weisz, *Proceedings of the 8th World Petroleum Congress*, Vol. 4 (1971), p. 245.
30. V. W. Weekman and D. M. Nace, *AIChE J.*, 16:397 (1970).
31. P. B. Venuto and E. T. Habib, *Catal. Rev.-Sci. Eng.*, 18:1 (1978).
32. J. J. Blazek, *Oil Gas J.*, 69(45):66 (1971).
33. N. Y. Chen, J. Maziuk, A. B. Schwartz, and P. B. Weisz, *Oil Gas J.*, 66(47):154 (1968).
34. N. Y. Chen, W. E. Garwood, and R. H. Heck, *Ind. Eng. Chem. Res.*, 26:706 (1987).
35. N. Y. Chen and W. E. Garwood, *Catal. Rev.-Sci. Eng.*, 28:185 (1986).
36. C. D. Anderson, F. G. Dwyer, G. Koch, and P. Niiranen, presented at the *9th Iberoamerican Symposium on Catalysis*, Lisbon (1984).
37. D. M. Nace and H. Owen, U.S. Patent 3,894,931 (1975).
38. A. W. Chester, W. E. Cormier, and W. A. Stover, U.S. Patent 4,309,279 (1982).
39. E. J. Rosinski and A. B. Schwartz, U.S. Patent 4,309,280 (1982).
40. W. E. Garwood, *Am. Chem. Soc. Symp. Ser.*, 218:383 (1983).
41. W. O. Haag, *Proceedings of the 6th International Zeolite Conference* (D. H. Olson and A. Bisio, eds.), Butterworth, London (1983), p. 466.
42. W. O. Haag, P. G. Rodewald, *J. Mol. Catal.*, 17:161 (1982).
43. W. O. Haag, in *Heterogeneous Catalysis* (B. L. Shapiro, ed.), Texas A&M University Press, College Station, Tex. (1984), p. 95.
44. N. Y. Chen and T. Y. Yan, *Ind. Eng. Chem. Process Des. Dev.*, 25(1):151 (1986).
45. S. L. Meisel, *Philos. Trans. R. Soc. London*, A300:157 (1981).
46. C. D. Chang, *Catal. Rev.-Sci. Eng.*, 25:1 (1983); *Hydrocarbons from Methanol*, Marcel Dekker, New York (1983).

47. S. A. Tabak, *AIChE National Meeting*, Philadelphia (1984).
48. S. A. Tabak and F. J. Krambeck, *Hydrocarbon Process.* (Sept. 1985), p. 72.
49. R. J. Quann, L. A. Green, S. A. Tabak, and F. J. Krambeck, *Spring AIChE National Meeting, New Orleans, La.* (1986).

28 | Hydrogen Effects in Olefin Polymerization Catalysts

POONDI R. SRINIVASAN, SHASHIKANT, and SWAMINATHAN SIVARAM

Indian Petrochemicals Corporation, Ltd., Vadodara, Gujarat, India

I.	INTRODUCTION	723
II.	CHEMISTRY OF OLEFIN POLYMERIZATION CATALYSTS	724
	A. Classification	724
	B. Kinetic Features of Olefin Polymerization Catalysts	724
	C. Nature of Active Centers and Their Determination	726
	D. Factors Determining Catalyst Activity	726
	E. Effect of Catalyst on Polymer Properties	727
	F. Mechanism of Polymer Growth Reaction	728
III.	HYDROGEN EFFECTS IN OLEFIN POLYMERIZATION CATALYSTS	728
	A. Titanium Catalysts for Ethylene and Propylene Polymerizations	728
	B. Chromium Catalysts for Ethylene Polymerization	737
IV.	CHEMISTRY OF TRANSITION METAL/HYDROGEN BOND	739
V.	CONCLUSIONS	740
	REFERENCES	742

I. INTRODUCTION

Catalysts for olefin polymerization by low-pressure processes had its origins in the discovery of a revolutionary new chemistry by Karl Ziegler at the Max-Planck Institute [1] and workers at Phillips Petroleum Company [2] in the early 1950s. It was soon realized that polyethylenes produced by these catalysts were structurally different from polyethylenes produced by high-pressure free-radical processes. Ziegler's discovery was subsequently extended to propylene polymerization by Natta [3], who made the seminal discovery that not only could propylene be polymerized to high-molecular-weight polymers, but the polymer so produced had an ordered structure, resulting in its unique physical properties. Commercial production of polyethylene and polypropylene based on the foregoing catalysts began in the mid-1950s and has grown substantially since then.

The olefin polymerization catalysts are a combination of a transition metal salt of groups IV to VIII metals (Ti, V, Cr, Co, and Ni) and a base metal alkyl of groups I to III (alkylaluminums). The Phillips catalysts, comprising chromium oxide on an inorganic support, are limited to the polymerization of ethylene. The chemistry of these catalyst systems for olefin polymerization has been reviewed extensively [4-6]. A wide variety of commercial processes, operating in solution, slurry, and gas phase, have been developed to exploit fully the potentials of these catalysts [7,8]. Simultaneously, our understanding of how these catalysts perform has advanced significantly in the past three decades [9-11].

Molecular hydrogen was used together with olefin polymerization catalysts in the late 1950s to regulate the molecular weight of polyolefins. The implications of this discovery to the commercial development of the polyolefin processes can hardly be underestimated. Molecular weight is one of the fundamental properties that determines the utility and end applications of a polymer; and without a simple and economical method of its control during the polymerization process, it is unlikely that polyolefins would have attained commercial success and the seminal discoveries of Ziegler and Natta would probably have remained mere laboratory curiosities.

Despite its almost universal application in polyolefin processes, relatively little attention has been bestowed on the role of hydrogen and the nature of its interaction with the olefin polymerization catalysts. The present chapter is an attempt to collect and collate all published information on the effects of hydrogen in olefin polymerization catalysts. In Section II we discuss briefly the chemistry of olefin polymerization catalysts. The hydrogen effects as applicable to specific classes of catalysts are described in Section III. In Section IV we review briefly the chemistry of transition metal-hydrogen interaction, knowledge of which substantially contributes to our understanding of the hydrogen effects in olefin polymerization catalysts.

II. CHEMISTRY OF OLEFIN POLYMERIZATION CATALYSTS

A. Classification

The olefin polymerization catalysts can be broadly classified as heterogeneous or homogeneous systems, the physical states referring only to the catalyst, not to the polymer. In the heterogeneous system, the catalyst is in a dispersed state coexisting with a liquid phase and the polymerization occurs mainly on the surface of the solid phase. Most practical catalysts for polyolefins are heterogeneous. Typical examples of heterogeneous and homogeneous catalysts are shown in Table 1.

B. Kinetic Features of Olefin Polymerization Catalysts

The elucidation of kinetic features of transition-metal-catalyzed polymerization of olefins provides important information, such as the rate law, activation energy, number of active centers, and average lifetime of the growing chain, which are of importance in the formulation of mechanisms. The principal kinetic features of both chromium- and titanium-based catalysts are well described in the literature [12-17]. The kinetic results are interpreted in terms of the familiar steps in a polymerization reaction: initiation, propagation, termination, and transfer.

Initiation:

$$C^*-R + H_2C=CH_2 \rightarrow C^*-CH_2-CH_2-R \tag{1}$$

TABLE 1 Heterogeneous and Homogeneous Catalysts for Olefin Polymerization

Catalyst	State	Olefin	Structure
CrO_3/SiO_2	Heterogeneous	Ethylene	Linear polyethylene
$(\eta-C_5H_5)_2Cr/SiO_2$	Heterogeneous	Ethylene	Linear polyethylene
$TiCl_3/Et_3Al$	Heterogeneous	Ethylene	Linear polyethylene
$TiCl_3/Et_2AlCl$	Heterogeneous	Propylene	Isotactic polypropylene
$TiCl_4/Et_3Al$	Homogeneous	Ethylene	Linear polyethylene
VCl_4/Et_2AlCl	Homogeneous	Propylene	Syndiotactic polypropylene
$(\eta-C_5H_5)_2ZrCl_2/[Me_2AlO]_n$	Homogeneous	Ethylene	Linear polyethylene
$(\eta-C_5H_5)_2ZrMe_2/[Me_2AlO]_n$	Homogeneous	Propylene	Atactic polypropylene

$$C^*\text{-}H + H_2C\text{=}CH_2 \rightarrow C^*\text{-}CH_2\text{-}CH_3 \qquad (2)$$

Propagation:

$$C^*\text{-}(CH_2\text{-}CH_2)_n\text{-}R + H_2C\text{=}CH_2 \rightarrow C^*\text{-}(CH_2\text{-}CH_2)_{n+1}\text{-}R \qquad (3)$$

Transfer:

By β-hydride elimination:

$$C^*\text{-}(CH_2\text{-}CH_2)_n\text{-}R \rightarrow C^*\text{-}H + H_2C\text{=}CH\text{-}(CH_2\text{-}CH_2)_{n-1}\text{-}R \qquad (4)$$

With alkylaluminum:

$$C^*\text{-}(CH_2\text{-}CH_2)_n\text{-}R + R'_3Al \rightarrow C^*\text{-}R' + R'_2Al\text{-}(CH\text{-}CH_2)_n\text{-}R \qquad (5)$$

With hydrogen:

$$C^*\text{-}(CH_2\text{-}CH_2)_n\text{-}R + H_2 \rightarrow C^*\text{-}H + H_3C\text{-}CH_2\text{-}(CH_2\text{-}CH_2)_{n-1}\text{-}R \qquad (6)$$

With olefin:

$$C^*\text{-}(CH_2\text{-}CH_2)_n\text{-}R + H_2C\text{=}CH_2 \rightarrow C^*\text{-}CH_2\text{-}CH_3 + H_2C\text{=}CH\text{-}(CH_2\text{-}CH_2)_{n-1}\text{-}R \qquad (7)$$

Termination (spontaneous):

$$C^*\text{-}(CH_2\text{-}CH_2)_n\text{-}R \rightarrow C\text{-}(CH_2\text{-}CH_2)_n\text{-}R \qquad (8)$$

In Eqs. (1) to (8), C^* represents an active center on catalyst. The interpretation of kinetics according to Eqs. (1) to (8) is rendered more complex because of such factors as the creation and nature of active centers on the catalyst; the need for the reactant monomer to diffuse from the gas phase onto the solid catalyst surface, often through the intermediacy of a liquid

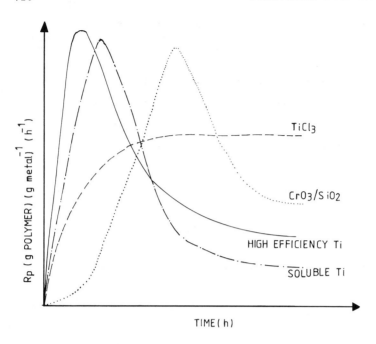

FIGURE 1   Typical kinetic profiles for olefin polymerization catalysts.

phase; followed by exothermic polymerization and the formation of a semicrystalline phase within the catalyst pores.

The kinetics of olefin polymerization are affected both by the nature of catalysts and the reaction conditions. Some typical kinetic profiles of olefin polymerization catalysts are shown in Fig. 1. The lifetimes of olefin polymerization catalysts are essentially determined by their kinetic profiles.

C.   Nature of Active Centers and Their Determination

The kinetics and mechanism of olefin polymerization are better understood in relation to the nature and number of active centers on the catalyst. The growth of polymers occurs on this active center, which can be created either by a physical process of activation (CrO_3/SiO_2) or by a chemical process of reduction by alkylation of the transition metal with alkylaluminum. The active centers are best described as a coordinatively unsaturated electrophilic site on the catalyst which can chemisorb an olefin specifically and selectively through donation of the π electron into the vacant d orbitals of the transition metal.

Methods have been described in the literature for determination of the number of active centers, which essentially consist of determining the concentration of transition metal/carbon bonds [18-24]. However, no direct methods exist as yet to characterize the structure of an active center during polymerization in terms of its oxidation states, ligand environment, and support-metal interactions. This is presently largely in the realm of speculation.

D.   Factors Determining Catalyst Activity

The activity of olefin polymerization catalysts is determined by (a) the nature, valency state, and type of ligands attached to the transition metal; (b) the

TABLE 2 Physical Characteristics and Activity of Olefin Polymerization Catalyst[a]

Catalyst	Surface area (BET) ($m^2 g^{-1}$)	Bulk density ($g\ cm^{-3}$)	Pore volume ($mL\ g^{-1}$)	Activity (kg polymer/g transition metal)
CrO_3/SiO_2	300-600	n.a.	0.4-1.6	1000
$(\eta-C_5H_5)_2Cr/SiO_2$	300-600	n.a.	1-1.2	1000
$TiCl_3 \cdot 0.3AlCl_3$	40-50	0.1	0.2-0.3	1
$MgCl_2/EB/TiCl_4$	200	0.4-0.5	1-2	200

[a]EB, Ethyl benzoate; n.a., not available.

nature of alkylaluminum compound; (c) the physical state of the catalyst; and (d) the conditions of polymerization. Recent studies indicate that Cr^{2+} is active in ethylene polymerization [22], Ti^{3+} and Ti^{2+} are active for ethylene polymerization, and only Ti^{3+} is active for propylene polymerization [23,24]. The nature of alkylaluminum compounds by virtue of their ability to reduce higher-valent transition metal salts to lower-valence states profoundly affects the activity of the catalyst. It has been found that for certain zirconium-based ethylene polymerization catalysts a novel organoaluminum compound consisting of an Al-O bond (called aluminoxane) gives the highest catalyst activity [25].

The physical state of the metal catalyst has a considerable effect on the kinetics of polymerization and product properties. These include the crystalline form of the transition metal salt, the nature of the support and its interaction with the transition metal compounds, particle size, surface area, bulk density, crystallite size, and crystallite surface area of the catalyst. The latter two appear to be especially important for crystalline $TiCl_3$ catalyst, where active centers are believed to be located on well-defined lattice defects [26,27]. Recent developments in the area of high-efficiency olefin polymerization catalysts are entirely due to the ability of catalytic chemists to synthesize new catalyst composition with very desirable physical characteristics. The exhaustive literature in this area has been reviewed in a number of publications [28-32]. In Table 2 we summarize some of the physical characteristics of olefin polymerization catalysts and their relationship to polymerization activity.

E. Effect of Catalyst on Polymer Properties

The important properties of polyolefins which are at least in part determined by the nature of catalysts are (a) molecular weight, (b) molecular-weight distribution, and (c) degree of crystallinity or stereoregularity. Molecular weight is controlled by the addition of chain-transfer agents. The multiplicity of active centers which differ energetically and structurally, and their differential rate of decay with time, have been ascribed as major reasons for the breadth of molecular-weight distributions observed with heterogeneous catalysts [33]. On the other hand, homogeneous catalysts with a single active center lead to a narrower distribution of molecular weights. Under stationary-state conditions with high rates of initiation and low conversions, this value conforms to the most probable distribution as defined by Flory [34]. Many of the high-efficiency catalysts supported on magnesium chloride as support give a narrower distribution of molecular weights, indicating the greater degree of homogeneity of the active centers in the supported catalysts [35]. Catalyst also determines the degree of linearity of a polymer as well as stereoregularity. The latter is shown in Table 1.

F. Mechanism of Polymer Growth Reaction

The mechanism of polymer growth reaction and the origin of regiospecificity and stereospecificity in olefin polymerization have been the subject of a number of reviews [9-11,36-38]. The accumulated literature now favors a mechanism in which polymer growth occurs on the surface of the transition metal, which first gets alkylated by the alkylaluminum compound. The active site is envisaged as a metal center with both an alkyl substituent and a chlorine vacancy located on the edges of the titanium trichloride crystal. Growth occurs by complexation of the monomer onto the vacant site followed by insertion of the polymer chain at the transition metal/carbon bond.

It is believed that energetically and structurally distinct sites exist on the catalyst which favor either an isospecific insertion (leading to isotactic polymer) or random insertion (leading to atactic polymer). Different models have been proposed to describe these sites. According to one model, the active sites located on the edge favor isospecific propagation, while those at the corner of $TiCl_3$ crystallite favor random insertion of propylene [4]. Soga and co-workers proposed that a center having two vacant sites gives atactic polypropylene, while that having only one vacant site gave isotactic polymer [38]. Pino proposed that a metal atom bound to solid surfaces possesses centers of chirality by virtue of which the active center of catalyst favors complexation of one prochiral face of propylene over the other [10]. The precise features of the catalyst that exercise such sharp selection is not known.

Comparatively little is understood about the mechanism of polymer growth on chromium catalysts. The precise initiation mode is still a subject of debate. Propagation is believed to occur by an insertion of ethylene into a Cr-alkyl bond. In the case of a chromocene on silica catalyst, the n-cyclopentadienyl ligand is associated with the active center and does not take part in monomer insertion.

III. HYDROGEN EFFECTS IN OLEFIN POLYMERIZATION CATALYSTS

A. Titanium Catalysts for Ethylene and Propylene Polymerizations

Titanium Trichloride/Alkylaluminum Catalysts

Molecular hydrogen has come to stay as the molecular-weight-regulating agent of choice in the polymerization of ethylene and other α-olefins catalyzed by titanium-based Ziegler-Natta catalysts. Hydrogen is an effective chain regulator for a wide range of catalyst compositions and reaction conditions. In addition, it is easy to use, inexpensive, and does not leave a residue in the polymer.

The use of hydrogen as a molecular-weight regulator was first disclosed almost simultaneously by Natta's school [39,40], workers from Farbwerke Hoechst [41], and Hercules Powder Company [42]. According to Natta, the chain termination involves the hydrogenolysis of the metal-to-polymer bond which would result in the generation of transition metal hydride. The hydride is expected to add to an olefin, which would then complete the catalytic cycle (Fig. 2). Based on the study of chain-transfer kinetics of ethylene polymerization with α-$TiCl_3$ and Et_3Al, Natta established the relationship

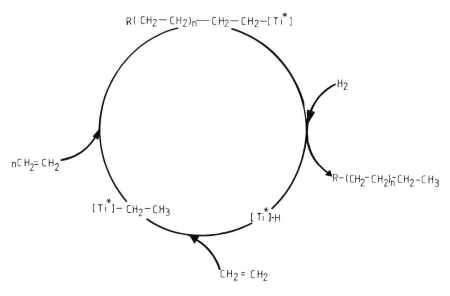

FIGURE 2 Catalytic cycle involving transfer of active catalyst center to hydrogen.

$$\frac{\bar{M}_n^0}{\bar{M}_n^H} = 1 + K(p_{H_2})^n \qquad (9)$$

where $\bar{M}_n^0$ is the molecular weight of the polymer in the absence of hydrogen, $\bar{M}_n^H$ is the molecular weight of the polymer in the presence of hydrogen, p_{H_2} is the partial pressure of H_2 and k and n are constants. The value of n for this catalyst system was found to be 0.5. Using data published by Keii [15] for propylene polymerization for the $TiCl_3/Et_2AlCl$ system with an isotactic index of 94%, the value of n was found to be 0.9, whereas for the $TiCl_3/Et_3Al$ system having an isotactic index of 64%, the value of n was 0.5. Keii reported a value of n = 0.88 with $TiCl_3/Et_3Al$ and n = 1.0 for $TiCl_3/Et_2AlCl$ for ethylene polymerization in toluene as a diluent. For polymerization in the gas phase the values of n were reported to be 0.35 to 0.45 for $TiCl_3/Et_3Al$ and 1.0 for $TiCl_3/Et_2AlCl$ [43]. The value of n = 0.5 is consistent with a hypothesis wherein a dissociative chemisorption of hydrogen onto the titanium surface takes place followed by the rate-determining reaction of the hydrogen adatom with the metal-polymer bond. This is analogous to the Horiuti-Polanyi mechanism for hydrogenation of olefin over metal surfaces [44]. In contrast, if the attack of molecular hydrogen on the metal-polymer bond is the slow step, the transfer reaction is expected to be first order with respect to $[H_2]$ (n = 1), analogous to the Twigg (or Eley)-Rideal mechanism for hydrogenation [45]. The observed rapid exchange of H_2 and D_2 over a $TiCl_3/Et_3Al$ catalyst system supports the hypothesis of a dissociative chemisorption of hydrogen on these catalysts [46,47].

The relative rates of chain transfer to hydrogen, olefin, and alkylaluminum for ethylene and propylene polymerization have been reported by Zakharov and co-workers [48] and Grievson [49]. The results (Table 3) showed that the rate of chain transfer to hydrogen was substantially higher relative to chain transfer to olefin or alkylaluminum. Also, at a given $[H_2]$, the molecular weight of polypropylene was suppressed to a greater degree than that of

TABLE 3 Rate Constants for Chain-Transfer Reactions

Olefin	Catalyst	Temperature (°C)	K_{tr}^M (liter mol^{-1} s^{-1})	$K_{tr}^{H_2}$ (liter$^{1/2}$ mol$^{1/2}$ s^{-1})	$K_{tr}^{AlEt_3}$ (liter$^{1/2}$ mol$^{1/2}$ s^{-1})	Ref.
Ethylene	δ-TiCl$_3 \cdot$ 0.3AlCl$_3$/Et$_3$Al	80	0.84	23	0.36	48
Propylene	δ-TiCl$_3 \cdot$ 0.3AlCl$_3$/Et$_3$Al	70	0.04	5.9	0.054	48
Ethylene	γ-TiCl$_3$/Et$_2$AlCl	40	0.01	3	0.003	49

polyethylene. This could be a reflection of the relative strengths of the metal-polymer bond.

Conflicting reports on the effect of hydrogen concentration on the overall rates of olefin polymerization have appeared in the literature (Table 4). For polymerization of ethylene, both in diluent and gas phase, the overall rate decreased substantially with hydrogen present [39,43,50]. Natta and co-workers [39] found that if the hydrogen-ethylene mixture was replaced by ethylene at the same pressure, the rate returned to its original value. This indicated that hydrogen has no poisoning effect on the catalyst. Similar observations were made by Okura and co-workers [51]. Other workers have observed an increase in the rate of polymerization in the case of propylene when hydrogen was present [52-54]. More recent studies indicate a negligible effect of hydrogen on the rate of propylene polymerization [55]. With higher α-olefins, butene-1, and 4-methyl-pentene-1, hydrogen enhances the overall rate of polymerization [56,57].

There appears to be no single explanation that accounts for all the observations. Assuming that both molecular hydrogen and olefin compete for the same active sites, hydrogen chemisorption should inhibit polymerization of olefin by causing a reduction in the number of active centers. This appears to be the case with ethylene, where Grievson [49] reported a reduction in the concentration of active centers with 50 mol % hydrogen ([C^*] = 6 × 10^{-3} mol per mole of TiCl$_3$) relative to a catalyst without hydrogen ([C^*] = 1.5 × 10^{-2} mol per mole of TiCl$_3$). Buls and Higgins [53] and Mortimer et al. [58] proposed that hydrogen must in some way help in the creation of new active sites on the catalyst to explain the rate increase observed by them for propylene polymerization in the presence of hydrogen. The possibility of additional new sites being created by reduction of Ti-Cl bond to Ti-H bond cannot be ruled out [43]. Pijpers and Roest [59] and Boucheron [60] proposed that in the absence of hydrogen, transfer occurs by β-hydrogen elimination, leading to unsaturation in the polymer. The chain-end double bond stays complexed to the active centers, thus inhibiting fresh complexation of olefin and causing a rate inhibition in the absence of hydrogen.

Sufficient data are not available in the literature to show which of the two factors, site activation in the presence of hydrogen or site deactivation by complexed unsaturated polymer, is responsible for the increase in the rate of propylene, butene-1, and 4-methyl-pentene-1 polymerizations. Precise determination of the number of active centers with and without hydrogen for α-olefin polymerization may help to resolve this question.

Other factors, such as the crystalline form of TiCl$_3$, the nature of alkyl-aluminum compound, the diluent, the ratio of Al/Ti, and the nature and concentration of reactive impurities also appear to have a significant role in determining the nature of response of the olefin polymerization rate to added

TABLE 4 Effect of Hydrogen on the Overall Rate of Polymerization

Olefin	Catalyst	Al/Ti	T (°C)	Pressure (kg cm^{-2})	Diluent	[H$_2$]	R_p [g polymer (g cat)$^{-1}$ h^{-1}]	Ref.
Ethylene	δ-TiCl$_3 \cdot$ 0.3 AlCl$_3$/Et$_3$Al	2	30	2	Toluene	0 9 × 10^{-4} M	23 12	49
	γ-TiCl$_3$/Et$_2$AlCl	1	50	3	Cyclohexane	0 0.8 atm	10 6	50
	δ-TiCl$_3 \cdot$ 0.3 AlCl$_3$/Et$_3$Al	3	35	0.4	Nil	0 0.2 atm	19 9	43
	δ-TiCl$_3 \cdot$ 0.3 AlCl$_3$/Et$_2$AlCl	3	35	0.4	Nil	0 0.02 0.2	8.9 10.5 7.5	43
	α-TiCl$_3$/Et$_3$Al	3	75	2	Heptane	0 1 atm	15 6	39
Propylene	δ-TiCl$_3 \cdot$ 0.3AlCl$_3$/Et$_2$AlCl	1	55	0.4 0.4	Heptane Toluene	— —	Dec. in rate with H$_2$ Inc. in rate with H$_2$	51
	δ-TiCl$_3$/Et$_2$AlCl	—	60	1	Heptane	0 1 atm	75 244	53
	δ-TiCl$_3 \cdot$ 0.3AlCl$_3$/Et$_2$AlCl	—	70	4	Nil	0 0.6 atm	1500 2000	54
	δ-TiCl$_3 \cdot$ 0.3AlCl$_3$/Et$_2$AlCl	3	70	5	Heptane	0 1.5 × 10^{-3} M	95 90	55
Butene-1	δ-TiCl$_3 \cdot$ 0.3AlCl$_3$/(iBu)$_3$Al	2	70	3.5	Isooctane	0 1 atm	111 226	56
4-Methyl-pentene-1	γ-TiCl$_3$/Et$_2$AlCl	3	54	1	Heptane	0 2 atm	75 118	57

hydrogen. It has been reported that site activation by hydrogen is more pronounced in the presence of diethylaluminum chloride compared to triethylaluminum. This has been attributed to the poorer alkylating (reducing) ability of the former and the consequent inability to activate all available sites [43]. These sites are then activated by conversion to Ti-H sites.

The nature of the alkylaluminum compound and the Al/Ti ratio determines the relative population of oxidation states on Ti (4+, 3+, 2+). It has been reported [15] that transfer by hydrogenolysis of metal-polymer bond occurs only on the Ti^{3+} center, whereas Ti^{4+} promotes only an exchange between added hydrogen and the hydrogen on the β-carbon of the polymer chain.

For polymerization of α-olefins, it is understood that at least two distinct sites exist on the catalysts, one leading to isospecific propagation resulting in isotactic polymers and the other leading to atactic polymer. It is becoming increasingly evident that the effects of hydrogen on these centers are not similar. The atactic polymer is much lower in molecular weight, and although the weight fraction of this in the total polymer is small, the number of atactic polymer molecules produced frequently exceeds the number of isotactic molecules. Consequently, a large proportion of hydrogen is consumed to hydrogenate an atactic polymer. This factor could also lead to anomalous observations on the influence of hydrogen on the overall rate of polymerization, which is the sum of rates of isospecific and nonspecific polymerization. The two rates may need to be separated from the point of view of hydrogen effects.

Buls and Higgins [61] also made an interesting observation that in the absence of hydrogen, the isotactic polypropylene polymer chain is attached to the metal catalyst and that other forms of chain transfer (to propylene, alkylaluminum, and β-hydrogen transfer) are not predominant. This led them to the remarkable conclusion that propylene polymerization in the absence of hydrogen is not truly catalytic.

For propylene polymerization, the isotactic index, measured as heptane insolubles, experiences a small initial drop upon the introduction of hydrogen, which remains relatively constant with increasing hydrogen concentration until very high levels of hydrogen are reached. The isotactic index then begins to decrease steadily (Table 5). This observation was made with both Et_3Al and Et_2AlCl. The reduction in isotacticity with increasing hydrogen concentration favors a site activation mechanism wherein new active centers are being created on the surface of titanium catalysts. A majority of the new sites being created are nonstereospecific sites, where only random propylene insertion is possible, leading to atactic polymer formation.

The mechanism of chain transfer by hydrogen as proposed by Natta for propylene polymerization (Fig. 2) will require the presence of an isopropyl

TABLE 5 Effect of Hydrogen on the Isotactic Index of Polypropylene

Catalyst	H_2 (bar)	R_p [g polymer (g cat.)$^{-1}$ h^{-1}]	I.I (%)	Ref.
δ-$TiCl_3$·0.3$AlCl_3$/Et_2AlCl	0	332	96.7	8
	0.6	380	95.7	8
	1.5	341	90.6	8
	0	449	91.1	55
	0.3	479	89.2	55
α-$TiCl_3$/Et_3Al	0	15	81.3	39
	1.0	6	76.5	39

TABLE 6 Reduction in Molecular Weight ($\overline{M}_w$) of Polypropylene and Polyethylene[a]

Polyethylene		Polypropylene	
$\overline{M}_w$	$[p_{H_2}]$ (psia)	$\overline{M}_w$	$[p_{H_2}]$ (psia)
700,000	0	480,000	0
270,000	0.5	110,000	0.2
210,000	1.0	72,000	0.5
170,000	2.0	60,000	1.0

[a] Conditions of polymerization: $TiCl_3$, Et_3Al, Al/Ti = 3, 75°C, heptane.
Source: Ref. 39.

end group, which indeed is found. Unsaturation at the chain end ($-\underset{\underset{CH_3}{|}}{C}=CH_2$) was found in the absence of hydrogen [61]. Using molecular tritium instead of hydrogen, Hoffman and co-workers [62] found that for 95% of the whole product, two tritium atoms per molecule of the polymer were incorporated. This was taken as evidence of a mechanism wherein each chain is initiated by a Ti-H bond and terminated by hydrogenolysis (Fig. 2). However, the authors did not rule out the possibility of H-T exchange at the β-carbon attached to the metal center. Such a process, which is known to occur, makes their experimental observation fortuitous.

π-Electron donors have been found to have desirable effects on the performance of titanium-based olefin polymerization catalysts [4]. Under optimum conditions, an electron donor causes an increase in polymerization rate, polymer molecular weight, and isotactic index. Consequently, to attain a desired molecular weight, a greater concentration of hydrogen is required for a donor-modified catalyst relative to an unmodified catalyst. No systematic study of hydrogen effects on the rates and stereospecificity with donor-modified titanium catalysts has appeared in the literature; thus no general conclusions can be drawn.

Some typical values of weight-average molecular weight obtained for polyethylene and polypropylene at different partial pressures of hydrogen are shown in Table 6. The fact that at very high hydrogen concentration the performance of titanium-based catalysts is adversely affected limits the use of this technique for the preparation of very low molecular weight resins, which have practical applications because of their high flow properties and ability to be shaped into intricate shapes and forms.

Diverse effects of hydrogen on the molecular-weight distribution of polyethylene and polypropylenes are reported in the literature [55,63-66] (Table 7). It is reported that hydrogen narrows [63] or broadens [65,66] the molecular-weight distribution of polyethylene or polypropylene or causes negligible changes [56,64] in it. The data available in the literature are inadequate to draw any general conclusions. Most authors have used a batch mode for hydrogen addition. This is expected to cause a broadening of the Q value due to the changing $[H_2]/[olefin]$ ratio over the course of the polymerization as hydrogen is consumed. In the case of propylene, the data reported in the literature are for the entire polymer (isotactic + atactic). However, as shown by Buls and Higgins [61], $\overline{M}_w$ of atactic polymer is drastically reduced by as much as 50% in the presence of hydrogen, and unless this effect is separated, any interpretation of hydrogen effect based on the Q value of the whole polymer could be highly misleading.

TABLE 7 Effect of Hydrogen on Molecular-Weight Distribution (Q)

Olefin		T (°C)	Pressure (bar)	[H_2]	Q	Ref.
Ethylene	$TiCl_4/(i-Bu)_3Al$	80	2	0	18	63
				6 vol %	9	
	γ-$TiCl_3/Et_2AlCl$	50	3	0	15	64
				20 vol %	13	
	γ-$TiCl_3/Et_3Al$	60	2	0	5	65
				1 atm	15	
Propylene	γ-$TiCl_3/Et_3Al$	75	n.a.	0	6	66
				1 atm	11	
	δ-$TiCl_3 \cdot 0.3AlCl_3/Et_2AlCl$	70	5	0	5.8	55
				2 vol %	6.4	

High-Efficiency Titanium-Based Olefin Polymerization Catalyst

The efficiency of olefin polymerization catalysts based on titanium has dramatically improved within the past 10 years, bringing in its wake new industrial processes for polyolefins. The underlying chemistry of these processes has been well reviewed [16, 28-32, 67]. The key to the improved performance of these catalysts is the use of an activated magnesium chloride as a high-surface-area support for titanium chloride which probably bonds to the coordinatively unsaturated magnesium ions located at the side surfaces and crystal edges, the (110) and (101) faces, through chlorine bridges, forming stable surface complexes [67]. The patent literature is replete with examples of such catalysts and a substantial amount of industrial research is being invested in the preparation, characterization, and evaluation of high-efficiency catalysts. Consequently, much of the information regarding these catalysts is highly proprietary, emphasis being on the commercial exploitation of the catalyst.

Nevertheless, a number of recent publications have been addressed to a fundamental understanding of these catalysts [68-71]. Guastalla and Giannini [72] have reported on the influence of hydrogen on the polymerization of ethylene and propylene with a $MgCl_2$-supported catalyst. Using a $TiCl_4/MgCl_2/Et_3Al$ catalyst they found that the initial rates of polymerization of propylene increased by 150% when the partial pressure of hydrogen was changed from 0 to 0.6 bar. At p_{H_2} > 0.6 bar no further increase in rate could be observed. These authors also observed that the productivity of both atactic (ether-soluble) and isotactic (heptane-insoluble) fractions increases with increasing partial pressure of hydrogen. However, beyond a p_{H_2} of 0.6 bar, the productivity of the isotactic fraction dropped. The reduction in molecular weight of heptane-insoluble fractions was proportional to the square root of hydrogen concentration [n = 0.5 in Eq. (9)]. Interestingly, the activity of the same catalyst toward ethylene polymerization was substantially reduced in the presence of hydrogen. A similar observation was made by Böhm using a high-activity magnesium hydroxide-supported titanium catalyst [73]. Soga and Siono [74] studied the effect of hydrogen on the molecular weight of both atactic and isotactic polymer produced during polymerization of propylene using a $TiCl_4/MgCl_2/ethylbenzoate/Et_3Al$ catalyst at 40°C and atmospheric pressures. They found that the catalyst activity, as well as the fraction of isotactic polymer, gradually decreased with an increase in hydrogen partial pressures. This was attributed to the decrease in molecular weight of the isotactic fraction, because of which it went into solution. No loss of isotacticity was observed with increasing hydrogen concentration. The value of exponent n in Eq. (9) was found to be 0.8 for isotactic fraction and 0.7 for atactic fraction. Although these values deviated from n = 1 or 0.5, Soga and Siono interpreted their result in terms of two distinct polymerization centers, one (C-1) having two vacant sites, giving atactic polypropylene, while the other center (C-2), having only one vacant site, giving isotactic polypropylene [38]. Dissociative chemisorption of hydrogen proceeds only on the C-1 center (n = 0.5), whereas molecular hydrogen is involved on the catalyst with the C-2 center (n = 1). In a related study Keii and co-workers [75] found the value of the exponent n to be 0.5 for polymerization of propylene using a supported catalyst (41°C, atmospheric pressure). Although the composition of the catalyst used in this study was similar to that used by Soga and Siono [74], different values of n were reported. In a more recent study, Chein and Ku [76] found that upon the addition of hydrogen, the number of active centers on titanium for ethylene polymerization increased by 50% and the number of isospecific active centers on titanium for propylene polymerization increased by 250%. The catalyst used was prepared by ball milling $MgCl_2$ with ethylbenzoate followed by reaction with p-cresol, triethylaluminum, and titanium chloride. Using a magnesium chloride-supported titanium catalyst of undisclosed composition, Jacobson [77] reported marginal increase in catalyst activity during propylene polymerization in liquid propylene. The reported effects of hydrogen on the rate, isotacticity, and molecular weight distribution of olefin polymerization using high-efficiency titanium catalysts are shown in Table 8.

In view of the complex structural features of $MgCl_2$-supported titanium catalysts, it is doubtful whether any general conclusions on the effect of hydrogen on the rates of polymerization of ethylene or propylene can be drawn. While interpreting results, it is necessary to recognize the following features of the supported catalysts: (a) surface characteristics such as surface area, crystallite size, porosity, titanium dispersion, and method of activation of support; (b) the presence or absence of an electron donor such as ethylbenzoate, which tends to weakly adsorb on the coordinatively unsaturated titanium sites; (c) the relative population of oxidation states on the catalyst surface and their stability, which is, in turn, controlled by the manner of reduction; and (d) the nature of the external electron donor generally used as a third component with propylene polymerization catalysts, which controls, by an as yet ill-understood manner, the relative rates of isospecific and nonspecific propagation.

Nevertheless, results available so far suggest that hydrogen suppresses the rate of ethylene polymerization, whereas the isospecific rate of propylene polymerization is enhanced by hydrogen. This observation is generally in agreement with results obtained using conventional titanium catalysts.

The isotactic index of polypropylene obtained with these high-efficiency catalysts is not generally affected by the addition of hydrogen [67,72,73]. The molecular-weight distribution of polypropylene and polyethylene obtained using a high-activity catalyst also did not change with increasing hydrogen concentration [78].

Supported catalysts also appear to show a more sensitive response to hydrogen. It has been reported that one-half as much hydrogen is needed to achieve the same polymer molecular weight as with a conventional catalyst system [76]. This could be a consequence of the higher concentration of active centers on supported catalysts (also capable of hydrogen adsorption) and the decrease in formation of the atactic fractions.

TABLE 8 Hydrogen Effect on High-Efficiency Catalyst

Olefin	Catalyst	Solvent	Pressure (bar)	Temp. (°C)	[H_2] (bar)	Yield [kg polymer (g Ti)$^{-1}$h^{-1}]	I.I. (%)	Q	Ref.
Propylene	$TiCl_4/MgCl_2/Et_3Al$	Hexane	3	70	0	36.6	36	n.a.	72
					1.2	112.0	35.9	n.a.	
Ethylene	$TiCl_4/MgCl_2/Et_3Al$	Hexane	3	60	0	177	—	n.a.	72
					1.0	171	—	n.a.	
Propylene	$TiCl_4/MgCl_2/Et_3Al/EB$	Heptane	0.5	40	0	3.02	90.2	12.0	74
					0.2	2.5	85.9	10.8	
Propylene	Mg-Ti Catalyst/Et_3Al	Liquid pool	31	70	0	565	95.5	7.3	77
					1.0	647	92.2	7.4	
Ethylene	$Mg(OR)_2/TiCl_4/Et_3Al$	Diesel oil	6	85	0	611	—	9.1	73
					7 vol %	499	—	9.0	

Soluble Organometallics as Olefin Polymerization Catalysts

Despite the fact that soluble organometallics are more amenable to unequivocal mechanistic studies, very little is known about the hydrogen effects on homogeneous olefin polymerization catalysts. The reason could be the unfortunate lack of commercial interest in these catalyst types, which has also inhibited fundamental studies.

Doi and co-workers [79] studied propylene polymerization at -78 C with a vanadium triacetylacetonate/diethylaluminum chloride soluble catalyst system. In the absence of hydrogen, the polymerization showed all the features of a transfer-termination free-living polymerization with molecular-weight distribution approaching unity. Addition of hydrogen caused the molecular-weight distribution to increase to 2.0 [34]. The catalyst system thus showed predictable kinetics. The value of n in the exponent of Eq. (9) was reported to be 1.0 for a homogeneous catalyst. Kaminsky and co-workers studied the influence of hydrogen on the polymerization of ethylene using a homogeneous bis(cyclopentadienyl)zirconium dichloride/aluminoxane catalyst [80,81]. The addition of hydrogen suppressed catalyst activity. However, the loss of activity was reversible when hydrogen was removed from the system. The exponent of n in Eq. (9) was also found to be 1.0, in agreement with earlier studies [78,79]. Homogeneous catalysts required less hydrogen to effect a given reduction in molecular weight compared to that required by heterogeneous catalysts.

A value of 1 for the exponent n in Eq. (9) for homogeneous catalysts is indicative of a mechanism wherein the rate-determining step is the reaction of alkylated transition metal with molecular hydrogen. Studies of kinetics of homogeneous hydrogenation of olefins with similar catalysts appear to substantiate this view (see Section IV).

B. Chromium Catalysts for Ethylene Polymerization

Chromium on Silica

Chromium catalysts for ethylene polymerization are generally prepared from an amorphous silica gel and CrO_3 (Cr ca. 1%) followed by activation with dry fluidized air at 500 to 1000°C for a few hours. Surface properties of silica determine polymer properties such as bulk density, molecular weight, and its distribution [2,30].

One of the major limitations of the Phillips catalyst in industrial practice has been its inability to produce low-molecular-weight polyethylenes. It was recognized very early that CrO_3/SiO_2 showed a negligible effect of hydrogen as a chain transfer agent in comparison with the Ziegler-Natta type of catalyst. It has been reported that 20 mol % hydrogen, when used with CrO_3/SiO_2 catalyst, depressed polyethylene molecular weight by only 50% [82]. However, even when hydrogen was used, the polymer contained predominately vinyl unsaturation, indicating that hydrogen acts in some way other than hydrogenation to shorten the chain length [83]. The use of deuterium instead of hydrogen resulted in both -CHD and -CH_2D groups. Exchange reactions led to the former, whereas the latter was believed to form as a result of the addition of Cr-D bond to an olefin [83].

The inability of hydrogen to control molecular weight with these catalysts has led to other methods of molecular-weight control, such as incorporation of titania in chromia [84,85], and by control of the hydroxyl population in the support [86]. With a given catalyst, the reaction temperature and ethylene pressure offer a further degree of molecular-weight control.

The lack of hydrogen effects on the chromium catalyst can be understood in terms of the active valence states of the chromium as well as strong metal-support interactions. Accumulated evidence now favors Cr^{2+} as active in polymerization. Catalyst activity and termination rate are inversely related to the hydroxyl population on the silica surface, as a surface hydroxyl group (silanol) can interfere with the reduction to Cr^{2+} [87]. It is conceivable that when hydrogen is added to the system, either fresh silanol groups are created or part of the existing silanol groups form water. Both of these can reoxidize Cr^{2+} to Cr^{3+}, the latter being less active for polymerization. On the contrary, for reactions such as hydrogenation of olefins, Cr^{3+} has been found to be the preferred valency [88]. An exhaustively reduced species where Cr^{2+} is the predominant valence state is inactive for hydrogenation. Cr^{3+} with a coordination number of 4 has two empty ligand sites, one on which the olefin is bound and the other which can accept a hydride ion by heterolytic dissociation of hydrogen. The following mechanism of hydrogenation over Cr^{3+} has been proposed by Burwell et al. [89, 90].

$$Cr^{3+}O^{2-} + H_2 \rightarrow [Cr^{3+}H^-]OH^- \qquad (10)$$

$$[Cr^{3+}H^-]OH^- + C_2H_4 \rightarrow [Cr^{3+}C_2H_5^-]OH^- \qquad (11)$$

$$C_2H_5^- + H^+ \rightarrow C_2H_6 \qquad (12)$$

The distinctness of the two sites, one active for polymerization and the other for hydrogenation, and their relative stability to surface hydroxyl groups could account for the poor hydrogen response of chromium/silica catalyst in ethylene polymerization.

More recently, aluminum phosphate has been used to replace silica in the Phillips catalyst [91, 92]. Interestingly, this support shows a greater sensitivity to hydrogen, and a wide range of molecular weights for polyethylene is accessible using hydrogen as a chain-transfer agent. In spite of this, polymers contained only vinyl unsaturation, indicating that β-hydrogen elimination is the preponderent chain transfer step [Eq. (4)]. In some unknown way, hydrogen must be assisting this process. The polymerization rate was not reduced by hydrogen. The reason for this major difference in hydrogen effect between the two supports, silica and aluminum phosphate, is not clear at present.

Another class of well-known chromium-based ethylene polymerization catalysts is that of the silica-surface-anchored chromium organometallics, discovered in the laboratories of Union Carbide Corporation in the late 1960s. A number of arene [93, 94], allyl [95, 96], and bis-cyclopentadienyl [97, 98] chromium compounds have been reported.

The organochromium compounds are usually more sensitive to hydrogen than are chromium oxide catalysts. In a study of the hydrogen effects of silica-supported chromocene catalysts, Karol and co-workers reported an order-of-magnitude higher hydrogen response than even $TiCl_3$-Et_2AlCl [99]. As expected, the polymer was fully saturated, indicating that hydrogenation of a Cr-polymer bond was the major transfer step. Polymers prepared from deuterium showed -CH_2D groups, confirming this view [100]. Substitution of cyclopentadienyl with other π ligands such as indenyl or fluorenyl led to catalysts with poorer response to hydrogen [101]. It has been reported that unlike chromocene, diarene chromium and diallyl chromium produce vinyl unsaturation when hydrogen was used to control molecular weights [102]. Obviously, hydrogenation of the polymer-chromium bond is not the only prevalent mechanism; β-hydrogen elimination can also be important. Hydrogen caused a loss of catalyst activity. The molecular-weight distributions of polyethylene produced by chromocene on silica were relatively narrow compared to the oxide catalyst and did not change with hydrogen concentration.

On the contrary, chromium catalysts, which transfer by β-hydrogen elimination, produced broad-molecular-weight-distribution polyethylene.

Any explanation of the high hydrogen response of such catalysts should consider the role of both the η-cyclopentadienyl ligand and the support which is linked to the chromium atom through a Si-O-Cr bond. In the absence of silica, chromocene does not catalyze polymerization of ethylene. The presence of coordinatively unsaturated chromium is responsible for both efficient polymerization and hydrogenolysis. Propagation is believed to occur by repeated insertion of the olefin into a Cr-alkyl σ bond formed initially by an as-yet-unknown mechanism. In the mechanism originally proposed [99], the chromium-to-silica bonding was envisaged as chemisorption, and hence the active valence state of Cr was considered as 2+. Polarographic evidence was presented on recovered catalyst after polymerization to show that Cr^{2+} was the dominant valence state (85 to 95%), with 5 to 15% Cr^{3+}. The latter was attributed to inadvertent oxidation.

However, the proposal of an oxide-ligated chromium species [35] with a Si-O-Cr bond would require a formal oxidation state of Cr^{3+} for the active propagating center. Such a center can undergo hydrogenolysis by either (a) an oxidative addition/reductive elimination process involving $Cr^{3+} \rightleftharpoons Cr^{5+}$ states, (b) by a heterolytic dissociation process similar to that proposed for hydrogenation on Cr^{3+} [89,90], or (c) by direct interaction of molecular hydrogen with the metal center via a vacant orbital without net oxidation of the metal center. The relatively unstable Cr^{3+} could be stabilized by the presence of a hard Si-O ligand. The soft η-cyclopentadienyl ligand could offer a degree of stability to the Cr-carbon or Cr-hydrogen bond. Experimental data currently available cannot distinguish among mechanisms (a) to (c). Knowledge of the order of reaction with respect to hydrogen can distinguish between (a) and (b) or (c). The drop in catalyst activity with increasing hydrogen concentration can best be explained by the reaction of H_2 with surface hydroxyl or oxygen, generating water or hydroxyl groups, both of which can poison the active sites.

IV. CHEMISTRY OF TRANSITION METAL/HYDROGEN BOND

The chemistry of metal-hydrogen bond plays an important role in both heterogeneous [103] and homogeneous catalysts [104]. Examples of homogeneous catalytic processes involving transient transition metal/hydrogen bonds are hydrogenation [105], hydroformylation [106], olefin isomerization and H-D exchange [107], hydrocarbon activation [108], and dimerization-oligomerization of olefins [109,110]. Of these reactions, hydrogenation bears the closest resemblance to the mechanism of chain transfer in olefin polymerizations. In fact, transition metal acetylacetonates and alkoxides in conjunction with trialkylaluminum at high H_2/olefin ratios are effective olefin hydrogenation catalysts [111]. The rate of reaction was found to be first order with respect to hydrogen and zero order with respect to olefin. The activity of the metal decreased in the order Co > Ni > Fe > Cr > Ti > Mn > V, approximately the order of decreasing value of Pauling's d character. Bis(η-cyclopentadienyl)-titanium dichloride in conjunction with metal alkyls is a hydrogenation catalyst [112,113].

The activation of molecular hydrogen on transition metal complexes can occur by three elementary mechanisms: oxidative addition [114], heterolytic or homolytic dissociation [89,90], and direct interaction of dihydrogen with the metal [35,115]. Oxidative addition, which requires both a vacant site and a lone electron pair, involves an increase in the formal oxidation state of the metal by two units. The oxidative addition of hydrogen has been demonstrated in a number of low-valent d^8 transition metal complexes. Heterolytic dissociation is known for a variety of Lewis acid complexes and metal oxides and is facilitated by the presence of a base to stabilize the proton liberated [90]. Homolytic dissociation normally accompanies the chemisorption of hydrogen on metal surfaces. The direct interaction of H_2 with metal has been proposed more recently. It has been proposed that for d^0 metals with no acceptable higher oxidation states, oxidative addition is unlikely and dihydrogen can directly interact with the metal via a vacant orbital without net oxidation of that metal. Such reactions show a first-order dependence on $[H_2]$. Ligands that donate electrons (Cl, alkoxy) diminish the activity of metal toward hydrogenation. The recent isolation of a true complex between a Mo (O) and dihydrogen and its structural elucidation lends further credence to this process of hydrogen activation [116].

It is therefore reasonable to believe that such a process also operates during the hydrogenation of olefins with groups IVA and VA metal salts in conjunction with alkylaluminums as well as hydrogenation of the metal-polymer bond. The key steps in these reactions are

$$R\text{-}Mx_{n-1} + H_2 \rightleftharpoons R\text{-}Mx_{n-1}(H\cdots H) \rightarrow HMx_{n-1} + RH \tag{13}$$

$$Pn\text{-}Mx_{n-1} + H_2 \rightleftharpoons Pn\text{-}Mx_{n-1}(H\cdots H) \rightarrow HMx_{n-1} + PnH \tag{14}$$

where R is an alkyl and Pn a polymer chain.

The hydrido-metal complexes can further react with an olefin by an insertion reaction. In a few cases the hydrido complexes can be isolated. Bercaw and Brintzinger showed that reaction of solid bis(η-cyclopentadienyl)titanium dimethyl with gaseous hydrogen gave two molecules of methane and an isolable μ-hydrido complex [117]:

$$(\eta\text{-}C_5H_5)_2Ti(CH_3)_2 + 2H_2 \rightarrow \begin{matrix} \eta\text{-}C_5H_5 \\ \eta\text{-}C_5H_5 \end{matrix} Ti \begin{matrix} H \\ H \end{matrix} Ti \begin{matrix} C_5H_5\text{-}\eta \\ C_5H_5\text{-}\eta \end{matrix} + 2CH_4 \tag{15}$$

Similarly, reduction of bis(η-cyclopentadienyl)titanium dichloride with a variety of reducing agents gave an intermediate species with a μ-hydrido bridge. This intermediate has been implicated in the reduction of molecular N_2 by low-valent titanium species [118].

V. CONCLUSIONS

The literature information on hydrogen effects in olefin polymerization catalysts is scattered and at times contradictory. Nevertheless, some general conclusions can be drawn. In the case of ethylene polymerization with either a conventional or a high-efficiency titanium-based catalyst system, hydrogen causes a reduction in the polymerization rate. However, in the case of propylene and other higher α-olefins, the literature evidence suggests an increase in catalyst activity with added hydrogen. Our current understanding of the valence states responsible for ethylene and propylene polymerizations suggests

TABLE 9 Value of Exponent n in the Chain-Transfer Equation, Eq. (9)

Olefin	Catalyst	I.I. (%)	n	Refs.
Ethylene	α-$TiCl_3$/Et_3Al	–	0.5	39
	δ-$TiCl_3$/Et_3Al	–	0.35-0.45	43, 15
	δ-$TiCl_3$/Et_2AlCl	–	1.0	43, 15
	$(\eta$-$C_5H_5)_2ZrCl_2$/aluminoxane	–	1.0	81
	Silica/Mg/$TiCl_4$/$(i$-$Bu)_3Al$	–	1.0	119
Propylene	$TiCl_3$/Et_2AlCl	94	0.9	15
	$TiCl_3$/Et_3Al	64	0.5	15
	$MgCl_2$/$TiCl_4$/Et_3Al/EB	90.2	0.8/0.7	74
	$MgCl_2$/$TiCl_4$/Et_3Al	37	0.5	72
	$MgCl_2$/$TiCl_4$/EB/Et_3Al	70	0.5	75
	$V(acac)_3$/Et_2AlCl	Syndiotactic	1.0	79

that only Ti^{3+} is active for propylene polymerizations, whereas the lower-valence states Ti^{2+} and Ti^{1+} are also active for ethylene polymerizations [24]. It is apparent that the effects of hydrogen on these two centers are distinct. However, the detailed mechanism that leads to activation or deactivation of specific active centers needs to be elucidated further.

The order of reaction with respect to hydrogen, as indicated by the value of exponent n in Eq. (9) as a function of catalyst type, gives further insight into the nature of hydrogen interaction with olefin polymerization catalysts. Literature data are collected in Table 9. Although comparisons should be made with caution, the results indicate that the order of reaction with respect to hydrogen is invariably 0.5 for ethylene or atactic polypropylene; for isotactic polypropylene the value approaches 1.0. Fundamentally different molecular mechanisms seems to operate on different sites of the catalyst. Sites responsible for polymerization of ethylene and nonspecific polymerization of propylene appear to have a similar interaction with hydrogen, which is distinctly different from its interaction with stereospecific sites. The mechanism proposed by Keii and co-workers [75] appears conceptually plausible, wherein a catalyst with two vacant sites can dissociatively chemisorb hydrogen by a homolytic mechanism (n = 0.5), whereas a center possessing only one vacant site can interact with molecular hydrogen (n = 1.0). Ti^{3+} (d^2s^1), with one low-energy vacant orbital and responsible for isospecific propagation, appears to directly interact with molecular hydrogen [115], whereas lower-valent Ti^{2+} or Ti^{1+} (d^2 or d^1), with more than one easily accessible vacant orbital, interacts with hydrogen by a dissociative mechanism. Whether an oxidative addition or a simple homolytic dissociation process is involved cannot be answered at the moment.

With homogeneous catalysts, both for ethylene and propylene polymerizations, hydrogen reduces the polymerization rate and shows a reaction order of 1.0 with respect to H_2. Although the data are very limited, it appears that in these cases direct interaction of molecular hydrogen with the metal center (possibly in its highest valence state, d^0) is the predominant mechanism. The data available with organochromium catalysts are too few to draw any general conclusions at the present time.

It thus appears that all the mechanisms proposed for hydrogen interaction on transition metals, namely, oxidative addition, homolytic dissociation, and direct interaction of molecular hydrogen, have a role to play in determining the nature of hydrogen effects in olefin polymerization catalysts. The fact that hydrogen can respond selectively and specifically to different polymerization centers point out to its applicability as a probe for the nature of active centers. This feature of hydrogen effects in olefin polymerization catalysts, which has not been explicitly recognized in the literature, is both intriguing and alluring.

ACKNOWLEDGMENT

The authors express their appreciation to the management of Indian Petrochemicals Corporation Ltd. for permission to publish this work.

REFERENCES

1. K. Ziegler, E. Holzkemp, H. Breil, and H. Martin, *Angew. Chem.*, 67: 541 (1955).
2. J. P. Hogan, *Appl. Ind. Catal.*, 1:149 (1983).
3. G. Natta, *Science, 147* (Jan. 15, 1965), p. 261.
4. J. Boor, *Ziegler-Natta Catalysts and Polymerizations*, Academic Press, New York (1979).
5. M. P. McDaniel, in *Transition Metal Catalysed Polymerization of Olefins and Dienes* (R. P. Quirk, ed.), MMI Press (1983), p. 713.
6. F. J. Karol, in *Encyclopedia of Polymer Science and Technology*, Supp. Vol. 1 (H. F. Mark, ed.), Interscience, New York (1976), p. 120.
7. J. N. Short, in *Transition Metal Catalyzed Polymerization of Olefins and Dines* (R. P. Quirk, ed.), MMI Press (1983), p. 651.
8. K. B. Triplett, *Appl. Ind. Catal.*, 1:177 (1983).
9. V. A. Zakharov, G. D. Bukatov, and Yu. Yermakov, *Adv. Polym. Sci.*, 51:61 (1983).
10. P. Pino and R. Mulhaupt, *Angew. Chem. Int. Ed. Eng.*, 19:857 (1980).
11. P. Pino and B. Rotzinger, *Makromol. Chem. Suppl.*, 7:41 (1984).
12. D. G. H. Ballard, *Adv. Catal.*, 23:163 (1973).
13. Yu. I. Yermakov and V. A. Zakharov, *Adv. Catal.*, 24:173 (1975).
14. F. J. Karol, G. L. Brown, and J. H. Davison, *J. Polym. Sci. Polym. Chem. Ed.*, 11:413 (1973).
15. T. Keii, *Kinetics of Ziegler-Natta Polymerization*, Chapman & Hall, London (1972).
16. P. Galli, L. Luciani, and G. Cechin, *Angew. Makromol. Chem.*, 94:63 (1981).
17. T. Keii, E. Suzuki, M. Tamura, and Y. Doi, *Makromol. Chem.*, 183: 2285 (1982).
18. G. D. Bukatov and V. A. Zakharov, *Makromol. Chem.*, 183:2657 (1982).
19. V. A. Zakharov, N. B. Chumyerskii, and Yu. I. Yermakov, *React. Kinet. Catal. Lett.*, 2:329 (1975).
20. G. D. Bukatov and V. A. Zakharov, *Makromol. Chem.*, 183:2657 (1982).
21. N. Kashiwa and J. Yoshitaka, *Makromol. Chem., Rapid Commun.*, 3:211 (1982).

22. R. Merryfield, M. McDaniel, and G. Parks, *J. Catal.*, 77:348 (1982).
23. K. Soga, T. Sano, and S. Ikeda, *Polym. Bull.*, 2:817 (1980).
24. K. Soga, T. Sano, and R. Ohnishi, *Polym. Bull.*, 4:157 (1981).
25. H. Sinn and W. Kaminsky, *Adv. Organomet. Chem.*, 18:99 (1980).
26. Yu. I. Yermakov, in *Coordination Polymerization* (J. C. W. Chein, ed.), Academic Press, New York (1975), p. 15.
27. A. Shiga and T. Sasaki, *Sumitomo Kagoku (Osaka)*, 2:15 (1984).
28. S. Sivaram, *Ind. Eng. Chem. Prod. Res. Dev.*, 16:121 (1977).
29. F. J. Karol, *Catal. Rev.-Sci. Eng.*, 26:557 (1984).
30. H. L. Hsieh, *Catal. Rev.-Sci. Eng.*, 26:631 (1984).
31. C. Dumas and C. C. Hsu, *J. Macromol. Sci. Rev. Macromol. Chem.*, 24:355 (1984).
32. K. Y. Choi and W. H. Ray, *J. Makromol. Sci. Rev. Macromol. Chem.*, 25:1 (1985).
33. U. Zucchuni and G. Cecchin, *Adv. Polym. Sci.*, 51:101 (1983).
34. P. J. Flory, *Principles of Polymer Chemistry*, Cornell University Press, Ithaca, N.Y. (1971), p. 334.
35. J. Schwartz, *Acc. Chem. Res.*, 18:302 (1985).
36. A. Zambelli and C. Tosi, *Adv. Polym. Sci.*, 15:32 (1974).
37. S. Sivaram, *Proc. Ind. Acad. Sci. Chem. Sci.*, 92:613 (1983).
38. K. Soga, T. Sano, K. Yamamoto, and T. Shiono, *Chem. Lett. (Jpn.)*, 425 (1982).
39. G. Natta, G. Mazzauti, P. Longi, and F. Bernardini, *Chim. Ind. (Milan)*, 41:519 (1959).
40. B. Ettore and L. Luciani, Italian Patent, 554,013 (1957) to Montecatini.
41. G. Seydal, C. Beermann, E. Junghanns, and H. J. Bahr, German Patent 1,022,382 (1958) to Farbwerke Hoechst A.G.
42. E. J. Vandenberg, U.S. Patent 3,051,690 (1962) to Hercules Powder Co.
43. Y. Doi, Y. Hattori, and T. Keii, *Int. Chem. Eng.*, 14:369 (1974).
44. J. Horiuti and M. Polanyi, *Trans. Faraday Soc.*, 30:1164 (1934).
45. G. H. Twigg and E. K. Rideal, *Proc. R. Soc. London*, A171:55 (1939).
46. A. Schindler, *J. Polym. Sci.*, C4:81 (1964).
47. A. Schindler, *Makromol. Chem.*, 118:1 (1968).
48. V. I. Zakharov, N. B. Chumyerskii, Z. K. Bulatova, G. D. Bukatov, and Yu. I. Yermakov, *React. Kinet. Catal. Lett.*, 5:429 (1976).
49. B. M. Grievson, *Makromol. Chem.*, 84:93 (1965).
50. M. N. Berger and B. M. Grievson, *Makromol. Chem.*, 83:80 (1965).
51. T. Okura, K. Soga, A. Kojima, and T. Keii, *J. Polym. Sci. Part A-1*, 8:2717 (1970).
52. L. S. Rayner, *Comm. J. Polym. Sci.*, C4:125 (1964).
53. W. W. Buls and T. L. Higgins, *J. Polym. Sci. Polym. Chem. Ed.*, 11:925 (1973).
54. J. F. Ross, *J. Polym. Sci. Polym. Chem. Ed.*, 22:2255 (1984).
55. H. G. Yuan, T. W. Taylor, K. Y. Choi, and W. H. Ray, *J. Appl. Polym. Sci.*, 27:1691 (1982).
56. C. D. Mason and R. J. Schaffhauyer, *J. Polym. Sci.*, B9:661 (1971).
57. E. M. J. Pijpers and B. C. Roest, *Eur. Polym. J.*, 8:1151 (1972).
58. G. A. Mortimer, M. R. Ort, and E. H. Mottus, *J. Polym. Sci. Polym. Chem. Ed.*, 16:2337 (1978).
59. E. M. J. Pijpers and B. C. Roest, *Eur. Polym. J.*, 8:1162 (1972).
60. B. Boucheron, *Eur. Polym. J.*, 11:131 (1975).
61. W. W. Buls and T. L. Higgins, *J. Polym. Sci. Part A-1*, 8:1025 (1970).
62. A. S. Hoffman, B. A. Fries, and P. C. Condit, *J. Polym. Sci.*, C4:109 (1963).
63. M. Gordon and R. J. Roe, *Polymer*, 2:41 (1961).

64. M. N. Berger, G. Boocock, and R. N. Howard, *Adv. Catal.*, *19*:24 (1969).
65. W. C. Taylor and L. H. Tung, *Polym. Lett.*, *1*:157 (1963).
66. M. Pegoraro, *Chim. Ind. (Milano)*, *44*:18 (1962).
67. P. Galli, P. C. Barbe, and L. Noristi, *Angew. Macromol. Chem.*, *120*:73 (1984).
68. J. C. W. Chein and C. I. Kuo, *J. Polym. Sci. Polym. Chem. Ed.*, *23*:731 (1985).
69. J. C. W. Chein, *Catal. Rev.-Sci. Eng.*, *26*:613 (1984).
70. P. Galli, P. C. Barbe, G. Guidetti, R. Zaneti, A. Martorna, M. Marigo, M. Bergozza, and A. Fichera, *Eur. Polym. J.*, *19*: (1983).
71. S. A. Sergeev, G. D. Bukatov, and V. A. Zakharov, *Makromol. Chem.*, *185*:2377 (1984).
72. G. Guastalla and U. Giannini, *Makromol. Chem., Rapid. Comm.*, *4*:519 (1983).
73. L. L. Böhm, *Makromol. Chem.*, *182*:3291 (1981).
74. K. Soga and T. Siono, *Polym. Bull.*, *8*:261 (1982).
75. T. Keii, Y. Doi, E. Suzuki, M. Tamura, M. Murata, and K. Soga, *Makromol. Chem.*, *185*:1537 (1984).
76. J. C. W. Chein and C. J. Ku, presented at the *Symposium on Recent Advances in Polyolefins, 190th National ACS Meeting*, Chicago (1985).
77. F. I. Jacobson, *Proceedings of 28th IUPAC Makromolecular Symposium*, Amherst (1982).
78. L. L. Böhm, *Polymer*, *19*:562 (1978).
79. Y. Doi, S. Ueki, and T. Keii, *Makromol. Chem.*, *180*:1359 (1979).
80. J. Dubchke, W. Kaminsky, and H. Luker, in *Polymer Reaction Engineering* (K. H. Reichert and W. Geisler, eds.), Hanser, Munchen (1983).
81. W. Kaminsky and H. Luker, *Makromol. Chem., Rapid Comm.*, *5*:225 (1984).
82. Belgian Patent 570 (1959) to Solvay et Cie.
83. J. P. Hogan, *J. Polym. Sci. Part A-1*, *8*:2637 (1970).
84. M. P. McDaniel and M. B. Welch, *J. Catal.*, *82*:118 (1983).
85. T. J. Pullukat, M. Shida, and R. E. Hoff, in *Transition Metal Catalyzed Polymerization of Olefins and Dienes* (R. P. Quirk, ed.), MMI Press (1983), p. 697.
86. M. P. McDaniel and M. B. Welch, *J. Catal.*, *82*:98 (1983).
87. H. L. Krauss, B. Rebenstorf, and U. Westphal, *Z. Anorg. Chem. Org. Chem.*, *33B(11)*:1278 (1978).
88. P. P. M. M. Wittgen, C. Groeneveld, P. J. C. J. M. Zwaans, H. J. B. Morgenstern, A. H. VanHeugten, C. J. M. Van Heumen, and G. C. A. Schuit, *J. Catal.*, *77*:360 (1982).
89. R. L. Burwell, A. B. Littlewood, M. Cardew, G. Pass, and C. T. H. Stoddart, *J. Amer. Chem. Soc.*, *82*:6272 (1960).
90. R. L. Burwell and C. J. Loner, *Proceedings of the 3rd International Congress on Catalysis*, Amsterdam (1964), p. 804.
91. R. W. Hill, W. L. Kehl, and T. J. Lynch, U.S. Patent 4,219,444 (1980) to Gulf Research and Development Company.
92. M. P. McDaniel and M. M. Johnson, U.S. Patent 4,364,842 (1982) to Phillips Petroleum Company.
93. D. W. Walker and E. L. Czenbusch, U.S. Patent 3,123,571 (1964).
94. Y. Tajima, K. Tani, and S. Yuguchi, *J. Polym. Sci.*, *B3*:529 (1965).
95. R. N. Johnson and F. J. Karol, Belgian Patent 743,199 (1969) to Union Carbide Corporation.
96. D. G. H. Ballard, E. Jones, T. Medinger, and A. J. P. Pioli, *Makromol. Chem.*, *148*:175 (1971).

97. G. L. Karapinka, U.S. Patent 3,709,853 (1973) to Union Carbide Corporation.
98. F. J. Karol and G. L. Karapinka, U.S. Patent 3,709,954 (1973) to Union Carbide Corporation.
99. F. J. Karol, G. L. Karapinka, C. Wu, A. W. Dow, R. N. Johnson, and W. L. Carrick, *J. Polym. Sci.*, *10*:2621 (1972).
100. F. J. Karol, G. L. Brown, and J. M. Davison, *J. Polym. Sci. Part A-1*, *11*:413 (1973).
101. F. J. Karol and R. N. Johnson, *J. Polym. Sci.*, *13*:1607 (1975).
102. F. J. Karol, W. L. Munn, G. L. Goeke, B. E. Wagner, and N. J. Maraschin, *J. Polym. Sci. Polym. Chem. Ed.*, *16*:771 (1978).
103. K. M. Mackay, *Hydrogen Compounds of Metallic Elements*, E. and F. N. Spon, London (1966).
104. E. L. Muetterties (ed.), *Transition Metal Hydrides*, Marcel Dekker, New York (1970).
105. R. S. Coffey, in *Aspects of Homogeneous Catalysts*, Vol. 1 (R. Ugo, ed.), Carlo Manfredi, Milan (1970), Chap. 1.
106. L. Marko, in *Aspects of Homogeneous Catalysts*, Vol. 2 (R. Ugo, ed.), Carlo Manfredi, Milan (1970), Chap. 2.
107. M. Orchin, *Adv. Catal.*, *16*:1 (1966).
108. G. W. Parshall, *Acc. Chem. Res.*, *46*:264 (1977).
109. G. Lefebvre and Y. Chauvin, in *Aspects of Homogeneous Catalysts*, Vol. 1 (R. Ugo, ed.), Carlo Manfredi, Milan (1970), Chap. 3.
110. S. M. Pillai, M. Ravindranathan, and S. Sivaram, *Chem. Rev.* (1986), *86*:353 (1986).
111. M. F. Sloan, A. S. Matlack, and D. S. Breslow, *J. Am. Chem. Soc.*, *85*:4014 (1963).
112. I. V. Kalechits, V. G. Lipovich, and F. K. Shmidt, *Kinet. Catal.*, *9*:16 (1968).
113. P. E. M. Allew, J. K. Brown, and R. M. S. Obaid, *Trans. Faraday Soc.*, *59*:1808 (1963).
114. A. Nakamura and M. Tsutsui, *Principles and Applications of Homogeneous Catalysis*, Wiley, New York (1980).
115. K. I. Gell, B. Posin, J. Schwartz, and G. M. Williams, *J. Am. Chem. Soc.*, *104*:1846 (1982).
116. G. J. Kubas, R. R. Ryan, B. I. Swanson, B. J. Vergamini, and H. J. Wasserman, *J. Am. Chem. Soc.*, *106*:451 (1984).
117. J. E. Bercaw and H. H. Britzinger, *J. Am. Chem. Soc.*, *91*:7301 (1969).
118. G. Henrici-Olive and J. Olive, *Angew. Chem. Int. Ed. Eng.*, *8*:650 (1969).
119. R. E. Hoff, T. J. Pullukat, and R. A. Dombro, presented at the *Symposium on Recent Advances in Polyolefins, 190th National ACS Meeting*, Chicago (1985).

Index

A

Absorption of hydrogen on metals, 40-45, 157, 303-304, 513
Adsorbed forms of H:
  as observed by IR, 76-78, 186-192, 567
  as observed by neutron scattering, 75-76, 167-180, 513, 591-592
  on Pt, from TPR, 127, 427, 429
  phase diagram, 19-23
  vibrational spectroscopy of, 58-57
Adsorption:
  activated, 5, 89-91
  of CO and H_2, 36, 190-192, 349-357
  competitive, 34-35
  cooperative, 34, 35, 37
  delocalized, 68-72
  energetics of, 140-153
  enhancement by spillover, 314-317
  heats of, 5, 72, 85-113, 141, 148
  of hydrogen:
    on metal films, 95-99, 149, 187, 213-221
    on oxides, 567-571
    on single crystals, 3-34, 59-75, 141, 148, 187, 426-427
    on sulfides, 588-604
    on supported metals, 75-78, 99-113, 139-162, 186-192
  kinetics of, 140-160, 399-421, 548
  reversibility of, 154-163

[Adsorption]
  sites, 59-60, 65-68, 399-418, 599-601
  slow, in determining selectivity, 637-639
  stoichiometry, 119-121, 158-163, 315
AES (Auger), 95, 196, 426, 433, 618-619
Alkylation, 703
Alloy(s), 8, 15, 119-123, 236-241, 388-390, 437, 463, 504, 653-674, 682-690
Amino alcohols, 529
Amination, 526
Ammonia synthesis, 38, 387
  hydrogen effects in, 616
Aromatic/C_5-cyclic ratio, hydrogen effects on, 456-458, 466, 487-488
Aromatization, 359, 449-450, 459-463, 480-482, 630, 660-669, 705, 715-717
  bifunctional mechanism of, 450, 655, 660-669
  of alkanes, 459-463, 660-667
  of cyclopentanes, 476, 488-489, 667-669
  of cyclohexane, 456
  of cyclohexene, 321, 432
  of isoalkanes, by dehydroisomerization, 463, 468-470
  on Pt single crystals, 442, 463
  promotion by high-temperature-adsorbed H, 481-482
  stepwise mechanism of, 459-461

[Aromatization]
  of unsaturated C_6-hydrocarbons, 461, 462, 478
Asymmetric hydrogenation, 625
Atomic migration, 205, 259, 262, 264, 266
Au black, sintering of, 299

B

BET equation (surface area), 93
Bimetallic catalysts, 119-123, 236, 239, 388-390, 437, 463, 504, 659-674
  surfaces, 15
Biohydrogenation, 628
Block walls, 199
Bragg maxima, 44
  reflection, 108, 169
Bronsted acid sites, 698-699
Bulk hydride, 195

C

Carbenium (carbonium) ions, 665, 696-699
Carbon monoxide (*see also* CO hydrogenation):
  adsorption for metal surface characterization, 349-357
  coadsorption with H_2, 36
Carboxylic acids, 530
Catalytic cracking, 613, 700, 711-715
Catalytic reforming, 119-121, 450, 614-617, 630, 653-674
C_1-chemistry, 118, 543-560, 615
C_5-cyclization, 449, 464-466
  hydrogenative and dehydrogenative routes, 463-466
  hydrogen effects on, 464-467
  of unsaturated hydrocarbons, 465-466
Cells, electrochemical, 228-229
  three-electrode, 228-229
Channeling, 276
Charge transfer, 7
Chemisorption, 4-23, 63, 71, 119, 127, 136, 142, 163, 426-429, 510-512, 568-571, 690-691
Chemoselectivity, 622
Chromium catalysts for polymerization, 737-739
Cluster compounds/catalysts, 626, 630
Coadsorption, 34-39
  of H_2 and CO, 36, 66, 190-192

[Coadsorption]
  of H_2 and N_2, 38, 616
  of H_2 and O_2, 39
Coalescence, 264, 267, 273
  crystallite, 306
Coal hydrogenation, 632
Cobalt sulfide, 596
CO hydrogenation, 363-364, 387, 543, 562, 630-631
  effect of promoters in, 560-561
  effect of support in, 555-560
  hydrogen effects in, 543-561
  kinetics of, 544-549
  product distribution in, 552-554
  SMSI in, 363, 364
Coke on catalysts, 131-136, 276-277, 306-307, 386, 387, 391, 442-444, 477-480, 605, 609, 614-619, 639, 656-669, 705
Conjunct polymerization, 704, 718
Constraint index, 708
Crackability of a paraffin, 701
Cross-flow catalyst/reactor, 645
Crystallite migration, 259, 262, 266, 281
Crystal shape, equilibrium, 293, 296-297

D

Deactivation of catalysts, 134-136, 276-277, 442-444, 477-480, 656-660
Decohesion, 46
Defect(s), 14, 81
  crystallographic, 11
  in MoS_2, 590
  sites, 12, 586
Dehydration, 525
Dehydrocyclization (*see* Aromatization)
Dehydrogenation:
  of alkanes, 390, 454-457
  of cyclohexane, 358, 440, 456
  of cyclohexanol and cyclohexanone, 523
  of cyclohexene, 432-433, 437
Dehydroisomerization (*see* Aromatization)
Deposition precipitation, 106
Desorption, slow, in determining selectivity, 637-639 (*see also* TPD)
Dewaxing, 713
Diastereoselectivity, 623
Diffusion, 265, 706

Index

[Diffusion]
  of H through metals, 206, 334, 390-391, 682-686
  of H through Pd and its alloys, 681-685
  of H on surfaces, 33-34
Diols, 527
Dioxacycloalkanes, 538-539
Dipole scattering, 60-62
Direct ripening, 262
Disproportionation, 699-700, 710
Dissolution of metals during hydrogenation, 513-514

E

EDAX, 387
EELS, 15, 175, 427
Electrical conductance of metals, effect of H adsorption on, 213-221
Electrocatalysis, 627
Electrochemical potential of metal catalysts, 225
Electron-hole-pair mechanism, 24
Electron microscopy of:
  H_2PtCl_6 in beam reduction, 298
  Pd-black, 300
  Pd/mica, 298
  Pt/Al_2O_3, 297-300
  Pt-black, 299, 301-303
  Pt/SiO_2, 297-298
  Rh/SiO_2, 297
  supported metals, 270-275
Enantiomeric excess in hydrogenation, 510
Enantioselectivity, 623, 625
Ensemble of metal atoms, 15, 418
Epimerization, 522-523
EPMA, 136
Equilibrium crystal shape, 205, 293-297
ESCA (XPS), 366, 590, 618
Ethylidyne, 430, 434-438
EXAFS, 102, 112-113, 336, 377
Excess surface free energy, 226
  measurement of, 230

F

Faceted particles, 205
Faceting of crystallites, 46, 267, 296
Faraday effect, 201
Fermi-Dirac distribution, 198
Ferromagnetic anisotropy, 210-213
Ferromagnetic coupling, 207
Ferromagnetic resonance, 201
Ferromagnetism, 201-210
Fick's law, 33, 681
Field emission measurements, 33, 89, 96
Field emission tips, sintering of, 296
Film growth, 205, 295
Film transport, 641-642
Fischer-Tropsch synthesis, 118, 133, 543-561, 615-617, 630, 643
Fourier-transform infrared spectroscopy (FTIR), 183-192, 333
Frequency factor for H adsorption, 30

G

Galvanostatic curve, 227
Globular morphology, 277-279
Grain boundary grooving, 201
Grain growth, 295, 302
  kinetics of, 295

H

Half-hydrogenated state, 461, 501-503, 506, 523, 529, 633-636
H_2-D_2 exchange, 13, 94, 98, 381, 430, 574, 591, 593
He diffraction, 426
He ion scattering, 13
H-H interactions, 170, 174
Homologation, 450, 472-473
Horiuti-Polanyi mechanism, 500-506
H_2-O_2 titration, 119-124, 136, 279, 355, 482
HREELS, 38, 57-68, 186, 303-304, 366, 426-433, 436-438
Hydride(s), 12, 40, 45-47, 66-67, 86-88, 196, 221, 304, 373-380, 483, 680-681
  as catalysts, 381-390
  magnetic properties of, 197-205
"Hydride" hydrogen, 384-385, 387-392
Hydrocracking, 697, 700
Hydrodenitrogenation, 631
Hydrodesulfurization, 163, 599-604, 631
Hydroformylation, 629-630
Hydrogen:
  availability (on surface), 500-509

[Hydrogen]
  delocalized, 68-72
  diffusion in metals, 206, 334, 679-693
  effects on polymer properties, 730-739
  as fuel, 611-612
  general kinetics of, 399-420
  generators, 689-694
  in olefin polymerization, 723-724
  solubility in metals, 680, 683
  types of adsorbed, 4-23, 75-78, 127, 167-180, 427, 429, 500-517
    "multibonded," 14, 19-21, 63, 68, 73, 187-188
    "on-top," 73-74, 187
    residual, 127-129
    strongly bonded, 127-129, 231-239
    subsurface, 42-44, 129, 157, 294, 303-304
    weakly bound, 188, 192, 231-241
  world production of, 611
Hydrogenation, 378-386, 430-432, 437, 499-515, 622-629
  of acetylene(s), 381-384, 507, 508, 514, 643
  of aromatics, 247-252, 358, 437, 502, 510, 624-625
  of butene, 573-575, 579-580
  of carbonyl compounds, 502-504, 508, 510
  of cyclohexene(s), 439-440, 624
  of ethylene, 321-323, 358, 430-432, 435, 623
  of fats, 625-626, 639
  solution of metals during, 513-514
  stereochemistry of, 500-503
Hydrogen bronzes, 318, 327-329
Hydrogen chemisorption/sorption:
  by metals (see Absorption of hydrogen on metals, Adsorption, Chemisorption)
  by oxide catalysts, 175-176, 568-570
  by sulfide catalysts, 176-177, 588-589
Hydrogen-deuterium exchange, 13, 94, 98, 381, 430, 574, 591, 593
  in hydrocarbons, 416-417, 439, 441, 483-484, 574
  in oxygenated compounds, 522, 527, 529, 537

[Hydrogen-deuterium exchange]
  in sulfur compounds, 589-590, 603-604
"Hydrogen fog," iv, 78
Hydrogenolysis, 129, 360-363, 385 389, 399-401, 409-412, 418-420, 449, 469-476, 525, 530-539
  coke formed during, 617
  pattern:
    hydrogen effects on, 471-473
    SMSI effect on, 360
  suppression by high-temperature-adsorbed H, 129, 490-492
Hydrogen spillover (see Spillover of hydrogen)
Hydrogen transfer in catalysis, 613-614, 628, 695-720
Hydroisomerization, 696, 703 (see also Isomerization)
Hydrotreating catalysts, 584

I

IES, 58
IETS, 58
IINS, 58, 75-76, 168-176
  of zeolites, 178
Impact scattering, 62-65
Inelastic neutron scattering, 167-178 (see also IINS)
Infrared spectroscopy, 183-192
  of Pt-H band, 189
Interaction:
  metal-support (see SMSI)
  physisorptive, 4
  van der Waals, 4
Intercalation of hydrogen, 592, 595
Interfacial free energy, 260, 281
  process, 265
Intrinsic selectivity, 622
Ion scattering, 92
Ising model, 2-dimensional, 20
Isomer/C_5-cyclic ratio, hydrogen effects on, 468, 489
Isomerization:
  of hydrocarbons, 359, 361, 385-389, 441-444, 466-468, 476-477, 572-578, 696-700, 710-711
  of oxygen-containing heterocyclics, 530-539
  of unsaturated alcohols, 503, 527-529
Isomerization/hydrogenolysis ratio, hydrogen effects on, 471
Isotactic index, 732-737

K

Kisliuk model/function, 25-26
Knudsen diffusion, 100

L

Landing site, 413, 418-419, 452, 453
Langmuir-Hinshelwood mechanism/reaction, 41, 399, 633
Laser Raman spectra of coke, 660
Lateral interactions, 16, 18
Lattice expansion upon H sorption, 44, 305
Lattice gas model, 22
Lattice strain in Pt-black, 302, 305-306
LEED, 11, 14, 16, 20, 23, 37, 44, 47, 49, 95, 196, 203, 215, 426
Lennard-Jones potential, 6
Lindlar catalyst, 508

M

Magnetic anisotropy, 102
Magnetic measurements, 100, 110, 113, 196-213
Magnetic properties of bulk metal hydrides, 198-213
Magnetic susceptibility, 199, 377, 380
Maximum yields, as function of H_2 pressure, 454-456, 460, 464, 469, 475, 487, 491, 500-515, 531, 533, 535, 616, 655-670
Membrane catalysts, 390-391, 461, 511, 691-693
Metal dispersion, 118-124
Metal hydrides [see Hydride(s)]
Metal hydriding process, 378-379
Metal-hydrogen bond, 7, 14, 21, 739-740
Metal membranes, 373, 390, 511, 614, 645-646, 679-693
Metathesis of propene, 575, 578
Methanation, 543-561
Methanol-to-gasoline (MTG) process, 717-718
M-forming process, 713
"Mischmetal," 386
MOGD process, 718
Molecular adsorption of H_2, 4-6, 186, 567-568, 636
Molecular beam measurements, 13
Molybdena-alumina, 566-570, 576-580
Molybdena-titania, 575
Molybdenum sulfide, 588-595
Monolayer adsorption, 97, 160-162, 170-175
Monte Carlo calculations, 22
Mossbauer spectra/studies, 207-209
  of Fe-Ni alloys, 200
  of sulfide catalysts, 588
Most abundant surface intermediate (MASI), 401-404, 412, 419
Multiparticle interaction, 16
Multisite effect, 411

N

Neel's theory, 213
Neutron scattering (see also IINS), 167-182, 591-592
Nickel, 8-14, 22, 28, 46-50, 72-74, 77, 97, 104-113, 140-150, 200-206, 244-247, 352, 403, 474, 512, 522-539, 544-562, 632

O

Octane booster/boosting, 614, 713-714
Ostwald ripening, 262
Oxanes, 537-538
Oxetanes, 530-537
Oxiranes, 530-537
Oxo compounds, 529
Oxolanes, 537-538

P

Palladium black, 76, 174-175, 490
PdH phases, 303-304, 375-386
Periodic operation (chemical process), 643
Permeable membranes as catalysts, 390-391, 461, 511, 645-646, 691-693
Phase diagram, 20, 87, 375, 681
Phase transitions, 16, 20, 87, 304, 375-386
"Pill-box" morphology, 277, 282
Platinum black, 76, 127, 173, 174, 296-308, 452-477, 487-490, 524
  sintering in electron microscope, 299
Poiseuille flow, 100
Poisoning of catalyst (see also Coke on catalysts):
  in catalytic reforming, 657-658
  in hydrocarbon reactions, 477-480

Polarizability of hydrogen, 92
Polarization measurements, 227-228
  galvanostatic methods for, 227, 232
  potentiodynamic methods, 227, 229, 232
Polymerization, 704, 718-720, 723-741
  kinetics, 724-726
Pore collapse, 267
Pore transport, 639-642
Potential energy curves, 5, 86-90
Potentiodynamic curve, 229, 231-234, 236-237, 240
Powder growth, 295
Promoters, 560
Proton band model, 34
Pt-H band:
  IINS spectra of, 173-175
  IR spectra of, 189-191
  vibrational spectra of, 73

Q

Quasielastic neutron scattering, 169, 172-173

R

RAIRS, 58, 77
Raman spectroscopy, 76-78, 187
Raney nickel, 76, 170-172, 378, 502, 512-513, 522, 525, 636
  hydrogen on/in, 513
  TPD of H_2 from, 513
Reactor:
  cross-flow, 645
  monolithic, 646
  trickle bed, 646
Reconstruction in H_2, 9, 23, 26, 46, 51, 376, 381
  of supported metals, 269-276, 297-300, 484, 486
  of unsupported metals, 300-308
Redhead model, 27-28
Redispersion, 260, 271
Regioselectivity, 506, 532-533, 623
Relaxation, 9, 23
  surface, 44, 51
Repulsive interaction, 16, 17
Reoxidation treatment, 355-356
Research octane number (RON), 653, 658, 673
Reverse spillover, 324, 337-338
Ring opening:
  of cyclopentanes, 474-476, 489-490
  of cyclopropanes, 483, 484-486

[Ring opening]
  of oxygen-containing heterocyclics, 530-539

S

Saturation magnetization, 200, 207
Selectivity:
  chemo-, 622
  enantio-, 623, 625
  intrinsic, 622
  isomerization and hydrogenolysis, 360, 471, 475, 482, 490
  regio-, 506, 532-533, 623
  shape, 622, 710-716
Selectoforming, 713
SIMS, 39-40, 307, 366, 618-619
Sintering, 45, 259-288, 293-307
  in electron microscope beam:
    of Au-black, 299
    of Pt-black, 299
  in ethylene, of Pt-black, 306
  in hydrogen:
    of Pd-black, 300
    of Pt-black, 300, 302, 487-490
    of Pt/carbon, 300-305
    of Pt/zeolite, 300-305
    of single crystals, 296-297
    of supported metals, 269-276, 656-658
  types of, of supported metals, 268
Site, subsurface, 43
Site blocking, 17
  ratio, 16
Skeletal rearrangement (see Isomerization)
SMSI, 107, 124, 127, 136, 146-149, 157, 279, 347-368, 385, 514, 615
  metals, catalytic properties of, 357-364, 514
  origin and nature of, 259, 285, 287, 364-366
  and selectivity, 359-364, 514
Soft X-ray emission (SXES), 306-307
Spatial constraint, 707-709
"Specialties," 539
Spillover of hydrogen, 102, 157, 311-341, 384, 491, 500, 509-510, 636, 667
  adsorption enhancement by, 314-321
  alumina activated by, 322-323
  bronze formation by, 318, 327-328
  and coke removal, 323-324, 337-338, 615

Index

[Spillover of hydrogen]
  double, 510
  magnesia activated by, 523
  mechanism of, 325-329, 340-341
  quantification of, 329-330
  rate of, 331-335
  reactions induced by, 321-322, 509-510
  retention of activity, 323-324
  reverse, 321, 509
Spin-orbit coupling, 210
Step sites, 9, 91
Stepped surfaces, 13-14, 426-444
Steric inhibition, 707-711
Sticking coefficient, 413
Sticking probability, 5, 13-14, 23-26, 47, 51
Stoichiometry of H sorption, 119-120, 229, 314-315
"Streak" phase, 48, 50
Stretching frequency, 67, 187-191
Structure-insensitive reactions, 357
Structure-sensitive reactions, 357, 359, 418, 485-486
Subsurface hydrogen, 42-44, 129, 294, 303-304
Sulfide catalyst activity, theories of, 585-588
Superactivity, 709-710
Superparamagnetic, 200
Surface anisotropy, 211-213
Surface area determination, 105-113, 119-124, 160-162
Surface chemistry of oxides, 566
Surface dipole selection rule, 61-62, 185
Surface energy, anisotropy of, 92
Surface free energy, 285, 293
Surface migration, 33-34, 91
Surface species in equilibrium, 616
Surface reconstruction, 296-297
Synergy in HDS catalysts, 584
  role of H spillover in, 587

T

TAIRS, 58, 76
TDS (see TPD)
Terraces, 8, 426-444
TPD, 117-118, 125-130, 134, 139-158, 197, 227, 231, 315-318, 331, 352-354, 426, 443, 512-513, 570-571, 591, 597-598, 616, 690
  kinetics of, 27-33

TPR, 117-118, 130-131, 136, 317
TPSR, 117-118, 136, 459, 616
TPT, 121-122
Tunneling, 33
Titania-supported catalysts, 227-280, 330, 349-368, 385, 555-556
Titanium catalysts, for polymerization, 728-737
Tungsten sulfide, 595-596
Turnover frequency (number), 385, 433, 440, 463, 483, 484, 490, 550-560

U

Ultrapure hydrogen, 679
  in manufacture of silicon chips, 679, 686
Unsaturated alcohols, 527-529
UPS, 10, 366, 427

V

Vibrational spectroscopy, 57-80, 168-178
Voltammogram, 232-233, 238

W

Wavefunctions, hydrogenic, 4, 10, 45
Weiss domains, 199
Wettability, 286
Wetting, 259, 279, 281, 294
Wigner-Polanyi equation, 27
Work function, 7-10, 379-380
Wulff plot, 294-297
Wulff point, 293

X

XPS (see ESCA)
X-ray diffraction, 301, 303, 304, 380, 383, 387, 483-484
X-ray line broadening, 101, 304

Z

Zeolites, 177-180, 187, 696-720
  classification, 705-707
  hydrogen transfer reactions on, 707-710
Zero coverage limit, 6
Ziegler-Natta polymerization, hydrogen effects in, 615, 728-737
ZSM-5 (also HZSM-5), 706-710